Pressure-Driven Microfluidics

For a listing of recent titles in the
Artech House Integrated Microsystems Series,
turn to the back of this book.

Pressure-Driven Microfluidics

Václav Tesař

Library of Congress Cataloging-in-Publication Data
A catalog record for this book is available from the U.S. Library of Congress.

British Library Cataloguing in Publication Data
A catalogue record for this book is available from the British Library.

ISBN-13: 978-1-59693-134-3

Cover design by Igor Valdman

685 Canton Street
Norwood, MA 02062

10 9 8 7 6 5 4 3 2 1

Contents

Preface

The subject of this book is handling and managing fluid flows. To an outsider, this may sound like a dull subject, but it is actually exciting and also very important from a practical standpoint. Fluids (a term covering both liquids and gases) are encountered almost everywhere, in the atmosphere, in rivers and seas, and indeed in our own bodies. In particular, they are almost omnipresent in technology. There are very few technological processes that can claim avoiding working with fluids. Even in what is seemingly a purely electric system, such as a computer microprocessor, it may be found that the limiting factor in further development is circuit cooling by the surrounding atmospheric air. If the passive and not very effective cooling by natural convection is replaced by controlled coolant flows, the subject of fluidics becomes a part of the project. Admittedly, the cooling and similar effects are areas where fluidics is not the star of the show and plays only a supporting role, but even there progress in the fluidic side of the development may bring a substantial overall step forward. There are other applications, like implanted devices for control of body fluids in medicine, where fluidics is the key factor.

Microfluidics is the outgrowth of fluidics—the technical field of working with fluids, mainly controlling their flows in a system of channels. Microfluidics is characterized by the small size of the channels. It emerged together with the more general area of microelectromechanical systems (MEMS), which became possible by applying the microfabrication techniques originally developed for microelectronics. Typical for MEMS are microfabricated moving mechanical components and among them there are also components acting on fluid flows. There are many existing applications, and even more proposed ones, taking a distinct advantage from the small size of microfluidic devices. The trauma of invasive implanting is lessened if the implanted device is very small. There are, however, still other applications where the small scale is not the decisive factor. There the devices may take over the operating principles of the earlier larger-scale fluidic (or perhaps mesofluidics). As the complexity of the actions performed by fluidics tends to increase, so does the complexity of the fluidic circuits and with this comes the general requirement to make the systems physically smaller. If not at present, fluidics is likely in the foreseeable future to move gradually toward microfluidics.

At the microscale, there is a general trend of the surface forces acting on the fluid becoming more important than the other, more conventional force types. Much interest in present-day development of microfluidics is devoted to the technique of driving the fluid by the surface-type electro-osmotic effects. While such aspects are not avoided and there is an ample discussion of such driving phenomena (often limited to special types of the fluids), this book tends to concentrate on the classical manner of driving the fluid flows, mainly by the action of a pressure difference. Not only does this driving mechanism have a much more general applicability, but it has also made it possible to include here among the treated subjects some interesting and little known principles used in the classical larger-scale fluidics.

However new is the subject of microfluidics, there are already in existence other books on this subject. Many of them naturally concentrate on the novel ideas associated with the microfabrication technology, producing the tiny devices, and with the mechanical actions made possible by the development in the field of MEMS. A typical book of this type is *Fundamentals and Applications of Microfluidics* by Nam-Trung Nguyen and Steven T. Wereley, Second Edition. The present book was carefully planned taking into consideration the existence of such literature. It aims at avoiding direct competition, and concentrates on those aspects of microfluidics that are outside the main interest in this and similar books, discussing the complementary aspects and issues.

Microfluidics can lead to profound changes in the world as we know it. It may have a large influence on transportation and even producing food and similar very essential aspects of human existence. In particular, its interest focuses discernibly toward collaboration with and influencing living organisms at the cellular level. This may bring many benefits in improvement of health care and various therapies, but may be also abused. One conclusion, however, seems to be clear. Fluidics—and its outgrowth microfluidics—is here and is going to stay with us. Engineers must learn its weaknesses and the advantages it offers.

Chapter 1

Introduction and Basic Concepts

The subject of this book, microfluidics, is the technology of handling small fluid flows in small devices. It is a new technology, existing roughly for two decades—a time too short for full recognition of its potential. According to the rather arbitrary definition in [1], microfluidic devices are characterized by dominant (smallest) channel width (or analogous transverse dimension of fluid flow paths) smaller than 1 mm. The general development trend is to make microfluidic devices gradually smaller, sometimes even so tiny that they are hardly visible with the naked eye (which can discern features larger than ~10 μm). Present-day microfluidics originated by using the manufacturing technology developed in microelectronic engineering for making fluid flow channels. It is possible—and advantageous—to make both electric and fluidic circuits on the same silicon chip (Figure 1.1) so that they can mutually collaborate. The task of fluidics is to handle fluid flows through the channels while a typical task of electronic sensors is to generate signals carrying information about the fluid presence and/or its properties so that the other parts of the electronic circuitry can process this information, perhaps influencing the fluid flows in a feedback action.

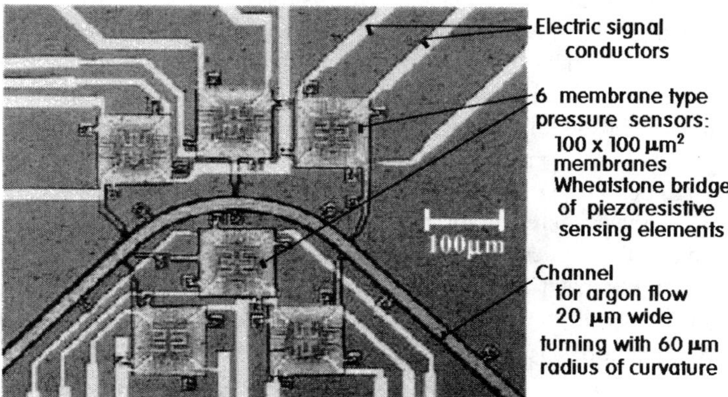

Figure 1.1 Detail of a silicon chip with typical coexistence and collaboration of microfluidics and microelectronics. The microfluidic part is extremely simple here – just a 90° bend of a constant width channel. Microelectronics converts the pressure readings at 6 locations along the channel into the electronic output. (*From* [2]. © 2001 Institute of Physics. Reprinted with permission.)

1.1 MEANING AND USE OF MICROFLUIDICS

Often, though not always, microfluidics is a part of MEMS technology [3, 4]: micro-electromechanical systems, characterized by electronics cooperating at the small scale with miniature mechanical devices implied by the name (Figures 1.2 and 1.3), but also with devices of other characteristics, such as fluidic [5], optical, thermal, acoustic, or chemical.

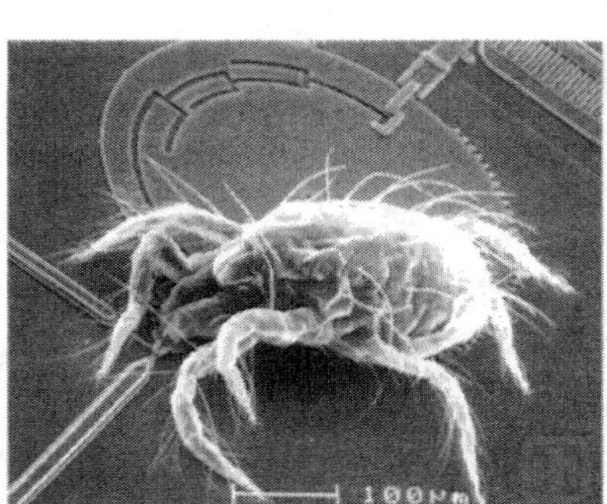

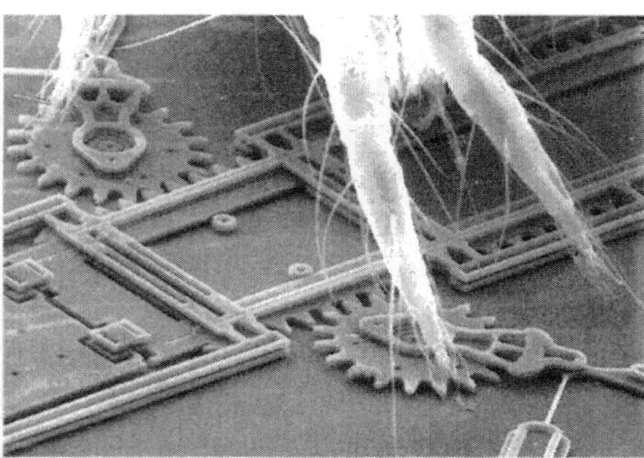

Figures 1.2 (Left) and 1.3 (Right) MEMS mechanism made by etching in silicon, compared with a mite, a creature normally hardly recognizable with the naked eye. (Courtesy of Sandia National Laboratories SUMMiTTM Technologies.)

Typical current applications of microfluidics are in biomedical sciences. An application that may serve as a characteristic example is an implanted device for diabetes sufferers, delivering insulin into the blood flow. It is easy to visualize—though at present not yet so easy to develop—a device for continuous glucose monitoring in a patient's blood, using this information for feedback-controlled insulin delivery (Figure 6.151). Both blood samples as well as the pumped insulin solution are liquids so that the need for fluidics is obvious. The small scale of microfluidics minimizes the problems associated with implanting the devices.

Another example where much current effort is directed is DNA "fingerprinting" [6], the exact identification of organisms. This has a vast potential not only for forensic purposes but also for objective diagnosis of illnesses by identification of the bacterial DNA and tailoring the medical treatment to the genetic makeup of the patient (see Section 6.7.2). This involves the polymerase chain reaction and analysis of the reaction product, again processes requiring precisely controlled handling of liquids. The small scale makes it possible to perform simultaneously a huge number of such analyses, perhaps in a portable "lab-on-chip." Typically, the background of biomedical research leads to the preference for microfluidic devices and circuits built on glass substrate (Figure 1.4) rather than silicon. Mass-produced final versions, still mostly under development, plan to use yet another material: plastics. So cheap as to be discarded after use, they eliminate any cross-contamination between the samples.

An emerging area of vast application potential for microfluidics is also microchemistry (e.g., [7]). All handling of the reactants (e.g., dosage and mixing) as well as of the reaction products (perhaps involving separation from product mixture) involves fluid motion. Performing chemical reactions in tiny microreactors has several distinct advantages; the obvious one is the possibility to perform simultaneously a huge number of slightly differing reactions in a drug or catalyst discovery activity. This saves the time needed for finding the best-performing drug.

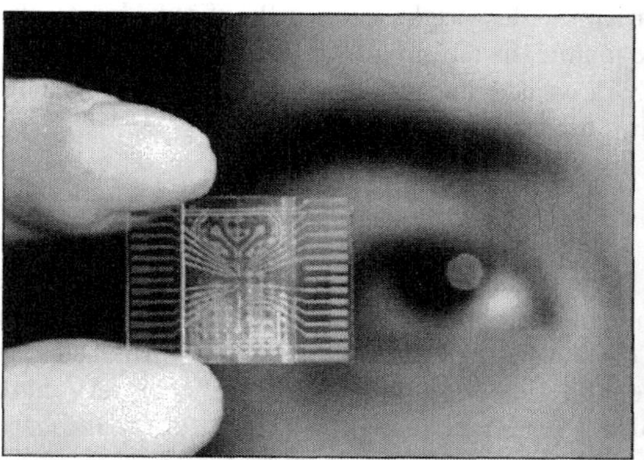

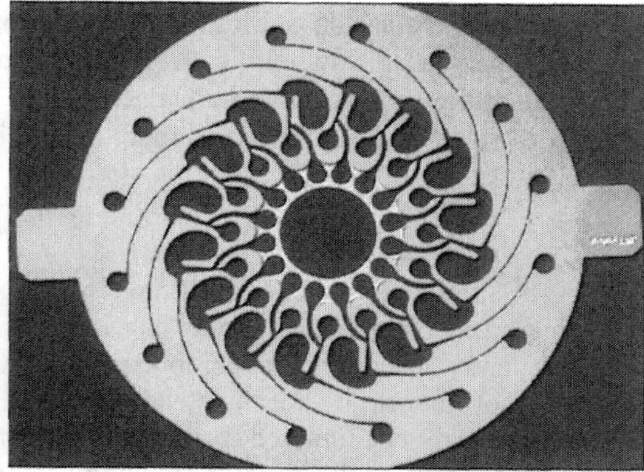

Figure 1.4 (Left) Microfluidics and electronic sensors on a glass chip for biological application: counting and classification of living cells (*From*: [4]. © 2004 Elsevier. Reprinted with permission.)

Figure 1.5 (Right) A typical high-temperature microfluidics: an array of 16 valves taking reactant samples from microchemical reactors that perform the Fischer-Tropfsch reaction, the synthesization of liquid hydrocarbon fuels.

Instead of silicon, the background and experience of chemical engineers has led to the common trend of using stainless steel (Figure 1.5) as their preferred material for microfluidic devices in this application. Especially in the case of exothermic reactions, the microchemistry is an example of non-MEMS microfluidics: the electronic circuitry is placed outside the chip, at a safe distance from the high temperature.

This book is mainly devoted to a particular area of microfluidics, characterized by the fluid flow in nonmoving part devices driven by the action of applied *pressure differences*. This is actually an approach building on past successes of large-scale pressure-driven fluidic devices. It should be said that in present-day microfluidics there is often an interest in using other driving mechanisms. A typical case would be electrokinetic or electroosmotic phenomena dominated by processes in an extremely thin near-wall layer. These are attractive due to their novelty; the small scale with resultant large surface-to-volume ratio leads to dominant effects of surface phenomena, impossible to use and mostly actually negligible at the large scale. The more traditional pressure-driven flows may be less spectacular, but they are more widely applicable. No special fluid properties are required, while electrokinetic flows are usable only with liquids; in fact, requiring a polar character of the liquid. Though quite special from a physical point of view, this generally means water and aqueous solutions, encountered very often in many applications. Nevertheless, in gas flows such as in the microchemistry of processing gaseous fuels, pressure-driven flows are the only possibility. Being applicable even at large scales, the pressure-driving in microfluidics can employ useful ideas developed earlier in large-scale fluidics, of which microfluidics is a direct descendant.

Many microfluidic devices were demonstrated to operate successfully using moving parts acting on the fluid [1]. As already mentioned, the attention here is focused on *purely fluidic* devices, operating without moving or deformed mechanical components. The absence of mechanical parts makes pure fluidics easier to manufacture, more robust, and more reliable, evading the problems caused by seizure or breakage. On the other hand, admittedly, generation and control of fluid flow is in principle easier with mechanical components. Fluid simply cannot enter a channel when its entrance is blocked by an inserted mechanical part in a mechano/fluidic valve. Similarly, a mechano/fluidic micropump is easier to design because fluid has no option other than to move when displaced by a moving piston. Pure fluidics employs more subtle phenomena. It relies on setting up special, sometimes downright exotic, flowfields inside the devices, such as a flowfield with some hydrodynamic instability. A complete change of the character of the flow is therefore triggered even by a weak input disbalancing the instability. This is more difficult to design, but also more interesting.

1.1.1 Why fluids?

The term "fluid" is a generalization, covering both liquids and gases (Figure 1.6), the latter differing from the practical point of view mainly in their much larger specific volume (lower density) and much higher compressibility. Compressibility can significantly influence the character of the flow, but in steady flows the effect becomes important only at very high velocities, rarely encountered in microfluidics. Also covered by the term "fluid" are multiphase mixtures, such as bubbly liquids (including foams), suspensions of solid particles or living cells, emulsions of mutually immiscible liquids, and so on. These cases, which may all lead to quite complicated fluid mechanics, are currently used more and more. A typical property of small scales is increasing importance of surface forces, usually negligible in large-scale flows. This is the case of surface tension on the interface between liquid and gas, which becomes the dominant factor in some microfluidic devices, sometimes even handling individual drops in gas (air) or individual air bubbles in liquids

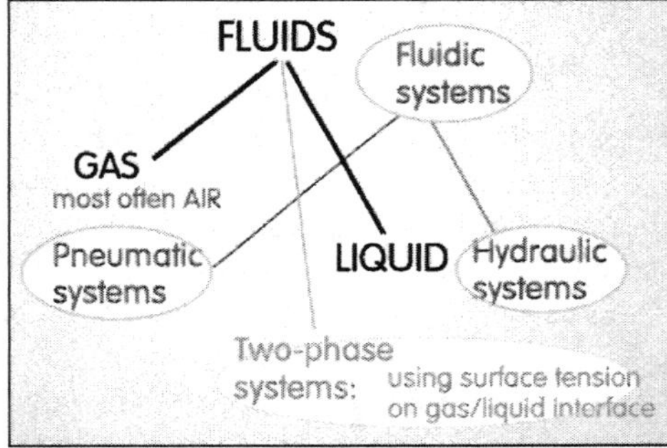

FLUIDS ARE NEEDED :

- Hydraulic power transfer
- Signal transfer
- Heat transfer
- Chemical reactions
- Biomedical material
- Food
- Lubricants, paints, solvents
- etc.

Figure 1.6 (Left) Just as the term *fluid* involves liquids as well as gases, *fluidics* is a general term encompassing both hydraulic systems working with liquids, and pneumatic systems working with air or another gas. Of increasing importance in contemporary microfluidics is handling two-phase flows, both gas and liquid (e.g., to generate and use air bubbles).

Figure 1.7 (Right) Some of the many reasons why handling fluids is indispensable.

The objects of interest in this book are direct descendants of classical hydraulic and pneumatic systems. Again, the term "fluidic" system (Figure 1.6) is a generalization covering both of them. Such systems usually involve generation of fluid motion, perhaps by a pump, distribution and transport of fluid into various destinations, and finally their use in the destination location. In classical large-scale hydraulics the fluid is just a working medium used to transfer power or, in lesser degree, transfer the control signals. The fluidic character of the transfer is in principle not very important; it may be thought of as replacing a mechanical transmission. The use of fluid brings some practical advantages. It may make it easier to transfer the power into a moving final destination by a flexible hose, but the properties of the fluid itself are of secondary importance. The fluid, for example a hydraulic oil, is perhaps chosen on the basis of its lubricating properties, preventing seizure of the moving parts.

On the other hand, the transport and handling of a fluid in modern fluidics is often chosen because a particular fluid is needed at the output destination. Figure 1.7 presents a list—far from exhaustive—of such typical uses that are the actual reasons why handling fluids is useful and in fact indispensable. Even in the classical application of fluids used to generate a force (piston force-generating drives have a number of advantages over electric drives for slow linear motions) and to perform energetic conversions in thermal machines, the fluid is not just a passive medium. Nevertheless, today these are rather exceptional motives for applying microfluidics – even though, for example, there were successful demonstrations of combustion microturbines (Figures 1.8, 6.19, and 6.20) made by etching in a chip.

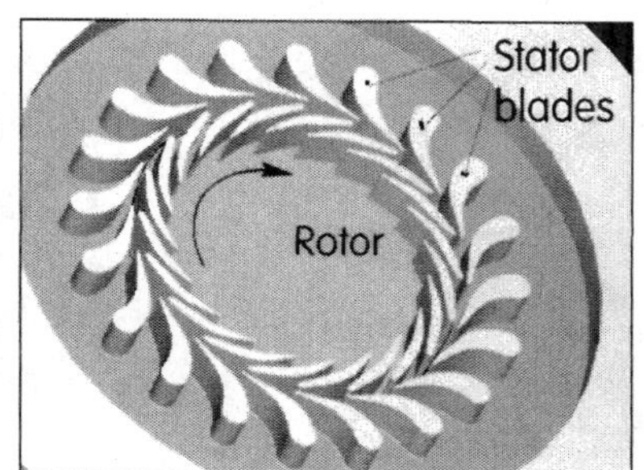

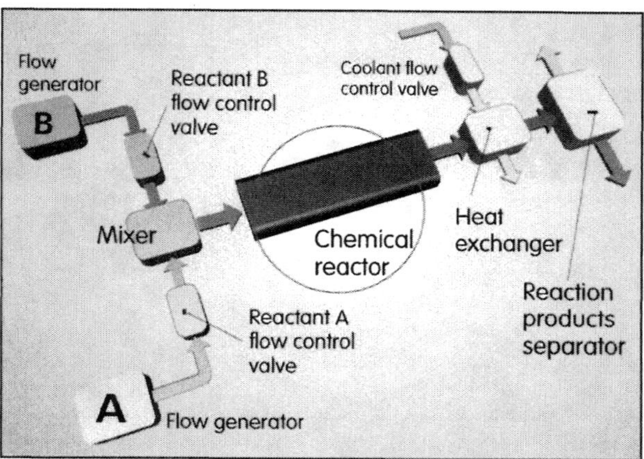

Figure 1.8 (Left) Even radial microturbines of millimeter overall size were made by etching. Such scaling down of large machines is rather exceptional in present-day microfluidics as the efficiency at Reynolds numbers is generally low.

Figure 1.9 (Right) An example of a more promising application: schematic representation of fluid transport used to perform a binary chemical reaction between two reactants A and B.

The idea is to use the extremely high density of energy storage in hydrogen as a fuel. The turbine driving an electric current generator should provide a more compact power package than energy storage in chemical batteries.

Fluid flows are also used to carry *signals* and even to process the information they carry. Again, despite several successful demonstrations (for example fluidic arithmetic units were demonstrated to perform mathematical operations even when heated to glowing red and hit by a hammer) this is not an important application in present-day fluidics. Electronic signal transfer and processing is generally more effective, mainly because fluid flows are slow and an almost universal requirement in information processing is fast operation. The limit to propagation speed in electronics is the speed of light, while with fluids the limit is the much lower velocity of sound.

Returning to Figure 1.7, it should be said that an increasingly often encountered reason for bringing a fluid into microsystems is *cooling*, carrying away generated heat. This is currently an area of intensive development activity. One of the main limits in further development of computer microprocessors is the increasing dissipated heat per surface unit. According to Moore's exponential law governing the progress in microelectronics, the thermal power density in the next generation of microprocessors will be comparable with the values existing in nuclear reactor cores. Increasing cooling effectiveness is indispensable for further miniaturization. As well, exothermic reactions in microchemistry (Figure 1.10) may require quite sophisticated cooling techniques.

Quite often, the reason for handling fluids in a mechanical MEMS device is *lubrication*: to prevent a friction contact between moving components (example: Figure 6.22). This need not be the traditional oiling of bearings; perhaps more important nowadays is data storage on a moving medium (such as a floppy disk) with creation of the supporting air layer between the medium and the read/write head past which it moves.

Fluids may be also simply consumed at the end of their flow path through the system: evaporated, discarded after use as tested sample, or used as food for captive biological organisms. In fact, the human food processing field has found many successful ways to use microfluidics, in particular, for producing an oil-water suspension, the basis for mayonnaise or salad dressings.

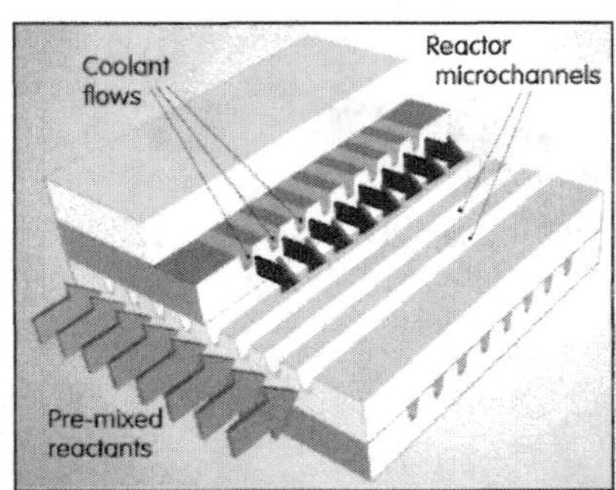

Figure 1.10 (Left) Chemical microreactors are basically just more or less simple channels - unless complicated by a requirement of large exposed surfaces coated by solid catalyst – so that they are usually arranged in large numbers in parallel. In the case of exothermic reactions, the reactor body may require intensive cooling.

Figure 1.11 (Right) An example of a typical contemporary microchemistry: a reformer for converting liquid fuels into hydrogen, needed for fuel cells. Devices developed from laboratory models like this are expected to cause a revolution in cars (© 2002 IMM - Institute for Microtechnology, Mainz, Germany, Printed with permission.)

One particular case belonging to this category, and of large and increasing importance, is the fluid consumed in a chemical conversion—burned as a fuel or undergoing more complex chemical reactions. Most industrial chemical reactions actually take place in the fluid phase. This is because the molecules of the reactants must be mobile so that they can meet and interact. The schematic block-diagram example in Figure 1.9 shows that performing even a simple binary chemical reaction may require a number of fluidic peripheral devices handling the flow of the reactants as well as of the reaction products. Analytical chemistry and biochemistry, where the small size makes it possible to perform a large number of simultaneous analyses, is a field where the advantages of microfluidics are particularly obvious. The microreactors are often made on a silicon chip together with electronic sensors, needed for example to monitor the reaction. The tiny amounts of samples needed for the analysis are an advantage in themselves, especially when samples are taken from a living organism. On the other hand, it is also advantageous in synthetic chemistry to perform the processes in microreactors, "numbered up" rather than "scaled up" to meet the requirement of a large production rate (Figure 1.11). The large surface-to-volume ratio makes possible much closer control of the reaction, the short residence time makes possible fast changes of the product composition, and there is less danger if the reaction or its product should explode. Obviously, fluidic no-moving-part technology has advantages in the supporting devices, such as sensors for sensing the reaction conditions, flow control valves, mixers, and separators.

Processing fluid samples is essential for one very important application of microfluidics: anti-terrorist warfare. Mass-produced—and therefore cheap—detectors taking molecular amounts of samples, for example, from the air and processing them to detect trace amounts of explosives, drugs, and other illicit substances may deprive terrorists of their present huge economic advantage: the crude bombs they use are cheap while current methods of detecting them are expensive.

1.1.2 Why devices without moving parts?

The devices mentioned above may be also made with moving components acting on the fluid. This is the standard way that fluid flow manipulating devices are designed at large scales. In a typical mixer, the pair of mixed fluids is contained in a vessel and agitated by a rotating impeller. There are pure fluidic mixer versions, accomplishing the mixing task by generating a flowfield in which the two fluids perform large-scale relative motions deforming the contact interface between them. Typically the two flowing fluids impinge on inclined or curved fixed plates protruding into the flowpath. The final mixing step takes place by diffusion across the interface. Obviously, such pure fluidic version is more suitable for scaling down and manufacturing at the microscale. To make, typically by etching, a free-moving component that can rotate inside an etched cavity and be driven from outside by a micromotor is possible (similar devices were actually demonstrated), but the process is far from simple and the resultant device is likely to be rather delicate. Making a small-scale version of the pure fluidic counterpart without the moving components is certainly much easier. The device is likely to be more robust, withstanding adverse operating conditions, and its operation life and reliability are generally much better. It may be discovered, however, that scaling down may lead to a different flow regime in which the mixing is far less effective.

Experience with other devices, such as the valves shown in Figures 1.12 and 1.14, is similar. The pure fluidic solutions are more suitable for being made at a small scale and offer operating advantages evident from comparing Figures 1.13 and 1.14 – provided the scaling down does not result in a very low Re flow regime with degraded efficiency. The term *fluidics* was actually originally coined just to describe the devices operating without moving components. It was introduced in the 1960s in an analogy to *electronics* and was initially meant to be applied to signal processing systems, where the fluid carried the signals just like electrons do in electronics. At about the same time, the progress in semiconductors in electrical engineering led to replacement of relays with mechanically moved contacts in industrial automation by no-moving-part switching and amplification of electric currents. The analogous switching of fluid flows by the no-moving-part devices seemed at that time to be similarly promising. The key element then was the no-moving-part fluidic signal amplifier, invented in 1959 by Bowles, Warren, and Horton at Ordnance Fuze Laboratories in Washington, D.C. The absence of mechanical components and their inertia made accessible unprecendently high operating frequencies, approaching the frequency band of early transistors. Serious attempts were undertaken to compete with electronics in information processing by building fluidic binary logical and computation circuits (binary fluidic flip-flop device, based on the Coanda jet-attachment effect, was described by Warren in 1962). They were demonstrated to withstand extreme conditions at a time when electronic devices were delicate and sensitive to variations in temperature.

In signal and information processing, however, fluidics lost and electronics won the competition. One of the principal reasons was the relative slowness of signal transfer by fluid flows by several orders of magnitude, which leads to inevitable slowness of the information processing. Another disadvantage was the size, larger than the size of comparable microelectronic integrated circuits because fluid flow channels needed larger cross-sections without abrupt area and direction changes.

With the demise of the applications in complex control systems, the interesting operating principles of no-moving-part fluidic devices were not forgotten. They found use in the "process fluidics" [8, 9] handling of fluids in applications where fluids have to be transported and processed anyway, so that there was no competition from electronics. The devices in which this branch of fluidics competed were the mechano-fluidic valves and other devices with moving components. Handling the reactant and reaction products flows in chemical plants were typical application examples. Pure fluidics without the bearings supporting the moving components and without the sealing of the mechanical input shafts is well suited to be made from, for example, refractory materials resisting high temperature and chemical aggressivity of the handled fluids. It is typical for the devices of process fluidics to handle large flowrates. That is why this field is also known as "power fluidics." In fact, many process fluidics devices were originally developed from those used as final output power stages of the earlier fluidic control systems. The advantages sought were the low cost and reliability [10]. The operation speed of pure fluidic devices was usually not the most important advantage, even though this was used in novel ways of modulations of the large flows, such as pulse-width modulation (PWM) in a valve operating in the regime of sustained oscillation. Despite some remarkable successes, process fluidics has not been adopted universally or without problems. To design the fixed geometry cavities in a device capable of generating a flowfield performing sometimes quite complex operations is by no means easy. A source of much frustration is then the fact that the device operation depends on the conditions existing in other parts of the fluidic system: a successful no-moving-part valve may fail to operate properly or at all in a different fluidic circuit in which it is connected to a mismatched load. Mechano-fluidic valves dictate the fluid flow paths without such limitations.

When microfluidics emerged, the spectacular achievements of mechanical MEMS and the newly developed capability to build using micromanufacturing techniques also the movable or deformable device components have initially led to a strong tendency to carry out also the handling of fluids by the mechanical actions. However, lessons learned with large-scale fluidics and the obvious advantages offered provided a challenging opportunity for using the no-moving-part pure fluidic principles also at the microscale.

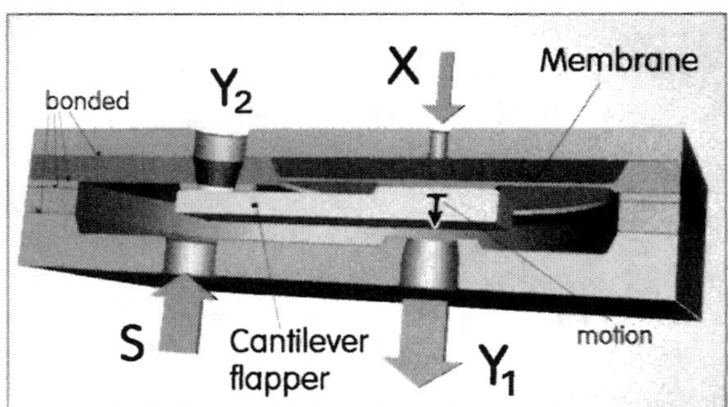

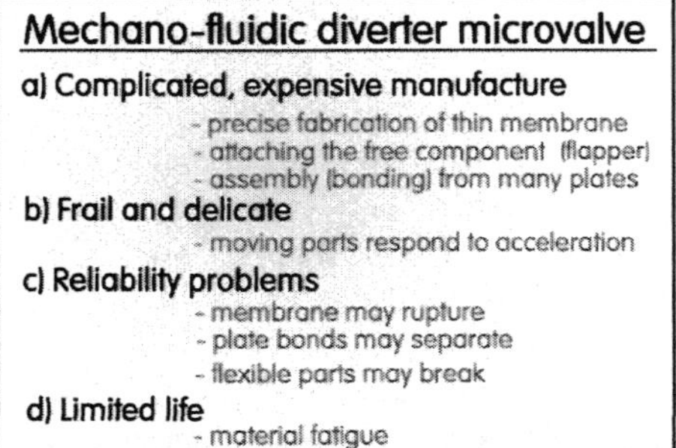

Figure 1.12 (Left) A typical example of a mechano-fluidic diverter flow control valve. Fluid supplied into S leaves either through Y1 or through Y2, depending on the position of the flapper that opens or closes the entrances into the two outputs. The flapper is attached to the thin membrane deflected by the control signal delivered into X.

Figure 1.13 (Right) Based on simple operating principles, the valves using an action of moving or deformed components have a number of disadvantages.

Weaknesses of typical mechano-fluidic devices and the advantages offered, on the other hand, by their purely fluidic counterparts are compared in the two examples of typical devices, flow control valves, in Figures 1.12 and 1.14. Both are intended to perform the same flow-diverting task controlled by the fluid-transmitted signal, brought to the input terminal X.

In the mechano-fluidic case shown in Figure 1.12, the mechanical motion is not the classical translation of a free mechanical component (perhaps supported by sliding bearings). Instead, the mechanical motion is made possible by deformation of one of the valve components (the membrane). The deformation moves the cantilever flapper bonded to membrane central boss. This closes the initially open output Y_1 and opens the initially closed Y_2. Obviously, this is not a fail-safe solution: the membrane may burst due to pressure overloading or material fatigue. One of the essential limitations of this design is that the only usable orientation of the membrane is with its plane parallel to the chip base plane. The membrane motion is perpendicular to this base plane and so is the associated opening and closing motion of the flapper. The opened and closed orifices facing the flapper, as well as the channels leading away from these orifices, can be reasonably arranged only in additional plates placed above and below. This necessitates a complicated assembly, with the valve consisting of a number of plates bonded above and below the base plate. Precise mutual positioning and bonding of the plates and of the flapper is an expensive operation. There is always a danger of the bonds becoming loose or perhaps not sealing perfectly. Other commonly encountered problems are listed in Figure 1.13.

The jet-type valve for the comparable operation in Figure 1.14 is based on the mechanisms of jet deflection and on the Coanda-effect attachment of the jet to the preferred attachment wall. The fluidic amplification principle is apparent: the large flow from S issuing as the jet from the main nozzle is diverted by the action of a much weaker control flow coming from X. The valve consists of a system of constant depth cavities and channels made by a single etching operation in the substrate plate. Of course, in addition there is a cover plate (not shown in Figure 1.14) bonded on top to close the cavities. This, however, is flat and needs no precise positioning. The result is manufacturing simplicity and robustness of the device, suitable for mass production.

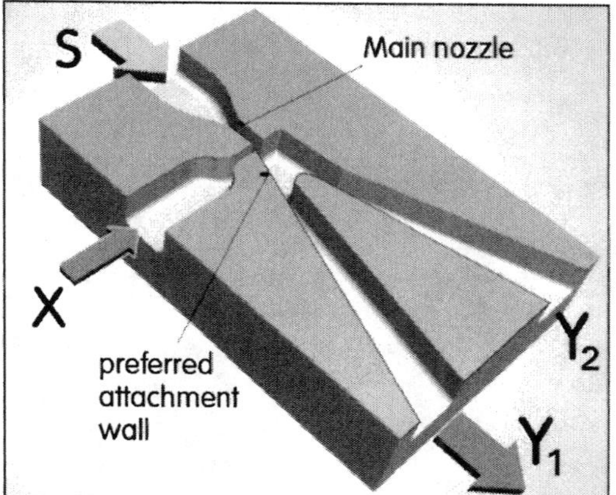

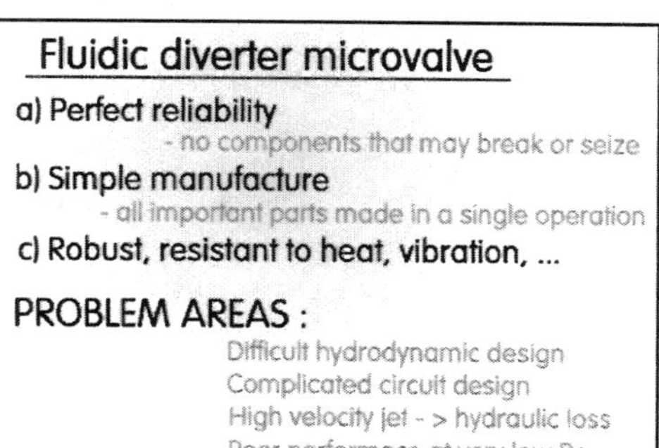

Figure 1.14 (Left) A typical example of a no-moving-part fluidic diverter flow control valve. The diverting action (fluid from S allowed to leave through either Y_1 or Y_2, depending on the control signal X) is the same as of the mechano-fluidic counterpart shown in Figure 1.12. Here, the operation is based on attachment of the jet leaving the main nozzle to the wall leading it into Y_1.

Figure 1.15 (Right) Advantages—but also problems—of the no-moving-part pure fluidic devices.

In all fairness, it should be said that similar to the problems encountered earlier in power fluidics, the no-moving-part solutions also bring potential disadvantages, which are listed at the bottom of Figure 1.15. The standard valves operating by mechanical motions have no such problems. Any mechano-fluidic valve of sufficient flow path cross-section and withstanding the operating pressure will divert the flow properly. Its complete closure of a particular flow path by the moving body means that the hydrodynamic properties of the connected other devices are irrelevant. This is in contrast to the difficult design of the pure fluidic devices. Not only do they have to be individually designed, but they must be matched to a particular fluid flow circuit, which may require changes to the other circuit components. Moreover, additional problems arise by the specific requirements of operation at the small scale.

1.1.3 Why the small size?

Common present-day microfluidic devices use flow channel widths in the region from 0.1 to 1 mm. Smaller widths, in the range of 10 to 100 μm, are likely to be used in the foreseeable future because smaller size accentuates some of the typical advantages listed in Figure 1.16. The advantages may be specific to a particular application, those listed here apply mainly to the use in microchemistry. The smaller devices may be often generally cheaper, especially if their design makes them amenable to the mass production methods used in microelectronics. Needing only tiny amounts of material makes it economically acceptable to use even very expensive materials; some current microfluidic devices use components made of, for example, gold (due to its excellent resistance to corrosion).

The advantages stemming from the possibility to perform a huge number of various reactions simultaneously in a huge number of reactors are obvious: more and varied performed analyses increase, for example, the probability of detecting another dangerous substance handled by a suspected terrorist. Also obvious is the decreased danger of handling explosives or chemically extremely aggressive substances: an accident at the milligram scale is easily contained and does not result in fatal consequences. Most importantly, chemistry is a field where a qualitative step in performance is obtainable by the decreased scale of the operation. The increased surface area per unit of the microreactor internal volume (Figure 1.17) means faster response to changes in control variables, increasing yield and selectivity of the reaction. In particular, improved controllability of chemical microreactors results from improved response to changes in temperature (which is the

MICROFLUIDICS :

- Better control of processes
  - high surface/volume ratio:
    fast response to control actions
    better selectivity, higher yield
- Less sample needed for analysis
- More analyses simultaneously
- Portability; Implantability
- Dangerous substances safely handled
- Approaching cellular size
  adressing individual cells
  in biomedical applications

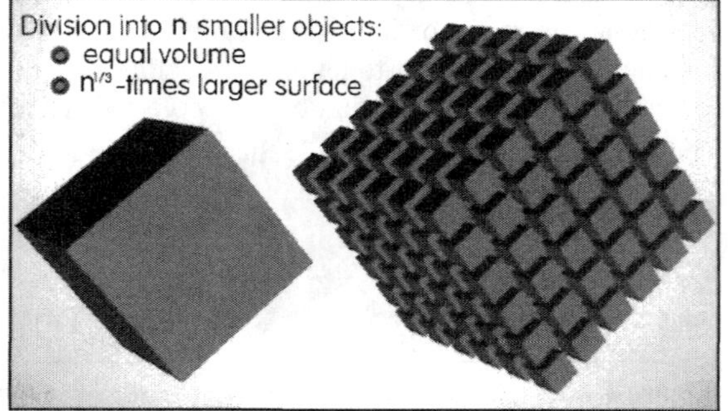

Figure 1.16 (Left) Several of many reasons why the small scale devices are advantageous.

Figure 1.17 (Right) The key factor behind several advantages of small size of microfluidic devices is their surface to volume ratio increasing in inverse proportion to the decreased linear size (easily demonstrated for cube shapes). This results in high effectiveness of transfer phenomena, such as heat transport across the outer surface of the device.

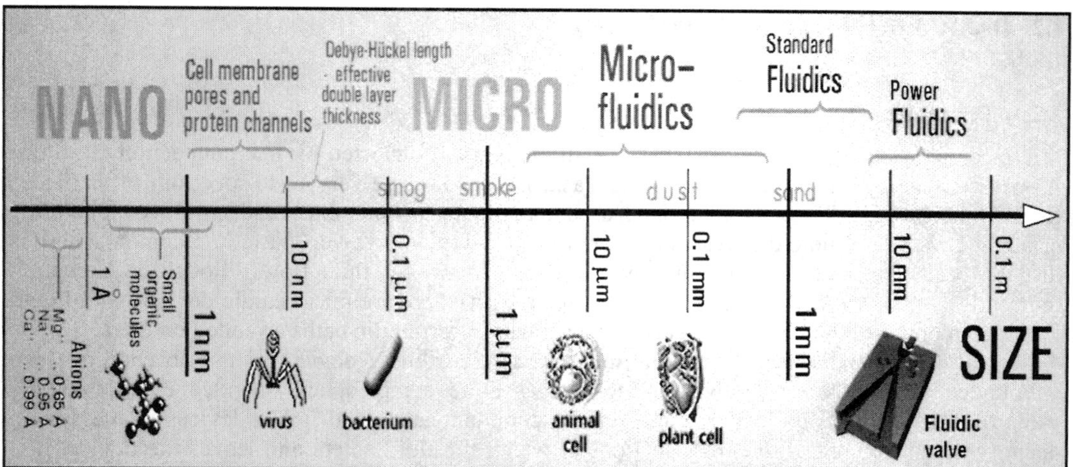

Figure 1.18 Microfluidic devices, characterized by the submillimeter size of the smallest transverse dimension of flow channels, are typically a decimal order of magnitude larger than typical animal body cells and four to five orders larger than the objects (molecules) handled in nanotechnology.

usual control variable in the control of chemical reactions). The faster response to thermal effects is, however, also of importance for some physical processes, such as diffusion. It makes some effects, hopelessly slow at a large scale, quite useful: for example turning down the flow in a channel by a component that swells in the presence of a substance contained in the fluid.

The smaller channel widths, on the other hand, may bring operational problems due to a seemingly trivial cause: their clogging. This is not a problem in the cases where the fluid is generated by a chemical reaction on the spot. Most other cases require filtering to remove solid (or semi-solid) particles carried with the fluid. This is unavoidable if the devices are to process sample fluid brought in from the outside world, for example. Filtering systems can protect the circuits located downstream from them, but may clog themselves.

Another problem, specific to pressure-driven devices, is that obtaining the desired flowrate may be found to necessitate very high driving pressure differences. These cause problems with leakages and even stresses in the device body.

Typical size range of microfluidics is shown in a wider context in Figure 1.18. It should be noted that the range of interest is quite far away from the realm of nanotechnology, with which it is sometimes mentioned together. The latter's aim is handling individual molecules (or complexes consisting of a countable number of molecules). At such molecular scales, the concept of fluid in its traditional sense—substance that can be modeled as continuum—may lose its sense. The fluid itself has to be treated as consisting of individual molecules. In microfluidics, the standard approach to fluid flow computations disregards the actual molecular structure, in common with the standard approach of general fluid mechanics. The fluid is treated as a continuum, with properties identical everywhere even if the investigated volume is decreased to an infinitesimally small size.

The fact that the range of dimensions handled in microfluidics, Figure 1.18, covers the scale of living cells is not unimportant. An interest in living objects and in the mechanisms they use (like transport of ions across semipermeable membranes through protein channels) is a distinctly recognizable trend in contemporary microfluidics. Indeed, a large number of perspective applications are currently seen in biology and medicine.

1.2 BASIC PROPERTIES OF DEVICES

1.2.1 Terminals

Microfluidic systems consist of devices, usually mutually connected by interconnection channels. A device is designed to fulfill a particular task. Because of the very wide spectrum of the tasks, fluid properties, device body materials, used manufacturing methods, and employed operating principles, various fluidic devices can differ widely and may have very little in common. There is just the basic fact that they contain cavities and channels for fluid flows, though an exception (currently not very important) exists even in this aspect: there are microfluidic devices with liquid flowing on open solid hydrophobic surfaces, following hydrophilic paths arranged on them.

A really universal characteristic of fluidic and microfluidic devices is the absence of open unbounded flowfields. The flowfields, sometimes quite complex, are bounded by surrounding walls (again with possible exceptions). Fluidic communication with other devices in the system occurs through terminals, the cross-sections at which the fluid enters and leaves the device. It is useful to characterize the properties of a particular fluidic device by describing the mutual interactions between the flows passing through the various device terminals.

Perhaps somewhat surprisingly, there are useful—and not uncommon—fluidic devices having just a single terminal. This is the case of an accumulating chamber, an example of which is shown

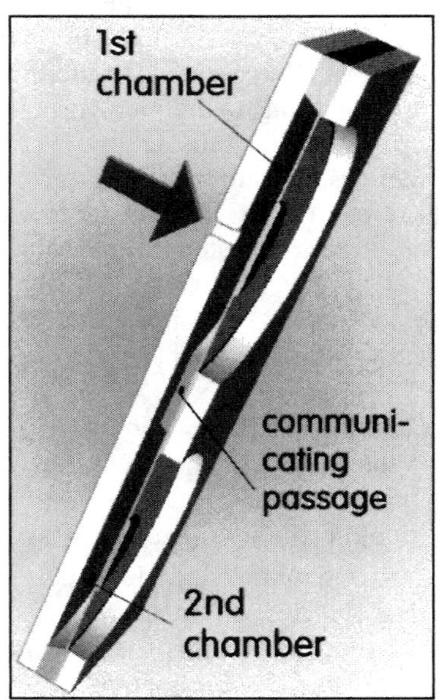

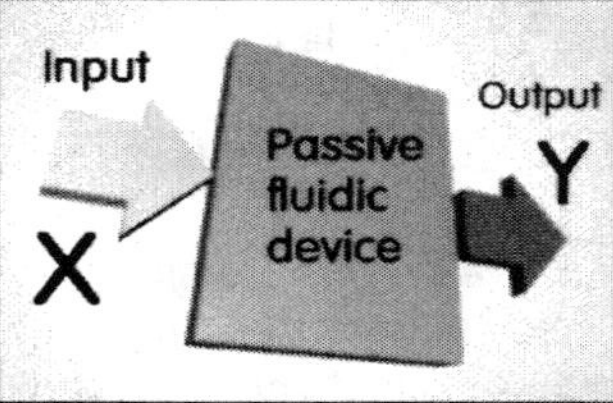

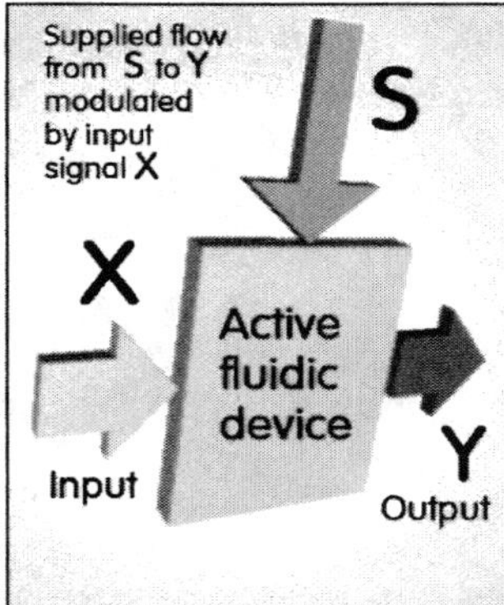

Figure 1.19 (Left) An example of a microfluidic device possessing only one terminal; a capacitance chamber for accumulating gas in a cavity. In this case, to increase the accumulation capability there are actually two mutually connected circular volumes. The gas compressibility is augmented by deformation of the thin membrane walls.

Figure 1.20 (Right, top) Schematic representation of a two-terminal, passive flow-through fluidic device with the upstream input inlet X and the output outlet terminal Y as the downstream exit.

Figure 1.21 (Right, bottom) Schematic representation of an active fluidic device. There are three terminals. The power available at the output Y may be higher than the weak control signal at X because of the powerful flow supplied through S, the progress of which into Y is controlled by the input signal.

in Figure 1.19. The fluid enters the device through the terminal to be accumulated inside. When it is later released, the fluid leaves through the same terminal.

Much more common are devices with two terminals, shown in Figure 1.20. This is the case of the most simple through-flow channels or cavities and also of microreactors operating with pre-mixed reactants. It is usually possible to call one of the terminals—the one through which the fluid enters—as an inlet or input, labeled X. The other, exit terminal is the output Y. Such devices are only exceptionally symmetric so that the role of the two terminals is mutually exchangeable. Devices are commonly designed for a particular orientation of the flow: a nozzle (such as the one in Figure 1.14) is designed to produce a fluid jet at its exit and would lose any meaningful purpose if the flow were reversed. Fluidics uses the alternating, direction-changing flows, in which this distinction is also not useful, much less often than electronic circuits.

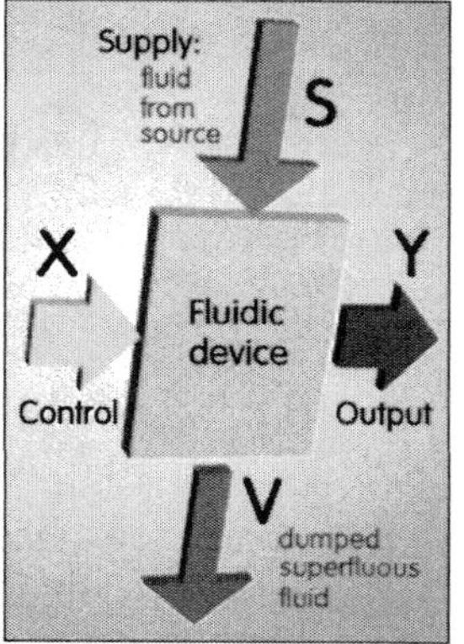

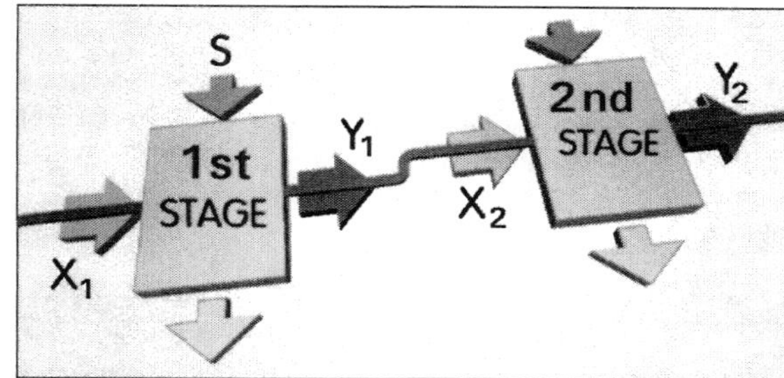

Figure 1.22 (Left) Schematic representation of a fluidic device possessing terminals of all four basic types: some fluid supplied from S is not progressing into the output Y and leaves through the vent V.

Figure 1.23 (Right) In a system with several devices, the fluid flow passes through so that the output Y_1 of the upstream device is connected to the input X_1 of the next device farther downstream.

A really interesting behavior is found in devices possessing more than two terminals. It is necessary to stress that the number of terminals in itself is not essential as long as they do not play a different role. It is possible, for example, to bifurcate the path in a simple restrictor so that the fluid leaves through two output terminals Y_1 and Y_2, without this increased number of terminals bringing a qualitative change to the wholly passive behavior of the restrictor. It is the addition of a terminal of a different type, the supply S (Figure 1.21) which makes the change. Its presence is essential for *active* character of the device. In a passive device (Figure 1.20), because of the energy conservation, the power available at device output Y cannot be higher than the value at the input X. This is different if the energy is supplied from some other available source. In pure fluidics, the basic case (not considering fluidic sensors and transducers, which handle energy in other forms, and also excluding chemical reactions), the supplied energy has the form of pressure energy or kinetic energy of the supplied fluid flow. The active device (Figure 1.22) capable of delivering a higher output power at Y than is the input power at X may be described as a fluidic amplifier. What takes place in the amplifier is the control action. This is brought in through the input X and acts on the flow from the supply S into the output Y so that the Y flow is varied. Generating the control effect is not easy and requires setting up a flowfield with some sensitive spot in which the weak input action can produce substantial changes in the character of the whole flowfield.

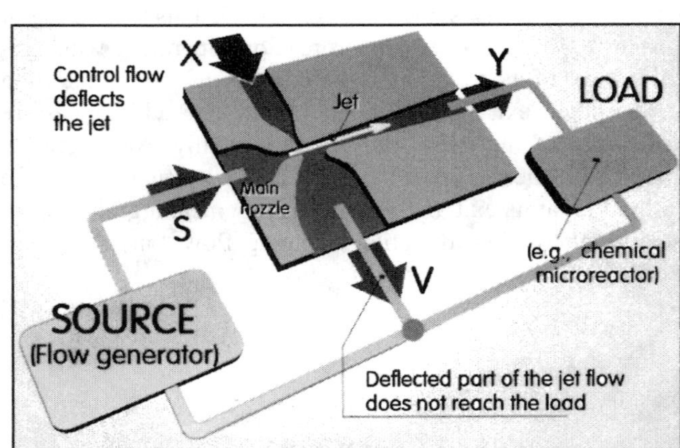

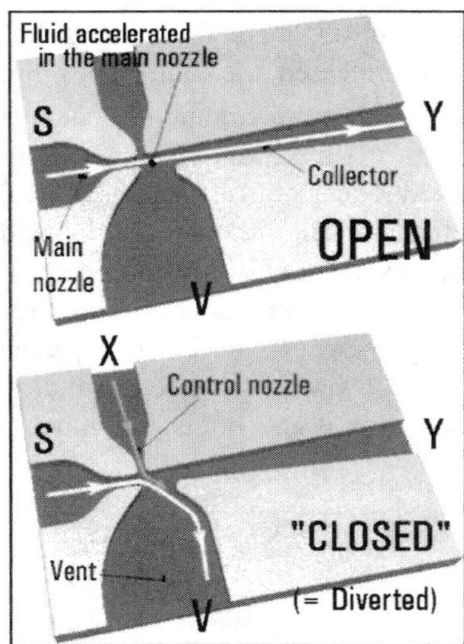

Figure 1.24 (Left) Jet-type diverter valve as a generic example of a device with all four terminal types (Figure 1.22). It is used to control, in the bypass mode, how much of the flow from the upstream source passes to the downstream load.
Figure 1.25 (Right) Mechanism of the jet-deflection diverting control: depending on the control flow intensity, a part of (or all) the supplied flow is prevented from getting into the output outlet and is forced to leave the valve through the vent.

The amplification property is nowadays hardly ever used for purposes analogous to those of an operation amplifier in electronics. Nevertheless, it is of importance in fluidic flow control valves. To control the operation of the device connected downstream to the valve (to the valve output terminal Y) requires modulating the output flow by the input action. As an example, the valve may be used to control the chemical reaction in a downstream positioned reactor by changing the flowrate of one of the reactants in response to the control flow signal. It is an advantage if the valve control does not require applying a large power action.

A typical feature of fluidic circuits with amplifiers is the often used connection into a cascade (Figure 1.23). It makes possible obtaining a higher amplification gain: the overall gain value of the cascade is the product of the gains of individual stages. The reason for using the cascade connection is much more general, however, and need not be guided by the amplification idea. It may be simply necessary to expose the processed fluid to various actions applied in series; the sequence of the mixer, reactor, heat exchanger, and separator in Figure 1.9 may provide a more typical reason for using such a throughput flow chain.

Typical pressure-driven flow control valves often use the sensitivity of a fluid *jet* produced by fluid flow from a nozzle. In the example of the jet-type valve shown in Figure 1.14, the jet leaving the main nozzle is directed to either one of the two output terminals Y_1 and Y_2. The simplest case, of course, is a valve with only one output terminal Y, as shown in another example in Figure 1.24. The control performed by this valve is of the bypass mode: the output flow passing through the controlled device downstream is decreased by the amount deflected away from Y by the action of the control signal. Obviously, there must be a way out from the valve for the fluid that does not reach the output terminal Y. The outlet used to dump this superfluous fluid is termed the vent V. As shown in the schematic representation in Figure 1.19, in this most general case there are in total four terminal types:

- Input X;
- Output Y;
- Supply S;
- Vent V.

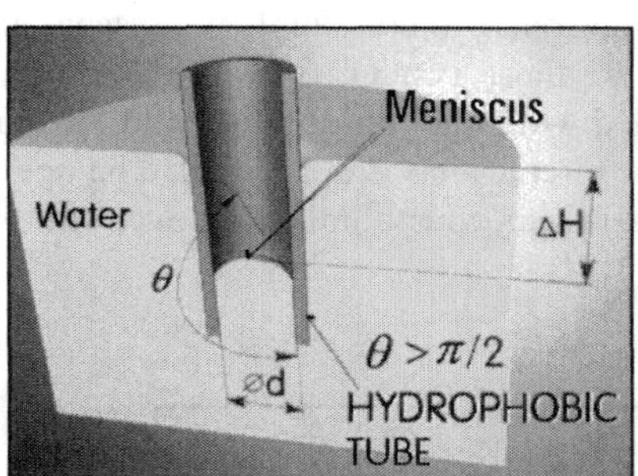

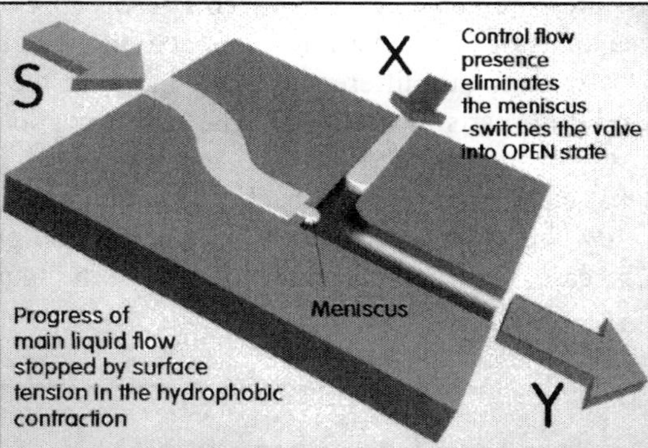

Figure 1.26 (Left) The surface tension mechanism of impeded liquid entry into a capillary tube with hydrophobic surface. The surface tension will resist the meniscus motion even if the fluid is acted upon by the pressure difference corresponding to the head ΔH.

Figure 1.27 (Right) The hydrophobic surface tension used in a valve controlling the propagation of a liquid into cavities originally filled with air. The meniscus in the area contraction resists the applied pressure difference between S and Y. This stopping action is eliminated when the control liquid inflow from X engulfs and removes the meniscus.

To present another example of a microfluidic valve operation, a valve based on a completely different operating principle than the jet-type valves of Figures 1.24 and 1.25, the liquid flow control valves based on surface tension effects may be mentioned. In a hydrophobic valve body, or at least a local hydrophobic surface in an area contraction, (Figure 1.26) the liquid encounters a sudden increase in the required driving pressure needed for the liquid to continue moving. According to Figure 1.27, this phenomenon is used for two-state flow control. The driving pressure applied in the supply terminal S is adjusted so that the liquid column motion stops in the contraction and cannot progress to the output Y. The valve switches into the OPEN state when a liquid column coming from the control terminal X removes the liquid/gas phase boundary downstream from the contraction. In this case, there is no vent V; the layout corresponds to the unvented three-terminal active device of Figure 1.21. Somewhat unusually, in contrast to the jet-type valves that may control the flow permanently, the valve of Figure 1.27 is designed to operate in a one-time way. It can control the gradual propagation of the liquid through the channels initially filled by a gas, but once the valve becomes OPEN, the subsequent changes in the control X have no effect. It is, of course, possible to repeat the opening effect if the valve is operated in a periodic manner (with retreat of the liquid columns into the initial state). This example documents the wide variety of possible operating principles that the pressure-driven microfluidics can offer.

1.2.2 Providing the driving pressure difference

In a typical fluidic system, presented in Figure 1.28, the paths through the devices from the supply terminals S to the vents V (and to the exits from the loads, as shown in Figure 1.24), form closed loops passing through the supply source. The source is usually a pump delivering the fluid at the required pressure. There is what seems to be an exception: in systems with air as the working fluid, the vents of the devices may be open into the atmosphere (as shown schematically in Figure 1.29). Of course, the difference is unimportant; the atmosphere there simply represents all the common return parts of the loops. The real exceptions are those systems operating in a one-time mode, which are supplied from a tank or vessel filled with the fluid. The fluid, perhaps a disposable one, is not returned back to the pump and the circuit loop is not closed. The operation stops when the tank is depleted. The tank may be pressurized, thus eliminating even the pump; this is the case of the external source in the list in Figure 1.31.

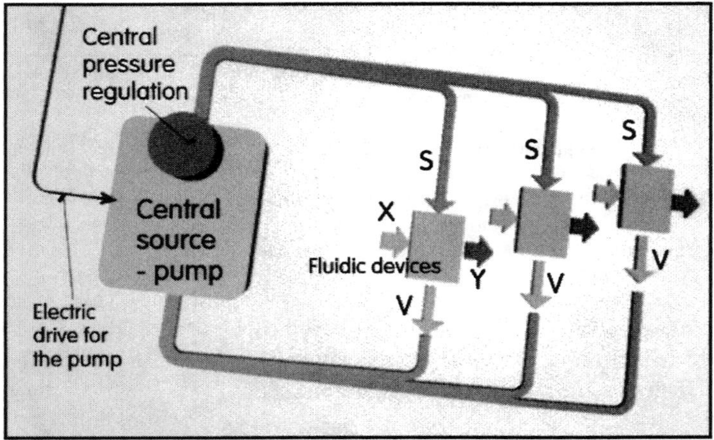

Figure 1.28 Supply fluid flowpaths from S to V through the active fluidic elements may be seen as perpendicular to the throughput paths from X to Y of Figure 1.23. The supply loops close through the supply source – in the most common layout shown here they pass through the common flow generator.

The delivered supply flow has to be regulated so that the devices can work under constant operating conditions as much as possible. The regulation task is typically fulfilled by a central regulator, usually a pressure-keeping device affecting all the supply flows at the source. The ideal of perfectly constant pressure (or constant flow in the case of flowrate regulator) is difficult to achieve, mainly due to practical spatial and power limitations. Some pressure variations are considered acceptable. They may be, however, found disappointing, especially in microfluidics, where the fluidic resistances of small devices are high and so that the acting pressure differences also have to be high. Even small changes of internal conditions during operation of some devices then have considerable consequences. The resultant mutual cross-influencing between the devices supplied from the same source is inconvenient. A solution may be sought in using many distributed regulators. In the extreme case the idea may be to govern the operating conditions of each device by a local regulator (Figures 1.29 and 1.30). This, of course, would complicate the whole system to a degree that makes this idea acceptable only in quite exceptional cases, perhaps

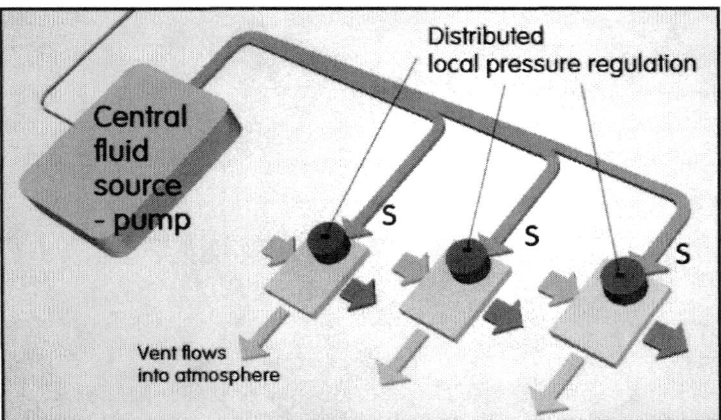

Figure 1.29 The extreme case of distributed pressure regulation. A regulator at each fluidic device would provide ideal operating conditions for the device. In this extreme form, however, the idea is usually impractical, not so much because of the required number of the regulator circuits (in the context of MEMS it is possible and relatively easy to make the large number of electronic controllers on the same chip), but because of the need to have at each location a dedicated additional pressure-control turn-down valve .

like that of the pressure-driven flows in some chemical microreactors. If the controlled device were just a flow control valve, the local regulator would be too much of a complication, unless, of course, it is an exceptionally important valve. This somewhat unusual case is shown in Figure 1.30. The pressure controller itself, operating in a feedback loop, is not a problem in the context of MEMS, especially if the electronic control circuit can be made on the same silicon chip. Neither is the sensor needed for detecting the pressure variations (Figure 1.1 demonstrated the ease with which there may be several pressure sensors made on the chip). The really questionable item is the turn-down valve operating as the actuator of the pressure controller. A compromise is sought in

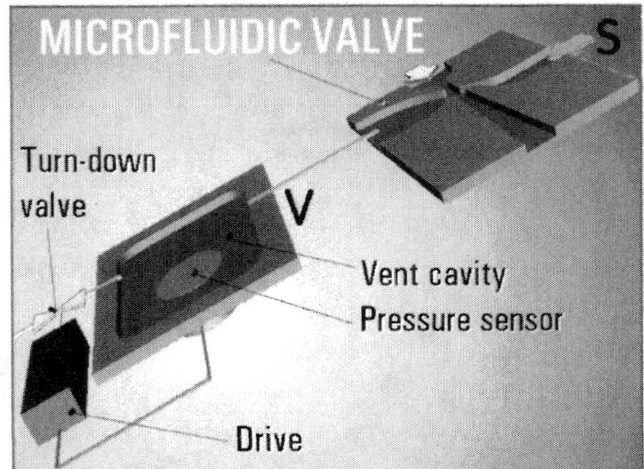

Figure 1.30 An example of a localized distributed pressure regulation. The driving pressure drop pushing the fluid through the valve at the right (or through another device; the valve shown here corresponds to the example from Figure 1.25) is kept constant. In contrast to Figure 1.29, the pressure regulation action is applied at the vent (and not supply) side of the device.

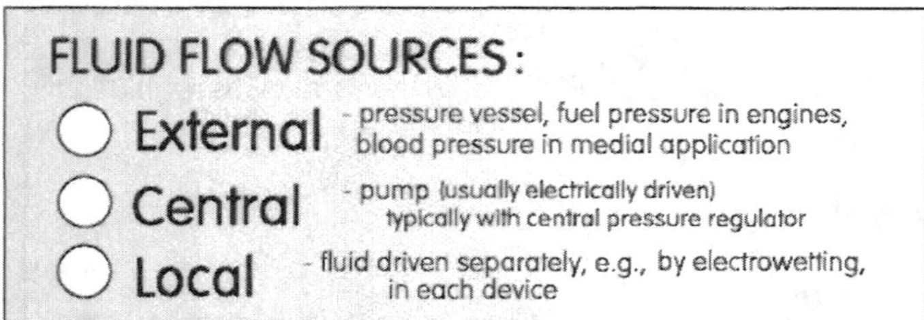

Figure 1.31 Alternative variants of arranging the supply source. Most often used are layouts with the central driving pump according to Figure 1.28.

controlling the supply pressure in a group of similar valves operating under similar conditions, as shown schematically in Figure 1.32. A practical example of an application of this approach is the sampling unit shown in Figure 1.6. Its task is delivering in a sequence a sample of reaction products into the common composition analyzer [11]. The unit consists of an array of identical no-moving-part valves. These operate at Reynolds numbers so low that the sample flows have to be driven by an applied pressure difference. In this case the pressure is relatively easily kept constant because of the prevailing nearly equal operating conditions in all the valves.

Rather than to regulate locally the pressure provided by a central supply pump, it may be found more convenient to generate the pressure locally at or inside each device. This decentralized approach is shown schematically in Figure 1.33. Again, there are no generally valid rules for applicability of any of these approaches and it is necessary to consider each case individually. This sometimes results in various compromise solutions, the local driving applied to only a group of some critical devices. A survey list of operating principles applicable to generate locally the flow in the devices is provided in Figure 1.34. Recent development in microfluidics is characterized by a considerable interest in unusual principles. Some of the principles are discussed in Chapter 5 on transducers and sensors. The direct driving by an electric voltage acting on charged particles in the fluid is perhaps the most straightforward method, but its use, as already discussed, is limited to a special sort of fluids, polar liquids with ions of dissolved salts generated by dissociation. As well, the methods dependent on surface tension can convert the driving electric power directly, due to

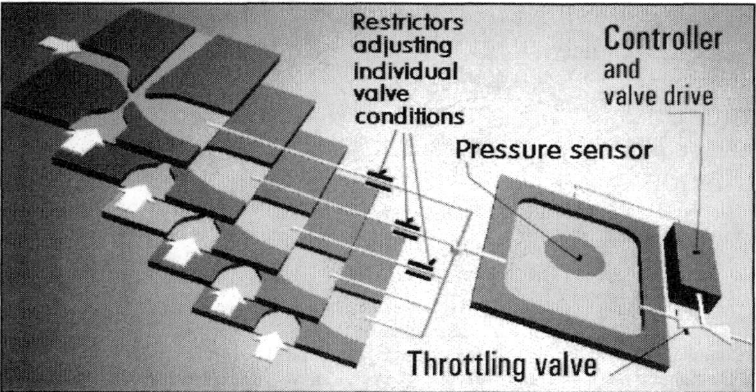

Figure 1.32 A compromise distributed pressure regulation applied to a group of similar valves operating under (nearly) the same conditions. The driving pressure levels in individual valves of the group may be adjusted by the restrictors.

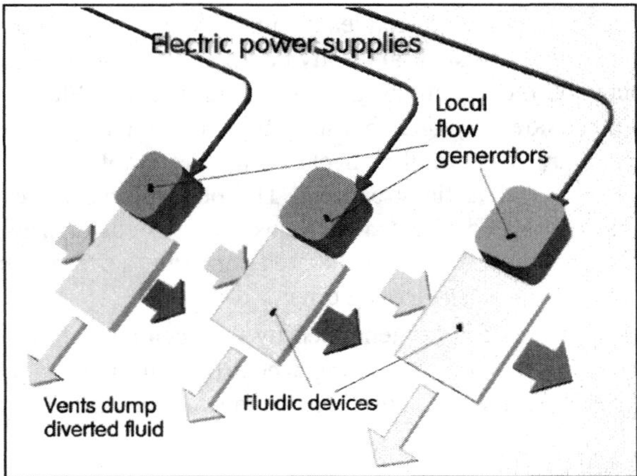

Figure 1.33 The third case from Figure 1.31: distributed pressure generation. Each device is provided with what in effect is its own small pump converting the supplied electric power into the required driving effect.

the electric field being able to change the surface tension. Again, this requires a somewhat special situation, two-phase flows with a liquid/gas interface.

The surface tension may be also varied by temperature changes (the Marangoni effect) and this has been also successfully tested as the driving effect in microfluidics. Converting the electric input into thermal effects is, in general, extremely easy and simple (the Joule heating) and this fact has ensured vivid interest also in other thermal actions on fluids, such as thermal expansion effects. The complication is these effects usually require operation in a periodic, reciprocating manner, requiring some sort of fluidic rectification. Of course, the favorable surface/volume ratio (see Figure 1.17) makes the thermal phenomena useful at operating frequencies which would be out of question at macroscale. Particularly large volume changes (Figure 1.42) are obtainable by

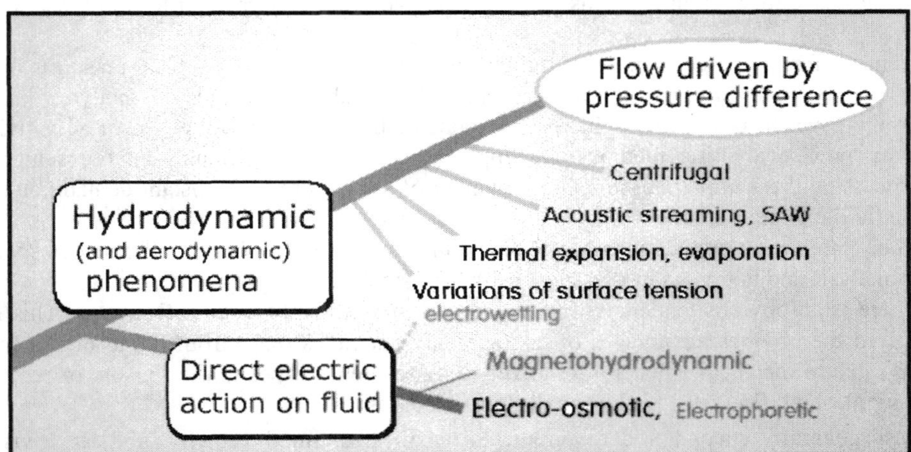

Figure 1.34 There are many ways to drive the fluid flow through a microfluidic device in the arrangement according to Figure 1.33. In this book, the interest is focused on the use of the more traditional hydrodynamic effects. A number of rather unusual principles has been investigated recently, for example, [1], mainly used as the central micropump supplying fluid into more complex multidevice circuits, but some principles may be applicable in a decentralized manner.

using phase change, evaporation and condensation. Fluidic rectification—the conversion of alternating flow into a flow with substantial steady component—may be realized quite simply due to the nonlinear character of the equations governing fluid flows. Flows in forward and return directions usually differ considerably and this may be sufficient for the rectification effect. The periodic driving effect may be simply mechanical, preferably at high frequencies. The phenomenon used is known as acoustic streaming. The periodic mechanical action often utilizes piezoelectric phenomena. A related effect, which has been tested for driving the flow through a microfluidic device, is traveling waves generated on the surfaces and also rectified by aerodynamic phenomena to provide the driving effect.

Finally, the driving pressure may be generated by the "centrifugal" forces acting on the fluid if the microfluidic circuit is placed on a rotating wheel or arm. This is actually a very effective method of pressure generation; the obvious disadvantage is the limited accessibility during the rotation.

1.3 FLOW CHARACTERIZATION PARAMETERS

1.3.1 Character of the flow and the Reynolds number Re

The initial approach to microfluidics was quite naturally characterized by attempts to scale down successful designs of the valves and other devices known and used earlier in the large-scale world. However, as the developing manufacturing techniques made possible progressive decreases in size, it soon became apparent that at the microscale some of the operating principles tend to be less effective or cease to work at all. This problem has been often aggravated by three characteristic factors:

- Demand for increased number of fluidic circuits on the chip. The typical reason is the trend to obtain more information: for example, the number of evaluated analytes in testing drinking water is known to increase exponentially with increasing sensor sensitivity. As more precise analysis methods are applied, using smaller amounts of the samples and processing them in parallel in a larger number of tests, there is a constant demand on making the fluidic circuits smaller.
- Many interesting applications require handling fluids (such as biological mucous specimens in analysis and hot gases in fuel synthesis) characterized by high viscosity.
- There is an obvious tendency of working with progressively smaller flowrates. This is both due to the obvious advantages of taking smaller samples (e.g., from living organisms) and also due to the requirements of relatively long residence times in the sensors or reactors. In general, small flowrates lead to small flow velocities.

These generally encountered requirements (small size, small velocity, and large viscosity) combine in microfluidic devices to result in generally small values of an important dimensionless parameter, the Reynolds number

$$\mathrm{Re} = \mathrm{w}\,\mathrm{b} / \nu \tag{1.1}$$

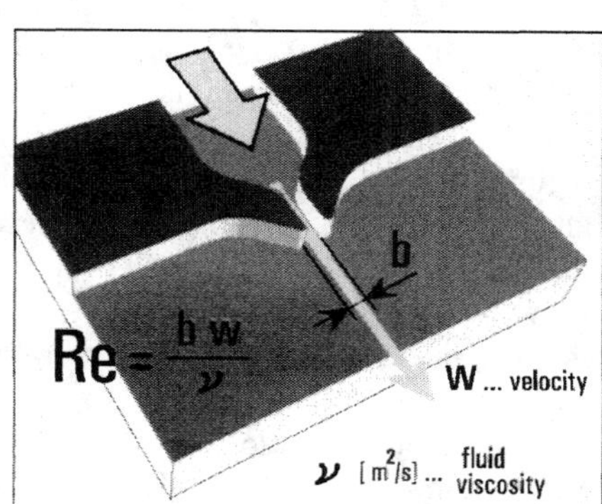

FLUID	SPECIFIC VOLUME v [m³/kg]	VISCOSITY ν [10⁻⁶m²/s]
air	0,83	15,13
H_2	11,93	104,96
O_2	0,75	15,22
N2	0,86	15,01
CO	0,86	15,12
CO_2	0,55	7,99
Ar	0,6	13,37
water	1.002 10^{-3}	1,01
Hg	0.073 10^{-3}	114,72
typical oil	1.148 10^{-3}	14,98
at T=20 °C P =101.3 kPa		

Figure 1.35 (Left) The standard definition of the Reynolds number in fluidics (and microfluidics); it is evaluated from the conditions in the main nozzle – the nozzle width b and the bulk (mean) exit velocity w.

Figure 1.36 (Right) Properties of some important fluids: specific volume and viscosity. Note that the values vary (sometimes significantly) with temperature and pressure.

where w [m/s] is the characteristic flow velocity, b [m] is characteristic dimension, and ν [m²/s] is kinematic viscosity of the fluid (Figure 1.36). As shown in Figures 1.35 and 1.39, it is customary to choose as the characteristic dimension of fluidic and microfluidic devices the width b of the main nozzle – the smallest cross section through which passes the permanent flow in the device.

Reynolds number value Re is of crucial importance for design of fluidic and microfluidic devices because it determines the character of the fluid flow. An important fact is that flows

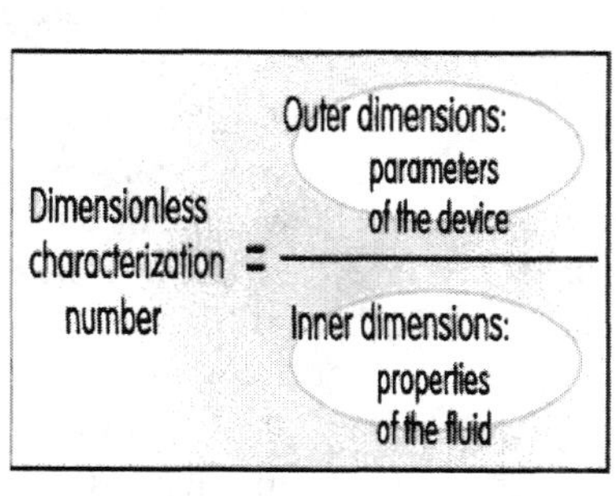

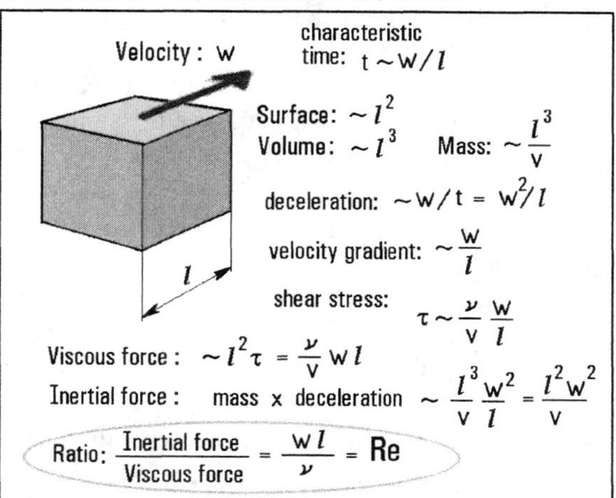

Figure 1.37 (Left) From several alternative interpretations of the meaning of characterization numbers Re, Ma, Kn , .. (compare with Figure 1.18), the meaning useful for study of flow regimes is the ratio of the outer characteristic variable—decided upon by the device designer—to the physical quantity, having the same dimensions, characterizing the fluid.

Figure 1.38 (Right) The meaning of the Reynolds number according to Figure 1.37; it is the ratio of inertial and friction forces acting on an element of fluid [11].

change their character at a critical value of Reynolds number. The critical value is not universal since the transition depends on the details of the geometry and also on the presence of disturbances. Very rough guidance is the transition taking place in straight circular cross-section pipes where the critical value (evaluated from the pipe diameter d) is $R_{ecrit} = 2\ 300$. The laminar flow at lower Re is characterized by well-organized concurrent streamlines. The turbulent flow above Re_{crit} is chaotic, with a whole spectrum of different size vortices present, usually pressure dissipating much more fluid energy. The two regimes differ substantially in the resultant pressure losses, heat and/or mass transfer intensity, and other important aspects. As already mentioned, turbulent pressure loss is generally higher, but turbulence is not wholly unwelcome. More intensive mixing caused by turbulent vortices is an obvious example, another is entrainment into a jet and the consequent jet-pumping effect, much more intensive if outer fluid is dragged along by the vortices from the jet.

The Reynolds number may be demonstrated to represent the ratio of dynamic forces acting on fluid particles to the viscous damping forces, as shown in Figure 1.38. As Re values decrease, the friction effects become relatively more important. Viscous friction usually adversely influences and finally inhibits the behavior of fluidic elements. At low Re, the effect may become overwhelming to the degree of stopping the oscillatory processes in a fluidic oscillator, for example.

1.3.2 Scaling down and Re

The easy part of scaling down is the task of maintaining similar geometry. If it is desired to utilize the working principle successful in a large-scale fluidic device, the scaled-down microdevice is required to possess identical value of the ratio k (Figure 1.39) of all linear dimensions. This may be not without problems, since the different fabrication technique used for manufacturing the microdevices may not for example permit the same aspect ratio $\lambda = h/b$ (relative depth h of the cavities).

The next consideration in scaling down is maintaining hydrodynamic similarity of large and small flows. Both have to be in the same hydrodynamic regime characterized by the Reynolds number. An attempt to scale down a successful fluidic valve developed for operation in the turbulent regime is bound to fail if the smaller Re of the microfluidic version will be subcritical so that the flow is laminar. The hydrodynamic similarity of the scaled-down device is secured if its Re value is the same as in the original large device, in other words, if the ratio r in Figure 1.39 is $r = 1$.

In standard large-scale fluid mechanics, at large Reynolds numbers the influence of friction effects is generally small. A scaled-down version of a device may therefore operate reasonably well even if its Reynolds number is not exactly the same. As long as the flow remains turbulent in both original and scaled-down devices, it may be acceptable to have the ratio r considerably different from the theoretically correct $r = 1$. In the laminar regime, there is less freedom. Even there, however, some Re mismatch may be accepted without grave consequences.

Where the requirement of the theoretically correct $r = 1$ results in problems is the required higher flow velocities in a small device. With the same working fluid (so that the viscosity ν is identical), the characteristic velocity in the microdevice (Figure 1.39) should be

$$w_{micro} = \Lambda\, w \tag{1.2}$$

that is, increased in proportion to the length scale Λ.

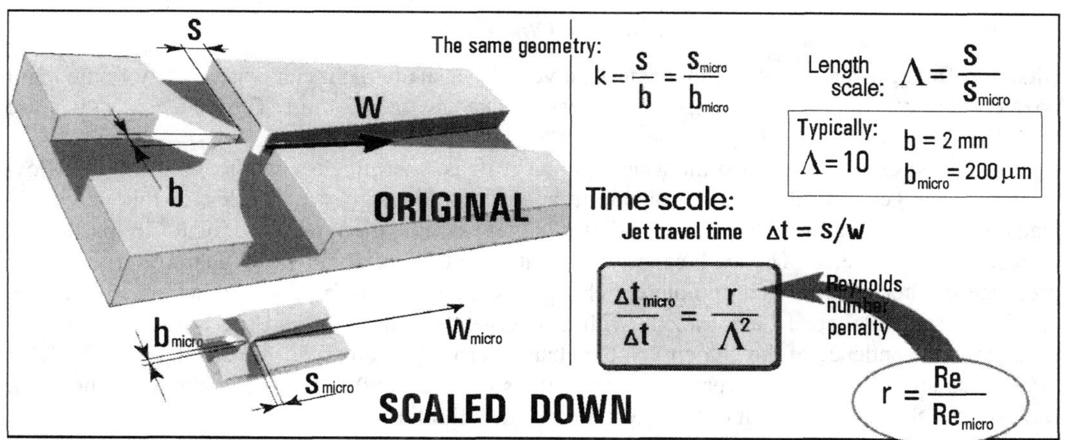

Figure 1.39 Scaling down from fluidics to microfluidics. The small size of the microdevices means faster operation. However, maintaining the same Re to retain an identical flow regime may result in impossibly high flow velocities [12].

The Reynolds number similarity is often put to use in developing a microfluidic device in an aerodynamic laboratory by using a scaled-up model, investigated more conveniently. The larger-size device is easier to build. Moreover, as stressed in the right-hand part of Figure 1.39, scaling up results in very slow flows in the model. This is useful for studies of unsteady phenomena, which in the scaled-up model take longer in proportion to Λ^2 (the square of the linear scale). The slower processes are more amenable to measurements and, for example, video recording.

Another transition to be considered, less widely known and perhaps less dramatic than the laminar/turbulent transition, takes place at extremely small Reynolds number values of the order 1.0. The regime entered below this critical value is the fully viscosity dominated *subdynamic regime*, where inertial forces become very weak in relation to the viscous friction forces. The flows there become self-similar and do not change with a further decrease of Re. The flows are of creeping character. Fortunately, this transition takes place at lower Reynolds numbers than encountered in most present-day microfluidics. Nevertheless, there is a gradual decrease of the dynamic effect relative intensity as this limit is approached. Already this may be of serious consequences, because the basic principles of the large-scale no-moving-part fluidics are critically dependent on inertia. In the absence of mechanical closure of the unwanted flow path, large-scale fluidics prevents the fluid from entering it by using the inertia of fluid accelerated into a different direction. Approaching the subdynamic regime may suffice in making this principle inapplicable. There are ways of circumventing this limitation (e.g., by using a direct localized pressure driving), nevertheless a careful consideration of this effect is necessary, especially as future trends are unequivocally moving toward smaller Reynolds numbers.

1.3.3 Compressibility and the Mach number Ma

Apart from the equal Reynolds number values, there are other requirements for perfect hydrodynamic similarity. One of them, in principle, is the theoretical requirement of the same values also for Mach and Knudsen numbers. The Mach number Ma characterizes the effect of compressibility in steady regimes (Figure 1.40). It is defined as

$$Ma = w / w_a \tag{1.3}$$

where w [m/s], as before, is the characteristic velocity, usually the mean exit velocity in the main nozzle, and w_a is propagation speed of pressure changes in the fluid. Typical w_a values are shown for several fluids in Figure 1.44.

As in the case of Re, also the Mach number values determine transitions into different flow regimes. As a general rule, it is reasonable to avoid high Ma values in fluidics, as this inevitably leads to large energetic losses and makes the device operation uneconomical. In particular, increasing the value to $Ma > 1$ causes a transition into supersonic flow, characterized by the presence of shock waves that completely change the character of the flow. Reaching this regime requires a special shaped Laval nozzle with a divergent downstream end (because in supersonic regime the dependence of the velocity on the channel cross-section area is reversed, the increase of w requires area decrease, the very opposite to the subsonic flow). The hypersonic and other high Ma regimes are practically out of the question for microfluidics.

The Mach number remains low and its influence may be neglected if the propagation velocity w_a is high, which happens if fluid specific volume as well as deformation stiffness, the measure of which is coefficient K (Figures 1.41 and 1.43), are both high. K is the proportionality factor in local linear approximation to the dependence of pressure rise on the relative decrease of the volume that the pressure causes. Its magnitude may be estimated from the local slope of the surface shown in Figure 1.42. The slope is steep for liquids and as a result, their propagation velocity w_a is typically very high, Figure 1.44. In a vapor or gas the compressibility is higher and this means much lower w_a. The dependence is complicated by the influence of temperature, bringing an additional degree of freedom. The standard derivation of the speed of sound assumes the changes of gas state to be so fast that there is no time for heat transfer between the gas and the surrounding walls. The change is then expected to follow the adiabatic law (Figure 1.43). However, the large surface/volume ratio (compare to the argument of Figure 1.17) typical for microfluidics can cause rather rapid thermal equilibrium, in which case the change follows the isothermal law, represented in Figure 1.42 by the sections T = const through the shown surface.

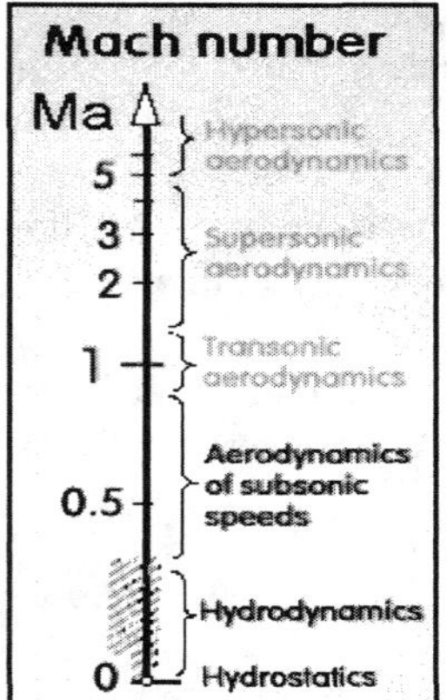

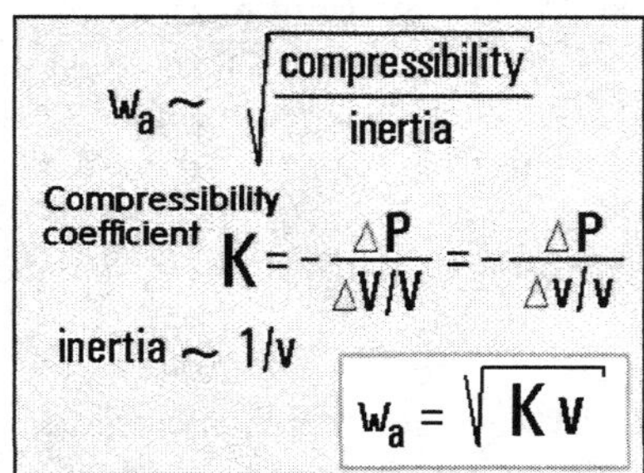

Figure 1.40 (Left) Flow regimes at different Mach numbers: the values of the ratio of flow speed to the acoustic propagation velocity w_a. Especially at low velocities the influence of Ma is weak and a device may be scaled without considering the Mach number similarity. However, the change in Ma must not result in getting into a different regime.

Figure 1.41 (Right) The "speed of sound" w_a —the velocity of pressure changes propagation in unbounded space— becomes low and can cause problems if the fluid has low compressibility characterized by the coefficient K.

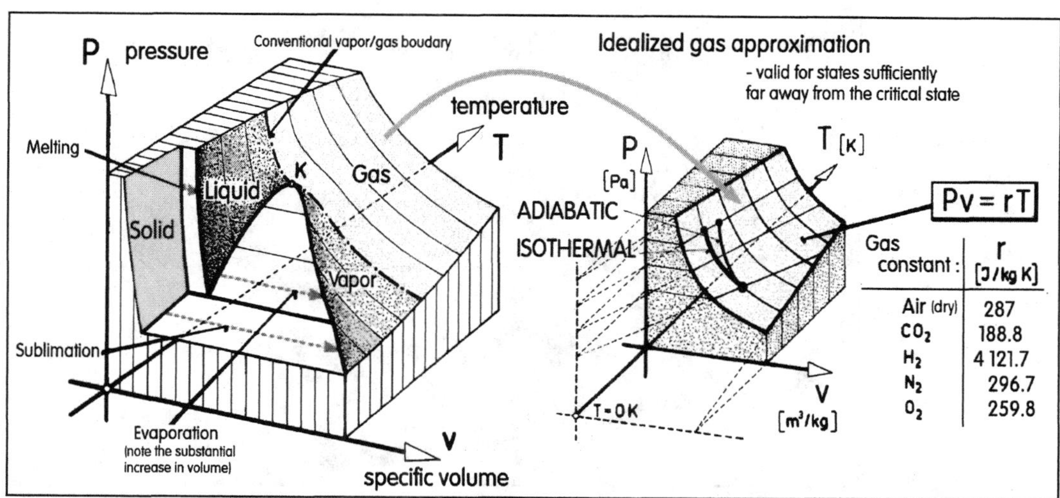

Figure 1.42 Fluid compressibility —variations of specific volume v due to the acting pressure—is complicated by the dependence on temperature T. The diagram at left represents typical shapes of the surface of equilibrium states including phase transitions. On the right-hand side is the usual simplified shape used for the description of the gas behavior.

The use of two-phase flow also may bring unexpected compressibility effects, especially in liquid (of high inertia, low v) containing bubbles (and hence exhibiting high, practically isothermal compressibility). Values w_a there are much lower than in gas (often as low as $w_a < 90$ m/s).

Mach number values may become a serious problem on the scaling-down way to microfluidics if the factor Λ (Figure 1.39) is chosen too large. As an exaggerated example, if a fluidic valve with b = 2 mm wide main nozzle and exit velocity w = 100 m/s were scaled down Λ=100–times to the microfluidic device with b_{micro} = 20 μm, the condition r = 1 would require, (1.2), the exit velocity Λ=100–times higher, inaccessibly high hypersonic w_{micro} = 10 000 m/s.

In gas governed by $Pv = rT$ undergoing fast
(and therefore adiabatic)
change of state

$$Pv^{æ} = \text{const}$$

$$K = æP$$

$$w_a = \sqrt{æ r T}$$

Gases	w_a [m/s]
Air at 20°C	343
He at 0°C	972
H_2 at 0°C	1 286
Liquids at 25°C	
Glycerol	1 904
Water	1 493
Mercury	1 450
Kerosene	1 324
Methyl alcohol	1 143
Carbon tetrachloride	926

Figure 1.43 (Left) The compressibility factor K for gas may be derived for the idealized approximation (in the right-hand side of Figure 1.42) using an assumption about the character of the state change. The standard approach is to use the adiabatic (= zero heat transfer) relation.

Figure 1.44 (Right) The acoustic propagation velocity w_a values for several most important fluids.

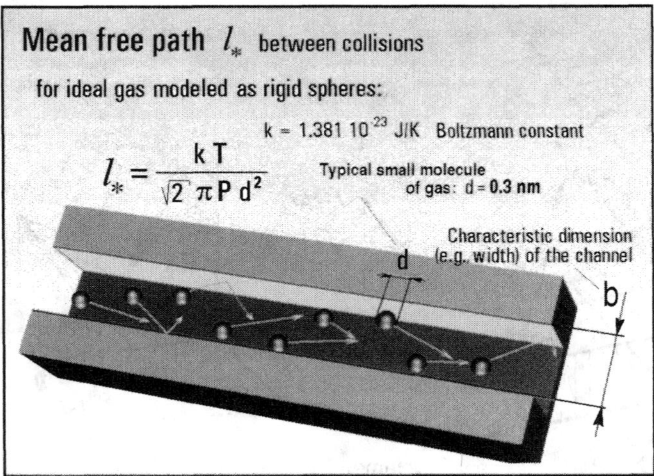

Figure 1.45 Fluids actually consist of individual molecules. This character is commonly neglected (so that fluid is handled as a continuum) as long as the size of the fluidic device is much larger than mean free path l_* of thermal motion of molecules.

1.3.4 Relation to molecular scale: Knudsen number Kn

Yet another dimensionless parameter which—in principle, at least—should be considered to ensure full hydrodynamic similarity is the Knudsen number Kn. It relates the size of the device internal cavities to the characteristic scale of the motion of fluid molecules (Figure 1.45). The definition in Figure 1.46 should be used with caution; some authors use an inverse definition violating the principle of dimensionless parameter formation according to Figure 1.37.

Maintaining simultaneously identical values of all three parameters Re, Ma, and Kn when scaling down a fluidic device is obviously impossible. Fortunately again, the effect of Kn is usually negligible, the exception being rarefied gases or extremely small devices. Most devices, even the smallest in current microfluidics, operate in the continuum regime (Figure 1.25), a pre-requisite for using standard CFD software for numerical computations of the internal flowfields.

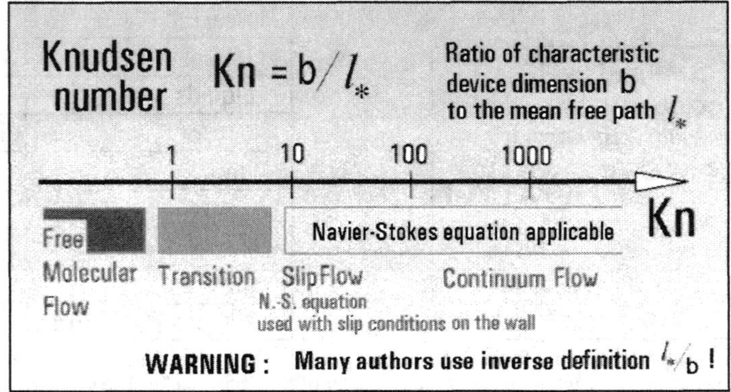

Figure 1.46 The Knudsen number Kn is the essential similarity parameter in rarefied flows in extremely small device sizes or at a low pressure P (as seen from the expression for the mean free path in Figure 1.45).

To conclude, of the various limiting factors to be considered in the scaling down, the most important is the Reynolds number Re. The general trend is toward operating the microfluidic devices at low Reynolds numbers, lower than in their large-scale counterparts. Since the dynamic forces used in the pure fluidics become less effective at lower Re in relation to friction forces (Figure 1.38), the device design is more difficult.

1.3.5 Periodic unsteady flows: Stokes and Strouhal numbers

Special similarity parameters become necessary when the device operates in reciprocating, periodically repeated regime. The basic similarity parameter, analogous to the Reynolds number of steady flows, (1.1), is the Stokes number

$$Sk = f\, b^2 / \nu \tag{1.4}$$

where f [Hz] is the repetition frequency, and other quantities are the same as in (1.1). Also the interpretation of the meaning is analogous, again the relative influence of inertial and viscous friction forces. The Stokes number is the essential parameter for characterization of mechano-fluidic reciprocating pumps. Figure 1.47 presents an example of the deviations from adiabatic propagation velocity of oscillatory motions, also dependent on Sk.

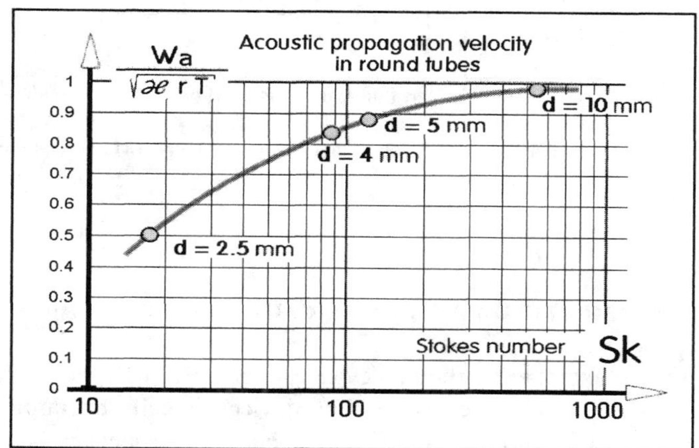

Figure 1.47 An example showing the importance of the Stokes number Sk, the similarity parameter of periodic unsteady flows. In this case it determines the acoustic propagation velocity w_a, which decreases with decreasing channel diameter d due to more intensive heat transfer effects. These effects make the adiabatic assumption in the derivation of the propagation velocity in Figure 1.45 untenable (experimental results from [13]).

If the oscillation is generated by hydrodynamic instabilities in a steady flow passing through the main nozzle of the device by the mean velocity w, the relation between Sk and Re is determined by the relation between w and f b. The ratio of these two quantities is the Strouhal number

$$Sh = f\, b / w = Sk / Re \tag{1.5}$$

This is usually constant across a wide range and, as a result, the Strouhal number is typically an invariant of many pure fluidic oscillation generators (i.e., with no mechanical moving parts). A typical example of such a self-excited fluidic oscillator is shown in Figure 1.48. Its dependence of the Strouhal number on the Reynolds number of supplied steady flow is presented in Figure 1.49. Of course, the Sh is constant only for a constant oscillator geometry, in this example a constant feedback loop length. In Figure 1.49 there are results for 7 different loop lengths *l*, in fact representing seven different oscillators. Their generated oscillation frequency, and therefore Sh, is inversely proportional to the length *l*.

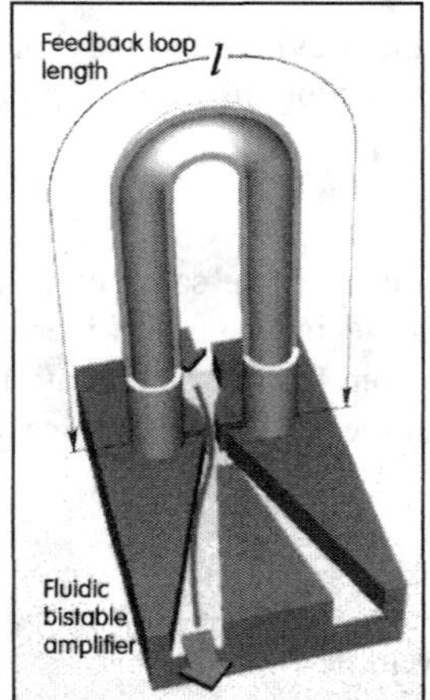

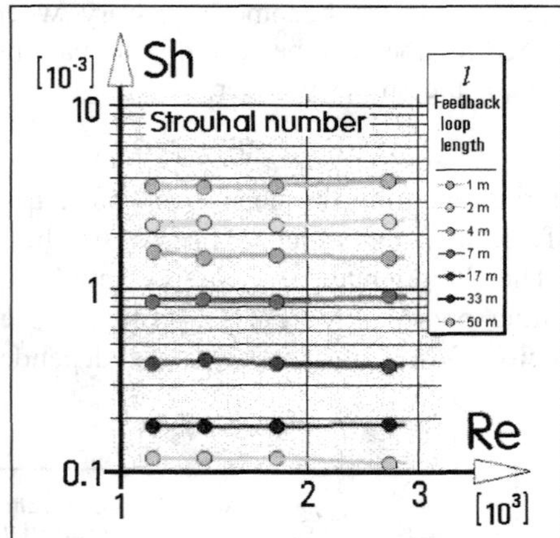

Figure 1.48 (Left) An example of a no-moving-part fluidic oscillator [14] obtained by providing a jet-type diverter amplifier with the feedback loop.

Figure 1.49 (Right) Constant Strouhal number Sh values found experimentally [14] for the family of oscillators of the type shown in Figure 1.48.

1.4 REGIONS OF OPERATING PARAMETERS IN MICROFLUIDICS

Although it is generally true that microfluidic devices are characterized by low Reynolds numbers, this is not valid universally. There are microfluidic devices actually operating in turbulent flow regime. Of course, the wide spectrum of fluidic device types makes general statements inaccurate. Nevertheless, the small size as one of the important factors in the Reynolds number definition is typical. The interest in turbulent regime stems on one hand from the effectiveness of mixing (such as the premixing of the reactants in microchemistry). On the other hand, this is the regime where it is possible to use the scaled-down versions of large-scale fluidics. To get beyond the laminar/turbulent transition limit, it is necessary to work with high flow velocities. This regime is found in the upper right-hand corner of the shaded field in Figure 1.50, which presents the various limits delineating the operational regimes of steady-flow microfluidics. The limiting lines presented in Figure 1.50 were evaluated for a particular case of air at standard laboratory conditions. Nevertheless, this may give an idea of quite general validity about the operating regime limits under which microfluidic devices work with a gas. Present-day microfluidics tend to operate under the conditions corresponding to the upper right corner of the diagram. The general development trend is shifting the operating regimes toward the lower left-hand side. Evidently, the problem with obtaining turbulent flows at the small, submillimeter size (characterized in Figure

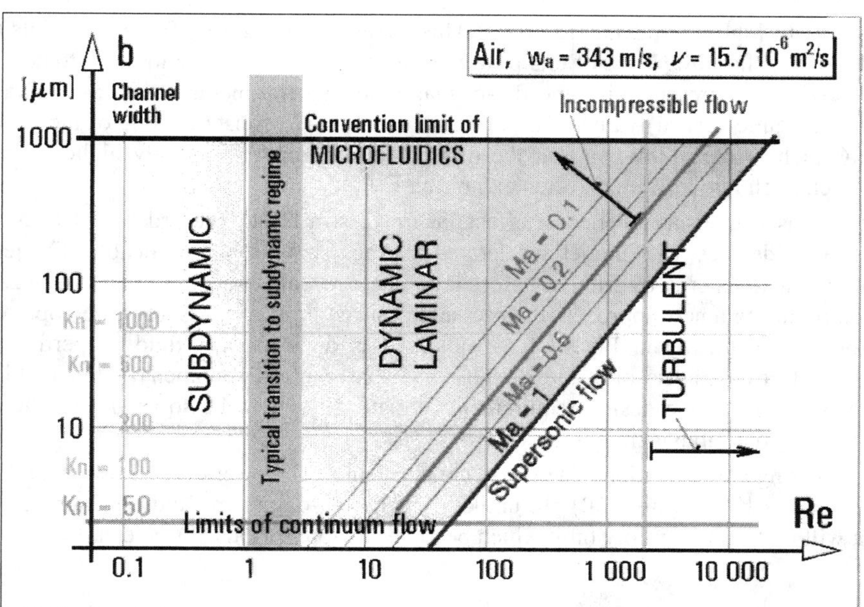

Figure 1.50 Range of Reynolds numbers Re and the nozzle widths b in which microfluidic devices usually operate (darker shade indicates more common conditions). The numerical values presented were computed for air at standard laboratory conditions.

1.50 by the main nozzle width b) is that it results in high Mach numbers. High velocity also means high necessary driving pressure values, which may easily lead to mechanical problems (leakages and delamination of the cover plates). The real limiting factor there, however, is the deteriorated

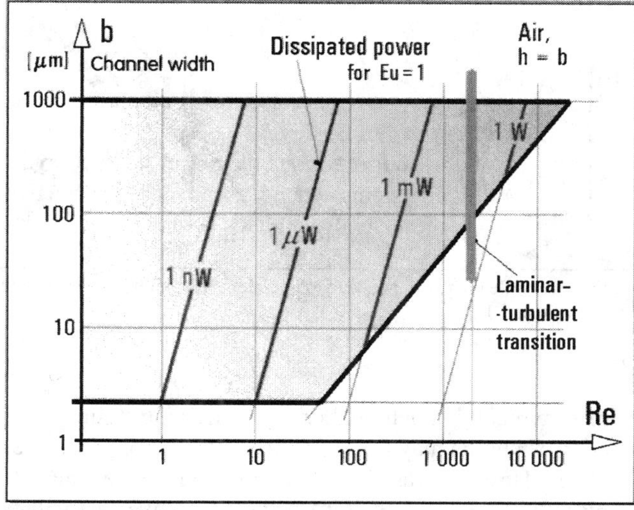

Figure 1.51 Another decisive factor limiting the applications of microfluidics is the power dissipated inside the device—plotted here again for air in the same coordinates as used in Figure 1.50.

economy due to high hydrodynamic losses. This is shown in Figure 1.51, which plots the power required to drive the flow. Most of this power is dissipated by hydraulic losses into heat inside the device. The diagram presents just an order-of-magnitude general picture as it was evaluated for the special (albeit quite typical) value of aspect ratio $\lambda = 1$ and a constant value of the Euler number $Eu = 1$ (which is rather low, so that the dissipated heat may be actually higher in more lossy devices, such as fluidic restrictors, often exhibiting $Eu > 1$).

The deceased intensity of inertial effects at lower still Re of the order of 10 can make the idea of jet-type devices, with the jet accelerated in a nozzle, no more applicable. The jet loses too much of its kinetic energy by friction before it reaches the collector opposite to the nozzle. This is a situation requiring a new approach, the pressure-assisted or even pressure-driven operation of the type shown schematically in Figure 1.29, unless it is decided (and fluid properties permit) to employ the electro-osmosis or some of the somewhat exotic phenomena listed in Figure 1.34. Only exceptionally do the design requirements dictate the size to be so small that the conditions are approaching the subdynamic limit.

Encountering any limitations due to the noncontinuum character of the flow (note the levels of Knudsen number Kn; Figure 1.50) are unlikely in present-day microfluidics. Even in the laminar regime it will be compressibility limit which becomes a more serious obstacle earlier.

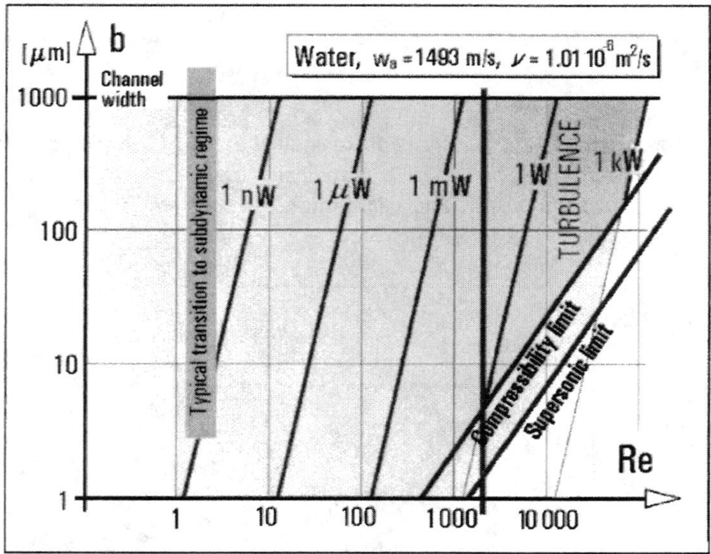

Figure 1.52 Diagram of the accessible range of operating parameters computed for water as the working fluid – otherwise analogous to Figures 1.50 and 1.51.

In liquids, with their typically much higher acoustic propagation speeds w_a (compare to Figure 1.44) the Mach number limits may be less restrictive. The opportunities for operating in the turbulent regime are wider. However (as shown for the example evaluated for water in Figure 1.52), at the typical microfluidic sizes $b < 1$ mm the restrictive factor there may be again the relatively high dissipated power.

References

[1] Nguyen, N.-T., and S. Wereley, *Fundamentals and Applications of Microfluidics*, Norwood, MA: Artech House, 2nd edition 2006.

[2] Lee, S. Y. K., M. Wong, and Y. Zohar, "Gas Flow in Microchannels with Bends," *Journal of Micromechanics and Microengineering*, Vol. 11, 2001, pp. 635-644.

[3] Hsu, T.-R., *MEMS and Microsystems: Design and Manufacture,* Boston: McGraw-Hill, 2002.

[4] Erickson, D., and D. Li, "Review, Integrated Microfluidic Devices," *Analytica Chimica Acta,* Vol. 507, 2004, p.11.

[5] Tesař, V., et al., "Subdynamic Asymptotic Behavior of Microfluidic Valves, " Journal of Microelectromechanical Systems, Vol. 14, April 2005, pp. 335–347.

[6] Burke, T., ed., *DNA Fingerprinting: Approaches and Applications*. Papers from the First International Symposium on DNA Fingerprinting, Bern, Switzerland, October 1990, Boston, MA: Birkhauser Verlag, 1991.

[7] Ehrfeld, W., ed., *Microreaction Technology: Industrial Prospects*, Berlin: Springer-Verlag, 2000.

[8] Tesař V.: "Großmaßstäbliche fluidische Ventile für die Durchflußsteuerung" (in German), *Messen-steuern-regeln*, Berlin, Germany, Vol. 26, No. 4, 1983.

[9] Tesař, V. "Valvole fluidiche senza parti mobili", (in Italian), *Oleodinamica–Pneumatica*, Torino, Italy, Vol. 39, No. 3, March 1998, p. 216.

[10] Tesař, V., "Fluidic Valves for Variable-Configuration Gas Treatment, " *Chemical Eng. Research and Design, Trans. Inst. of Chem. Eng*, Part A, 83(A9), p. 1, September 2005.

[11] Tesař, V., et al., "Development of a Microfluidic Unit for Sequencing Fluid Samples for Composition Analysis," *Chemical Engineering Research and Design*, Vol. 82 (A6), p. 708-718, Transactions of the Institution of Chemical Engineers, U.K., Part A, June 2004.

[12] Tesař, V., "Microfluidic Valves for Flow Control at Low Reynolds Numbers," *Journal of Visualisation*, Vol. 4, No. 1, 2001, pp. 51–60.

[13] Beskok, A., et al., "Rarefaction and Compressibility Effects in Gas Microflows, " *Trans. ASME, Journal of Fluids Eng.*, Vol. 118, No 3, p. 448.

[14] Tesař, V., C.-H. Hung, and W. Zimmerman, "No-Moving-Part Hybrid-Synthetic Jet Actuator, " *Sensors and Actuators A (Physical),* Vol. 125, Issue 2, January 2006, p. 159.

Chapter 2

Basics of Driving Fluid by Pressure

Of all possibilities on how to move fluids in devices (as listed in Figure 1.34), this book concentrates on the driving by an action of a pressure difference, applied between an upstream and a downstream terminal. This is a classical fluid driving method, compared with the principles recently introduced into microfluidics [1], which employ phenomena such as electrokinetic or capillarity forces. The pressure action uses the fact that an integral of the pressure over a given area is the mechanical force pushing the fluid. Using this principle can capitalize on previous developments in hydraulic and pneumatic systems. Device designers easily acquire the intuitive guidance or "feel" for the employed effect. The reason why there is a search for new principles is due to a practical problem: the increasing magnitude of required pressure differences due to the general trend of fluidic systems being designed progressively smaller and exhibiting therefore a higher hydraulic resistance. With the very small cross sections to pass through, the necessary fluid entrance pressure may reach inconvenient levels, perhaps even endangering the mechanical integrity of the device.

2.1 PRESSURE AND VELOCITY

As a matter of fact, the essential underlying relation that governs the necessary magnitude of the acting pressure does not involve the channel cross section. The high pressure levels are due to a secondary effect, the hydraulic losses that cause resistance to the flow. At large scales, the losses are of subordinate importance, sometimes to the degree of being commonly neglected in the first approximations to a device design. It is the increasing role of viscous forces at the decreasing Reynolds numbers, discussed in Chapter 1, which makes the losses more important.

Pressure is defined as the mechanical force per unit area of a surface (often imaginary) chosen inside the fluid-occupied volume. In fluid at rest these elementary forces act in directions normal to the surface. It is useful to decompose the force action into components parallel to the directions of the coordinate axes. All the three components are equal; this is the Pascal law. Pressure magnitude is measured by manometers, as described for example in [2].

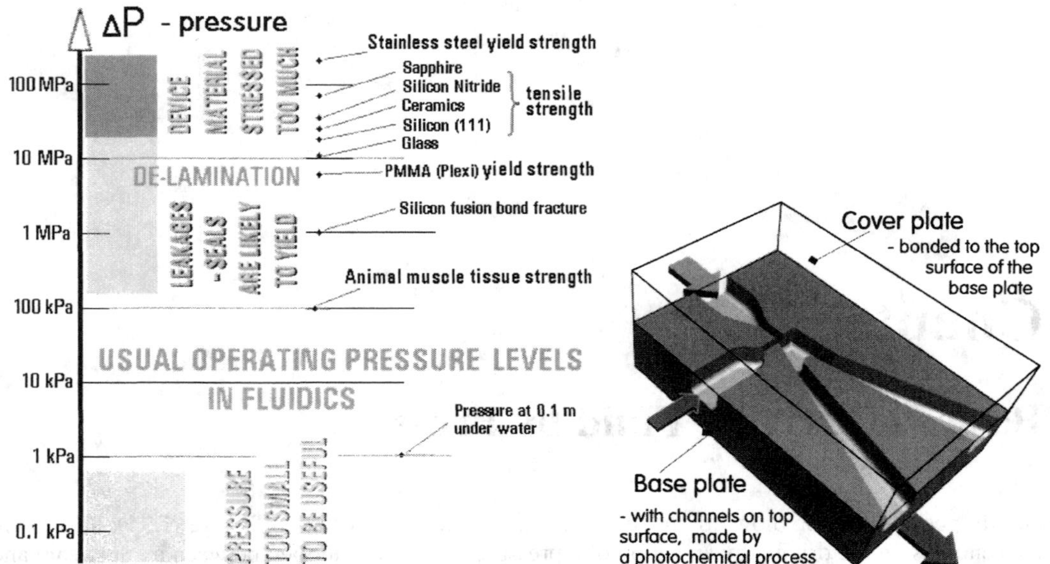

Figure 2.1 (Left) Pressure scale: the usual operating pressure range in microfluidics is between 1 kPa and 100 kPa. Lower pressure levels do not produce reasonably fast fluid motion (Figure 2.6). On the other hand, pressure levels above ~ 0.5 MPa stress the device bodies to levels causing leakages and increasing the danger of cover plate delaminations.

Figure 2.2 (Right) Characteristic build-up of a present-day microfluidic device. The layout is planar: the cavities for fluid flow are made (usually by a chemical micromachining) in a flat base plate and then covered by bonded flat cover plate. Making a single device as shown here is unusual – more typical is applying the integrated circuit idea with devices etched simultaneously and covered by a common cover plate.

Pressure propagates in the fluid by finite speed and its changes do not occur everywhere at the same instant. The propagation speed is limited by the speed of sound (Figure 1.41). In channels of a fluidic system, pressure actually propagates at a somewhat slower speed than shown in Figure 1.44, which is valid for an unbounded space. Nevertheless, even in the channels where it is influenced by thermal interaction with the walls (Figure 1.47), the propagation is fast enough for the Mach numbers to be in the deep subsonic regime, perhaps with the (very unusual) exception of supersonic flows in the limited regions downstream from specially shaped Laval nozzles. Pressure is actually transmitted in the fluid by intermolecular forces and/or (especially in a gas) by momentum exchange taking place in molecular collisions. This nano-scale viewpoint considering the molecular structure is actually rarely, if at all, adopted in contemporary microfluidics, because molecular interactions are very complicated and difficult to compute. More common is the simplifying continuum assumption. Even so, computing the total force by integrating the pressure distributions over a surface of interest may be not an easy task, especially in flowing fluids where in addition to the normal component there is also a tangential (shear) friction force essentially due to viscosity (in a turbulent flow, modified by momentum exchange of turbulent eddies).

Fluidic devices, especially the pure fluidic no-moving-part ones, are very reliable and failure-free. For making use of this property, it is imperative not to stress the devices too much with internal pressure. Devices must be operated at pressure levels substantially lower than the nominal strength of the body material (Figure 2.1) and in particular below the usually lower strength, of the order of 1 MPa, of the bonding between the individual plates of the usual multilayered

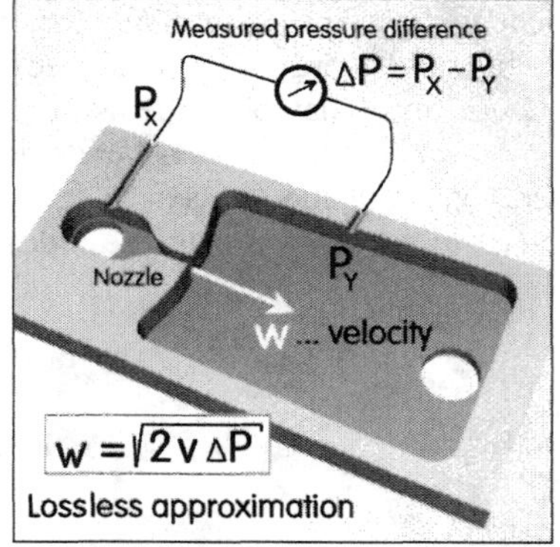

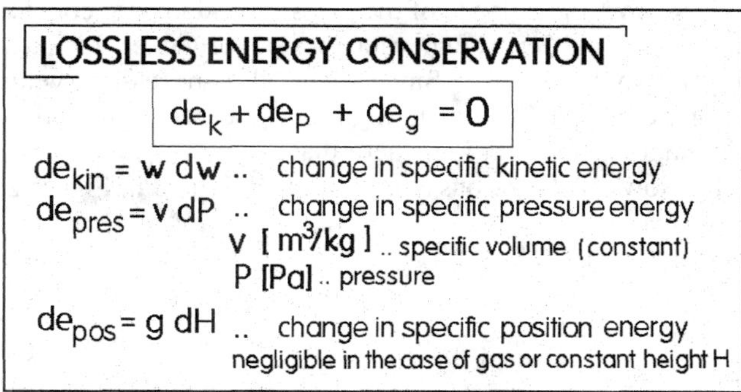

Figure 2.3 (Left) The required pressure difference for driving the fluid in a fluidic device is essentially (neglecting the dissipative effects) given by the required velocity of fluid flow.

Figure 2.4 (Right) The usual starting point in calculations of fluid flow: the Bernoulli equation in the differential form, here neglecting the loss term.

construction (Figure 2.2). Devices and circuits are typically built from at least two layers (note the number of bonds shown in Figure 1.12). Even if the cavities are distributed in a single plane, they are usually closed on top by the bonded cover plate. The bonding between the plates rarely keeps them together equally well over the whole theoretical contact area. There are weaker spots. It is not enough for the pressure to not remove the cover plate; even local delaminations are detrimental, causing a nuisance of leakages or contaminations between neighboring cavities or channels.

Approximately, neglecting dissipative losses, the pressure may be evaluated from the required velocity w of fluid motion using the simple relation presented in Figure 2.3. The velocity is usually determined by the requirements of a system's response to input changes and by such

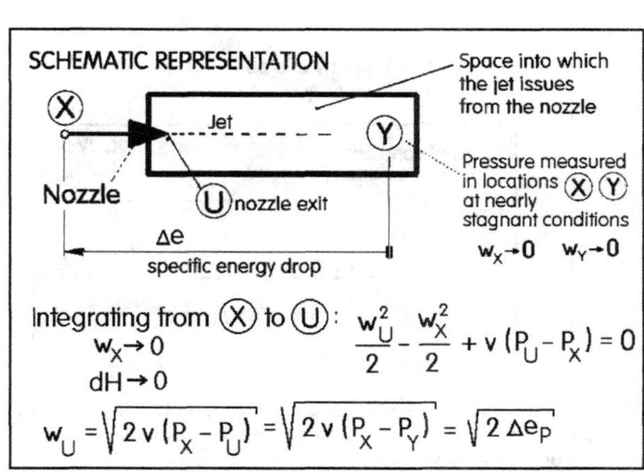

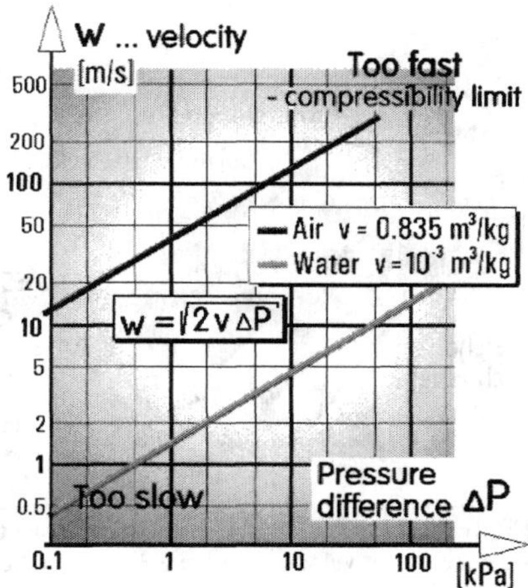

Figure 2.5 (Left) The equation presented in Figure 2.4 is integrated along the flowpath to derive the velocity–pressure relation as plotted in Figure 2.6.

Figure 2.6 (Right) Diagram for estimating the magnitudes of pressure and velocity. Velocities are usually much lower when working with liquids, where the usual reasonably chosen flow speeds are within ~ 1 – 10 m/s, while gas flow speeds are usually chosen in the range of 10 – 100 m/s.

aspects as residence time (of importance for example in microchemistry). The driving pressure is derived from the balance of the pressure and kinetic components of fluid energy as represented in Figures 2.4 and 2.5. Rough estimates of the values for liquid and gas flows may be made using the diagram in Figure 2.6. Since higher-than-necessary velocity inevitably deteriorates the device energetic budget (it leads to high loss), velocities should not be chosen outside the central part of this diagram. It is obvious there that the corresponding necessary pressure differences are indeed well below the dangerous stress levels presented in Figure 2.1.

2.2 FLOWRATE AND CHANNEL CROSS SECTIONS

2.2.1 Integral state parameter

The velocity, discussed in the previous section, is a parameter providing a useful idea about the fluid motion in the device, but it is not a suitable quantity for characterization of the state in a particular location of a fluidic circuit. This characterization is an essential basic step in device evaluation and design and also in solutions of circuits. What is actually needed for this purpose is (1) an integral quantity, and (2) a quantity obeying a conservation law. The location in the circuit (e.g., in an input terminal of a device) in physical terms actually is a cross section of a fluid flow channel, such as the entrance channel leading into the device. The integral character means that in contrast to the velocity the value is not defined locally at a point but obtained by integrating the local values over the whole channel cross section area. This leads to the flow rates shown in

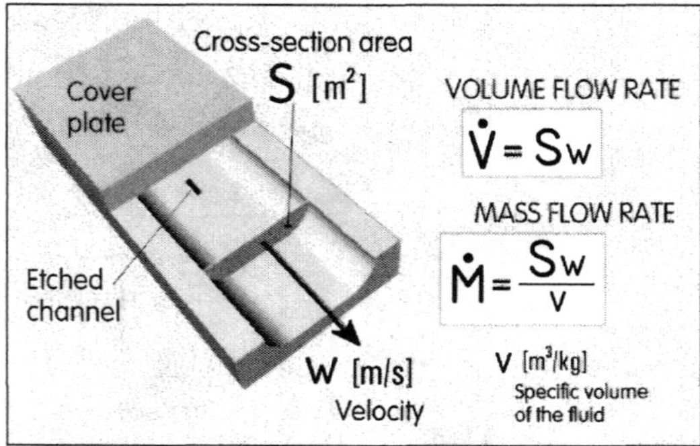

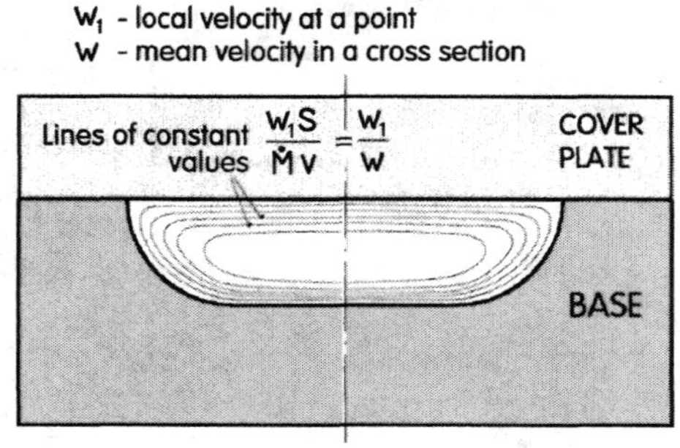

Figure 2.7 (Left) The state parameter used to specify a state at a location in a fluidic circuit (especially in a terminal of a device) is not velocity (Figure 2.3), but the flow rate - either volume or mass flow rate—which provides integral information. The simple expressions given here are valid for a one-dimensional approach, assuming constant velocity across the whole of the cross section.

Figure 2.8 (Right) Local velocity w_1 defined at a location like a device exit is in fact almost never constant across whole cross section of the exit or a channel. Sometimes its distribution may be very intricate. The example here shows the distribution in a typical channel made by one-sided isotropic etching. What is actually meant when a discussion mentions "velocity" in a particular cross section is the mean velocity w.

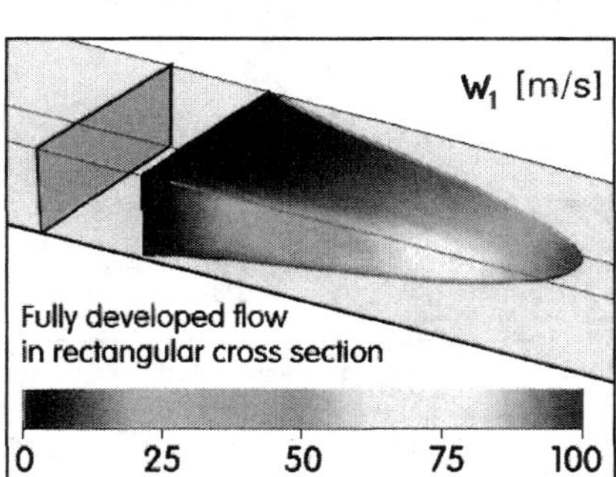

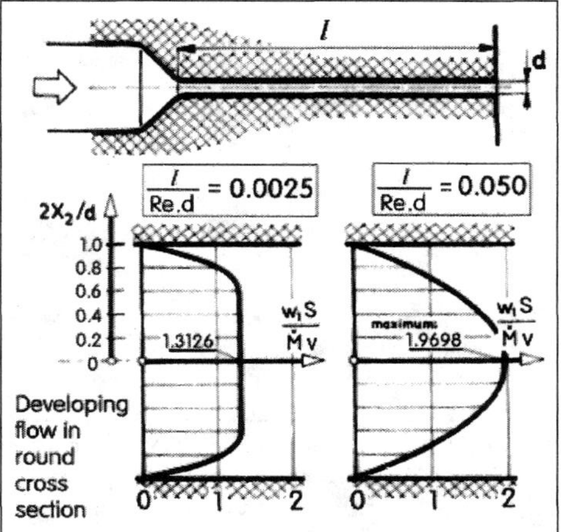

Figure 2.9 (Left) Another example of computed distribution of the local velocities – here in an ideal rectangular cross section. Channels in fluidic circuits are, however, rarely long enough for the flow reaching this fully developed state.

Figure 2.10 (Right) Numerical solution of developing velocity profiles in a circular cross-section tube.

Figure 2.7. Perhaps often used, but suitable only for handling those situations where there are no important effects of fluid compressibility, is the volume flow rate. A more general character possesses the mass flow rate, which is a quantity universally conserved along a flow path. The instruments used to measure the flow rates are flowmeters [2, 3].

There are, in fact, several approaches to the integration of the velocity field across the channel area to obtain the through state parameter. Theoretically a different result is obtained from integration of the local elementary mass flow rates, momentums, and specific energy. In engineering, knowledge about the minute details of the flowfield, obtained either computationally or experimentally, is rarely exact enough for considering these aspects. Because it is the mass that is generally needed for evaluation of the fluid mass balance in the system, it is the mass flow rates integration approach that is used almost exclusively. More importantly, the usually imperfect knowledge causes problems due to the character of the velocity distributions of the kinds shown in Figures 2.8 and 2.9 actually vary with the flow rate (due to changes of friction with the Reynolds number). The distribution also depends on the character of the device connected upstream from the investigated input cross section. The interconnecting channels are usually not long enough for the velocity profiles to develop fully before entering the downstream device. As a result, a certain degree of imprecision has to be generally accepted, though the possible consequences must be then kept in mind.

2.2.2 Implications of manufacturing technology

The main influence on the relationship between the velocity and the flow rate as the state parameter has the shape of the channel cross section. Most rules of hydrodynamics in textbooks are formulated for the round cross section pipes, with other shapes converted to this basic case by the use of the concept of hydraulic diameter. Unfortunately, this concept can usually lead to an imprecision acceptably small only for turbulent flows, with most friction effects taking place in rather narrow regions near the pipe walls. In fluidics in general—and particularly in microfluidics—the conditions are far away from those under which this approach may be recommended.

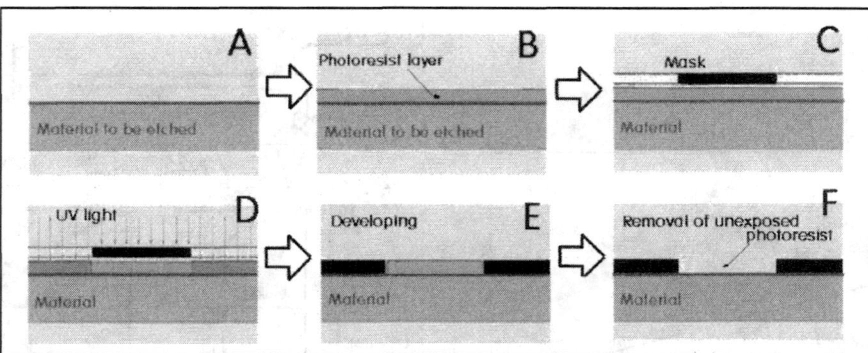

Figure 2.11 Preparatory stages in the standard etching processes used in manufacturing channels and device cavities. The cavity shape is determined by the mask that protects some parts from irradiation by UV light that polymerizes the photoresist. Nonpolymerized photoresist is then removed, producing unprotected surface areas accessible for the etchant in the next stages.

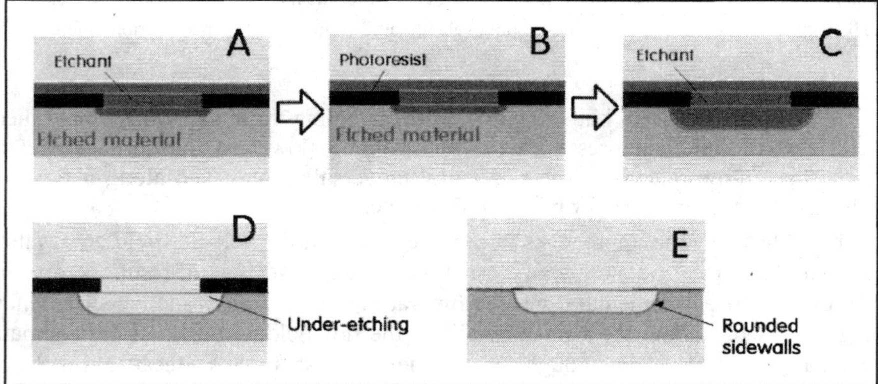

Figure 2.12 Manufacturing the channels and cavities on the top side of the base plate by isotropic etching. An important effect for the cross-sectional shape of the resultant channels is the underetching below the edges of the photoresist.

Current manufacturing methods, such as etching (Figures 2.11 and 2.12) used for making fluidic devices, favor the planar design of cavities, with depth the same everywhere. Nominally, the cross sections of the channels in the terminals are therefore rectangular, of different aspect ratios (Figure 2.16). Different widths b are selected for obtaining different cross-sectional areas needed to keep the velocity within the desirable range. The dependence between the mean velocity (Figure 2.8) and the flow rate is much more complicated for the rectangular shape (Figure 2.9) than is the case for the circular-section pipes and may be even more so for the shapes resultant from various etching methods (Figures 2.13, 2.14, and 2.15).

The aspect ratio $\lambda = h/b$ of depth to width (Figure 2.16) is of essential importance for device performance. The planar, essentially two-dimensional device shape in plan view may give rise to a wrong impression that it is possible to compute the flowfields in the devices as two-dimensional. This would mean a substantial reduction of the necessary computational effort. In general, this impression is wrong because the influence of the friction on the top and bottom surfaces of the cavities makes the flow strongly three-dimensional, with influential secondary flows and often even helical character of the flow in the channels. This is true even if the aspect ratio λ is much *higher* than 1.0, the condition sometimes applied as enabling a crude applicability of the two-dimensional solution approach. Apart from making the design more difficult, the low aspect ratios λ are inevitably detrimental to device performance. The friction on the top and bottom of the cavities necessarily dissipates kinetic energy of the fluid.

What makes the situation particularly unpleasant is the fact that producing shapes with the aspect ratio λ higher than 1.0 requires more expensive manufacturing methods. The standard manufacturing factors often actually limit the aspect ratio λ to values much *less* than 1.0. This is the case with one-sided isotropic etching, the mainstay method. The reasons for the limitation are obvious from Figures 2.11 and 2.12, which show the succession of the basic steps of which this method consists. This is commonly used to make the microfluidic device cavities in materials not have a uniform crystalline structure – such as in glass or stainless steel. These two materials are favored by chemical engineers and are used widely whenever there is no requirement of electric circuits made on the same chip.

The usual way of evading the aspect ratio limitation in these cases is two-sided etching: the same photoresist layer mask allows etching from both opposing sides. This progresses until the etched cavities meet. Further improvement of the aspect ratio is then possible by stacking many plates with cavities etched through on top of one another. Unfortunately, two-sided etching is more difficult, more expensive, and prone to errors due to imperfect mutual placement of the masks on both sides. More importantly perhaps, etching the plates through severely limits the freedom in the device and circuit design – plates with the isolated "islands" surrounded on all sides by the cavities fall apart and it is difficult, expensive, and sometimes impossible to put them back and hold them in their proper positions.

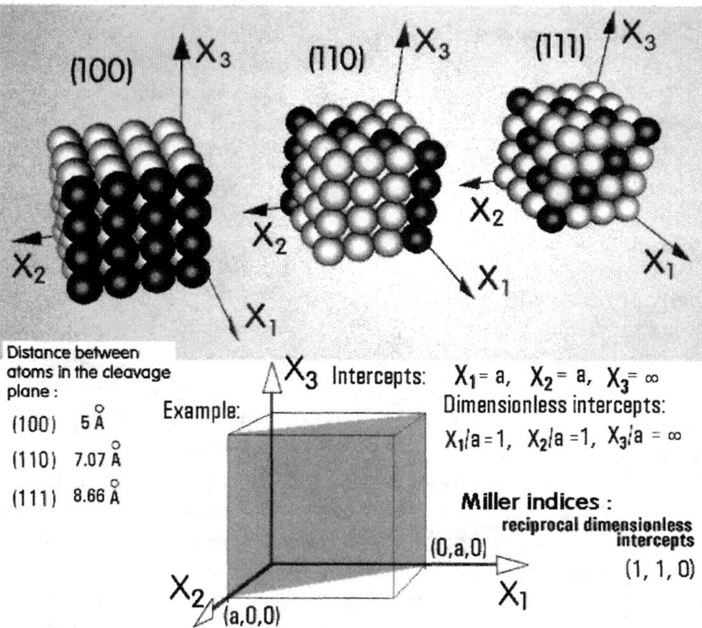

Figure 2.13 Top: the three cleavage planes of a silicone crystal and their Miller indices. Bottom: Example of evaluating the Miller indices.

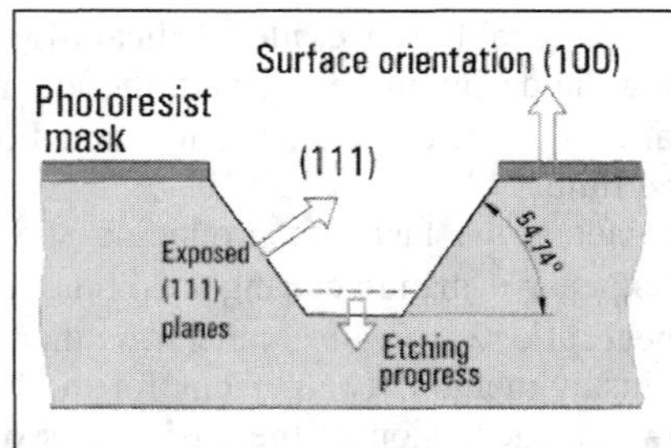

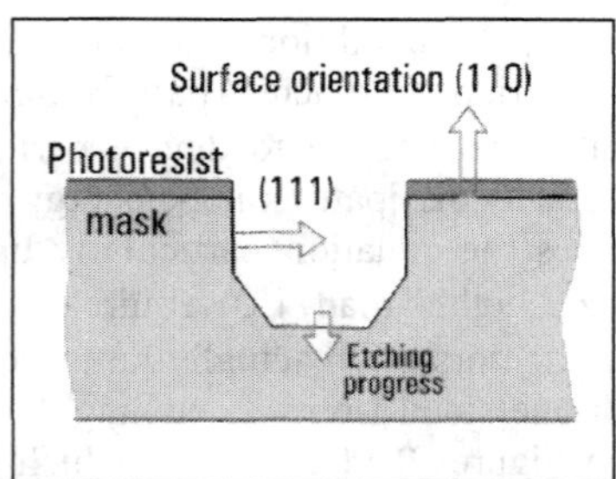

Figure 2.14 (Left) Cross section of a channel formed by anisotropic wet etching of (100) silicon base plate. If the open area in the photoresist mask were square, the resultant shape would be a pyramidal pit.

Figure 2.15 (Right) Cross section of the shape formed by anisotropic wet etching of silicon base plate having the (110) surface orientation.

Anisotropic etching can produce more suitable channel cross sections at reasonable economy – but cteates its own problems. This etching is applicable to monocrystals. The progress of the material removal is slowed down substantially at planes in which there are small distances between the atoms forming the crystal grid. These planes are called cleavage planes, since they become exposed when some crystalline substances are split. In silicon, with its cubic crystalline structure (Figure 2.13), there are three systems of parallel cleavage planes, usually identified by their Miller indices. These specify the orientation of a crystal plane by considering how the plane intersects the main crystallographic axes. The three values are reciprocals of the intercepts expressed in dimensionless coordinates, as is also shown in the example in Figure 2.13. The coordinates of the intercepts are a , a , and ∞. The dimensionless relative coordinates are obtained by dividing the intercepts by the cell edge length a. The reciprocals of 1 and ∞ are 1 and 0, respectively, and the Miller indices are these reciprocals.

Silicon monocrystals are usually supplied in the form of wafers 4 in diameter, with Miller indices of the flat top surface either (100) or (110). The resultant channel section shapes produced by the etching in these two cases are shown in Figures 2.14 and 2.15. The former explains the characteristic 54.74° inclination of walls seen on many surfaces of microfluidic devices.

Using monocrystals with the (110) surface orientation, shown in Figure 2.15, one can obtain nearly perfectly rectangular channels because the etching rate in the (110) direction is much higher relative to the other crystallographic planes. Of course, only two resultant opposing sides are parallel; the perpendicular sides are convergent.

Finally, an important manufacturing method that deserves being mentioned in this context is the process shown schematically in Figure 2.17. The method is called LIGA, a German acronym. It has the potential of making channels of excellent aspect ratios of the order $\lambda \sim 10$. The process does not involve etching; the cavities are essentially made in a thick layer of photoresist. This requires illumination into considerable depth. For this purpose, it is common to use a more penetrating form of electromagnetic radiation than light. Very energetic radiation also has the advantage of not producing significant sideways illumination due to scattering. Mechanical properties of photoresists are not acceptable for most purposes, however, so that it is used only as a mold for a metal "negative," made by electroplating. This (unless the original photoresist mold is "negative" itself) is not usable (the concave and convex shapes are inverted) but serves as a master for series production of copies from plastics. This method requires a complex and expensive facility, often including a synchrotron as the source of the penetrating radiation. Thus the method is obviously suitable for microfluidic systems made in large numbers, perhaps in true mass production.

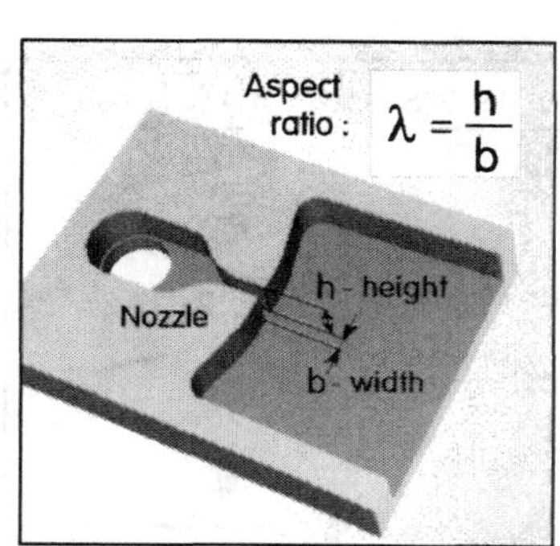

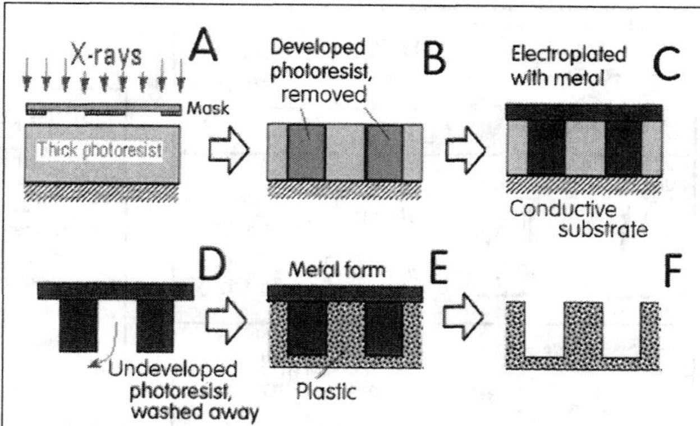

Figure 2.16 (Left) Performance of fluidic jet-type devices as a rule improves with increasing aspect ratio λ of the nozzle exit. In general, increased λ usually also makes possible operation at a lower Re.

Figure 2.17 (Right) The LIGA manufacturing method: two stage process producing in the first stage a metal form suitable for series production of microfluidic devices from plastics.

2.3 STATE PARAMETERS

For description of fluidic device properties, it is necessary to introduce and use suitable state parameters (Figure 2.18). Behavior of devices with several interacting flows coming through several terminals may be rather complex. As a suitable, not too complex case to start with, the attention here is first focused on devices having only two terminals. At this stage, we can neglect the actual internal constitution of the device, which may be simply a connecting pipe or may be a quite complex assembly of many components. To emphasize the generality of the approach, the device is represented in Figure 2.19 by the symbolic *black box.* This is the basis of the zero-dimensional representations: the actual physical locations of the devices and the shape of the individual connecting pipes are irrelevant; of importance is only the topological character (what is connected to what). When drawn as circuit diagrams, the results of using this approach are similar to electric circuit diagrams. Even the basic governing laws are analogous, at least to a degree. One of the dissimilarities between the fluidic and electric circuits is actually just the basic concept of the state parameters.

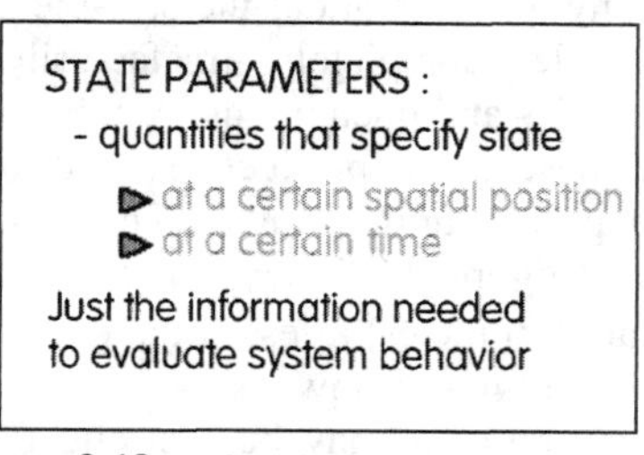

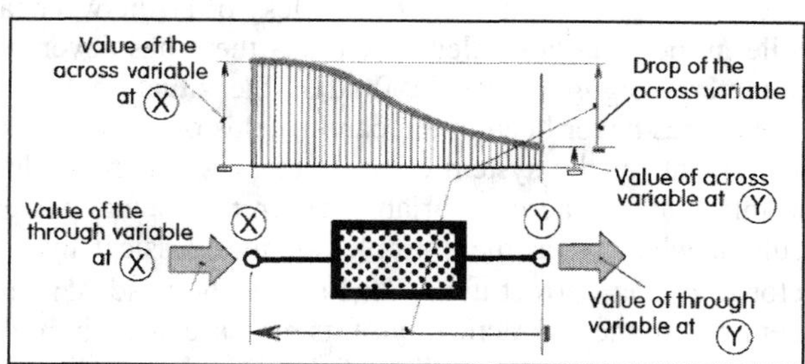

Figure 2.18 (Left) The local state, usually specified for a terminal of a fluidic device, is determined by two state parameters.

Figure 2.19 (Right) Of the two state parameters, one is the *through variable* indicating the extent of the flow through the device while the other one is the *across variable*, useful to characterize the decrease of the flow driving capability due to dissipation, energetic conversions, and other processes taking place inside the device.

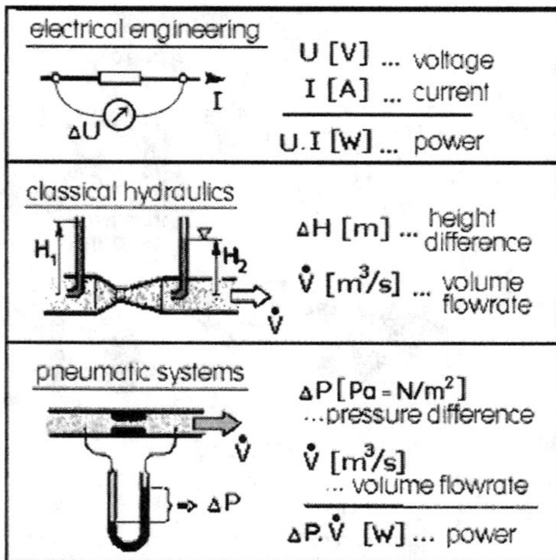

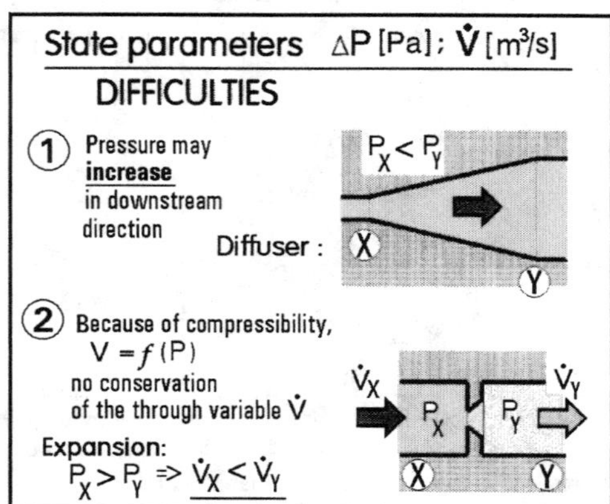

Figure 2.20 (Left) The use of the state parameters is quite analogous to the use of voltage and current in electronic circuitry. A very desirable property of the two parameters is that their product should be the (dissipated) power in proper units, watts.

Figure 2.21 (Right) The classical state parameters, pressure and volume flow rate, cause several problems not known in the analogous electrical quantities. Even if the devices are passive, the exit pressure may be higher than at the inlet (if the device contains a diffuser) and the outlet flow rate may be larger than the inlet one due to volume expansion.

There are two basic types of state parameters, the *across* type and *through* type. For specification of a state of the two-terminal device, two parameters are needed, one of each type. Section 2.2.1 has already introduced the requirement of the integral character of the quantity used as the *through parameter*. The actual conditions in a device terminal are often a quite complicated spatial field of quantities defined at a point, such as velocity. Though perfect for detailed studies of the flowfield in the device, specification of the velocity field in a terminal carries so much information that its use for fluidic circuit design would be an insurmountable task. Integral quantities, such as the volume flow rate $\dot{V}$ [m³/s], the velocity integral over the terminal cross section, (Figure 2.7) - renounce the information about the details. They are suitable for the through state parameter role since they contain just the necessary information in the form of a single scalar numerical value. More problems are associated with the *across parameter*. It should be noted that what is actually used is always just the difference of its values, mostly the difference between the device terminals (Figure 2.19) and less often between a terminal and some suitable reference. While in the analogous electric circuits there has never been any question about the suitable state parameters (Figure 2.20), in fluidics the situation has not settled completely satisfactorily, as indeed it has never been in the parent fields of traditional hydraulics and pneumatics. A reasonable requirement for the system of parameters is coherence in the sense of the product of the through variable and the across variable difference being power in proper units (watts). The product resultant when using the variables of the classical hydraulics (Figure 2.20) needs a conversion factor to obtain correct units. This makes the head ΔH (measured in meter) a less suitable choice, together with the impracticality of its measurements in high-pressure liquid flows.

Of course, the amenability of the quantities to direct measurements may be a strong factor from a practical point of view. Direct measurability by manometers makes pressure favored across state parameter, used quite universally—though never without profound questions being asked. This has led to the classical pneumatic system of state parameters (Figure 2.20). As shown in Figure 2.21, there are two essential problems. The first obvious difference when compared with

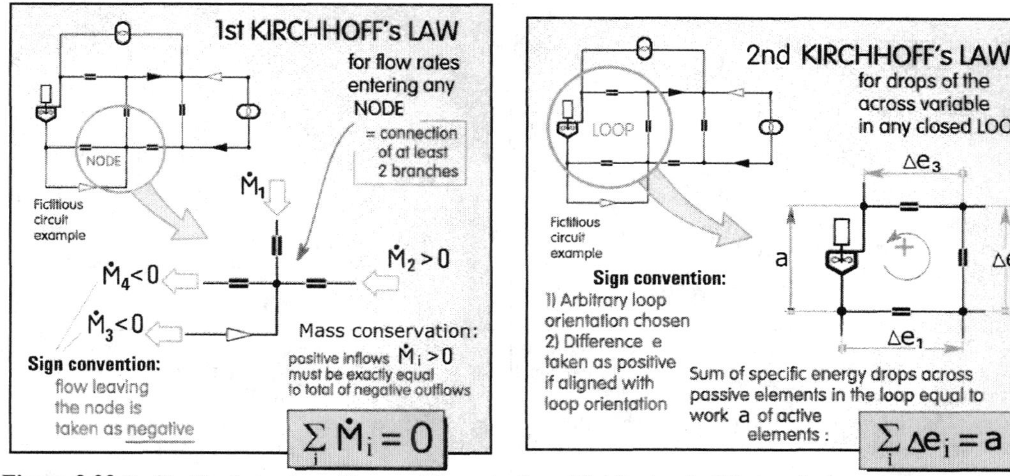

Figure 2.22 (Left) The first universal law used for evaluation of fluidic circuits. Whatever happens in other parts of the system, at each of its nodes the sum of mass flow rates (with proper orientations) is zero. The law as written here is valid for steady states – in unsteady regimes the mass balance may be invalidated by some fluid being stored inside the node.

Figure 2.23 (Right) The other basic law of fluidic circuits. Without regard to processes taking place in other parts of the system, it is possible to take any closed loop of devices and evaluate the state in the loop from the energy conservation condition.

the electric circuits is the existence of passive components, diffusers, in which the across parameter value actually increases in the flow direction without there being any energy supplied from outside. This is possible due to the multicomponent nature of fluid energy: in contrast to voltage, pressure energy is not the only component and may be increased at the expense of the other components. In particular, in the diffusers the reason for the deviation is the kinetic energy component. This varies in a manner making it uniquely dependent on the size of the channel cross section. If this increases in the downstream direction in the diffuser, the kinetic energy component must decrease – and this is done by conversion into the pressure component. Of course, there is also (in a gravity field) the positional energy component that may be converted – but this is usually much smaller than the other two, at least if the working fluid is gas.

The most important complicating factors, however, are the consequences of fluid compressibility. As also shown in Figure 2.21, the volume flow rate $\dot{V}$ is not a conserved quantity. This seriously impairs the fluidic circuit solution procedures, as these are based on the two Kirchhoff conservation laws (Figures 2.22 and 2.23).

The problem of the through parameter variations due to the compressibility may be solved by using the mass flow rate $\dot{M}$ [kg/s]. The resultant system of state parameters, shown in Figure 2.24, is the most popular choice in contemporary fluidics. Nevertheless, there are several drawbacks it does not solve, as already indicated by its incoherence (a in Figure 2.24). The main problem stems from the fact that pressure is not a conserved quantity and cannot be simply and universally used in formulating the condition valid for the loops in fluidic circuits (Fig. 2.23). In fact, this law assumes use of the differences in the across parameter. However, in thermodynamic expressions governing the dependence of the through parameter $\dot{M}$ on pressure values, the latter actually appear as ratios rather than the differences (b in Figure 2.24).

The suitable quantity in this respect is the specific energy e [J/kg] (Figure 2.25). This is a quantity for which there exists a universal conservation law in the proper form of the differences. Using it together with the mass flow rate leads to a state parameters system that is satisfactory

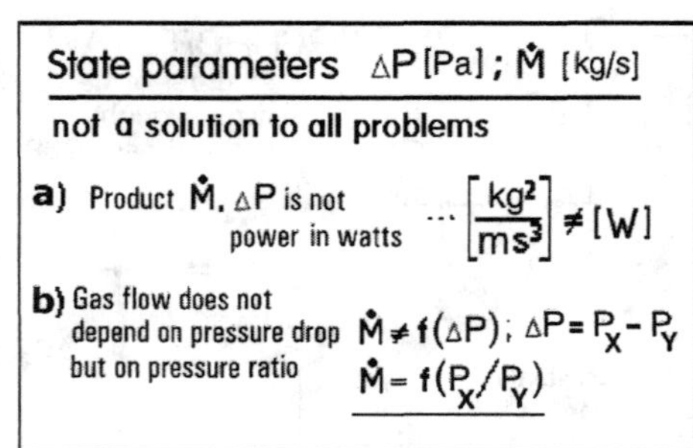

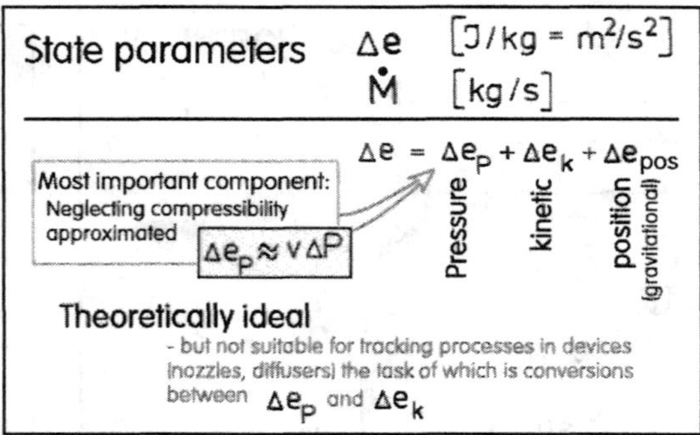

Figure 2.24 (Left) The system of state parameters most often used in practice, in spite of its obvious theoretical weaknesses. The mass flow rate is conserved along the flowpath and pressure is useful for practical purposes as well as directly measured quantity.

Figure 2.25 (Right) Though theoretically the best choice, the state parameter system using specific energy did not become very popular in practice.

from a theoretical point of view. Unfortunately, this system (Figure 2.25), failed to gain popularity in practice. One of the reasons is perhaps the fact that specific energy is not a directly measurable quantity and has to be computed from measured pressure (and other measured quantities). The computation—unless there is complete information available about some rather elusive effects—is not straightforward and brings several problems of its own. Pressure is also more instructive and working with it offers a better "feel" for its magnitudes. That is why pressure as a state parameter often remains in use despite its theoretical shortcomings.

In this book, the system of parameters from Figure 2.25 is used whenever practicable in an attempt to promote this less known but theoretically better variant. The pressure-based system of parameters will be, however, used in parallel, especially in the cases where the compressibility effect can play no significant role.

The compressibility effects, especially in steady states, are really important only at high flow velocities. These are generally avoided in microfluidics, being associated with undesirably high energy losses. If the compressibility may be neglected, then the specific pressure energy difference is simply proportional to the pressure difference:

$$\Delta e_P = v\,\Delta P \tag{2.1}$$

the energy and pressure as state parameters differ in the latter being simply multiplied by specific volume as the proportionality constant. The more complicated expressions valid for flows with significant compressibility effects are discussed in Section 2.5.

A complicating fact is that the resultant pressure energy difference Δe_P is just a component of the total specific energy drop Δe across a device. Differences of the other energy components (kinetic and gravitational) must be added to Δe_P to obtain the total drop applicable in the energy conservation law of Figure 2.23. Unfortunately, if this is done, then the most important components used in making fluidic devices, nozzles and diffusers, appear in the circuit solution expressions only as if they were mere resistors. Their pressure conversion effects—the very reason why they are used—and the resultant pressure distributions we wish, after all, to evaluate, do not appear directly in the used equations. Circuit solutions with the components performing conversions between the components of fluid energy have to be performed in two steps: first evoking the total energy balance and then computing the pressure conditions.

2.4 DISSIPATION OF FLUID ENERGY

The dissipative effects discussed in this part are, after all, the real reason why the pressure differences driving the flow in microdevices can often become very high. These effects can cause the pressure levels to be much higher than the values obtained from the lossless expressions given in Figures 2.2 and 2.3 and discussed in the initial part of this chapter.

It should be noted that however satisfactory the first-approximation description by the lossless approach may be, applied for example in Figure 5.5 to evaluate the energetic conversions associated with generation of a jet in a nozzle, this approximation in fact means that the drop Δe across a device is identically zero. The evaluated difference brings no contribution to the solution procedure based on the second Kirchhoff law (Figure 2.23). Evidently, it is the dissipation that is of interest in the circuit evaluation and design.

2.4.1 Conversion $e_K \rightarrow e_T$

The general expression for the energy balance across a two-terminal device with the energy dissipation term included is presented in Figure 2.26. The dissipation actually causes a misbalance; some energy escapes in the form of heat. This is transported into the surroundings and therefore not available for subsequent energy conversion processes. To emphasize this misbalancing character, the budget in Figure 2.26 is written with the dissipated part replacing the zero on the right-hand side of the earlier lossless case in Figure 2.4. Also written on the right-hand side is the other possible misbalancing term, this time a positive one: if the device is an active one (perhaps a pump), this term represents the specific work **a** input into the device by the pump or any external action.

The energy dissipation is a consequence of a very general tendency in nature, the trend of organized motion to be converted into the disorganized, chaotic thermal motion of fluid molecules. The increase in chaos corresponds to the rise in entropy, characteristic for these processes. The

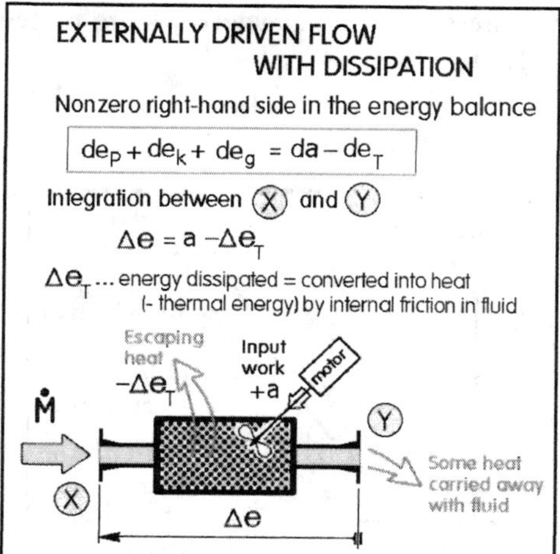

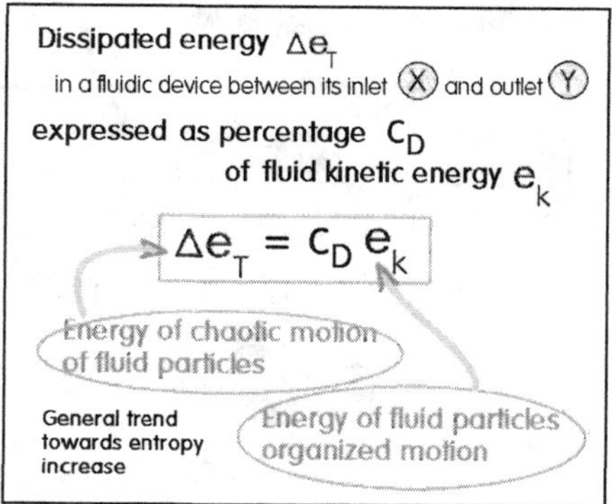

Figure 2.26 (Left) General expression for the specific energy change in steady-state flow through a two-terminal device. The mutual conversions between the pressure component, kinetic component, and gravitational component are governed by a condition valid for the sum of their differences across the device. In the lossless passive case, the sum was zero. In the general case, the sum may equal to work input minus the dissipation.

Figure 2.27 (Right) Drag coefficient c_D is introduced to indicate how a large part of organized kinetic energy of the fluid is converted into heat — disorganized energy of thermal motion.

kinetic energy—energy of the organized motion—is the source from which the disorganized motion is supplied, so the magnitude of the conversion into heat is usefully expressed as a percentage of the kinetic energy. The percentage magnitude is the loss or drag coefficient c_D in Figure 2.27 (it is equivalent to the drag coefficient in flow past bodies and indeed may be useful in evaluation of the drag force with which fluid acts on a device through which it is passing). There is nothing paradoxical, in fact not really extraordinary, if a situation arises where $c_D > 1$, although this means the converted percentage is larger than 100%. What this means is the converted kinetic energy is resupplied from some other components of fluid energy, most importantly from the pressure energy. This is the reason behind the "pressure loss" leading to the potentially dangerous very high driving pressure levels.

2.4.2 Steady-state characteristic and the characterization parameter Q

The kinetic energy of fluid is proportional to the square of velocity. This, in turn, is proportional to the flow rate passing through the device, both to the volume flow rate as well as mass flow rate (Figure 2.7). If the proportionality constant c_D were invariant, then this chain of proportionalities would lead to quadratic dependence between the across and through state parameters,

$$\Delta e \sim e_k \sim w^2 \sim \dot{M}^2$$

The constant of proportionality between them, the *quadratic dissipance* Q [m^2/s^2], then characterizes the device (Figures 2.28, 2.29, and 2.30) from the point of view of its behavior in a circuit – in the same way as the resistance magnitude is used to characterize steady-state behavior of components in electronic circuits. The dissipance is a very important concept. Hydrodynamic properties of all devices in steady state for the purpose of circuit analysis and design may be fully characterized by the single scalar value Q.

In reality the coefficient c_D varies with the Reynolds number. The variation is small enough for there to be a mass of literature on hydraulic losses that presents the values of hydraulic loss coefficients as constants (similar to the example of the losses in bends in Figure 2.31). However, the deviations from the quadratic behavior are generally known to exist and provide one of the reasons why, in the absence of a simple algebraic expression, the characteristics are preferably presented in the diagrammatic form, similar to Figure 2.29.

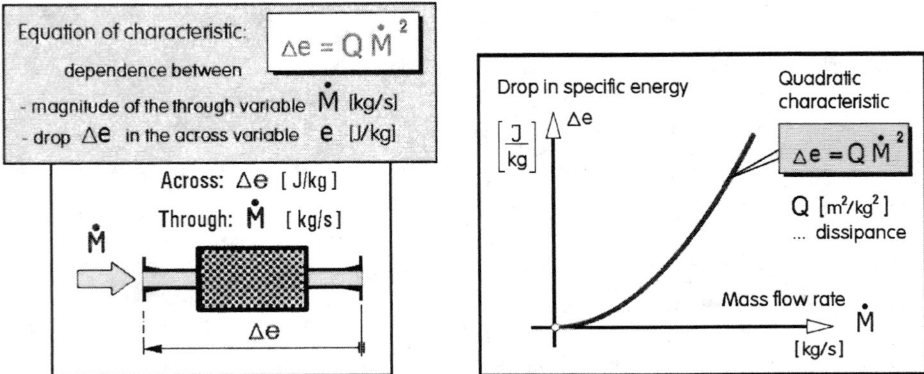

Figure 2.28 (Left) Characterization of hydrodynamic properties of a two-terminal fluidic (or microfluidic) device by the mutual dependence between the two state parameters.

Figure 2.29 (Right) Schematic example of the quadratic characteristic obtained for constant value of the loss coefficient c_D. Also useful in this context is the diagram in Figure 2.33, which presents (in logarithmic coordinates to cover a wide range) numerical values of the quantities used for the steady-state characterization of devices.

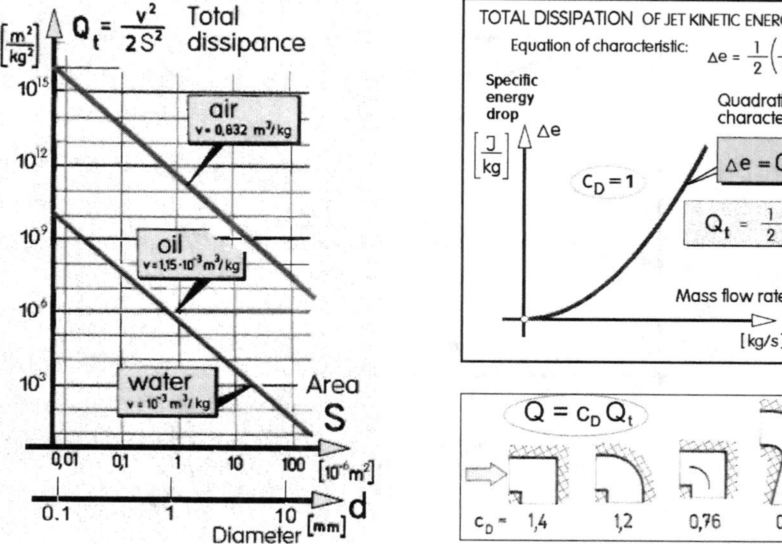

Figure 2.30 (Left) The total dissipance Q_t depends only on the characteristic cross section S of the device (usually the dominant section (the smallest through which the fluid passes in the device) and on the specific volume v of the fluid. This diagram provides an opportunity to estimate the order of magnitude of these quantities, with an auxiliary scale for the diameter d of an equally large circular section.

Figure 2.31 (Right, top) The relation between the dissipance Q and the total value Q_t and an example of the loss coefficient values for a number of channel bends. The values were found experimentally at very large Reynolds numbers.

Figure 2.32 (Right, bottom) The characteristic of a device that dissipates totally the whole kinetic energy the fluid of specific volume v has in the (usually smallest) cross-sectional area S in the device.

The two concepts, the dissipance Q and the coefficient c_D, do not compete but are mutually complementary. The dimensionless coefficient c_D depends solely on the geometry—and not the size—of the device. If a device is scaled up (or down) then its c_D is the same as the value for the smaller (or larger) original. This is very useful for laboratory investigations, made at a convenient size using scaled models. For example, the values given in Figure 2.32 are applicable to channel bends of any size, provided that they are operated at the same Reynolds number. On the other hand, if a designer has to make a decision about the proper size of a device, c_D is not useful. It is the dissipance Q that has to be decided upon, since it involves the information about the absolute size as well as about the used fluid (its specific volume).

Mutual relation between dissipance Q and coefficient c_D is shown in Figure 2.31 to be mediated by the concept of the *total dissipance* Q_t. This is the dissipance of a device in which the kinetic energy of fluid it has in the dominant cross-section S is completely dissipated. The value depends only on the area of the dominant cross section and the fluid specific volume. The expression in Figure 2.31 follows from the expression for $\dot{M}$ in Figure 2.7 together with the expression for the specific kinetic energy $e_k = w^2/2$. To gain some familiarity with the not very common concept of Q_t, the diagram in Figure 2.30 may be useful. It makes possible rough estimates of the order of magnitude values of the total dissipance Q_t for a given channel cross-sectional area with the three fluids. Note the ratio v/S appearing there as well as in Figure 2.31. It is found again in many other expressions for characterization quantities (e.g., Figure 2.89). Another useful diagram providing an idea about typical numerical values is in Figure 2.33.

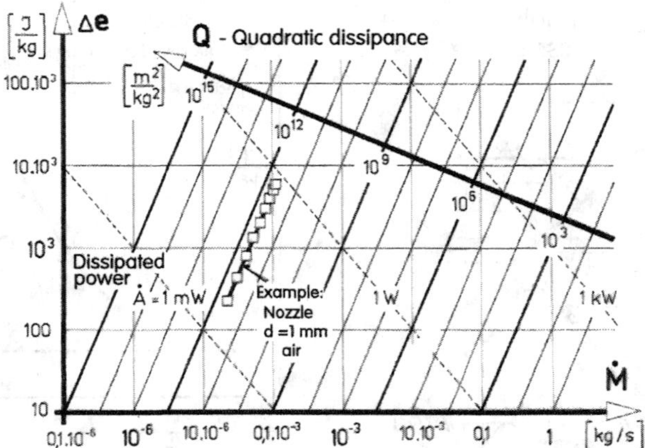

Figure 2.33 To provide an idea about the orders of magnitude of the quantities appearing in the expression for dissipance as the characterization quantity in Figure 2.29, in this diagram it is possible to find the specific energy drop for a given value of the quadratic dissipance Q and the mass flow rate passing through the device. Note also the system of dashed lines of constant dissipated power.

2.4.3 Total dissipation of jet energy

Many fluidic devices—at least those not operating at extremely low Reynolds numbers—are based on use of fluid jets. Some idea about jet generation by fluid issuing from a nozzle was already presented at the beginning of this chapter, in Figures 2.3, 2.4, and 2.5. This was useful to demonstrate the relationship between pressure and velocity using the lossless model of the flow. In fact, to make the discussion more simple, the derivation of the relation there was not fully flawless

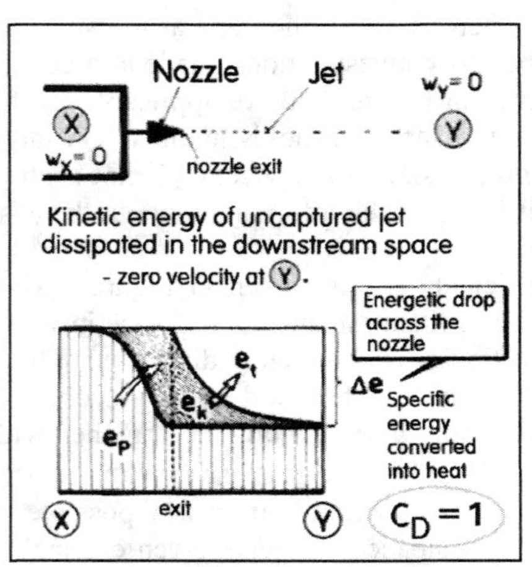

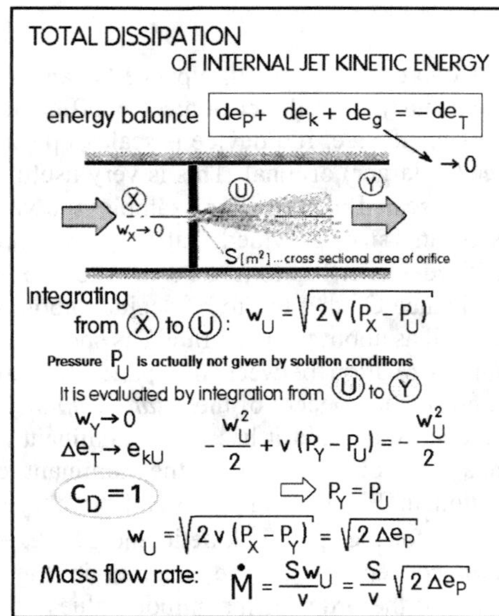

Figure 2.34 (Left) Energetic changes in the nozzle, jet, and the downstream cavity involving the dissipation.

Figure 2.35 (Right) The equation from Figure 2.26 integrated along the flowpath to derive the expression for the mass flow rate. The hydraulic losses inside the nozzle are not considered.

from a theoretical point of view. Without giving any deeper justification, Figure 2.5 assumed that the pressure in the two locations, U and Y, was the same. In terms of the lossless theory, with the governing energy conservation equation $w\,dw + v\,dP = 0$, this cannot be true. According to this equation, the decrease of velocity (and kinetic energy) between U and Y should be compensated by a pressure rise. It is the presence of the dissipation term in the equation that provides an explanation for the equal pressure $P_Y = P_U$. The jet flow, in fact, is a case of the total, 100% loss of all available kinetic energy, $c_D = 1$.

The integration of the equation with the dissipation term on the right-hand side (but no work input $a=0$) from Figure 2.26 is presented in Figure 2.35. It may be usefully compared with the previous analogous but lossless version in Figure 2.5. Figure 2.34 shows the corresponding schematic distribution of the specific energy component along the flowpath graphically. Note that this is still a simplified case; the graphical presentation reveals a neglect of (in fact small) hydraulic loss taking place inside the nozzle. This is done because the nozzle loss is of a different, frictional character to be discussed later.

Of course, what Figures 2.34 and 2.35 represent is the case of jet flow into an unbounded space. Real downstream cavities into which a jet issues from a nozzle in microfluidic devices are of finite dimensions, certainly not so large as to fully justify the condition $w_Y = 0$. The fluid leaves the cavity with some, however small, exit velocity.

2.4.3 Dissipation in separated regions

A useful (and actually rather surprisingly successful) method of evaluation of hydraulic losses in internal flows with separation from the wall—such as the examples in Figure 2.36—is associated with Borda's 1766 study of conditions in a pipe with sudden area expansion. This is a case of the fluid leaving a cavity with smaller but nonzero velocity. It is useful to consider first an application (Figure 2.37) of the approach from Figure 2.4, the one that neglects the loss. The net effect of the

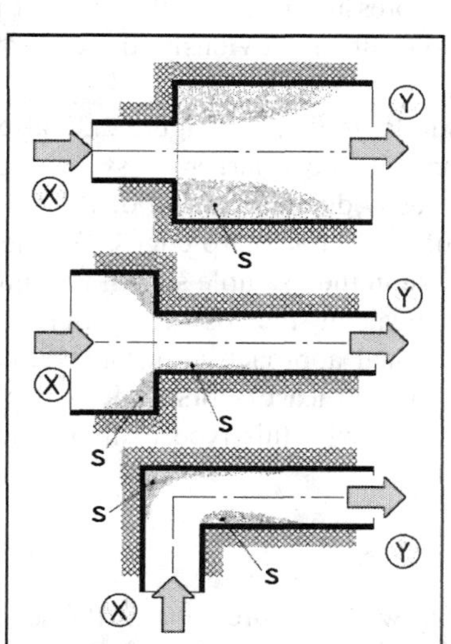

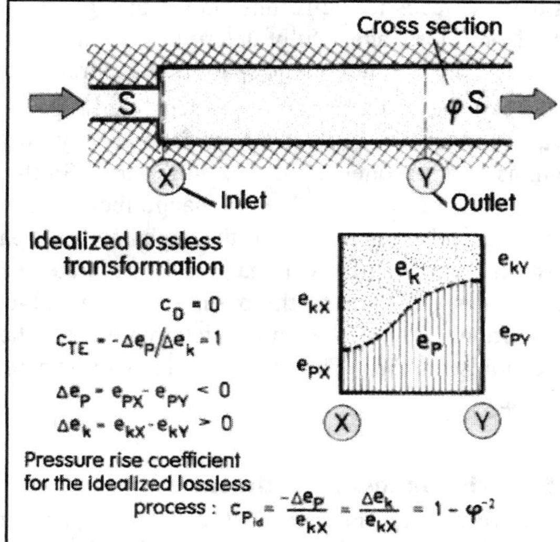

Figure 2.36 (Left) Of the two types of hydraulic losses, the one associated with flow separation from the walls, shown here in three typical examples, may be evaluated applying the expression for the Borda loss in a sudden channel expansion.

Figure 2.37 (Right) The lossless approximation applied to the sudden expansion: in the absence of the loss, pressure rises at the expense of the kinetic energy just like in a diffuser. The rise is determined by the area ratio φ.

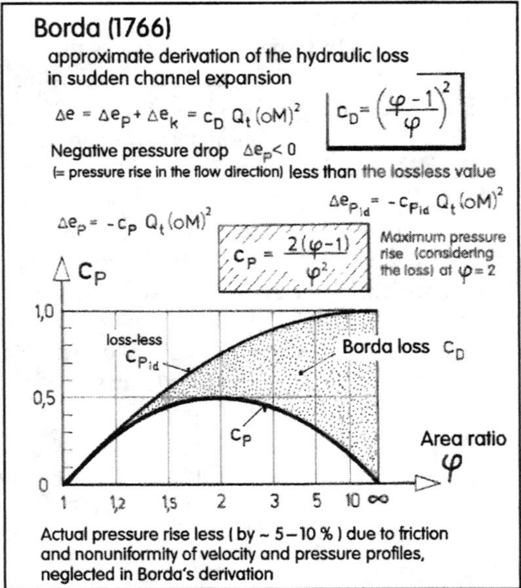

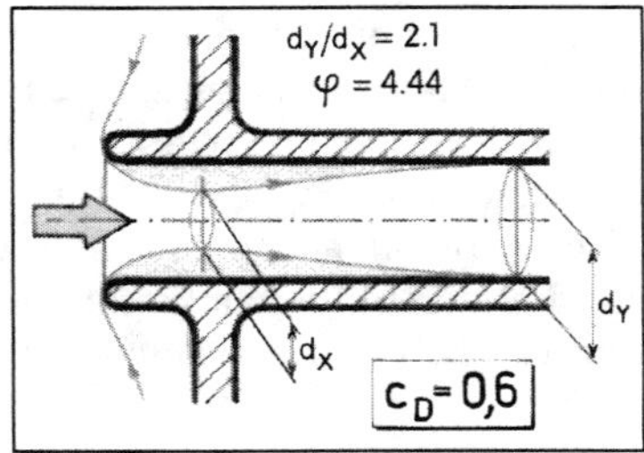

Figure 2.38 (Left) The Borda loss in sudden increase of channel cross section in the streamwise direction is derived directly from basic principles. It is a useful concept for evaluating hydraulic losses arising due to flow separation from the wall (Figure 2.36).

Figure 2.39 (Right) An example of finding the loss coefficient c_D for the separation loss in an unfavorably shaped pipe entrance. Determining the smallest cross section (here the diameter d_X), however, is not a simple task.

energy conversions in that case is the pressure rise. Applying the three available conservation conditions, for mass flow rate, momentum rate, and specific energy, for one-dimensional flow with the loss leads to the results shown in Figure 2.38: the pressure rise is still there, but reduced by the Borda loss, again fully determined by the area ratio φ alone. Obviously, the case of the jet issuing into the infinitely large space is the extreme case $\varphi \rightarrow \infty$.

The flows with formation of the separated regions S similar to those in Figure 2.37 also lead to producing a local area contraction followed by an expansion into a larger cross-sectional area. Again, as in the sudden expansion in Figure 2.36 this is mostly due to inability of fluid flowpaths (due to fluid inertia) to suddenly change their direction of motion at sharp edges. Whenever it is possible to find the area ratio of the expansion flowpath, like in the example shown in Figure 2.39, the Borda formula offers a reasonably accurate way to evaluate the loss coefficient c_D. The inaccuracy is only due to the assumptions of one-dimensional approach (constant velocity in a cross section) and the neglect of frictional effect. There is no—at least to first order—dependence on Reynolds number. The resultant characteristic of a device with this type of hydraulic loss is quadratic.

2.4.5 Friction loss mechanism

Dissipative mechanism acting without flow separation from walls (Figure 2.40) is the second type of hydraulic losses. These are caused by *friction,* in fact, by transversal momentum transport between coflowing layers having different flow velocity. Outwardly, this transport is manifested by presence of shear stress slowing down the faster layer. Velocity of the slowest layer at the wall is zero, unless there is the slip effect (Section 1.3.4). In laminar flows, the momentum transport is

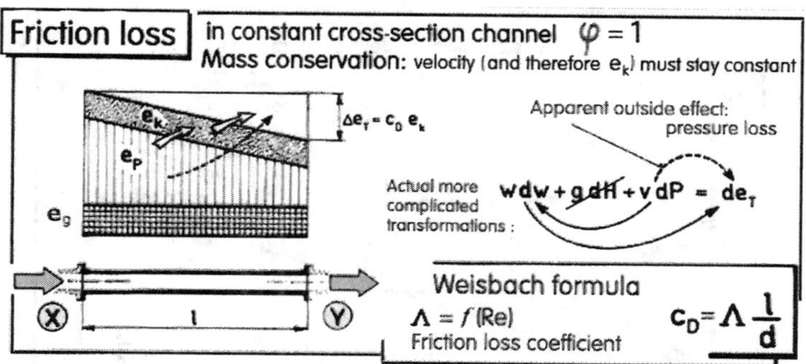

Figure 2.40 Friction type loss in its most typical form, the dissipation of kinetic energy in a constant section channel. The main practical difference from the separation type loss discussed above is the strong dependence of the loss coefficient on Re.

due to thermal motion of fluid molecules while in turbulent flows the motion perpendicular to the flow direction is due to turbulent eddies. In entrance parts of channels (Figure 2.10), this friction mechanism takes place in the boundary layers at and near to the walls while there may be no shear stress in the core part of the flow. Boundary layer thickness grows in a streamwise direction, gradually diminishing the extent of the core until the opposite boundary layers meet. After a further accommodation distance, the velocity profile filling the whole cross section becomes self-similar. Literature (e.g., [4]) provides data for the loss coefficients in fully developed channels of various cross sections. Figure 2.41 shows some data for the fully developed flow in circular cross section pipes. Note that in the laminar regime, the Hagen-Poiseuille formula $\Lambda = 64 / Re$ represents a linear relation between the state parameters. Linearity may be a desirable property, especially in dynamic regime where it prevents spectral distortions of the transferred flow dependence on time.

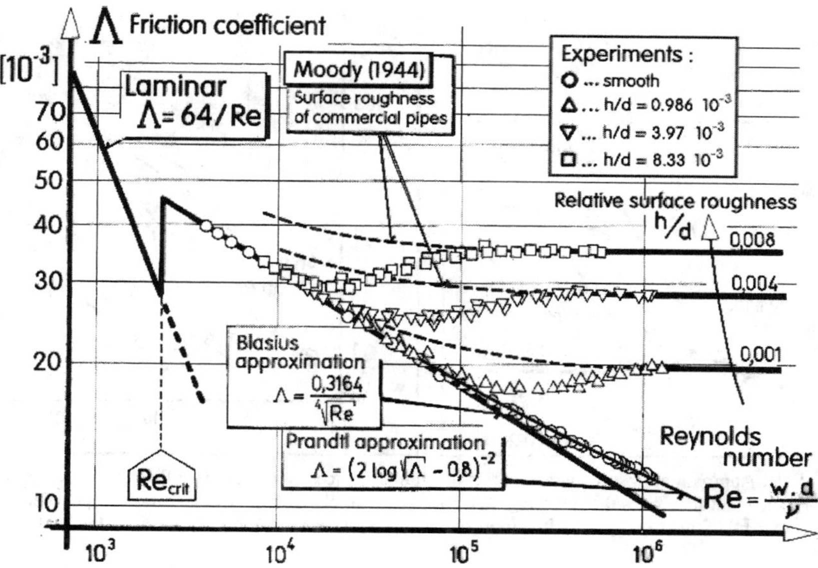

Figure 2.41 Friction coefficient dependence on Reynolds number for fully developed flow in a circular cross section pipe.

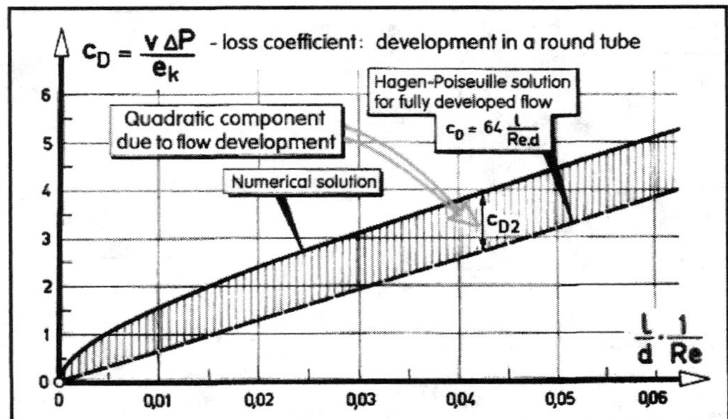

Figure 2.42 Failure of linearity of the laminar friction type loss: in addition to the linear Hagen-Poiseuille component there is the quadratic loss component due to incomplete flow development, associated with the changes of the velocity profile as shown in Figure 2.10.

Unfortunately, channels in fluidic circuits are seldom long enough for the full development. The developing entrance flow contributes by a quadratic additional term (Figure 2.42) in the equation of channel characteristic (and there is often also the similarly quadratic exit loss, for example, of the Borda type if the channel exits into a larger cavity). As a result, really linear characteristics are practically nonexistent in fluidics. Nevertheless, at very low Reynolds numbers the approximate, nearly linear behavior may be found to the degree of justifying the introduction of the corresponding characterization quantity, the pneumatic resistance R shown in Figures 2.43

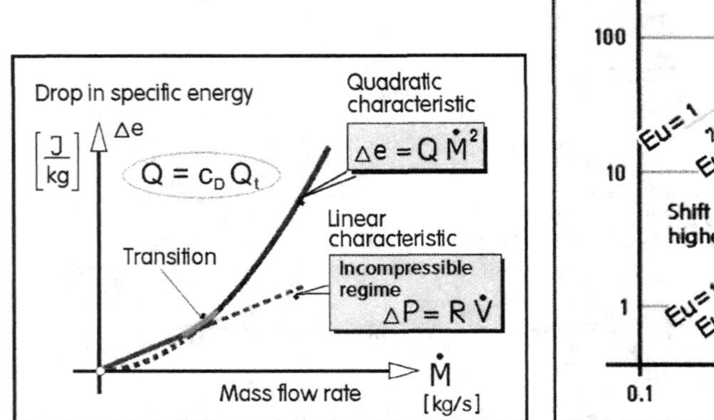

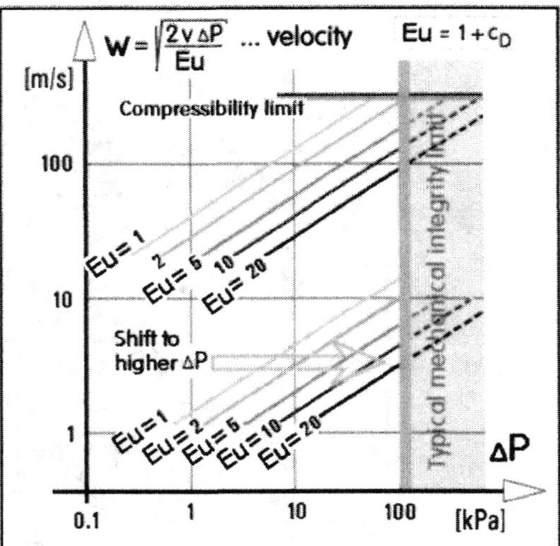

Figure 2.43 (Left) Approximations to characteristics of fluidic devices. While at large flows the best approximation is usually quadratic, at small flow rates there is a region of nearly linear characteristics that may be specified by the pneumatic resistance R. In this usually fully incompressible regime it may be practical to use the traditional state parameters.

Figure 2.44 (Right) Dissipation causes a shift toward the higher pressure differences in the diagram for estimating the magnitudes of pressure and velocity as shown in Figure 2.6. It is because of this shift that the pressure needed to obtain the desired flow velocity may be beyond the acceptable limits imposed by material and bonding stress.

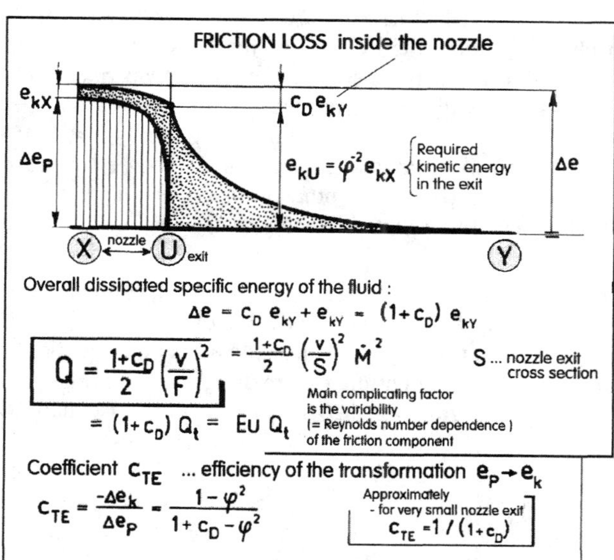

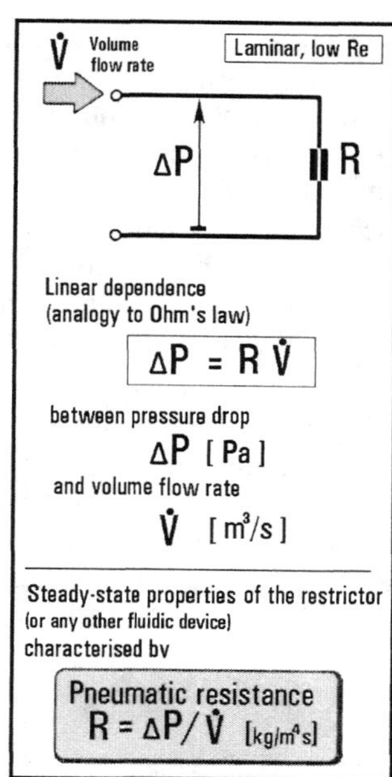

Figure 2.45 (Left) After the simplification of Figure 2.5 and then the better approximation of Figure 2.34, this is finally the actual course of energetic conversions in a nozzle. It includes the friction-type loss in the nozzle, characterized by its loss coefficient c_D. The overall pressure drop across the nozzle is then given by the equation from Figure 2.31 with overall loss coefficient $c_D + 1$, here called the Euler number Eu of the total loss.

Figure 2.46 (Right) If flows in a microfluidic device (at low Re) are laminar and without separation from the walls, the mutual dependence between the two state parameters may be acceptably approximated by the linear analog of Ohm's law. The corresponding velocities are necessarily low so that it may be safely expected there is no supersonic flow and perhaps no compressibility effects. This justifies the use of the classical pneumatic system of state parameters (bottom of Figures 2.20 and 2.21).

and 2.46. A theoretically interesting fact seen in Figure 2.43, which shows the quadratic and theoretical linear characteristics superimposed, is that of the two possibilities, nature always selects the one leading to the higher dissipation (and to faster entropy increase with time).

In most fluidic devices, the two loss mechanisms—friction and separation—are present simultaneously. If the walls past which the fluid flows are short, the friction effect is small and the quadratic separation loss dominates. As well, losses associated with flow direction and cross-section changes are of the quadratic nature. While the separation type loss can lead to maximum value of the loss coefficient $c_D = 1$, in long narrow channels and at low Re (note that c_D increases as Re goes down) may be much higher. The value $c_D > 5$ seen in Figure 2.42 is not particularly exceptional. It is this existence of large c_D that may lead to the very high necessary driving pressures (Figure 2.44), the notoriously reported disadvantage of pressure-driven microfluidics. Of course, there is the current tendency to make the microfluidic systems more complex, with a larger number of devices. The pressure drops across each of the series-connected devices are summed, due to the second Kirchhoff law (Figure 2.23), so as to make the total that is to be overcome by the central pressure source.

The sum of the loss coefficient of the two kinds is the resultant overall nondimensional loss parameter of the device, Euler number Eu. As a result of the presence of the frictional component, the Euler number also varies with the Reynolds number (increases when Re decreases, in accordance with the trend in Figure 2.42). The magnitude of the change with Re depends on the ratio of the two summed loss components summed to form the Euler number. An example of the dissipation with two separable loss coefficients is shown in Figure 2.45. It is the case of the jet formation in a nozzle, following closely the similar Figure 2.6 for the "lossless" case and the Figure 2.34 which gives more attention to the actual processes in the jet downstream from the nozzle, where the initial jet kinetic energy is totally dissipated but the friction losses are neglected. The schematic diagram of the distribution of the components of fluid energy in Figure 2.45 also shows the additional friction effect. This is shown there taking place in the nozzle (in most real fluidic devices there is much more friction taking place in other components, in particular in diffusers in the collector that capture the jet). The Euler number in this case is

$$\mathsf{Eu} = 1 + \mathsf{cD}$$

the unity value for the separation loss is due to $\varphi \rightarrow \infty$ assumed here for the uncaptured jet. Assuming again the device, nozzle is designed for a given required value of the initial jet velocity (or, equivalently, for a given nozzle exit kinetic energy e_{kU} – Figure 2.45) it is now the diagram in Figure 2.44 that presents an order-of-magnitude estimation of the necessary driving pressure difference. When compared with the analogous diagram for the "lossless" model (actually with

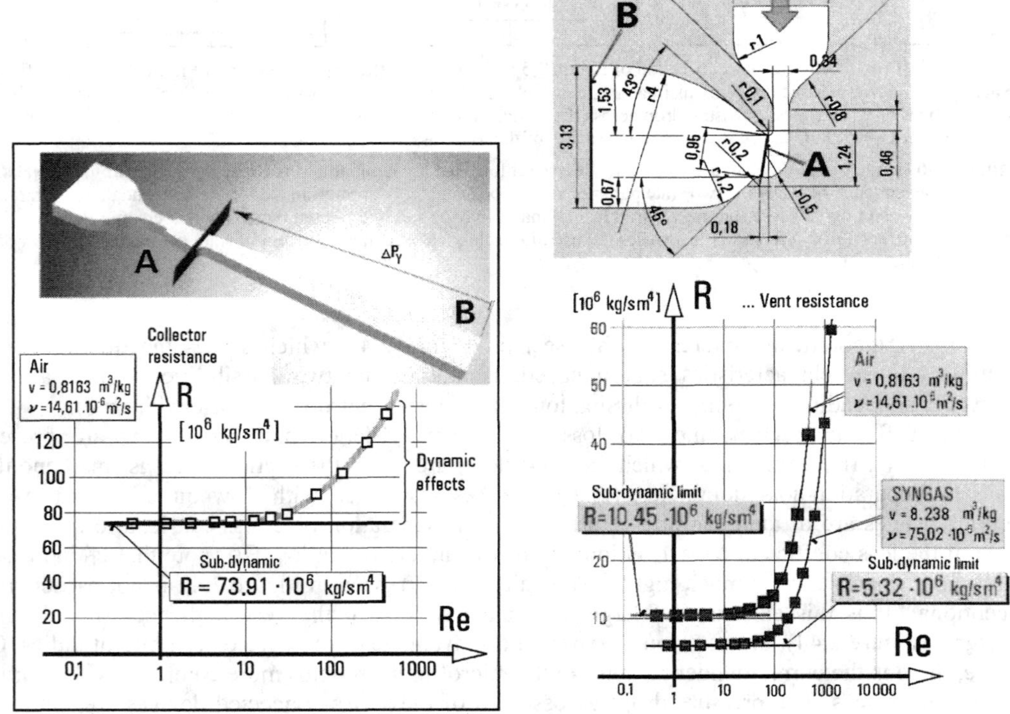

Figure 2.47 (Left) An example of the pneumatic resistance R (defined in Figure 2.46) evaluated for a simple straight collector downstream from a nozzle. The pressure drop is evaluated between the sections A and B. It is apparent that to obtain R reasonably constant, the Reynolds numbers must be extremely low, [16].

Figure 2.48(a) (Right, Top) Geometry of an investigated example of a component (vent channel of a no-moving-part valve) with rather strong quadratic loss components due to changes in cross section and flow direction.

Figure 2.48(b) (Right, Bottom) Values of the pneumatic resistance R evaluated for the flowpath between the sections A and B of the vent channel shown in Figure 2.48(a) with two different gases.

Eu = 1) in Figure 2.6, it is apparent how strongly the presence of the losses shifts the conditions towards the much higher driving pressure difference levels.

2.4.6 Asymptotic subdynamic regime

At extremely low Reynolds numbers, the increase of the laminar friction coefficient c_D with diminishing Re (Figures 2.40 and 2.41) may overcome other mechanisms and lead to Eu ~ 1/Re. The resultant linearity of the characteristics would have several attractive features (thanks to the linear principle of superposition) and make circuit solutions particularly easy. However, a really linear behavior takes place at much lower Re than generally thought, so low, in fact, that operation in this *subdynamic* regime is for most purposes practically not useful. With the exception of extremely long constant cross-section channels, even at the lowest Re the regimes encountered in practice in current microfluidic devices are not in the fully subdynamic region. The devices still exhibit considerable quadratic components in their behavior. The quadraticity is due to dynamic effects (associated with kinetic energy e_K, which is proportional to the square of velocity and hence to the square of the flow rate). In the linear regime, behavior of devices and their components would be characterized by the constant resistance R of Figure 2.46. Practical examples of evaluated resistance in Figures 2.47 and 2.48 show that the constant R is approached only in the creeping flows below Re = 10. Often the constancy is encountered only at values even less than Re = 1, especially if the shape of the device cavities are complicated, with cross-section and flow direction changes leading to dynamic effects. The subdynamic regime therefore may be useful only in the role of an asymptotic reference that provides an idealized guidance for hydrodynamic computations but is not really attained in practice.

As opposed to the absolute magnitude of the resistance to flow at low Reynolds numbers, expressed by means of R, the relative properties are characterized by a dimensionless parameter. This is analogous to the uses of Q for absolute magnitudes and Eu (or c_D) for the dimensionless relative characterization in large Re flows. For low Re the use of a Euler number is not a good

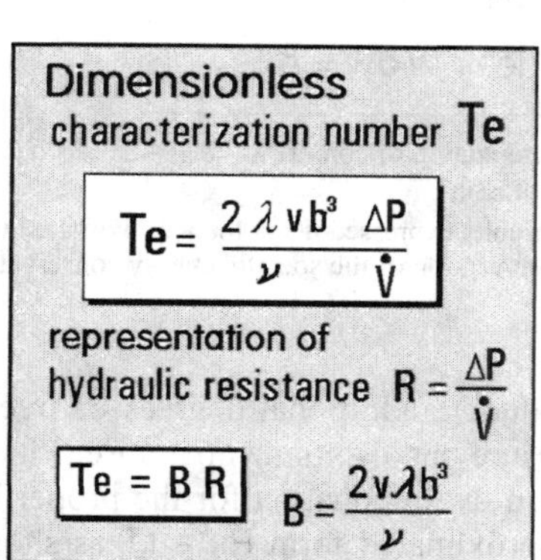

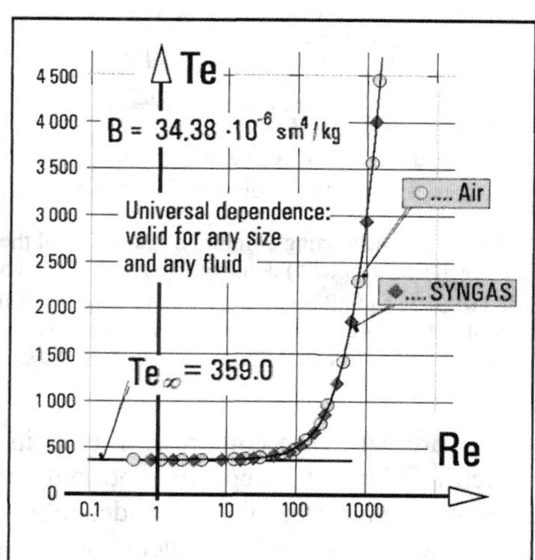

Figure 2.49 (Left) Dimensionless parameter characterizing the relative magnitude of the pneumatic (or hydraulic) resistance R. A particularly useful advantage of using this parameter is the resultant independence on the fluid properties.

Figure 2.50 (Right) The dependence of the effective pneumatic resistance on the Reynolds number from Figure 2.48(b) replotted in terms of the parameter Te.

idea since instead of approaching a value suitable for use as the characteristic quantity of the device, it keeps increasing with increasing Re. A better choice is to use the dimensionless parameter Te defined in Figure 2.49, which does approach a constant asymptotic value Te_{∞}. Figure 2.50 shows it to be independent of the properties of the fluid.

2.5 STATE PARAMETERS FOR COMPRESSIBLE FLOWS

It is also useful to now turn our attention to the other, high end of the spectrum of flow velocities. Most practical fluidic devices operate in incompressible, low Ma regime. High velocities, perhaps even supersonic ones, are rare. They are avoided for the practical reason of leading to high energetic losses, which in almost all applications it is desirable to minimize. Mentioning the details of this region may be nevertheless useful for clarifying the very basic problem of the proper choice of state parameters in fluidics. The quantities $\dot{M}$ and Δe recommended here for this role, and bring forth the question of the more exact evaluation when in the compressible regime the simple expression (2.1) ceases to be applicable.

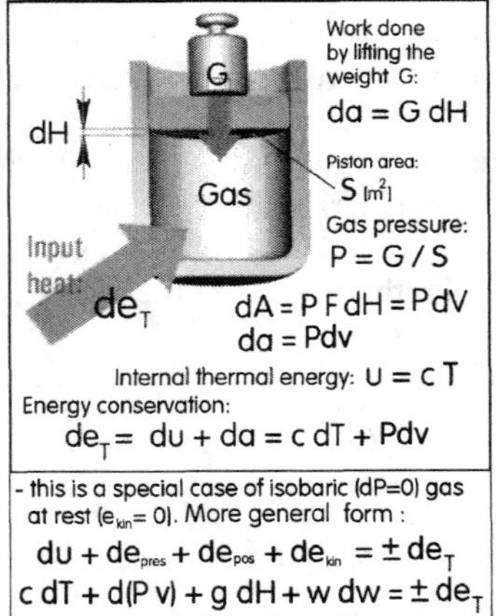

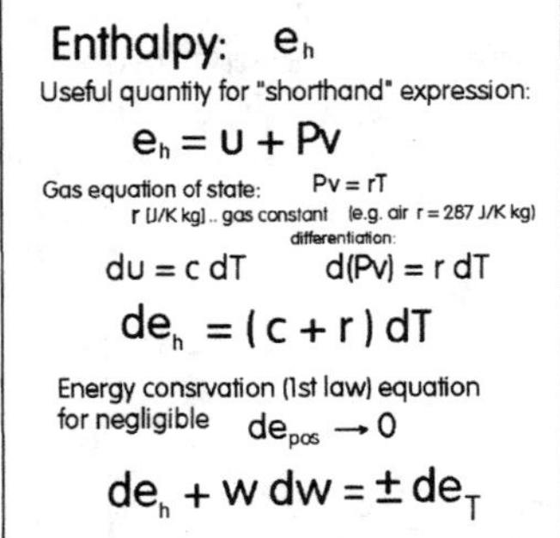

Figure 2.51 (Left) The starting point in investigation of the characterization of compressible gas flow is consideration of the work A (or specific value a) done by a gas heated at a constant pressure.

Figure 2.52 (Right) Definition of enthalpy as a useful auxiliary variable, expressed using the approximated version of the surface from Figure 1.42 in the gas region. The result is a simplified version of the specific energy conservation law from the bottom part of Figure 2.51.

The basic problem with compressibility is its introducing additional degrees of freedom: the varying specific volume v is dependent not only on pressure but also on temperature (Figure 1.42). One degree of freedom from this dependence is removed by an equation for the properties of the fluid. This is the equation of state, here used in the approximated form $Pv = rT$ as shown at the right-hand side of Figure 1.42. This simplified version is in common usage for engineering computations. Yet another degree of freedom has to be then removed. This is done by specifying the dependence between the temperature T and pressure P, using the basic conservation concepts derived in Figures 2.51 and 2.52, the latter introducing the simplifying concept of enthalpy.

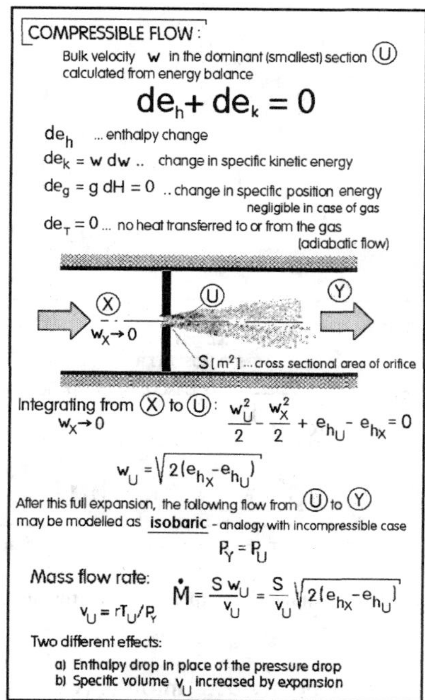

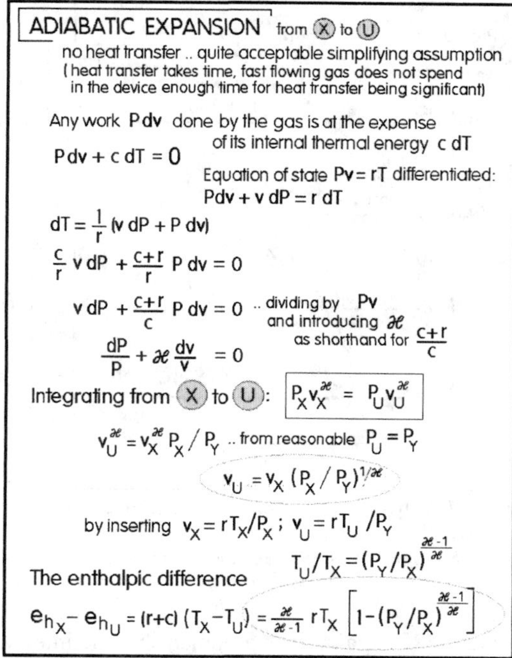

Figure 2.53 (Left) Further development of integrating the energy conservation from the simple incompressible case in Figure 2.35 to the compressible flow.

Figure 2.54 (Right) The enthalpic drop and expansion evaluated between the locations **X** and **U** in Figure 2.53 for adiabatic isentropic flow of idealized gas.

This dependence is complicated by the fact that temperature changes involve heat transfer between the gas inside the fluidic device and the walls, and perhaps even with whole surroundings. In theory the heat transfer problem unfortunately depends on minute details of the temperature field not only inside the device but also outside. Investigation of the fluid flow through a fluidic device with generation of an internal jet to follow the energetic changes in the incompressible case, as presented before in Figure 2.35, is for the compressible flow replaced by the model shown

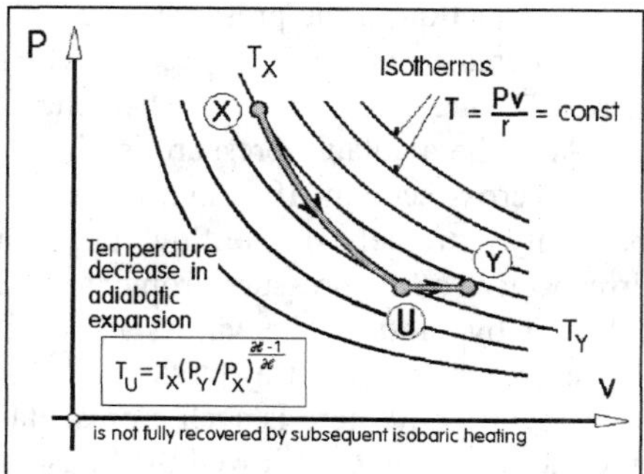

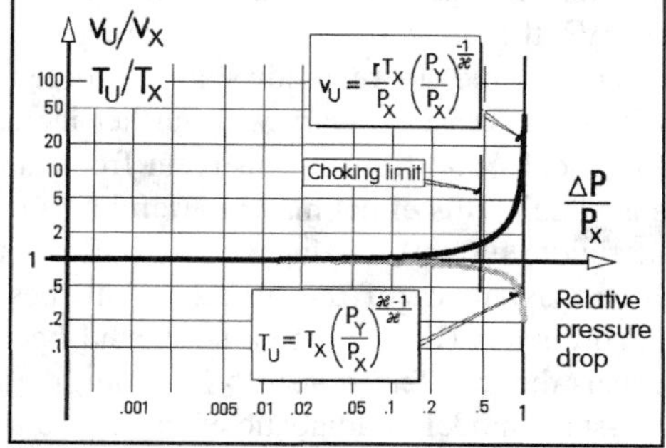

Figure 2.55 (Left) The trajectories of the adiabatic expansion and isobaric dissipation between the locations X, U, and Y (Figure 2.53) plotted on the surface from Figure 1.42 observed along the temperature axis.

Figure 2.56 (Right) Volume increase and temperature drop in the adiabatic expansion. Computed for $\varkappa = 1.4$ gas.

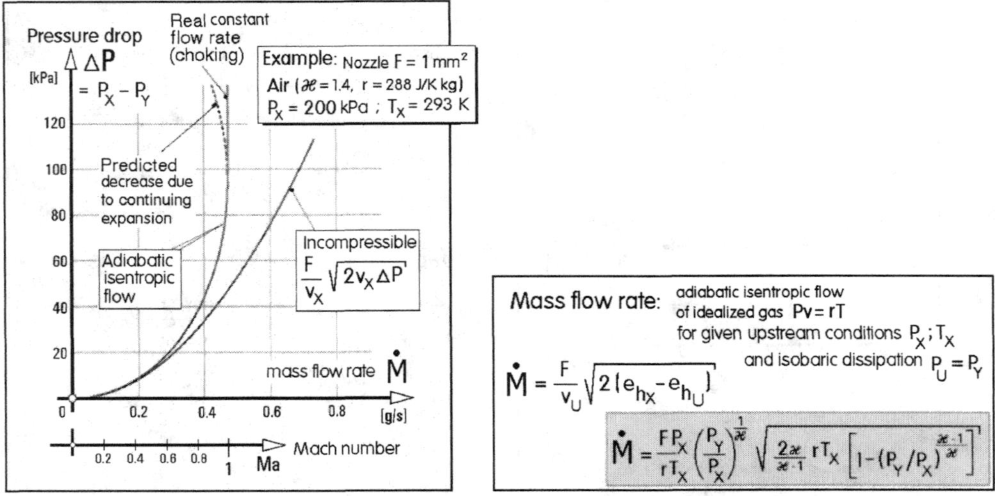

Figure 2.57 (Left) Mass flow rate computed for different acting pressure differences for incompressible and adiabatic incompressible flows.

Figure 2.58 (Right) Resultant expression for the mass flow rate with the enthalpic drop expressed from the given upstream and downstream pressure values.

in Figure 2.53. Again, it is assumed that the internal flow may be divided into the two segments: first an acceleration of the fluid from the inlet X up to the smallest cross section in the exit location U, followed by the dissipation of the kinetic energy between U and the outlet Y. In the acceleration segment it is an acceptable first approximation to neglect the dissipation, conversion into heat and heat transfer, so that the consideration of the thermodynamics of the gas state changes (Figure 2.51) leads to the expression derived in Figure 2.54. This is inserted into the mass flow rate formula in Figure 2.53. In the second, dissipation segment of flow between U and Y the process may be reasonably simulated by the isobaric state change.

Before discussing the result, two observations may be helpful. First, Figure 2.56 emphasizes again that the pressure drop ΔP across the device must be indeed a very large proportion of the inlet pressure P_X for the deviations due to the expansion effect being noticeable. Second, Figure 2.57 brings a warning that blindly following the formula may lead to physically impossible conclusions. It presents the computed mass flow rate evaluated from the resultant expression in Figure 2.58 for the assumed adiabatic isentropic process as a function of the pressure drop ΔP. At very large ΔP the computed values $\dot{M}$ are predicted to decrease with increasing ΔP. This is physically impossible. In real flows it is avoided by the choking phenomenon: when the flow reaches $Ma = 1$ it ceases to vary with further increase in ΔP. The available pressure is dissipated in standing shocks that form downstream from the smallest cross section. If a high-pressure gas source is available, this effect may be useful as a mechanically extremely simple flow regulator to supply the fluidic system, keeping a constant flow rate irrespective of the pressure drop changes.

The thermodynamics of the processes are best followed by plotting the values in the heat diagram coordinates (Figure 2.59), so named because the area below the trajectory of the state change equals the transferred heat to (or from) the gas. In Figure 2.60 there is such a presentation for the two-stage model of adiabatic expansion to the intermediate state U followed by the isobaric dissipation. Also shown there is a curve representing typical real processes: real expansion does not take place with absolutely no heat transfer and also the subsequent heating of the gas by the dissipated heat is not perfectly described by the reversible isobaric model. After all, real processes

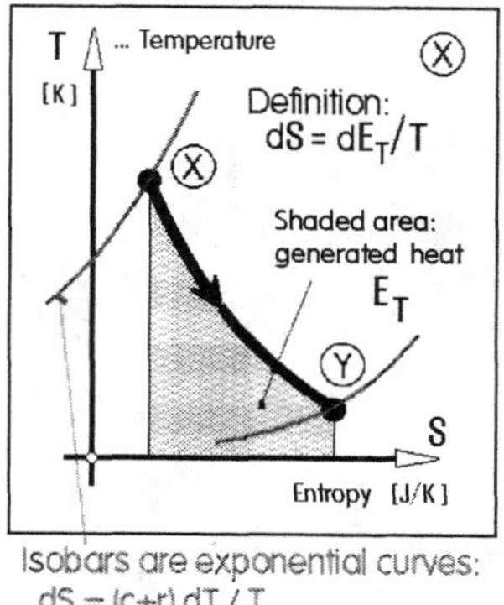

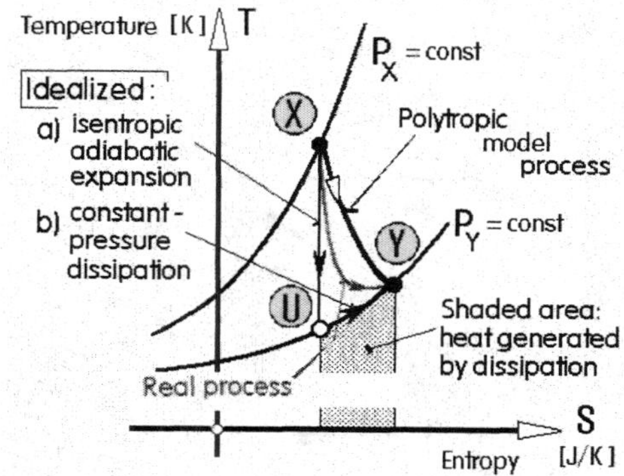

Figure 2.59 (Left) Polytropic model of thermodynamic changes in the course of the compressible flow between the inlet X and outlet Y, plotted in the heat diagram coordinates.

Figure 2.60 (Right) The model of flow between X, U, and Y (Figure 2.53) plotted in the heat diagram together with the polytropic relation and a typical real state change.

are not modeled well by textbook reversible changes. A useful way to describe the complete change between the inlet X and the outlet Y (without introducing the auxiliary point U, which may be illusory) is provided by the polytropic law (Figure 2.61). The polytropic expression for the energetic drop has the advantage of being written in the same form as for the adiabatic isentropic process, which actually is one of its two extreme limit cases, and differs from it solely in replacing the ratio of specific heats $\mathcal{H}$ by the polytropic exponent k. This may assume, in dependence on the available opportunity for heat transfer, values between $k = \mathcal{H}$ for zero dissipation on the one hand and $k = 1$ on the other hand for full temperature recovery (or, perhaps, for heat transfer from the device walls to the gas, cooled by the expansion).

The expression in Figure 2.62, based on the polytropic law, is the recommended way of evaluating the drop of the energy across-variable, including the effects of compressibility in a gas. As shown in the example Figure 2.63, which compares the pressure difference with the specific energy difference evaluated according to Figure 2.62 for a particular case (air flow with inlet pressure 200 kPa), the energy across variable differences is practically identical to the incom-

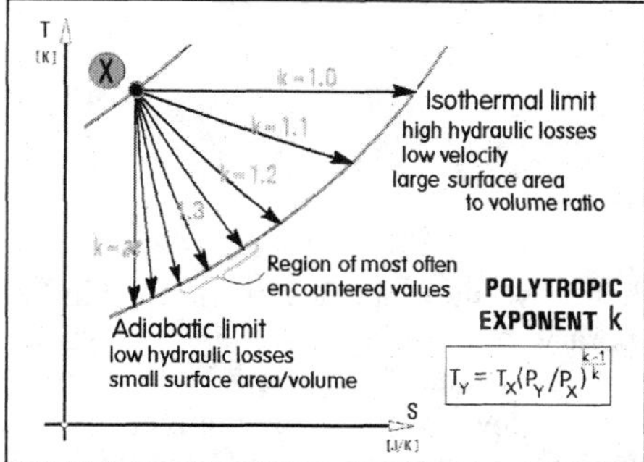

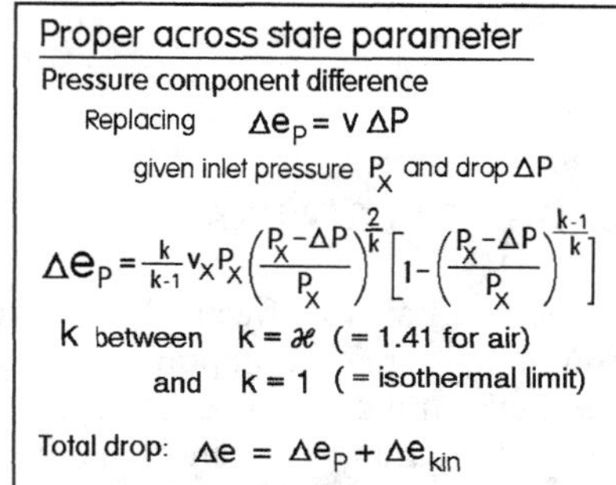

Figure 2.61 (Left) Trajectories of the change expressed by the polytropic model in the heat diagram at different values of the polytropic exponent.

Figure 2.62 (Right) Sum of the pressure component and the total energy drop across the device is the suitable across parameter for Kirchhoff's law (Figure 2.23) in compressible flows.

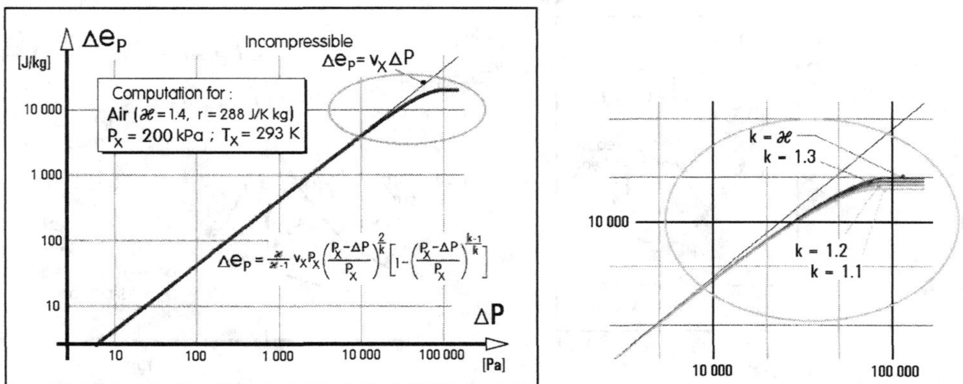

Figure 2.63 (Left) Dependence between the pressure drop and the drop of the pressure component of the specific energy compared with the incompressible model (2.1).

Figure 2.64 (Right) Enlarged end of the diagram from Figure 2.63 showing the different results obtained with different values of the polytropic exponent.

pressible result (2.1) until the pressure drop across the device becomes extremely large. Then the pressure energy becomes lower than obtained by the incompressible (2.1) value. Finally, it levels off as a manifestation of the aerodynamic choking. In the logarithmic scales of Figure 2.63 (and the detail in Figure 2.64) only at this leveling end of the curve does the value of the polytropic coefficient become important.

Computations of the pressure energy drop as the across variable in practice, according to the recommended expression in Figure 2.62, of course, are made somewhat uncertain by the necessity to insert the values of the polytropic exponent k in situations where there are no experimental or computational data. Experimental investigations actually indicate that k is not really constant during the flow through a fluidic device, usually decreasing from the nearly adiabatic initial value, near to $k = \mathcal{H}$, in the locations upstream from the smallest cross section to a value much lower, perhaps nearly isothermal $k = 1$, in the separated flow region. Estimation of the mean value and of heat transfer effects, which determine it, is not easy. On the other hand, the exponent value is seen in Figure 2.63 to be not very influential. Inserting $k = 1.3$ in the case of air may be the best choice - unless there is a very slow flow in a cavity having very large surface relative to the internal value, when a lower value of k is to be inserted.

2.6 LAWS OF FLOW BRANCHING

The problem of how to effectively distribute a fluid flow into branches, perhaps as in Figures 2.65 and 2.66, is of increasing importance in contemporary microfluidics. After all, the final objective of using a microfluidic system is very often some large-scale effect. Microdevices "numbered up" rather than scaled up to obtain such a gross effect, are operated in large numbers in parallel. The chemical synthesis may be a characteristic example. The supply inflow, brought in by a single large pipe, has to be distributed into all the paths through the microdevices. The essential requirement placed on this distribution is the uniformity of conditions in all microdevices. Usually, a reverse process of summing up the flows takes place on the exit side (Figure 2.65).

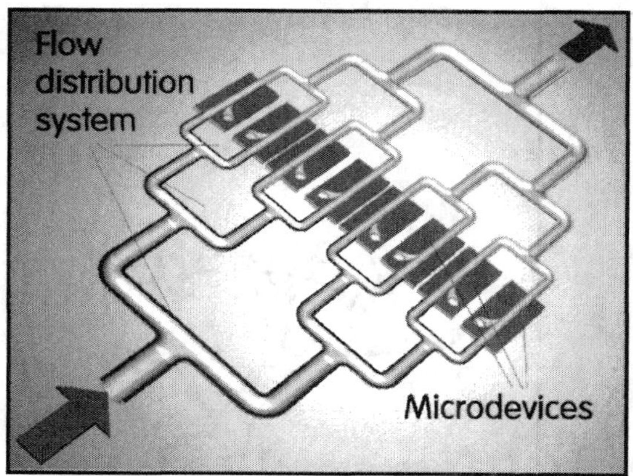

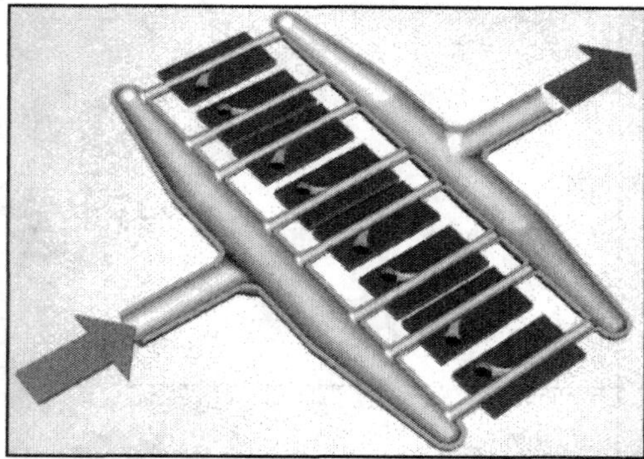

Figure 2.65 (Left) A typical multilevel branched distribution system supplying fluid into fluidic devices operated in parallel. The mirror arrangement of branches on the exit side performs the reverse task of flow summation to produce the desirable large output.

Figure 2.66 (Right) The branching may be replaced by a common manifold chamber open in parallel into all microdevices. A similar common collector vessel is again on the exit side. This layout may save space, but is obviously not favored by natural selection in living beings. It fails, unless the hydraulic resistance of the microdevices is very high, to secure equal flowpaths needed for equal distribution of flows.

Apart from the flow distribution role, there may be other good reasons for using the flow distributing networks. As an example, the two systems of inlet branches in Figure 2.68, bringing close together two fluids, is one of the several possible solutions of the fluid mixing. The task there is to bring the two mixed fluids into close mutual neighborhood so as to shorten the diffusion path of the molecular gradient transport performing the final equilibration.

In practice, the flow distribution systems often occupy more volume than the microdevices themselves. In planar designs (as will be shown below in Figures 2.76 and 2.77), the necessity to accommodate the branching channels may critically restrict the freedom of placement of the micro-devices. Effective design of the branching therefore presents a considerable challenge, becoming more important as microfluidic systems tend to become progressively more and more complex with a larger number of devices.

This problem is not new: Nature had to solve essentially the same task to sustain cells in higher living organisms by supplying them with fluid nutrients and oxygen. The cells are also "numbered

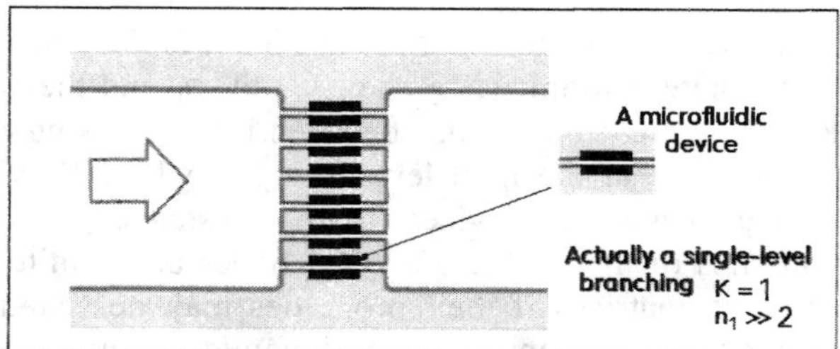

Figure 2.67 The principle of a membrane, evolved from the common manifold principle (Figure 2.66) by decreasing the size and increasing number of microdevices so that they occupy a surface perpendicular to the (local) flow direction. To save space, the membrane surface may be convoluted. The problem is again to achieve an equal distribution of flow into the devices, which must exhibit high hydraulic resistance to approach this goal.

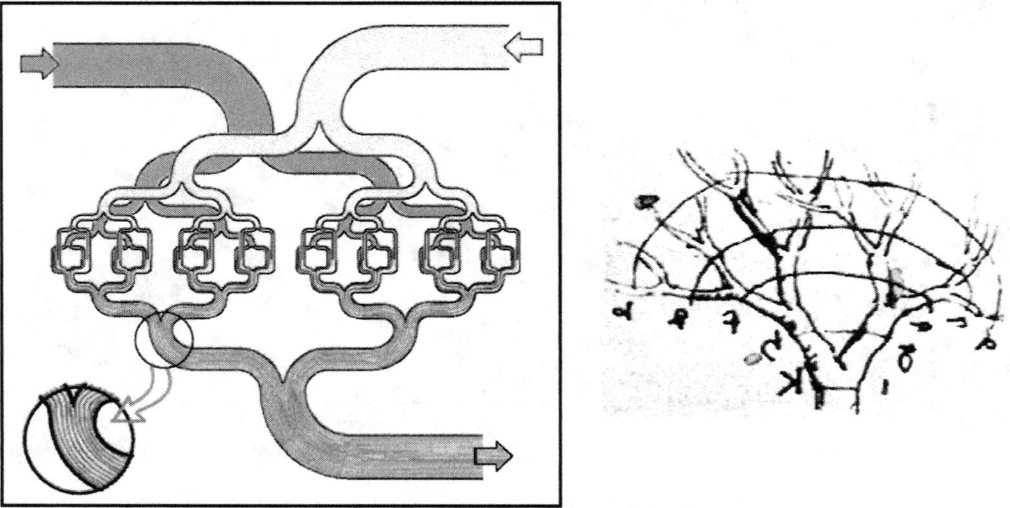

Figure 2.68 (Left) Principle of one type of microfluidic mixers. There are two independent branched channel systems (in two planes of a multiplane arrangement) at the inlet side that produce the interleaved strips of the two mixed fluids. These are typical for most microfluidic (no turbulence) mixers. Strips are thin and this reduces the distance to be traversed further downstream by Fickian low Re diffusion which finally removes the concentration gradients.

Figure 2.69 (Right) A sketch from Leonardo da Vinci's notebook [5] depicting his theory of tree branching, according to which the total cross section of branches at each branching level is constant and equal to the cross section of the trunk.

up" rather than scaled up. Of many conceivable arrangements, natural selection found and promoted as the most effective principle the multilevel branching, with tube size progressively decreasing at each branching level. Mammalian respiratory (Figure 2.70) and cardiovascular circulatory systems are an example. Also botanical organisms, plants, use the branching principle (Figures 2.69, and 2.71). The desirable identity of conditions at the cell level is achieved by maintaining sameness of hydrodynamic conditions at each branching level, where the small number of branches makes the task easier than in multiexit manifolds of Figure 2.66. Designers of microfluidic systems can learn many useful lessons from these natural optimum solutions.

2.6.1 Branching factors

A typical case of a multilevel branching is shown in Figure 2.72. Channels at the same k-th level are of the same size, here determined by their width b_k and their length l_k. A channel belongs to the level k if there are k branchings between it and the single large input channel (level 0). The microdevices are at the final level $k = K$, where K is the total number of branchings between the inlet and the devices. In real systems, one of the most difficult tasks is evaluating properties of the nodes. Since the nodes are multiterminal devices, their characterization needs using matrices if their properties may be stated as linear, which is usually not the case since they are dominated by dynamic effects associated with changes of flow direction, otherwise, it is necessary to work with quadratic tensor equations, too complicated to be handled here. To reduce the complexity, the pressure conditions in the nodes are simplified to the very extreme of neglecting their hydrodynamic properties at

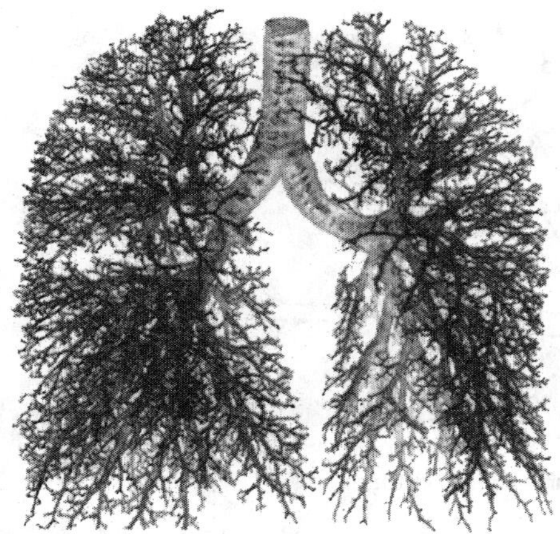

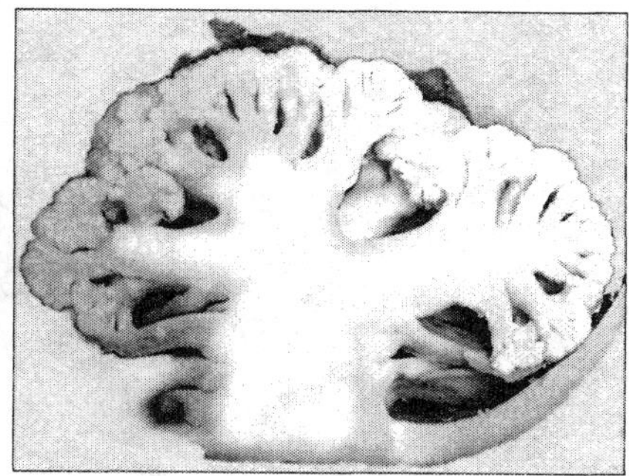

Figure 2.70 (Left) System providing useful lessons for microfluidics: human respiratory system up to and including the $k = 7$-th branching level (i.e. without the smallest-level branches, the inclusion of which, since they approximate a space-filling fractal, would make the branching design impossible to follow).

Figure 2.71 (Right) Branched supply system of cauliflower. Even though at the final level the "microdevices" are effectively on a surface, the branching principle is used rather than the principle of the membrane Figure 2.69.

all. In the branched system, as a result, the pressure conditions are assumed to be determined only by the pressure drops due to friction in the channels. Geometry of the branching network is characterized by the distal-to-proximal channel size factors, the width ratio β_k and the length ratio γ_k (Figure 2.72). The kinematic factor, which characterizes the distribution of the flows into the branches, is

$$u_k = w_k / w_{k-1} \tag{2.2}$$

the velocity ratio, where velocity w_k is averaged over the channel cross section and, in pulsatile flows, over time. The ratio of pressure drops across the branches is

$$p_k = \Delta P_k / \Delta P_{k-1} \tag{2.3}$$

The total number of branches at a level k (assuming a single aorta, $n_0 = 1$) is

$$N_k = n_1 n_2 \dots n_k = \prod_{j=1}^{j=k} n_j$$

where

$$n_k = N_k / N_{k-1} \tag{2.4}$$

It should be noted that the most popular alternative in man-made systems (e.g., Figure 2.65) are the *bifurcation* networks with $n_k = n = 2$ at all k levels. Figure 2.72 presents the trifurcation case $n_k = n = 3$. When available space is to be used efficiently, the branching pattern resembles a space-filling fractal – of course not sharing its theoretical property $K \to \infty$. Fractal objects as well as known efficient branchings are typically self-similar, with not only $n_k = n$, but also the values of the geometric factors $\beta_k = \beta$ and $\gamma_k = \gamma$ the same at all k levels. In the self-similar case, the expression for the number of all branches at the level k simplifies to

$$N_k = n^k \tag{2.5}$$

The essential requirement is the equal distribution of the flow into the branches, the discussion may be limited to *equipartition* branching, with all daughter branches at a level k identical.

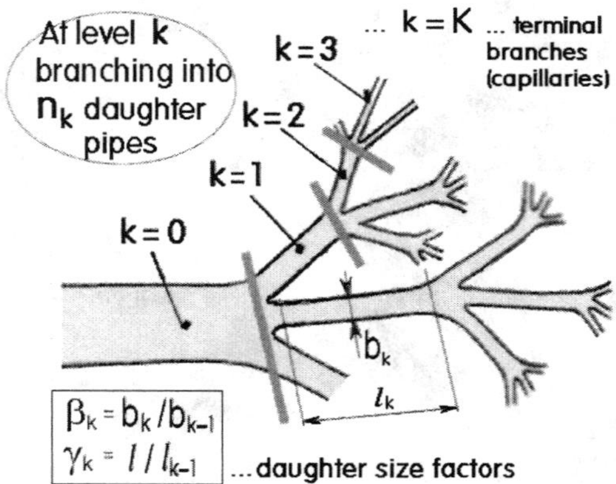

Figure 2.72 At a level k between the "aorta" channel ($k = 0$) and the "capillary" channels ($k = K$) there are N_k channels or pipes. The branching factors n_k,, β_k, γ_k ,... are the ratios of the distal to proximal quantities. Very often —but not necessarily—the system is self-similar, with identical values of the factors at each branching level.

The isokinetic alternative

The most popular version of engineering branching layouts is the *isokinetic* (or da Vinci's) branching $u_k = 1$. In particular, most often encountered is the bifurcating ($n = 2$) self-similar ($u_k = u = 1$, $n_k = n$) isokinetic case. However, data for naturally developed fluid flow branching networks do not support the opinion that this alternative is the best choice. For example, in cardiovascular systems the mean velocity w_K of blood flow in the capillaries ($k = K$) is much slower than in the aorta. In human arterial system the ratio of capillary/aorta velocities is $w_K / w_0 \approx 10^{-3}$

2.6.2 Comparison with data for biological branchings

With millions of years eliminating less successful solutions and much deeper branching levels due to the number of terminal microdevices much higher—of the order of 10^{10} (= the number of capillaries) in the human cardiovascular system—the biological flow distribution systems are more advanced than contemporary engineering microfluidics. Despite the dissimilarities, it is useful to consider the mammalian arterial systems as a reference. With data available from investigations by a number of researchers, they provide useful numerical background data to verify the derived branching laws.

It is, of course, necessary to keep in mind the different conditions in man-made and natural branching systems:

(a) The manufacturing techniques of microfluidics prefer planar designs. The space-filling tree of Figure 2.70 cannot be directly copied. The round cross sections of the tubes are also very exceptional in present-day microfluidics.

(b) In plants, especially trees, studied around 1500 by Leonardo da Vinci [5] (Figure 2.69), the branching laws are strongly influenced by the different underlying principle. The long plant cells pass through several subsequent branching levels so that the cross-sectional equivalence is inevitable.

(c) Plants grow continuously and the branching system must be arranged so as to cater to the necessary presence of budding branches.

(d) In pulmonary systems, the diffusion across the pipe walls becomes progressively important as the size decreases. Optimality criteria valid for convective transport at large scales cease to be applicable at the diffusion dominated levels.

(e) The mammalian distribution trees operate in unsteady periodic flow regimes. In respiratory systems the flow direction is reversed at each half of the operating cycle. This double (forward and backward) use of the same branching tubes, as opposed to the two separate systems in Figure 2.65, saves space (and tissue) but makes the hydrodynamics more complicated. Perhaps somewhat strangely, this reversing principle is not used in the cardiovascular circulatory system. Nevertheless the hydrodynamics is also complex there, due to the operation in pulsatile regime.

Mammalian respiratory and arterial systems are of equipartition bifurcational type with round pipe cross sections. The channel widths and their ratio β_k are here replaced by diameters d_k and

$$\beta_{dk} = d_k / d_{k-l} \qquad (2.6)$$

Assuming the self-similarity, the goal is to identify the dependence of the four parameters β_d ; γ ; u ; p on the branching number n. Some conditions mutually relating these variables are obvious:

(A) The first one is a straightforward consequence of incompressible (constant volume flow rate) version of the mass conservation law. Under the simplifying conditions of incompressibility $v = = \text{const}$ and no accumulation in the network, the proximal and distal total volume flow rate at all branching levels are equal

$$N_k \dot{V}_k = N_k \pi d_k^2 w_k / 4 = N_{k-1} \dot{V}_{k-1} = N_{k-1} \pi d_{k-1}^2 w_{k-1} / 4$$

so that from (2.2), (2.4), and (2.6) $\qquad n_k \beta_{dk}^2 u_k = n \beta_d^2 u = 1 \qquad (2.7)$

(B) With the simplification of the pressure drops due only to frictional losses in the pipes, it may be plausible (even though somewhat more forced) to assume that at the microfluidic scale the loss may be computed using the Hagen-Poiseuille linear dependence on the volume flow rate:

$$\Delta P = R \dot{V} \quad \text{with} \quad \dot{V} = \pi d^2 w / 4 \quad \text{and} \quad R = \frac{128 \nu}{\pi v d^4} l_k$$

Of course, pipes in pulmonary or arterial systems are usually not long enough for the flow being fully developed, as the Hagen-Poiseuille law assumes. Nevertheless, this may be perhaps at least a suitable starting point.

The pressure drop at level k is proportional to $\Delta P_k \sim l_k w_k / d_k^2$, which means

$$p \beta_d^2 = \gamma u \qquad (2.8)$$

Note that in the self-similar isokinetic bifurcating case $u = 1$, $n = 2$, (2.7) reduces to

$$\beta_d = n^{1/2} \qquad (2.9)$$

and the numerical values of the branching factors are:

$$\beta = n^{-1/2} = 0.707, \quad \gamma = n^{-1/3} = 0.7937, \quad p = n^{2/3} = 1.587$$

(C) If the system is not isokinetic, other assumptions are needed. An important assumption was introduced by Bengtsson and Edén [6]. It states that arterial systems are built with constant power dissipated per unit pipe wall area. This secures an equal stress distribution among all the constitutive cells, keeping an equilibrium in the body.

The constant power loading (disspated power $\Delta P \dot{V}$ per unit mantle area $\pi\, d^2 l/4$) in both proximal and distal branches at the k-th level meets the condition

$$4\, N_k\, \Delta P_k\, \dot{V}_k / (N_k \pi\, d_k\, l_k) = 4\, \Delta P_{k-1} \dot{V}_{k-1} / (\pi\, d_{k-1}\, l_{k-1})$$

which means

$$\frac{p\, u\, \beta_d^2}{\beta_d\, \gamma} = 1 \quad \text{and, in view of (2.8) leads to} \quad u^2 = \beta_d$$

Inserting this relation into (2.8) and (2.9) results in

$$u = n^{-1/5}$$
$$\beta_d = n^{-2/5} \tag{2.10}$$

The relation between the two remaining undetermined factors is

$$\frac{p}{\gamma} = n^{3/5}$$

(D) The final condition to be met in biological organisms follows from the requirement of the volume filling character of the network. For the assumption $d < l$, necessary for validity of Hagen-Poiseuille law,

$$N_k\, l_k^3 = N_{k-1}\, l_{k-1}^3$$
$$\gamma = n^{-1/3}$$
$$p = n^{4/15}$$

Indeed, this means $\beta/\gamma = n^{-1/15}$, the arteries become more elongated as the branching level progresses. If this were not the case, this final condition would become $N_k\, d_k^2\, l_k = N_{k-1}\, d_{k-1}^2\, l_{k-1}$, resulting in

$$\gamma = n^{-1/5} \tag{2.11}$$

The derived branching laws may be compared with available data. The dimensionless (2.10), is satisfied in the self-similar network with $d_k \sim N_k^{-2/5}$ $d_{k-1} \sim N_{k-1}^{-2/5}$ at each level k.

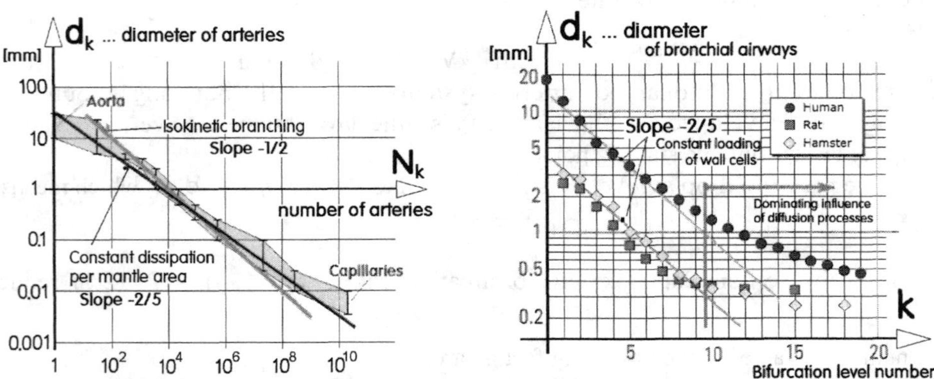

Figure 2.73 (Left) Verification of the used branching laws using dependence of the diameter d_k of human arteries on the branching level k, which is here characterized by the number of branches N_k at the particular level. The slope for constant dissipated power loading of the arteries wall fits available data [6, 7] (the shaded area indicates typical uncertainty) visibly better than the isokinetic branching law.

Figure 2.74 (Right) Measured [8, 9] diameters d_k of mammalian bronchial tubes at various branching levels k. For size large enough for diffusion across the walls being insignificant, the slopes correspond well to the dependence derived for the constant dissipation per tube mantle area.

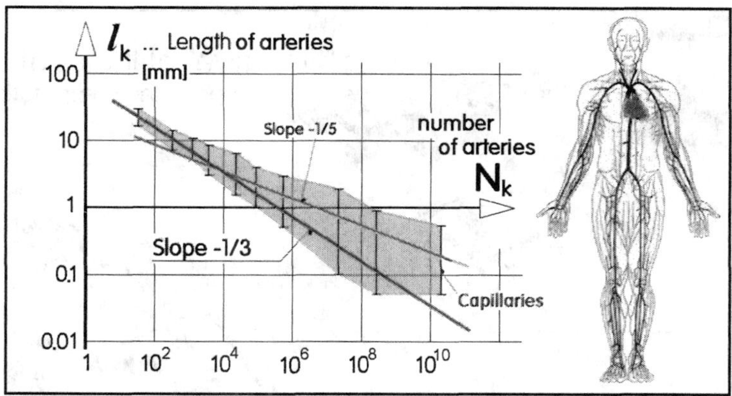

Figure 2.75 Measured lengths l_k of human arteries at a branching level k (again characterized by the number of branches N_k) [6, 7] compared with the slopes corresponding to volume filling by long tubes, slope -1/3, (2.15), and by tubes of larger diameter-to-length ratio, slope -1/5 (2.16).

This is shown to hold both in Figure 2.73 for the human arteries and in Figure 2.74 for mammalian lungs - in the latter case, of course, only in the convection dominated region of interest here, at small $k < 9$.

Similarly, the distribution of human arterial lengths in Figure 2.75 supports reasonably the space filling law $\gamma = n^{-1/3}$ for narrow long tubes, even though the scatter does not completely exclude validity of short-tube law (2.11), except at small k, where filling the space by short tubes is out of the question anyway.

2.6.3 Optimality criteria dictated by manufacturing technology

Microfluidic flow distribution networks have to meet criteria similar to A, B, C, and D, but the latter two are dictated by technological rather than biological reasons. The contemporary planar manufacturing methods lead to different geometric conditions. Typically, the channels are etched to a constant depth h everywhere (Figure 2.76). The geometric branching factor β is now the ratio of channel widths b (Figures 2.72, and 2.77). The criteria A, B, and D, adapted accordingly to these geometric constraints are then:

(A) $n\,\beta\,u = 1$

(B) $p = \gamma\,u \qquad \alpha = p\,\beta\,u = \gamma\,\beta\,u^2$

(D) Plane filling property $n\,\beta\,\gamma = 1$

The criterion C, justified for biological organisms, is replaced by a different constraint. Practically more important is the overall *pressure drop* across the branching. Alternatively, this criterion may be formulated as desirability of small power dissipated by friction.

This, however, cannot be the decisive criterion alone as it would simply lead to the channel cross sections as large as possible together with short channel lengths. This would jeopardize the paramount requirement of equal distribution of the flows, because it would tend to lead to the membrane case (Figures 2.66 and 2.67) with its problematic flow distributions into the channels.

In fact, the large width of the distribution channels is likely to cause design problems. This is particularly obvious for the microdevices distributed in the base plate plane, shown in Figures 2.76 and 2.77 rather than in line (as was the case in Figure 2.65). The devices in Figure 2.76 are cooling micronozzles, generating impinging jets. Ideally, they should cover

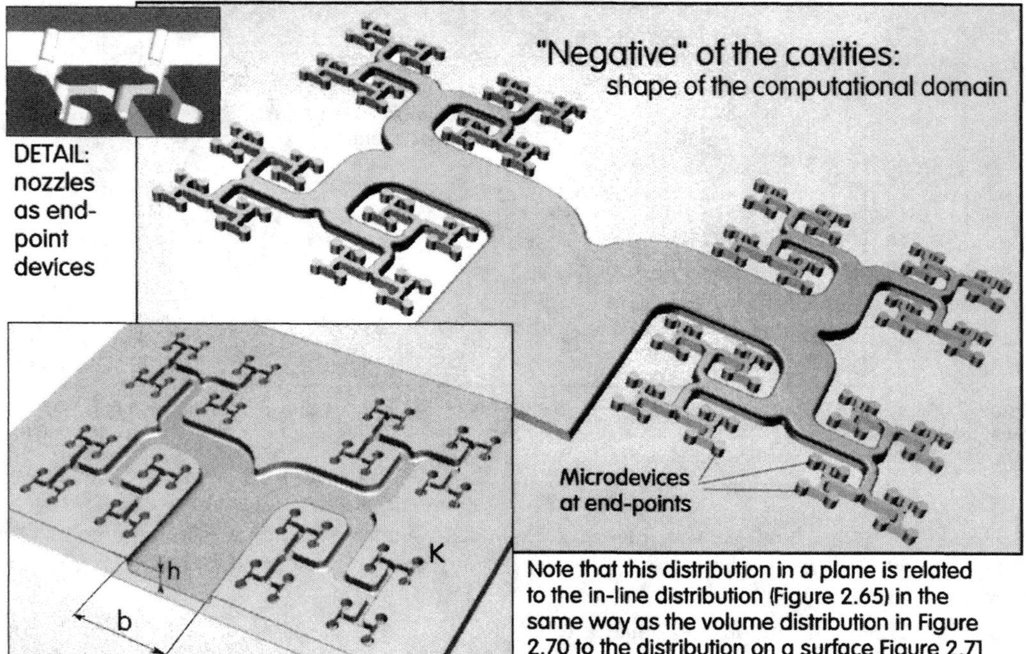

Figure 2.76 The H-scheme of bifurcating supply channels into microfluidic devices at end-points K distributed across the plane (rather than in-line as in Figure 2.64). If the fluid is not stored, consumed, or left to leave (irrigation, impinging microjets), and the design is complicated by the necessity of multilayer configuration providing in another parallel plane the space for the exit network.

the plane uniformly. However, if the widths of the supply channels are large, it is difficult to fit them between the nozzles. The designer is forced to concentrate the nozzles in clusters with voids between the clusters – if the distribution in the plane is to be regular and the condition of equal path lengths to the nozzles is to be met. The problem is more obvious for larger number n of daughter channels (Figure 2.77). A nonisokinetic branching law $\beta > 1/n$, making the placement of the supply channels easier, is welcome even if it means increased total dissipative loss.

Another reasonable optimality criterion is the small total *volume* of the distribution network. It may be a useful design goal because of the usual imperative requirement placed on microfluidic systems - their small overall dimensions. Also, under the usual condition of incompressibility, the small total volume is equivalent to minimum *residence time*. This may be an important criterion especially for transitional regimes like start-up, rinsing, change of reagents in microchemistry, or depressurization. Characteristically, an optimization with respect to this criterion alone would lead to extremely narrow channels, the very opposite to the consequences of applying the pressure loss criterion.

What is needed is a compromise choice. This may seem to be the power dissipated per unit volume, leading to the criterion C in the form

$$n\,\alpha / (n\,\beta\,\gamma) = u^2 = 1$$

Unfortunately, this is equivalent to the da Vinci's isokinetic $u = 1$, $\beta = 1/n$, dismissed as leading to the excessively large substrate area occupied by the low k level channels.

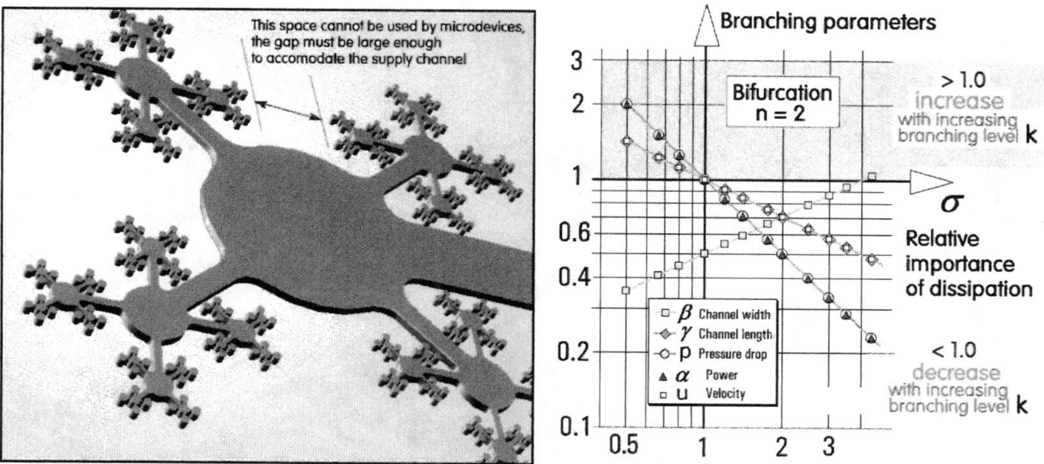

Figure 2.77 (Left) A layout with end-points across a plane: quadrifurcation n = 4 scheme of alternating " + " and "x" crosses (shown here again in the "negative"). Similar to Figure 2.76, the necessary space for the wide supply channels makes achieving an equidistant end-points distribution practically impossible if the flowpaths are to be of the same lengths.

Figure 2.78 (Right) Values of the branching parameters evaluated with the compromise criterion C with different dissipation weights σ for the bifurcating branchings.

A more general alternative is to place a different importance weight on the dissipation by

$$\sigma\, \alpha / (\beta\, \gamma) = 1 \qquad (2.12)$$

with adjustable weight factor σ. This permits covering the whole range between the minimum volume $\sigma=1$ (or even $\sigma<1$) and the more emphasis placed on minimal losses with $\sigma>1$. The latter is more important in devices spending most of their operational time in steady regime where the minimum residence time is of secondary concern. The corresponding values of the branching factors u, β, γ, α, and p are plotted as a function of the chosen σ in Figure 2.78, valid for the bifurcating network n = 2. Note that the factor u of velocity distribution into the branches varies with σ equally as the length ratio factor γ. Also the pressure and power change curves coincide. The velocity change in the branching

$$u = 1/\sqrt{\sigma} \qquad (2.13)$$

means lower velocity in the daughter channels for $\sigma>1$, corresponding to the larger channel widths factor

$$\beta = \sqrt{\sigma}\,/\,n \qquad (2.14)$$

Different requirements are used if the task is to distribute the flows into a device array positioned along a line, Figure 2.65. A detailed example of such branching network with channel bends at the inlet as well as at the outlet is shown in Figure 2.79. Contrary to the plane case, there is no strong motive for the non-isokinetic branching law $\beta > 1/n$. It is easy to derive the expressions for the channel length l_k decrease with progressing branching level k. If the relative pitch a/b_o of the endpoints is within reasonable limits, the rounded geometry (surprisingly perhaps) has no effect on the bifurcation factors γ and β, the values of which are the same as derived above for the optimality condition of filling the plane.

If Reynolds numbers of flows are sufficiently high for significant dynamic effects, behavior of the downstream summation circuits may be usefully influenced by shaping the summation nodes, Figures 2.80, 2.81, and 2.82.

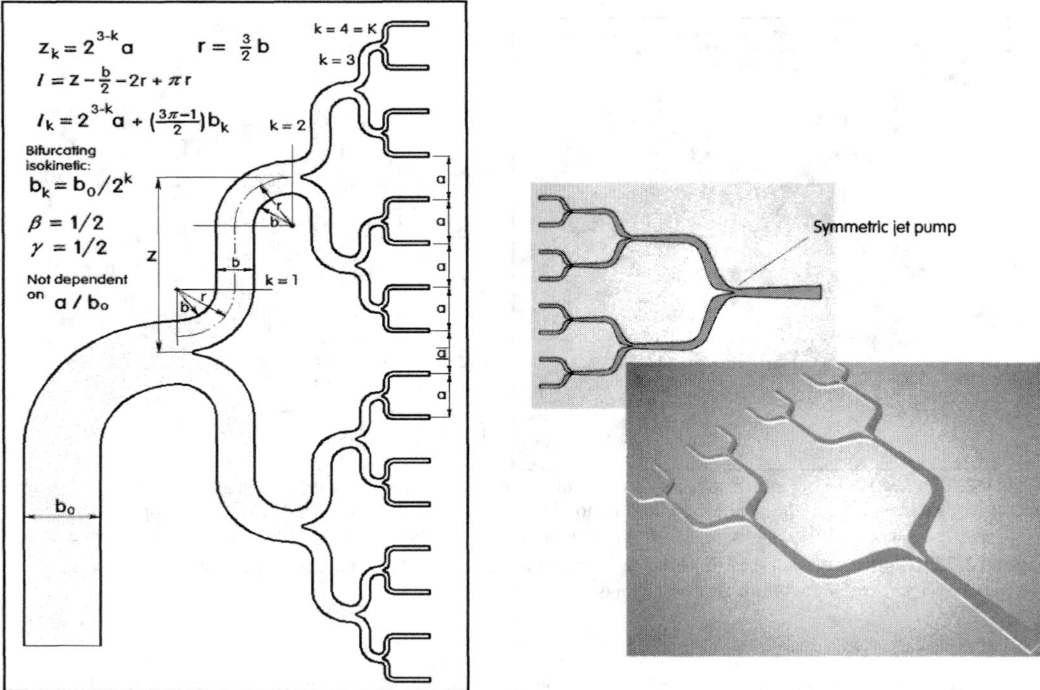

Figure 2.79 (Left) Planar microfluidic flow distribution network with endpoints distributed along a line: bifurcating isokinetic n = 2, u = 1, K = 4 self-similar geometry complicated by the rounded channel entrances and exits. The derived expressions may be easily generalized to nonisokinetic cases with various channel shapes and parameters of the geometry.

Figure 2.80 (Right) Jet-pumping effect, see Section 3.8.2, used in the summation nodes can generate cleaning flows used to remove previous samples from the parallel paths in a fluidic sampling system [10]. This is required to prevent cross-contaminations between samples led sequentially into a single composition analyzer.

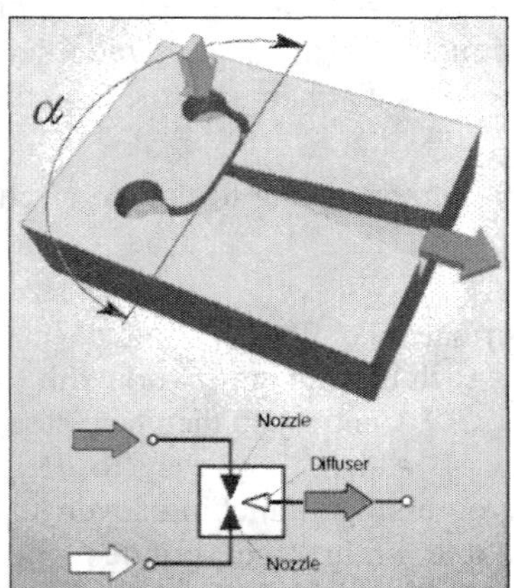

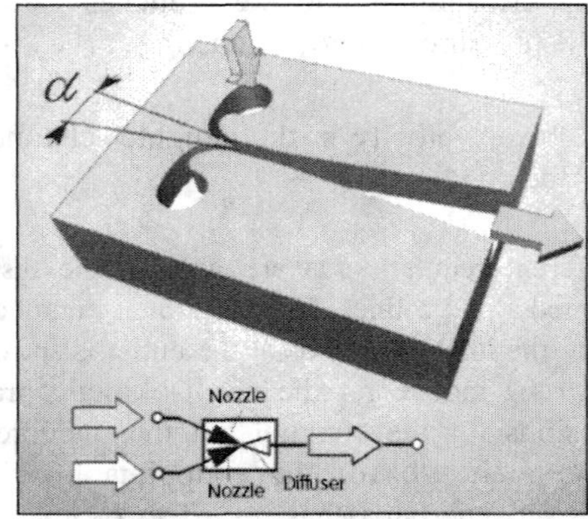

Figure 2.81 (Left) A simple planar fluidic device. The interaction of the incoming flows may be useful in the device's role as a summation node: head-on collision with the opposing flow accelerated in the nozzle inhibits the other flowpath.

Figure 2.82 (Right) Another case of a device with two nozzles and a single collector. The two flowpaths meet at a small angle α and the resultant mutual entrainment leads to jet-pumping effect. It is essentially this device that is used in the example shown in Figure 2.80.

2.7 UNSTEADY FLOW EFFECTS: INERTANCE

The problem discussed in previous parts of the present chapter was characterization of steady-state properties of fluidic devices. In unsteady regimes, with time-dependent mass flow rate $\dot{M}$, the steady-state characterization parameters fail to characterize the device properties completely. In particular, there are two phenomena that have to be taken into the consideration.

One of them, discussed in this part of the chapter, is the additional pressure (or specific energy) drop across the device—positive or negative depending on the sign of the time change (Figure 2.83)—appearing due to fluid inertia. Overcoming the inertial forces when flow velocity varies spends some of the available fluid energy. The other effect, discussed later, is changes of fluid flow. In the unsteady regime, mass flow rate in the device exit may be different from what is supplied to the device inlet. This phenomenon is the consequence of fluid accumulation inside the device.

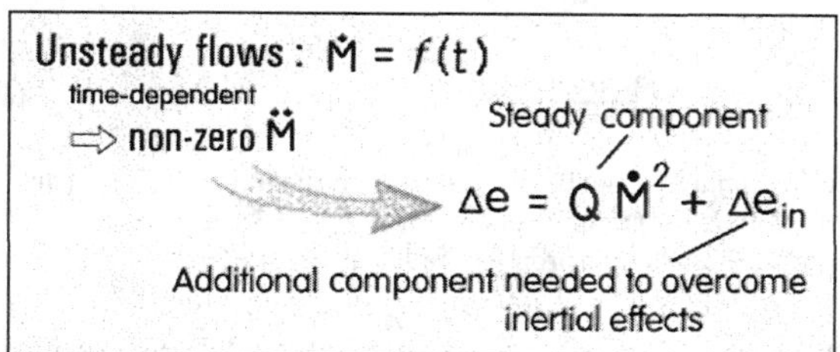

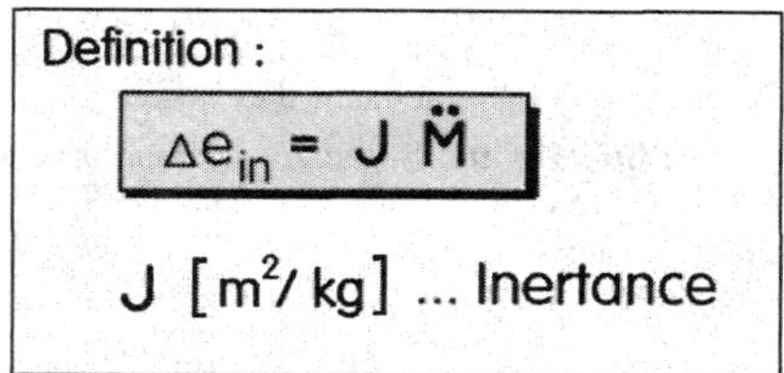

Figure 2.83 (Left) The additional difference in specific energy of the fluid across the device, due to inertia, has to be taken into the account in unsteady regimes.

Figure 2.84 (Right) The concept of inertance, the characterization quantity introduced to quantify the added energy drop, (Figure 2.83). This drop (at least for not very fast processes) is a reasonably linear function of the time rate of change of the mass flow rate through the device.

The inertial effects may become very complex if the change of the flow rate through the device takes place very fast (e.g., at very high repetition frequency of periodic processes). It may lead to complete restructuring of the velocity profiles. Generally speaking these phenomena are less understood than other parts of fluid mechanics – mainly because of the difficulties associated with experimental investigations of rapid change processes. Nevertheless, for most cases in fluidics, it suffices to assume the simple proportionality relationship of Figure 2.84. The proportionality constant, inertance J, then suitably characterizes the behavior of a fluidic device.

Evaluation of inertance is usually done by using the one-dimensional approach: by integrating the acceleration $\dot{w}$ (= velocity derivative with respect to time) along the flowpath

$$\Delta e_{in} = \int_0^l \dot{w}\, dX_1 \tag{2.15}$$

The integration is particularly easy in the case of the constant cross-section channel (Figure 2.85), where velocity, evaluated from the expression for the mass flow rate Figure 2.7, and its derivative are constant along the whole channel length. An important fact in the resultant expression is the proportional increase of inertance with increasing channel length l. It may be often neglected in fluidic elements of short length. On the other hand, in long channels inertial effects may be damped by the inevitable frictional dissipance. The length is stressed in the circuit-diagram symbol, Figure 2.86. Interesting is the appearance of the ratio S/v that appeared already (as a reciprocal value) in the expressions for dissipance in Figures 2.30 and 2.31.

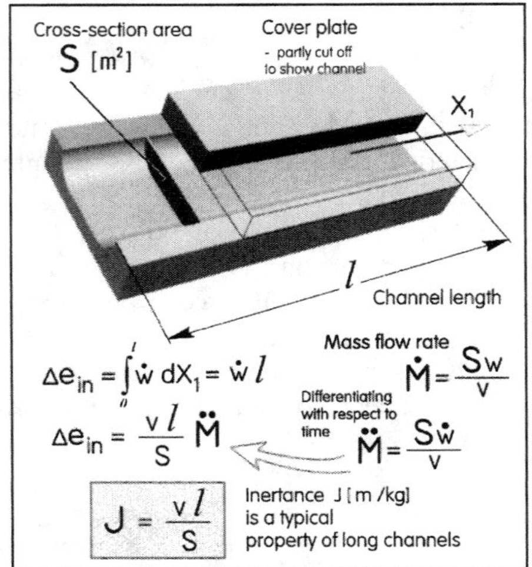

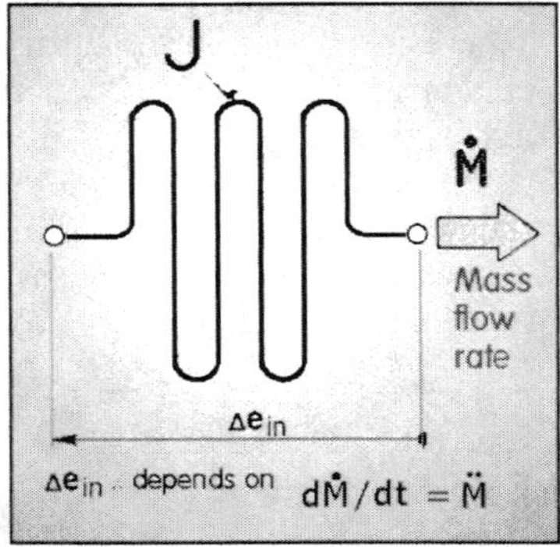

Figure 2.85 (Left) Inertance J evaluated for a channel of constant cross section.

Figure 2.86 (Right) Schematic symbol for an inertance dominated channel in fluidic circuit diagrams. It may really correspond to a very long channel, which are usually folded (Figure 3.7) or coiled (similar to Figure 5.157). Also, it resembles a symbol for coil in electric circuits, which exhibits the analogous property, inductance.

It is useful to be accustomed to typical inertance values, perhaps by means of the diagram in Figure 2.87. What should be noted is that using the *mass* flow rate in the definition (Figure 2.84) causes the magnitudes for liquids to be much lower than for a gas—the very opposite of the general idea that accelerating heavy fluids is associated with spending more energy.

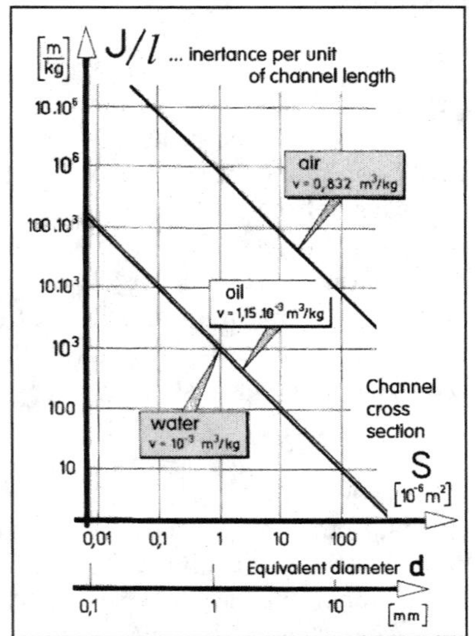

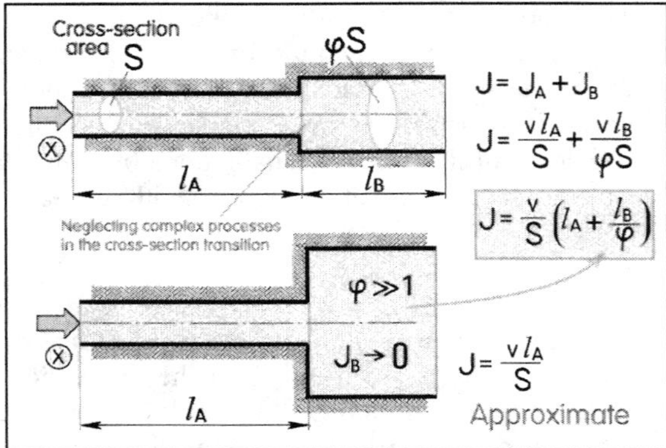

Figure 2.87 (Left) Useful ideas about the order-of-magnitude of inertance values may be obtained using this diagram.

Figure 2.88 (Right) Integration used to evaluate the inertial specific energy drop in Figure 2.85 was particularly easy in the case of the constant channel cross section. It may be extended to the case of the sudden cross-sectional change shown here. Integral of a sum is evaluated as the sum of two integrals, both of which have here, with constant acceleration $\dot{W}$ in each segment, the simple form of Figure 2.85.

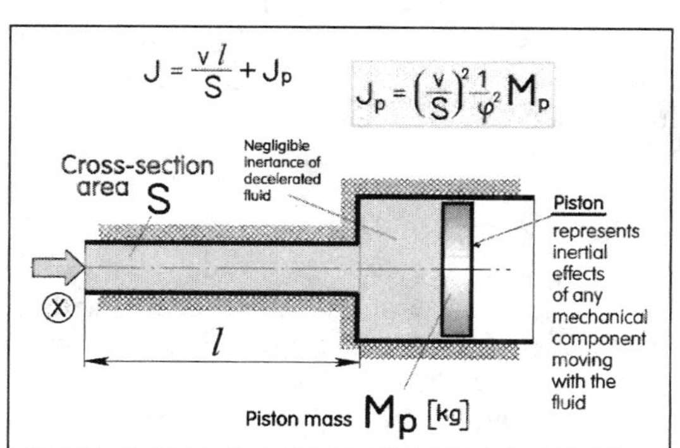

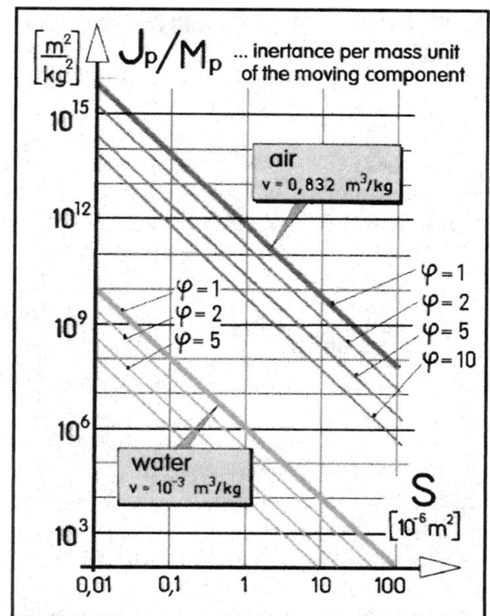

Figure 2.89 (Left) Inertial effects are extremely prominent in the classical mechano-fluidic devices where the fluid motion is accompanied by spatial translations of moving mechanical components, here represented by a piston.

Figure 2.90 (Right) Inertance per kilogram of moving component mass may be usefully found in this order-of-magnitude diagram.

Equation (2.15) may be also easily integrated in the cases of the channel consisting of several parts with constant cross-section in each segment (Figure 2.88). This, of course, is an extreme simplification that neglects the very complex unsteady flow phenomena in the location of the area change. The individual segments enter the expression for the overall inertance with their respective lengths divided by the area ratio φ. If the cross-sectional area of one of the segments is large, as shown in the bottom part of Figure 2.88, then its inertance may be neglected. An important fact seen in the results there is the general rule for evaluating the total inertance of two series-connected devices: the overall inertance is the sum of the component inertances J_A and J_B.

Inertial effects in pure fluidic, no-moving-part devices are generally small, of course, with the exception of devices with long channels such as some mixers. It is a property usually of importance in the interconnecting channels between devices. High values of inertance, however, are encountered in devices with moving components, the consequent slowness of operation (together with more difficult manufacture and worse reliability) being the very reason why the moving parts present a disadvantage. Figure 2.89 shows the inertial properties of moving (or deformable) mechanical components represented by a piston of mass M_p. The acceleration forces acting on the piston are shown in this example (as is often the case) decreased by the piston moving in a space of φ-times larger cross-section area and correspondingly lower velocity. This, however, does not eliminate the effect, as may be found in the following typical case corresponding to Figure 2.89.

Air in a channel 6 mm long, of 0.8 mm diameter ($S = 1.13 \cdot 10\text{-}6$ m^2) drives in an enlarged end chamber of 1.2 mm diameter ($\varphi = 2.25$) a stainless steel piston 2 mm long. Figure 2.90 indicates $J_p/M_p = 0.54 \cdot 10^{12}$ m^2/kg^2. Piston mass is $M_p = 17.8$ mg, so that $J_p = 9.6 \cdot 10^6$ m^2/kg. This inertance value is practically 1 000 times larger than the inertance $v\,l\,/\,S = 9.9 \cdot 10^3$ m^2/kg of the channel.

2.8 FLUID ACCUMULATION: CAPACITANCE

2.8.1 Accumulation mechanisms

Temporary accumulation of fluid (since the process obviously cannot go on forever in a device of finite dimensions) in a device is responsible for the second one of the two phenomena that in unsteady flows upset Kirchhoff's laws (Figures 2.22, and 2.23) balance of mass and energy, the basis of the fluidic circuit solutions. If the acting pressure rises, some of the input fluid is accumulated in the internal cavities and does not leave the device. The accumulated fluid, of course, is missing in the mass budget of the circuit evaluated according to the steady-state rules. Similar mass unbalance is also found during the reverse process of pressure decrease, when some of the accumulated fluid is released from the device.

The characterization parameter used to specify the magnitude of these effects is the capacitance C [kg^2/J]. Most accumulation processes being nonlinear, their progress is best followed on storage characteristic, the dependence of accumulated mass M on the rise of the fluid specific energy Δe with which the accumulation is associated (Figure 2.91). Capacitance C is the slope of this characteristic. The definition of C used here and presented in Figure 2.91 operates with the mass of the accumulated liquid and specific energy drop. This brings the advantage of operating with conserved quantities but provides a somewhat unfair advantage to storage of liquids. Because of their higher density (= smaller specific volume v) the change of mass of a liquid stored even in a relatively small volume is much higher than in the case of a gas.

Already mentioned in Section 1.2.1 is the fact that contrary to analogous electric capacitors, the fluid accumulation is described by a single terminal model, shown in Figure 2.92. This illustration also presents the general schematic symbol used to represent accumulation properties when drawing fluidic circuit diagrams—unless it is desirable to stress a particular accumulation mechanism, like the elastic mechanism schematically represented by the piston and spring (e.g.,

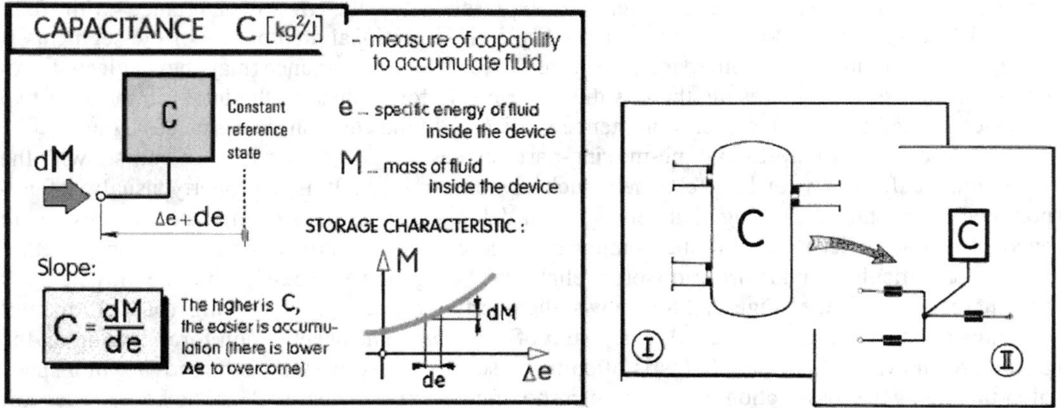

Figure 2.91 (Left) Fluidic capacitance C [kg^2/J] is a measure of accumulating capability of the fluidic device. The accumulation effect is of importance only for unsteady regimes, where it accounts for a temporary "loss" of some fluid (or its emergence), upsetting the steady-state Kirchhoffian mass balance in the circuit.

Figure 2.92 (Right) The single-terminal character of the accumulation effect. Physically the accumulation cavity shown in I possesses three inlet (or outlet) pipes, here shown as containing fluidic restrictors. However, the device behavior is fully described by the presence of the single-terminal accumulator chamber, as shown in the schematic representation II connected to the STAR connection of the three restrictors.

Figure 2.95). The absolute value of fluid specific energy (which would include all the energy associated with molecular and atomic structure) is never used. We work with differences, which in the case of the single terminal can raise the slightly embarrassing question of where is the actual location of the reference state (Figure 2.91) to which the difference Δe is related. In practice, the energy in question is almost always only the pressure component, evaluated using the isochoric model $\Delta e = v\, \Delta P$, with the atmosphere being the reference "terminal".

Even though this chapter describes the mechanisms rather than devices, it is useful to mention here the specialized devices the task of which is to generate a specified—and often also large—accumulation effect. Their specialized purpose makes easier the discussion of the fluid accumulation mechanisms. Four cases representing the four most important mechanisms are presented in the following sections as follows: the use of fluid compressibility (Figure 2.93), liquid gravitational energy when raising in a vertical column (Figure 2.94), compliance due to volume change made possible by elasticity (Figure 2.95), and compliance in a volume confined by surface tension (Figure 2.96).

The mass stored due to compressibility is most pronounced if the working fluid is gas, due to its lesser slope of the pressure-volume dependence (Figure 1.42) than found in liquids. The design of the "fluidic capacitor" employing this mechanism is extremely simple; any cavity will do. A larger cavity can accumulate more gas, but this is just the question of how large capacitance C is asked for. Liquids, being much less compressible, in the same cavity can store transitionally a much smaller volume, but their compressibility factor K (Figure 1.41) is by no means negligible. Since it is defined in terms of stored mass, the resultant capacitance may be surprisingly large. Extreme values may be then achieved with bubbly liquids, with their large mass (of the liquid component) and yet large differential compressibility (of the gas in the bubbles).

If the task is to store large amounts of a liquid, an excellent (and simple to manufacture) solution is the gravitational mechanism of Figure 2.94. If the cavity is prismatic, of constant cross section along its height, the storage characteristic is linear, which may be a desirable property in dynamic systems where, after all, the capacitance is of importance. On the other hand, there is the disadvantage of requiring an alignment with respect to the external acceleration (mostly gravitational, but centrifugal action may also be an option). In some applications the very presence of the vertical columns may be impractical. Also, an inconvenient aspect at small scales typical for microfluidics may be the operation complicated by the inescapable influence of capillary effects on the surface.

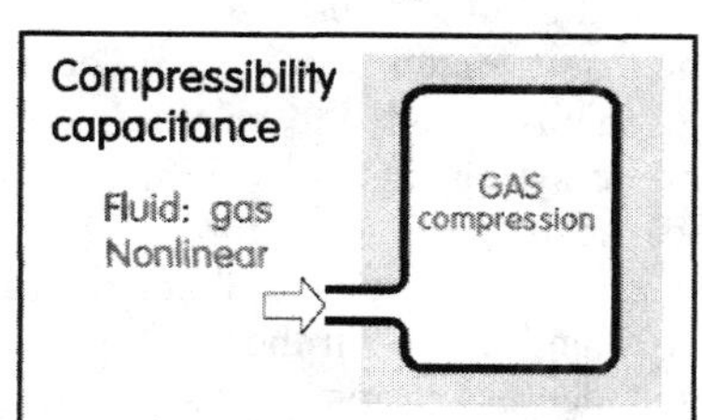

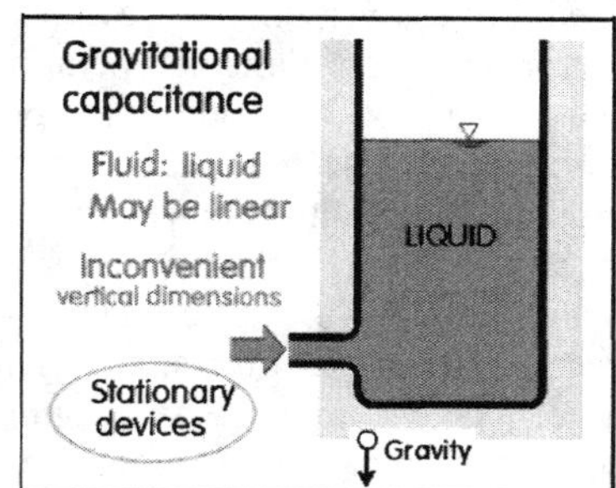

Figure 2.93 (Left) The simplest design of an accumulation device is used if the storage is due to compressibility of the fluid. This storage mechanism, of course, is most effective if the stored fluid is gas.

Figure 2.94 (Right) When working with liquids, the simplest form of the accumulation is obtained by an open vessel in which the liquid surface is left to rise (and fall). This is the typical mechanism used in stationary large-scale fluidic systems.

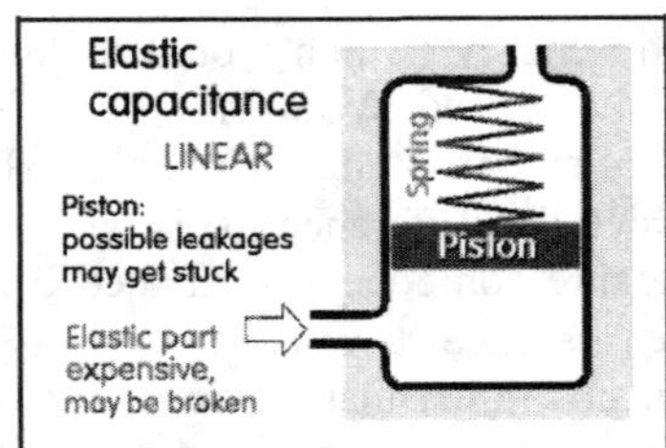

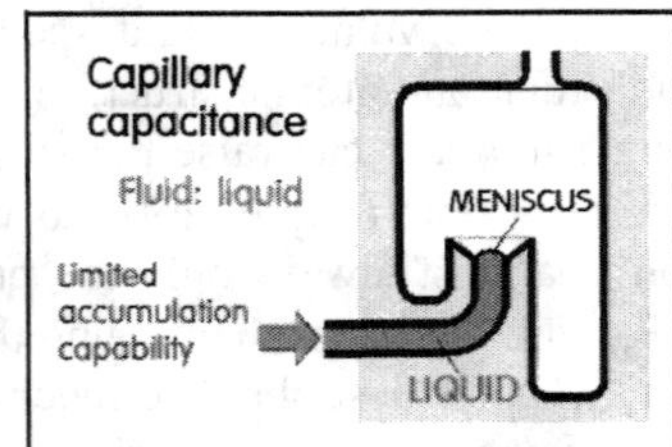

Figure 2.95 (Left) Another commonly used accumulation mechanism uses expansion of the device's available internal volume. The piston and spring version shown here is mostly used as a representative of this form of the fluid storage. More practical for storing small amounts is thinning of some walls so that they can easily deform.

Figure 2.96 (Right) The last of the four basic accumulation mechanisms utilizes the surface tension increase associated with changing the surface curvature of the liquid surface, which acts in a manner similar to deformation of an elastic membrane.

Especially when operating at high pressure levels, it is necessary to take into account the accumulation due to deformation of the device body, schematically represented in the case of elastic deformation in Figure 2.95. The elasticity may be, of course, enhanced by making some walls easy to deflect as a diaphragm (Figure 2.98), either inserted and fixed or made integrally with the remaining parts of the body by a localized thinning.

Finally, somewhat related in principle though made in a wholly different manner are the devices (Figure 2.96 and 2.97) employing the surface tension effect on the interface between a liquid and gas. Two immiscible liquids—perhaps oil and water—may be also used but this version is very rare, as working with gas as one of the fluids brings the increase of the storage effect by using the additional compressibility effect.

This brings us to the theme of various combinations of the four basic mechanisms. This is so common that, in fact, some principles are very rarely used in their pure form. Three basic principles in their various combined forms are shown in Figure 2.99. Some of the combinations come with an additional idea, such as e.g. that of the separation component in Figure 2.99: D (the component is represented in this schematic picture by the piston).

What is the purpose of the specialized accumulation devices? The simplest case of use of the storage capability is in the form of an external driving source (Figure 1.31) of pressurized fluid. The running time of the system is, of course, limited by the capacitance and total volume of a reasonably sized storage vessel. However, the flow rates in some microfluidic systems are so small that the corresponding times of operation may be surprisingly long. The storage vessels are also often filled by the processed samples to be analyzed in a "lab-on-chip" at the beginning of the operation, especially if it is not known in advance how much sample will be needed. Some samples, such as biological fluids taken from living organisms, have to be stored somewhere once they are taken and a storage vessel at the microfluidic system input is the natural choice. More sophisticated are the uses of (usually small-sized) accumulation chambers for adjustment of dynamic properties in fluidic circuits, especially those operating on oscillatory periodic regimes. Two examples of this sort are shown in Figures 2.100 and 2.101. In the former one, the point of interest is a contradiction to the statement that capacitance is a property of single-terminal devices (or their representations). As seen there, capacitive fluidic devices with two terminals do exist – but they may be shown to consist in principle of two basic one-terminal components (in this case corresponding to Figures 2.94 and 2.98) with some form of connection between the motions with which the accumulation is associated. Of course, two mechanically coupled diaphragm devices (in the basic form of Figure 2.98 the diaphragm is exposed to fluid on one side only) may

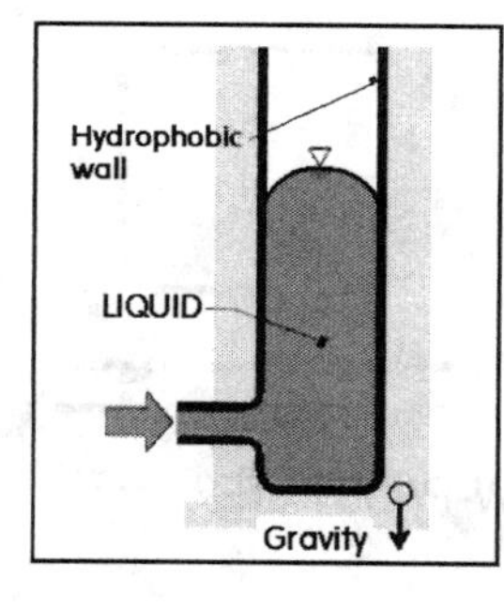

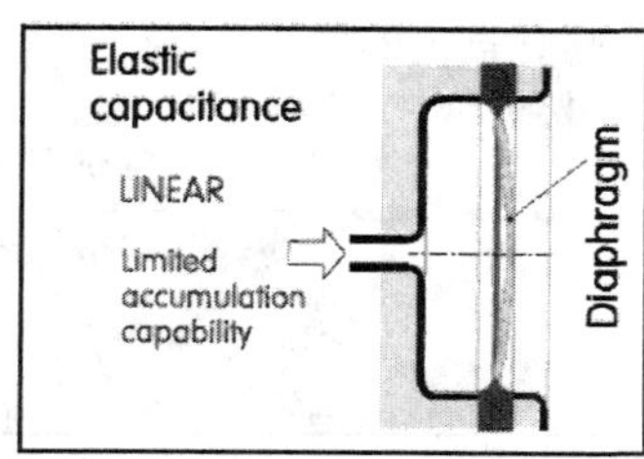

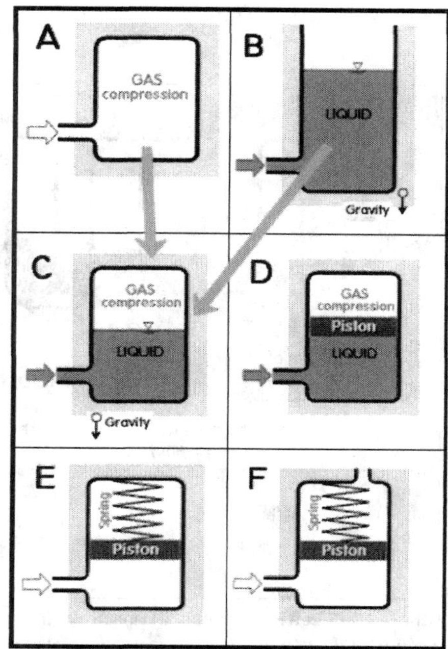

Figure 2.97 (Left, top) A typical example of combined accumulation mechanisms: in the small scales typical for microfluidics it is actually nearly impossible to employ the gravitational capacitance mechanism alone without a considerable effect of capillary "spring" acting on the surface meniscus.

Figure 2.98 (Left, bottom) The piston with supporting spring shown schematically in Figure 2.95 (and in **D**, **E**, and **F** of Figure 2.97), is mostly just a representation of various deformable rather than moving components, such as the diaphragm, easier to manufacture and less prone to sticking, leakage, and other malfunction.

Figure 2.99 (Right) Combined accumulation mechanisms. Fluid accumulation due to compressibility **A** is usually not very effective unless it is possible to provide a very large accumulation cavity. The gravitational accumulation **B**, with rising liquid surface in the vertical column, is more effective, but brings the inconvenience of keeping orientation of the device with respect to gravity. The two mechanisms may be combined according to **C**. Freedom to move the device, without keeping its vertical alignment, is then obtained by placing the separating piston **D**. The piston is usually replaced by a diaphragm, the deformation of which requires overcoming the elastic force, in the model **E** represented by the spring support of the piston. Note also that with the piston (or diaphragm) separation the input fluid need not be a liquid. The other alternative **F** operates without the gas compression on the other side of the piston.

be replaced in Figure 2.100 by the equivalent device with a single diaphragm on which fluids act from both sides. Steady flow is prevented from passing through, but oscillatory flow (of limited amplitude) is allowed to get across. A more simple use of this in principle double device is transfer of motion (e.g., in the form of an alternating flow) between two different fluids. When working with a gas that may contain dangerous (e.g., radioactive or microbially contaminated) aerosols, energy in the oscillatory flow form may be transported through with either one of the two versions shown in Figure 2.100 serving as an aerosol stop or trap.

The two-element module (capacitor – restrictor) shown in Figure 2.101 has two terminals due to the two terminals of the restrictor. The module properties are less influenced by the presence of inertance, which limits the usefulness of the high-pass filters of Figure 2.100 to only low frequency range. The low-pass combination of Figure 2.101, on the other hand, may be useful up to ultrasonic frequencies. It finds use in fluidic oscillators consisting of a fluidic amplifier and feedback lines, in which it is used to invert the phase of the feedback alternating flow signal.

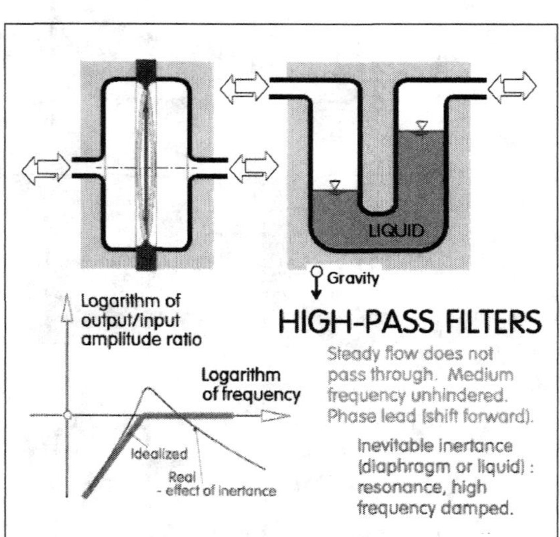

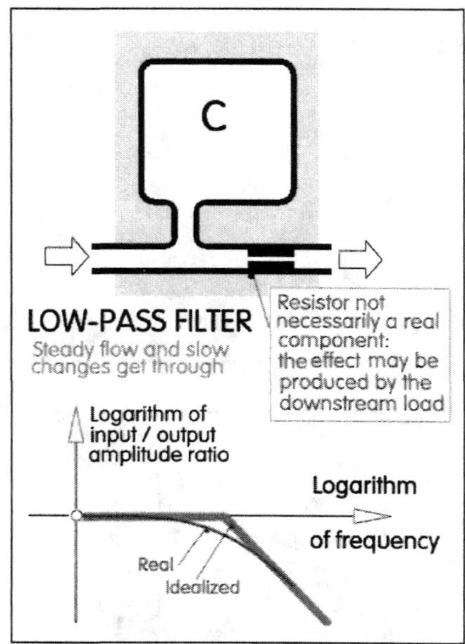

Figure 2.100 (Left) Exceptional capacitive fluidic devices with two terminals, useful as high-pass filters. The connection of the two motions is always associated with high inertance, the presence of which deteriorates the high-frequency behavior.

Figure 2.101 (Right) There are also the opposite fluidic circuits, those generating the low-pass effect. They are more useful and more often used in practice, especially for their phase-shifting property.

2.8.2 Gravitational capacitance

This is a principle more typical for large scale fluidics, though its use in liquid processing microdevices is not uncommon. Common at small scales is the use for storing the supply of a liquid before its entering the functional cavities. The storage cavities are sometimes simply made by drilling a vertical hole at the entrance part of the body, into which the liquid is input by means of a syringe. Also the processed liquid at the output side may be stored in a similar manner.

In its most common version, that of constant section (prismatic) vertical accumulation cavity, the storage characteristic (Figure 2.91) is linear and very easy to evaluate, as shown in Figure 2.102. Obviously, a larger cross section S of the storage cavity leads to a larger magnitude of the resultant capacitance, an expected fact with the capacitance being the measure of the capability to accumulate the working fluid. The presence of the ratio S/v in the resultant expression for the capacitance is very typical for all characterization quantities (it was already met in the expression for the steady-state characterization by dissipance and in the inertial effects in Figures 2.30, 2.88, and 2.89). The gravity acceleration g in the expression should not obscure the fact that a very effective variant of this storage principle is offered in rotating microfluidic circuits, with flows driven by the centrifugal action. The acceleration magnitudes may be many times higher than the Earth surface gravity acceleration value $g = 9.81\ \mathrm{m/s^2}$.

With a given liquid (which, with the exception of mercury, tend to have specific volumes mutually very similar, see also Figure 2.130) and the usual stationary gravity action, the only variable to select when designing this type of accumulation is usually the cross-section area S.

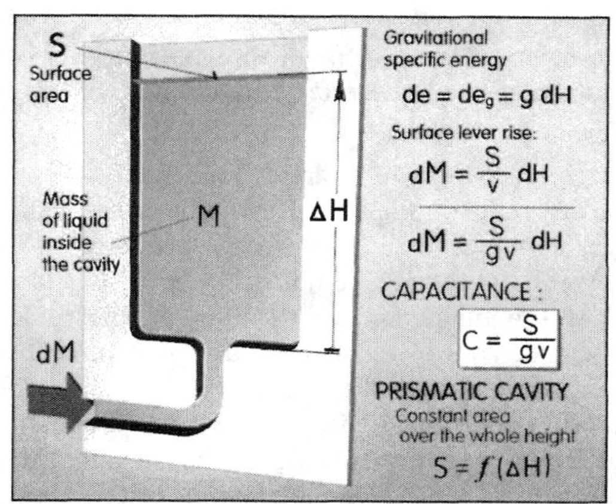

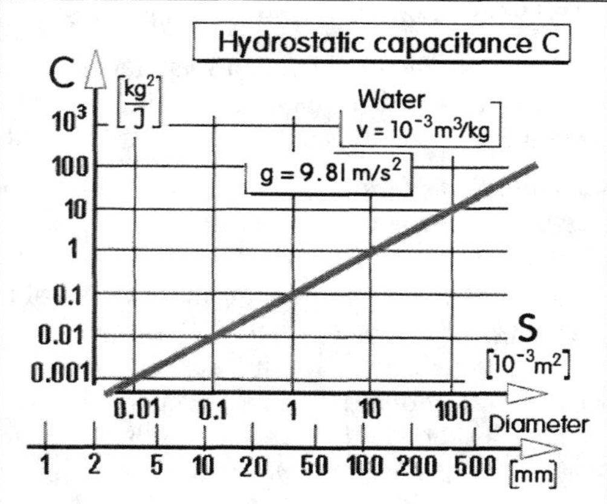

Figure 2.102 (Left) Gravitational type of fluid accumulation, simple, inexpensive and offering large capacitance values —but applicable only to liquids and to systems, usually stationary, operating at a given orientation with respect to vertical.

Figure 2.103 (Right) Rough estimation of the magnitude of gravitational capacitance C of vessels with constant cross-sectional surface area S over the whole height, filled with water.

Typical smallest cross section areas S in microfluidic devices are of the order of $1 \cdot 10^{-3}$ m^2. Smaller sizes are possible, but the result is the combined principle of Figure 2.97, with overwhelming influence of capillarity. The diagram in Figure 2.103 is included to provide an idea about the order of magnitude values for this storage principle. Typical magnitudes of C are of the order of 0.1 J/kg^2. This is by many decimal orders more than offered by other principles, with the nearest competitor being the surface tension principle.

Existing variations of devices using this storage principle are the nonprismatic vessels (Figure 2.104), and a system of several cavities filled in parallel, in which case their surface areas S are simply summed to obtain the resultant overall capacitance value.

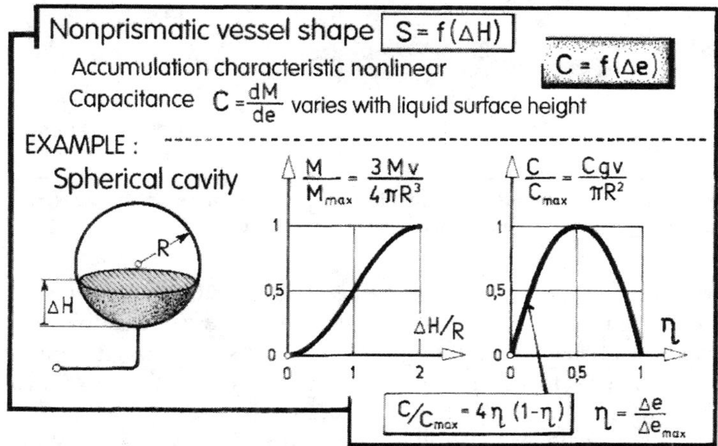

Figure 2.104 With nonprismatic storage vessels the gravitational capacitance device becomes nonlinear. The diagrams show variations of the stored liquid mass M and the corresponding variations in the value of the capacitance C. This may be perhaps used to compensate an opposite nonlinearity of some other simultaneously used capacitance principle.

2.8.3 Fluid compression capacitance

Although extremely simple to manufacture, this principle is quite difficult to evaluate. This is a nonlinear mechanism, its computation (as stated for compressibility effects in another context in Section 2.5) made difficult due to the additional degrees of freedom.

Accumulation of gas: The variations of the fluid specific volume v are dependent not only on pressure but also on temperature (Figure 1.42). The derivation in Figure 2.105 applies the equation of state $Pv = rT$ to the total mass M of the gas in the cavity, $PV = M\,rT$, and uses as the constraint the condition of constant cavity volume V (obviously, any combinations with the elastic mechanism, Figures 2.98 and 2.99E, calls for a corresponding adaptation of this approach). The resultant conclusion is the dependence of the evaluated capacitance on the interconnection between variations dT of gas temperature and the pressure rise dP. This is determined by the opportunity the gas is given for exchanging heat with its surroundings. The simplest situation is states as a) in Figure 2.106. If the processes in the cavity are very slow (and there is ample wall surface within reach), any change in gas temperature gives rise to gradient transport of enthalpy in the direction of reaching the equilibrium state of constant temperature T everywhere. The resultant expression for the capacitance is $C = M / rT$. The nonlinearity is due to the accumulated mass change dM being dependent on the mass M already stored. The specific-energy (and pressure) difference raises exponentially, the storage characteristic is of the "stiffening" type.

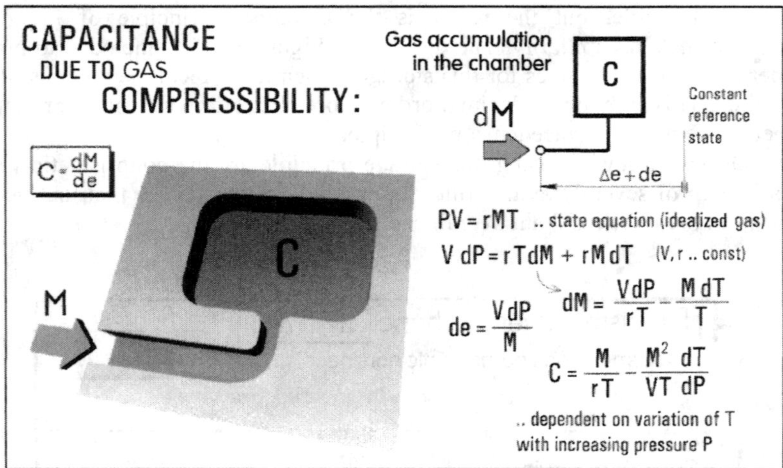

Figure 2.105 Gas accumulation in a constant-volume chamber evaluated on the basis of the equation of state $Pv = rT$ as shown in Figure 1.42. The capacitance depends on the slope of temperature rise dT/dP.

Accumulation by processes so slow is rather exceptional. More often, fluidic capacitive elements are used for adjusting dynamic processes, where the gas state changes tend to be fast. The extreme case is the adiabatic state change $Pv^{\varkappa} = \text{const}$ with no heat transfer at all. This, as derived in Figure 2.106, leads to a smaller value of the capacitance $C = M / \varkappa rT$.

Obviously, most actual state changes in the accumulation chambers are somewhere between these two extremes and—analogously to the discussion in Section 2.5—may be characterized by polytropic exponent polytropic exponent k, assuming values between $k = 1$ and $k = \varkappa$. Available experimental data demonstrate that the effective instantaneous values k vary during the processes in the cavity. For example, during the expansion after sudden pressure decrease, the

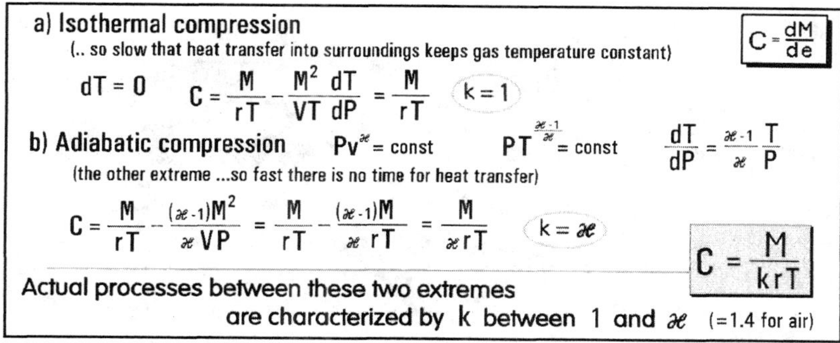

Figure 2.106 Two extreme cases of the gas accumulation: (a) a very slow state change so that the gas temperature T remains equal to that of the cavity walls, and (b) and the adiabatic state change so fast the gas cannot exchange heat with its surroundings.

value is at first near to $k = \varkappa$ in the initial, fast response to the change, but then decreases as the change gradually slows down toward attaining the final equilibrium. Evaluating the effective values, even by sophisticated numerical computations of temperature and flow fields, is far from reliable. Estimated mean values are used instead. The diagram for finding the corresponding magnitudes of the capacitance in Figure 2.107 may suggest that the errors due to wrong estimates are not significant, but in some more detailed investigations the resultant loss of precision may be unpleasant. Some guidance as to the polytropic exponent values is provided by the diagrams in Figure 2.108 for periodic processes and in Figure 2.109 for a single transition change. Noteworthy is the dependence on the shape of the cavity: a flat space with large surface/volume ratio leads to processes nearer to isothermal ones while the extremely small surface/volume ratio for spherical cavity makes them nearer to the adiabatic limit at the same frequency (and hence same Stokes number). This is consistent with the concept of the isothermal thermal boundary layer at the walls

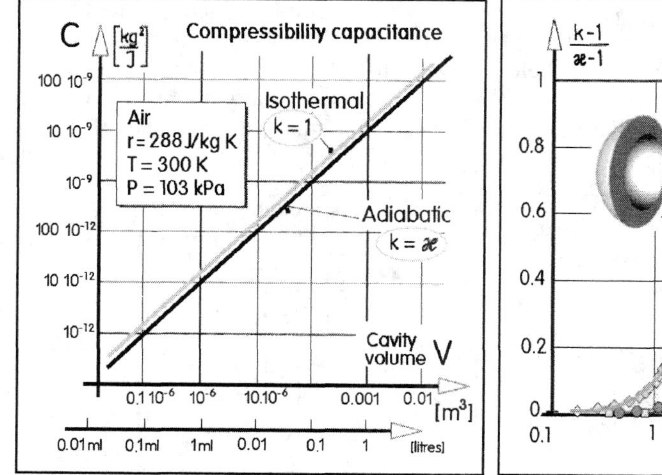

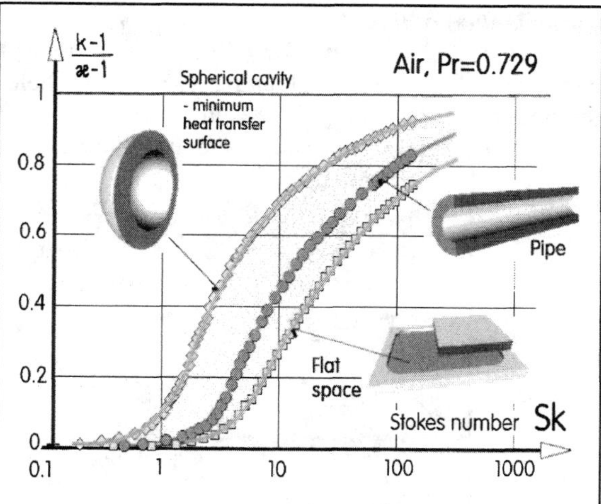

Figure 2.107 (Left) Useful diagram for estimating order of magnitude values of compressibility capacitance C for air in cavities of a given volume V.

Figure 2.108 (Right) Variation of the polytropic exponent k between the limits k = 1 and $k = \varkappa = 1.4$ for periodic processes: the value increases with increasing frequency nondimensionalized to the Stokes number Sk. The increase depends on the shape of the cavity, which determines the opportunities for heat transfer to the walls.

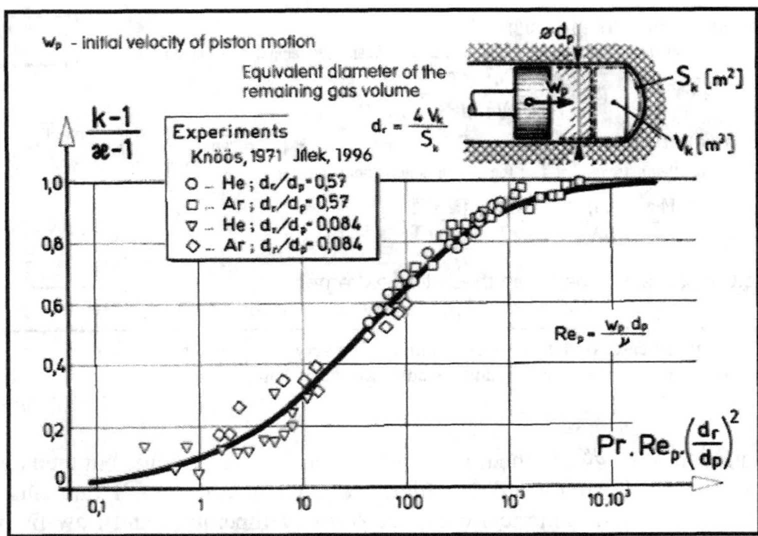

Figure 2.109 Variation of the polytropic exponent k for single-stroke compressions by a piston moving with the initial velocity w_p nondimensionalized into the Reynolds number. Again, k approaches the adiabatic limit $\varkappa$ if the process is fast so that there is no time for considerable heat exchange between the gas and the cavity walls.

and adiabatic state change in the core of the cavity, shown schematically in Figure 2.110. The flat cavities, typical for microfluidics with its characteristic planar patterns manufactured in flat substrate plates therefore tend to exhibit higher-than-usual values of the compressibility capacitance C. Providing the cavity walls with thin ribs (or filling it with very thin flat objects of negligible volume) may help in increasing C in critical situations with lack of space for a larger volume, although Figure 2.107 reminds us that the improvement cannot be substantial.

Accumulation of liquids due to their compressibility: The absolute volume of accumulated liquid may seem to be nearly negligible, due to their large volume stiffness characterized by the coefficient K (Figures 1.41 and 2.111). Nevertheless, due to the mass (and not volume) in the definition Figure 2.112, the resultant values of the capacitance are actually higher than those exhibited by gases of the same volume (Figure 2.113). Also shown in this diagram is the effect of

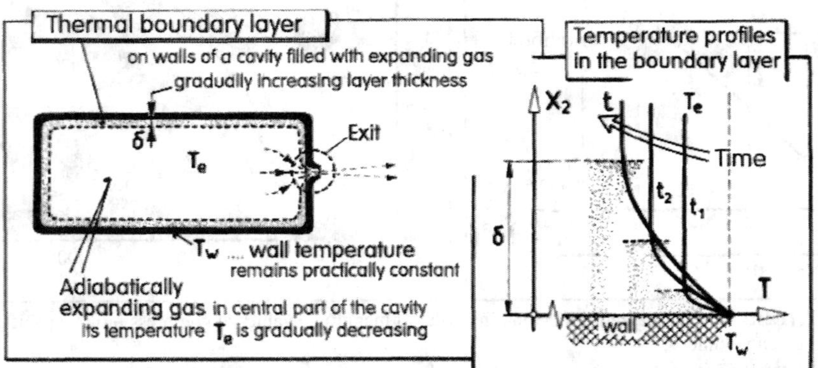

Figure 2.110 A simple model for the processes of Figures 2.108 and 2.109 based on the idea of an isothermal boundary layer of gradually increasing thickness and an adiabatic core.

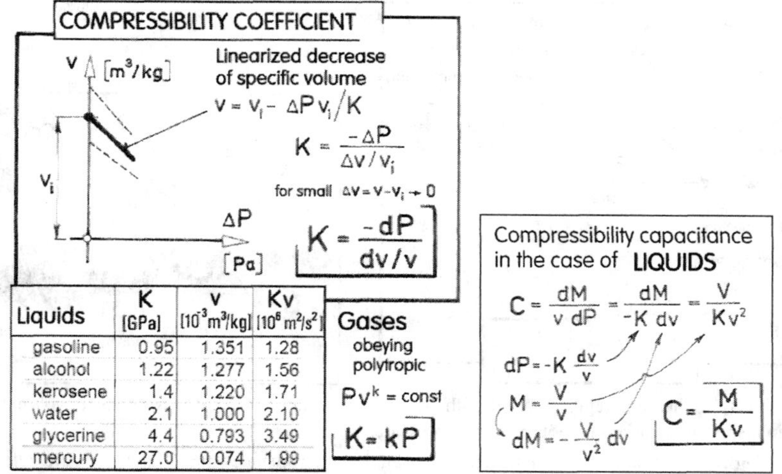

Liquids	K [GPa]	v [10^{-3} m³/kg]	Kv [10^{6} m²/s²]
gasoline	0.95	1.351	1.28
alcohol	1.22	1.277	1.56
kerosene	1.4	1.220	1.71
water	2.1	1.000	2.10
glycerine	4.4	0.793	3.49
mercury	27.0	0.074	1.99

Figure 2.111 (Left) Compressibility capacitance for liquids is evaluated on the basis of their compressibility coefficient K.

Figure 2.112 (Right) The corresponding magnitude of the liquid compressibility capacitance.

pressure rise on the gas compressibility expressed by means of the coefficient $K = k\,P = k\,r\,T$ in Figure 2.111 (k – polytropic exponent). The increase in P increases K in the same way as an increase in k, but the effect on C at a constant temperature is different in Figure 2.113 because of the pressure dependence of the mass M of fluid present in the cavity: increasing pressure and the consequent increase of gas density $1/v$ brings the conditions nearer to that of liquids.

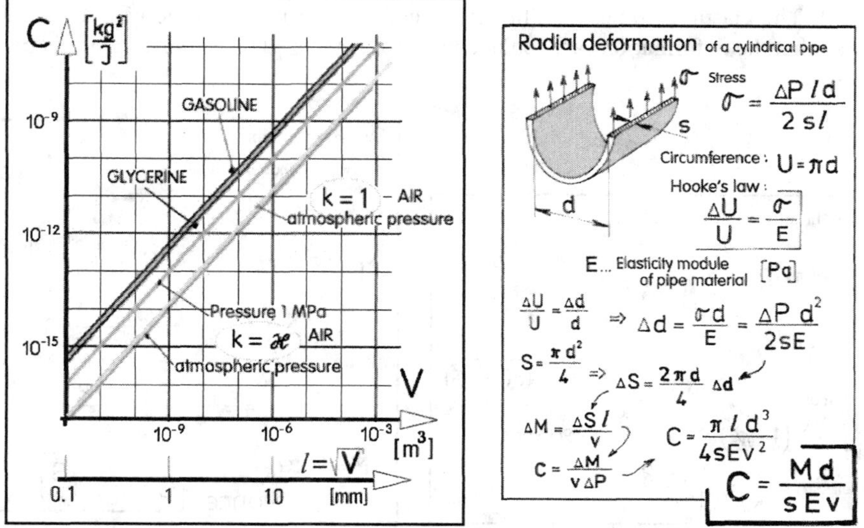

Figure 2.113 (Left) Order-of -magnitude values of the capacitance C due to compressibility of liquids plotted as a function of cavity volume (with the corresponding characteristic length *l* on the auxiliary scale) in Figure 2.79.

Figure 2.114 (Right) Elastic capacitance of circular cross-section pipes due to their radial expansion in response to pressure rise.

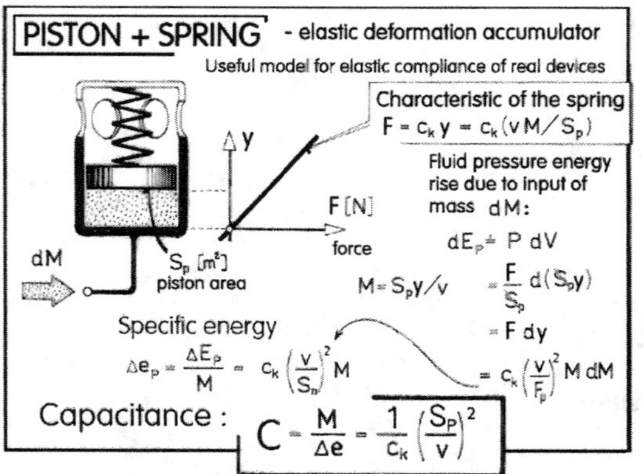

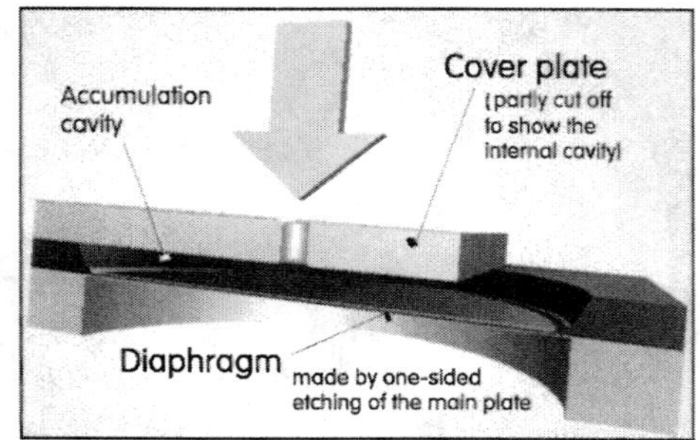

Figure 2.115 (Left) Schematic representation of the fluidic capacitance associated with elastic volume change of fluidic devices in response to variations of the internal pressure.

Figure 2.116 (Right) An example of a dedicated accumulation microdevice employing the elastic deformation mechanism: one wall of the accumulation chamber is thinned (by etching)to form an easily deformable diaphragm. Some additional capacitance is also due to the compressibility of the fluid in the chamber.

2.8.4 Capacitance due to wall elasticity

Walls of the cavities containing fluid are never perfectly stiff, a fact particularly apparent at the high pressure levels typical for the small size end of the range of pressure-driven microfluidics. They expand under the action of the internal pressure. As long as the strain levels remain in the elastic, Hooke's law domain, these volume changes provide another fluid accumulation mechanism. The accumulated amount is usually small, unless the device is specially adapted for the accumulation purpose, though elastic capacitance of some connection tubes may be significant, Figure 2.114. The elastic member may be separated, as a spring, from the wall a part of which is made free so that the deformations are replaced by movements. This is the case of piston accumu-

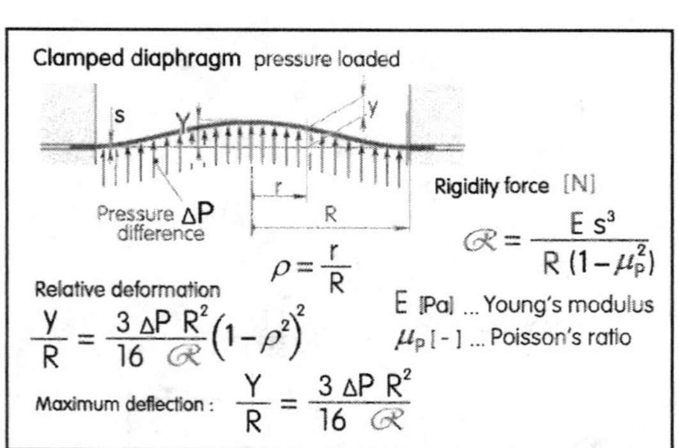

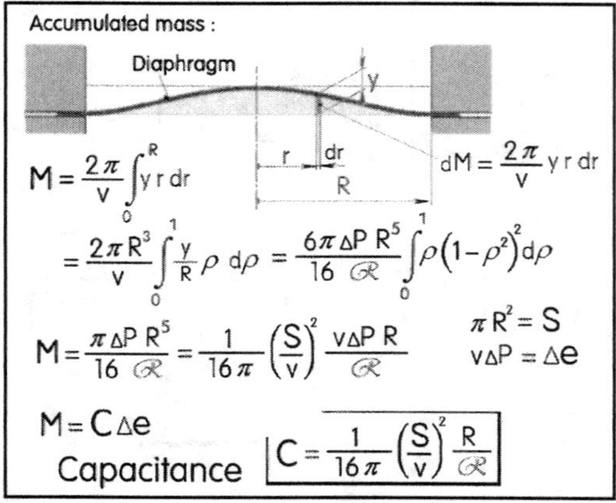

Figure 2.117 (Left) The basic relations for deformation of a pressure loaded (i.e., with continuously distributed equal load)diaphragm tightly clamped on the periphery (so that the deflection curve there has a circle of zero-derivative points).

Figure 2.118 (Right) Evaluation of the mass of the fluid accumulated in the space provided by the deformed diaphragm, and the corresponding expression for the capacitance. Note again the appearance by the ratio v/S.

	E	μ_P	$\frac{E}{1-\mu_P^2}$
	[GPa]		[GPa]
Rubber	0.007	0.5	0.0093
Polyethylene	0.7	0.42	0.849
Parylene	3.2	0.4	3.81
Photoresist	3.71	0.375	4.32
Polyamide	36	0.42	43.7
Aluminium alloy	70	0.34	79.1
Silicon (110)	161	0.279	174.6
Polycristalline silicon	169	0.22	177.6
Silicon (100)	169	0.278	183.2
Stainless steel	205	0.28	222.4
Silicon nitride	317	0.23	334.7

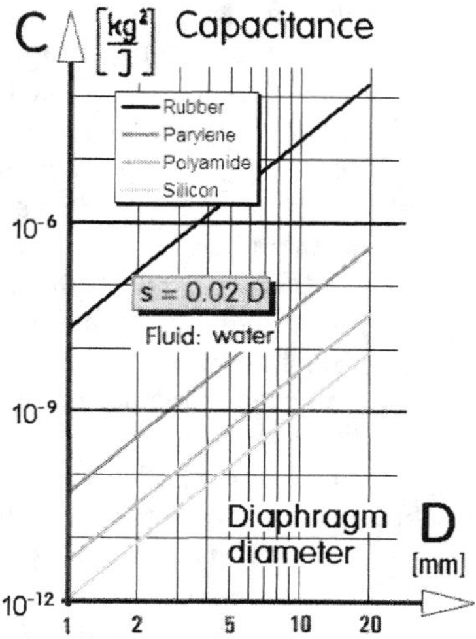

Figure 2.119 (Left) Elastic properties – the Young's modulus E and Poisson's ratio μ_P – for several materials of particular importance for making diaphragms for microfluidics.

Figure 2.120 (Right) Another of the presented diagrams for order-of-magnitude estimations of the capacitance values – here presented for the deformed diaphragms corresponding to Figure 2.117. As derived in Figure 2.118, the capacitance is critically dependent on the diaphragm thickness **s**, here assumed to be in certain proportion to the diameter.

lators, schematically represented in Figure 2.115. A prismatic cavity (cylinder) path has to be provided for piston movements and the piston has to be sealed. Such devices can be arranged to store of large fluid volume (an application example: Figure 6.146). The necessity of an expensive assembly operation, problems with the sealing, and the never eradicable danger of spring breakages make it unpopular in microfluidics.

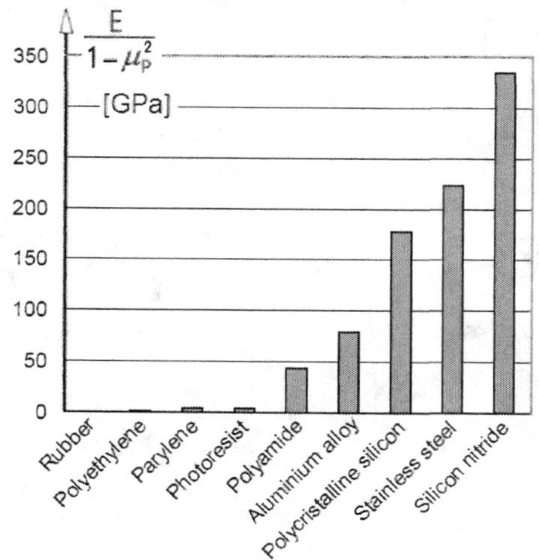

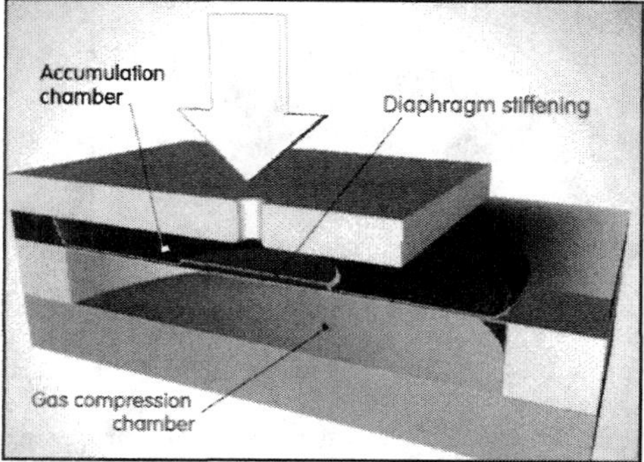

Figure 2.121 (Left) Elastic properties defining the rigidity force (Figure 2.117) for various diaphragm materials.

Figure 2.122 (Right) An example of a microfluidic capacitor combining diaphragm elasticity with gas compression.

In the other adaptation for accumulation, the deformation of the wall(s) is facilitated by making it very thin and thus converted into a diaphragm. The total volume change may be increased by stacking several diaphragms on top of each other, with a central common communication hole – this is the idea of bellows. Bellows are sometimes made by plastic deformation of an initially smooth tube, a process with such high demands on material ductility that its other properties, such as elasticity, are rather poor. Similarly as when the bellows are made from a fully compliant material, its elastic character of the accumulation must be provided by a supporting elastic element, a spring. All this can be made for microfluidic purposes by applying the microfabrication techniques, but the manufacture is obviously exceptionally demanding. This is why the version with a simple, flat diaphragm, especially having its own elasticity (Figures 2.116 and 2.122), is almost universally preferred in the no-moving part fluidics.

The equations of equilibrium of a plate under uniform lateral loading are a traditional problem solved by Hencky in 1913 [11]. The main results for the deflection y at a radius r are summed in Figure 2.117. In Figure 2.118 the deflections are integrated to obtain the mass M of the additional fluid stored under the deformed diaphragm. This is then used to evaluate the expression for the capacitance. Its value depends on the properties of the membrane expressed by the reference quantity—the rigidity force $\mathcal{R}$. This depends on diaphragm thickness s, its diameter $D=2R$, and elastic properties of the diaphragm material, presented for a number of important materials in Figures 2.119 and 2.121. Figure 2.120 again provides a diagram for estimating order-of-magnitude values, for a particular relative diaphragm thickness, for a range of materials.

2.8.5 Capillary capacitance

This fourth accumulation mechanism, the storage of liquid inside the drop, or of a gas in a bubble, Figure 2.123, is typical for microfluidics. The surface of the drop can extend to accommodate the additional stored liquid and the extension is associated with rise in the internal pressure energy. Most often, the actual accumulation "chamber" is actually just a part of such a drop (or bubble): it is the meniscus (Figure 1.26) that forms due to capillarity on the interface between immiscible

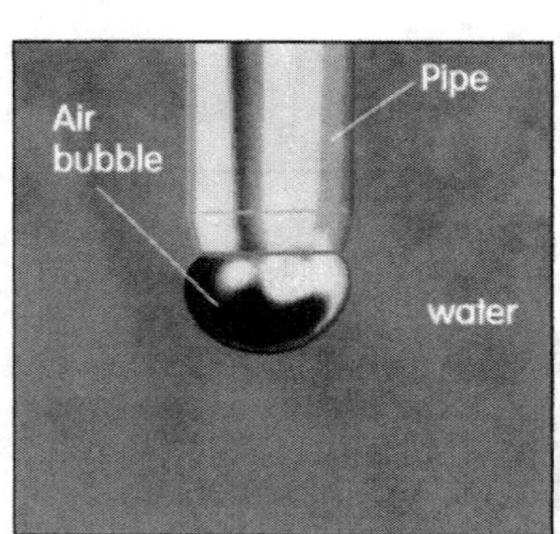

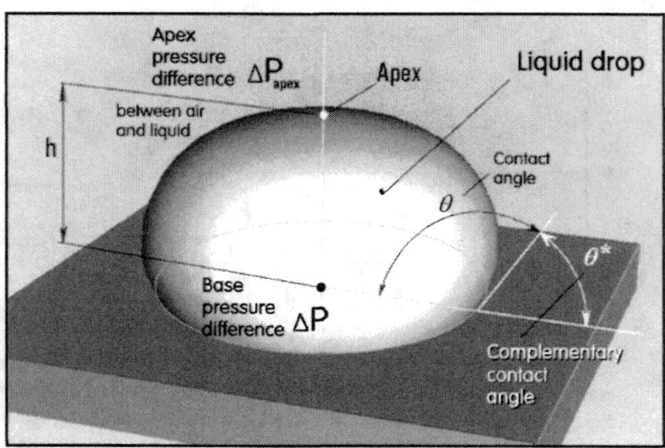

Figure 2.123 (Left) Photograph of the rather exceptional, inverted case of the capillary accumulation mechanism: storage of gas (air) temporarily inside the sessile bubble that forms at the bottom end of a small pipe immersed into liquid. The shape of this bubble is identical (for the same properties of fluids) to the shape of the "normal" liquid storing bubble.

Figure 2.124 (Right) Details of sessile liquid drop geometry. The surface tension provides the "elastic" force holding the accumulated liquid brought into the drop from below.

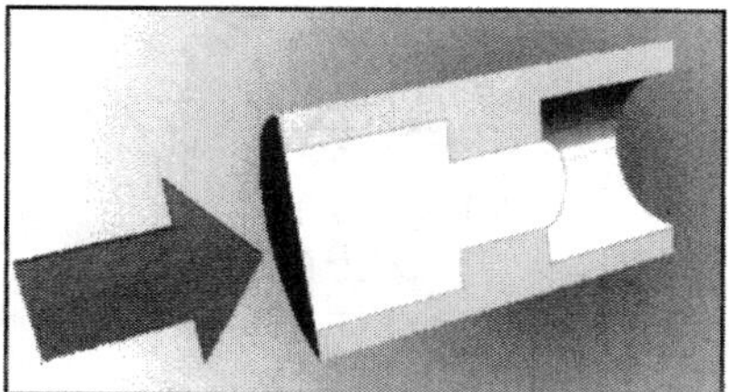

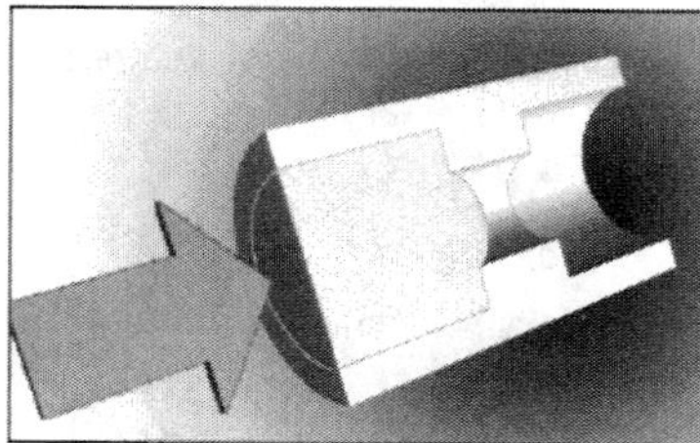

Figure 2.125 (Left) Formation of the liquid accumulation meniscus at the exit end of a small constant cross-section channel in hydrophilic material.

Figure 2.126 (Right) Formation of the capacitance meniscus at the channel entrance in hydrophobic material.

Contact angles for different materials with water/air meniscus		
Glass, [12]	HYDROPHILIC	51.0 deg
Perspex (PMMA), [13]		73.5 deg
Teflon (PTFE), [14]	HYDROPHOBIC	115 deg
Superhydrophobic silane coating, [15]		150 deg

Figure 2.127 The stationary position of the meniscus at the channel end is based on the singularity (with hysteresis) of the driving pressure as the contact angle traverses the 90° change.

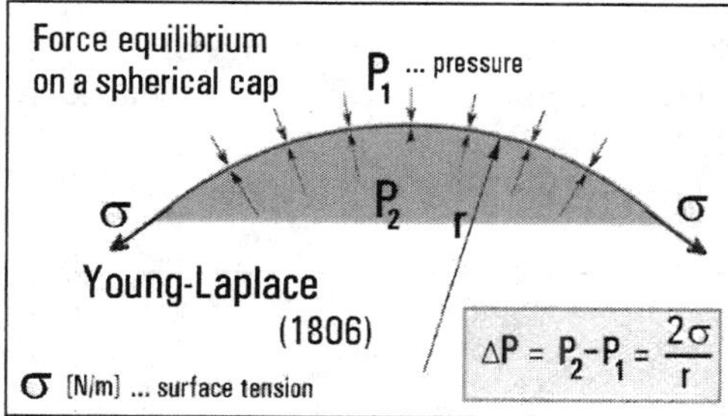

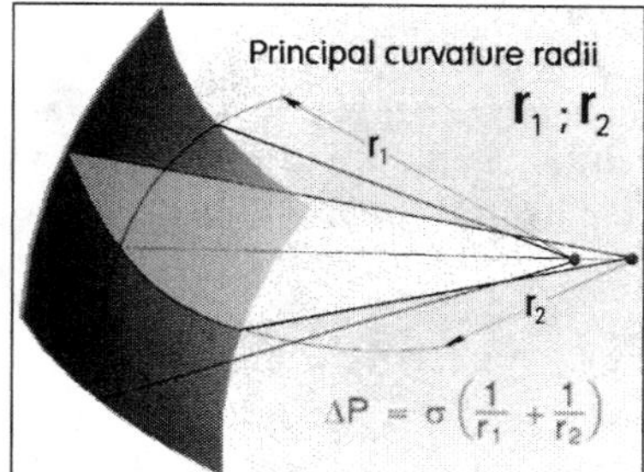

Figure 2.128 (Left) The Young-Laplace law of the force balance on the liquid surface: the higher pressure inside the drop is inversely proportional to the surface radius of curvature. The liquid properties are characterized by the proportionality constant, the surface tension σ.

Figure 2.129 (Right) On the surface of the drop, keeping the accumulated liquid, the conditions are complicated by the nonspherical character of the surface: the law from Figure 2.128 is to be adapted by considering the two principal curvature radii.

fluids at an end of a communicating channel. The fluids may be both liquids, such as oil and water, but in typical cases the storage is provided by the interface between a liquid and a gas, most often air. Typically, it is the liquid working fluid of the microfluidic system that is in unsteady states accumulated under the convex surface intruding into an air-filled space. This surface behaves similarly as the diaphragm of the elastic storage discussed in Section 2.8.4. The capillary storage effect is much more effective. The other advantage is it is generally much cheaper to manufacture (what is manufactured is just the inlet channel; the extensible surface forms spontaneously). There is no danger of destroying the diaphragm irreparably by an overpressure (even though such an event may, hopefully reparably, upset the operation of the device). On the other hand, the utilizable size of the surface and consequently the absolute amount of stored fluid is very limited. Also, working with two fluids and keeping them both inside a fluidic circuit has its practical disadvantages.

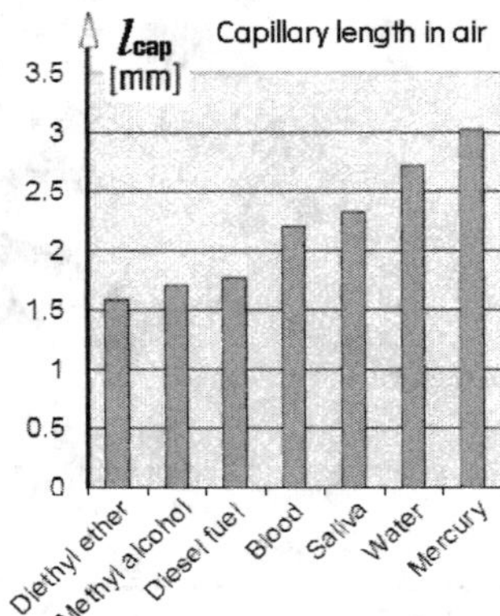

	σ [mN/m]	v [10^{-3} m^3/kg]	l_{cap} [mm]
Acetone	24.0	1.336	1.81
Benzene	28.9	1.144	1.84
Blood	50.0	0.945	2.20
Diethyl ether	17.6	1.402	1.59
Diesel fuel oil	26.2	1.170	1.77
Mercury	483.5	0.184	3.01
Methyl alcohol	22.6	1.263	1.71
Phenol	39.3	0.933	1.93
Saliva	53.0	0.993	2.32
Water	72.3	1.000	2.72

Figure 2.130 Surface tension material constants of several liquids important for possible use in microfluidic devices.

The drop or bubble, Figure 2.124 used for the storage is usually formed at an end of a small cross-section channel (Figures 2.125 and 2.126). The standard layout uses the exit end of a channel formed in a hydrophilic material. Typical materials of both sorts are in Figure 2.127. Initially, the liquid/gas interface is pushed by the driving pressure and passes through the channel. When it reaches the exit edge, the movement stops and the pressure has to rise by a large amount before the contact angle (Figures 2.124, 2.131, and 2.132) increases its size by 90° to be reconciled with the different direction of the wall and can continue moving. It is in this transitory state of stationary location, with the drop at the end of the inlet channel and fixed to its exit edge that the drop is

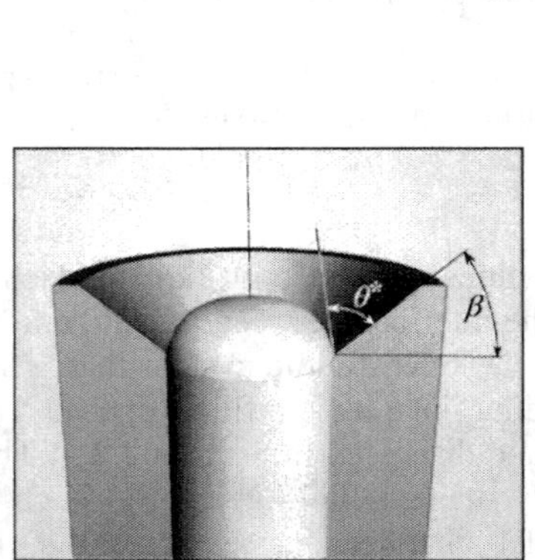

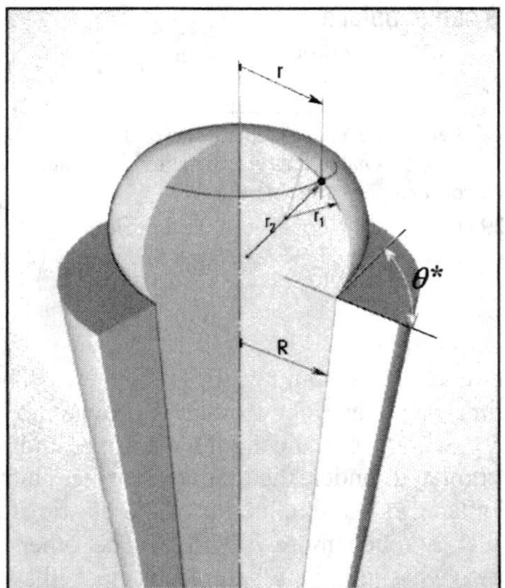

Figure 2.131 (Left) The meniscus, a top part of liquid drop used to accumulate the fluid in unsteady regimes, is formed at the end of a channel, here shown in the axisymmetric version. Because of the importance of the conditions in the contact location, it may be useful to adjust the properties by choosing a nonzero angle β of the channel end rim.

Figure 2.132 (Right) The two principal radii at a point of the drop surface. They are the starting point in derivation of the differential equation (not discussed here in detail) that defines the shape of the interface.

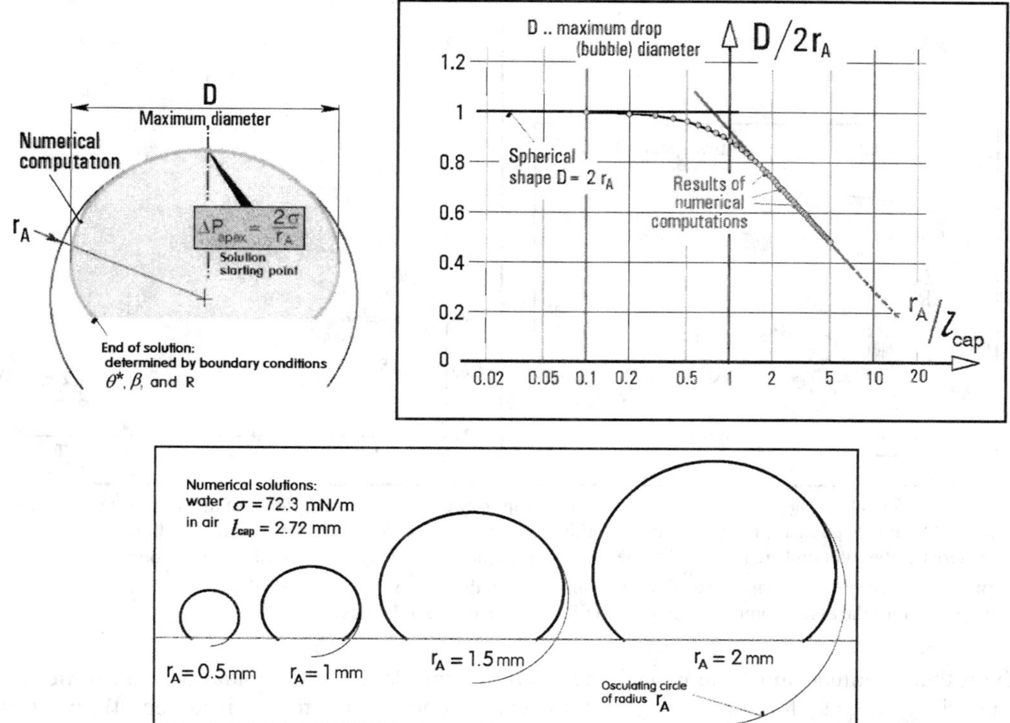

Figure 2.133 (Top left) The computed shape of the surface tension storage volume. The computation starts in the apex point where the conditions are simple, since the two principal curvature radii there are the same and equal to the apex radius r_A is easily evaluated (Figure 2.128) from the local pressure difference across the drop surface.

Figure 2.134 (Top right) The ratio maximum drop diameter to the apex radius is a useful indicator of the deviations from the spherical shape. The deviation depends on the ratio of the drop size to the capillary length.

Figure 2.135 (Bottom) Computed shapes of the sessile water drops in air for different values of the apex radius.

employed in its liquid the accumulation role. The storage characteristic is nonlinear, capacitance varies in the process of filling the drop. In the typically planar shape of cavities in microfluidics, the drop is deformed by the different in-plane and out-of-the-plane conditions for its shape changes, making the description rather complicated. It is useful to follow the basic ideas on the simpler axisymmetric case, with the drop formed at the end of a circular cross-section channel (Figures 2.131 and 2.132). Even then, the shape of the drop is governed by a nonlinear differential equation that has to be solved numerically. Examples of the solution are shown in Figures 2.133 to 2.135. The shape varies with the absolute size (characterized by the apex radius r_A) of the drop in relation to the capillary length l_{cap} (Figure 2.130)

$$l_{cap} = \sqrt{\frac{\sigma}{g/v_w - g/v_a}} \tag{2.16}$$

where g is the acting external acceleration (in gravitational field $g = 9.81 \ m/s^2$), v_w is the specific volume of the liquid, and v_a is the specific volume of the surrounding gas. For water and air, the shape may be taken as spherical if the apex radius r_A is less than ~ 1 mm.

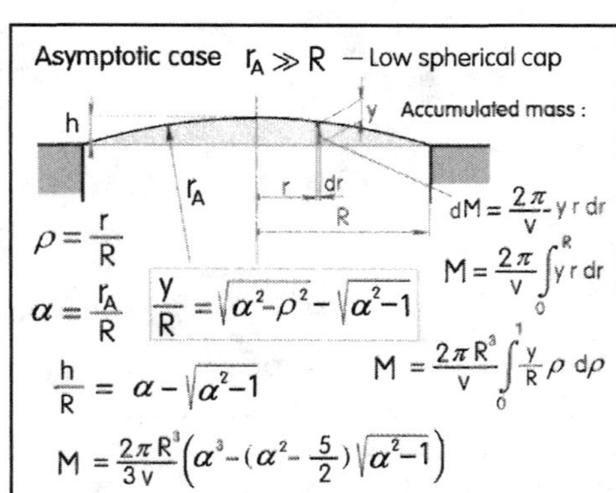

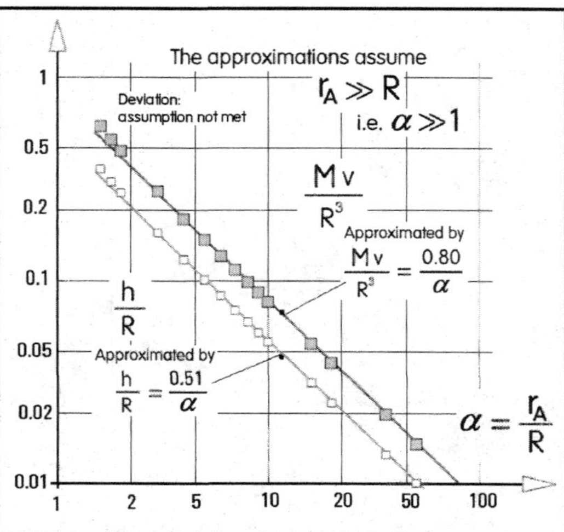

Figure 2.136 (Left) The first step in derivation of the capillary capacitance: integration of the drop surface to find the amount of liquid stored inside the drop. Due to limiting the interest only to the initial part of the nonlinear storage characteristic, the solution here is limited to the conditions in the nearest neighborhood of the apex point.

Figure 2.137 (Right) Perhaps somewhat surprisingly, the rather complex expressions derived in Figure 2.137 may be approximated for the assumption of large α by very simple reciprocal relations.

Even then, evaluation of the capacitance is not a simple task. To obtain at least some useful numerical values for comparison with capacitances of other discussed fluid storage mechanisms, Figures 2.137 and 2.138 present a derivation for the very beginning of the storage characteristic where the bubble is still extremely low (for a very large ratio α of

Approximated $\frac{Mv}{R^3} = \frac{0.8}{\alpha}$ $\quad \pi R^2 = S$

$$M = \frac{0.8\, R^4}{v\, r_A} = \frac{0.4\, S^2}{\pi^2 v^2 \sigma} v\, \Delta P_{apex}$$

$$v \Delta P = \Delta e$$

$$v\, \Delta P_{apex} + gh = \Delta e$$

The approximation Fig. 2.137 makes it easy to evaluate the hydrostatic term

$$gh = g\frac{0.51\, S}{2\pi\sigma v} v \Delta P_{apex}$$

but it is negligible under the $r_A \gg R$ assumption

$$M = \frac{0.4}{\pi^2 \sigma}\left(\frac{S}{v}\right)^2 \Delta e \qquad M = C\Delta e$$

Capacitance

$$C = \frac{0.4}{\pi^2 \sigma}\left(\frac{S}{v}\right)^2$$

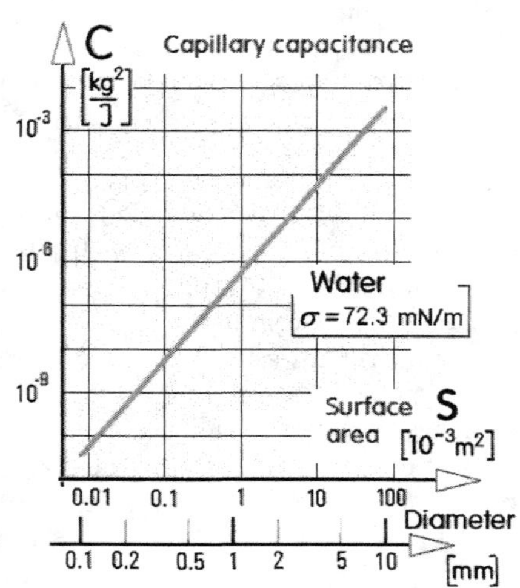

Figure 2.138 (Left) Initial capacitance (at the beginning of the accumulation curve) evaluated for the approximate reciprocal relations from Figure 2.137.

Figure 2.139 (Right) Diagrammatic presentation of the relationship from Figure 2.138 may be used for rough estimates of the accumulation properties of liquid (water) drops and for comparisons with similar relations for the other capacitance mechanisms: Figures 2.103, 2.113, and 2.120.

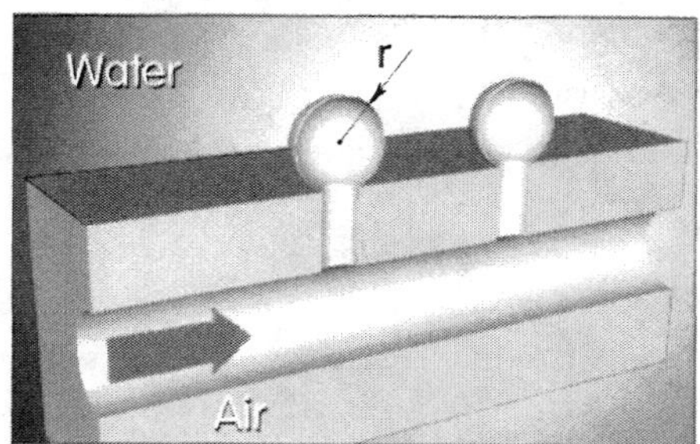

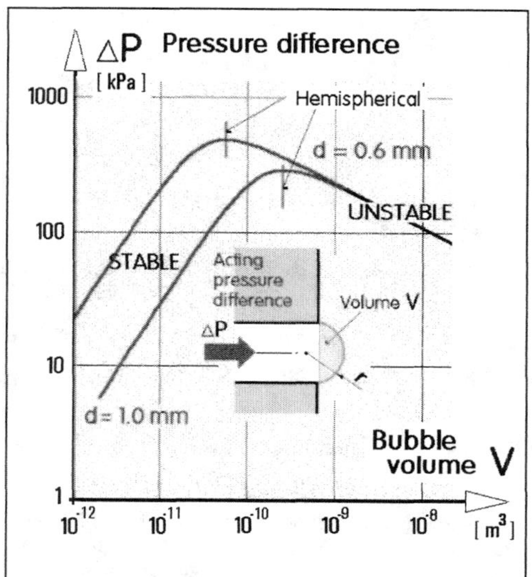

Figure 2.140 (Left) The standard method of increasing the capacitance beyond the limit imposed by the acceptable size of a single storage volume is using several storage mechanisms in parallel—here in the inverse case of air bubbles formed in a water-filled space. Unfortunately, this parallel operation is unstable.

Figure 2.141 (Right) The doubling (or, in general, multiplying) the storage mechanisms according to Figure 2.140 is stable only in the initial part of the characteristic, where, of course, the stored amount is not large. Quite easily one of the bubbles can get beyond the stability limit and then grows suicidally at the expense of air taken from neighbor bubbles.

the apex radius r_A ratio to the channel exit radius R). The results are presented in Figure 2.139 in a form corresponding to the earlier similar plots for other cases. The comparison for water as the working fluid shows the capillary mechanism can achieve very high capacitance (second to the gravity storage) at the same diameter of the active cavity. On the other hand, the amount of the stored fluid is very small. When using other mechanisms, a safe method of storing more fluid is to use several accumulation cavities in parallel, with the resultant capacitance equal to the sum of the capacitances of all the devices. This, unfortunately, is not recommended in the case of the capillary capacitance due to the instability once the drops (or bubbles in Figure 2.140) get beyond the hemispherical shape (Figure 2.141), whereby the increase in size and of the curvature radius leads, according to the Laplace-Young law (Figure 2.128), to a decrease of the pressure opposing the input of additional fluid.

Some useful ideas about unsteady flow processes, their character and computation in fluidic circuits with devices exhibiting inertance and/or capacitance properties are available in [16].

References

[1] Nguyen, N.-T., and S. Wereley, *Fundamentals and Applications of Microfluidics*, Norwood, MA: Artech House, 2002.

[2] "Process Measurement and Analysis, " In *Instrument Engineers' Handbook, Vol. 1*, B. G. Liptak, (ed.), 4th edition, Butterworth-Heinemann, 2003.

[3] Baker, R.C., *Flow Measurement Handbook*, Cambridge, U.K.: Cambridge University Press, 2000.

[4] Tesař, V., et al., "Subdynamic Asymptotic Behavior of Microfluidic Valves, " Journal of Microelectromechanical Systems, Vol. 14, April 2005, pp. 335–347.

[5] Richter, J. P., The Notebooks of Leonardo da Vinci (1452–1519), *Compiled and Edited from the Original Manuscripts*, New York: Dover, 1970, p. 499.

[6] Bengtsson, H.-U., and P. Edén, "A Simple Model for the Arterial System," *Journal of Theoretical Biology*, Vol. 221(3), 2003, p. 437.

[7] Schneck, D.J., "An Outline of Cardiovascular Structure and Function," *The Biomedical Engineering Handbook*, Bronzino J. D. (ed.), New York: CRC Press & Springer, 2000.

[8] Weibel, E.R., *Morphometry of the Human Lung*, New York: Academic, 1963.

[9] Liu, T., "Optimum Bifurcating-Tube Tree for Gas Transport," *Trans. ASME, Journal of Fluid Mechanics*, Vol. 127, 2005, p. 550.

[10] Tesař V., "Sampling by Fluidics and Microfluidics", *Acta Polytechnica - Journal of Advanced Engineering*, Vol. 42, No. 2, 2002, p. 41.

[11] Hencky, H., "Über den Spannungszustand in kreisrunden Platten mit verschwindender Biegungssteifigkeit," *Zeitschrift für Mathematik und Physik*, Bd. 63, 1915, p. 311.

[12] Sklodowaka, A., M. Wozniak, and R. Matlakowska, "The Method of Contact Angle Measurements and Estimation of Work of Adhesion in Bioleaching of Metals," *Biological Procedures Online*,1: doi:10.1251/bpo14, November 1998, p. 114.

[13] *Computer Image of a Water Drop Profile on a Polymethylmetacrylate Surface*, http://www.mie.utoronto.ca/labs/last/kwok/drop.html.

[14] Harkema, S., "Self-Cleaning Surfaces," http://www.phy.cam.ac.uk/steiner/project_information.php/project-self_cleaning.

[15] Zhai, L., et al., "Stable Superhydrophobic Coatings from Polyelectrolyte Multilayers, " *Nano Letters*, Vol. 4, No. 7, 2004, p. 1349.

[16] Tesař ,V., "Fluidic Circuits," Chapter 13 in *Microfluidics: History, Theory, and Applications*, ed. W. B. J. Zimmerman, Springer, Wien, New York, 2006.

Chapter 3

Simple Components and Devices

A characteristic feature of fluidics shared by microfluidics is building units from individual elements. This takes place at several levels. Figure 3.1 introduces three levels discussed here: fluidic *systems* built from *devices*, which in turn consist of *components*. The division into the building blocks is conceptual rather than really tangible: systems today are mostly made in the form of a single mechanical object, following the principle of integrated circuits: all the devices of a circuit are etched simultaneously in a single base plate. If a system is actually built from parts manufactured separately, the division into the parts is dictated by manufacturing convenience—there is, for example, usually a separate base plate and a cover plate—rather than functional differences between the constituent devices.

Nevertheless, the concept of a device from which systems are built is not only useful for a discussion but is actually applied in the development stage. What is individually designed and tested in the laboratory before declared suitable for use as a building block are fluidic devices. In the earliest days of fluidics, manufacturers even tended to deliver individual devices, such as fluidic signal amplifiers, separately. The users were expected to connect them by means of short lengths of flexible tubing. Indeed, even today, several manufacturers offer kits with separate reactors, mixers, and so on for experimentation (mainly educational) in microchemistry. In the development stage anyway, having an isolated device is indispensable for tests, usually requiring free access to device terminals, where the flowmeters and manometers are connected during the investigations.

There used to be a distinct development trend in fluidics to not aim at universality of devices and their uses. Devices tended to be individually designed for a particular task. With the typical nonlinear characteristics and capability to perform complex nonlinear interactions between the flows coming from several terminals, fluidics perform complex tasks using very simple circuits, consisting at most of a handful of devices, if not just a single, very specialized device. To perform the task, the devices were often made with rather extraordinary characteristics, usually laboriously

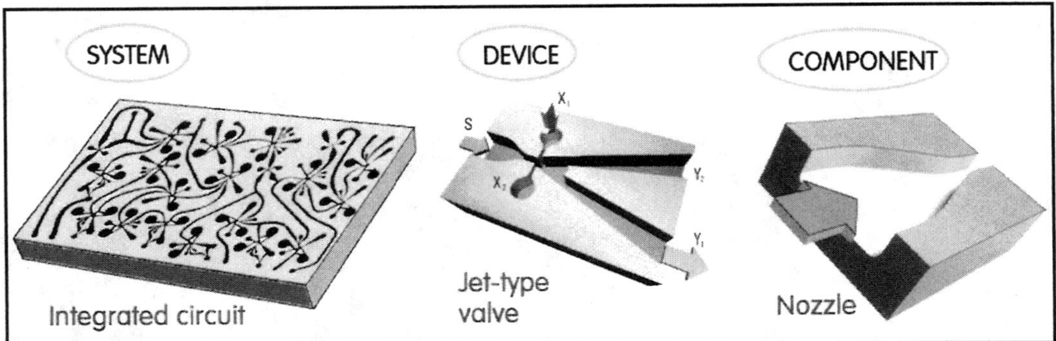

Figure 3.1 A fluidic *system* can perform an economically useful operation and is sold as a self-contained unit. Typically, it consists of a number fluidic *devices* interconnected by channels. Each of the devices can perform a technologically useful operation such as diverting the fluid flow. Devices are designed using *components* as (perhaps merely conceptual) building blocks.

developed ad hoc for the particular purpose. This approach is in a direct contradiction to what is typically done in electronic circuit design. There the designers use a very limited number of standard device types, usually with rather simple linear properties, made in a range of sizes (or parameter values). Developing specialized devices is, of course, expensive. The current growing market supports the trend of standardization even in fluidics. Also the typical number of fluidic devices in a circuit tends to increase [1]. The earlier trend of circuitry with a very limited number of special devices gradually disappears.

3.1 CONNECTING CHANNELS

Not really devices themselves, channels perform the essential but simplest of tasks: transferring fluid from one device into another, usually as a continuous fluid flow. In contrast to electronic circuitry, alternating, direction sign changing flows are rather exceptional. The distinction between

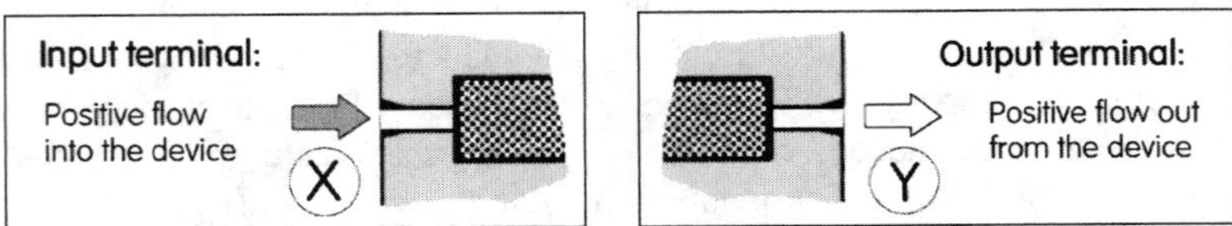

Figure 3.2 (Left) Figure 3.3 (Right) Sign convention: fluid flowrate is taken as *positive* if it enters an input terminal of a device and/or if it leaves the output terminal.

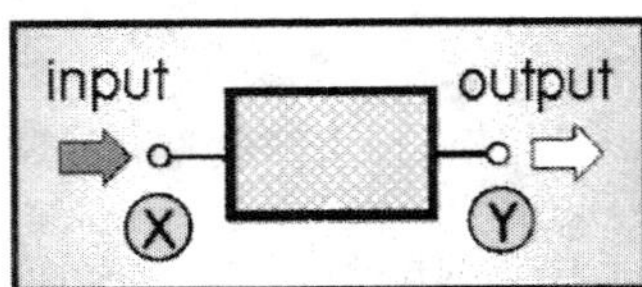

Figure 3.4 Schematic representation of a device as a "black box." As a result of the two conventions from Figures 3.2 and 3.3, the input and output flows in the two terminals are equal (in steady-state regime there no accumulation due to device capacitance).

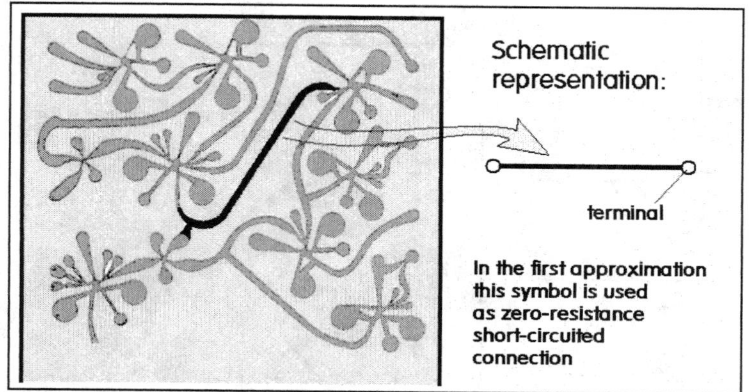

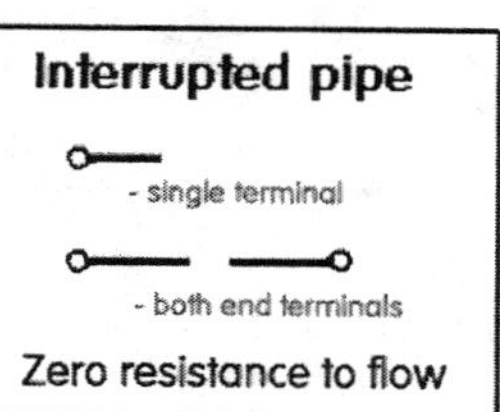

Figure 3.5 (Left) A typical example of a connecting channel in a fluidic integrated circuit and its schematic representation used in circuit diagrams.

Figure 3.6 (Right) In contrast to an electric conductor, which exhibits (in air) a practically infinite resistance when interrupted, the interrupted fluidic connecting channel behaves in the very opposite manner; it lets the fluid leave into the surrounding space (usually the atmosphere).

the input (Figure 3.2) and output (Figure 3.3) terminals therefore agrees with the directions of the fluid flow (Figure 3.4). The channels (Figure 3.5) are almost invariably closed conduits. Guiding liquids in grooves open from above or by strips of wetting material (Figure 2.127) on nonwetting surfaces, though possible are extremely rare, mainly because of sensitivity to external accelerations. This makes the channels (Figure 3.7) slightly more complex to manufacture than simple electric conducting strips, though wires with an insulation can be comparably expensive (and heavier). Other interesting differences in comparison with electric conductors are the different consequences of simple interruption (Figure 3.6), due to electric circuits operating in a nonconducting environment. Electric current does not escape if a conductor is interrupted, while in fluidics it is necessary to block the channel (Figure 3.8) to prevent leakage into the atmosphere.

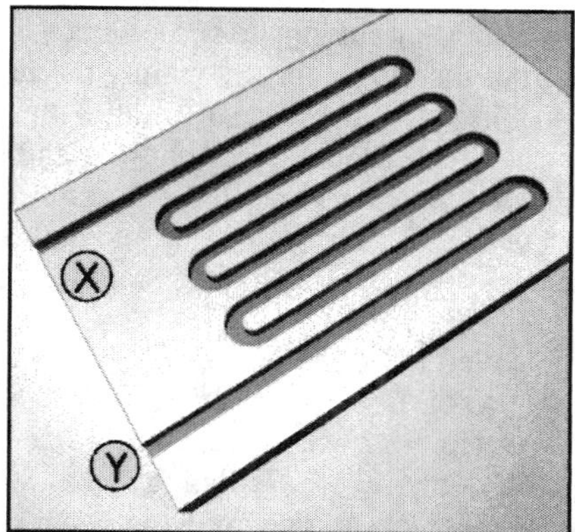

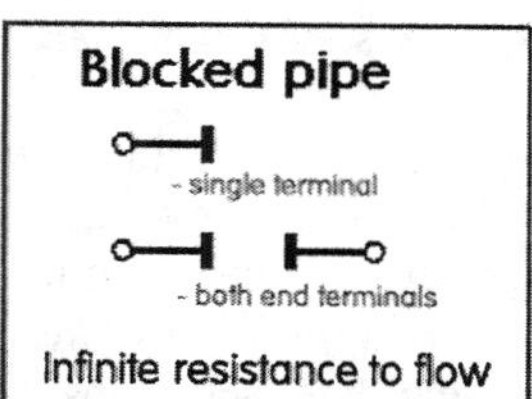

Figure 3.7 (Left) Connecting channels in fluidics are usually of constant cross-section. A change in cross-section performs energetic conversions. The changes of flow direction are common, though not welcome (they increase the hydraulic loss) but often inevitable. Channels as long as the one in this example may be made with some special intentions, perhaps for their inertance (Figure 2.86).

Figure 3.8 (Right) Symbol for blockage of a channel, necessary to prevent fluid flow out from the device or circuit.

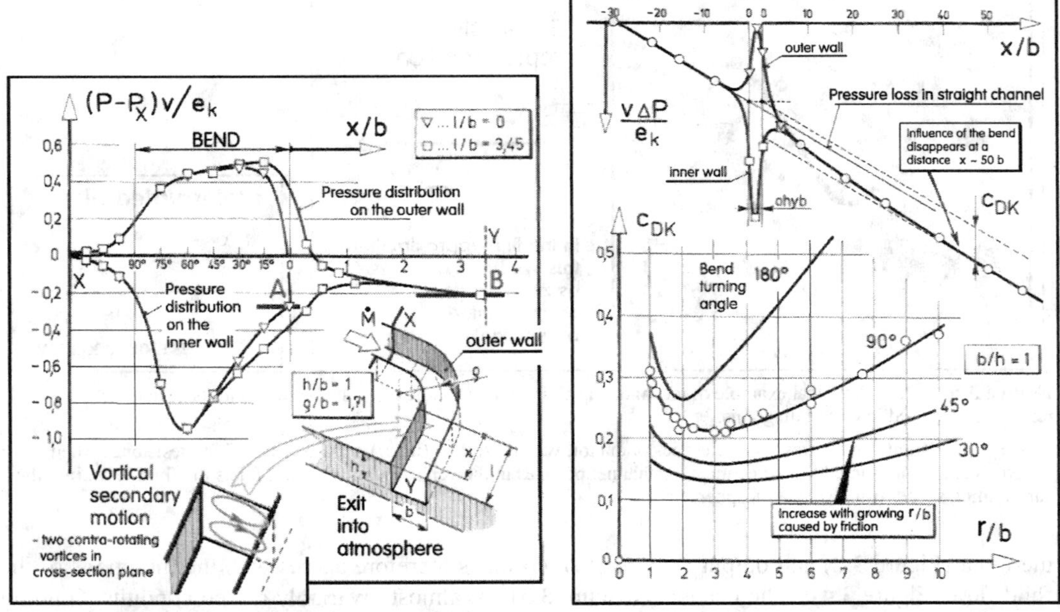

Figure 3.9 (Left) Pressure distributions on the inner and outer walls of a channel bend. The large differences give rise to secondary motions. Note the larger pressure loss at A than in B for the zero length of the exit.

Figure 3.10 (Right) Pressure distribution in a typical channel bend with a long straight inlet and outlet straight channels. Values of the bend loss coefficient c_{DK} should be noted as well as the method of its evaluation, taking into account and subtracting the straight channels friction losses.

Unless of eceptionally small cross section, electric connecting conductors are considered to be of practically negligible electric resistance. Terminals become short-circuited when connected. In fluidics, there is no real danger of short-circuiting. Reasonably sized channels exhibit nonnegligible pressure loss due to friction (Figure 2.41), especially if they span distances more than 10 or 20 times their widths. Rather than short-circuiting, it is a leakage from improperly closed conduit into the surrounding space that is of the most concern, mainly because of the loss of the fluid. Bending an electric conductor is inconsequential while in fluidics the friction energy losses in the channel are exacerbated by the usually inevitable bends causing additional quadratic losses. They are mainly caused by the complicated secondary motions (Figure 3.9). The pressure changes in a bend are relatively very large but usually cease to be influential a short distance downstream (though some disturbances caused by the bend remain recognizable even at large distances; top part of Figure 3.10).

Very long channels, sometimes folded or coiled as in the example shown in Figure 3.7, are used for special purposes. They may actually act as friction-type restrictors. Another reason to use them may be a desire to introduce into the circuit a significant inertance (see schematic representation in Figure 2.86). This changes the phase of oscillating flow, which may be a desirable property in feedback loops of fluidic oscillators. As mentioned in connection with Figure 1.48, such long channels may be also used to generate time delay due to finite propagation velocity (caused by compressibility). Yet another reason for very long channels is in mixers for microfluidics, where at low Reynolds numbers it takes considerable residence time for the reactants to mix properly before entering the reactor. Finally, extremely long channel lengths are needed in gas chromatography detectors.

3.2 AREA CONTRACTIONS AND NOZZLES

In a channel the cross-sectional area of which decreases in the streamwise direction, mass conservation demands an increase in flow velocity. Because of energy conservation, this has to be associated with decrease in pressure as the pressure component of fluid energy is converted into the kinetic component. Such a channel with the area contraction is a two-terminal fluidic device. Its schematic representation is in the top part of Figure 3.12. It is not used often, as there are not many situations demanding increase of dynamic action of the fluid in a channel. A reason for its use may be pressure effects: the pressure decrease depends on magnitude of the flowrate, which may be thus evaluated from the pressure difference output signal. This device is usually, as soon as the increased velocity is not needed anymore, followed by a diffuser converting the kinetic energy back to pressure rise to avoid large dissipative losses associated inevitably with fast flows.

More often, the accelerated fluid leaves the channel downstream from the contraction into a more or less open space where it forms a jet (Figure 3.11). The corresponding single-terminal devices for this particular purpose are *nozzles* (second schematic symbol in Figure 3.12). Nozzles may be used alone as devices performing, for example, agitation of sedimenting particles in fluid, cooling by impinging cool fluid, heating by hot fluid, deposition of bacteria on a test surface, or perhaps needleless hypodermic injections. More often, nozzles are used in fluidics as components of more complex devices.

Strictly speaking, the acceleration of the fluid in the contraction part of the nozzle is not necessary. There are nozzles in which the contraction is absent. Generally, however, there are good reasons which make the acceleration desirable, such as increase of the jet striking distance. The other useful, though not necessary, part of a nozzle is the exit channel of constant cross section immediately upstream from the exit. It stabilizes the jet flow and secures reliable directional orientation of the jet. Its usual length is mostly two to three channel widths (or diameters in the axisymmetric case). Some designers tend to make it short in an attempt to avoid the loss due to friction between its walls and the fluid accelerated there to the highest velocity.

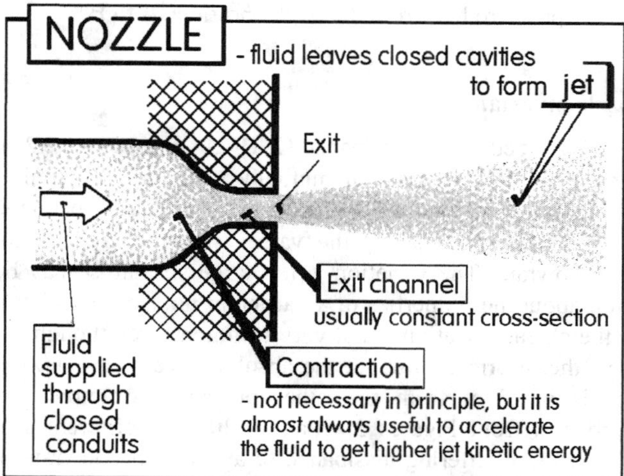

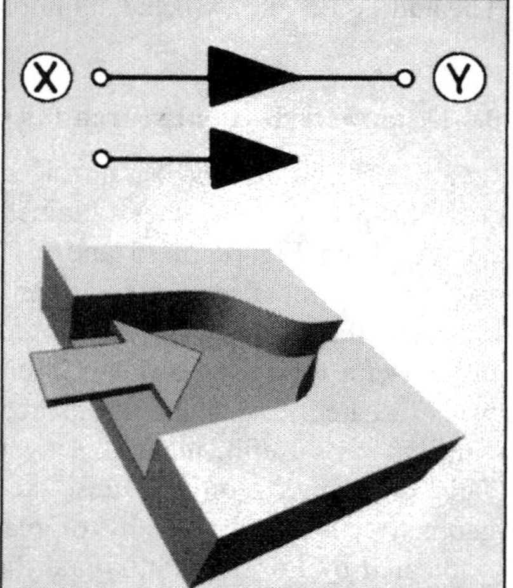

Figure 3.11 (Left) Nozzles are essential components of many fluidic devices, especially those operating at higher Reynolds numbers. They are used to generate jets.

Figure 3.12 (Right) Schematic representation of area contraction and a nozzle, with the contraction upstream from the exit, and an example of a typical planar nozzle.

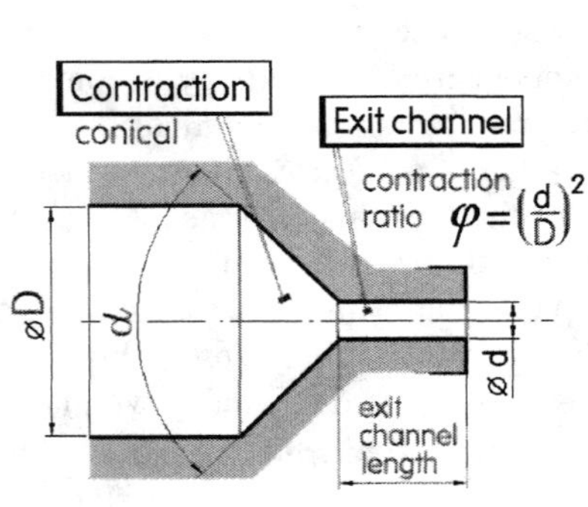

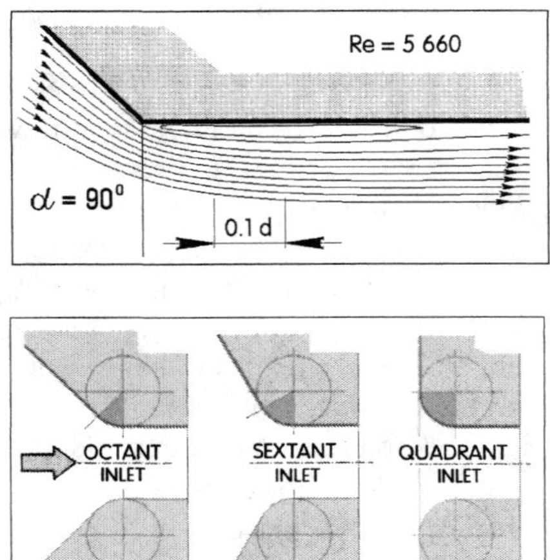

Figure 3.13 (Left) Basic parts of a nozzle—here of the axisymmetric shape—are its area contraction, accelerating the flow, and the exit channel, stabilizing the flow and directing it. Only the cheapest versions are made (perhaps by drilling) with sharp edges.

Figure 3.14 (Right, top) Computed pathlines at the entrance into the exit channel of an axisymmetric nozzle revealing the typical separation of the flow and formation of the recirculation bubble downstream from the sharp-edged entrance.

Figure 3.15 (Right, bottom) Usual shapes of the contraction and the rounding of the entrance into the exit channel to suppress the separation from Figure 3.14. Smaller vertex angle of the conical contraction permits choosing a smaller rounding radius.

Experience shows, however, that an exit channel, Figures 3.13 to 3.15 (provided it is not too long) actually decreases the overall loss. This is particularly effective if the fluid experiences a large change of flow conditions in the nozzle. Essentially this is effect is demonstrated in Figure 3.9 in the case of fluid experiencing upstream from the exit channel a change of flow direction. The addition of the exit channel improves the pressure loss from the exit conditions A to B.

3.2.1 Characterization: search for a nozzle invariant

Steady state properties of a nozzle are characterized by its dissipance Q (Figures 2.28 and 2.29). An example of the values obtained experimentally for a set of mutually geometrically similar nozzles from Figures 3.16 and 3.17 is presented in Figure 3.18. It is obvious that due to the variation of the frictional dissipance (similar to Figure 2.41) the values shown there are not constant; they vary slightly with the mass flowrate. The variation is not large so that Q may be quite useful for giving a rough information about the properties of a particular nozzle for which it was evaluated. Use of the dissipance for the characterization is not very common, even though the value is very useful, including, as it does, the information about the absolute size of the nozzle. This dependence on the absolute size is apparent from the other presented example, the geometrically similar family of planar nozzles from Figures 3.19 and 3.20 with different size (indicated by nozzle widths) and therefore widely differing dissipance Q at the same flowrate in Figure 3.21. More common is use of the dimensionless characterization by the Euler number Eu (derived in Figures 3.22, 3.23, and 3.24). Its variations with the Reynolds number (and hence flowrate), schematically shown in Figure 3.25, are the same as found for Q.

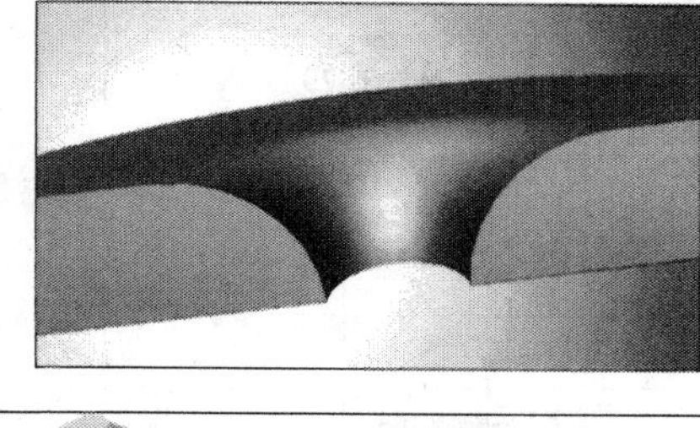

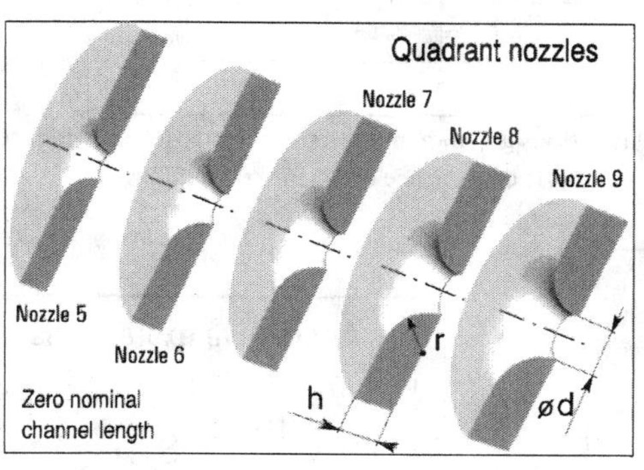

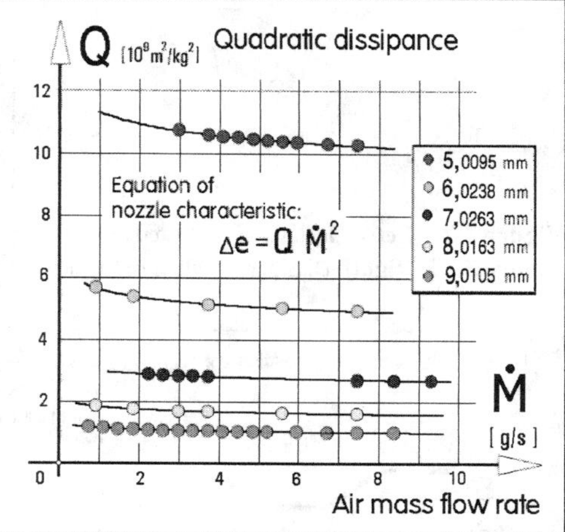

Figure 3.16 (Left, top) Meridian plane section through a micromachined nozzle for generating jets perpendicular to the device plane. Laser drilling produces nozzles practically equivalent to the quadrant-contoured shape.

Figure 3.17 (Left, bottom) A family of scaled-up mutually similar quadrant-shaped models used to investigate properties of nozzles having no exit channel.

Figure 3.18 (Right) Experimentally evaluated dissipance values of the nozzles from Figure 3.17, plotted as a function of the flowrate.

Variations of Eu with the Reynolds number Re for the quadrant nozzles in Figure 3.26 provide an the obvious explanation: the typical variation of the frictional dissipance. Since the nozzles in this family are mutually similar, the dependences $\mathrm{Eu} = f(\mathrm{Re})$ all collapse into the single curve (in the logarithmic coordinates of Figure 3.26 resembling the straight line, which is not really true). This hydrodynamic similarity is an important tool for testing nozzle designs experimentally using scaled up laboratory models, which are more convenient to work with. The universality of this presentation (independence on fluid properties) is also useful for recalculation of nozzle properties

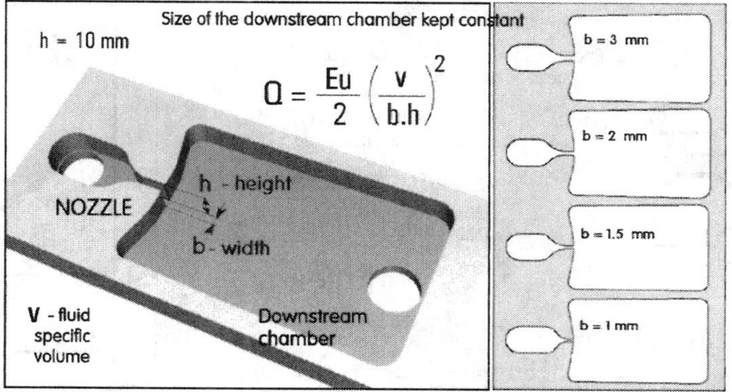

Figure 3.19 A set of nozzles of the planar shape, more usual in fluidics, tested with the integral downstream chamber. Nozzle geometry is presented in Figure 3.20. The scaling up in this case did not involve a change of the chamber size.

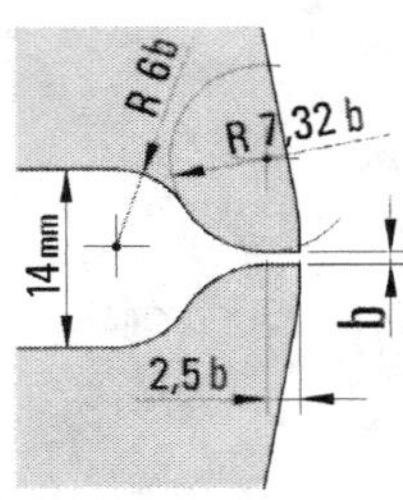

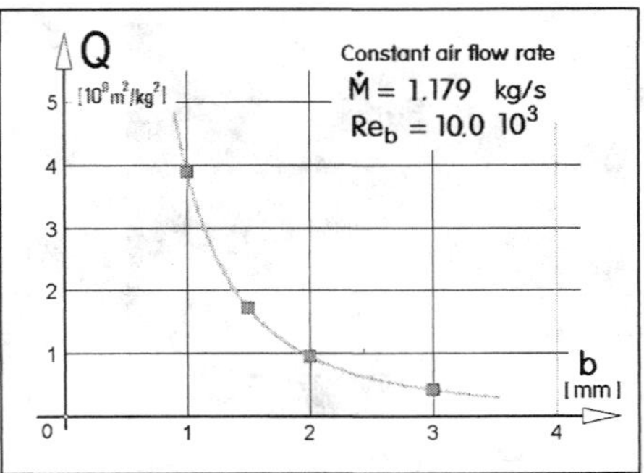

Figure 3.20 (Left) Details of the tested nozzle geometry. Dimensions given in multiples of the nozzle exit width b.

Figure 3.21 (Right) Dissipance values of the nozzles from Figure 3.1: dependence on the nozzle width.

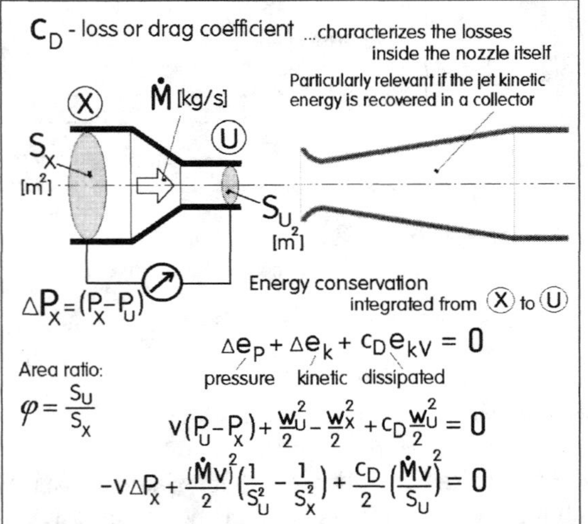

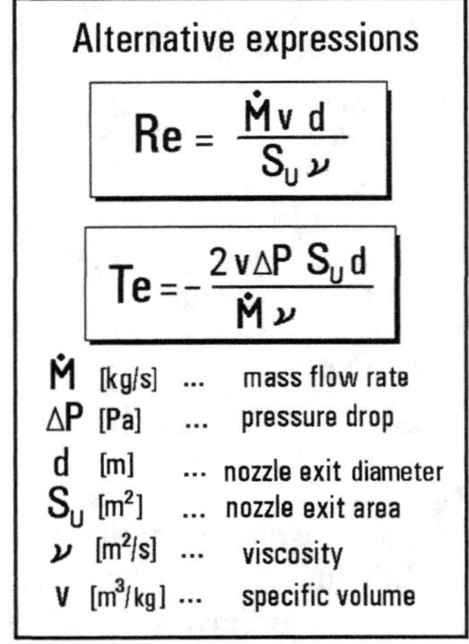

Figure 3.22 (Left) Energy conservation equation integrated between the inlet and exit of a nozzle.

Figure 3.23 (Right) Alternative versions of the two parameters that convert the energy conservation equation from Figure 3.22 into dimensionless form.

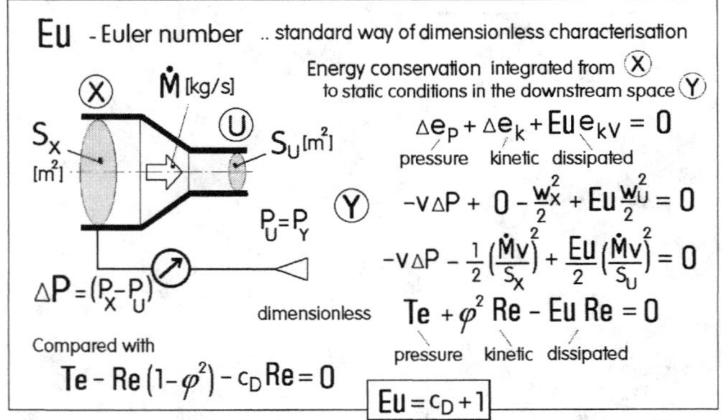

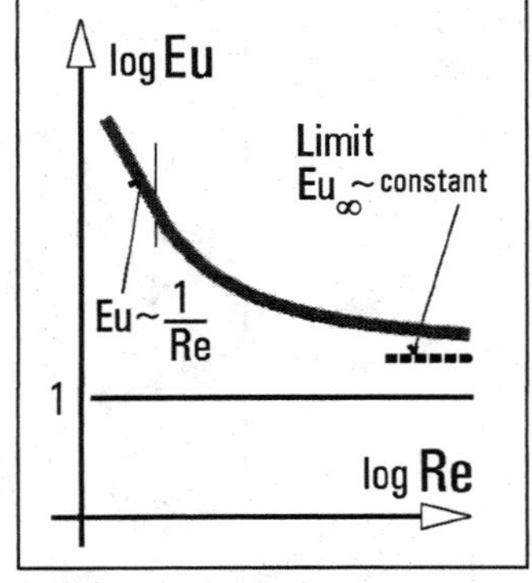

Figure 3.24 (Left) Energy conservation equation integrated up to the stagnant conditions in the downstream space.

Figure 3.25 (Right) Schematic dependence of Euler number of a loss in a nozzle on Reynolds number.

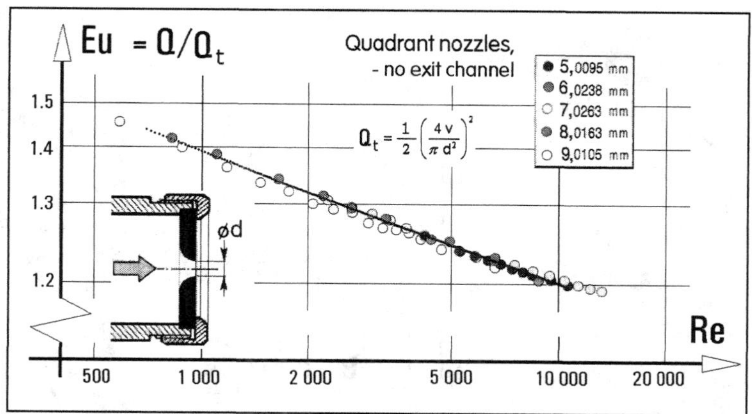

Figure 3.26 Experimental results: Euler number for the nozzles from Figure 3.17 as a function of Reynolds number.

for use with different fluids. The nonidentical values of Eu in Figure 3.27, despite the same Re, may seem to be in contradiction to this identity of properties in dimensionless coordinates – but it is explained in this case by the models incorporating the downstream space in which the kinetic energy of the jet is dissipated and the fact that the size of this space did not vary when the nozzle size was changed.

Yet another way of characterizing nozzle properties by dimensionless parameters is the use of the parameter Te (Figures 2.42 and 3.23). In contrast to Eu, the value of Te rises with increasing Re (Figure 3.28), which presents a better picture about the losses, which actually increase with increasing flowrate. This parameter Te may be especially recommended for use at low Re where it approaches a constant asymptotic value (an example is in Figure 2.50).

The losses in nozzles may be decreased by careful shaping. In particular, always recommendable is rounding of sharp edges to suppress possible separation of the form (Figures 3.14 and 3.15). In the axisymmetric shapes, often made by drilling, the choice may be significantly influenced by manufacturing economy—designers tend to favor the cheap sharp-edged shapes

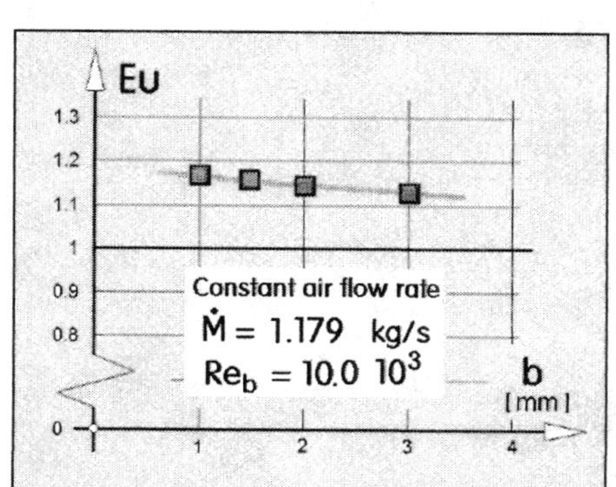

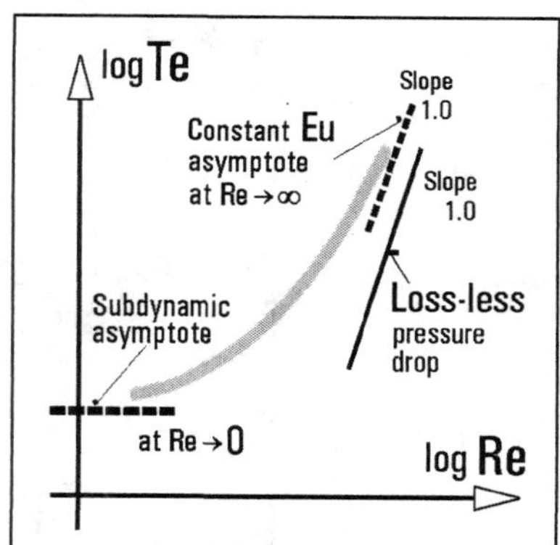

Figure 3.27 (Left) Euler numbers evaluated for the set of planar nozzles from Figure 3.19 at the same Reynolds number.

Figure 3.28 (Right) The other dimensionless parameter Te for characterization of nozzle losses (Figure 3.24) provides a more logical presentation (rising losses with rising flow) and is particularly suitable for very low Reynolds numbers of microfluidics, where it tends to be constant.

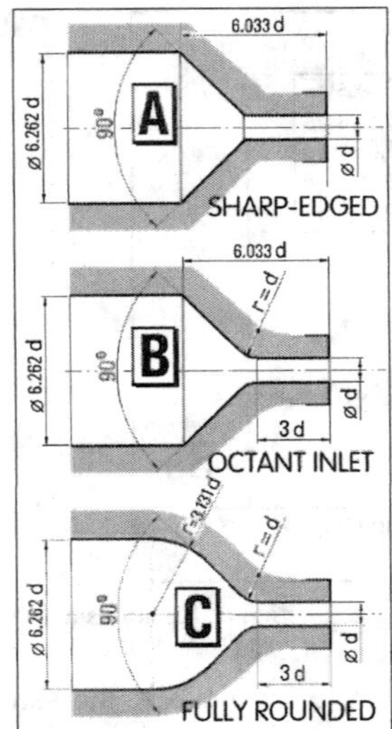

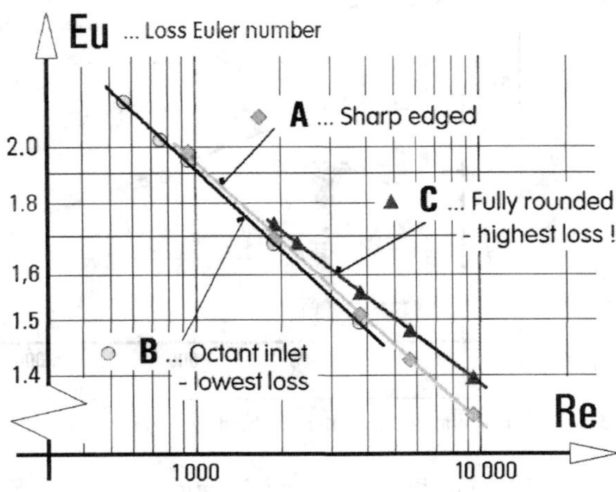

Figure 3.29 (Left) Nozzles of progressively more rounded shapes used in the study of the effects of the rounding.
Figure 3.30 (Right) Euler number for the nozzles from Figure 3.17 as a function of Reynolds number – showng the surprising failure of the complex (and most expensive) shape.

simply made by standard drill bits. In fact, the details created by smooth rounded shapes may bring unpleasant surprises. Figures 3.29, 3.30, and 3.31 show an example of the most elaborate shape actually exhibiting a higher loss than the less sophisticated variants.

Ideally a nozzle would be characterized by a single number, a constant. Unfortunately, even Te is not really very much better than Eu from this point of view, as shown in the schematic representation of a typical dependence in Figure 3.28. The subdynamic asymptote is approached at

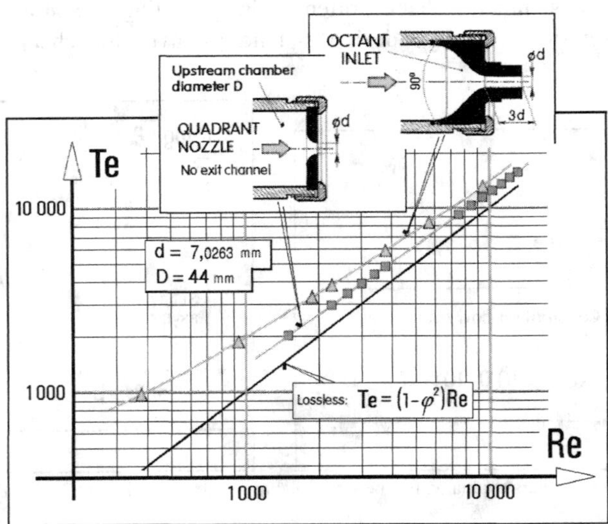

Figure 3.31 Nozzle losses presented using the alternative parameter Te. Contrary to common beliefs, this study has found for the fully rounded smooth nozzle shape from Figure 3.29 actually higher loss than for the simple quadrant nozzle of the same area ratio φ.

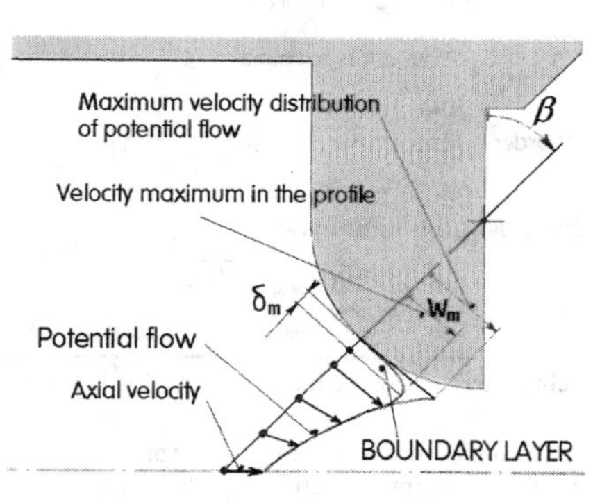

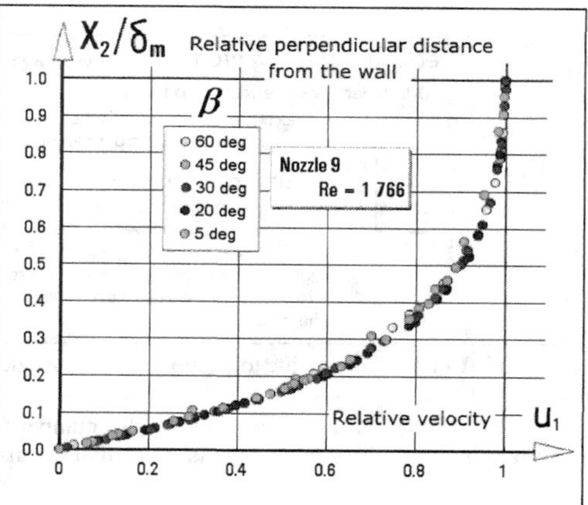

Figure 3.32 (Left) Geometry of the potential flow and the boundary layer inside the quadrant nozzle.

Figure 3.33 (Right) Velocity profiles in the boundary layers in quadrant nozzles exhibit similarity: they all collapse into a single curve when plotted in relative coordinates. This example shows profiles inside the nozzle 9 from Figure 3.17.

a too low Re, where the nozzle ceases to be practically useful. The search for the elusive characterizing invariant therefore had to go deeper, to basic aerodynamics of nozzle flow. This is shown (Figure 3.32) to be dominated—in smooth, separationless shapes—by the potential lossless flow in the core and the boundary layer formed on the walls. Figure 3.33 documents that character of the velocity profile shapes does not change along the surface. A special nozzle contouring is necessary if there is a demand for constant velocity in the core. Otherwise, with simple shapes, the potential flow often exhibits velocity increasing towards the walls. Thickness of the boundary layer (Figure 3.34) grows with decreasing Re. Since the thickness determinates the hydraulic loss

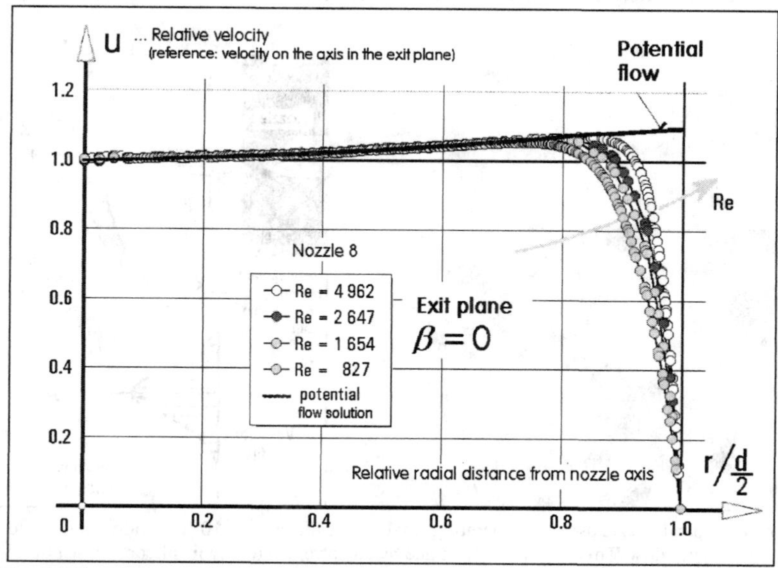

Figure 3.34 Flowfield inside the quadrant nozzle. Lossless potential flow predictably agrees with the actual velocity distribution in the core flow away from the walls, near to which the flow is dominated by the boundary layer.

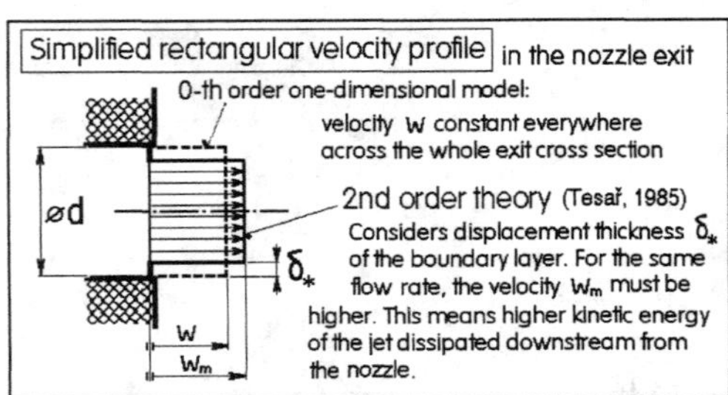

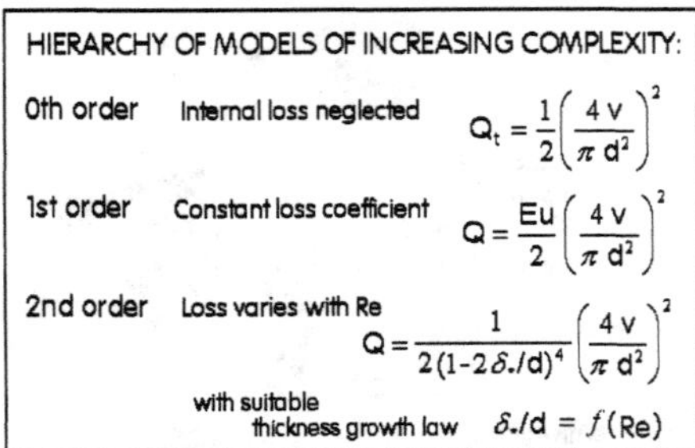

Figure 3.35 (Left) The loss due to friction inside a nozzle may be characterized by the magnitude of displacement thickness.

Figure 3.36 (Right) The details of the Reynolds number dependence of nozzle properties may be surprisingly simply characterized by coefficients in the laminar growth of the displacement thickness introduced in Figure 3.35.

in the nozzle (Figure 3.35), it is this growth that determines the variations of dissipance Q with flowrate. The real characterization constant of a nozzle may be defined as the one determining the thickness growth law (Figure 3.36) [2].

3.2.2 Generation of free jets and droplets

Free jets are liquid jets issuing into air or another gas. Their applications tend to concentrate on the small-size end, the very small size of the nozzle made by micromanufacturing methods ensuring the often desirable small droplet size or small dosage in various dispensers. Typical is pipeting (with frequent change of the liquid), for example at the input end of biochemical analyses, and dosage of glue to connect optical fibers. Free jets may reach further than the submerged jets—gas in a gas or liquid issuing into liquid—since they lose less momentum by mixing with the surroundings, but they also have their limiting mechanism: it is the surface tension induced

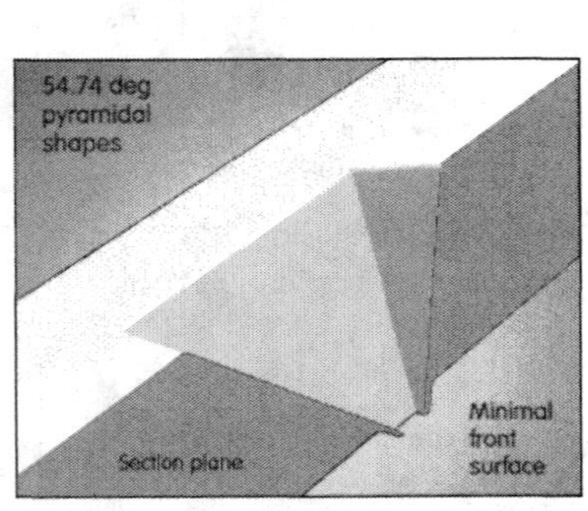

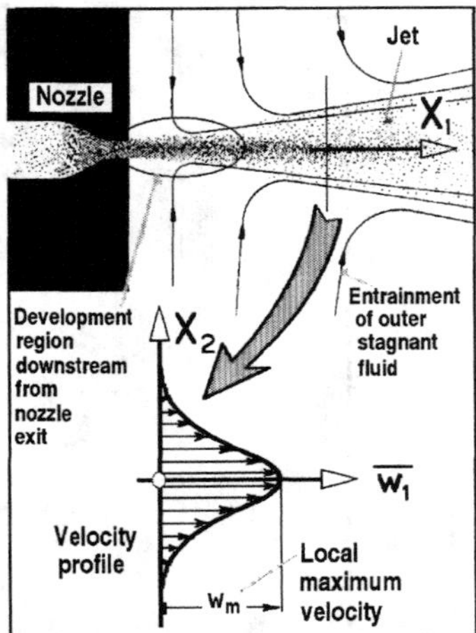

Figure 3.37 (Left) Typical nozzle used for generating smallest (of the order ~ 10 μm) individual liquid droplets by pressure pulses in the supply flow. This is an example made by anisotropic etching in silicon, shown in mid-plane section.

Figure 3.38 (Left) Schematic drawing of a submerged jet, issuing from the nozzle into a fluid of comparable specific volume. By entraining the outer fluid, jet increases its width (and, to conserve momentum flowrate, decreases its velocity).

instability, that, enhanced in the later stages by pressure forces on the bulging interface, results in breakup of the jet into a train of drops (often accompanied by smaller satellite droplets). In microfluidics, it is the formation of the drops at the nozzle exit, obtained by short driving pressure pulse, which is of particular interest. Indeed, the ink jet printer nozzles doing exactly this were one of the economically most important applications in the beginning of microfluidics. Current interests concentrate on microdipensing, handling samples down to the picolitre range. The droplets are usually generated to move perpendicularly to the plane of the base plate. The example of a nozzle for this purpose, shown in Figure 3.37, features the typical inclined walls and pyramidal shapes made by the anisotropic wet etching in <100> oriented silicon (Figure 2.14). An important design feature is etching off a layer also at the exit side, leaving low outer pyramid around the protruding exit. The minimal surface of the nozzle lip is important at these smallest scales. Drop formation is strongly influenced by not completely predictable wetting of the outer parts of the lip, especially after operational pauses, during which the wetted lips fast become dry, leaving remains of uncertain composition in this sensitive area.

3.2.3 Generating submerged jets

Various forms of submerged jets—the accelerated flow that is left to leave fully closed conduits in a nozzle to issue into a more or less open space filled with the same fluid (schematically presented in Figure 3.38)—are very common in no-moving-part fluidic devices. Even in its basic from, which may be treated as two-dimensional (Figure 3.39), submerged jet flows are rather complicated. Full understanding requires numerical computations of complete (and preferably three-dimensional) Navier-Stokes equation. Each such computation result, however, is valid only for the particular set of boundary conditions. To obtain general insight, it is useful—like in other flows dominated by shear stresses in the fluid—to use the simplified Prandtl equations (Figure 3.40) and preferably transform it to an even simpler form by similarity transformation. This is possible due to the property of jets (after negotiating an initial distance, where they develop) to retain along their flowpath a certain invariant character of the distribution of variables. This enables the partial differential Prandtl equation to be transformed into the much simpler set of ordinary differential equations (as shown in Figure 3.51). Even though the jet width increases in the direction of the flow (due to entrainment of outer stagnant fluid) and maximum velocity

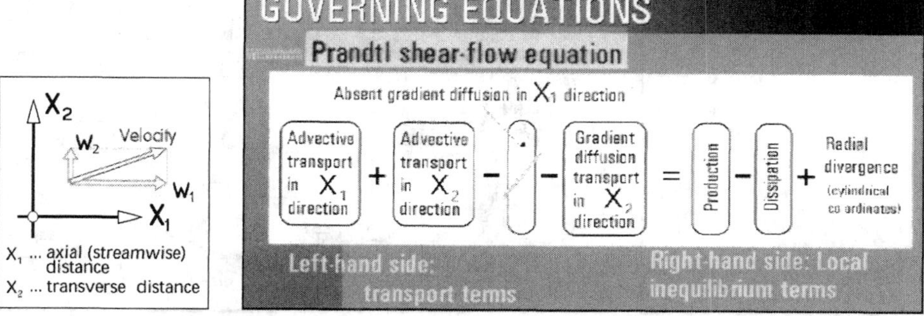

Figure 3.39 (Left) The two coordinates used in the similarity computations of jets. Planar jets may be computed as two-dimensional only in the first approximation; the top and bottom cover plates cause three-dimensionality.

Figure 3.40 (Right) Classical solutions of jets are based on the Prandtl equation, which gives almost exact solutions in shear flows like the jet.

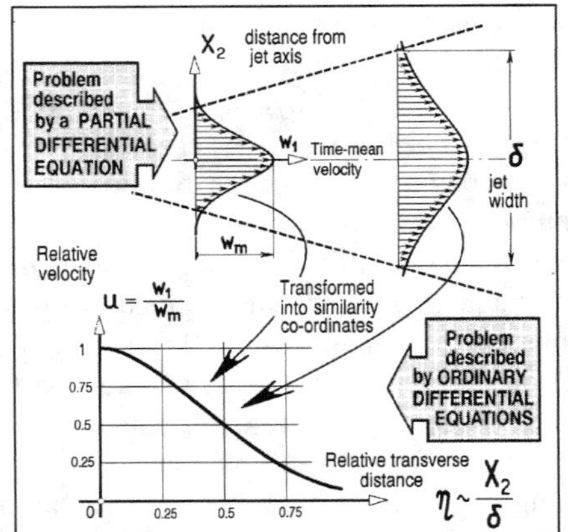

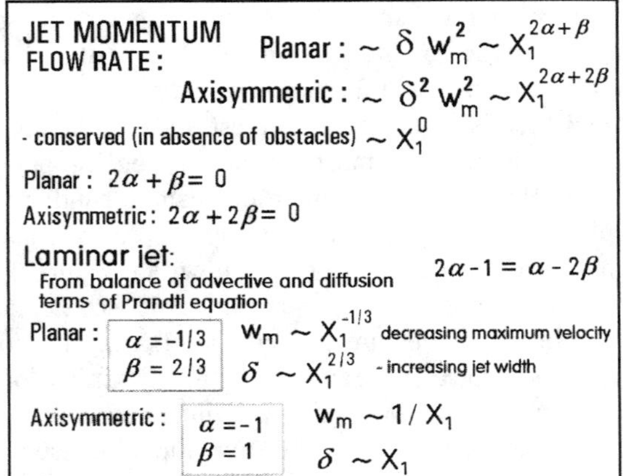

Figure 3.41 (Left) The principle of similarity solution: because profiles of quantities are similar, the dependence on the longitudinal variable may be removed by suitable similarity transformation of coordinates.

Figure 3.42 (Right) The basic relations for dependence of jet widths and axial velocity in a developed jet may be obtained from simple considerations of friction and inertial forces as well as momentum rate on a fluid element.

consequently gradually decreases (to keep constant momentum flowrate), all the velocity profiles (Figures 3.38 and 3.41) are mutually similar. They collapse into a single curve (Figure 3.41) when plotted in dimensionless similarity variables: transformed velocity and transformed transversal coordinate. The reference quantities, to which the absolute velocity and transversal distance are related, are derived for the laminar case in Figure 3.42 and the solution is in Figure 3.43.

More complicated (note the typical differences in Figure 3.44) is the turbulent jet, the basic features of which are presented schematically in Figures 3.45 to 3.47. Fortunately, jet turbulence reasonably meets the simplifying assumption of isotropy, no preference for a particular orientation in space. It may be, as a result, fully described by a scalar quantity, the turbulent viscosity ν_t (Figure 3.48) used as a replacement for the fluid molecular viscosity ν of the laminar case. Unfor-

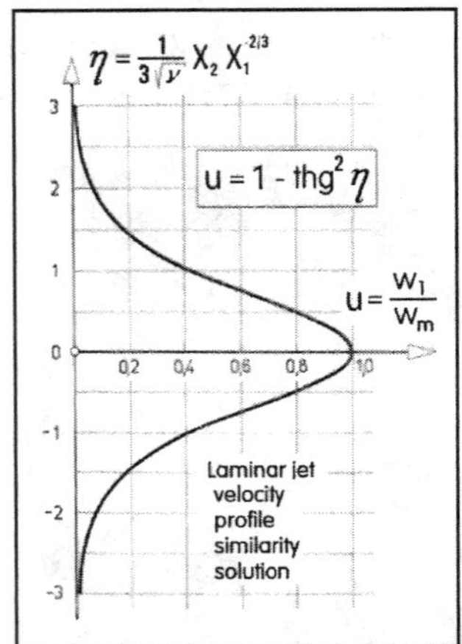

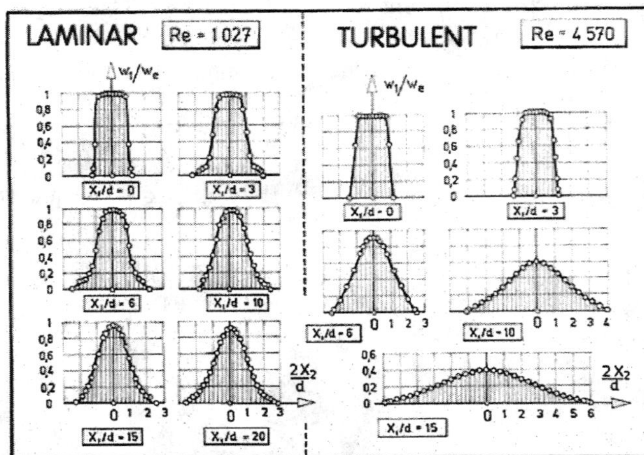

Figure 3.43 (Left) The velocity profile in the transformed coordinates obtained by the similarity solution for the laminar planar jet.

Figure 3.44 (Right) Experimental velocity profiles of the axial velocity (related to the nozzle exit velocity w_e) of a helium round jet (Figure 3.57) in the laminar and turbulent regimes at different axial locations.

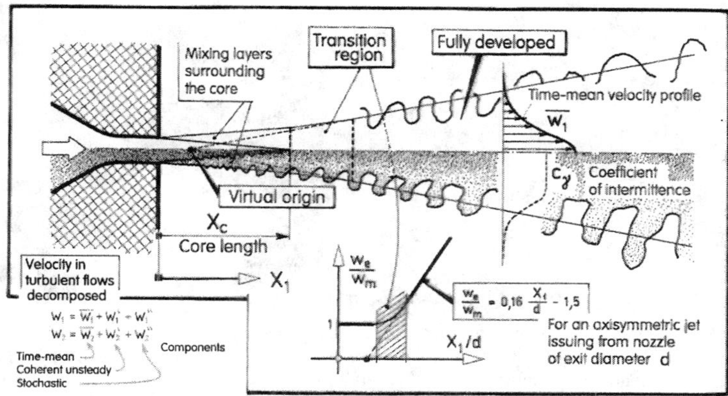

Figure 3.45 Basic features of turbulent submerged jets. Note the necessity to decompose the steady and unsteady velocity components to apply the statistic approach to turbulence and also the intermittent character on the edges; entrainment of the outer nonturbulent fluid leads to alternating turbulent and nonturbulent character.

tunately, unlike ν, the turbulent counterpart is not constant and its distribution in space is evaluated rather laboriously, from the spatial distributions of velocity and size scales of turbulence. These are evaluated by solving, in addition to the transport equations for the two velocity components (not for pressure, since this in a developed straight jet may be taken as constant) the simultaneous transport equations for parameters of turbulence. The first of them is

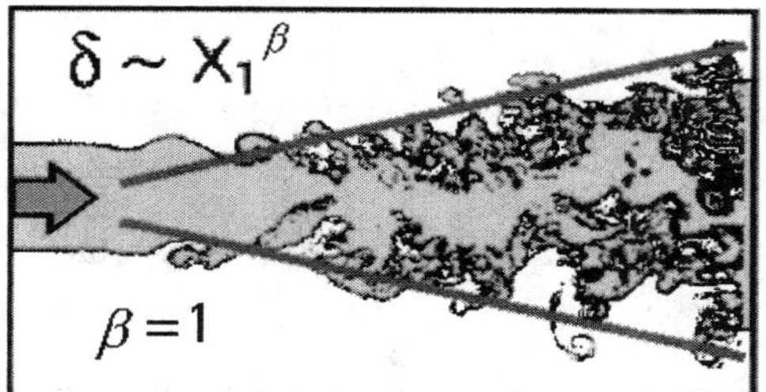

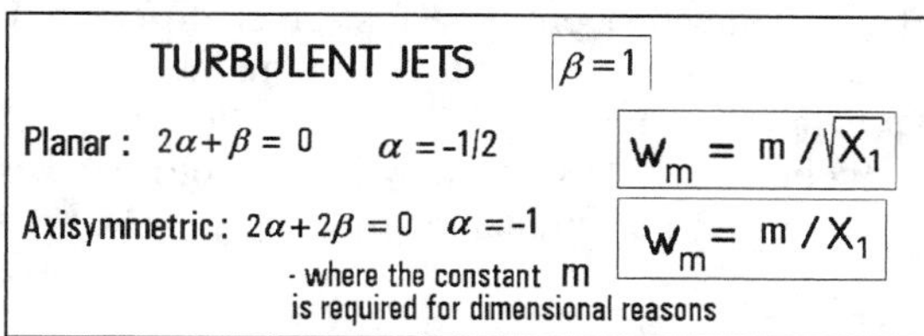

Figure 3.46 (Left) Photograph of an actual turbulent jet with superimposed straight lines of turbulent jet nominal edge.

Figure 3.47 (Right) Basic laws of the velocity variation of turbulent jet, using the momentum rate conservation (top part of Figure 3.41), which remains valid also in the turbulent case.

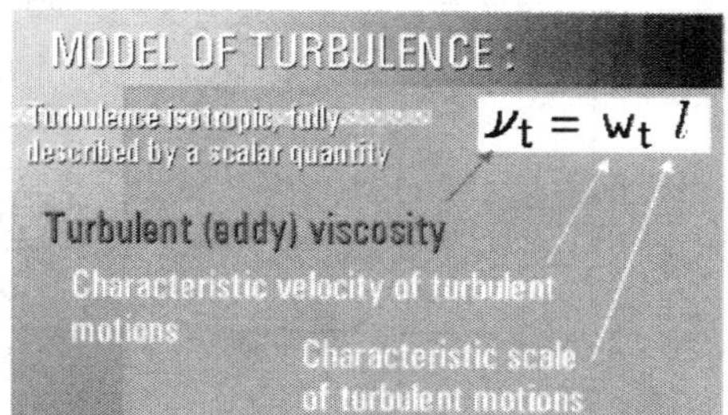

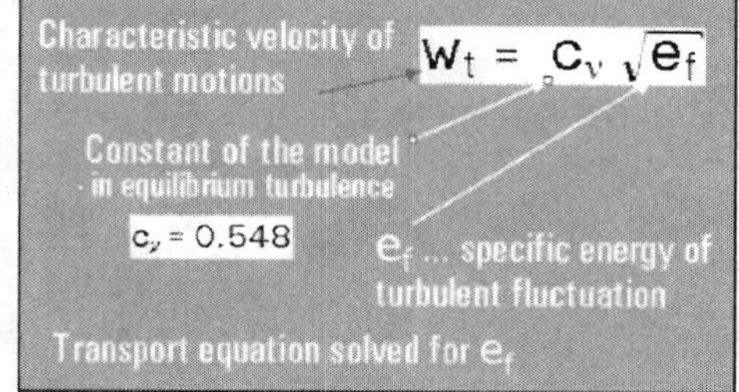

Figure 3.48 (Left) From computation point of view, turbulent flow differs by the constant viscosity ν of the laminar case replaced by the turbulent viscosity, a quantity varying in a complicated manner across the flowfield. It may be shown to depend on the product of characteristic velocity and scale of turbulent motions.

Figure 3.49 (Right) Velocity of turbulent motions is evaluated by solving a transport equation (with terms analogous to those of Prandtl equation, Figure 3.40) for specific energy e_f of turbulent fluctuations.

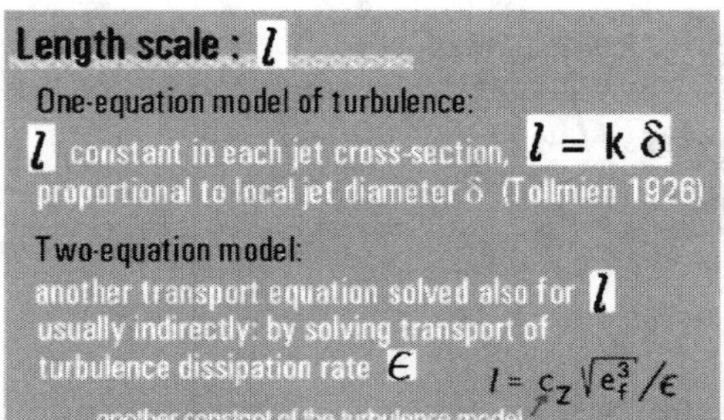

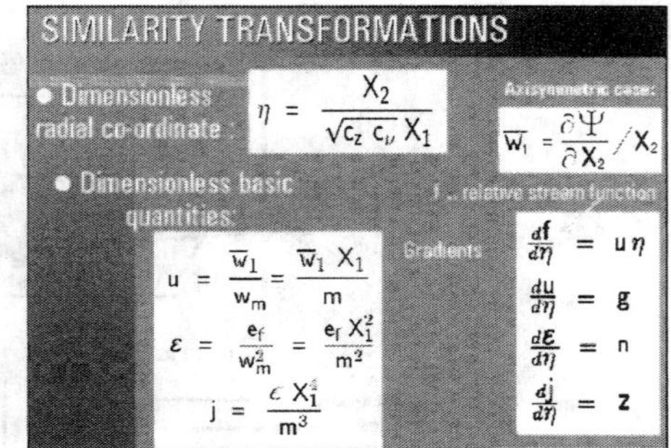

Figure 3.50 (Left) The characteristic size of turbulent vortices, needed for evaluation of turbulent viscosity in Figure 3.47, may be evaluated either by an algebraic expression – unlikely to be successful in general, but surprisingly good for computing jets – or by solving an additional simultaneous equation.

Figure 3.51 (Right) The transformations of coordinates used in the similarity solutions of turbulent jets – here in the axisymmetric version.

the equation for the specific energy e_f of turbulent fluctuations. This is used to evaluate the characteristic velocity of turbulent motions (Figure 3.49). In the one-equation model of turbulence it is used alone. The expression for the length scale – the characteristic size of turbulent eddies – is the very simple (Figure 3.50) hypothesis due to Tollmien [3]. In the 2-equation model of turbulence, an analogous transport equation is solved for turbulence dissipation rate ε, from which the length scale l is found using the expression in the bottom part of Figure 3.50. It is the large number of the partial differential equations of the rather complex transport type that makes general solution of turbulent flows so difficult. Nevertheless, for jets [4, 5] these equations may be converted by the similarity transformations (using expressions from Figure 3.51 and/or similar) also into a set of ordinary first-order differential equations. Figure 3.52 shows them for the 2-equation planar case (the first four are mere definitions, which for the axisymmetric case are of the form shown in Figure 3.51). The solution of the main set in Figure 3.52 (the laminar set, included for comparison, has an analytic solution) needs, of course, 7 values of the (boundary) starting conditions, the determination of which may be not straightforward [6, 7].

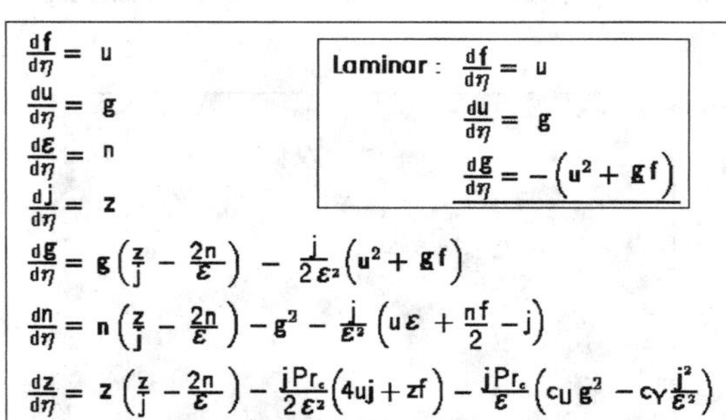

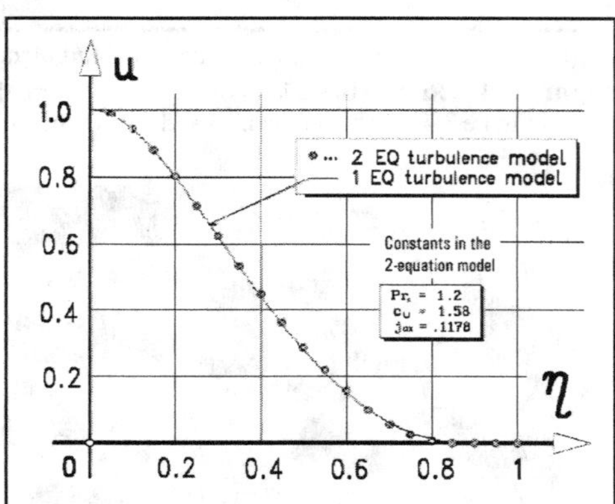

Figure 3.52 (Left) Set of ordinary differential equations obtained by the similarity transformations of the Prandtl-type transport equations describing fully developed laminar (inset, top left) and turbulent planar jet, the latter evaluated using the 2-equation model of turbulence.

Figure 3.53 (Right) Similarity solutions of turbulent planar jets: there is no discernible difference between the results obtained with the 1-equation and 2-equation models of turbulence.

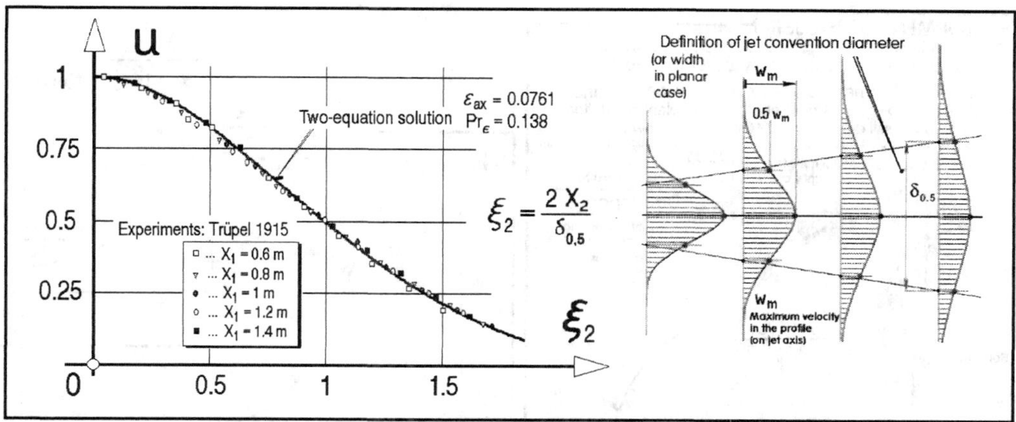

Figure 3.54 Similarity solution of an axisymmetric turbulent jet using the two-equation turbulence model, compared with classical experimental data. Note the use of relative transversal reference dimension $\delta_{0.5}$. This is better suited to processing experimental data than the definition in Figure 3.51,which is dependent upon a particular theory.

The simpler 1-equation model leads to a set of only 5 equations, needing 5 starting conditions. This model, of course, is less versatile in general, but for jets the Tollmien's simple expression is surprisingly successful, as documented in Figure 3.53. Both models agree very well with experimental data (Figure 3.54 documents this for the 2-equation model).

While the similarity solutions of jets are useful for gaining familiarity with this important type of flow and may provide useful ideas, especially about distributions of some more esoteric quantities (such as those used in turbulence theory), for practical design of fluidic devices it represents just an approximation.

First, most submerged jets in fluidics are oriented parallel to the base plane, issuing from a slit nozzle into a space bounded on top as well as bottom. They may be near to the two-dimensional similarity solution only if the nozzle exit aspect ratio (Figure 1.35) is extremely high. This, unfortunately, is not obtainable with the usual, less expensive manufacturing methods (recall the

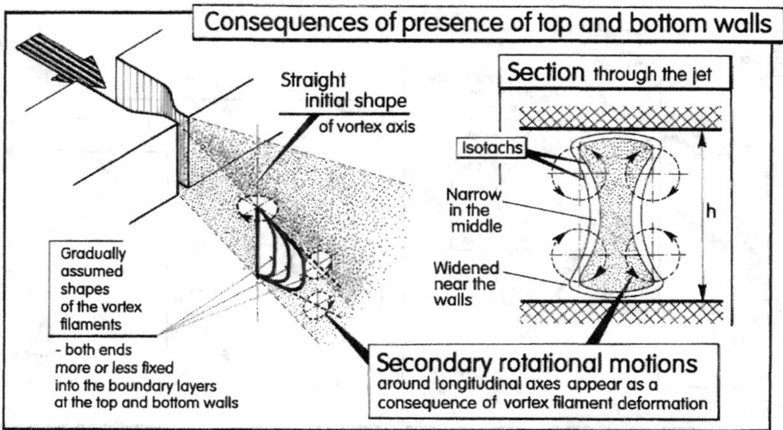

Figure 3.55 Due to the limitations of manufacturing methods, most jets issue into a bounded space and are therefore influenced by the friction on the bottom and the top cover plate. The vortices, characteristic for the outer parts of jets, are deformed into roughly U-shaped configurations, the rotational motion of which makes the jet thinner in the middle and wider near the top and bottom.

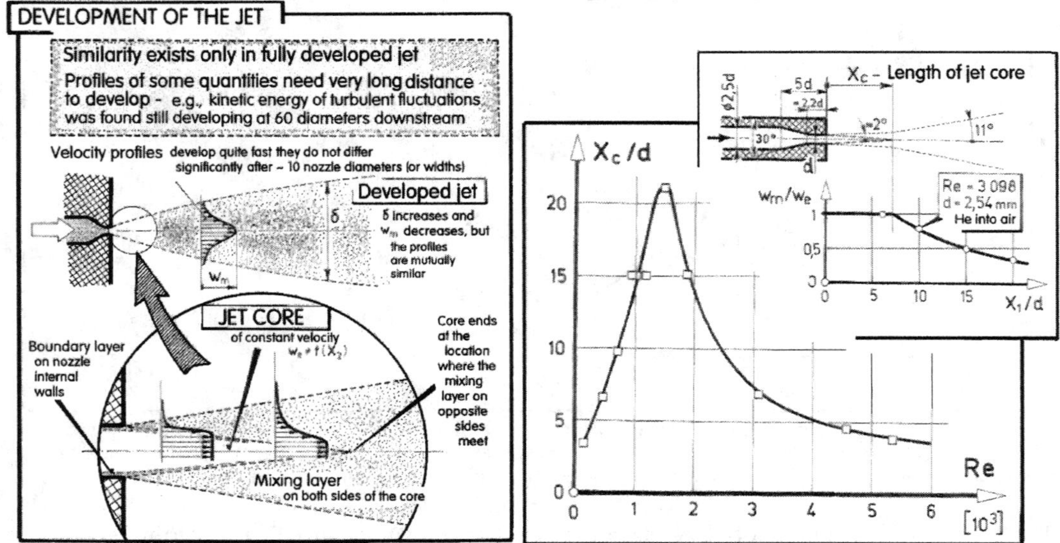

Figure 3.56 (Left) Usefulness of the similarity solution is limited by the presence of the initial development segment. In most applications in fluidics the jets is captured before it becomes fully developed.

Figure 3.57 (Right) Length of the jet core may be very large, in particular at low Reynolds numbers before the jet transition into turbulence.

aspect ratio less than 1 obtained with the isotropic etching, Figure 2.12). The in-plane jets are therefore bounded. This leads, as shown in Figure 3.55, to distinctly three-dimensional effects.

Second, jets need quite a long distance to develop to the (at least nearly) self-similar shape. When the jet leaves the nozzle, its velocity profile has features quite opposite to the developed jet shape with local maximum on the axis. Unless the boundary layer at the wall is very thick, the potential flow in an area contraction tends to have a minimum on axis and maxima at the wall (as

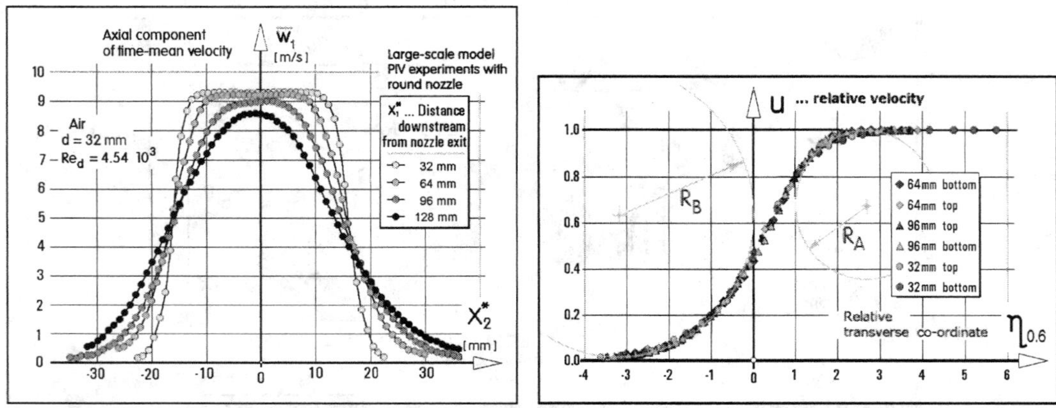

Figure 3.58 (Left) Velocity profiles evaluated by PIV (particle image velocimetry) in the development region of a large-scale laboratory model jet.

Figure 3.59 (Right) In relative coordinates (evaluated using analogous principle of conventional reference length as shown in Figure 3.53) the velocity profiles from Figure 3.53 demonstrate a remarkable mutual similarity.

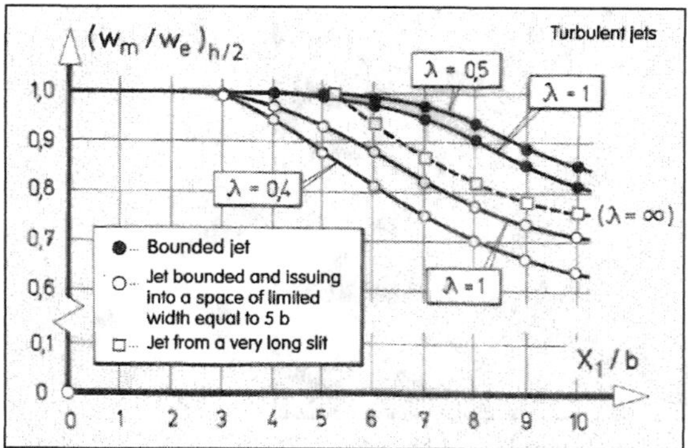

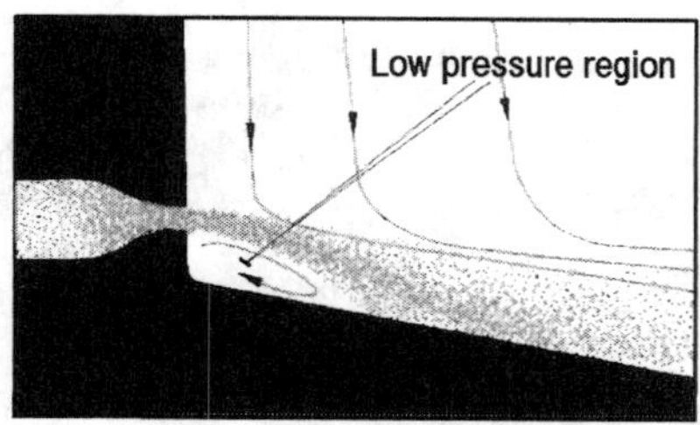

Figure 3.60 (Left) Jet core length may be significantly influenced both by the jet being bounded (i.e., by small aspect ratio λ) and confined.

Figure 3.61 (Right) Confined jet, not having enough available space in its sides for inflow of fluid replacing the entrained fluid, tends to change its flow direction and attach to the confinement walls. This is the Coanda jet attachment effect, particularly strong if there is only one wall on one side.

documented in the example presented in Figure 3.34). This is gradually eradicated as the jet develops by the mixing layer of growing thickness, which surrounds the potential core. The core length (Figures 3.56, 3.57, and 3.60) may be unexpectedly long, especially in low Re laminar jets of interest in microfluidics. From a fluidic device's point of view, the presence of the core is a favorable phenomenon – it increases the reach of the jet. In many devices, the jet fluid is recaptured and converted back into a channel flow in a collector. It is desirable to place the latter as near to the nozzle as is compatible with operational requirements (e.g., if the jet is to be deflected, a collector placed too near may require a too-large deflection effort because of the large resultant deflection angle). At any rate, the jet is preferably captured while still before the core disappears, which obviously makes the similarity theory not really useful. On the other hand, as documented in Figure 3.59 on the velocity profiles from Figure 3.58, there is a reasonable similarity (of a different sort, but also governed by the Prandtl equation, though with different boundary conditions) found for the velocity profiles in the mixing layer surrounding the core. This may be used for developing another approach to description of the jets in their developing part.

In many jet-type fluidic devices it is useful to utilize the Coanda effect of jet attachment to a wall or alternatively to one of two opposite walls (Figures 3.63 and 3.64). Of interest for the

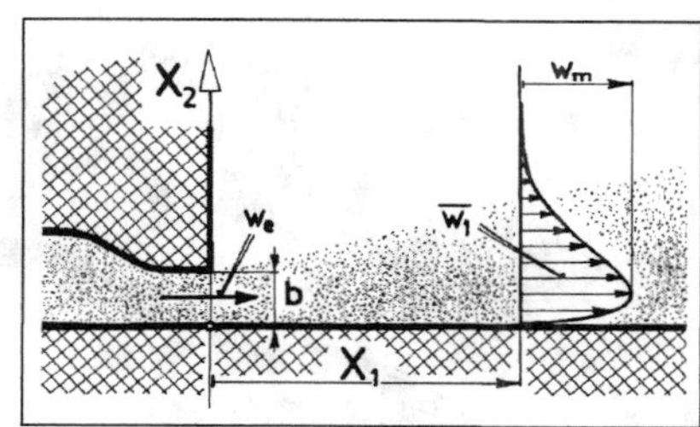

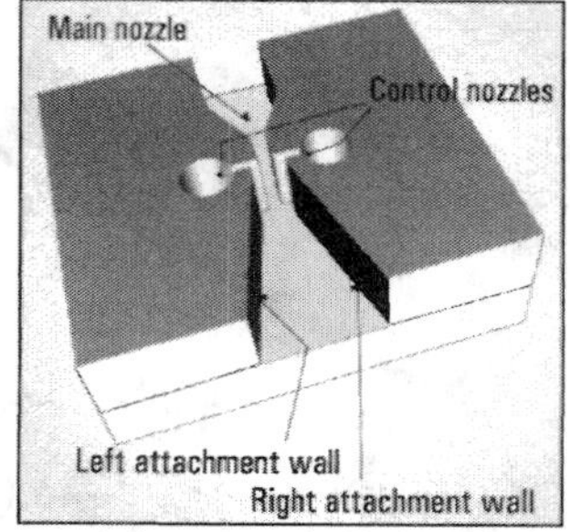

Figure 3.62 (Left) Once the jet becomes attached, the conditions change and the proper similarity solution used as the reference is the wall-jet—here shown in its straight-wall form.

Figure 3.63 (Right) With two attachment walls, arranged in a symmetric manner on both sides of the space in front of a nozzle exit, the Coanda effect leads to an interesting and useful bistable behavior.

Figure 3.64 The two stable states of a bistable jet-type device. The jet may be switched from one side to the other by a quite week action of the control flows brought in an alternating sequence into the two control nozzles (Figure 3.63).

design of jet-capturing collectors are the profiles of the jet following a wall, near which it acquires the character of a boundary layer. This is the case of the wall-jet. In the laminar case there is also a similarity solution, set up along analogous lines to what was discussed above about the submerged jet solution, actually also available in the analytic form.

3.3 DIFFUSERS AND COLLECTORS

Diffusers are devices performing a task opposite to that of the nozzles. Essentially, they are channels the cross-sectional area of which increases in the streamwise direction. The mass conservation demands an increase in flow velocity as the fluid moves downstream. If the process were governed by the simple lossless energy conservation (Figure 2.37), this would be associated with an increase in pressure as the kinetic component of fluid energy is converted into the pressure component. Positive axial pressure gradient, however, is unfavorable for behavior of the near-wall flow that (Figure 3.67) it may cause to separate. The formation of the separated recirculation region is always associated with huge rise in the (Borda-type; see Figure 2.38) hydraulic loss, which may completely destroy the pressure rise mechanism. To avoid this catastrophic consequence, the diffusers have to be very carefully shaped, with very gentle changes of the cross section so as to keep the pressure gradient small. Short diffusers accept a larger divergence; this

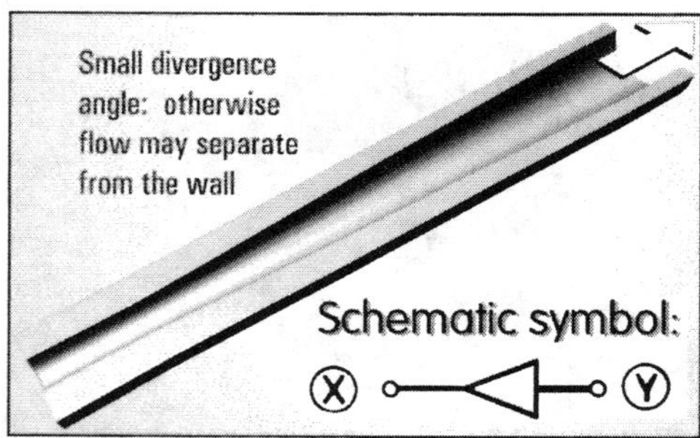

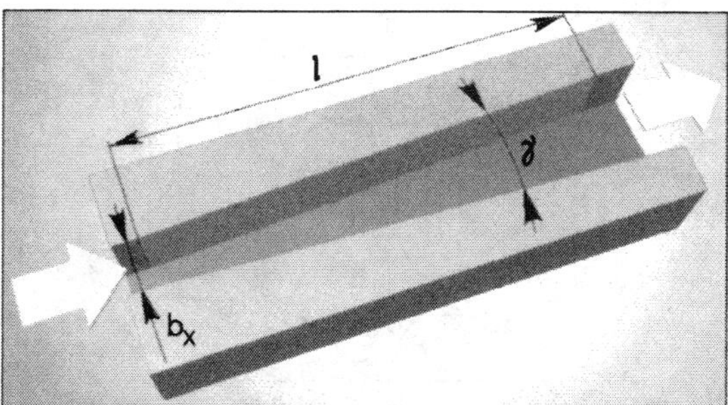

Figure 3.65 (Left) Example and symbol of a diffuser: cross-sectional area increases in the flow direction. Because of the acute danger of flow separation from the wall, the area change must be very gentle.

Figure 3.66 (Right) Of more importance for fluidics are diffusers of rectangular cross-section.

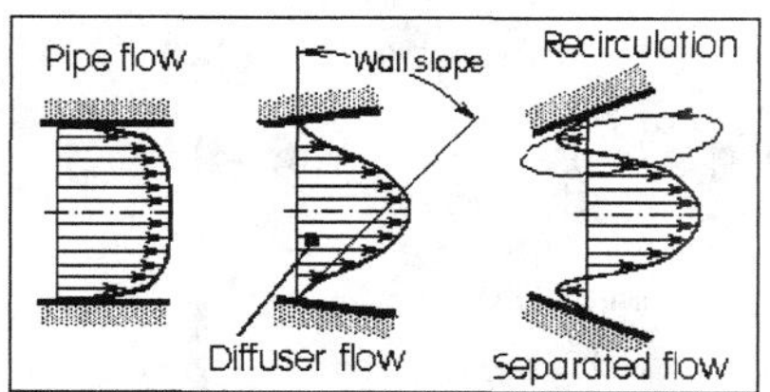

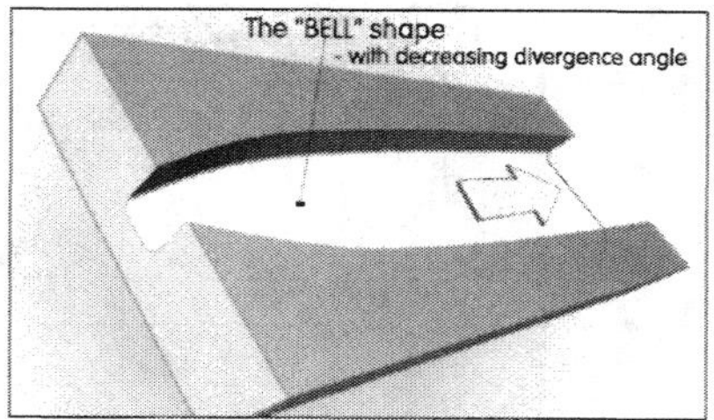

Figure 3.67 (Left) Characteristic shapes of velocity profiles in a constant cross-section channel or pipe, a properly operating diffuser, and a diffuser with too-large divergence angle, leading to separation of the flow from the walls.

Figure 3.68 (Right) Sophisticated shapes, like the one shown here, did not prove to be better than the simple straight-wall wedge-shaped diffuses.

nonlienar dependence on axial position has led to tests with a number of contoured diffuser geometries with nonconstant divergence angle. The example in Figure 3.68 shows an example of the short wide-angle diffuser with gradual transition into the narrow angle. In general, such schemes did not bring a convincing advantage over the linear (constant angle) geometry. The small optimum divergence angles lead to diffusers that are inconveniently long. Lack of available space for them often leads to attempts at "bending" the diffuser, curving its axis. Whenever possible, this has to be avoided as it more often than not leads to one-sided separation.

Of course, the separated flow in Figure 3.67 is not likely to remain in the indicated symmetric position. The central flow, actually a jet, is unstable and even a weak disturbance is likely to

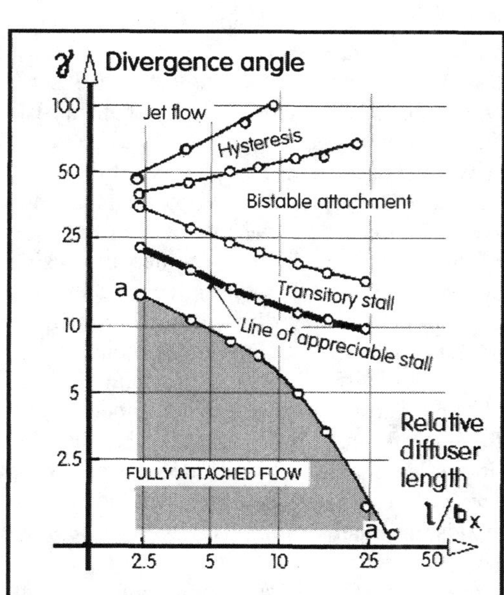

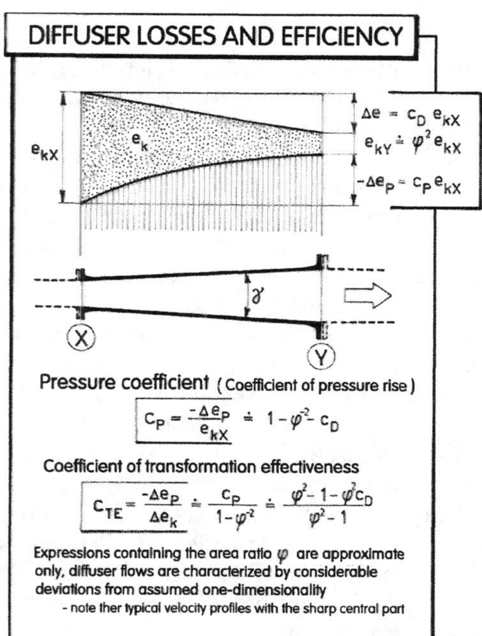

Figure 3.69 (Left) Operating regimes in rectangular diffusers having different divergence angles. The best performance may be obtained near the line of appreciable stall, but the danger of separation may dictate a design staying cautiously behind line a – a.

Figure 3.70 (Right) There are actually several different criteria of what is optimal: note that a minimum loss diffuser requires high coefficient c_{TE}, but this does not secure the highest output pressure at the diffuser outlet.

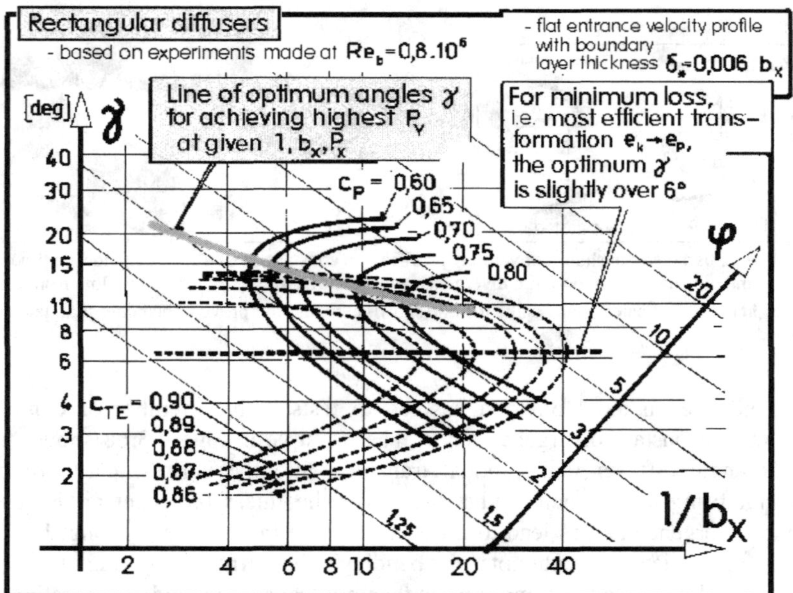

Figure 3.71 A design diagram like this one, for selecting the best diffuser for a given area ratio φ or alternatively for a given relative available length, needs much experimental test data. This is, unfortunately, mainly available for high Reynolds numbers. Using this diagram in microfluidics may be done with discretion.

switch it toward one wall. At a large divergence angle γ (Figure 3.66) the diffuser flow becomes equivalent to the jet flow with attachment to one of the walls (Figure 3.64).

In spite of the simplicity of the geometry, design and experimental studies of diffusers is complicated, mainly due to the rather sophisticated character of the friction loss mechanism. Even numerical solutions are not particularly reliable. The usual steady-state models of turbulence fail to predict the friction reliably, because really effective diffusers are to be operated in the regime (Figure 3.68) of incipient stall and transient separations from the wall – phenomena which are strongly unsteady even if the basic input flow is nominally steady. As the interest in fluidics and microfluidics turns to the low Reynolds numbers, turbulence modeling becomes increasingly dependent on the low-Re modifications of the models. The modifications have to predict the transitional regime between laminarity and turbulence, which is very difficult and responsive to minute disturbances. Even in steady flows this is the least reliable part of turbulence modeling.

With the theoretical predictions bogged down by these difficulties, the practical design of diffusers is mostly guided by correlations obtained from a large number of experiments – like, using the performance parameters of Figure 3.70, the one presented in Figure 3.71. Note that this particular case—unfortunately for rather too large Reynolds numbers than needed for small scale fluidics—is dependent on the particular entrance conditions, characterized by a very thin boundary layer surrounding a potential, constant velocity flow. Use in other conditions may be necessitated by unavailability of better data, but must be accompanied by warnings of circumspection.

Typical for fluidics is the use of diffusers in *collectors*. These are device components that capture an incoming jet and convert it into a flow in a connected closed conduit. Functionally, the collector is therefore the very antithesis of a nozzle. Like exceptional nozzles with constant cross-section shape, the collector may also be in principle built without any diffuser at all, but this would be a rare case indeed, mostly when the collector output is connected to a manometer admitting no

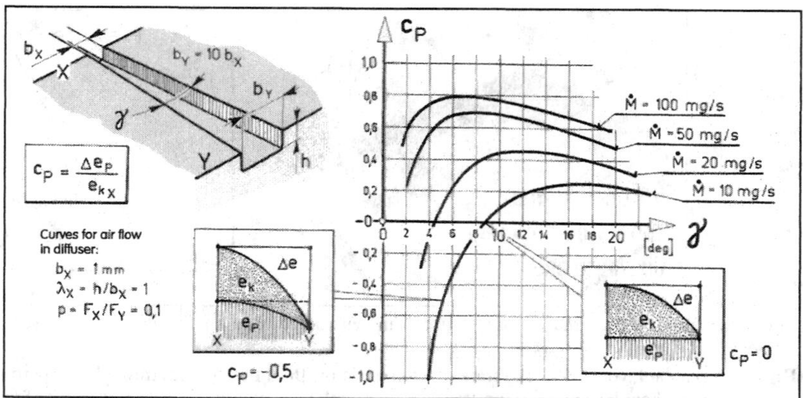

Figure 3.72 Some results available for very small diffusers. Avoiding disappointingly low (perhaps even zero or negative!) pressure rise in the presence of the strong viscous friction may require larger divergence angles than is common usage in the large-scale devices. The indicated two cases of the axial distribution of the components of fluid specific energy correspond at these low Re situations to the analogous top diagram in Figure 3.69.

fluid passage through it. Such a collector is, therefore, an equivalent to the Pitot impact tube as used for investigations of local pressure in a fluid flow. The local character is due to the small cross section and the small entrance as there is no reason for making the tube large. Even the smallest cross section suffices to transfer the pressure signal. The small cross section, of course, damps by friction all changes in time; an advantage, perhaps, if the fluctuation is a mere noise.

Typical for fluidics are other situations, with nonzero output flow extracted from the collector. and a diffuser. Evaluating its properties is, however, difficult. The captured jet is likely to have rather complex velocity profiles, far from those with the thin boundary layer flow assumed for diagrams like the one in Figure 3.71. Of particular difficulty are the cases of a deflected jet, where the complex character of the input velocity profiles also varies in time. Also the trend towards low Re in microfluidics makes designing a diffuser difficult, as shown in Figure 3.72.

3.4 RESTRICTORS: OBSTACLES TO THE FLOW

At this stage the discussion progresses from elements, like nozzles and diffusers, used mostly as components in more complex devices, to elements that are used almost always as devices in their own right. The first devices discussed are the simplest ones: restrictors or fluidic resistors. Their task is to generate a pressure difference dependent on the flowrate passing through. The difference is simply a hydraulic loss, dissipation of fluid energy by converting it into heat. This occurs spontaneously in any channel or cavity. What the restrictors do is to amplify locally the dissipation mechanism and perhaps make it more stable and more reliable by ensuring reproducibility of conditions, in particular of the geometric dimensions. A schematic representation of the restrictors used in circuit diagrams is presented in Figure 3.73.

There are three restrictor types. The first one utilizes the dissipation in the regions of separated flows (Figure 2.36). The regions are usually generated in cavities placed downstream from a sudden change in the wall shape that causes the separation. Figure 3.74 shows a very typical example, with two symmetrically placed recirculation regions. The desirable reproducibility of the dissipation processes is achieved by stabilizing the flow separation on the well-defined sharp edges and by ensuring that the recirculation regions remain at their intended positions inside the

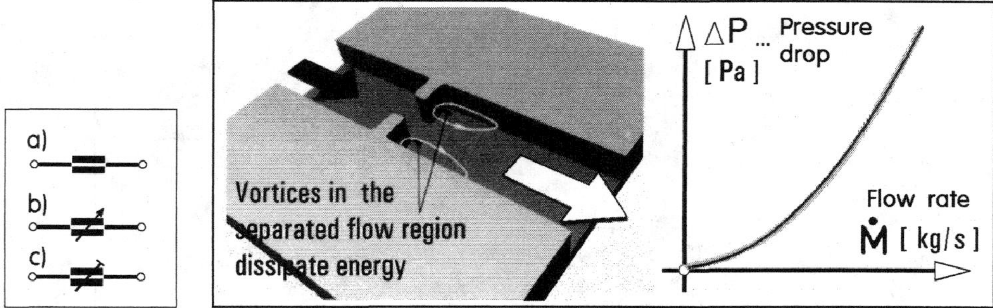

Figure 3.73 (Left) Schematic representation of a restrictor in general (a), a manually adjustable one (b) and one adjustable with a tool (c).

Figure 3.74 (Right) A typical layout of the separation type restrictor: fluid flow passes through an aperture forming a jet the kinetic energy of which is then dissipated in the stationary recirculation regions. The characteristic of this type of restrictors is more or less quadratic.

device. Typical for this type of restrictor is an almost perfectly quadratic characteristic (the dependence of the across state-parameter on the through state-parameter, Figure 3.74). The quadraticity is due to dissipation of kinetic energy of accelerated fluid, proportional to square of velocity (and hence of flow rate). Deviations from the perfect quadratic dependence—and from the constant value of dissipance Q—are due to remnant friction effects that are usually suppressed but cannot be eliminated completely. They arise in locations where the fluid velocity is highest. Decisive influence on the suppression of the friction has shaping of the apertures. In the example in Figure 3.74, the friction is suppressed by making the walls defining the aperture very short.

The other type of restrictors, *frictional* ones, are shaped to promote the friction and suppress the quadratic component of the behavior. Even though it is practically impossible to obtain a perfectly linear characteristic, numerous attempts at least try to approximate it. The large friction is obtained by arranging long parallel surfaces along which the accelerated fluid has to move.

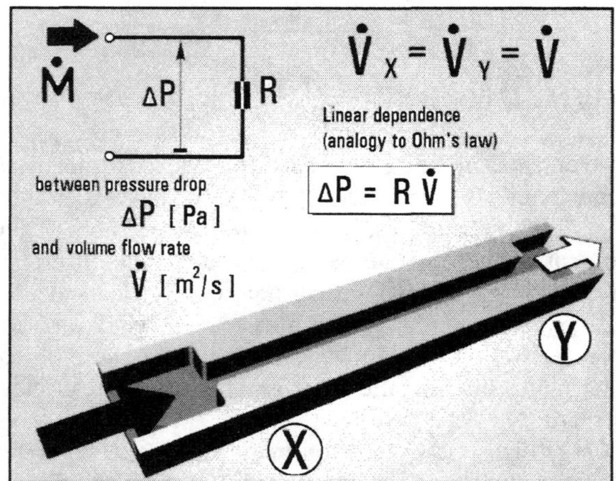

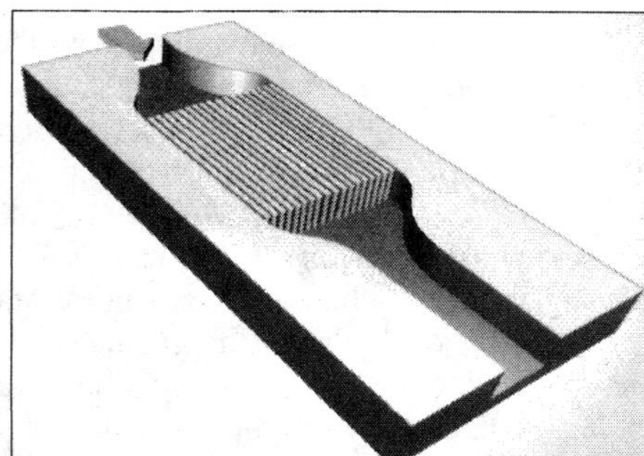

Figure 3.75 (Left) An example of a friction-type restrictor. This layout is built in an attempt (never completely successful) to obtain the linear characteristic with the constant pneumatic resistance R, in theory attainable with laminar, low Re flows.

Figure 3.76 (Right) Parallel channels are used if, for obtaining the linearity, the channel Reynolds number is to be kept low while it is necessary to accommodate a large flow rate.

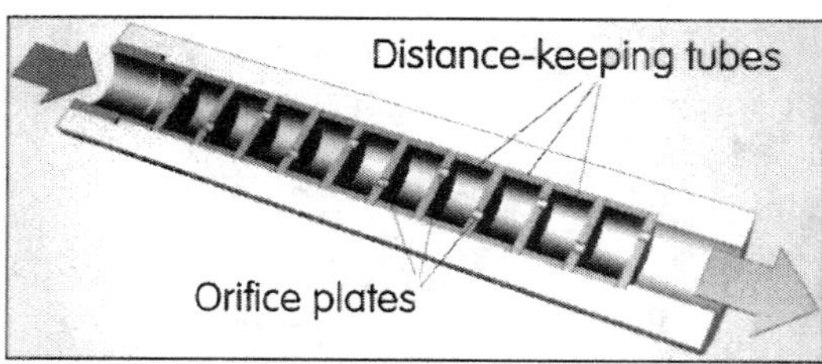

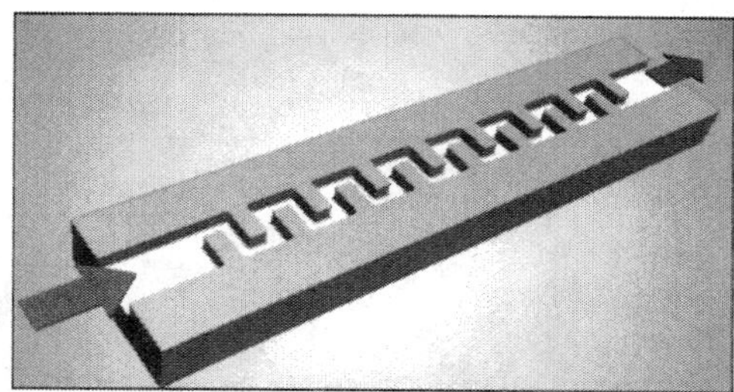

Figure 3.77 (Left) Series connected quadratic (separation type) restrictors for obtaining very high pressure drop without needing to use an extremely small orifice. Note the eccentric and alternating location of the orifices to avoid generating a long straight jet passing through all orifices.

Figure 3.78 (Right) Labyrinth restrictor – a planar version of the idea to use many separation-type restrictors in series in a single body.

Instead of the aperture there is a capillary tube or its planar equivalent shown in Figure 3.75. The narrow space has to be long and narrow for the friction to be large enough to make negligible the inevitable quadratic component losses at the entrance (compare to Figure 2.42) as well as the exit. Friction itself does not secure the linearity: the flow also has to be laminar, at a low Reynolds number. Why exactly should the ideal of linearity be so desirable is sometimes not easy to understand; probably the reason is the tendency to copy the analogous electronic circuits with their Ohm's law linear resistors. At any rate, fluidic friction restrictors are impractical, having a number of disadvantages. They tend to be very long, occupying precious space on the base plate. Due to the temperature dependence of viscosity, their resistance varies with temperature. This may cause serious unbalancing of adjusted pressure levels in the system when fluid temperature is changed.

The low Reynolds number limitation may be circumvented rather easily, by providing a large number of flowpaths in parallel (Figure 3.76). On the other hand, series connection (Figures 3.77 and 3.78) is used for the separation-type restrictors if the requested pressure drop is large and the corresponding single orifice would be extremely small – and therefore difficult to manufacture reproducibly and also sensitive to possible clogging by particles carried with the fluid.

The third restrictor type is less common. It is the *centrifugal* type, an example of which is presented (in meridian plane section) in Figure 3.79. Its main component is an axisymmetric, usually rather flat vortex chamber with central outlet and an inlet on the circumference. Because of the tangential inlet direction, the fluid entering the chamber rotates. The rotation speed increases (due to conservation of rotational momentum) as the fluid progresses toward the central exit and the radius of rotation increases. Near the exit the rotation is so fast that the centrifugal effects tend to prevent the fluid from further radial motion to the exit. It is only due to the boundary layers on the top and bottom flat wall, where the rotation is slowed down, that any fluid gets to the exit at all. The important fact is that the opposition to the flow may be high even if the smallest cross sections in the device are not exceptionally small. This makes this flow restriction mechanism particularly useful for generating high pressure drops in fluids that contain—or may occasionally carry—particles large enough to block the otherwise very small orifices.

The *adjustable* restrictors are almost always of the separation type to avoid the inconvenient temperature dependence and also because of the compactness obtainable with the separation type mechanism. The adjustment of hydraulic properties is done by mechanical motion which usually varies the gap width. Ideally, both opposing surfaces, one of them movable, should be shaped with protruding sharp edges (A). It is, however, often tempting and usually more convenient from the device design point of view to select a design with either the movable (B) or the fixed (C) part having its edge replaced by a flat surface, relative to which the edge moves. The design (B) in

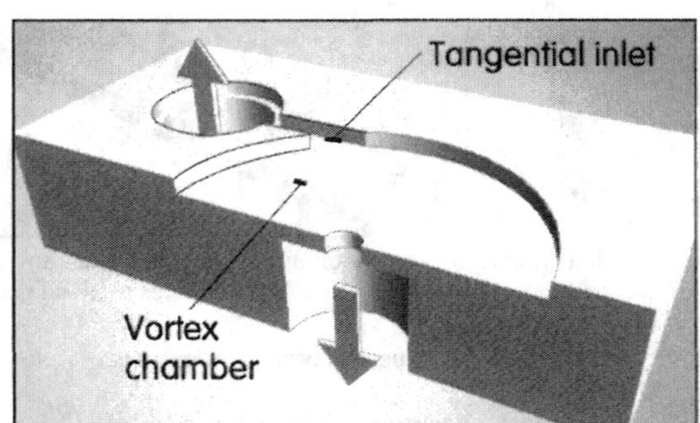

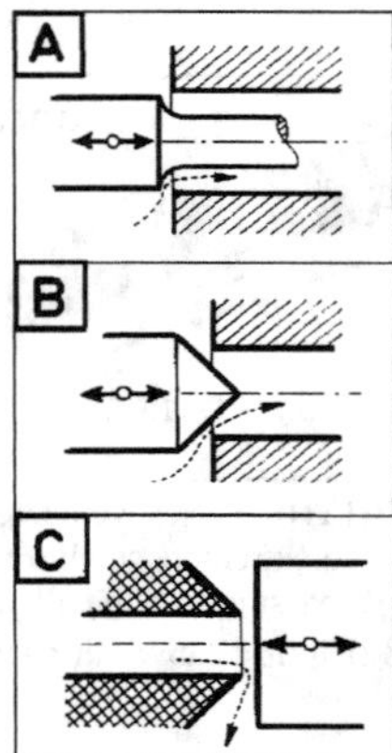

Figure 3.79 (Left) The third type of restrictors: by relying on the centrifugal acceleration in rotating fluid, opposing centripetal motion to the central exit from the vortex chamber, this device can generate a large pressure drop without very small cross sections.

Figure 3.80 (Right) The basic principles of adjustable restrictors. The basic requirement is suppressing frictional dependence on temperature by limiting occurrence of an accelerated flow past tangential surfaces.

Figure 3.80 is the principle of the often used "needle valve" (usually made with smaller vertex angle than shown here, to reduce sensitivity and increase precision of the adjustment). Both B and C, however, should be considered inferior as they unnecessarily expose what is a short but nevertheless influential frictional tangential surface to the flow.

3.5 DIODES

Diodes are essentially restrictors exhibiting different dissipances $Q_{forward}$ and Q_{return} in the two flow directions. Their main (but not sole) application is in rectification of alternating flows. The different properties in different directions is rather a rule than an exception in fluidics, both nozzle and diffuser geometries are asymmetric with respect to the flow direction and they hence exhibit such nonsymmetry of behavior. This, after all, is one of the reasons why the alternating flow is not popular in fluidics, as opposed to the alternating electric current. However, in devices not specially designed for the rectification the diodity

$$\mathscr{D} = Q_{return} \,/\, Q_{forward} \tag{3.1}$$

is not much larger than 1.2 or perhaps 1.6. Ideally the value should be near to ∞ and this is indeed obtained with moving-part devices, the nonreturn valves, an example of which is presented in Figure 3.81 (they gave rise to the circuit symbol for diodes as shown in Figure 3.82). Alternative versions with the closure body held on deformable "hinges" are also known and used. Inertia of the moving body, of course, limits the available range of operating frequencies.

Inertance (see Section 2.7) is not so much of a problem in the no-moving-part fluidic diodes. One of them, patented in 1920 by the famous American inventor N. Tesla (Figure 3.83) is one of the earliest known devices without moving parts. Unfortunately, even though it does exhibit a certain above average diodity, experience with this particular design has been somewhat disappointing.

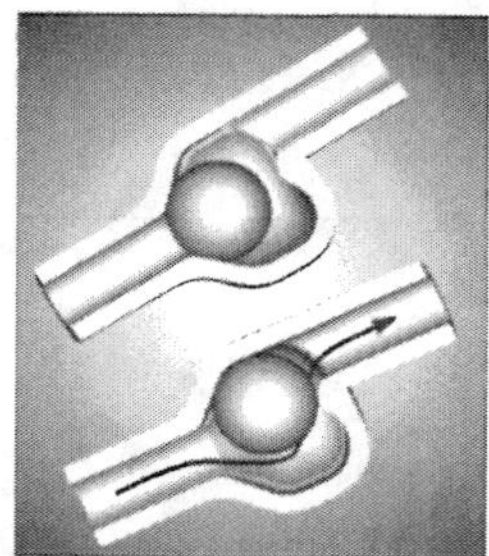

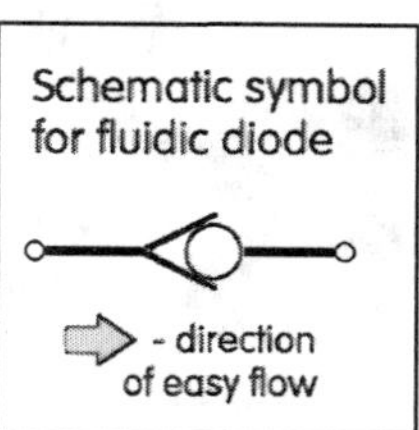

Figure 3.81 (Left) The nonreturn valve with free-moving spherical blocking body. This is a very reliable layout that may be relatively easily manufactured in submillimeter sizes. The only disadvantage is the limited frequency range due to the inertia of the moving body.

Figure 3.82 (Right) The schematic symbol of a fluidic diode used in circuit diagrams.

3.5.1 Labyrinth diodes

A similar problem is encountered in another quite early design of a no-moving-part diode, the labyrinth type, shown in Figure 3.84 in a version suitable for large size and in Figure 3.85 for micromanufacturing. Figures 3.86 and 3.87 present the idea upon which this type is based: the expected idealized flowpaths show the return flow to be much more difficult because of its entering into traps, cul-de-sac chambers located straight ahead in front of the channel a. Fluid accelerated in channel a is directed into the trap chamber, from where it has to return by following a rather tortuous path before it finds its way through the inclined channel b. On the other hand, the return flow is expected to be much easier since there is no difficulty in finding the correct way from the inclined channel b into a. In similarity to the Tesla diode, the effectiveness is increased by placing several rectification stages in series.

Of course, the rectification effects depends on fluid inertia [9]. It is an inertial effect acting on the fluid in channel a that forces it in Figure 3.38 into the trap chamber and prevents it from

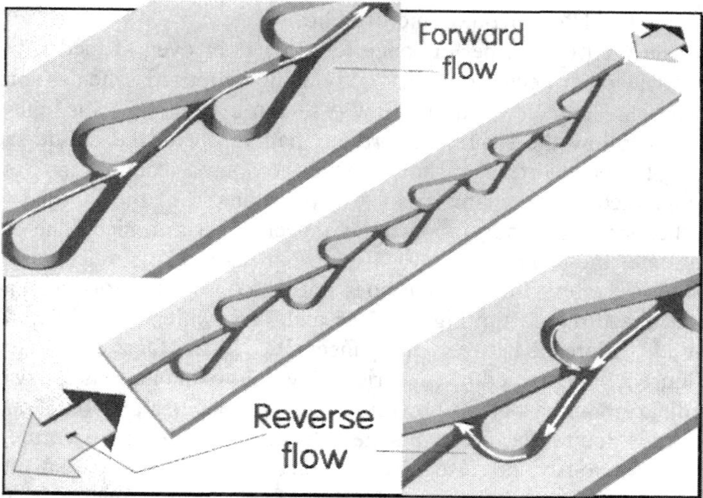

Figure 3.83 The Tesla diode: return flow is diverted into the more complicated side path by interaction of the two parts of the split flow. Increasing at each of the stages, the side flow finally becomes dominating even though the end of the side path is turned back, opposite to the flow direction.

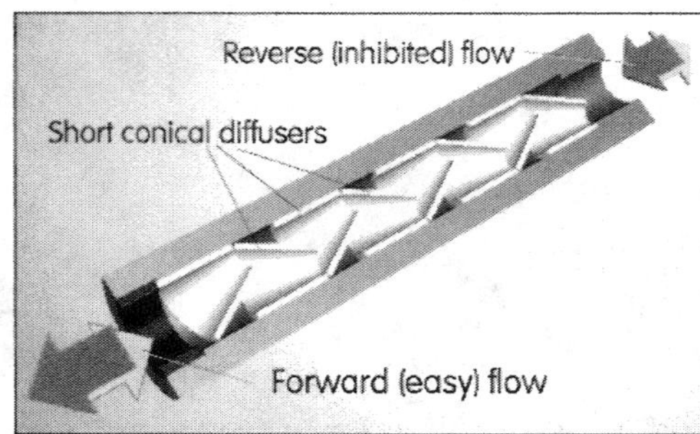

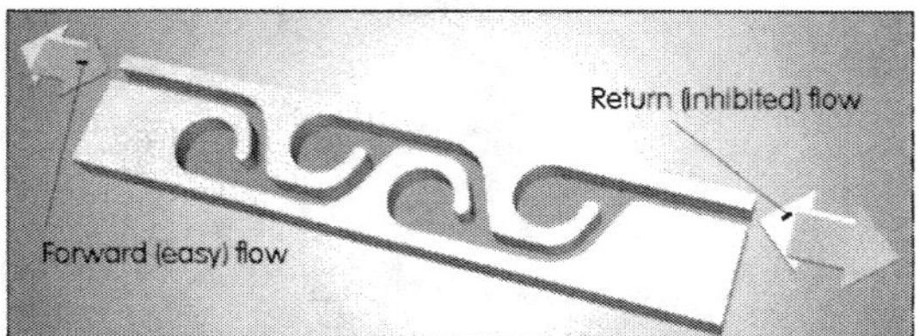

Figure 3.84 (Left) Assembled labyrinth diode. It consists of hollow cone bodies acting as nozzles in the forward flow direction by diverting most of the return flow at each stage into the cul-de-sac.

Figure 3.85 (Right) Etched labyrinth diode—the basic principle of Figure 3.84 adapted to a planar geometry.

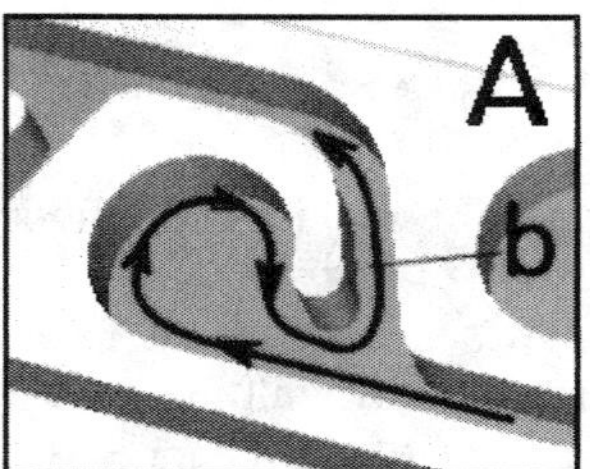

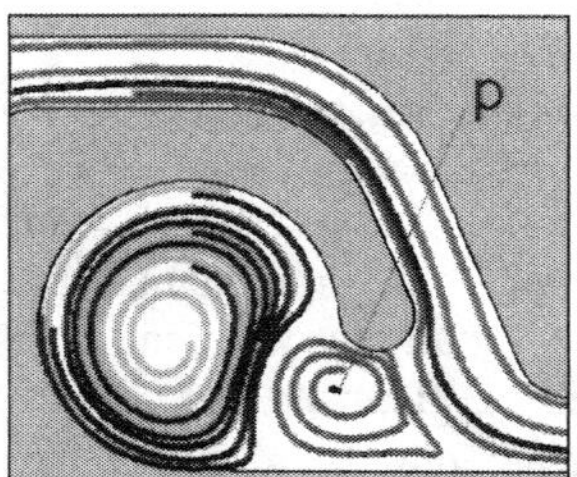

Figure 3.86 (Left) Idealized path of the return (inhibited) flow in the labyrinth diode from Figure 3.85.

Figure 3.87 (Center) Forward (easy) flow in the labyrinth diode from Figure 3.85.

Figure 3.88 (Right) Computed pathlines in the return flow regime in the diode from Figure 3.85. Note the chamber being filled by steadily rotating vortex and there being another small vortex p at the chamber entrance.

making the slight bend into b and the easier way out. As a result, the effectiveness of the rectification decreases with decreasing Reynolds number (the parameter indicative of the ratio of inertial and dissipative forces of fluid). No particular effectiveness cannot be expected if such a diode were used in the low Re domain of microfluidics.

Unfortunately, the rectification performance is rather poor even at medium Re. The reasons are similar as found in other seemingly very convincing proposed schemes and may deserve a more detailed discussion. As mentioned, the flow A shown in Figure 3.38 is highly idealized. Real flows are unlikely to follow the suggested tortuous path. Instead, the cul-de-sac is likely to be filled by a standing trapped vortex, as shown in the computed example in Figure 3.88. This is reflected in the computed actual characteristics of an example of the labyrinth diode shown in Figure 3.89. Another smaller vortex p occupies the chamber entrance. The actual return flow therefore does not enter the chamber at all and its progress into b is much easier than originally expected. The negative, return flow branch does not differ from the forward branch to a degree suggested by the idealized pictures in Figures 3.86 and 3.88 (which would lead to expectations of an order of magnitude difference between dissipances $Q_{forward}$ and Q_{return}). The actual values of the diodity (3.1) in Figure 3.90, obtained by numerical flowfield computations are perhaps higher than found in most ordinary devices but do not reflect any spectacular phenomenon taking place. In fact, the diode effect is due to the forced changes of rotation of the small vortex p located in the chamber entrances. Obviously, because of its contact with the flow passing through, it has to change its rotation sense at each change of the through flow direction. It is this change in the rotation that is the main cause of the observed enhanced diodity. Also apparent in the diagram in Figure 3.90 is the deterioration of properties associated with a decreasing Reynolds number.

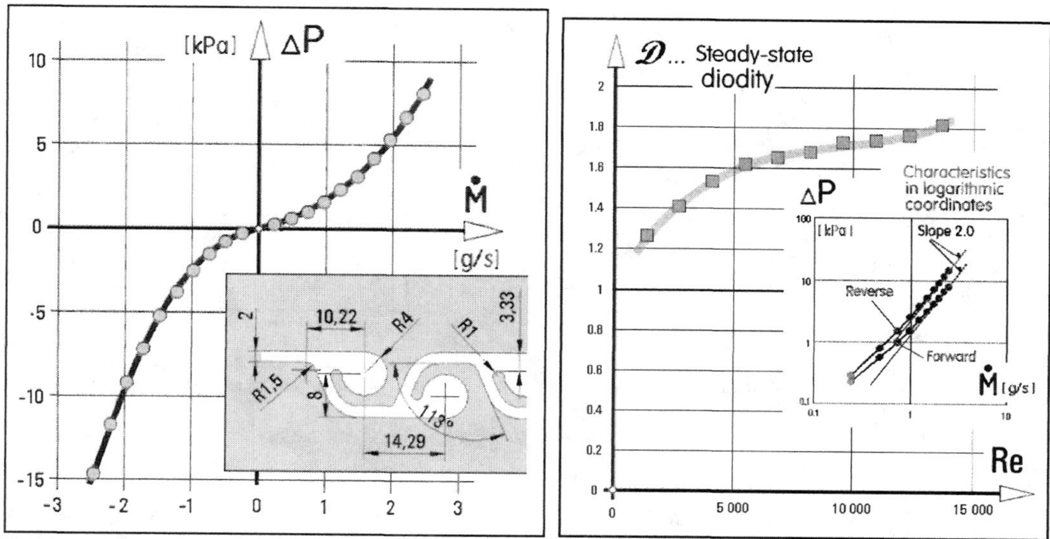

Figure 3.89 (Left) Characteristics of an example of a four-stage planar labyrinth diode (dimensions shown in the inset) – pressure drop dependence on the mass flowrate.

Figure 3.90 (Right) Diodity of the labyrinth diode evaluated from the results in Figure 3.98 and, in the inset, both branches of the characteristic plotted in logarithmic coordinates.

The change of rotation effect is particularly prominent and actually leads to a reasonable rectification effect in regimes with periodic change of the flow direction. This is, of course, the regime in which the diode operates when used for the rectification. Computations have shown that the large vortex occupying the chamber continues to rotate in the same sense. The small entrance vortex p, however, is forced to change its rotation sense periodically. The change takes place at

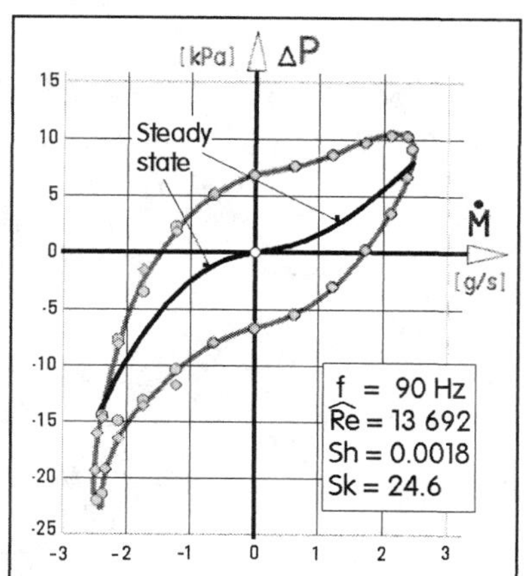

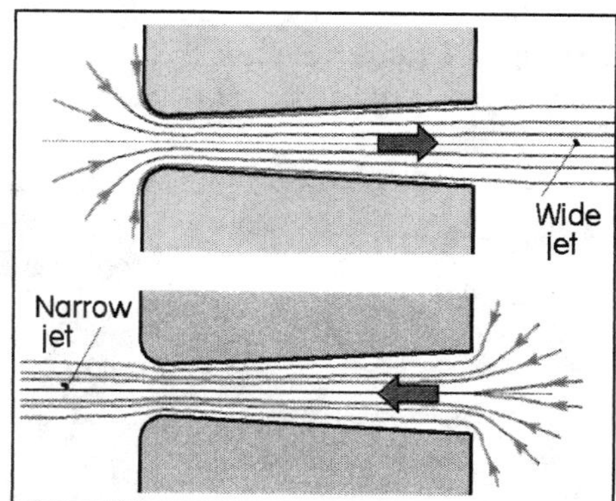

Figure 3.91 (Left) Dependence of pressure drop across the labyrinth diode from Figures 3.85 and 3.89 on a harmonic input mass flowrate. Because of the need to overcome the inertia of the small vortex p, the negative pressure drop is higher than expected on the basis of the steady-state characteristic.

Figure 3.92 (Right) The basic principle of the Venturi type diode. Different pressure drops in different flow directions are due to the difference in dissipated kinetic energy of the slow wide jet in forward flow and the fast narrow jet in the reverse flow. Pathlines computed for the diode from Figure 3.93.

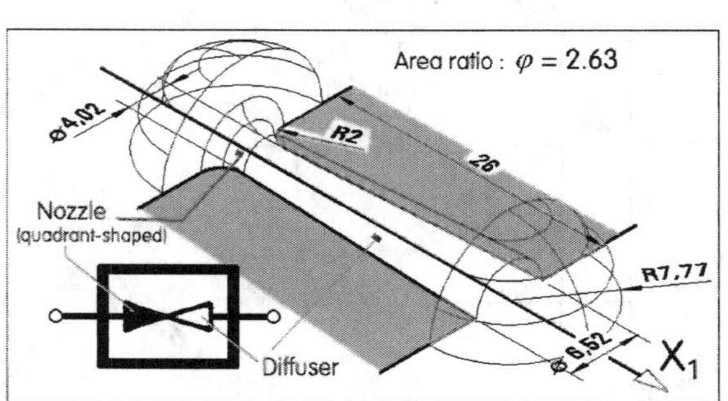

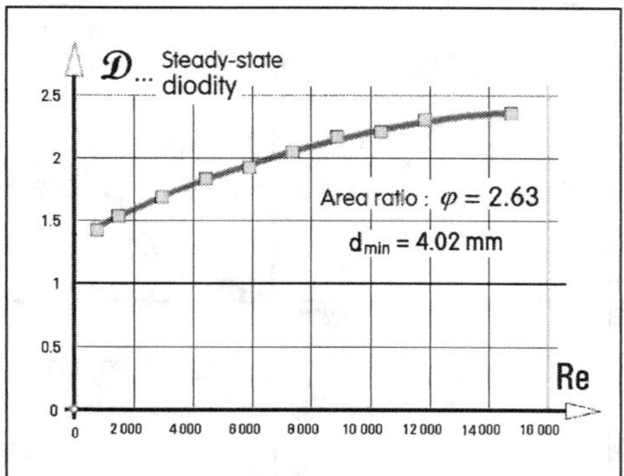

Figure 3.93 (Left) Nozzle and diffuser – the schematic symbol for Venturi diodes and the geometry of an example of a diode investigated in detail.

Figure 3.94 (Right) Values of the diodity evaluated as a function of the Reynolds number for the diode from Figure 3.93.

the negative peak of the dependence plotted in Figure 3.91. This explains why this peak is so sharp and tall. The change of the rotation absorbs work extracted from the channel flow; this explains the large hysteresis apparent in the diagram.

A popular vortex diode used for rectification at rather high, acoustic frequencies in air – and therefore suitable for the rectification task in the "valveless" pumps driven by loudspeakers and similar, very economical drivers using components of consumption electronics – is the Venturi type shown in Figure 3.92. As emphasized by its schematic representation in Figure 3.93, this device consists of two basic components, a nozzle (in this example a quadrant nozzle according to Figure 3.17) and a diffuser (Figure 3.65). They are connected immediately so that no jet is formed at the nozzle exit. The computed values of diodity in Figure 3.94 are higher than the four-stage labyrinth diode from the previous example.

3.5.2 Vortex diodes

Much higher values of the diodity $\mathcal{D}$, up to about $\mathcal{D} = 20$ (and even more) are obtained with another type of fluidic diodes: the vortex diodes. In their "return" flow direction, the flow is opposed by the centrifugal action as in the restrictor in Figure 3.79. The diodity is due to there

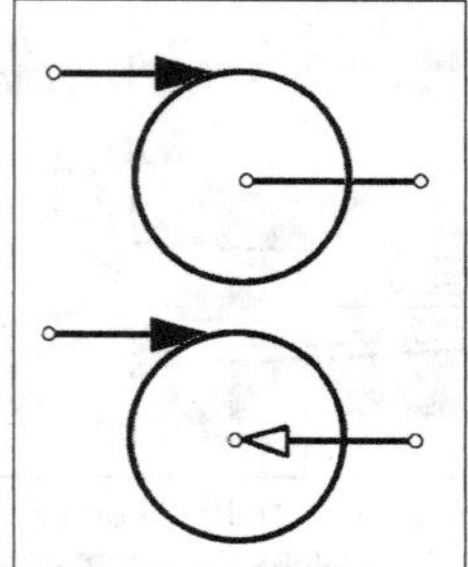

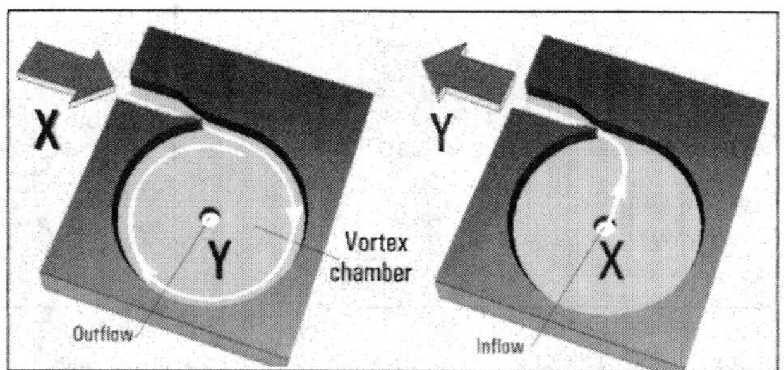

Figure 3.95 (Left) Nozzle and vortex chamber: schematic symbol of a fluidic diode. Bottom: high-performance variant with emphasized presence of the diffuser in the exit.

Figure 3.96 (Right) The basic principle of the vortex diode: the reverse flow (left) toward the central exit is opposed by centrifugal force acting on rotating fluid and exhibits much higher pressure drop than the forward flow (right).

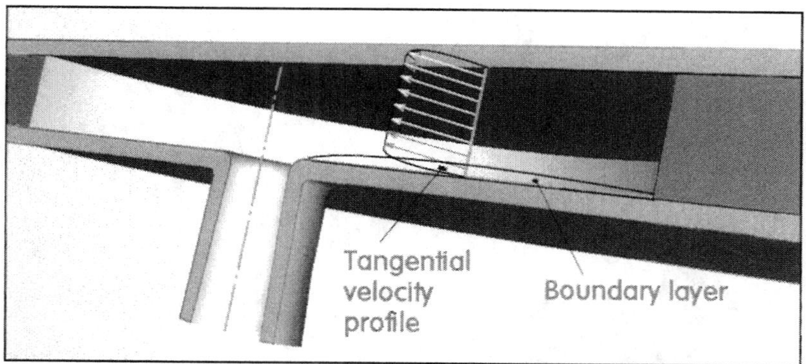

Figure 3.97 Velocity profile of the rotational, tangential motion in the vortex chamber. Similar slowed-down boundary layer as shown on the bottom wall is formed, of course, also on the top cover plate.

being essentially no rotation in the vortex chamber in the "forward" flow regime. In fact, the simple version shown in Figure 3.96 (schematic symbol: Figure 3.95) does not achieve such a high performance and (depending on geometry) usually exhibits values of about $\mathscr{D}$ = 4, still much better than the labyrinth type. The real high-performance versions have a more complex design, typically with a large number of inlets on the periphery of the vortex chamber, necessary for symmetry of the vortex motion. The inlets and the outlet have to be carefully shaped, the latter as diffusers. Opinions vary as to the desirability of contouring the vortex chamber cross-section; the shape with the flat top and bottom in Figures 3.96 to 3.98 may exhibit good performance.

There are two limiting factors. First, vortices take a considerable time to start up and then disappear when the flow stops. Vortex diodes in general are unsuccessful for rectification of alternating flows unless the operating frequency is extremely slow. They are used for pumping applications with very slow switching, the time scale dictated by slow filling of the displacement vessel. Second, the centrifugal force acting in rotating fluid being a typical dynamic effect, vortex diodes cease to be effective at low Reynolds numbers below about Re ~ 800. This is actually the level at which the microdiode from Figure 3.99 (characteristics: Figure 3.100) is operated. Also the favorable time scaling with decreasing size (processes at microscale are much faster) makes the microfluidic versions interesting.

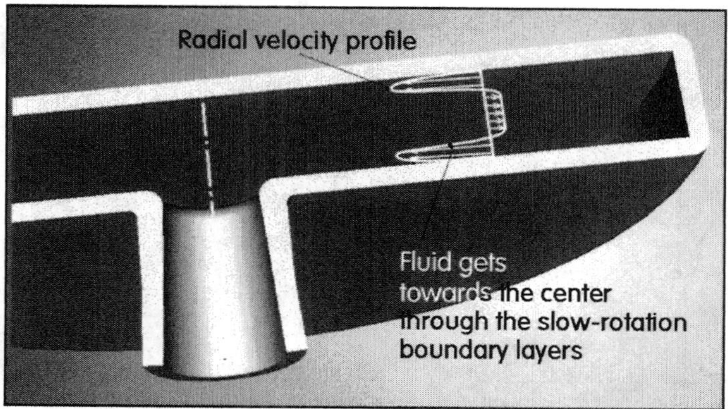

Figure 3.98 Velocity profile of the radial flow. The centrifugal action on fluid is so strong that the core flow outside the boundary layers is actually directed away from the axis.

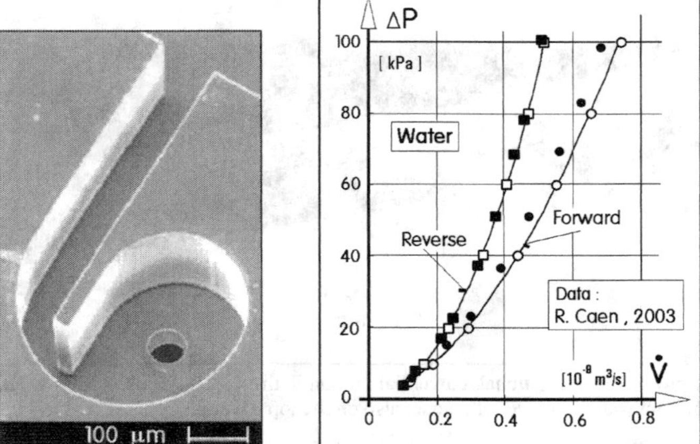

Figure 3.99 (Left) Microfluidic vortex diode. (*From*: [10]. © 2001 Institute of Physics. Reprinted with permission.)
Figure 3.100 (Right) Steady-state experimental characteristic of the microdiode from Figure 3.99. Despite the small relative diameter of the vortex chamber and the small Reynolds number, an appreciable diodity could be demonstrated.

3.5.3 Reverse flow diverters

Much more effective than turning the return flow down can be diverting it away. This is easily done with various jet type devices. Some of them, strictly speaking, do not really belong in the class of the simplest devices discussed here. They are near relatives of the diverter valves discussed in Chapter 4. It may be useful, nevertheless, to discuss them here because their operation has much in common with the diodes.

Perhaps the simplest of the reverse flow diverters is shown schematically in Figure 3.101. Like the Venturi diode in Figure 3.93 it also consists of a nozzle and a diffuser, except that here the two components are not connected and there is a gap left between the nozzle exit and the collector entrance. Both the exit and the entrance are on the same line, so that the jet formed in the nozzle enters the collector in the forward flow regime. The collector axis, however, is inclined. When in the reverse flow regime fluid leaves the collector, the inclination causes it to miss the nozzle. This way, the paths of the forward flow and the return flow are different.

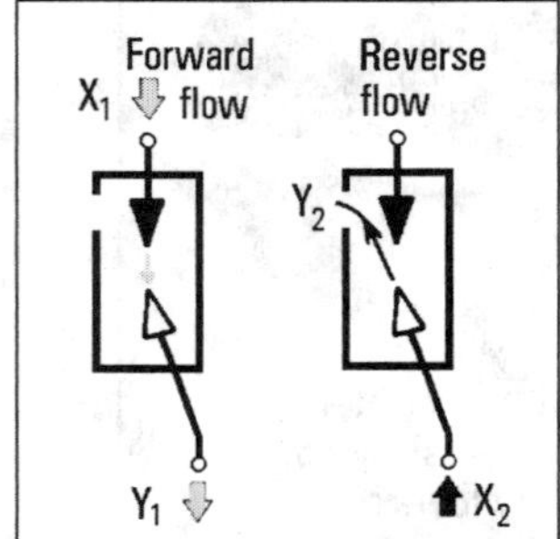

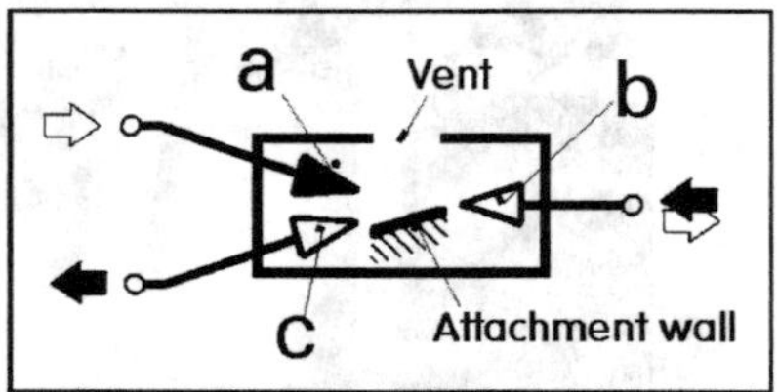

Figure 3.101 (Left) Schematic representation of one version of the reverse flow diverter based on a nozzle exit and a collector entrance positioned on the same line, with the collector axis inclined.
Figure 3.102 (Right) Schematic representation of another reverse flow diverter, employing the Coanda jet attachment.

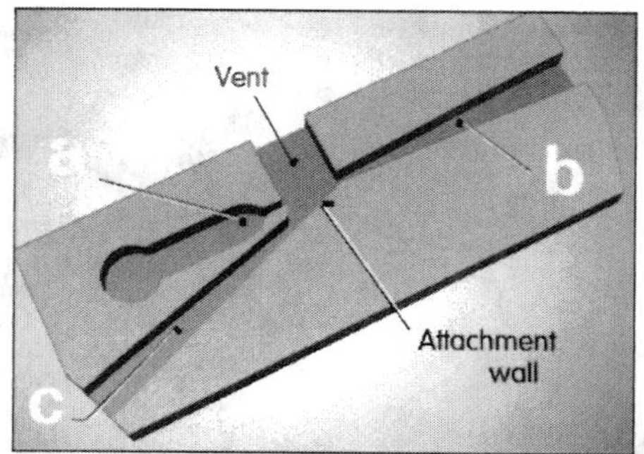

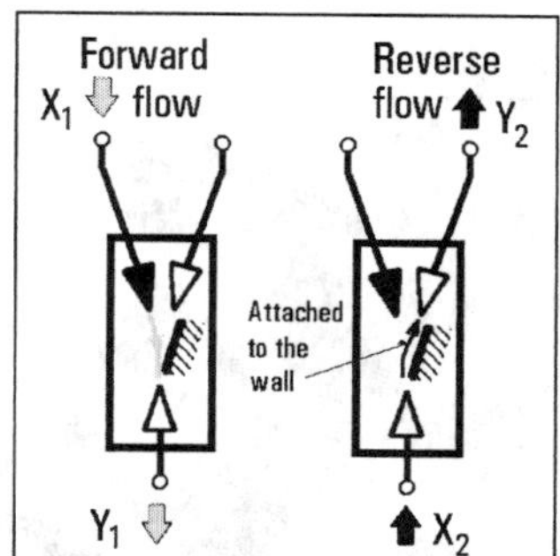

Figure 3.103 (Left) An example of etched planar diverter corresponding to the schematic diagram in Figure 3.102.
Figure 3.104 (Right) Explanation of the working principle of the Coanda-effect reverse flow diverter.

Similar disagreement of the two flowpaths may be also obtained by the action of the Coanda jet attachment, Figure 3.103. In the schematic representation (Figures 3.102 and 3.104), the forward flow leaving nozzle a forms a jet captured by the collector b. If, in the reverse flow state, an oppositely oriented jet leaves the collector b, its Coanda effect attachment diverts it into a different flowpath, to the second collector c.

3.6 REACTORS AND HEAT EXCHANGERS

Including microreactors in the list of simple devices may be surprising – but in principle they are very simple indeed, in particular those processing premixed reactants. In this case the reactors are little more than just a channel, mostly of constant cross section. The advantage of microdevices is precise controllability due to the high surface to volume ratio 10 000–50 000 m^2/m^3 (compared with mere 50–100 m^{-1} of standard batch reactors). To use it, the channels are usually provided with temperature control, consisting of a controllable heat source (or sink) and a temperature sensor (mostly a thermocouple). These, however, bring no complication from the fluidics point of

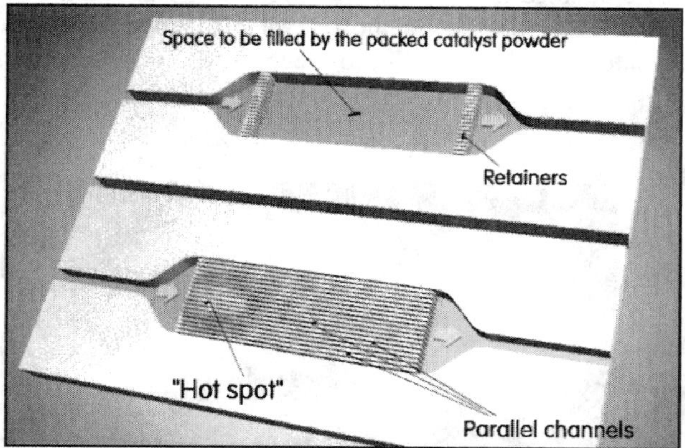

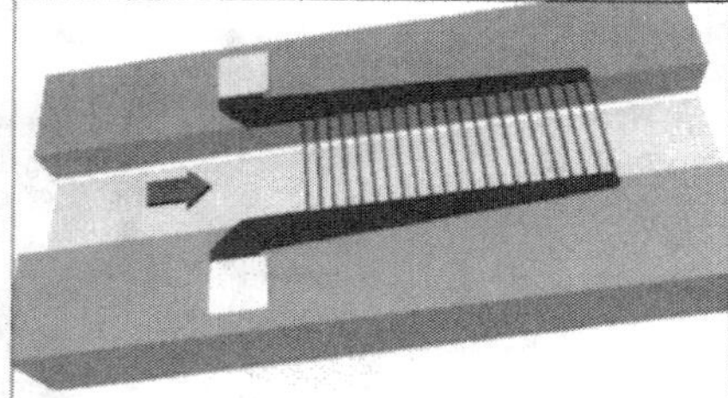

Figure 3.105 (Left) Parallel operation of microreactor channels or the packed bed should be designed with equal inlet conditions in mind – the narrow inlet shown here, though common, is wrong, causing higher flow in the centrally located channels, where exothermic reactions generate the hot spot.

Figure 3.106 (Right) An example of an electric heater with the heating resistor strips directly exposed to the flow of reactants and spread so as to cover a large area of the channel.

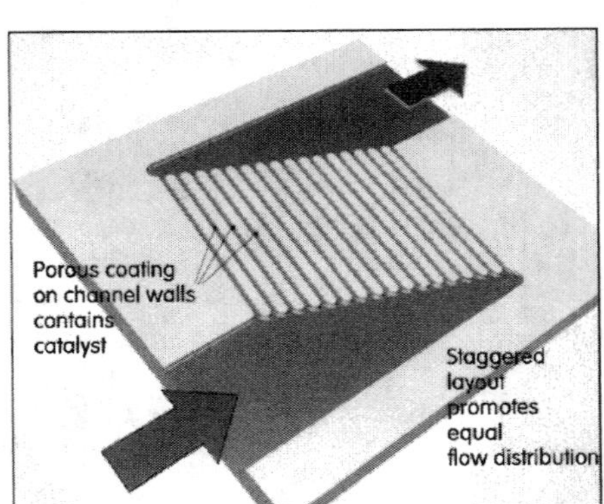

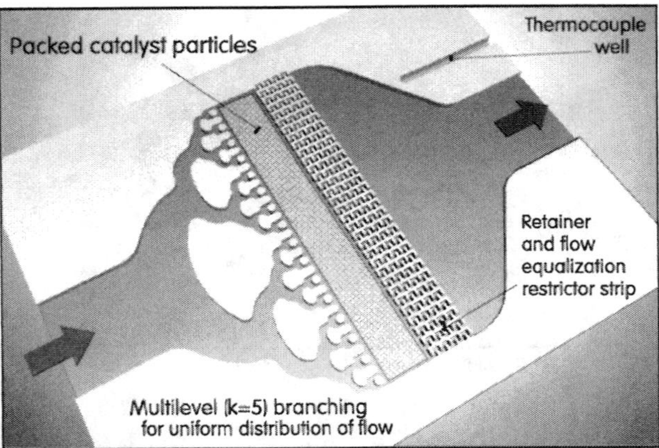

Figure 3.107 (Left) Staggered arrangement of the microreactor channels is recommended in place of the unsatisfactory designs of Figure 3.105 to eliminate the danger of the "hot spots" in channels with higher reactant flowrates.
Figure 3.108 (Right) An example of the packed bed reactor with carefully designed inlet flow distribution and a catalyst retainer strip generating high pressure drop for the same purpose of constant reactants flowrate everywhere.

view. Electric heating, perhaps by directly exposed electric conductors Figure 3.106, covering as large area of the flow as possible, is preferred at laboratory development stage. An economically more favorable source is an exothermic reaction (combustion) taking place in the same body. This is common in the fuel processing applications.

The "numbering up" to obtain the sizeable overall production rate is often already done at the reactor channel level, with a large number of parallel reactor channels in the same body (of course, the bodies are also used in large parallel numbers). The inlet into the parallel channels needs careful design for equal inlet conditions, important for maximum selectivity and yield. The bottom example in Figure 3.105 shows high temperature areas due to the improper flow distribution from the narrow inlet with an exothermic reaction. The packed bed, top part of Figure 3.105, may be prone to a similar effect. Layouts like Figure 3.107 offer an improvement. The other example in Figure 3.108 demonstrates the extreme care devoted to equal flow distribution by the multilevel branching (Figure 2.79). The retainers keeping the powder in the reactor are shown in Figure 3.108 arranged as a system of labyrinth restrictors (see Figure 3.78) to secure the even distribution.

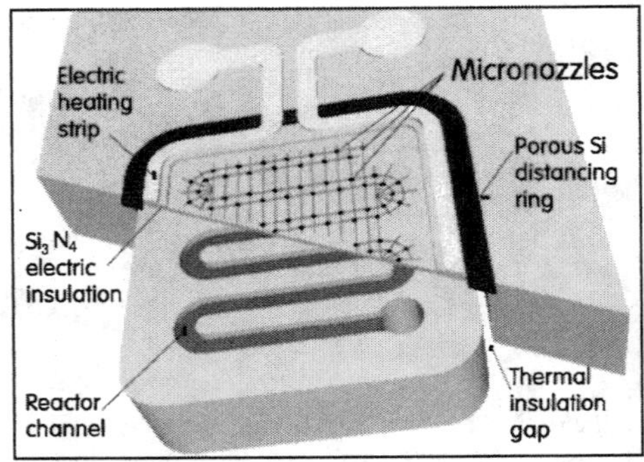

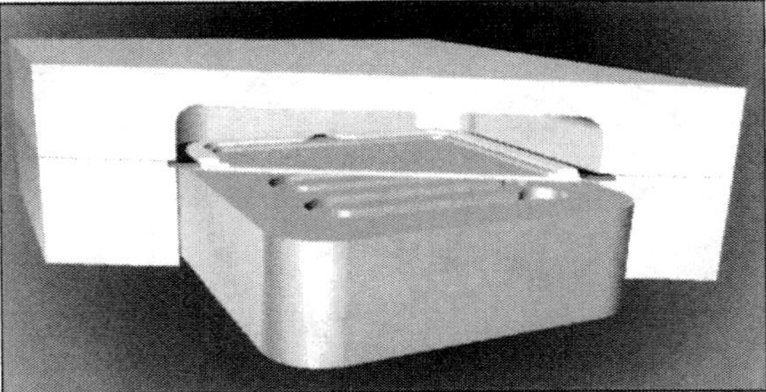

Figure 3.109 (Left) An example of a reactor for operation with reactant admixture through micronozzles.
Figure 3.110 (Right) Another section view of the reactor from Figure 3.109 showing the top chamber for inlet of the added reactant.

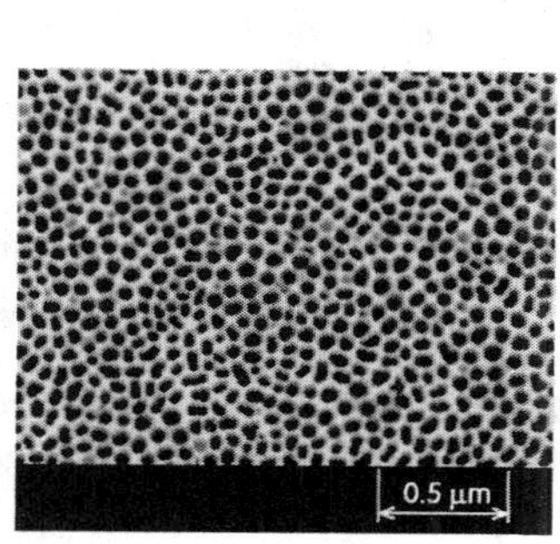

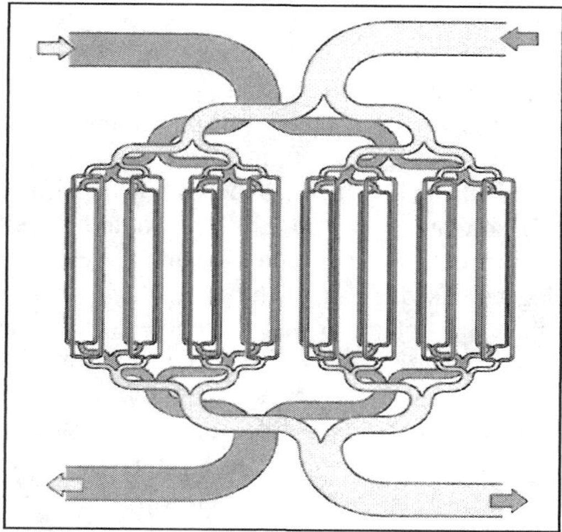

Figure 3.111 (Left) Typical porous layer used to increase the available surface for supporting immobilized catalyst. This is a layer made by anodic oxidation of aluminium. Similar and more regular structure is formed by porous silicon.
Figure 3.112 (Right) Basic principle of microfluidic parallel-flow heat exchanger.

bution. Reactor channels often have a porous coating on walls (Figure 3.111), to increase the surface area of deposited catalyst.

Compared with the plain channels of the premixed operation, fast mixing reactants, like hydrogen, which are input directly into the reactor, lead to more complex design. In Figures 3.109 and 3.110, hydrogen is injected into the channel through micronozzles. Another solution is hydrogen transport through selectively permeable palladium membrane, supported by a porous layer. Palladium also serves as a catalyst. Porous membranes made by the anodic oxidation (Figure 3.111) are also used to hold solid catalysts, distributed in a thin surface cover layer.

Heat exchangers are related to reactors. The difference, in the recuperative exchangers, is in their use of two mutually separated channel systems, one for each fluid (Figure 3.112). The separation walls between the channels are here thin, but their thermal resistance is negligible anyway. The limiting factor is the transfer between fluid and the wall. It is the large surface-to-volume ratio of channels—and the consequent short distance to be traveled by thermal gradient diffusion—that brings the advantages of the microfluidic versions. The cross-flow exchanger layouts, corresponding to Figure 1.10, may be simpler, since they avoid the complexity stemming from the need to separate the two fluids when brought in from the same side.

3.7 MIXERS

Used as separate devices upstream from premixed-reaction reactors, mixers are often encountered in microchemistry. As well, many biological processes, such as antibody-antigen binding and DNA hybridization, require rapid and effective mixing.

Mixing processes depend, at least at their final stages, on gradient diffusion: the spontaneous transport of species tending to gradually eliminate existing concentration gradients. This diffusion is very effective at large Reynolds numbers, where it is dominated by the vortical turbulent motions. In laminar microscale flows the transport has to rely on thermal motions of molecules, with the resultant classical gradient transport Fick's law. This mechanism is much slower. Generally, this is not much of a problem with gases, as is apparent from their values of diffusivity χ, the coefficient in the gradient transport relation. It is useful to compare it with the coefficient of momentum transport by evaluating Schmidt number Sc, the ratio of viscosity ν and diffusivity χ (both have the same dimension, m^2/s). Typical Sc values for gases are of the order of 1 or even less: in particular, H_2 diffuses into CO_2 with diffusivity $\chi = 60 \cdot 10^{-6}\ m^2/s$ (at T = 20°C = = 293.2 K), so that its Sc = 0.133. (It should be mentioned that diffusivity value is usually not a constant as it depends on concentration and therefore varies in the course of the diffusion process.) The mixing task is difficult in the case of liquids, with much lower diffusivity χ, making the molecular transport much slower. As a general rule, the diffusive propagation depends on the size of liquid molecules. It is faster if the molecules are small, such as is the case of water. Even so, diffusion of water into alcohol is characterized by $\chi = 2.2 \cdot 10^{-9}\ m^2/s$, which means very high value Sc = 820. Particularly difficult and slow is diffusive transport of large and long molecules, such as those of proteins. Their typical diffusivity values are around $\chi \sim 50 \cdot 10^{-12}\ m^2/s$.

To provide an idea about the slowness of the nonstirred molecular (Fickian) diffusion, Figure 3.113 presents the simplest case of gradient diffusion: propagation of species A, occupying the bottom half-space, into the upper half-space initially wholly occupied by another species B. This process is time-dependent and Figure 3.114 shows the distributions at three times after the start of the process—computed for the particular case of water diffusing into ethylalcohol. To make the solution simpler, several aspects in this problem are imposed artificially: in reality alcohol will also diffuse into the water below but here the alcohol concentration at $X_2 < 0$ is assumed to be zero, kept by a continuous inflow of fresh water. Also in this case the diffusivity χ is concentration dependent (the diffusive transport of the same two fluids at the opposite extreme of the concentrations, with a small amount of alcohol diffusing into water, is characterized by almost one half value $\chi = 1.13 \cdot 10^{-9}\ m^2/s$). In Figure 3.115, this relatively well-diffusing liquid pair is

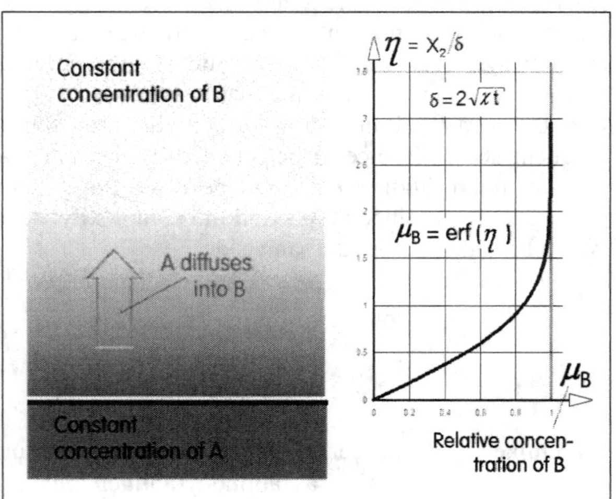

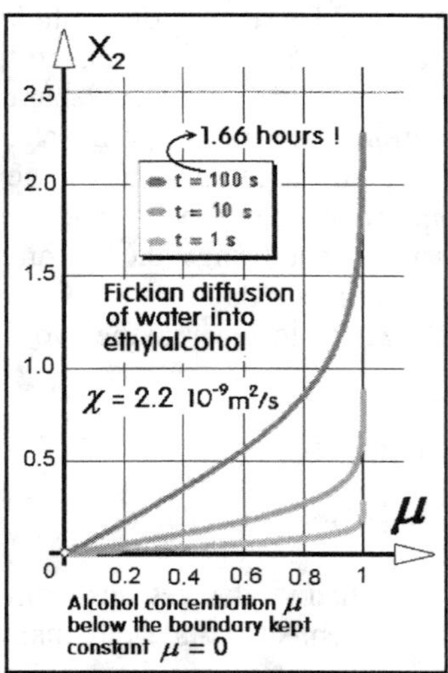

Figure 3.113 (Left) The simplest diffusion problem: propagation of liquid A from the constant 100% concentration in the lower half-space into liquid B. Note the universal solution of the concentration distribution in relative coordinates. The thickness of the diffusion layer δ is proportional to the square root of the product of diffusivity χ and time t.

Figure 3.114 (Right) Spatial distributions of concentration of alcohol diffusing into water according to Figure 3.113, evaluated for three instants of time after a sudden start.

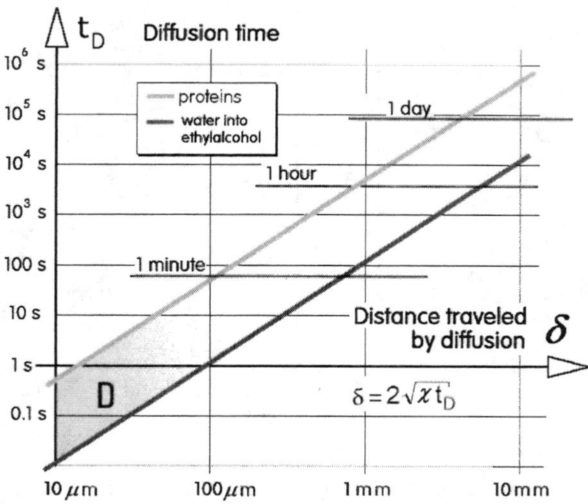

Figure 3.115 Diffusion time and corresponding characteristic thickness of the diffusion layer δ for the simple problem according to Figure 3.112, evaluated for easily diffusing small molecules and long organic molecules.

shown to need hours to cover by the diffusive propagation distances of the order of millimeters. With long organic molecules the mixing would take days! Relying on the molecular diffusion transport alone in a device that simply brings the two fluids side by side (Figure 3.116) requires a combination of conditions falling into the shaded region D (Figure 3.115). Even then it is quite likely that the necessary the mixing channel will be very long, perhaps as shown in Figures 3.117 and 3.118.

If the estimated diffusion times are longer, it is necessary to introduce some means that result in shortening the final distance to be traversed by the diffusion. The obvious possibility is continuing and translating into the microscale the idea of mechanical stirring, the standard at the large scales. Of course, the classical rotating or otherwise moving stirrers as known in large-scale devices are not suitable. They would be difficult to manufacture and also too frail. Instead, there are at least three principles that may be useful. Lu et al. [11] used a 3 x 3 array of magnetic components of 400 μm size powered by external rotary magnetic field. Moroney et al. [12] used agitation by ultrasonic traveling waves generated by a piezoelectric film located at the bottom of a mixing chamber. Perhaps most interesting is to use (Kiu and Grizdinski) as the actuators oscillating small

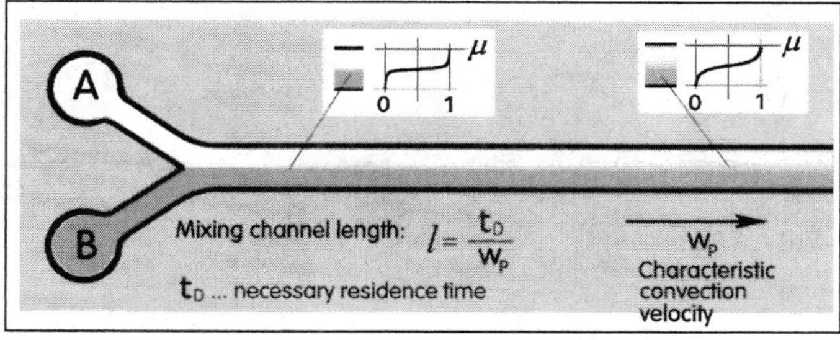

Figure 3.116 A contact mixer that simply brings the two fluids A and B together and relies on the molecular diffusion. The required residence times of the order shown in Figure 3.115 call for slow flow velocity and extreme mixing channel lengths downstream from the confluence point.

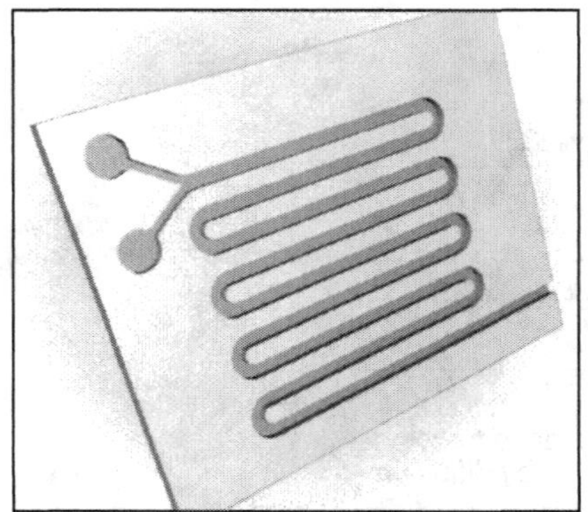

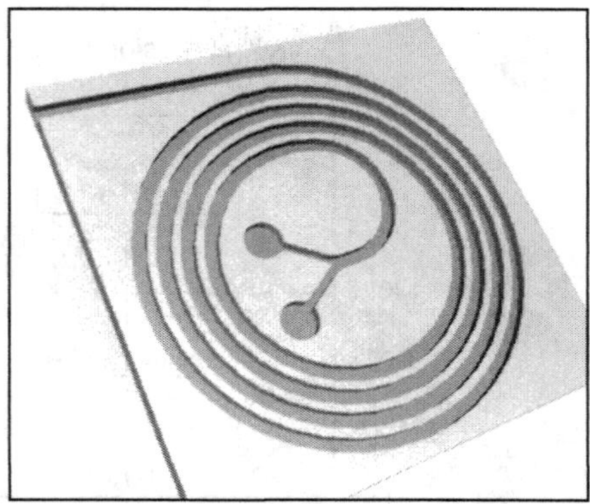

Figure 3.117 (Left) and Figure 3.118 (Right) Characteristic appearances of the liquid mixing channels: they are often folded or coiled to accommodate the usually necessary large lengths.

trapped air bubbles. The bubbles are driven by a piezoelectric actuator at the resonant frequency f [Hz] according to the equation:

$$\pi d f = \sqrt{3 \varkappa P_o v} \tag{3.2}$$

where d is the bubble radius, $\varkappa$ is the ratio of specific heats for the gas, P_o is the hydrostatic pressure, and v is the specific volume of the liquid. For $\varkappa = 1.4$, $P_o = 106$ kPa, and water $v = 1.0 \cdot 10^{-3}$ m^3/kg, with bubbles of roughly 400 μm diameter, the resonance is at f = 8 kHz.

Of course, pure fluidic pressure driven principles of generating smaller sizes are easier to implement. One principle of how to get nearer to the region D in Figure 3.115 is known under the (perhaps not very fitting) name of *hydrodynamic focusing* (Figures 3.110 and 3.120). The central layer of fluid B is thinned by the action of two incoming side flows of the other fluid A. The resultant thickness of layer B may be as small as commensurable with the size of living cells. Indeed, one of the common applications of this principle, other than the mixing, is actually in investigating and counting cells and particles. The obvious problem of the basic version shown in Figures 3.119 and 3.120 is that it is suitable for making only mixtures with a very small percentage of the central fluid B. The percentage may be easily increased by providing a larger number of inlets of the fluid B—the example in Figure 3.121 uses 3 inlets. An often-printed picture of the SuperFocus mixer developed by Löb et al. [13] with 138 feed nozzles is e.g., on p. 158 in [14].

Another no-moving-parts principle of generating the lamellae in the output channel is the principle with interdigitated "finger" exits, shown in Figures 3.123 and 3.124. The motion of the generated lamellae through the slit in the cover plate is its only inconvenient aspect – it may be

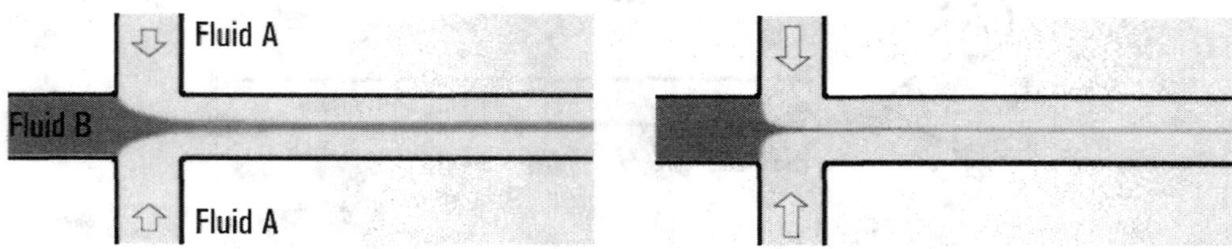

Figure 3.119 (Left) and Figure 3.120 (Right) The thinning of the central liquid flow by forced inflow of another fluid from both sides became known as *hydrodynamic focusing*. The higher the acting pressure of the side inflows, the thinner the central layer. Because of its thinness, the layer can diffuse into the outer liquid rather fast.

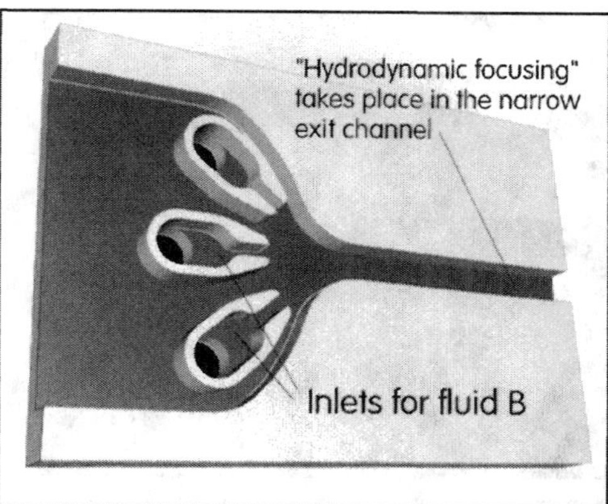

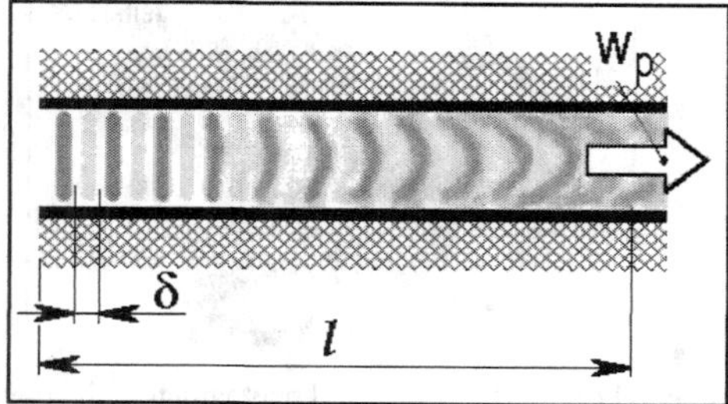

Figure 3.121 (Left) Use of the so-called "focusing" to generate seven thin longitudinal lamellae.

Figure 3.122 (Right) Schematic diagram of the transversal lamellar structure in the mixer exit channel.

difficult to get the exit flow somehow back into the usual position in the plane of the base plate. Also the walls of the "fingers" have to be exceptionally thin, which may cause manufacturing problems.

There are principles that produce the lamellae in the output channel oriented transversally – they are perpendicular to the direction of flow, as shown in Figure 3.122. This is a rather welcome feature since the lamellae are then elongated and thinned by the natural shear stresses occurring in the channel flow. An example of a device producing the transversal lamellae is the oscillator mixer [15, 16] shown in Figures 3.125 to 3.128. Its operation is based on hydrodynamic instability of colliding jets. Their oscillation is a dynamic effect and there is a limit of low Reynolds numbers below which the oscillator ceases to operate. Practical experience has shown, however, that this is no serious obstacle as the device laboratory models could be operated reliably at Re below ~ 100, especially if it were possible (by stacking plates with through cavities made by two-sided etching) to use a higher aspect ratio λ (Figure 3.128).

To show the wide spectrum of available possibilities, yet another principle of generating the lamellae was found in the vortex mixer [17] (Figure 3.129) operated [18] below the usual limit Reynolds number at which the rotation in the chamber ceases.

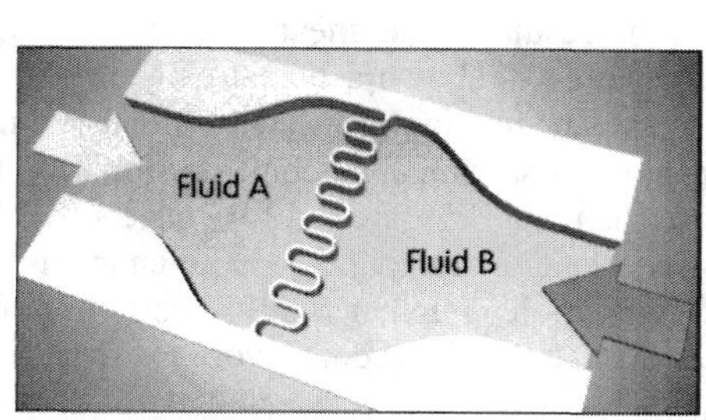

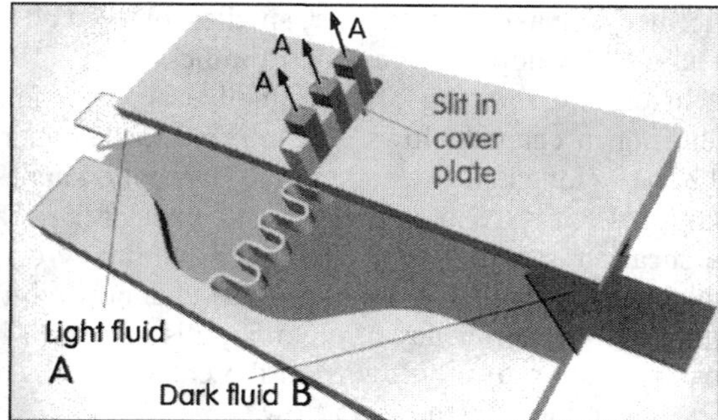

Figure 3.123 (Left) Mixer generating the lamellae by etched interleaved "fingers".

Figure 3.124 (Right) Disadvantage of the interleaved fingers mixer is the out-of-plane direction of lamellae motion.

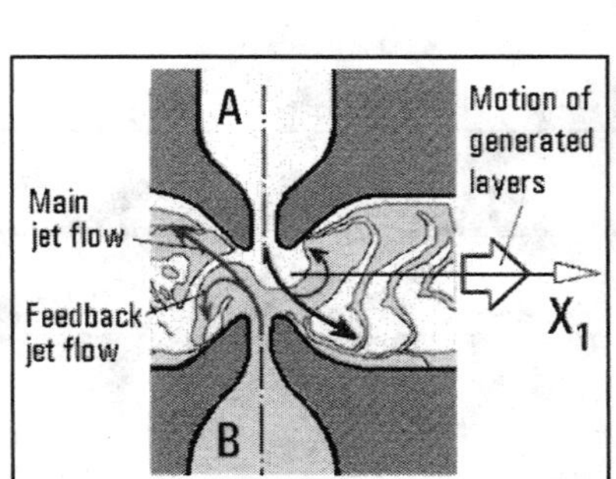

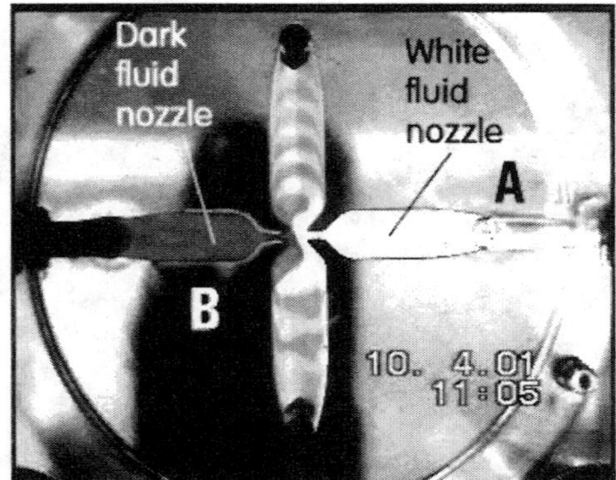

Figure 3.125 (Left) Posterized photograph of the flow in an earlier version of the colliding jets mixer.

Figure 3.126 (Right) Video frame (camera rotated by 90 deg) of a later colliding jet mixer.

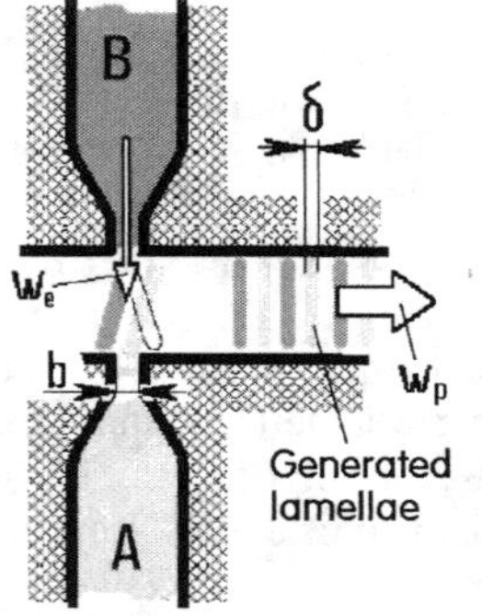

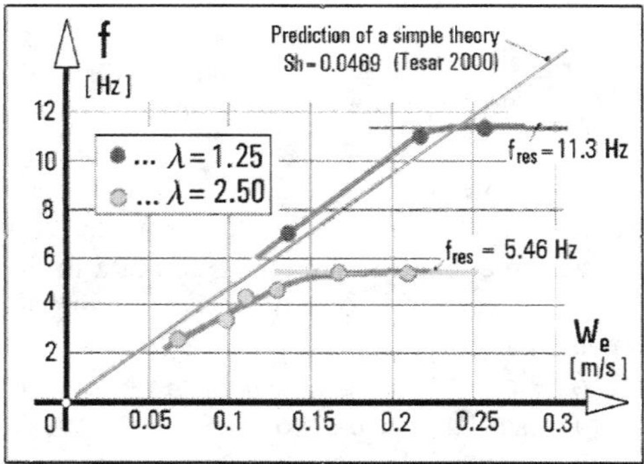

Figure 3.127 (Left) Schematic representation of the generated lamellar structure in one of the exit channels of the colliding jet mixer with interleaved lamellae of fluids A and B.

Figure 3.128 (Right) Experimental results—oscillation frequency as a function of nozzle exit velocity—obtained with the laboratory model of the colliding jet oscillator in Figure 3.126.

The generation itself of the lamellae in the exit channel of a mixer device is not all one can do to obtain the short diffusion path. Once the lamellae are available, they may be subsequently deformed to make them thinner, smaller, or even chaotically tangled. This is an area of particularly intensive development in microfluidics (e.g, [19–21]). The simplest processing leading to extension and thinning of the lamellae is very simply the result of the shear stresses naturally occurring in channel flows. The transversal lamellar structures as shown schematically in Figure 3.122 are elongated to produce the parabolic shapes (compare to Figure 2.9) with no particular additional effort. Also, at least in the center of the channel cross section a welcome deformation by the shear stresses is found for the longitudinally oriented lamellae (Figure 3.130). Instead of the straight channels, it is possible to use channels with wavy walls or the zigzag shape channels. In both cases, the efficiency of the mixing is increased if the Reynolds number is sufficiently high for local separation of the flow from the channel walls and formation of stationary vortical structures [22] interacting with the main flow.

The processing of the lamellar channel flow may be, however, more sophisticated. Of particular theoretical interest is the "split-and-recombine" principle, shown in Figure 3.131 [23]. If there were only two lamellae in the channel inlet, the dark one and the white one, a channel built

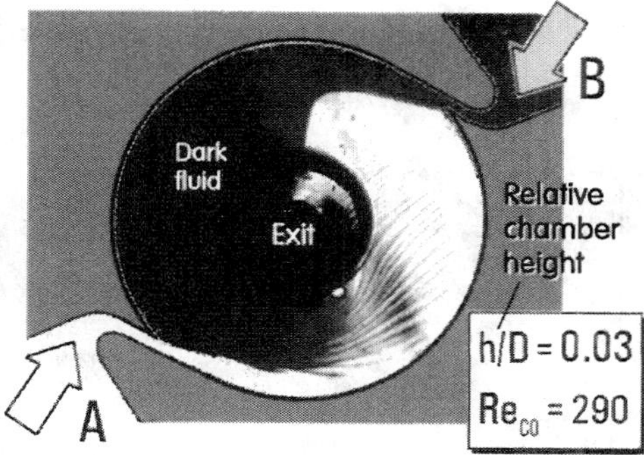

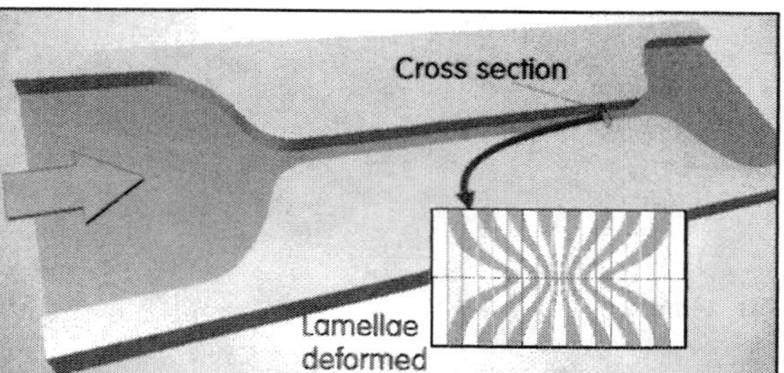

Figure 3.129 (Left) Interleaved lamellae of the two fluids were observed at very low Re in vortex mixer, which is normally used at higher Re with pronounced rotation in the vortex chamber generating the expected interleaved spirals.

Figure 3.130 (Right) The simplest case of deformation of the interleaved lamellae by shear in a straight channel.

according to this principle will convert it into four lamellae. In a multistage layout, the number of the lamellae is doubled at each stage, with each lamella made correspondingly thinner. Practical realization of the idea is not easy; each stage would need a very careful diffuser-type shaping to perform the theoretically expected change of the cross-section. The necessary complex spatial curvature of the walls makes the shape difficult to manufacture even at a large scale, not to mention the limitations of microfabrication with its preference of planar, non-3D configurations. Practically, the idea still remains a manufacturing challenge, Figure 3.132.

The conversions performed in the split-and-recombine devices are ordered and have predictable results (provided each stage works as expected in the theory). There is, however, also a class of channel shapes the action of which on the lamellar structure of the flow may lead—again theoretically, at least—to a chaotic disordered distribution of the two fluids in the channel cross section. In principle, these arrangements are based on succession of channel segments, each producing secondary flow rotation in a sense opposite to that of its predecessor. Aref [24] has demonstrated that such a procedure can lead to completely chaotic trajectories of fluid particles. The effect has been demonstrated with the herringbone pattern ribs on the channel floor with

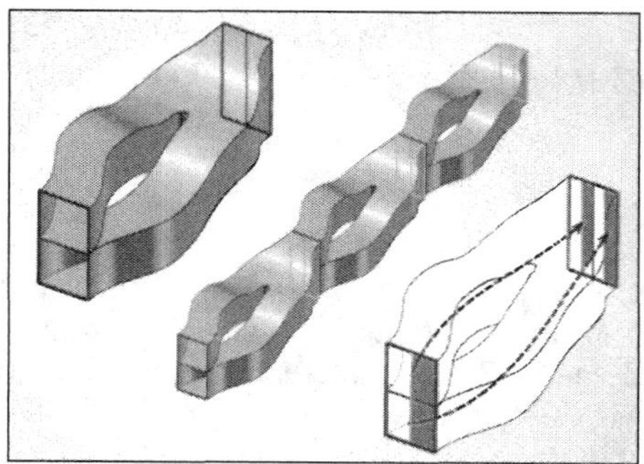

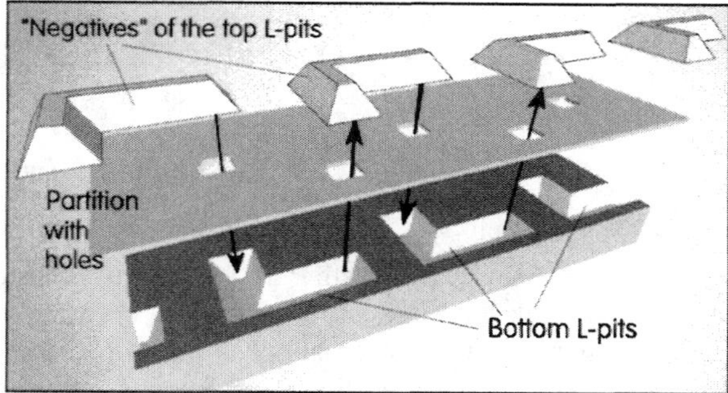

Figure 3.131 (Left) Often proposed "split-and-reconnect" processing of the longitudinal lamellar structure in a mixer exit channel. It is difficult to make by micromanufacturing.

Figure 3.132 (Right) Spatial deformation of the lamellar structure by a succession of L-shaped short channels (pits) etched in an alternating manner in two plates and interconnected by holes in a partition wall. Note that the cavities are sharp-edged approximation to a motion on a helix.

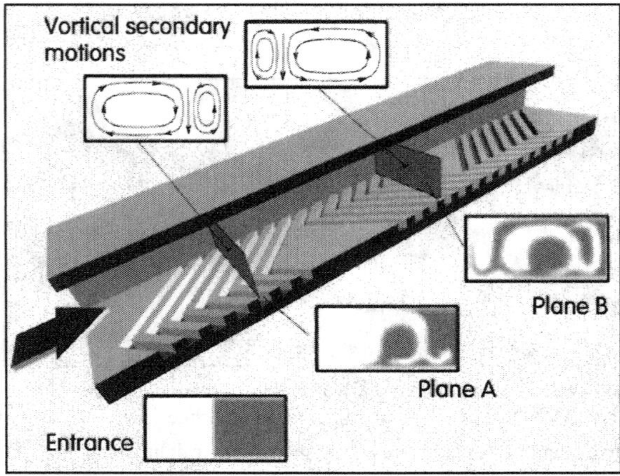

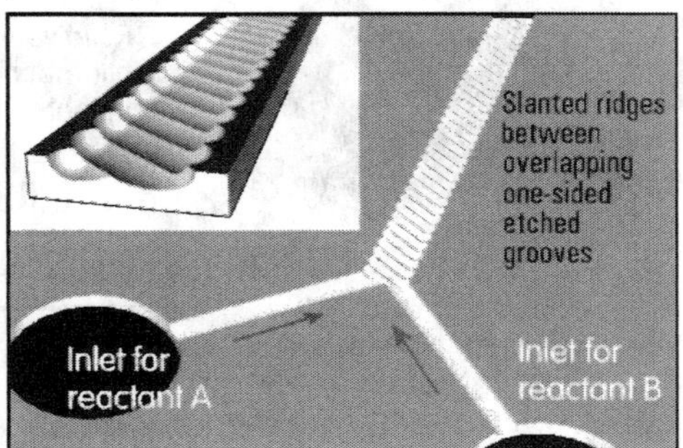

Figure 3.133 (Left) Popular "herringbone ribs" principle of deforming the lamellar structure. The inclined ribs at the channel bottom generate rotational secondary motion. The sections alternate so that the location of the downward secondary motion varies between the left and right side of the cross section.

Figure 3.134 (Right) The ribs on the bottom of very small channels may be made by the isotropic etching (Figures 2.11 and 2.12) through a mask consisting of isolated slanted access strips.

slanting of the ribs changing in repeated cycles (Figure 3.133). The rectangular cross section of the floor ribs may be difficult to make by micromanufacturing. An easier to produce version uses the ridges made by overlapping grooves made by the simple isotropic etching. In Figure 3.134 these ridges are shown to generate a single rotation sense of the secondary motion, but they may be also made in the repeated cycles with the opposite sense of the generated rotation. Of course, the floor ribs are not the only possibility for generating the flows with the secondary-flow rotation. In the example shown in Figure 3.132 there is an approximation to helical motion made in two stacked plates by the anisotropic 54.74° etching (Figure 2.14). here shown with constant chirality of the helical trajectory, the chirality may be also changed in repeated cycles to lead to the predicted complete chaotic distributions.

3.8 THREE-TERMINAL JET PUMP TRANSFORMERS

3.8.1 Venturi transformers: a nozzle and a diffuser

Very simple fluidic devices, consisting of just two components, a nozzle and a diffuser, can perform surprisingly interesting tasks. Let us consider again the Venturi device, discussed in association with its rectification properties (Figures 3.92 and 3.93). The interesting behavior that shows important information about multiterminal devices is obtained when the Venturi device is provided with a third terminal connected to pressure tap at the smallest cross section of the channel (Figure 3.135). In this form, the device is used as a flowmeter: the sum $\Delta P_X + |\Delta P_Y|$, dependent on evaluated flowrate, is larger than a pressure drop obtainable with a restrictor (which can at most produce an output pressure drop equal to ΔP_X). The flowrate can be thus measured with higher precision.

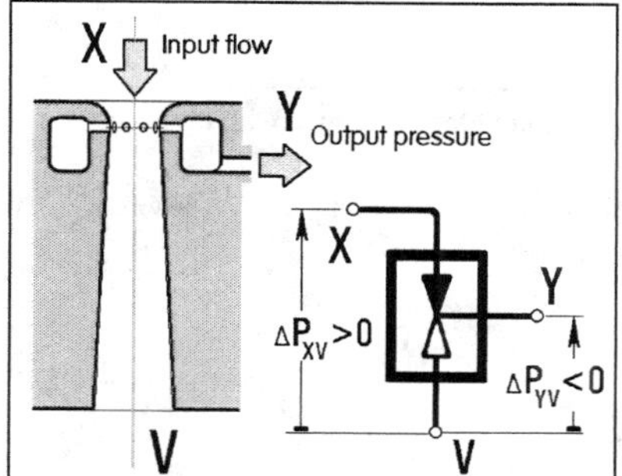

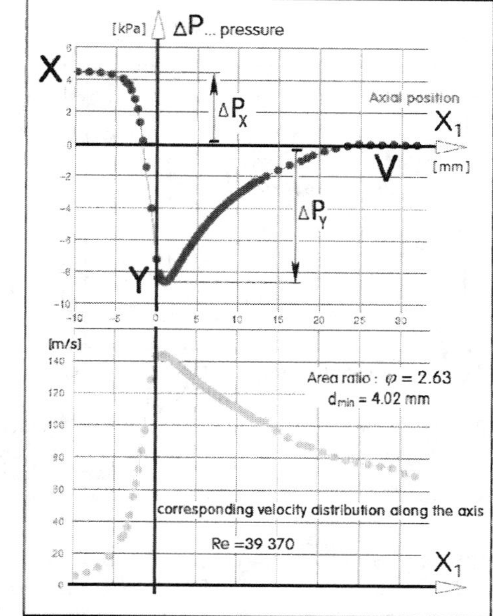

Figure 3.135 (Left) The "nozzle and diffuser" device with the pressure outlet at the smallest flow cross-section location can perform interesting transformation of the state variables.

Figure 3.136 (Right) Pressure (top) and velocity (bottom) distribution on the axis of the device from Figure 3.93 with the pressure taps in the smallest cross section (so that it corresponds to Figure 3.135).

It is less known that the same Venturi device can be used in a manner that is similar to an electric transformer: it can produce at its output a multiple of the input pressure difference, at the expense of a decreased flowrate. For simplicity, this pressure amplification effect was evaluated for the large-scale model device already shown in Figure 3.93, even though it was by no means optimized for this particular role. The addition of the pressure tap actually causes an essential change: it converts the device into a three-terminal one. This increases the number of possibilities of connecting the device into a fluidic circuit. These possibilities, rather than performance of the particular device, are the reasons this behavior deserves discussion.

One interesting aspect of the device is its capability to invert sign: in response to the positive input pressure difference ΔP_X there is, between the output terminals, available a negative, subatmospheric pressure difference ΔP_Y. This is useful in applications where there is a need for a suction – perhaps suction pads of a robot's hand carrying various objects without pressing them by grip force.

Another interesting possibility offered by this device follows from the fact that the absolute value $|\Delta P_Y|$ of the output pressure difference (Figure 3.136) is higher than the input value $|\Delta P_X|$. This passive pressure amplification is demonstrated in Figure 3.137, where the plotted values are those of the pressure gain: the output-to-input ratio of the pressure differences ΔP_Y / ΔP_X. Admittedly in this case the gain, with absolute value hardly more than 2, is not impressive, but even this may be useful. Higher gains, up to about ~ 10, are achievable in specialized designs and also the low Re performance may be improved. In Figure 3.137, the gain is characteristically seen to decrease with decreasing Reynolds number Re and even becomes less than unity (meaning an attenuation rather than amplification) at low Reynolds numbers, due to viscous losses. Of course,

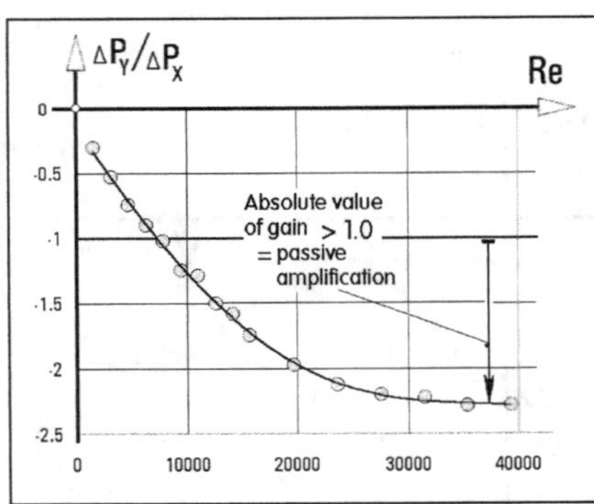

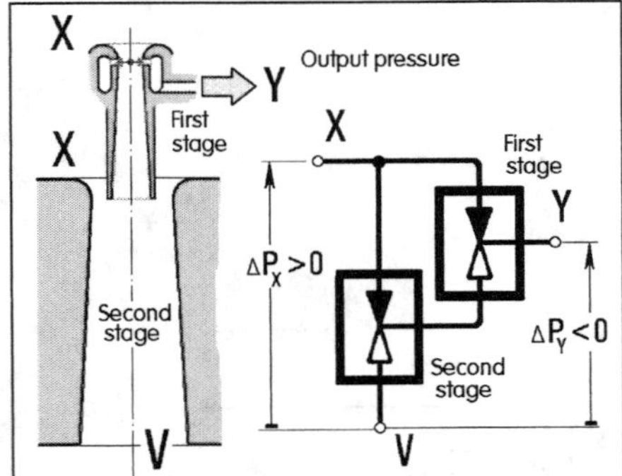

Figure 3.137 (Left) Ratio of the output-to-input pressure differences computed for the device with dimensions according to Figure 3.93 is Reynolds number dependent.

Figure 3.138 (Right) Staging the two Venturi devices to obtain a higher output resultant pressure difference; the overall pressure amplification is the product of the gain values of the two devices.

the power (product of across and through variables) cannot be increased in a passive device. One of them may be increased at the expense of decrease of the other one. The product, power, actually decreases due to the inevitable losses. As shown in Figure 3.138, the amplifier devices may be staged in a cascade to obtain higher amplification: values about ~ 100 obtainable with the two-stage cascade are certainly significant enough to improve pressure signal reading. Of course, the pressure gain is traded for a decrease in flowrates – the cascade needs very high available input flowrate.

3.8.2 Essential facts about jet pump transformers: two nozzles and a diffuser

A disadvantage of the Venturi transformer is its small loading capability. The output pressure may be conveniently read on a manometer that extracts no flow, but if the output is loaded by a connected device demanding some fluid flow, the pressure decreases considerably. This inability to provide a larger output flowrate stems from the fact that the output flow has to pass through the small pressure tap holes.

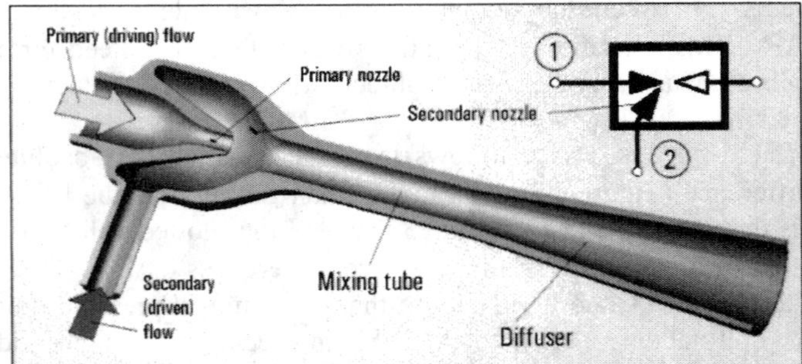

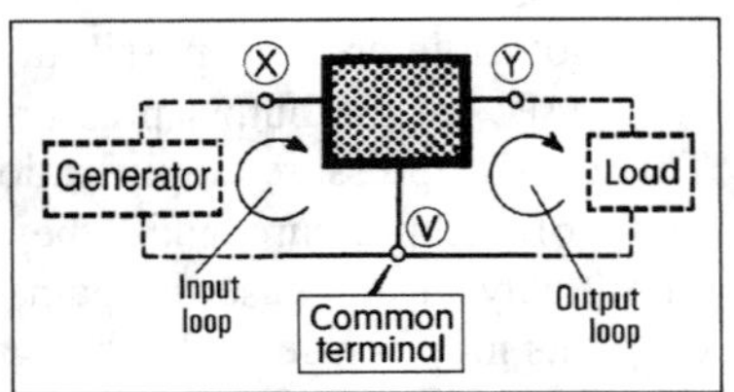

Figure 3.139 (Left) Basic components of an axisymmetric jet pump and its symbol for circuit diagrams.

Figure 3.140 (Right) General schematic representation of the transformer action performed by a three-terminal device placed between the generator loop and load loop, with one of its terminals belonging to both loops.

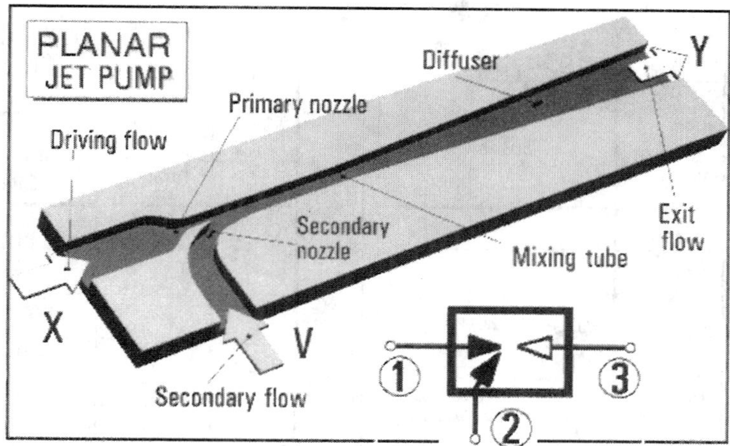

Figure 3.141 Planar design of a jet pump, essentially equivalent to the axisymmetric version in Figure 3.139, perhaps with slightly less effective suction (due to only one-sided admission of the secondary flow), but suitable for the progressive planar manufacturing techniques.

The device can be, of course, easily adapted to deliver a useful output flow. Instead of the output flow through the taps, it is to be delivered through a large area access to the main, primary flow. This is achieved by making a gap between the nozzle and the diffuser (as a matter of fact, such a large gap is seen in the first, power stage of the cascade in Figure 3.138). The flow into the gap has to be arranged through some smooth transition, which is effectively a nozzle. This way,

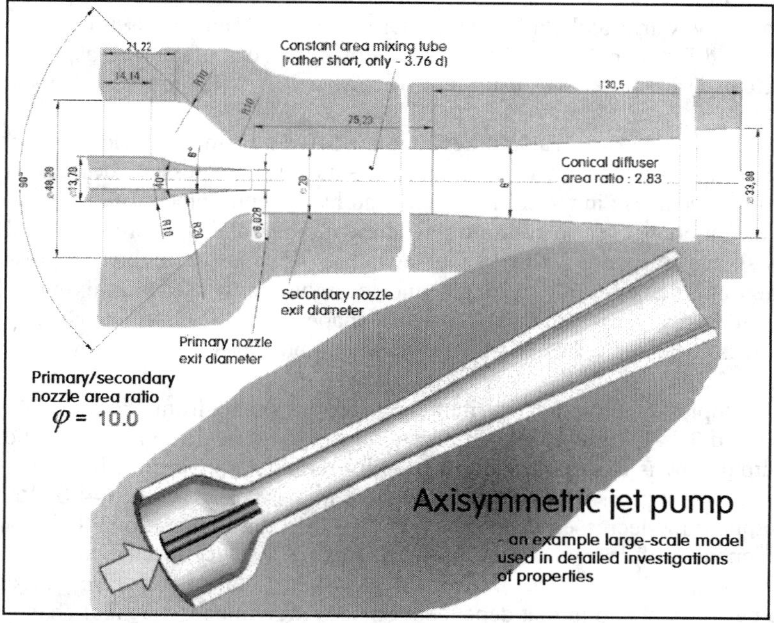

Figure 3.142 Example of a scaled-up model of a jet pump used for demonstration studies of jet pump properties.

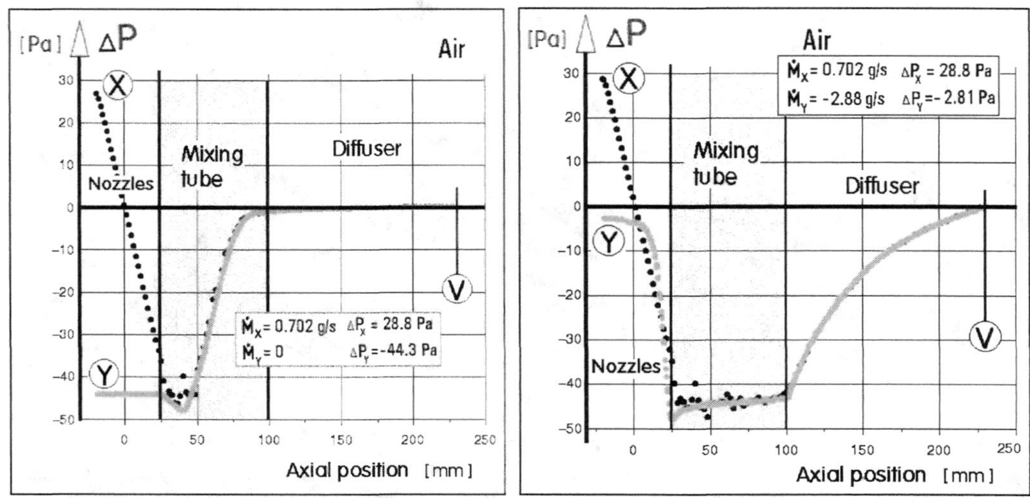

Figure 3.143 (Left) Pressure difference relative to the vent V along the primary and secondary flowpaths in the jet pump from Figure 3.142 when operated with zero output flowrate through the output terminal Y.

Figure 3.144 (Right) Distribution of pressure along the primary and secondary flowpaths in the same jet pump from Figure 3.142 at an output flowrate (through the output terminal Y) 4.1 times larger than the input flowrate.

the adaptation ends in a device having not two but three components: apart from the diffuser (Figure 3.65) there are two nozzles (Figure 3.111). The additional nozzle is called the *secondary* nozzle and the flow through it is the secondary flow. The schematic representation in Figures 3.139 and 3.141 emphasizes the additional nozzle presence. The resultant device is known as the jet pump (or, in Europe, as *ejector*). It is one of the oldest pure fluidic devices; the earliest variant was investigated by Vitrio and Philibert de l'Orme in 1570 and an essentially present-day form was patented by d'Ectot in 1818. In modern fluidics, the manufacturing aspects (planar design suitable for etching) often demand using the somewhat less efficient one-sided secondary nozzle (Figure 3.141).

The interesting transformer capabilities (Figure 3.140) of these devices are based upon the suction effect. This is generated by two mechanisms: by the pressure decrease in the primary nozzle with subsequent rise in the diffuser, and also by the entrainment of the secondary fluid into the jet. Also of importance is the pressure rise in the mixing tube. The presence of the mixing tube is crucial (as shown in Figure 3.143) if the secondary flowrate is small. The pressure rise in the diffuser is then almost negligible. Though other designs are in use, the mixing tube of constant diameter, about 8 diameters long, is the usual shape used in axisymmetric jet pumps. If the secondary flow is large, the diffuser is the essential component for the streamwise pressure rise (as shown in Figure 3.144).

The two examples of pressure distribution in the jet pump from Figure 3.142, presented in Figures 3.143 and 3.144, demonstrate an important but often neglected fact of fluidics in general: if a larger output flow is to pass through the output terminal Y, we must be prepared to accept a lower generated output pressure difference (negative in these cases). The –44.3 Pa between Y and V at zero output flow decreases to mere –2.9 Pa with the higher flow. The magnitude of the output flow depends, of course, on the properties of the load connected to the jet pump. Some newcomers to the field of fluidics find a dirty trick the failure of generated output pressure difference to be constant and in fact dependent on seemingly inconsequential changes elsewhere in the fluidic system.

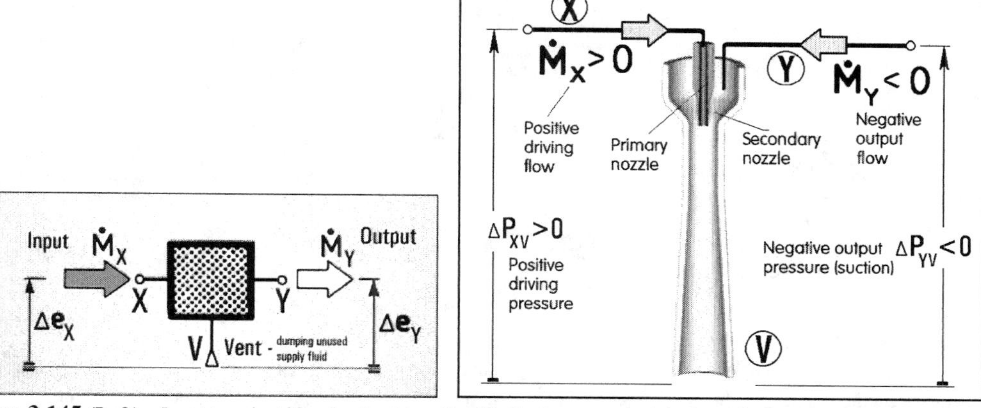

Figure 3.145 (Left) Input terminal X and output terminal Y of a three-terminal device. The input and output differences of the *across* state parameter are defined as the differences between the respective terminal and the vent.

Figure 3.146 (Right) The *common diffuser* connection of a jet pump as a transformer device between the input and output loops. It reverses sign of both pressure difference and the flowrate.

In fact, these distributions (Figures 3.143 and 3.144) show that the pressure level at the beginning of the mixing tube is nearly the same despite the loading by the output flow. Most of the pressure difference resulting in the decrease of the output pressure difference is due to the increased pressure drop across the secondary nozzle. This emphasizes the need of good second nozzle contouring. Also important is proper dimensioning of the jet pump. In particular, the loading properties depend on the choice of the area ratio φ of the secondary to primary nozzle exit cross-section areas. These aspects are of general importance and are discussed in Chapter 4.

3.8.3 Common terminal and different connections into the circuit

Another important fact of more general validity, of importance also for other devices, is the possibility of using the same multiterminal fluidic device in a number of different roles. Jet pumps possess three terminals and though this may seem to be surprising, either one of them may be used as the input terminal X or as the output Y (Figure 3.145). This is why the terminals in the schematic symbols in Figures 3.139 and 3.141 were labeled neither X nor Y, but neutrally as 1, 2, and 3. As shown in Figure 3.145, the remaining third terminal, vent V, is connected simultaneously to both input and output circuit loops. The device performs different tasks in the different connections, which may be characterized by indicating which of the terminals is common to the two loops.

The application in which the signs of the flow (through variable) as well as of pressure (or specific energy or other across variable) are reversed (Figure 3.146) uses the jet pump in the *common diffuser* connection. This is the most common usage of jet pumps. They in this role create a vacuum (subatmospheric pressure—or pressure negative with respect to the vent) from the positive input pressure. This may be useful for such tasks as pipetting fluid samples. Schematic presentation of this role of the device and its connection into the fluidic circuit between the source and a load is presented in Figures 3.147 and 3.148.

The same jet pump may be connected—however strange it may seem to be at first sight—also with the primary nozzle open as the vent to the atmosphere. To generate the primary flow in the

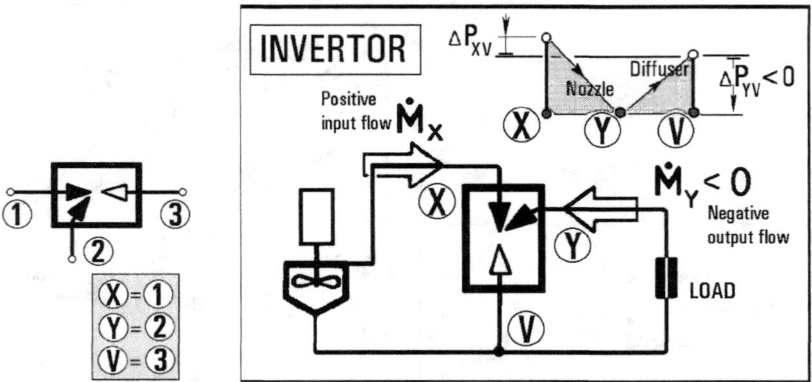

Figure 3.147 (Left) Relabeling (in fact change of their roles) of the terminals in the *common diffuser* connection of a jet pump. Here the primary nozzle inlet serves as the input terminal and the secondary nozzle inlet as the output.

Figure 3.148 (Right) Schematic diagram of the fluidic circuit with the jet pump in *the common diffuser* connection placed between the source (electrically driven pump at left) and a load (restrictor).

source loop the input pressure drop must be negative. The driving source therefore must generate a suction extracting air from the collector. This will generate a suction also in the secondary nozzle terminal, which in this connection has the role of the output Y. A fact of essential importance is this output pressure difference is larger than the one applied at the input side, as is seen in the schematic representation of the pressure differences in the upper part of Figure 3.150. Here the roles of the jet pump terminals are changed as indicated in Figure 3.149. The device performs *passive amplification* of the pressure differences. This may be useful if the pressure difference available from a suction source is not sufficient for a particular task, but its flowrate capacity is larger than needed—because, again, the pressure amplification is obtained at the cost of decrease in the output flowrate, which is again negative—it is (note the definitions in Figure 3.3) directed into rather than out from the device.

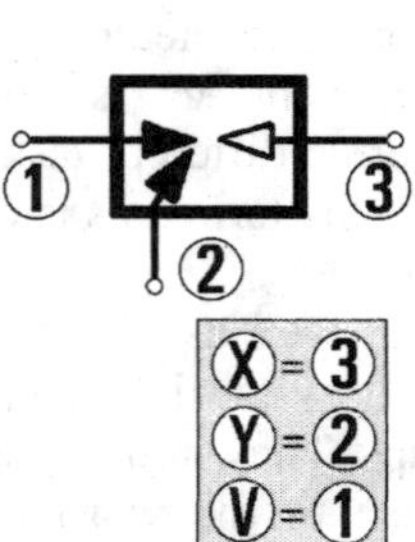

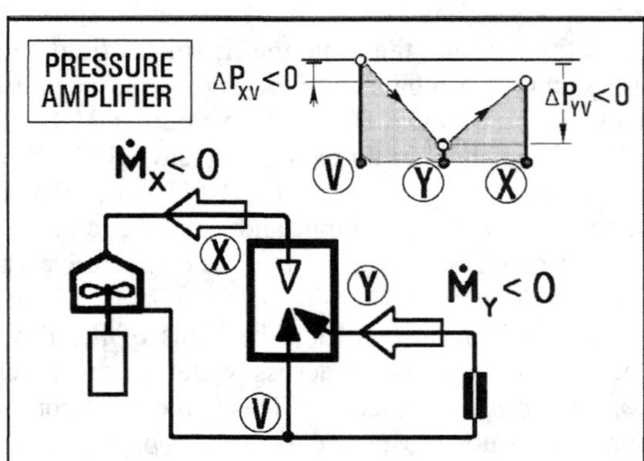

Figure 3.149 (Left) The *common primary nozzle* connection of the same jet pump as before. The (negative) input flow is applied to the diffuser; secondary nozzle is connected to the output terminal.

Figure 3.150 (Right) Use of the jet pump in the *common primary nozzle* connection between the source and load shown in circuit diagram. The schematic pressure distribution above shows how the load receives a much higher acting pressure difference than provided by the source.

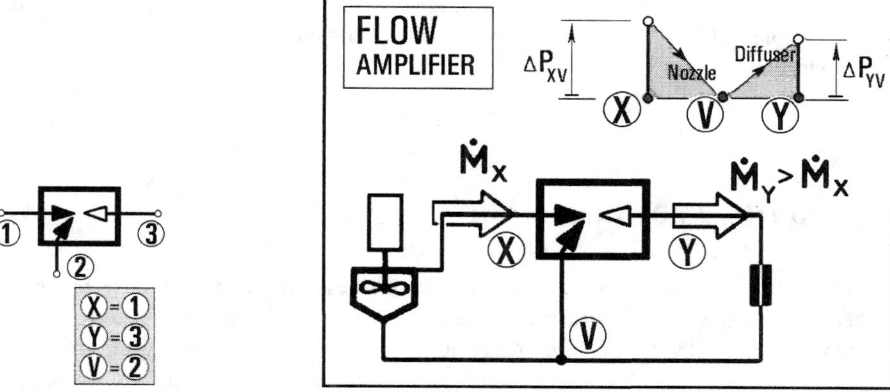

Figure 3.151 (Left) The last possibility, the "common secondary nozzle" connection of the same (at least in principle) jet pump to serve as transformer increasing the flowrate through the load, at the cost of decreasing the pressure difference. **Figure 3.152 (Right)** Circuit diagram of the "common secondary nozzle" connection, performing passive amplification of flowrate (at the expense of decreased pressure).

Finally, there remains the third possibility: to connect the jet pump with its secondary nozzle as the common terminal (Figures 3.151 and 3.152). This is the only case in which the signs of the input and output state parameters remain the same. The output flowrate is increased by the amount sucked in through the secondary nozzle. Such passive amplification of flowrate finds a number of practical uses; it should be kept in mind, however, that since the power is not increased (there is, in fact, a considerable power loss), it is necessary to accept a decreased available pressure difference. As will be shown in Chapter 4, to get significant flow gain, it is desirable to choose a large area ratio φ, the ratio of the secondary to primary nozzle areas. A simple decrease of the primary nozzle size, however, may be not the right way: it is difficult for the entrainment effect on

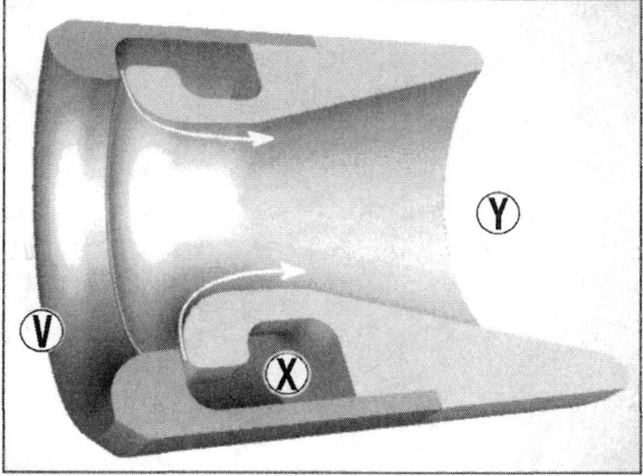

Figure 3.153 Coanda-effect jet pump with an annular slit primary nozzle. This is an example of a special version jet pump, in this case with large area ratio φ particularly suited for the connection with the common secondary nozzle. Effectiveness of entrainment into the jet is increased due to the deformation of turbulent structure by jet curvature.

the edges of a tiny jet to reach far enough into the surrounding secondary fluid. A special version, such as the example in Figure 3.153, was developed for this particular role, useful for such tasks as cooling by a large air flow.

3.9 TOWARD THE SUBDYNAMIC LIMIT

All these interesting properties of jet-type fluidic devices are made possible by the action of submerged fluid jets, which unfortunately become less effective as the nozzle exit Reynolds number decreases [26]. Typical jet Reynolds number values in present-day microfluidic devices are in the range from Re = 100 to Re = 2 000, with interest gradually shifting towards the lower values. As microfluidic systems become more sophisticated, they need more devices on the same chip – and the devices then tend to be smaller and this leads to lower Re. The jets themselves actually remain recognizable (as demonstrated in Figure 3.154) even at Reynolds numbers by at least a decimal order of magnitude lower than the above range. The problem is that at these conditions they entrain less outer fluid and possess less kinetic energy needed for the fluid passing through all the resistances represented by the components and devices further downstream. How far can we get down with Re while the jet-type devices are still useful? The answer depends on how poor performance may be still accepted. The lower, practically not attainable limit is the critical Reynolds number of the order 1.0, beyond which in the subdynamic regime the jets are not formed. Device properties there even cease to be dependent on Reynolds number. When approaching this regime it is inevitable to assist the operation or even drive it by different working principles.

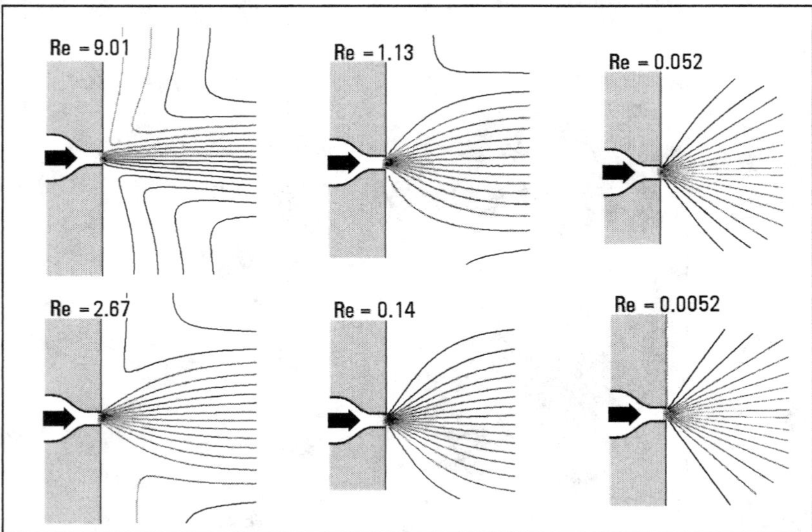

Figure 3.154 Computed planar jet flows at gradually decreasing Reynolds numbers. In the subdynamic limit the character of the jet is lost; the pathlines assume an invariant pattern, independent of Re, with indiscriminate fluid propagation into all available directions.

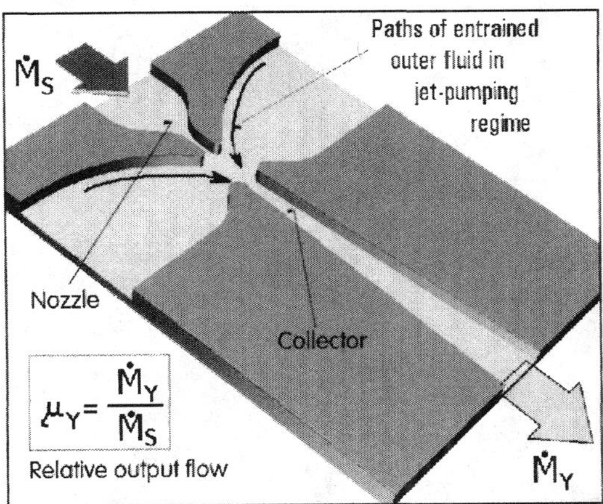

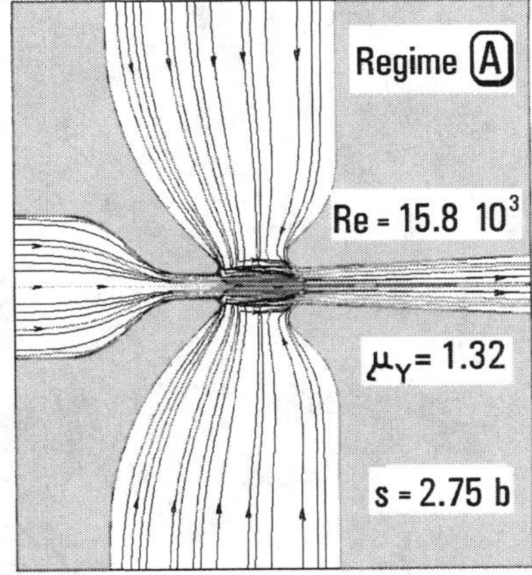

Figure 3.155 (Left) Planar jet pump device used in demonstration investigations of properties at very low Reynolds numbers. The secondary inlets are open as vents into the atmosphere.

Figure 3.156 (Right) Flow paths in the device from Figure 3.155 in the turbulent flow regime A (Figure 3.159) at high Reynolds numbers. The value s is the distance of the collector entrance from the nozzle exit. (*After:* [25].)

It is interesting to get some insight about the low Re operation from systematic investigations [20] of flow control valves. The simplest among them, shown in Figure 3.155, was the passive valve, the operation of which depends on the changes of properties with the Reynolds number. It possesses only the essential components: the nozzle generating the jet, the collector for capturing the jet, and a pair of wide vents decoupling the nozzle from the diffuser. The criterion of performance is the relative output flowrate μ_Y as defined in Figure 3.155. It is the percentage of the input air flow that is available at the output of the device. The overall schematic presentation of the results is in Figure 3.159. At the high Reynolds number end there is the turbulent flow regime A (Figure 3.156). The device there behaves as a proper jet pump: the jet entrains surrounding air brought in from the atmosphere through the vents (Figure 3.155). As a result of the inflow of additional air, the magnitude of the relative output flowrate is $\mu_Y > 1.0$. The entrainment is greatly enhanced by the presence of turbulent vortices (Figure 3.46) at the jet boundary.

As Reynolds number decreases, the vortices become less prominent. They remain present even in the laminar flow regime B, but they are more regular, missing the irregular surface formed by smaller vortices protruding into the air in the vents. As a result the jet entrain less. Also, the relatively large magnitude of friction slows the jet down and leaves less momentum for its passing through the relatively narrow collector. The relative output flowrate drops to values $\mu_Y < 1.0$. The flowpaths (substituting streamlines that do not exist in the three-dimensional flows like this one) shown in Figure 3.157 for a representative Reynolds number of this regime, indicate a change in the character of the flowfield in the devices. No air enters the device from the atmosphere. On the contrary, the flowpaths actually leave the device through the vents. There is still some entrainment into the jet, but it achieves little. It influences only the fluid near the jet, in the vicinity of the nozzle exit. This fluid, however, does not come from outside but is separated from the jet near the entrance into the collector. The entrainment turns some of this separated fluid back and prevents it from leaving the device. This return flow thus forms a stationary recirculation region. Boundaries of these regions are plotted in Figure 3.160 for several Reynolds numbers at which the flowfield was evaluated. The important fact is that as the entrainment effect decreases

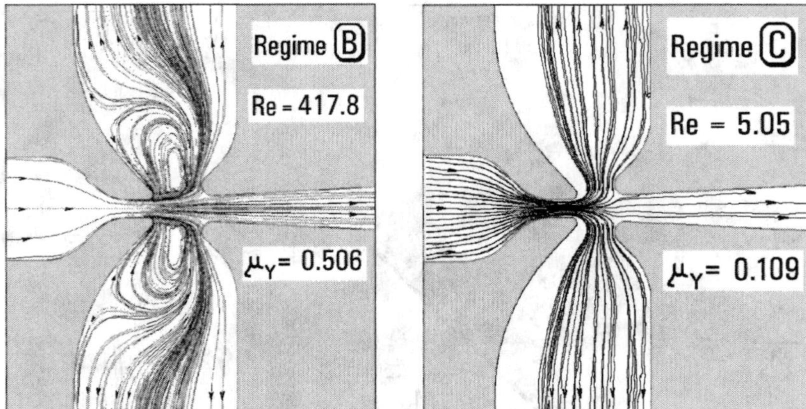

Figure 3.157 (Left) In laminar flow regime B (Figure 3.159) the friction in the collector of the device from Figure 3.155 presents such strong resistance that instead of jet pumping entrainment fluid tends to leave through the vents. The flowpaths there form recirculating regions. (*After:* [25].)

Figure 3.158 (Right) In the subdynamic flow regime C (Figure 3.159) the fluid leaves through the vents with no sign whatsoever of any tendency of entrainment into the jet. (*After:* [25].)

with decreasing Re, the size of the recirculation region also diminishes. One possible point of view is to consider the recirculation region as a sort of blockage that partly closes the vent outlet and thus keeps the fluid inside the device. As the region becomes smaller with further decrease in Re, more fluid is allowed to leave the device. This is in agreement with the fact that the relative

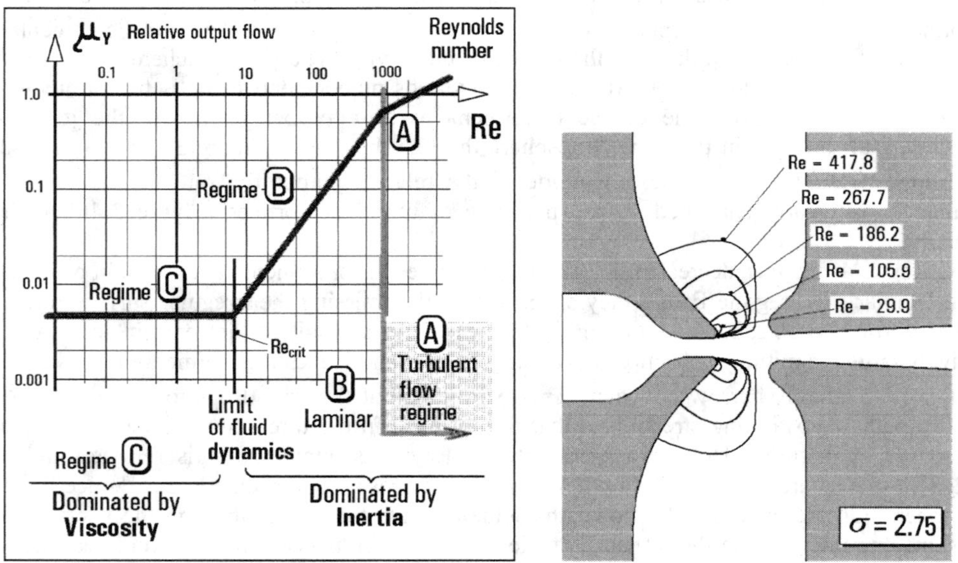

Figure 3.159 (Left) Schematic representation of Reynolds number dependence of the relative output flow (defined in Figure 3.155).

Figure 3.160 (Right) Boundaries of the recirculating region that partly blocks the vents of the device from Figure 3.155. The blocking effect decreases with decreasing Reynolds number and disappears in the final subdynamic regime. Together with the effect shown in Figure 3.154 this results in the loss of dependence on Re. (*After:* [25].)

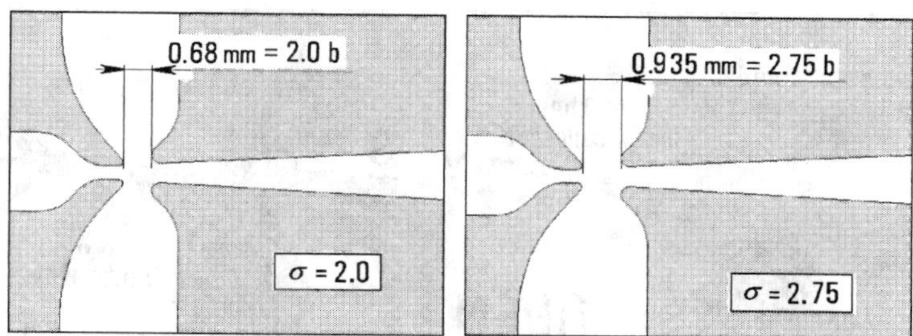

Figure 3.161 One of the investigated changes of the geometry of the device from Figure 3.155: change in the nozzle-to-collector distance. **Left:** the original geometry. **Right:** collector at a larger distance.

output flowrate μ_Y is seen to drop continually as Re decreases in the dependence presented in Figure 3.159. This development comes to an end when at Re_{crit} the recirculation region finally disappears. There is nothing then to stop fluid from leaving the device, the geometry of the flowpaths ceases to vary. This is why the further dependence on Re is lost.

This final stage means entering the subdynamic flow regime C, Figure 3.158. Of course, the absolute values of pressure drops and flowrate continue to become smaller with further decrease of Re. The invariance and loss of dependence on decreasing Re is an effect observable on the dependences between the evaluated relative dimensionless values. That this final, relatively invariant regime C is hardly of any practical importance is apparent from the fact that the relative output flowrate μ_Y is reduced to mere ~ 0.005. Only one half of a percent of the supplied fluid gets into the device output terminal. The remaining 99.5% is lost through the vents. The narrow collector represents a considerable resistance compared with the much easier way out from the device through the wide vents.

It may be interesting to consider the changes in the subdynamic behavior (and changes of the transition into it at the critical Reynolds number Re_{crit}) caused by the changes in the device geometry. As may be expected, placing the collector entrance farther from the nozzle (Figure

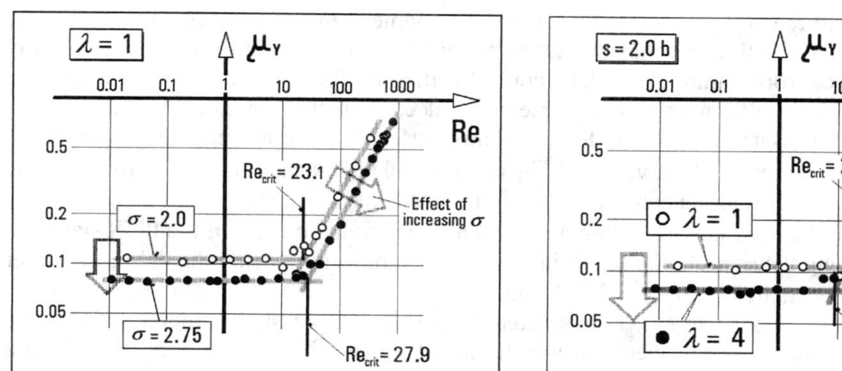

Figure 3.162 (Left) The dependence of the relative output flowrate μ_Y on Reynolds number for the two geometry variants shown in Figure 3.161. (*After:* [25].)

Figure 3.163 (Right) The change in the dependence of the relative output flowrate μ_Y on Reynolds number caused by increasing the nozzle aspect ratio λ. (*After:* [25].)

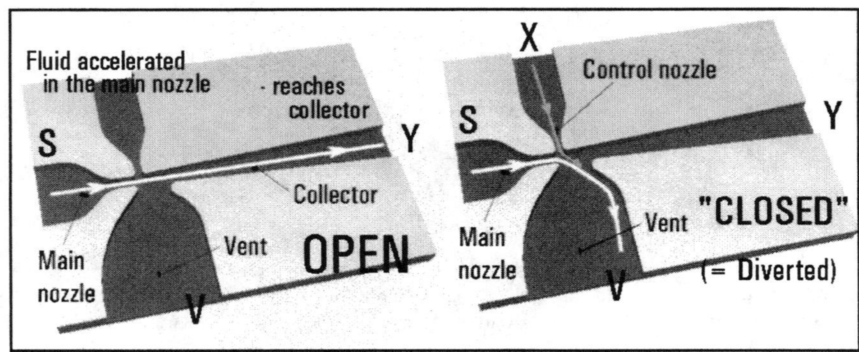

Figure 3.164 An example of an active device – a flow control valve in which the main flow may be denied access to the output by the action of a control signal that diverts it into the vent. The fact that this is a four-terminal device makes its behavior (and characterization of the behavior) much more complicated. From the manufacturing point of view, however, it is almost as equally difficult to make as the passive device from Figure 3.155.

3.161) decreases the relative output flowrate μ_Y, both in the dynamic as well as the subdynamic range (Figure 3.162). With the increased distance the fluid loses more kinetic energy by friction and is offered more opportunity to escape through the vents, which become wider. There is no significant change in the critical Reynolds number defined to lie at the intersection of the two lines (straight lines in the logarithmic coordinates) corresponding to the two regimes B and C. The other investigated geometric change was the increase in the aspect ratio λ (Figure 1.35) of the nozzle exit channel. The results shown in Figure 3.163 were obtained for the small aspect ratio $\lambda = 1$, characteristic for one-sided isotropic etching and also for the increased $\lambda = 4$, obtainable by stacking plates etched from both sides. This change improves the output flowrate μ_Y in the dynamic regime, which leads to a lower transitional critical Reynolds number.

While providing instructive conclusions thanks to its simplicity, the passive device from Figure 3.155 does not admittedly offer many useful applications. Much more important are active flow control valves controlled by an external control signal. These are the subject of Chapter 4. Nevertheless, is may be seen from the example in Figure 3.164 that even in the valves it is necessary for the jet to reach the collector with sufficient momentum. In the OPEN state at left, the control flow is zero and the fluid reaches the output terminal of the valve in exactly the same way as it does in the device from Figure 3.155. The control action, by diverting the main flow into the vent, then more or less—depending on its intensity—decreases the amount of fluid passing through the valve. The essential condition for the control effect is there is something available at the output that can be decreased. If at very low Re the available OPEN state relative output flow μ_Y is small, then the possibilities of the control action are reduced.

Among the possibilities of how to obtain a reasonably high value of μ_Y in the OPEN state, the obvious one is reducing the opportunity for fluid loss through the vent, placing the collector nearer to the nozzle (to get a smaller σ; Figures 3.161 and 3.162). Alternatively, the connection between the vents and the outer atmosphere may be restricted. This way, the pressure in the vent rises. The strategies of this pressure assisted operation are discussed in association with Figures 1.30 and 1.32 in Chapter 1.

In the extreme, the driving pressure effect may become more important than the momentum of the jet. This is the case in the valve shown in Figure 3.165 (designed by Dr. Tippets [25]). This is a valve designed for operation at extremely low Reynolds numbers. The collector (connected to the output Y) is not even placed in line with the fluid inlet from the supply terminal S. In fact,

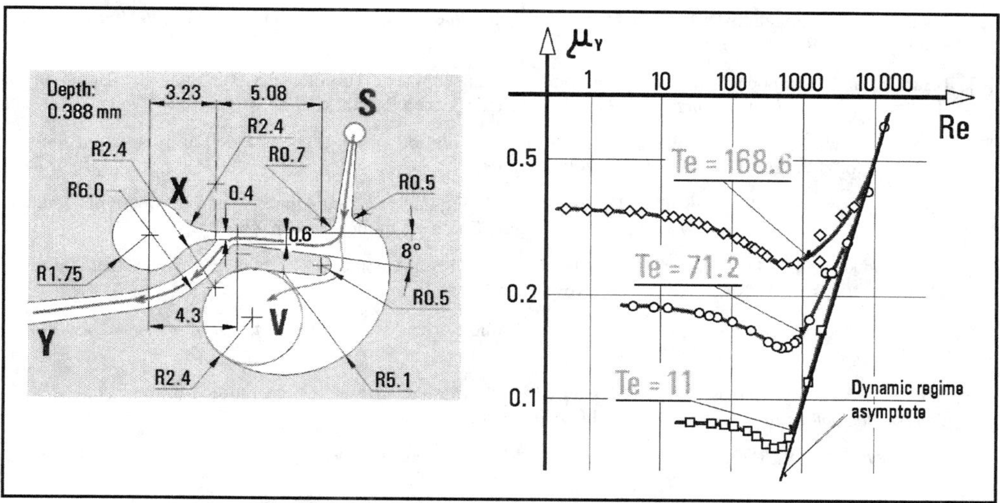

Figure 3.165 A practical example of a four-terminal active valve (actually an element of the sequential sampling unit from Figure 1.5) and its characteristic—dependence of the relative output flowrate on Reynolds number—in the pressure-driven operating regime, with the driving pressure characterized by the dimensionless pressure parameter Te. (*After:* [25].)

this inlet is here not even shaped as a nozzle with its characteristic area contraction. If the Reynolds number is increased, what momentum the fluid acquires in this inlet actually leads it away, into the vent V rather than into Y. As a consequence of this disregard for dynamic effects, the dependence of the output flow on Reynolds number shows an initial decrease of μ_Y. The resultant minimum on the curves provides a particularly distinct location of the critical Reynolds number. The operation of this device is wholly dependent on maintaining the pressure conditions between the device vent and the locations downstream from the load connected to Y. It is useful to characterize the driving pressure by the dimensionless parameter Te (Figures 2.49 and 3.28).

References

[1] Nguyen N.-T. and S. Wereley, *Fundamentals and Applications of Microfluidics*, 2nd ed., Norwood, MA: Artech House, 2006.

[2] Tesař, V., "Nozzle Characteristics - the Boundary Layer Model," *Hydraulika i Pneumatika*, Vol. XXIV, No. 3, Poland, May 2004, p. 35.

[3] Tollmien, W., "Berechnung turbulenter Ausbreitungsvorgänge," *Zeitschrift für Angewandte Mathematik und Mechanik*, Vol. 6, 1926, p. 468.

[4] Tesař, V., "The Solution of the Plane Turbulent Jet," *Acta Polytechnica*, Vol. 36, No. 3, Prague, Czech Republic, 1996, p. 15.

[5] Tesař, V., "Two-Equation Turbulence Model Solution of the Plane Turbulent Jet," *Acta Polytechnica*, Vol. 35, No. 2, Prague, Czech Republic, 1995, p. 19.

[6] Tesař, V., and J. Šarboch, "Similarity Solution of the Axisymmetric Turbulent Jet Using the One-Equation Model of Turbulence," *Acta Polytechnica*, Vol. 37, No. 3, Prague, Czech Republic, 1997, p. 5.

[7] Tesař, V., "Two-Equation Turbulence Model Similarity Solution of the Axisymmetric Fluid Jet," *Acta Polytechnica*, Vol. 41, No. 2, Prague, Czech Republic, 2001, p. 26.

[8] Tesař, V., "Similarity Solutions of Jet Development Mixing Layers Using Algebraic and 1-Equation Turbulence Models," *Acta Polytechnica*, Vol.46, No.1, Prague, Czech Republic, 2006, p. 40.

[9] Forster, F. K., and B. E. Williams, "Parametric Design of Fixed-Geometry Microvalves – The Tesser Valve," *Proc. of IMECE 2002*, ASME, New Orleans, LA, November 2002.

[10] Anduze, M., et al., "Analysis and Testing of a Fluidic Vortex Microdiode," *Journal of Micromechanics and Microengineering*, Vol. 11, 2001, p. 108.

[11] Lu, L.-H., K. S. Ryu, and C. Liu, "A Magnetic Microstirrer and Array for Microfluidic Mixing," *Journal of Microelectromechanical Systems*, Vol. 11, 2002, p. 462.

[12] Moroney, R. M., R. M. White, and R. T. Howe, "Microtransport Induced by Ultrasonic Lamb Waves," *Applied Physics Letters*, Vol. 59, 1991, p. 774

[13] Löb, P., et al., "Steering of Liquid Mixing Speed in Interdigital Micro Mixers—From Very Fast to Deliberately Slow Mixing," *Chemical Engineering & Technology*, Vol. 27, 2004, p. 340

[14] Berthier, J., and P. Silberzan, *Microfluidics for Biotechnology*, Norwood MA: Artech House, 2006.

[15] Tesař, V., J.R. Tippetts, and Y.-Y. Low, "Oscillator Mixer for Chemical Microreactors," *Proc. of 9th International Symposium on Flow Visualization*, Edinburgh, paper No. 298, August 2000.

[16] Tesař, V., and K. Peszynski, "Microfluidic Oscillator for Generating Oil/Water Emulsions," *Pneumatyka*, Poland, Vol. 34, No. 3, May—June, 2002, p. 8.

[17] Tesař, V. and Y.-Y. Low: "Study of Shallow-Chamber Vortex Mixers for Microchemical Applications," *Proc. of 9th International Symposium on Flow Visualization*, Edinburgh, paper No. 445, August 2000.

[18] Tesař ,V. and Y.-Y. Low, "Microfluidic Mixing Based upon Interfacial Instability," *Proc. of Colloquium Fluid Dynamics 2000*, Inst. of Thermomechanics, Academy of Sciences of the Czech Republic, Prague, October 2000, p. 59.

[19] Hardt, S., et al., "Passive Micromixers for Applications in the Microreactor and μTAS Fields," *Microfluidics and Nanofluidics*, Vol. 1, 2005, p. 108.

[20] Hessle, V., H. Löwe, and F. Schönfeld, "Micromixers – A Review on Passive and Active Mixing Principles," *Chemical Engineering Science*, Vol. 60, 2005, p. 2479.

[21] Campbell, C.J., and B. A. Grzybowski, "Microfluidic Mixers: from Microfabricated to Self-Assembling Devices," *Phil. Trans. Royal Society*, London, 2004.

[22] Mengeaud, V., J. Josserand, and H. H. Girault, "Mixing Processes in a Zigzag Microchannel: Finite Element Simulations and Optical Study," *Analytical Chemistry*, Vol. 74, 2002, p. 4279.

[23] Hardt, S., H. Pennemann, and F. Schönfeld, "Theoretical and Experimental Characterization of a Low-Reynolds Number Split-and-Recombine Mixer," *Microfluidics and Nanofluidics*, Vol. 1, 2005, p. 237.

[24] Aref, H., "Stirring by Chaotic Advection," *Journal of Fluid Mechanics*, Vol. 143, 1984, p. 1.

[25] Tesař,V., et al., "Subdynamic Asymptotic Behavior of Microfluidic Valves," *Journal of Microelectromechanical Systems*, Vol. 14, April 2005, p. 335.

[26] Zimmerman, W. B. J. (ed.), *Microfluidics: History, Theory, and Applications*, New York: Springer-Verlag, 2006.

Chapter 4

Valves and Sophisticated Devices

Chapter 3 discussed the simplest devices, consisting of only a few components and generally not exhibiting a particularly complicated behavior. Most of the devices were *passive*, just responding to the changes of flow conditions at their inlets. This chapter progresses to devices performing more interesting operations. Typically, many of these devices are *active*, able to deliver more power at their output than what is input into them at the inlet side since they are supplied by energy-containing fluid from a powerful supply source.

The discussion of these sophisticated devices needs introduction of two important concepts: the concept of *control action* and the concept of a response to *loading*.

The importance of the control action is obvious. In an active device, the input fluid flow is not directly led to the output and in fact it may not physically reach the output at all. What it does is modulation of the fluid flow from the supply source, deciding how much of it can get to the output. The task of the designer is to ensure that the changes of the output flow follow the input changes. Another important task is usually to ensure a high gain. This is the ratio of the output effect to the input effect. If, as usual, the gain is $> |1.0|$, then the device may be described as an amplifier. This is an important property, even though amplification of fluidic signals ceased to be a task of primary importance in present-day fluidics. It is necessary to distinguish between flow gain, pressure (or specific energy) gain, and the power gain, which is the product of the two former ones. Some control principles are more suitable for producing a higher flow gain while the amplification of the pressure changes is small. Others are suitable for amplifying the pressure differences. There are quite useful devices exhibiting a negative gain, with the output flow reacting by a decrease to a rise in the input control flow. The sign is usually unimportant what is really necessary is the absolute magnitude of the gain.

The other basic concept of this chapter, the response to varying the connected load, is much less generally known and appreciated. Some newcomers to fluidics actually find it an unexpected and almost perfidious property. Properties of a fluidic device are dependent on the properties of whatever is connected to it, mainly to the output outlet. Behavior of a device tested alone in the laboratory (and perhaps laboriously developed to be satisfactory) may be completely different from the performance when the device is connected into a fluidic circuit together with other devices. The most influential are the properties of the load, the device immediately connected to the output of the investigated fluidic element. Some devices react to the changes in their loading more, some less. The response may be, for example, a decrease of the amplification gain. The most obvious effect of the loading is the extreme case of a total loss of proper behavior if the output terminal is nearly blocked by a high dissipance load or blocked completely by a load allowing no output flow. There are ways how to decrease this dependence on the loading. One of them is providing vent outlets through which the fluid may leave if the flow path through the output terminal is closed.

However vast are the possibilities of what fluidic devices can do, their most important and most often performed immediate task is to control fluid flows. As a result, many of the devices described in this chapter have the character of flow control valves.

4.1 LOADING CHARACTERISTICS

Because of the usually complex, nonlinear properties of fluidic devices, the effect of their loading is best followed in a graphical representations. The most useful diagram for this purpose is the loading curve, which is usually a part of the output characteristics of the device: the dependence of all pressure drops (or specific energy differences) between the device terminals on the magnitude of the output flowrate.

In large-scale fluidics as well as in most cases of present-day microfluidics, the curves in this diagram are best approximated by quadratic parabolas. Of course, as the continuing trend towards the smaller scales leads to decreasing Reynolds numbers, the curves become nearer to linear dependences, though they practically never become exactly linear.

To make the basic concepts of the loading more obvious, it is useful to begin the discussion by considering the case of the very simple device, the jet pump, presented in Section 3.8.2 as a device used for transformations between the input loop and output circuit loop. Even though this device is very simple, its operation being based on properties of a jet makes its properties very near to those of the fluidic jet-type valves, which are of the ultimate interest in the present context.

4.1.1 Loading a simple jet-type device

It was stated in Chapter 3 that a jet pump in the circuit schematically represented in Figure 4.1 is useful as a passive flow amplifier, generating at its output terminal Y an output mass flowrate $\dot{M}_Y$ larger than the input flowrate $\dot{M}_X$, larger by the amount sucked into the device through the common secondary nozzle. This occurs mainly due to the entrainment into the jet that issues from the primary nozzle. The magnitude of the flow gain $\mu_Y = \dot{M}_Y / \dot{M}_X$ is, of course, dependent on the geometry of the jet pump, but it also depends on what load is connected to the jet pump output. The smaller are cross sections of the flowpath in the load, the smaller is the amplification, and if they are really small, the blocking load can change the character completely. Instead of the expected increase, we may discover that the fluid spills over into the secondary nozzle and the output flowrate is smaller than the input one, Figure 4.2. In Figure 4.3 the behavior of the jet pump

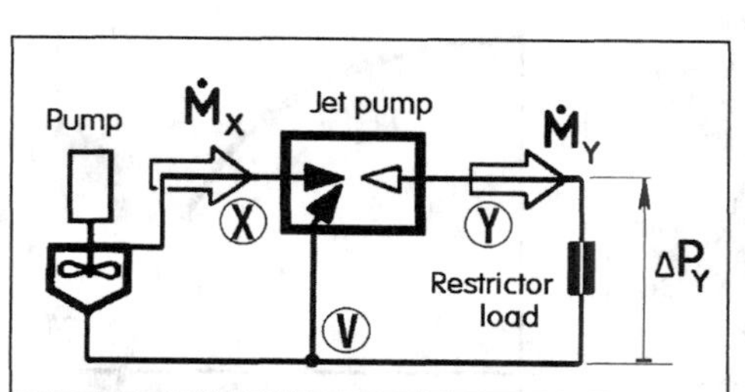

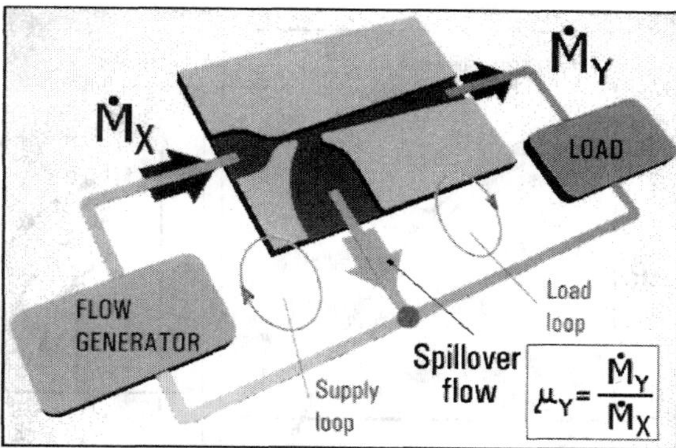

Figure 4.1 (Left) Circuit diagram of a simple jet-type device—a jet pump in the common secondary nozzle connection (Figure 3.151)—loaded by a simple load. The output flow $\dot{M}_Y$ passing through the load varies with the load dissipance.

Figure 4.2 (Right) A planar version example of the jet pumping device from Figure 4.1. If blocked by a high dissipance load, the expected flow amplification $\mu_Y = \dot{M}_Y / \dot{M}_X > 1$ by entrainment of additional fluid from **V** is replaced by some fluid being actually spilled over, so that $\mu_Y < 1$.

is characterized by the solid black line of dependence of the output pressure difference on the output flowrate (Figure 4.1). The load is there a simple separation-type restrictor. Its characteristic, corresponding to Figure 3.74, is one of the gray lines. The state in the output loop of this circuit is determined by the intersection point of the characteristics.

In Figure 4.3, there are three load characteristics for three different alternative loads. The slope of characteristics increases with decreasing size of the cross section in the restrictor. As long as the Reynolds number is not too small, the restrictors may be characterized by the magnitude of their dissipance Q (Figure 2.29). As expected, the characteristic for the highest Q and steepest slope has its intersection point with the jet-pump characteristics at the smallest output flowrate. Pushing the fluid through the small cross-section restrictor, requires higher pressure drop. Indeed, as might be expected, the intersection shows the pressure drop to be largest for the load of smallest cross-section (and highest dissipance). The problem with newcomers in the field of fluidics is that they do not expect these changes to take place and find it surprising that the output flowrate is not constant.

The output flow, of course, also depends on how much fluid is supplied into the jet pump at its input terminal X, through the primary nozzle. This may be not easy to determine in general, For jet-type devices, however, the dependence of the input flow on the loading at the output is rather weak, especially if the jet from the primary nozzle traverses a larger distance of vented space before it is captured by the collector. The space isolates the primary nozzle from conditions further downstream. In the present case, the distance is not very long (Figure 4.2), but we may assume that it is possible to keep the input flowrate constant (and hence a constant nozzle Reynolds number) during the loading process. The loading curves of the jet pump for three different supply flow conditions A, B, and C are presented schematically in Figure 4.4. For each loading curve, this picture shows an important "no spillover" point, the state in which the output flow is equal to what is supplied at the input.

Figure 4.5 then explains the transition of the jet pump performance from the normal flow amplification operation, with the small dissipance load A and the intersection point at $\dot{M}_Y > \dot{M}_X$, to the state with the fluid "spilled" over to the vent V, which takes place with the higher dissipance

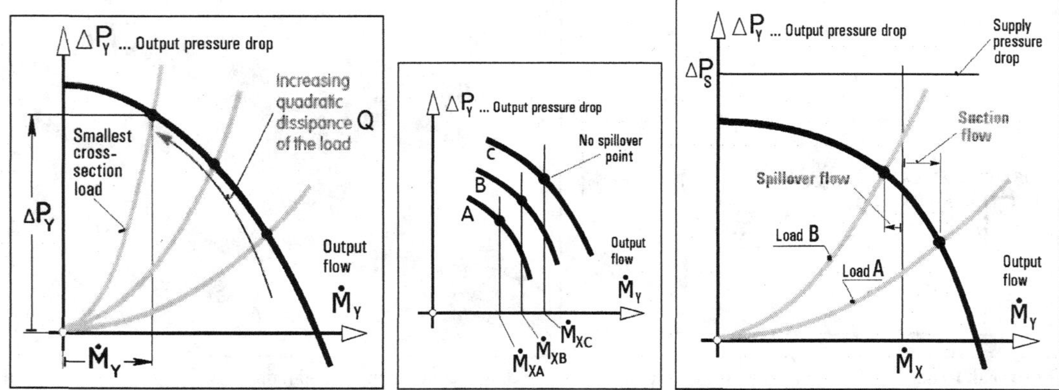

Figure 4.3 (Left) Typical quadratic shapes of the characteristics of the devices in Figure 4.1 at high Reynolds numbers. Black curve: loading characteristic of the jet pump. Gray curves: characteristics of the load.

Figure 4.4 (Center) Actual use of the loading characteristics is complicated by the fact that instead of the single loading curve there is a family of curves, each valid for a particular input flow.

Figure 4.5 (Right) Conditions near the no-spillover regime presented in the characteristics.

load B. Since the usual performance criterion for the flow amplifier is the magnitude of the flow gain, it is evident that the circuit in Figures 4.1 and 4.2 will perform best if operated with loads having the minimum possible dissipance.

The discussion above used schematic examples of the characteristics, but it may be useful to present an actual evaluated example of a jet pump, as shown in Figure 3.142. Rather than the typical planar layouts of microfluidics, the device in this example is axisymmetric. Moreover, to demonstrate the Eulerian similarity (which tends to be somewhat ruined by the friction effects at low Re), the behavior was evaluated at rather high Reynolds numbers. To show the seemingly different appearance that may arise in some cases, this example was evaluated for the common diffuser connection (Figure 4.6). In that case, the output flow (compare with the convention Figure 3.3) is negative and also negative is the output; note the opposite orientation in relation to the positive values shown in Figure 4.1). The task of the jet pump in this connection is just to invert the signs. It may be equally well operated with high dissipance load, thus generating a high (negative) output pressure difference (or specific energy difference in Figure 4.6) with low dissipance load to generate a large output flowrate. There does not seem to be, in contrast to the previous example, an obvious performance criterion telling which load is best. In fact, as we shall see, the criterion is usually transferring to the load the maximum available power.

If the comparison is not led astray by the negative values, it is apparent that the three loading characteristics in Figure 4.7 are very similar to what was schematically presented in Figure 4.4. The replotting in the relative coordinates in Figure 4.8 shows, that the curves are mutually similar. This jet pump (Figure 3.142), was designed in a somewhat haphazard way, without any optimization; in particular, the mixing tube is much shorter than the usually recommended length of 8 diameters. Yet the performance judged from the point of view of flowrates, is rather good: note in Figure 4.8 that with very small load dissipance the output flowrate is as much as more than four times the input flow ($|\mu_Y| = 4.4$). On the other hand, the suction effect obtainable in the output terminal Y is not particularly strong here. It increases with increasing dissipance of the load. However, even with the infinite dissipance (complete blockage of the output outlet) the absolute value of the pressure gain is a mere $|\pi_Y| = 0.155$ at the smallest Reynolds number in

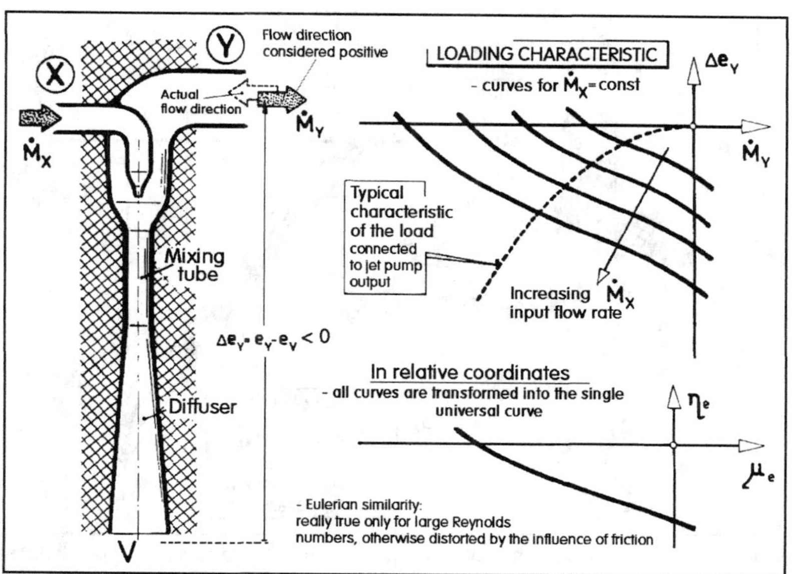

Figure 4.6 A typical axisymmetric jet pump in the common diffuser connection (Figure 3.146) together with schematic presentation of its loading characteristics.

these tests. A certain improvement with increasing Re, impairing the similarity, is recognizable in Figure 4.8 (as might be expected due to the decreasing influence of the friction losses). The pressure gain improves to $|\pi_Y| = 0.175$ at the largest input flowrate.

Another example of the common diffuser connection of a jet pump, and its corresponding loading characteristics in the relative coordinates, this time of really microfluidic dimensions, is presented in Figure 4.9. Apart from the improvement of performance with increasing Re evident again, what is immediately apparent in Figure 4.9 is the worse flow gain $|\mu_Y| \sim 2$, less than one

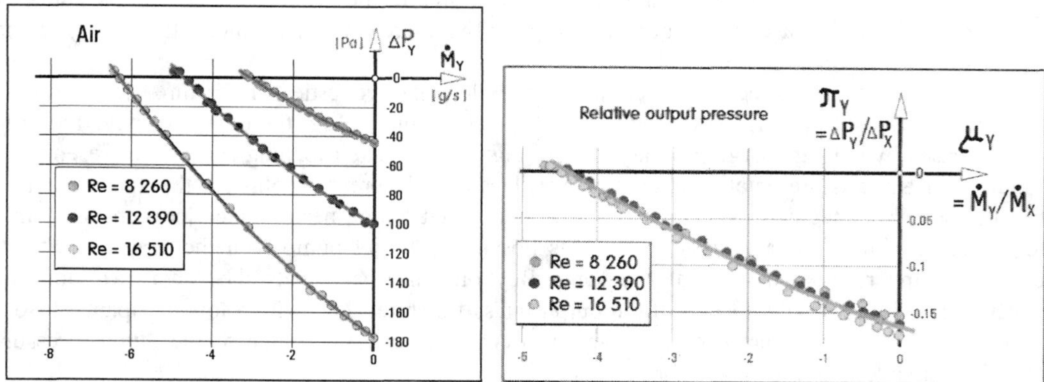

Figure 4.7 (Left) Loading characteristics evaluated for the 6.018 mm main nozzle diameter jet pump shown in Figure 3.142 with three different (rather large) constant input air mass flowrates $\dot{M}_X$, specified here in terms of the primary nozzle exit Reynolds numbers. The output pressure difference as well as the output mass flowrate are negative in the used common diffuser circuit connection (Figure 4.6).

Figure 4.8 (Right) Demonstration of the Eulerian similarity: all three characteristics in Figure 4.7 are converted into virtually single curve in the relative coordinates (output values related to the values of the same quantity in the input **X**).

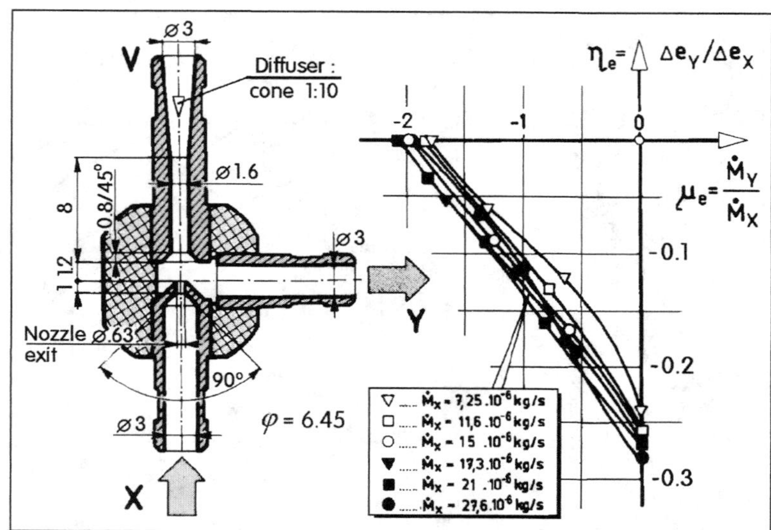

Figure 4.9 Although its axisymmetric components (for one-off manufacturing on a lathe) make it different from the current planar layouts, this jet pump is of true microfluidic size (submillimetre size of the primary nozzle). Despite the low Reynolds numbers (Re = 780 for the smallest input flowrate), the Eulerian similarity in the relative coordinates is quite convincing.

half of that in Figure 4.8. On the other hand, this jet pump exhibits a better pressure gain, $|\pi_Y| \sim 0.27$, roughly almost twice as much than before.

Of course, the geometries of the two jet pumps (Figures 3.142 and 4.9) are different and indeed, the differences in their characteristics are attributable to their different design. Also the smaller Reynolds number in the latter case has a nonnegligible effect on the characteristics. Nevertheless, there is a certain systematic dependence on an important geometric parameter, which deserves a closer look. As already mentioned in association with Figure 3.153, this parameter is the ratio φ of the mixing tube entrance area to the primary nozzle area. Its smaller value $\varphi = 6.45$ in the case from Figure 4.9 suggests that increasing φ promotes the flow gain and decreases the pressure gain.

This is, indeed, supported by the comparison of loading properties of the three jet pumps in Figure 3.10. One of them is the original jet pump from Figure 3.142; the other two have the ratio φ decreased by having a larger primary nozzle, Figure 4.10. Their loading curves are presented in Figure 4.11 and this presentation may, however, lead to a different conclusion: the diagram seems to suggest rather unequivocally that what is needed to get a high performance jet pump is simply to design it with the area ratio φ as large as possible. The jet pump with the largest $\varphi = 10$ evidently, under the condition of the same Reynolds number, beats the other two both by generating a larger absolute value of the output mass flowrate $|\dot{M}_Y|$ with a low dissipance load – as well as generating a larger output suction effect (high absolute value of the output pressure $|\Delta P_Y|$) with a load of high dissipance.

This conclusion, however, is valid only for the condition of the same input Reynolds number. Despite the design of the jet pumps in Figure 4.10 actually used the same outer body, with different primary nozzles, from the Re point of view, the situation used in evaluating the curves in Figure 4.11 corresponded to what is shown in Figure 4.12. Here the nozzles are of the same diameter and it is the outer body that is changed in each case so as to vary the area ratio values

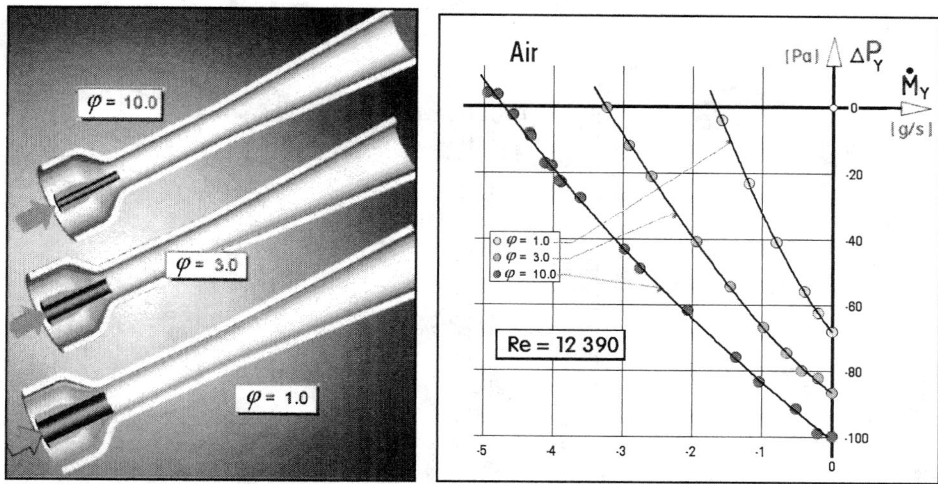

Figure 4.10 (Left) A study of the effect of area ratio φ was made by providing the same jet pump from Figure 3.142 with different primary nozzles.

Figure 4.11 (Right) Loading curves in the output characteristics of the three jet pumps from Figure 4.10.

as shown Figure 4.10. The advantage of the jet pump with the largest $\varphi = 10$ now becomes evident: it has a big an effective jet pumping mixing tube, diffuser, and all the other parts. In contrast to this, the poor performance of the $\varphi = 1$ device is now not surprising: it is just a small body around the large nozzle, which cannot produce any effective jet pumping effect.

This example shows clearly the wrong conclusions that may be drawn from improperly referenced comparisons. To get a real picture of the advantages and disadvantages obtained by varying the area ratio φ requires comparing the properties (loading characteristics or power characteristics, Figure 4.13) of the three designs replotted in the relative coordinates (Figures 4.6 and 4.8). This is done in Figure 4.14. Although the numerical values are valid only for the

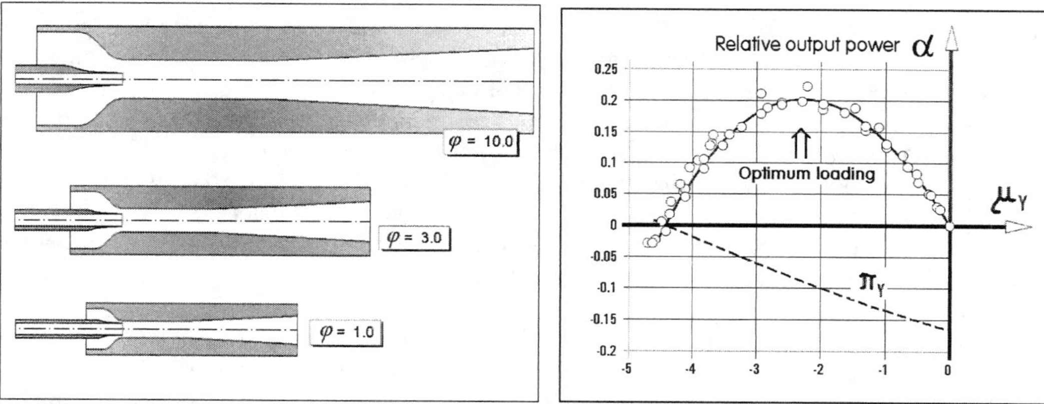

Figure 4.12 (Left) Jet pumps from Figure 4.10 redrawn so that the diameters of their primary nozzle exits are equal.

Figure 4.13 (Right) Relative output power transferred from the jet pump into the load connected to its output terminal computed from the results plotted in Figure 4.8 for the jet pump shown in Figure 3.142.

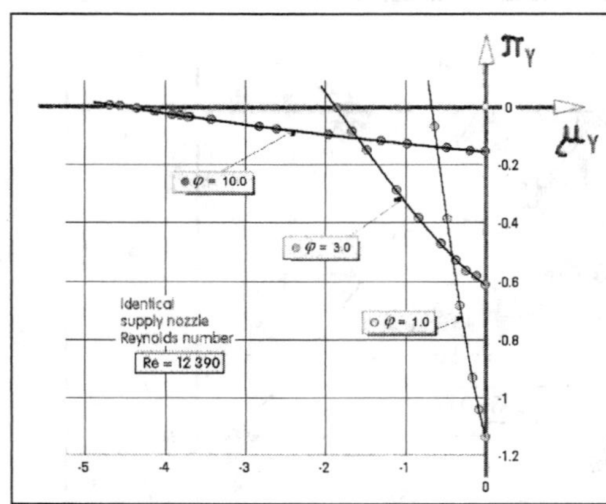

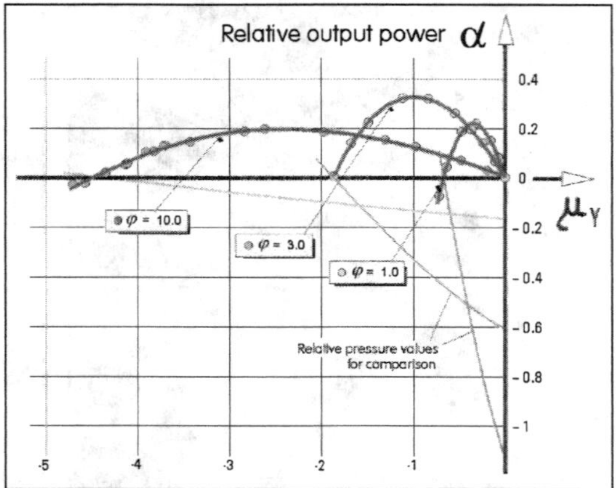

Figure 4.14 (Left) Loading curves (from Figure 4.11) of the three jet pumps shown in Figure 4.10 replotted in the relative coordinates: relative output mass flowrate μ_Y and the output-to-input pressure ratio π_Y.

Figure 4.15 (Right) Relative output power characteristics obtained as a product of pressure gain and volume flow gain of the three jet pumps shown in Figure 4.10.

particular example of the jet pumps of Figure 4.10, the generally valid conclusions are apparent: the jet pumps with small area ratio φ are suitable for the applications in which it is desirable to obtain very large suction pressure. Here for $\varphi = 1$ it was possible to get pressure amplification $|\pi_Y| > 0$ ... with the sign inversion characteristic for the common diffuser connection there is with the blocked output (infinite load dissipance) $\pi_Y = -1.13$. On the other hand, with the zero dissipance load (suction from the atmosphere), it is preferable to chose a large ratio φ.

Are there any limits to the choice of the area ratio φ? Generally speaking, choosing extremely small or extremely large area ratio (unless, in the latter case, there are special means undertaken to entrain more secondary air, for example, with several nozzles or the annular primary nozzle, Figure 3.153) leads to loss in efficiency. This is best presented in terms of the pneumatic (or hydraulic) power transferred from the jet pump to its load. This output power (Figure 2.20) is the product of the two output state parameters. Again, it is the relative value, related to the input power supplied into the primary nozzle (the input terminal), which is of importance for judging the device performance. It is evaluated as a product of the pressure and volume flowrate gains, for example in Figure 4.13 for the jet pump from Figure 3.142. The modest maximum achievable power transfer efficiency in Figure 4.13 shows that this jet pump was indeed not optimized. The top of the power loading curve there reaches the maximum at 20 % efficiency (there are optimized jet pumps reaching—usually, however, at large Re—efficiency values approaching 40 %).

The relative power loading characteristics were evaluated also for the family of jet pumps from Figure 4.10. The results are plotted in Figure 4.15. There is a value of the area ratio φ with which it is possible to get the best efficiency. Even though in this case there are not enough evaluated jet pumps in the family, it is apparent that the optimum is near to $\varphi = 3$, a value that agrees with the experience obtained also with other jet pump designs.

It may be interesting now to return to the common secondary nozzle connection of Figure 4.1 with its more usual, positive valued loading characteristics of Figure 4.3 and ask about the effect of the area ratio φ on the performance. In this case, the jet pump used as a passive flow amplifier, the performance criterion is usually the achieved maximum flow gain μ_Y. As may be expected from the experience with the other circuit connection in Figure 4.14, high flow gains are obtained

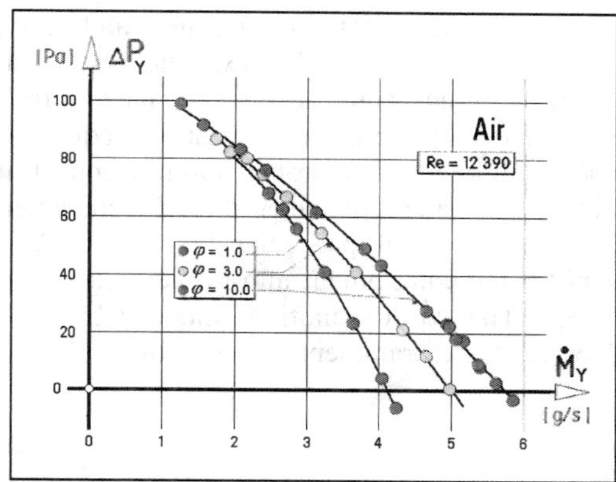

Figure 4.16 The influence of the nozzle area ratio φ on the loading curve of a jet pump in the common secondary nozzle (passive flow amplifier) circuit connection. The characteristics, corresponding to the one shown schematically in Figure 4.3, were evaluated for the family of three jet pumps from Figure 4.10.

with large values of the ratio φ. Of course, the load should possess the minimum possible dissipance so that the advantage of the high flow gain is employed. For operation with air, the secondary nozzle should be ideally supplied directly from the atmosphere. The larges entrained air floe is obtained if also the diffuser exit is open into the atmosphere as well (so that, apart from the inevitable but small entrance and outlet losses, the load represents a zero output pressure drop ΔP_Y, zero dissipance load). In fact, the results plotted in Figure 4.16 show that the loading curves tend to approach a single common curve as the output pressure drop ΔP_Y increases, so that the advantage obtained by choosing large φ becomes lost anyway.

4.1.1.1 Performance improvement by superimposed oscillation

It may be useful to note that the performance of jet pumps increases considerably if there are oscillations superimposed on the basic supplied input flow. An improvement is caused by the formation of large eddies which encompass surrounding fluid and entrain it more effectively, as

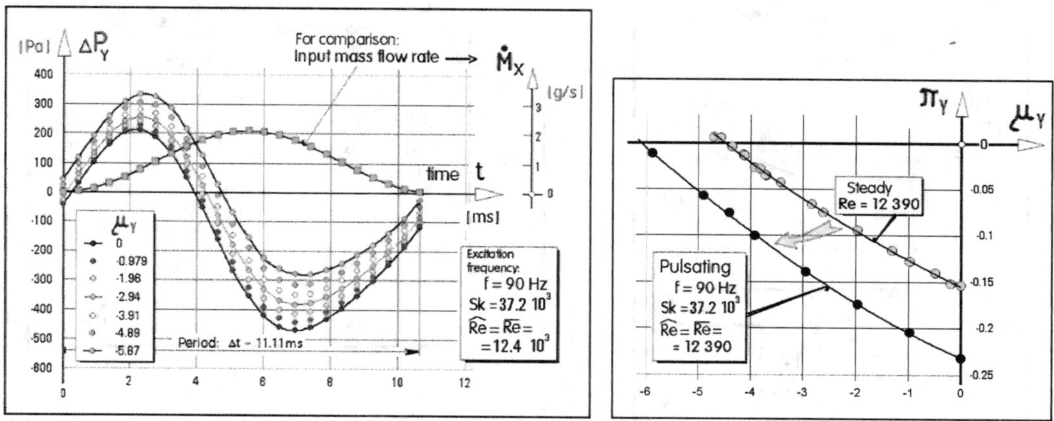

Figure 4.17 (Left) Time dependence of the output pressure difference ΔP_Y in the course of one period for the jet pump from Figure 3.142 with harmonically pulsating input flow.

Figure 4.18 (Right) The time-mean loading characteristic with the pulsation according to Figure 4.17 compared with the much worse efficiency of the same jet pump with steady input flow.

well as due to inertia of the rectified flow in the mixing tube and the diffuser. An example of the improvement is presented in Figures 4.17 and 4.18. The results shown here are for strong superimposed oscillation with amplitude equal to the time-mean value, so that the driving input flow comes to a complete halt in each period. The oscillation frequency $f = 90$ Hz was chosen by a chance; the results may be much better after proper optimization of the Stokes number Sk (1.4) here computed from the primary nozzle exit diameter d as the characteristic length.

This way toward improvement is of particular importance for small-scale microfluidics, where at low Reynolds numbers the entrainment into the jet tends to be much less effective than into turbulent jets at high Re. The vortical motions induced by the oscillation can effectively substitute turbulence and increase the entrainment considerably.

4.1.2 Passive flow control valves

The intention of the previous sections was to present the basic idea of multiterminal device characteristics, but they also attempt to show the usefulness of jet pumps, the rather neglected fluidic devices. In principle they act as passive (not controlled by signals brought from outside) fluid flow control valves. The discussion can now progress to related devices that are really called valves. The first are still passive ones that do not have a control terminal bringing in an external control signal. Nevertheless, they do exhibit a certain measure of a control action.

Before discussing the first example of such valves, it is useful to consider Figure 4.19. It shows the basic types of the flow control valves, using for this purpose the more obvious models with a moving part, the insertion of which into the flowpath forces the fluid to flow in the desired way. Of course, these mechano-fluidic valves are outside our focus of interest here, where we are inte-

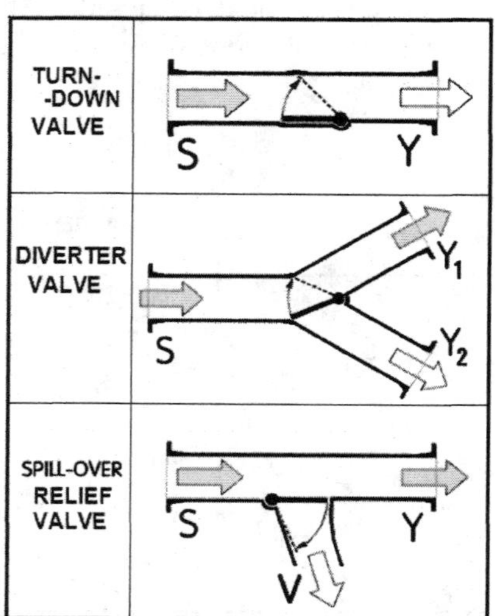

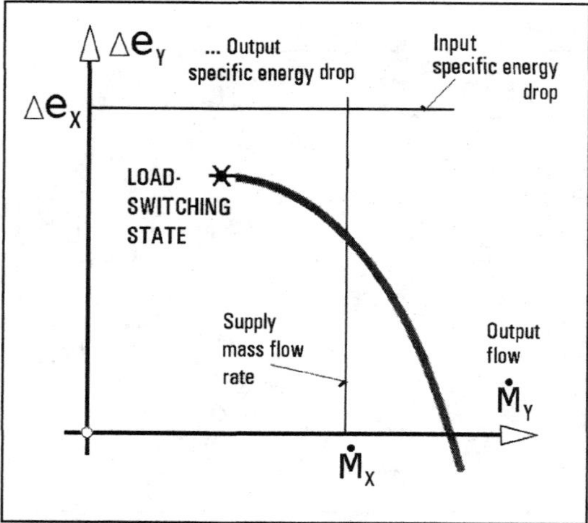

Figure 4.19 (Left) The basic operating modes of flow control valves. For easier discussion, they are here presented on mechano-fluidic examples, with the control action produced by motion of a moving component.

Figure 4.20 (Right) Schematic example of a loading characteristic ending in the load switching point.

rested in valves performing the tasks from Figure 4.19 by aerodynamic (or hydrodynamic) action in fixed geometry cavities. In the circuit connection shown in Figures 4.1 and 4.2, the jet pumps operated in the mode corresponding to the spillover valve model at the bottom of Figure 4.19. The attention is now turned to the flow diverting valves, in which both exit outlets may be used as the output terminals.

4.1.3 Load-switching in a passive Coanda-effect valve

While the jet pumps were in the previous chapter characterized as devices having two nozzles (the primary nozzle and the secondary one) and a single collector (with its diffuser), the valves of the present part form a class of devices with only a single nozzle, but two collectors. In addition – note the schematic representation in Figure 4.21 – there is a simple but important component: the attachment wall. The Coanda-effect attachment is necessary to avoid the trivial splitting of the flow into both outputs. In some valves there may be yet another attachment wall on the opposite side. Its influence has to be suppressed by being positioned at a larger setback distance from the nozzle exit or with a larger deflection angle to make the valve monostable. In the basic, unloaded state the fluid always leaves through the main collector.

In the basic case, the other collector is used as a vent V: the common terminal through which some fluid may be entrained into the jet in a similar manner as in Figure 4.1 while the load is connected to the output terminal Y. Across the largest part of the corresponding loading characteristic, the behavior is analogous to the one shown in Figure 4.3. The essential difference is in the presence of the *load-switching* point (Figure 4.20). As the dissipance of the load increases and the intersection point with the curve of the load characteristic is approaching the load-switching point, most fluid actually laves (spills over) through the other terminal (V) than the one

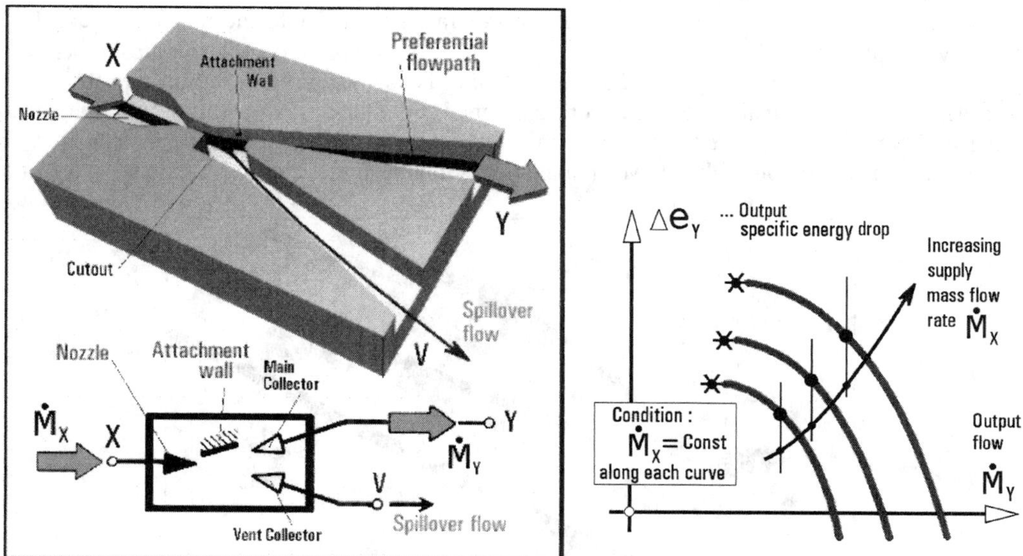

Figure 4.21 (Left) An example and schematic symbol of the monostable diverter valve. Coanda effect of attachment is utilized for guiding the jet into the preferred collector, as long as dissipance of the load connected to Y does not reach the switching level.

Figure 4.22 (Right) Family of loading characteristic curves, each for a different constant input mass flowrate.

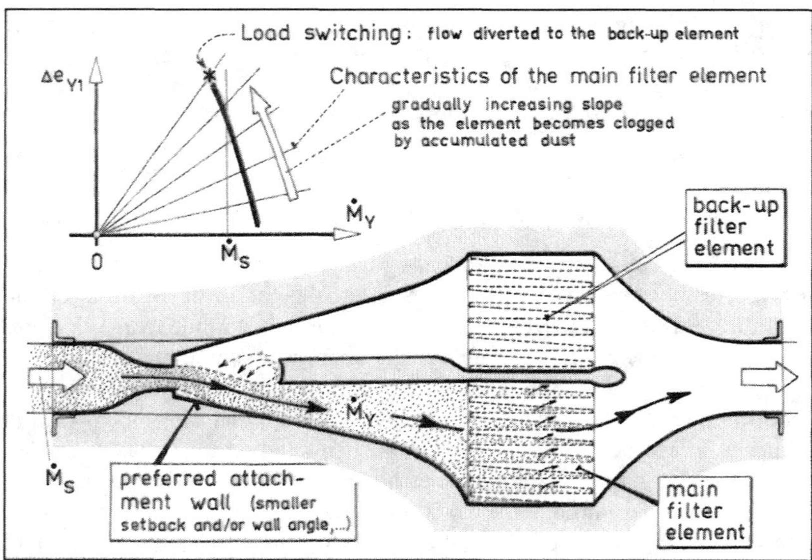

Figure 4.23 Application of the load-switching phenomenon in fluid filtering.

(Y) toward which it is directed by the Coanda-effect attachment. At the end of the attachment wall the jet is forced to turn away. This has a detrimental effect on stability of the attachment. When the load-switching state is reached, the jet jumps into the other collector. This load-switching effect is utilized in flow control in an open-loop mode, without flowrate sensors, the control signal channels and general simplicity and low cost. Of course, the exact load-switching state is different for different magnitudes of the flowrate supplied into the nozzle (Figure 4.22). This may complicate the circuit design, but may be utilized for a more complex control tasks.

In the example shown in Figure 4.23, the switching is used to direct the fluid into the backup filter element when the main element becomes clogged. Of course, the operation with the backup filter is only temporary, before the clogged main filter element is replaced. Nevertheless, it avoids endangering the operation of the downstream devices that need the cleaned fluid and would not

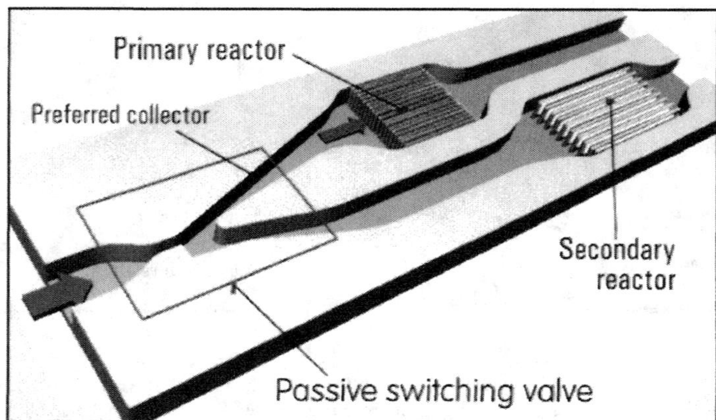

Figure 4.24 An example of a passive flow control: switching the gas flow into the secondary microreactor takes place when the reaction temperature in the primary reactor reaches a certain undesirable level.

get it in sufficient amount through the clogged main element. In the other example, Figure 4.24, the switching is used to protect the exothermic primary reactor from reaching temperature levels at which the catalyst ceases to be effective (or may undergo irreversible changes). The gas supplied into the primary reactor (attached to the preferred outlet Y) is heated there by reaction, which results in increasing the gas viscosity. This increase means rising effective dissipance of the reactor in its role of the load. The intersection point, similar to the top part of Figure 4.23, climbs up the loading curve of the valve. When it reaches the load-switching point, the gas flow is switched into the secondary reactor. Overheating and thermal aging of the catalyst in the primary reactor is avoided.

4.1.4 Passive jet-type pressure regulators

Another typical example of the elegant solutions made possible with fluidics are the open-loop controls performing a rather complex control task in a single device. In traditional solutions with closed control loops, this task would often require a quite complex circuit, consisting of controller, sensors, channels for the signa transfer, and an actuator. The whole task may be fulfilled in fluidics (perhaps with a somewhat larger final deviation of the controlled variable) by what is in effect just the actuator device – by using its special shape of the loading characteristic. An example is the characteristic in Figure 4.25, with the horizontal segment. Its intersection point with the characteristic of loads diagram remains at the same output pressure level whether the load dissipance is high (the CLOSED state) or low (the OPEN state in Figure 4.26). Of course, developing a geometry exhibiting the desirable behavior is by no means simple. In the example presented here, it was facilitated by the fact that the device used as the starting point, which was a bistable valve

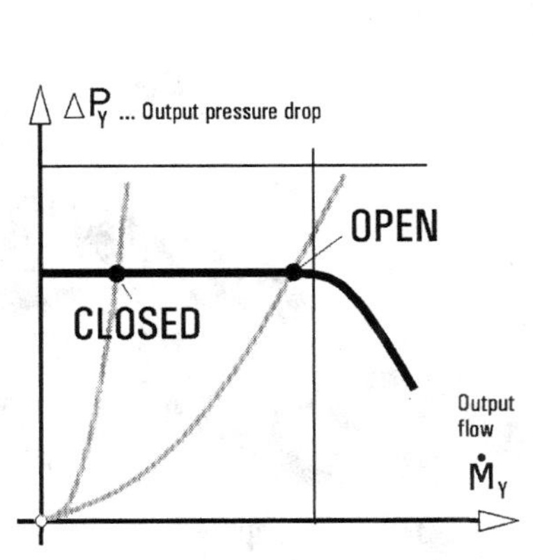

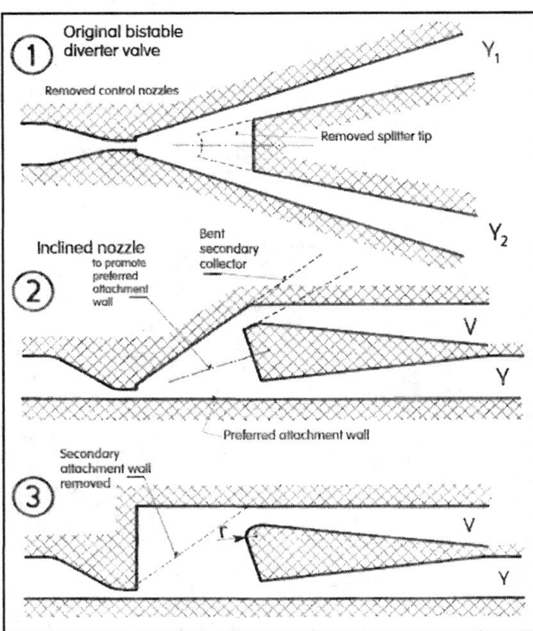

Figure 4.25 (Left) Special shape of valve loading characteristics: operated in the horizontal section the valve keeps a constant pressure at its output irrespective of the exit opening or closing.

Figure 4.26 (Right) Development of the passive pressure regulator geometry from a bistable diverter valve [1].

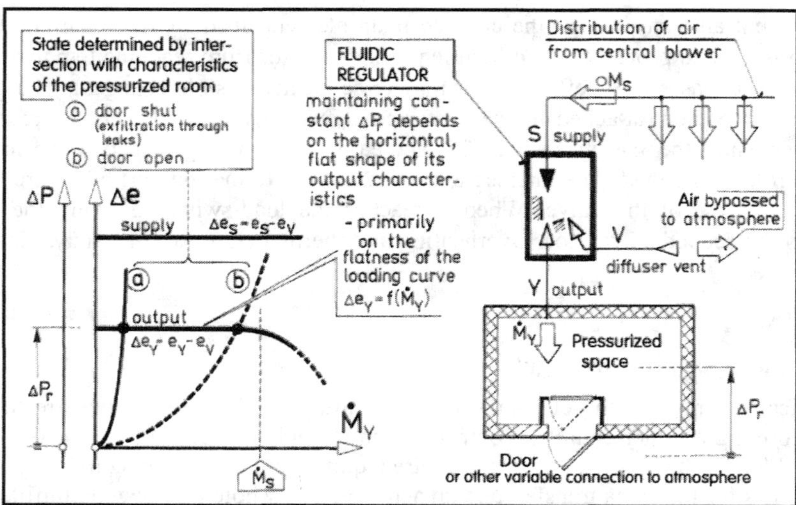

Figure 4.27 Use of the passive pressure regulator diverter valve for keeping a constant pressure in a space with an unpredictably variable opening into the atmosphere.

(1 in Figure 4.26) with a removed splitter tip, already exhibited a surprisingly flat spillover part of its loading characteristic. It also did not have any tendency to load-switching, an important property for the operation conditions with the output flowpath often fully closed. The development steps are shown in Figures 4.26, 4.28, and 4.31. They resulted in a valve with geometry presented in Figure 4.30. In addition to the proper behavior, the valve is also spatially very compact. It finds applications in circuits that require a constant pressure in a downstream space (Figure 4.27), irrespective of opening or closing the door that connects this space with atmosphere, or, according to Figure 4.29, in a space with a turn-down valve downstream from it. Apart from the loading curve, a complete diagram of the output characteristics contains also the dependence of the supply pressure (or energy) drop on the output flowrate. An important aspect of the device geometry

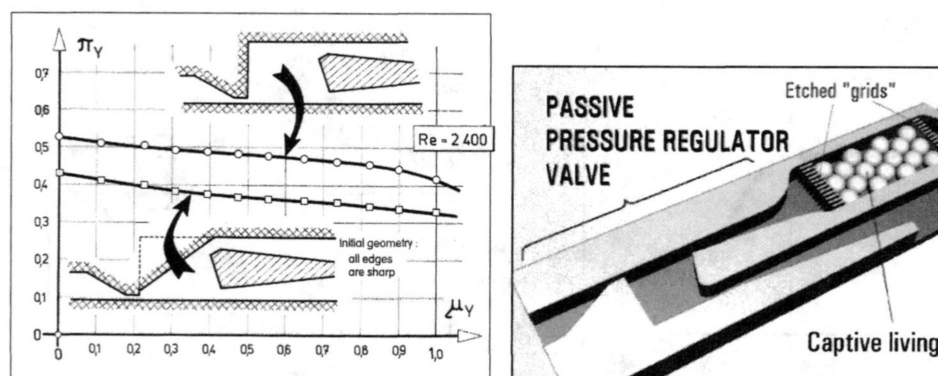

Figure 4.28 (Left) One of the steps in the development of the pressure regulator geometry (3 in Figure 4.26). Rather surprisingly, the pressure recovery was much better after the removal of the original secondary attachment wall.

Figure 4.29 (Right) An application of the regulator valve to keeping a well-being of biological objects that would otherwise react adversely to the pressure changes caused by variations of the culture medium inflow and metabolite removal flowrate (controlled here by the colliding-jets valve, Figure 4.35).

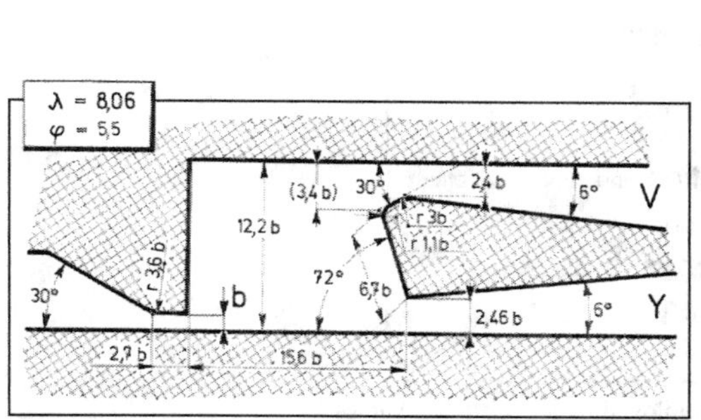

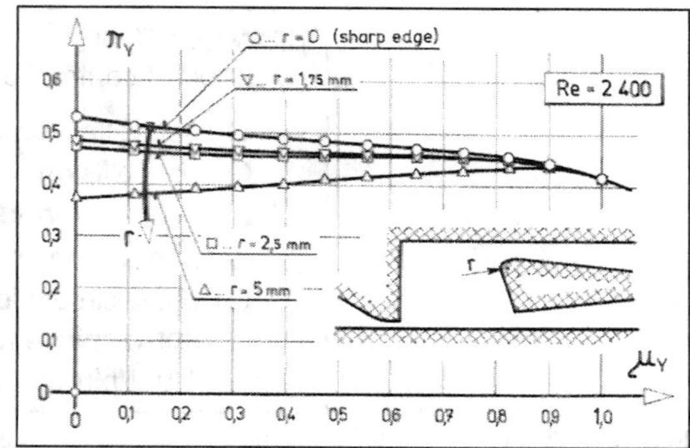

Figure 4.30 (Left) Geometry of the passive pressure regulator valve.

Figure 4.31 (Right) An important step in the development of the regulator was the adjustment horizontal, zero slope of the loading curve by rounding of the splitter nose.

from Figure 4.30 is that the supply curve is also in its spillover part almost ideally flat. This may be employed using this valve as an impedance isolator, insulating the upstream source from any changes taking place farther downstream. Again, as in Figure 4.23 this pure fluidic open-loop control is advantageous in its simplicity: the absence of sensors, drives, connecting lines and, for example, electric sources used in traditional control systems.

4.2 FLUIDIC CONTROL ACTION: ACTIVE VALVES

A very interesting behavior is obtained in active fluidic devices. In them, the strong main flow undergoes substantial changes in response to quite weak input stimuli. This chapter concentrates on those cases where the input is of a fluidic character (i.e. consists of some weak fluid flow). Of course, a similar effect may be obtained with mechanical, electrical, or other sort of inputs. These cases, of course, belong in a discussion of sensors and transducers in Chapter 5.

The fact that a large flow can react to a weak stimulus at all is an important property of fluid flows. Not all flows exhibit such behavior; on the contrary, identification and use of such "weak spots" with consequences for the whole flowfield are exacting activity requiring a lot of knowledge about hydromechanics.

There is a seemingly safe way toward this goal. It is the use of some hydrodynamic instability, where indeed, if the conditions are adjusted on the very verge before some spontaneous transition is about to take place, a feeble tripping activation suffices to produce momentous changes. A typical phenomenon requiring only a weak input is the transition from laminar flow into turbulence. If the flow is adjusted just below the critical transition Reynolds number, the change is known to take place in response to a weak acoustic disturbance. In fact, practical experience with such instabilities as the basis for the control action are rather disappointing. The flow in the critical state is often simply too sensitive, with the transition happening without the control signal in response, for example, to the vibration of the device support. This may be avoided by adjusting the state farther away form the critical condition, but these adjustments of the critical Reynolds number is generally rather tricky business sometimes with unpredictable results, especially because of the temperature dependence of fluid viscosity.

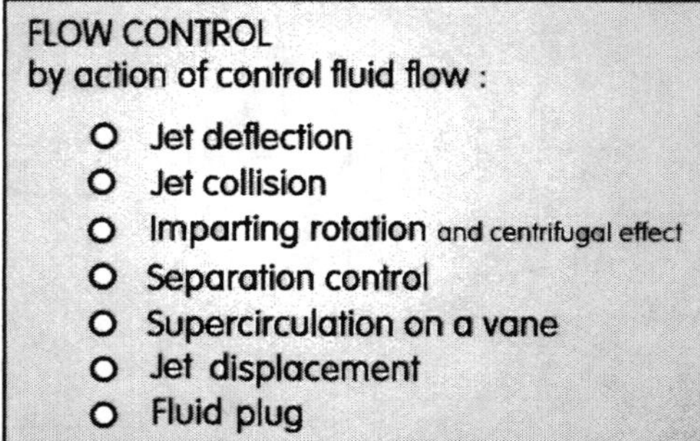

Figure 4.32 List of most important flow control principles employed in active fluidic valves.

The list of other, more recommendable mechanisms is in Figure 4.32. Not all the principles listed are equally useful. The jet deflection principle (Figure 4.33) is so important that it is presented in detail in Section 4.3. On the other hand, the last four principles are there more or less for completeness or for their usefulness in special situations. They are little used and perhaps little known. Some of the principles are suitable for continuous, proportional control while others, most characteristically the control by flow separation, are suited only for the binary, ON-OFF operation.

The criterion of performance is the amplification gain. It is generally not a constant value. Besides its dependence on the magnitude of the supply flowrate and the loading of the device, it varies due to the nonlinearity of the *transfer characteristic,* the dependence between the input (control effect) and output values. The quantity usually presented as characterizing these devices is therefore the differential gain (of which, naturally, the designers and manufacturers are inclined to present the maximum achievable value). It is necessary to distinguish the flow gain (compare with the definition of the values in Figure 4.34):

$$K_\mu = d\dot{M}_Y / d\dot{M}_X \tag{4.1}$$

the specific energy gain

$$K_\eta = de_Y / de_X \tag{4.2}$$

which, under the usual prevalence of enthalpic component in the specific energy and the often encountered incompressibility (or near-incompressibility), $v \sim \text{const}$ may be equated with the pressure gain

$$de_Y / de_X = d(v P_Y)/d(v P_X) = dP_Y / dP_X = K_\pi \tag{4.3}$$

and the power gain

$$K_\alpha = K_\mu \, K_\eta \tag{4.4}$$

If any of these (absolute) values is larger than 1.0, then the valve may be described as a fluidic amplifier. The resultant overall gain of two amplifiers in series is the product of the two component gains (will be shown later in Figure 4.77). The amplification property is essential for several very useful concepts, such as building self-excited fluidic oscillators. At any rate, high gain is usually desirable. Some of the principles listed in Figure 4.32 can achieve higher gains, others, such as the jet collision and displacement, are less effective from this point of view. The general problem of microfluidics is the usual tendency of all these gain quantities to decrease with decreasing Reynolds numbers. It is not uncommon to encounter microfluidic valves operating with gain values less than 1.0. Characteristically, more suitable for application at low Reynolds numbers are the principles that tend to exhibit lower gain.

4.2.1 Jet deflection

If a flow is converted in a jet in a nozzle and then captured and reconverted back into a channel flow by a collector a short distance downstream (Figure 4.33), it is easy to control the captured flowrate by jet deflection. Before the advent of no-moving-part fluidics, this principle had been used (e.g., by Askania in Germany) in pneumatic control systems with mechanical deflection using turnable or deformable nozzles. Warren, one of the founding fathers of present-day fluidics, reportedly thought of the idea while spraying his lawn. The advantages, which gained high popularity to this principle, are the very high achievable power gain and no particular demands on the control pressure. In fact, because of the conversion of pressure into kinetic energy in the supply nozzle, the pressure in the main jet may be lower than the atmospheric pressure. The valve may be then controlled by simply opening and closing an inflow into the control nozzle from the atmosphere. This is a welcome property when compared with the centrifugal vortex valves, collision, or plugging principle which in their basic form generally demand a control pressure higher than the pressure in the controlled fluid. The disadvantage of the diverting action is the inability to stop or turn down the flow.

A problem with this type of valve is the sensitivity to loading. The classical solution to this problem is *venting* of the cavity between the nozzle exit and the collector entrance: it means opening this cavity to the atmosphere when working with air or, when working with a liquid, connecting the cavities with the main oil tank. The vent provides a path out from the device for the fluid that the blockage by the load prevents from entering the collector. The minimum number of the basic valve terminals (Figure 4.34) is four. Two of them, through which passes the main fluid flow, are of large cross-section: the supply terminal S and the output Y. This flow is modulated by the small control flow through the small cross-section terminal X. This prevents some fluid from reaching the output Y by diverting it into the vent V. Not all jet-type valves are vented. Venting may be not really necessary in two-exit or multiexit diverter valves where there is another way out

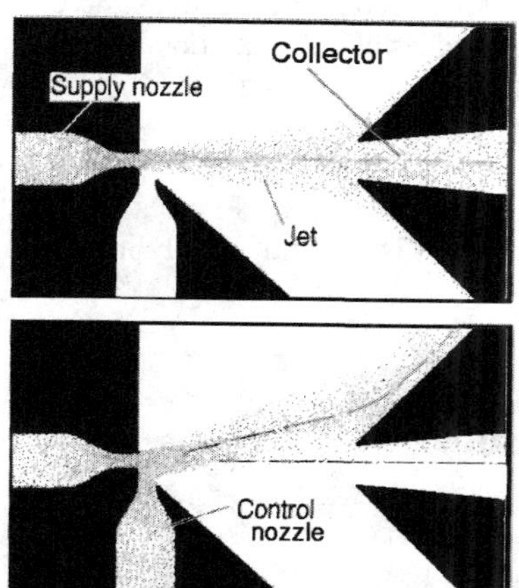

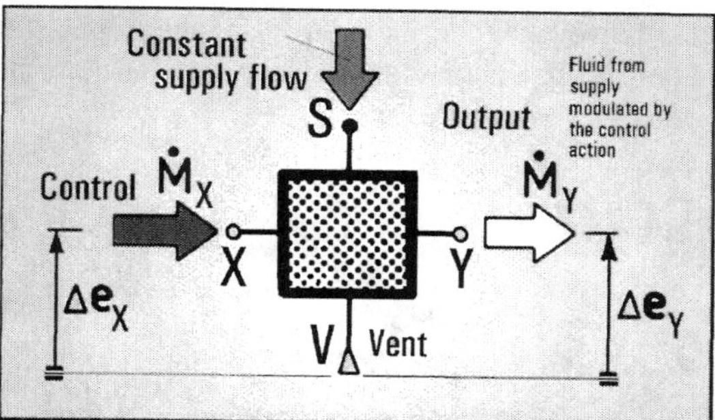

Figure 4.33 (Left) Control by deflecting the main jet by action of the control flow is by far the most important among the principles in Figure 4.31.

Figure 4.34 (Right) Schematic representation of the basic case of an active valve and its terminals. The supply S is connected to a source of pressurized fluid, X brings in the control signal, the output terminal Y is connected to the load. The vent V is used to discharge unneeded fluid. The main, high-power flow passes from S to V. The control action determines the admission of the fluid into the output or vent. A number of alternative variants exists differing in the number or existence of terminals.

from the valve. It may be also undesirable when working with different fluid species (e.g., in chemical applications) where it is demanded to avoid the leakage of the fluid through the vent and the indiscriminate mixing (resulting in possible uncontrollable mutual contamination) in the common vent space.

Because of their importance, Section 4.3 is devoted to the jet-deflection diverter valves.

4.2.2 Colliding jets

There are not many choices to select from if the valve is required to turn down the flow rather than divert it. The principle of the colliding flows is among the few possibilities. Its disadvantage in the basic form discussed here are the rather low values of achievable gains, both flow and energy ones. The advantage, on the other hand, is that this is the only principle capable of stopping the controlled flow completely. Also, it is a principle that operates well even at very low Reynolds numbers.

The flow blockage or turning down action removes the need for the vent. In the schematic representation of such valves (Figure 4.35), there are only three basic terminals. The main flow from S to Y is here varied by the control action input at X. Note that Figures 4.35 and 4.36 assume that the load is connected downstream from the valve, to the output terminal Y.

Visualization of the flow in an example the colliding-jets valve is shown in Figures 4.37 and 4.38. Although the differential gain may be quite good across a part of the operating range, as implied by experimental data in [2], it is apparent that to overcome the main flow, the opposing control flow has to posses a comparable momentum. This is not a principle exhibiting much amplification effect. In fact, to get the full closure as shown in Figure 4.38 the control jet has to be more powerful than the flow input on the control side. This may be improved by combining this valve in a two-stage cascade layout with some other flow amplification principle. Nevertheless, the valves with colliding flows are usually used in situations where the high gain is not an essential factor, especially in small-scale devices fulfilling the flow closure task. The small Reynolds number and the consequent viscous damping are here actually welcome to suppress the otherwise rather pronounced tendency towards instability and self-excited oscillation, as documented by the use of essentially identical configuration as a dynamic mixer (Figures 3.125 and 3.126).

The load in which this valve is to control the flowrate may be connected not only in the downstream location, but also upstream, as shown in the application example in Figure 4.39. In this case, however, the usual naming of the supply terminal S as used in Figures 4.35 and 4.36 obviously loses its appropriateness, not only because this terminal is then not supplying a constant

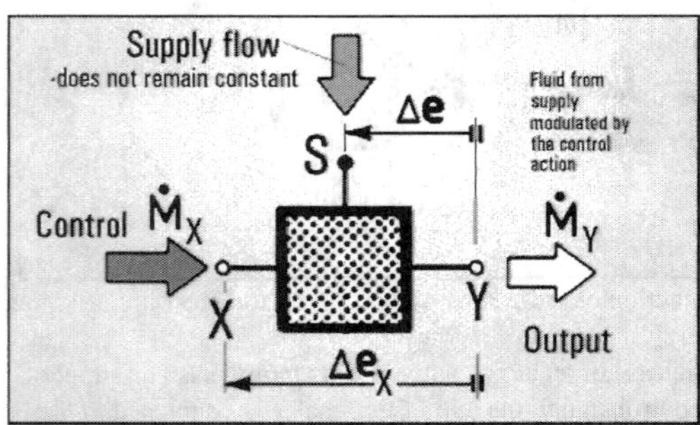

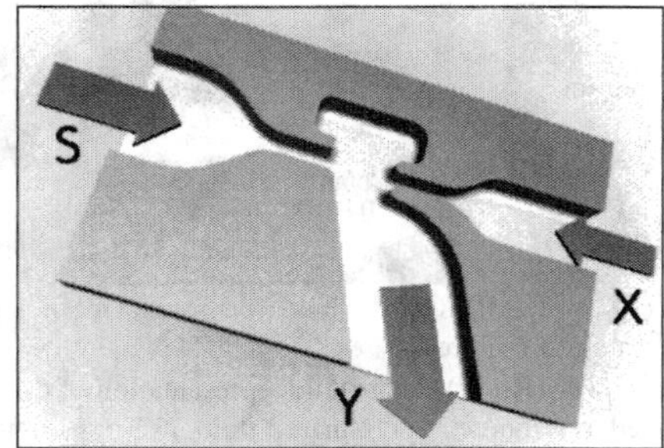

Figure 4.35 (Left) The simplified block diagram (compare with Figure 4.33) of unvented turn-down valve.

Figure 4.36 (Right) A three-terminal turn-down valve based on the principle of colliding jets.

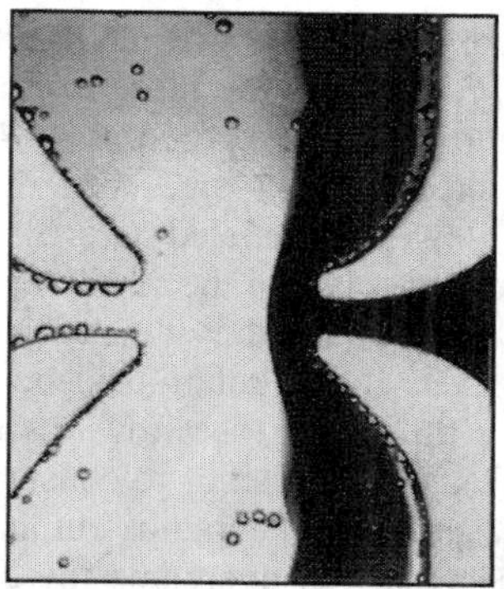

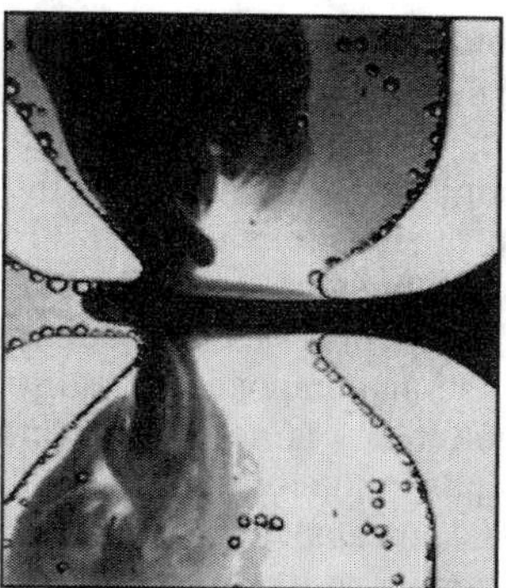

Figure 4.37 (Left) Water model tests of the colliding jets principle. With this weak control flow (visualized by addition of black dye) the valve allows an almost unrestricted flow of the (transparent) fluid from the supply nozzle at left.

Figure 4.38 (Right) A powerful control flow (black) can almost completely stop the flow of the transparent fluid from supply nozzle. It can do so even at very low Reynolds numbers. The only problem is the rather poor flow gain.

constant fluid flowrate, but because the terminal connected with the load should be the output Y. This different point of view is expressed by the different naming of terminals in Figure 4.39 and in the schematic representation in Figure 4.40.

4.2.3 Centrifugal action: vortex valves

The other available principle of turning down the flow in a no-moving-part fluidic valve is based on essentially the same principle as in the vortex diodes, discussed in Section 3.5.2. The basic configuration, shown in Figure 4.41 (and schematically represented in Figure 4.42), besides being

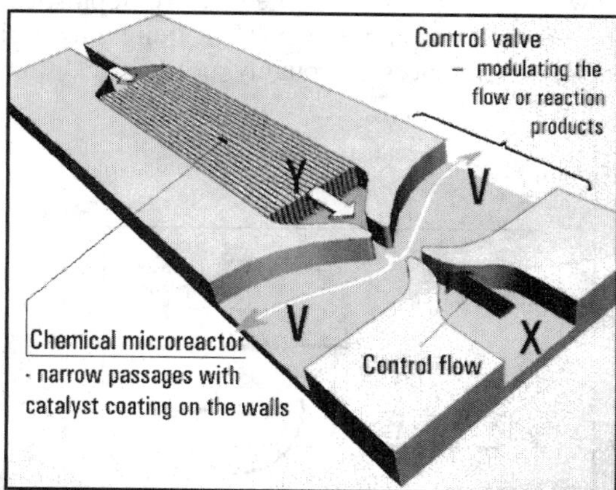

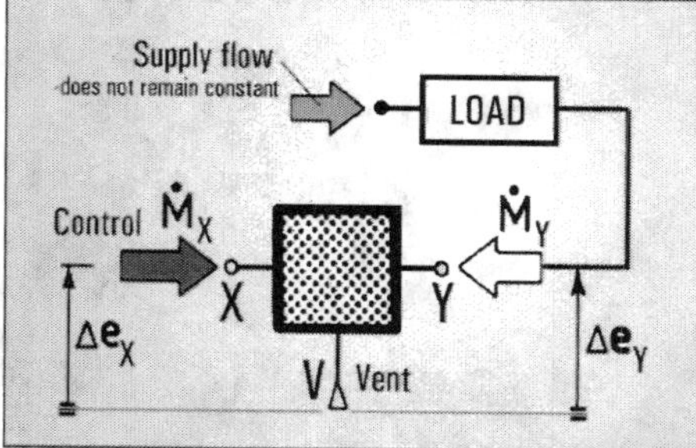

Figure 4.39 (Left) An example of the colliding jets valve used for control of reaction products flow at the exit of a microreactor. The fact that there are physically actually two vent channels—one on each side—is irrelevant. Both perform the same task as if there were just a single vent path out from the valve.

Figure 4.40 (Right) If the load (such as the reactor in Figure 4.39) is positioned upstream from the valve, then the basic schematic representation from Figures 4.35 and 4.39 ceases to be really fitting. The load should be always connected to the output terminal Y irrespective of the flow direction and it should be the vent V which is there for the discharge of unneeded fluid. The naming of terminals corresponds to Figure 4.36.

operated in only one flow direction, differs from the diodes in having three terminals, corresponding therefore to the schematic Figure 4.35. Apart from the more common "turn-down" version, Figures 4.41 and 4.42, there us also the "turn-up" valve, Figures 4.44 and 4.45. The naming od terminals again assumes the load connected downstream, even though it may be equally well in the locations upstream from the main flow inlet, labeled here S. This inlet injects the main flow into the vortex chamber. If the fluid does not rotate in the vortex chamber, the valve represents only a small resistance. Turning the flow down is caused by the control flow imparting a tangential motion component to the main flow. Even if the rotational speed at the vortex chamber circumference is not high, it increases as the fluid progresses towards the center of the chamber, since, as in the reverse flow state in the diodes, the fluid tends to conserve its moment of momentum. This means rapidly increasing rotation speed as the radius diminishes. The flow through the chamber is then more difficult as it is necessary to overcome the centrifugal forces acting on rotating fluid particles.

As in the diodes, this momentum conservation tendency is limited by the friction taking place in the boundary layers in the top and bottom flat walls of the chamber. The flow also cannot be turned down completely—first, because this would mean stopping the very motion on which the centrifugal effect depends, but also because some fluid (compare with Figure 3.98) succeeds in reaching the central outlet through the boundary layers. The deteriorating influence of the boundary layers on the valve performance becomes more apparent at low Reynolds number, where the boundary layers are relatively more thick. Although the vortex valve principle can be quite useful at Reynolds numbers of the order of hundreds where many present-day microfluidic valves operate, it is less useful for very low Reynolds number microfluidics.

The other disadvantage is the need for a high-pressure control signal required to generate the rotation in the vortex chamber. Usually, the pressure at the control inlet X has to be higher than the pressure in the supply terminal. There are designs that circumvent this drawback, at a price of a more complex layout, but in the simplest design corresponding to Figure 4.41, it is necessary to provide somehow the suitable higher-pressure source of the control fluid.

Performance obtainable in these simplest layouts, usually characterized by the achieved turning down ratio (OPEN to CLOSED output flow, Figure 4.43), is limited. More sophisticated designs can achieve the ratio of flows up to 16 or even more, which is quite impressive considering that the corresponding ratio of pressure differences is roughly proportional to the

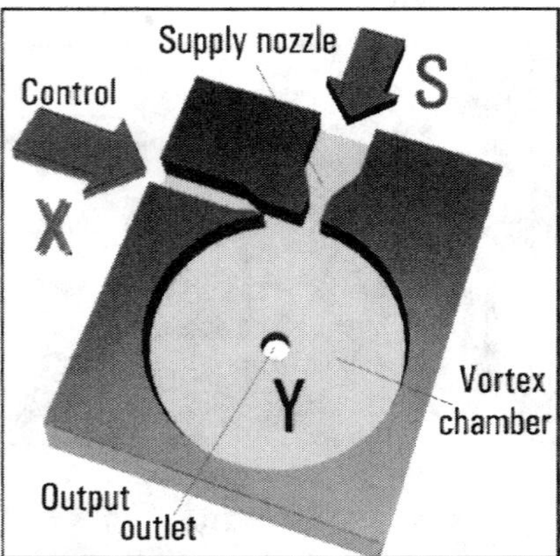

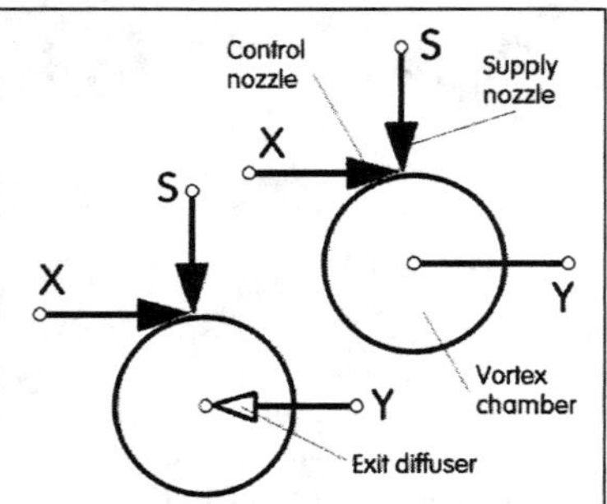

Figure 4.41 **(Left)** The vortex amplifier valve (or "vortex triode," in contrast to the diode, Figure 3.96). Relatively easy radial flow from S to Y becomes much more difficult if the control flow from X imparts some tangential momentum, since the rotational speed increases fast towards the central outlet and the flow has to overcome centrifugal effects.

Figure 4.42 **(Right)** Schematic representation of the vortex valves in circuit diagrams.

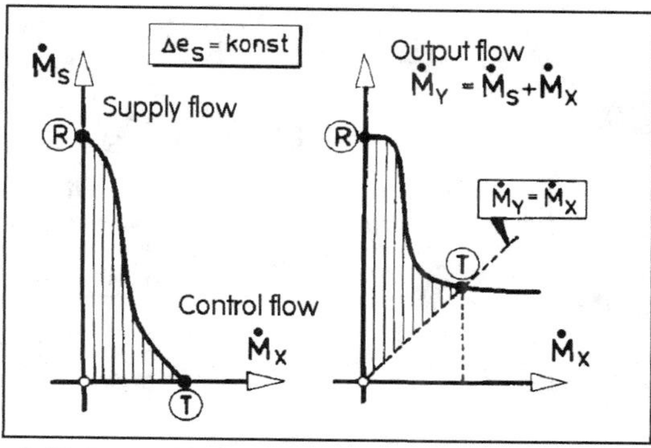

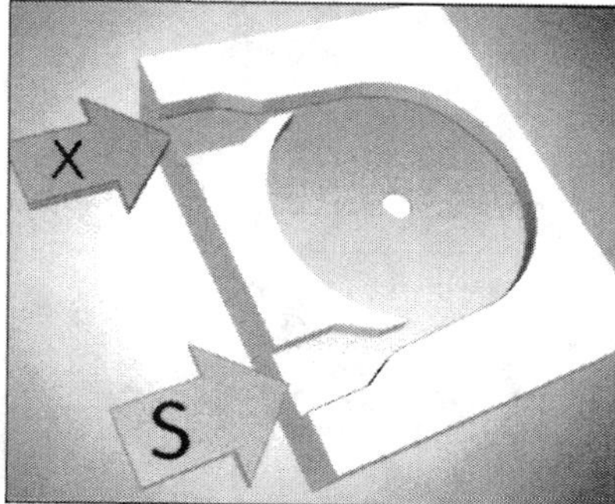

Figure 4.43 (Left) Schematic presentation of typical flow transfer characteristics of the usual (turn-down) vortex valve. The slope (= differential flow gain) is negative, the main flow decreases with increasing control flow $\dot{M}_X$.

Figure 4.44 (Right) The turn-up vortex valve: rotation in the vortex chamber is decreased by the control action.

square of this value. Typical features of such designs are the large number of inlets on the vortex chamber circumference (Figure 4.46) and the diffuser in the exit (Figures 4.42 and also 4.49). The latter is useful for decreasing the pressure drop across the valve in the **OPEN**, radial flow state if there is a small exit diameter d, which is otherwise a desirable feature in the (nominally) **CLOSED** state.

It is helpful to follow the action of the usual turn-down valve on the transfer characteristics; the dependence of the output changes on the input. Schematically shown in Figure 4.43 is the flow transfer characteristic $\dot{M}_Y = f(\dot{M}_X)$ together with the similar dependence of the flow in the supply terminal S. The two curves differ in the addition of the control flowrate to make the output value. Because of this addition, the output flow can never decrease to zero. Both curves pass through two important points, the radial state point R and the point of the tangential state T. The mass flowrate $\dot{M}_{RS}$ in the **OPEN** radial state R is the reference condition used in nondimensionalizing the flowrates

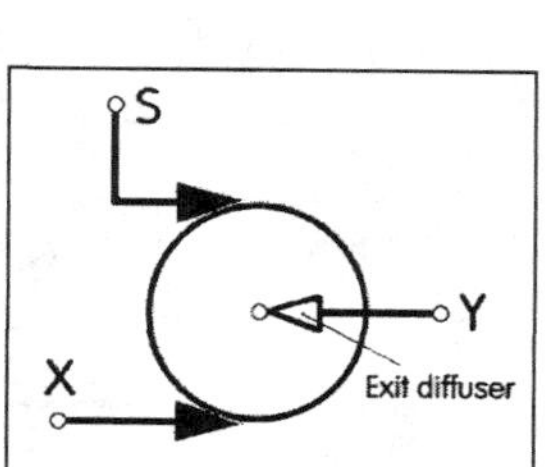

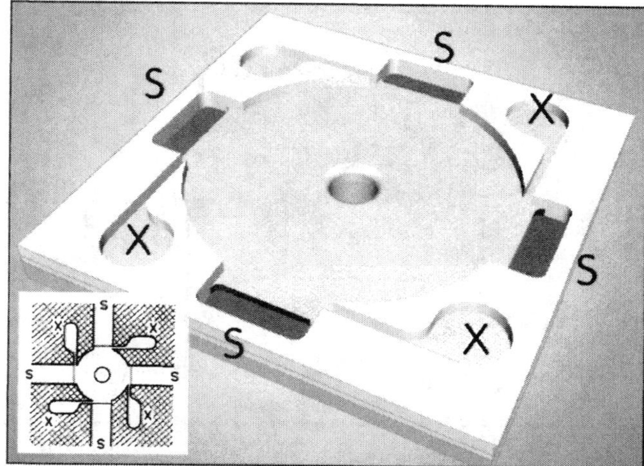

Figure 4.45 (Left) Schematic symbol of the turn-up vortex valve.

Figure 4.46 (Right) Multi-inlet vortex valve designs eliminate the flowfield asymmetry, which in the simple versions of Figure 4.39 (or Figure 4.42) has a detrimental effect on the performance.

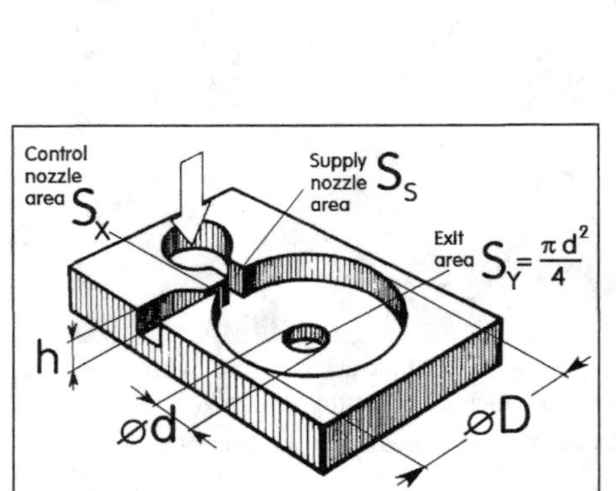

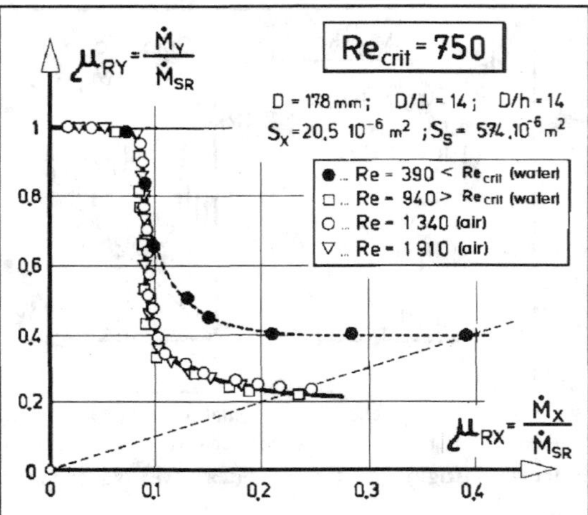

Figure 4.47 (Left) Basic dimensions of the simplest vortex valve.

Figure 4.48 (Right) Typical flow transfer characteristic of the simple valve layout: performance decreases significantly if the valve is operated below the critical Reynolds number (diagram using data from Wormley and Richardson [3], with permission of ASME).

rates. The usual Euler similarity (compare with Figure 4.8) applies: as shown for the valve from Figure 4.47 in Figure 4.48, all transfer characteristics of a particular vortex valve obtained at different Reynolds numbers collapse into a single universal curve in the coordinates. A significant deviation from the similarity is observed when the Reynolds number evaluated from the chamber diameter D (Figure 4.47) becomes lower than the critical value $Re_{crit} \sim 750$.

The tangential state T is used as the reference for the CLOSED regime. The turn-down ratio mentioned above is evaluated as the inverse $1 / (\mu_{RY})_T$ of the relative output flowrate in the

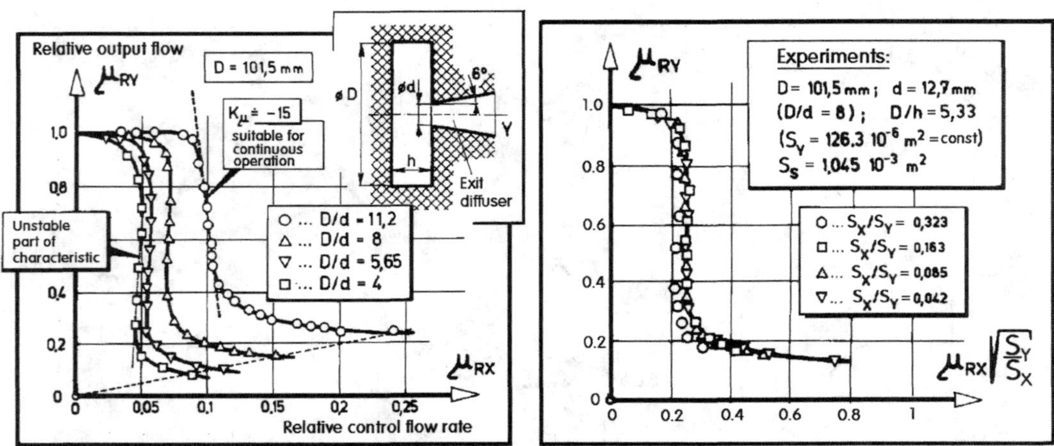

Figure 4.49 (Left) The effect of the exit diameter on the slope of the flow transfer characteristics. Very large, in fact infinite flow gains are attainable with very small outlet size, but this makes the valve prone to instabilities.

Figure 4.50 (Right) Universal presentation of the flow transfer characteristics of a vortex valve with different areas of the tangential control nozzle. (Both Figures 4.49 and 4.50 use data from [3], with permission of ASME.)

T state. For the valve for which the characteristics in Figure 4.48 were evaluated, the turn-down ratio was 4.5 in the supercritical regime. At Re = 390, a value that may be said to be typical for current microfluidics, the ratio decreased to 2.5. The data plotted in Figure 4.49 demonstrates that it is possible to get much higher flow turn-down ratio by choice of the diameter ratio D/d, though this can lead to slope of the transfer characteristic becoming too steep and unsuitable for continuous operation. The other diagram, Figure 4.50, documents that the effect of the control nozzle exit area is basically a horizontal dilatation of the transfer characteristic.

4.2.4 Separation and supercirculation

These two valve control principles are included here mainly to make the survey of control action complete. They are more typical for large-scale fluidics. In microfluidics they may be useful in situations where a very large flow to be diverted and small depth of cavities results in a very small aspect ratio. For deflecting the very wide jet in a shallow cavity, the classical perpendicular control jet action is very ineffective. Such wide jets may be effectively deflected by a mechanical, rotated vane positioned inside the jet, or, if the width is very large, perhaps by a whole louvre consisting of a cascade of vanes in parallel. The mechanical rotation of the vanes may be replaced by fluidic control that produces the same effect while the vane is a fixed, nonmoving vane.

One possibility is to set the vane at an angle of attack so that it deflects the jet if there is no control action. The control flow issuing, for example, from a slit on the suction side of the vane, similar to a spoiler on an aircraft wing, causes a separation of the jet from the inclined vane. In the example in Figures 4.51 and 4.52, only the suction side of the stub "wing" is exposed to the jet, which in the absence of the control signal is not deflected by the inclined pressure side.

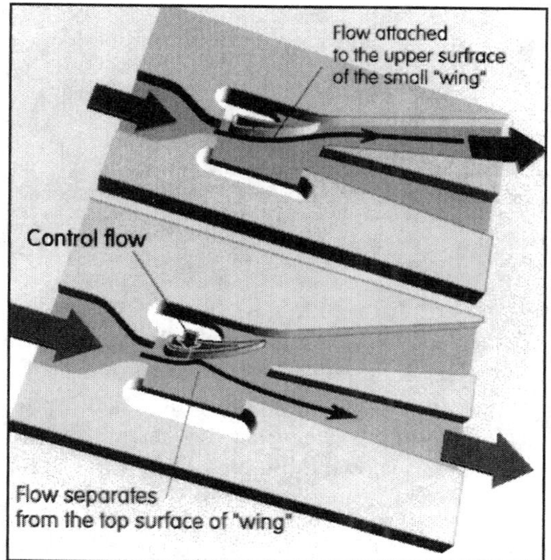

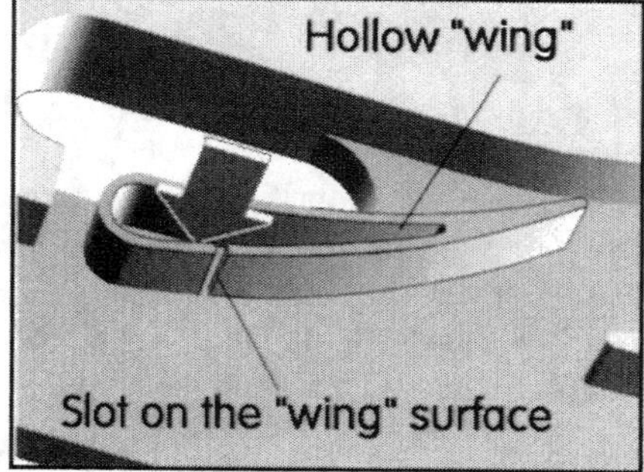

Figure 4.51 (Left) The separation control of the flow past an inclined little stub "wing" exposed on its suction side to the main jet which it deflects. The input signal issuing from the small slit causes separation of the jet from the wing surface so that the jet ceases to be deflected.

Figure 4.52 (Right) Detail of the "wing". The input control flow is admitted into its internal cavity.

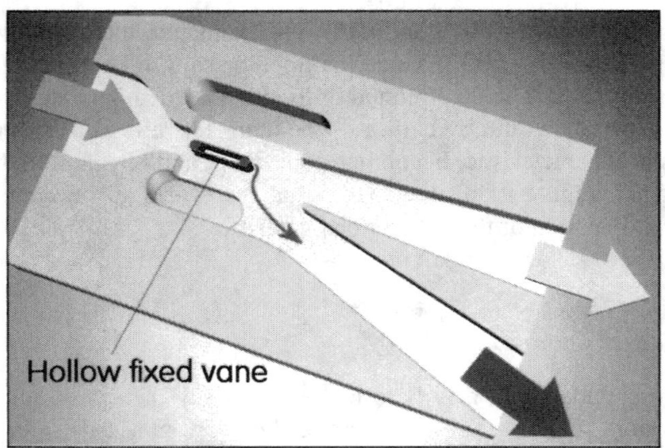

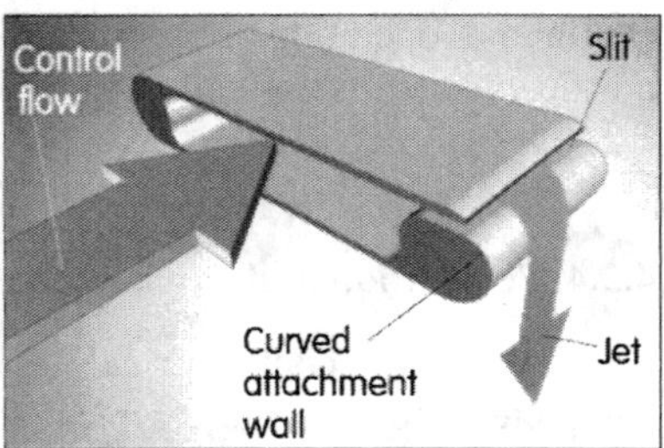

Figure 4.53 (Left) Deflection of the main jet flow past the fixed vane, caused by circulation.

Figure 4.54 (Right) The circulation on the vane is produced by the control jet issuing from the vane and attaching by the Coanda effect the round trailing edge of the vane.

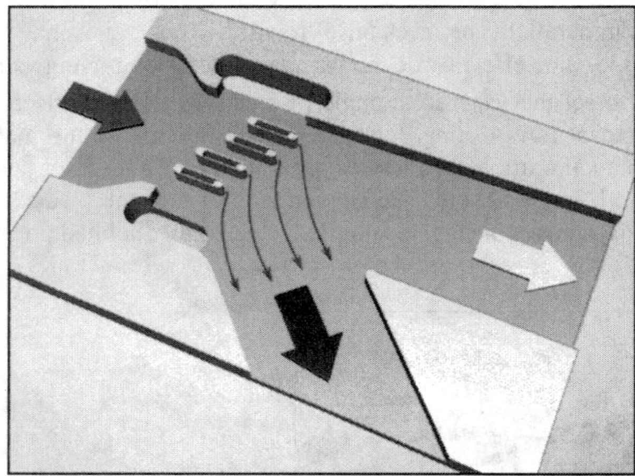

Figure 4.55 Cascade of fixed vanes with supercirculation control by the Coanda-effect jets at the vane trailing edges may deflect effectively even very wide flows.

The deflection of the jet by the inserted vane may be attributed to the circulation past the vane. This increases with vane inclination, or may be generated by the fluidic control action on a vane remaining at a zero attack angle. Particularly effective is the supercirculation produced by the control flow leaving the vane at a location and direction where it forcibly places the effective stagnation point. An example of this supercirculation control using the Coanda effect of the control flow attachment to the rounded vane trailing edge is shown in Figures 4.53, 4.54, and 4.55.

4.2.5 Displacement

In this also rather exceptional control mode, again based on jet type flows, the supply nozzle is positioned at a wall and directed so that it a wall-jet (Figure 3.62), following the wall toward the collector. The control flow is admitted between the wall and the wall-jet, which it forces to move

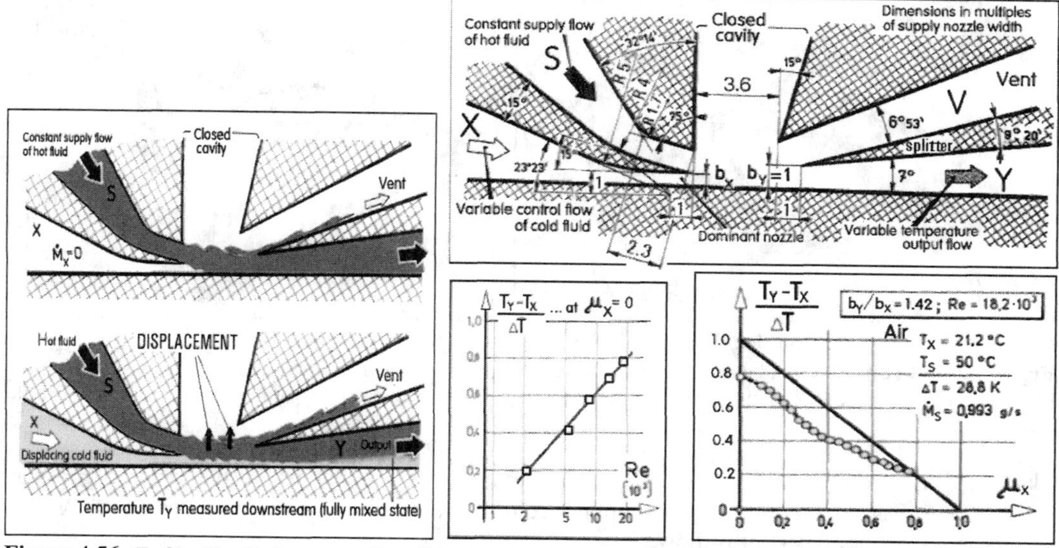

Figure 4.56 (Left) The displacement of a wall jet from its wall. In this particular application, the hot wall jet is dsiplaced by cold fluid. The result is a change of temperature in the output collector.

Figure 4.57 (Top) Geometry of the valve used in scaled-up experimental investigation of the control by displacement.

Figure 4.58 (Bottom center) A complicating factor: temperature in the exit decreases at zero control flow in dependence on the magnitude of the main flowrate, here presented in terms of the main flow Reynolds number .

Figure 4.59 (Right) Theoretical linear decrease of the exit temperature and an example of experimental results.

farther away from the wall. There are several possibilities how to utilize this displacement of the main jet. A collector may be placed with its entrance a certain distance away from the wall, so that it captures the displaced flow. In other cases, use is made of a physical or chemical difference between the main and the control fluids. The output collector is at the wall and the mixture in it, formed from the two fluids, changes is averaged composition in dependence on the magnitude of the displacement. As an example, Figures 4.56 to 4.59 present a test and its result for the case of hot wall jet displaced from its wall by cold fluid flow. Ideally, the increase in the control flowrate results in inversely proportional decrease of exit temperature (Figure 4.59), which the actual data only crudely approximated.

4.2.6 Fluid "plug"

The blocking of a channel by stationary fluid is a concept which was found to operate particularly at extremely low Reynolds numbers [4]. The control action (Figures 4.60 and 4.61) is based on blockage of the entrance into the collector by the control fluid, which may be for this purpose chosen to be of high viscosity. It forms inside the entrance a volume of nearly stagnant fluid held there by viscous forces. At very low Re, this actually works even with gas since the viscous effects then dominate. Operation with mutually immiscible control and main flow liquids is, of course, easier to set up. To switch the valve into the OPEN state, the control fluid may be sucked back into the control port. If for some reason the control fluid viscosity cannot be as high as to maintain it in place – and the Reynolds number is not very low, the fluid "plug" does not stay in

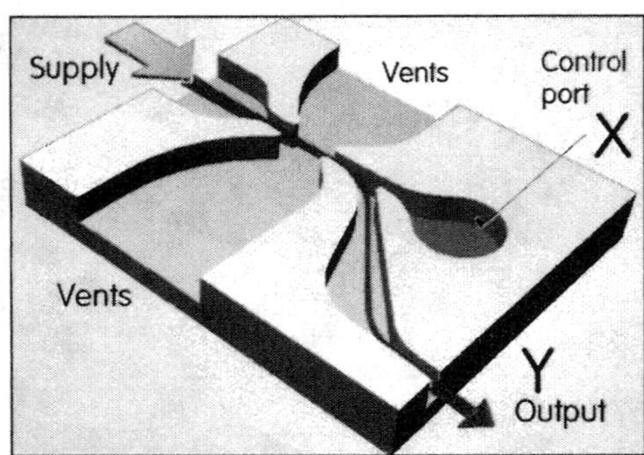

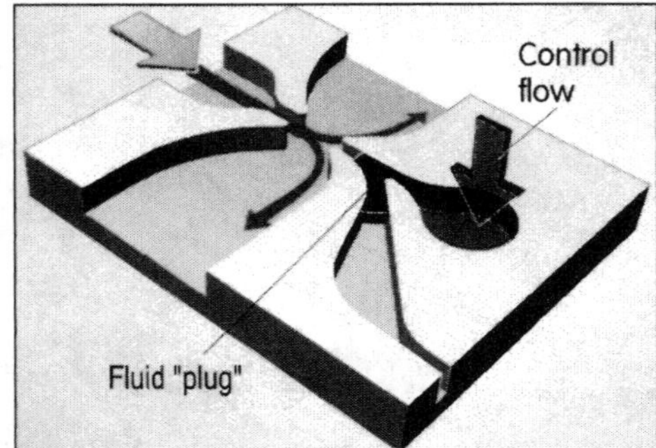

Figure 4.60 (Left) The fluid plug valve in the OPEN state with zero control flow.

Figure 4.61 (Right) The fluid plug valve in the CLOSED state. Very strong viscous effects at the low Reynolds number keep the control fluid in the collector entrance, which the main flow cannot enter so that it leaves the valve through the vents.

place and is continually removed into the output under the action of the driving pressure difference. It has to be continually resupplied from the control port. Switching into the OPEN state occurs if the resupplying control flow stops.

4.3 JET DEFLECTION

Jet deflection control of flow in fluidic valves, so important and so often used, deserves being discussed in more detail, starting from the basic facts about the deflection mechanism and their consequences for valve design. To be specific, a particular geometry of nozzles was chosen without attempting an optimization and investigated first on its own and then in connection with several layouts of the collectors. If, according to Figure 4.33, there is a single collector opposite to the supply nozzle, connected with the output terminal, the deflection of the main jet can result in a decrease of the amount of captured fluid and consequent reduction of the output flowrate. Early researchers discovered that this simple basic layout—with the single collector and the single control nozzle—does not operate in the expected straightforward manner, mainly due to the Coanda-effect attachment to nearby walls. A solution to this problem was found in symmetric layouts, with two opposing control nozzles and two collectors side by side.

Instead of being just a source of trouble, however, the Coanda effect was gradually found to be actually very useful in specialized valve designs with attachment walls arranged so that the jet remains in the deflected position without any control flow. These bistable valves have become so important that they will be discussed in a dedicated part of this chapter.

The original idea of valves not influenced by a jet attachment to the wall was mainly developed with an intention to use them as fluidic operational amplifiers. The task was to design them so that they could exhibit linear transfer characteristics and very high values of the gain. Most known designs operate in turbulent regime. Very special later laminar flow designs are known, however. Their main advantage is the absence of the turbulent noise, which, being amplified together with the useful signal, is otherwise the limiting factor to the overall gain in an amplifier cascade. It is, of course, very useful in low Reynolds number microfluidics.

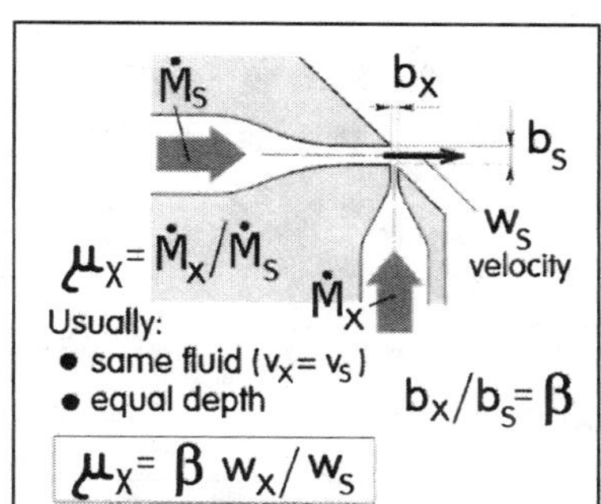

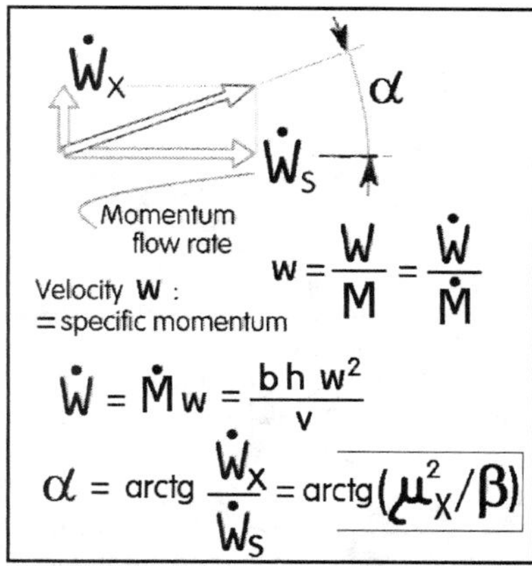

Figure 4.62 (Left) The interaction of two jets orthogonal flows. Conditions are characterized by the magnitude of the relative control flow μ_X.

Figure 4.63 (Right) Evaluation of the deflection angle by considering only the relatively easily evaluated effect of momentum flowrates of the two jets.

4.3.1 The deflection mechanism

The jet is deflected from its original direction by a sideways force produced by the control fluid. The conditions are characterized by the relative magnitude μ_X of the flow leaving the control nozzle (Figure 4.62). The deflection force may be decomposed into the dynamic and the pressure components. Of them, there are good chances of reasonable estimation of the dynamic force and hence evaluation of the deflection angle α. This is quite easy because the velocity profiles in nozzle exits (compare with Figure 3.34) are usually reasonably flat, so that a simple one-dimensional approach may be applied with reasonable accuracy. Jets are characterized by their momentum flowrate, an invariant quantity of a jet if far away from walls (and from the pressure effects the flow past walls tend to cause). If the main and control jets were interacting under the constant pressure conditions, the direction of the resultant jet formed by their coalescence could be computed from the final formula in Figure 4.63, where μ_X is the relative control flowrate and β is

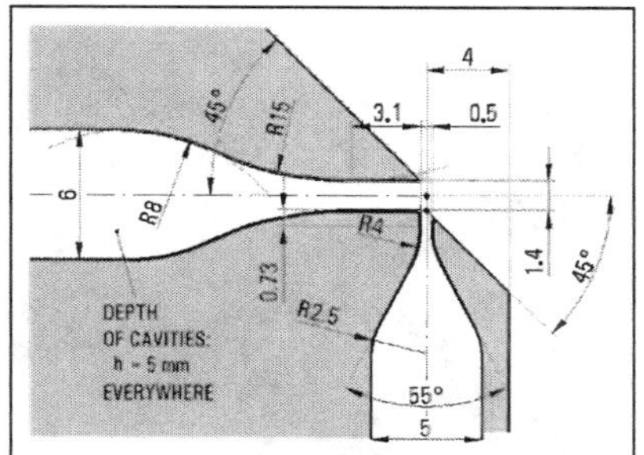

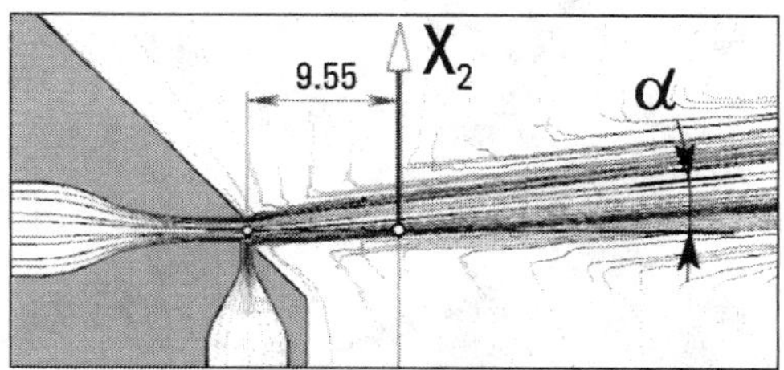

Figure 4.64 (Left) Geometry of the two nozzles (supply S and control X) in the investigated example of jet deflection. The size, large from the point of view of microfluidics, was chosen for easy laboratory measurements on a scaled-up model.

Figure 4.65 (Right) The jet deflection was evaluated from the position of maximum in the velocity profile measured at a distance (equal to 6.82 b_S) sufficiently far downstream from the complicated interaction region.

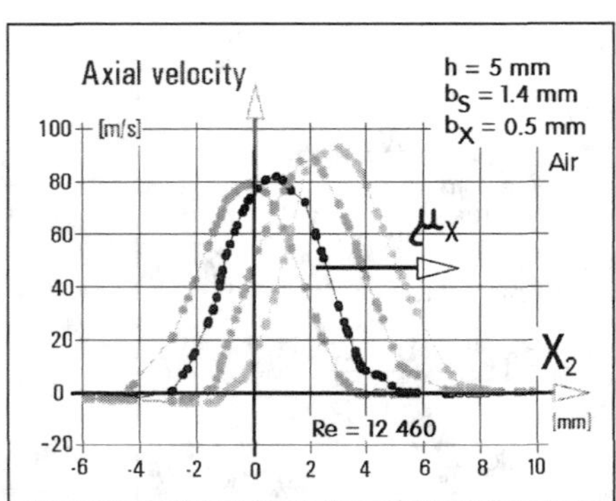

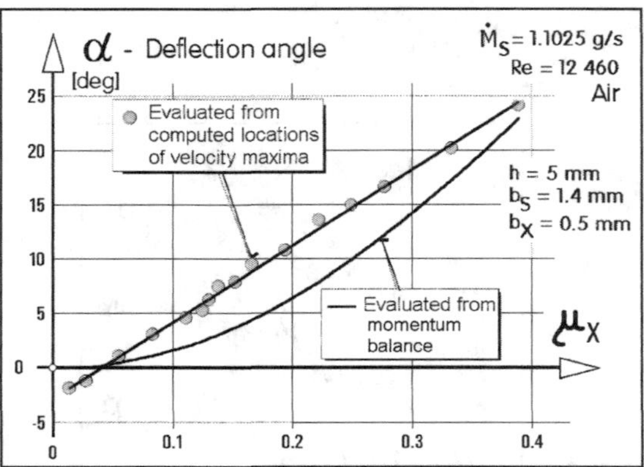

Figure 4.66 (Left) Examples of the velocity profiles along the line X_2 show the shift of the maximum when the relative control flowrate μ_X is increased. The Reynolds number chosen for suitability to scaled-up laboratory tests.

Figure 4.67 (Right) Comparison of the actual deflection angles α and the results of the momentum balance (Figure 4.63). The latter can provide only an order-of-magnitude estimate, especially at small angles α. The difference is due to the difficult-to-predict pressure field.

the ratio of b_X , the control nozzle exit width, and the main (supply) nozzle width b_S, assuming the same fluid and the usual equal height of the rectangular nozzle exits h (= the depth of the cavities).

Unfortunately, the actual deflection angles usually differ substantially from this simplified prediction. For the example pair of nozzles shown in Figure 4.64, using the evaluation method presented in Figures 4.65 and 4.66, the difference is seen in Figure 4.67. The evaluated profiles in Figure 4.66 are remarkable: even at the rather short distance from the two nozzles the two flows merge into what is effectively just a single jet; the profiles do not exhibit any tendency to having two velocity maxima corresponding to the two interacting components. Also remarkable is the visible increase of the maximum velocity with μ_X due to the addition of the control flow. The reason for the discrepancy in Figure 4.67 is the fact that the two flows (as is the case in typical fluidic devices) actually meet and interact before they become jets. The decisive role is taken over

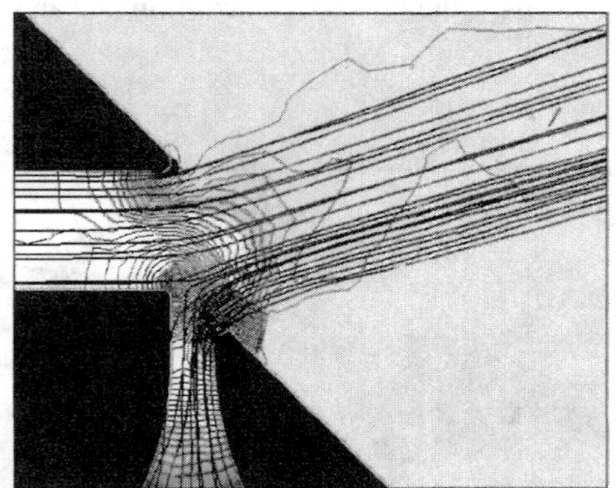

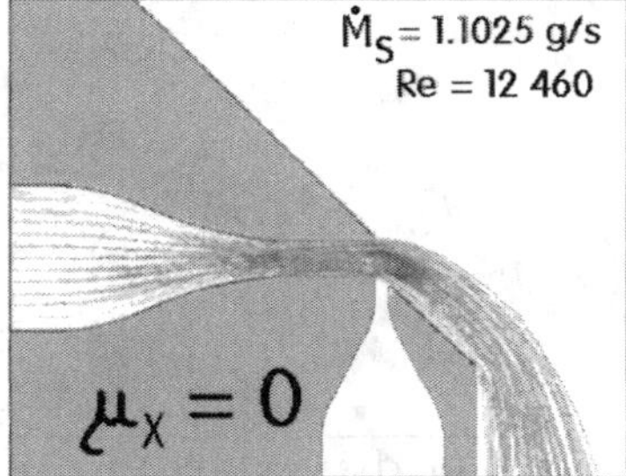

Figure 4.68 (Left) An example of computation results showing the complexity of the interaction flowfield: isobars and superimposed pathlines in the vicinity of the exits from the two nozzles at relative control flowrate μ_Y = 0.278.

Figure 4.69 (Right) Computed flow pathlines at zero control flow: despite the momentum of the jet, the Coanda effect is strong enough to deflect it and lead away from the valve.

by the complex interaction region in which, due to the presence of the nozzle exit walls, dominate the pressure forces (Figure 4.68). There is also the effect of the top and bottom wall.

Instead of the expected dependence from Figure 4.63, the actual deflection angle values are found in Figure 4.67 on practically straight line, which, perhaps at first sight surprisingly, does not pass through the origin. The negative deflections at zero control flow is caused by the Coanda effect attachment. Computation results in Figure 4.69 show a surprising effect of the attachment, considering the rather large momentum of the flow, which may be expected to resist changes of flow direction and pathline curvature by the pressure action (which is the force behind the attachment).

4.3.2 Simplest example of the jet-deflection valve

The two nozzles, retaining their geometry shown earlier in Figure 4.64 (including the constant depth of the cavities $h = 5$ mm), were used together with a single collector placed opposite to form a seemingly straightforward and guileless valve model corresponding to Figure 4.33. This valve is used here as an example for discussing the basic problems of this valve type. It can perform a useful task to control in a bypass mode the flowrate passing through a load. The load is connected to valve output Y, the flow in it is varied by diverting into the vent V a part of the supplied fluid flow coming through the supply terminal S at a constant flowrate. The collector, as usual, contains as its important part a diffuser (compare with Figure 3.66, note the change in the meaning of what is now the input X) for reconverting the kinetic energy of the captured jet into pressure energy. This conversion is necessary to avoid losses that would be otherwise caused by high velocity flow in downstream flow paths. Details of the geometry are shown in Figure 4.70. The entrance into the collector is wider than the main nozzle exit, to account for the jet width growth by entrainment of the outer fluid from the vent.

The flow controlling properties of such valves are presented by means of the flow transfer characteristic, the dependence of the output flow on the input flowrate. The slope of the curve is the differential flow gain K_{μ}, here negative since the jet deflection decreases the amount of fluid captured in the collector. Based on the idea presented above in Figure 4.33, there are several aspect of this transfer characteristic that may seemingly be reasonably expected. The relative output flow μ_Y is defined again (e.g., in analogy to the input flow in Figure 4.62) by relating the absolute output flow to the supply flowrate, held constant.

First, at zero control flow $\mu_X = 0$ it may seem natural that the undeflected jet reaches the collector entrance at its full strength – so that the relative output flowrate value should be near to $\mu_Y = 1.0$, depending on the choice of the collector entrance width. A smaller entrance would not capture the whole the jet, resulting in $\mu_Y < 1.0$. Evidently a better choice is, as in Figure 4.64, the entrance which is quite wide. This should encourage the jet-pumping mechanism (Figure 3.152) to entrain additional fluid, leading to $\mu_Y > 1.0$. However, a very wide collector is likely to cause problems when the valve is loaded by the connected load. Instead of continuing into the load, the captured fluid can find it easier to return inside the collector and leave through the low velocity parts of the entrance. A narrow collector would be therefore more suitable if it is expected that the load will possess a high dissipance.

Second, gradually increasing control flow may be expected to cause a gradual decrease of the relative output flow μ_Y. Because the collector entrance is at the downstream distance equal to $6.82\ b_S$, the same at which the velocity profiles were investigated in Figure 4.65, it seems quite easy to evaluate the flow transfer characteristic curve by integrating across the entrance width the profile at each relative input flow μ_X. The output flow, of course, depends also on the dissipance of the load as established already for the similar case of the jet pump in Figure 4.3. To simplify the investigation and make the isobaric results from Figure 4.65 applicable, the idealized zero

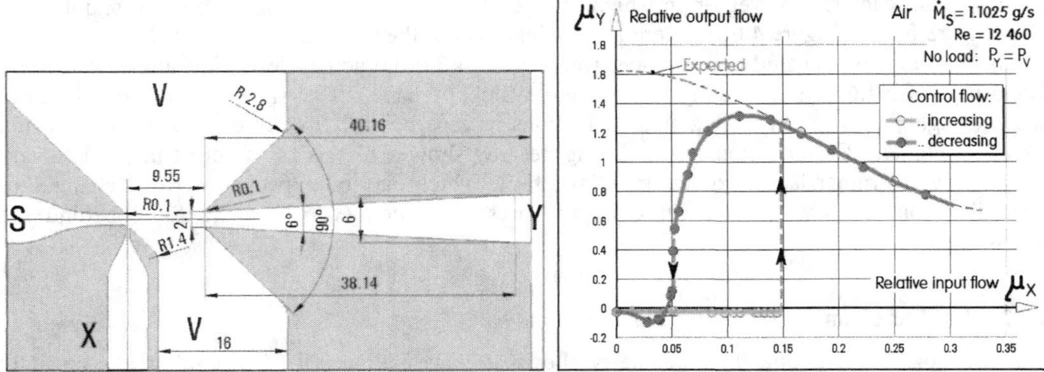

Figure 4.70 (Left) Valve with a single collector opposite to the supply nozzle built using the nozzles discussed earlier in connection with Figure 4.64. The collector entrance, 2.1 mm wide, is at the distance equal to 6.82 b_S from Figure 4.65. An important part of the collector is the 6 deg vertex angle diffuser.

Figure 4.71 (Right) The flow transfer characteristic, dependence of the relative output flow on the relative control (input) flow, of the simple jet-deflection valve from Figure 4.64 exhibits rather strange features, making it different from the expectations.

loading dissipance condition was now assumed. The valve behavior was investigated with the pressure in the output Y kept constant and equal to the pressure in the vents V. In other words, the output pressure difference ΔP_Y (measured between the output terminal Y and the vent V, the usual pressure reference) was maintained zero. The resultant flow transfer characteristic for this simplifying condition is plotted as the "expected" curve in Figure 4.71.

The actual results, also presented in Figure 4.71, are significantly different from this expectation. Instead of the simple monotonous decrease from maximum flow at $\mu_X = 0$, the output flow μ_Y is initially negative and remains constant. Only when the control flow reaches the value $\mu_X = 0.15$ does something happen. It is a very strong sudden response indeed: the output flow jumps to a value higher than the supplied flow. Thereafter the output flow slowly decreases. Only then, in this decreasing part, the characteristic follows the expected behavior, indicating that the idea on which the expectation was based was not completely wrong. What may seem to make the actual behavior even stranger is the different return path for decreased input flow. There is a strong hysteresis below $\mu_X = 0.15$. The response again fails to follow the expected line. At about $\mu_X = 0.1$ the output flow decreases and the return ends again by the negative output flow.

The reason for this complicated behavior is easy to identify: it is again, as seen in Figure 4.72, the Coanda effect of Figure 4.69. The jet is led by the attachment away from the collector. Moreover, as outer fluid is entrained into the jet from all available directions, a considerable negative flow is coming also through the collector. This is the reason behind the small negative values of μ_Y. This example of the differences between the real flows and the seemingly reasonable expectations is indicative of typical frustrations experienced by fluidic valve designers. It may be actually possible to find some useful application for the observed hysteretic behavior. Nevertheless in the early history of fluidics, when the interest was in signal processing and similar valves were developed to operate as fluidic signal amplifier, the hysteretic and poorly predictable behavior was a real frustration.

Another aspect of the hysteretic behavior is seen in the input characteristics presented in Figure 4.73. Simple expectation may be based on the idea of the nozzle flows issuing into the "open" vented space. The supply nozzle pressure drop ΔP_S would be then constant while the control

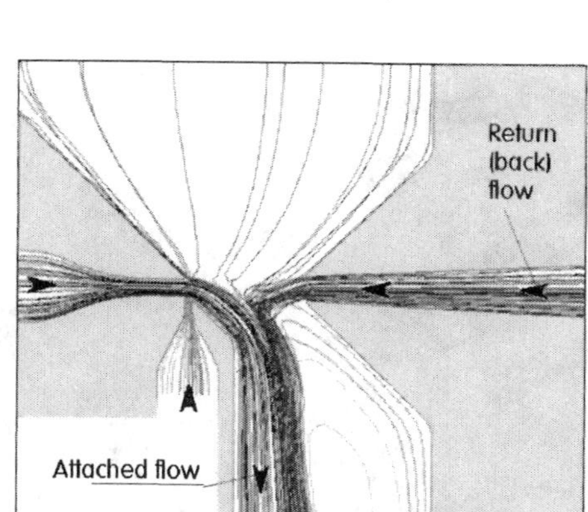

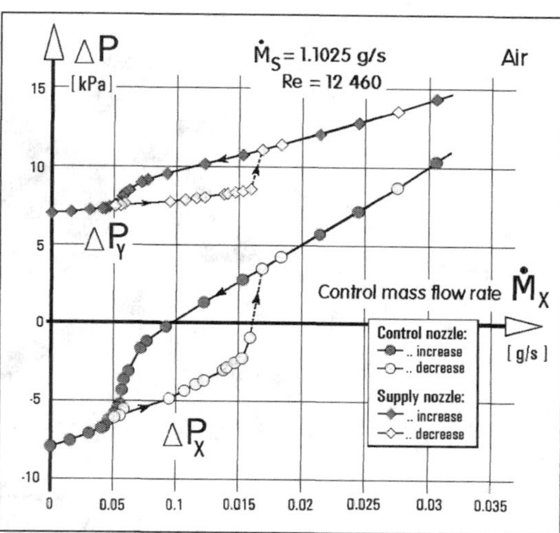

Figure 4.72 (Left) Pathlines inside the valve from Figure 4.70 at a small control flow $\mu_X < 0.05$: the control action does not suffice to overcome the Coanda effect that diverts the jet away from the collector. The jet entrainment generates a small negative output flow in the output terminal.

Figure 4.73 (Right) The input characteristics of the valve from Figure 4.70 – the control terminal and supply terminal pressure differences relative to the vent plotted as a function of the input flowrate into the control nozzle.

nozzle characteristic would follow the quadratic law of Figure 2.31. In fact, the real input behavior is dominated by the pressure field in the interaction region. This makes the characteristic more or less linear, though again with shifted origin. The shift is also here due to the Coanda effect, which makes the control pressure values ΔP_X initially (at small control flows) negative. Conspicuous in Figure 4.73 are the hysteretic jumps associated with the separation (and reattachment) of the jet.

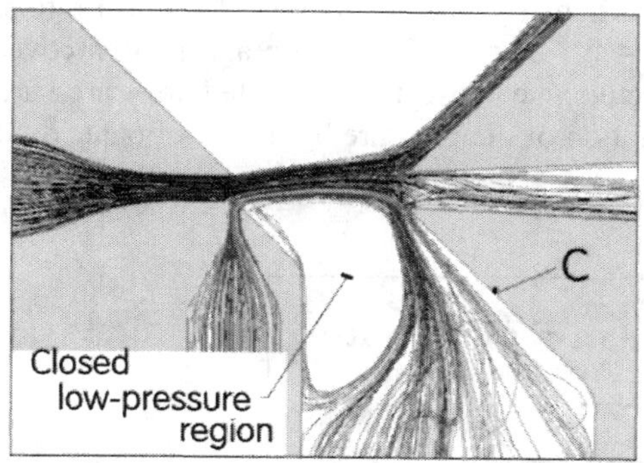

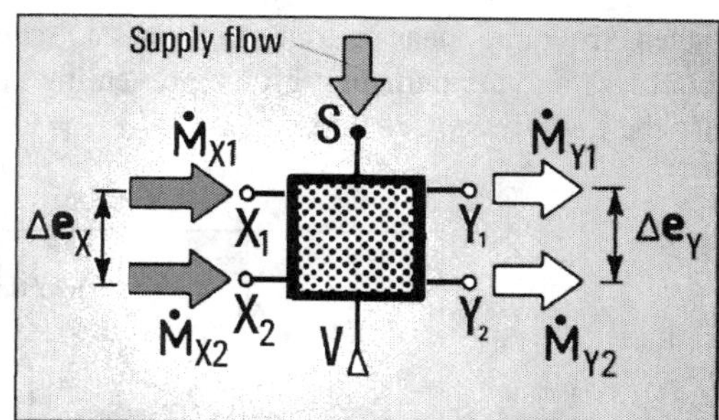

Figure 4.74 (Left) Pathlines inside the valve from Figure 4.70 at the control flow large enough to overcome the Coanda-effect deflection. They indicate the formation of the closed recirculating "vacuum bubble" with the low pressure due to fluid entrainment into the jet. The pressure deflection force on the jet opposes the momentum of the control nozzle flow.

Figure 4.75 (Right) Schematic "black box" representation of terminals of the proportional amplifier operating with the difference signals. The input is the difference between the specific energy and/or mass flowrate values in the two input terminals X_1 and X_2, similarly the output is the difference between the values in the two output terminals Y_1 and Y_2. In the incompressible regimes, it may be convenient to work with directly measurable pressure differences ΔP instead of the specific energy difference Δe, which differ simply by the multiplication constant, $\Delta e_P = v\,\Delta P$.

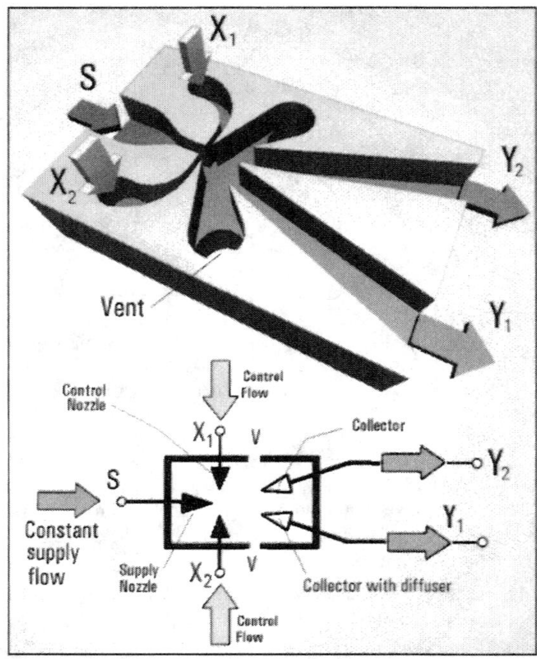

Figure 4.76 An example—and schematic representation in circuit diagrams—of the differential-signal proportional amplifier, with a pair of control nozzles as well as a pair of collectors.

Another complicating effect observed in some regimes in the flowfield of this type of valve is the formation of the closed recirculation "bubbles," as shown in the example in Figure 4.74. This also influences the resultant direction of the jet (notice the jet axis curvature towards the bubble in Figure 4.74). These examples show a remarkable capability of the jet to attach even to walls placed at extremely large divergence angles. This is particularly true at small nozzle aspect ratios, in the present case the main nozzle aspect ratio is $\lambda = h / b_S = 3.57$. The top and bottom cover planes are quite near to one another and can effectively enclose the generated low pressure "bubbles" and attachment regions, preventing equilibration of the pressure by inflow of outer fluid into the low pressure regions.

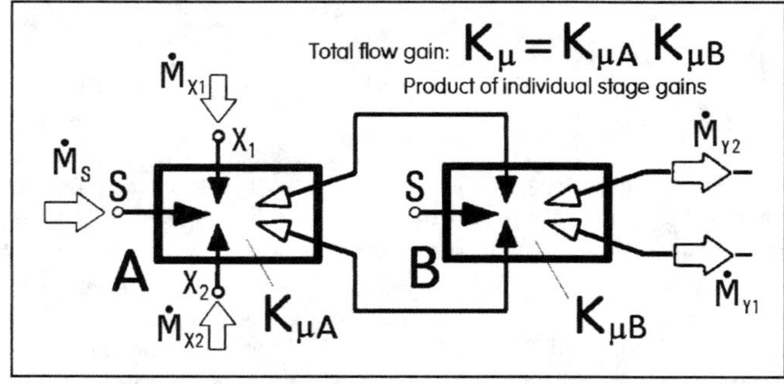

Figure 4.77 Schematic representation of an amplifying cascade of two amplifying valves.

4.3.3 Symmetric proportional control valves

An important development step that took place in the early history of fluidics was the idea of eliminating these complicating effects by using a symmetric layout of the valve. The two flows from the two opposing control nozzles can eliminate the problems by forcing the jet into the desirable straight, nondeflected direction. Intyroducing this modification required a significant change in mental attitude. Signals have to be carried not by a single channel but by a pair of channels, as shown schematically in Figure 4.75. There are also two signal-carrying channels leading from the device output. This is done by replacing the single collector from the basic device in Figure 4.70 by the layout with two collectors placed side by side – as shown in Figure 4.76. This makes possible a cascade connection (Figure 4.77) useful for obtaining a high overall gain when amplifying feeble signals available from some sensors. In the symmetric pair of collectors, a decrease of the captured flow in one collector when the main jet is deflected is accompanied by simultaneous increase of flow on the other side. These symmetric layouts exhibit—within a certain range, before the main jet is pushed beyond the collector—an excellent linearity the transfer characteristic. Linearity is a property of paramount importance for fidelity (lack of spectral distortions) of the amplified signals. Nonlinearities ale eliminated by their subtraction: they take place on both sides and since the output signal is the difference between them, they tend to cancel themselves.

To provide an idea how the symmetric approach improved the valve properties, it may be interesting to continue in the lines of previous Figures 4.64 and 4.70 and modify the original geometry of the two nozzles by placing symmetrically the second control nozzle of the same shape on the opposite side. Initially, the results as presented in Figures 4.78 to 4.80 do not show any improvement. They represent the situation with zero control flows from the two control nozzles. The Coanda effect is not suppressed; the jet clings to even very widely divergent walls. Figure 4.81 shows what nominally may be considered the same situation with zero differential input signal. The two control flows are equal, the jet is deflected so much that there is no doubt they do not reach to the collectors. Nevertheless, because they are nonzero, there is a visible change. There is a quite large recirculation low-pressure bubble between the jet and the wall to which it attached in Figures 4.79 and 4.80. All that is needed is the wall being not very long for the jet to become

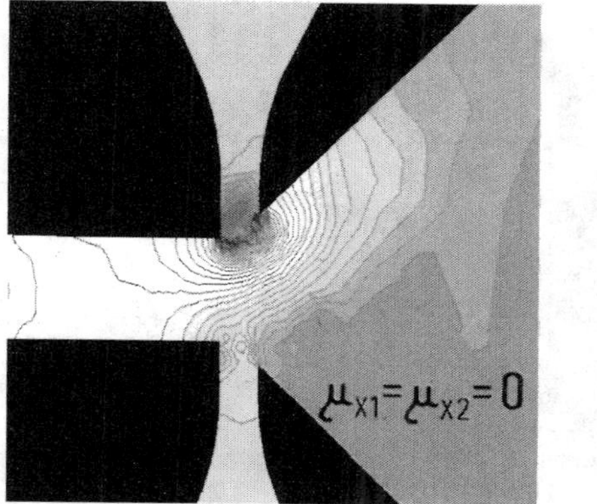

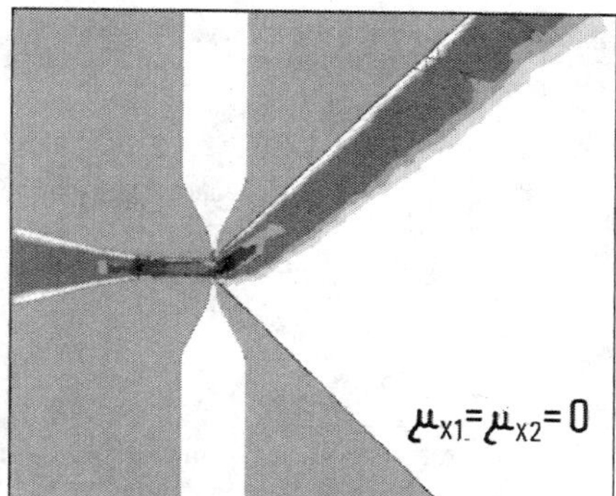

Figure 4.78 (Left) Isobars in the midplane (at $h/2 = 2.5$ mm from the bottom of the cavity) demonstrate the complexity of the pressure flowfields, nearly impossible to predict by simple analysis.

Figure 4.79 (Right) Contours of constant velocity magnitude computed for the symmetric layout with two control nozzles at zero control flows. The Coanda attachment is still effective.

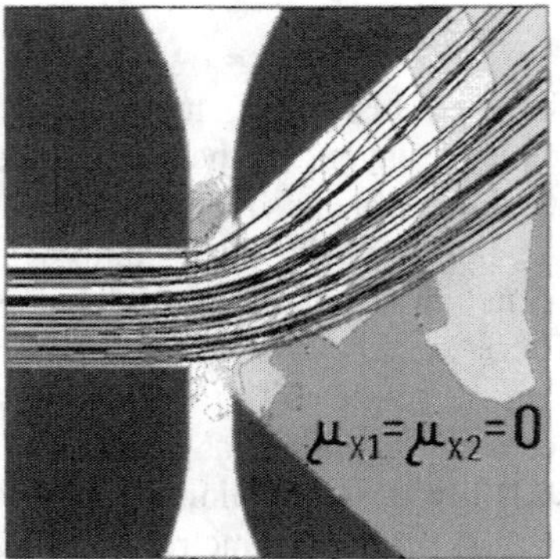

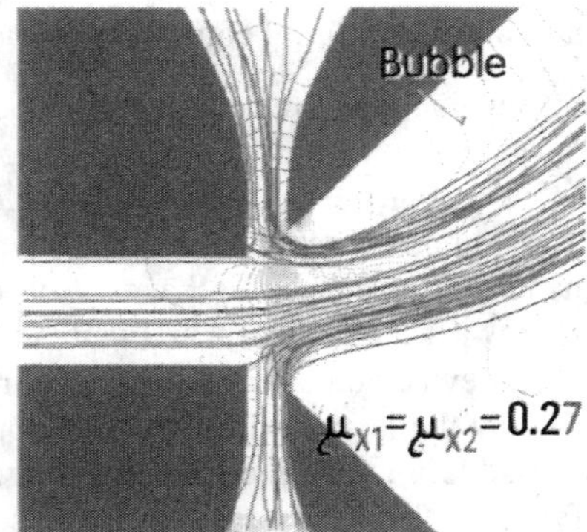

Figure 4.80 (Left) Pathlines in nozzle exits superimposed on the pressure field at zero control flow in the symmetric layout. There is a small separation bubble immediately downstream from the nozzle exits, but it is insignificant.

Figure 4.81 (Right) With equal flows from the two opposing control nozzles the value of the differential input signal is still zero; nevertheless, the pressure field and superimposed pathlines demonstrate the jet straightening tendency.

straight. As soon as the bubble is long enough to reach the end of the wall, the fluid from the vent enters the bubble and destroys the low pressure there. The deflecting pressure force ceases to exist.

The proportionality property of the symmetric amplifier is demonstrated on the design from Figure 4.82, retaining most features of the single-nozzle single-collector design from Figure 4.70 with the symmetric layout of the nozzles (Figure 4.79) and two collectors. The narrower entrances (1.9 mm in Figure 4.82 compared with 2.1 mm in Figure 4.70) were required to accommodate two collectors side by side. This results in slightly lower maximum output flows, but, capturing mainly the most energetic central part of the impinging jet, lead to higher pressure recovery efficiency.

The properties are judged by the shape of the differential flow transfer characteristic plotted in Figure 4.83, the dependence between the difference $\Delta\mu_X$ between the two control flows on the

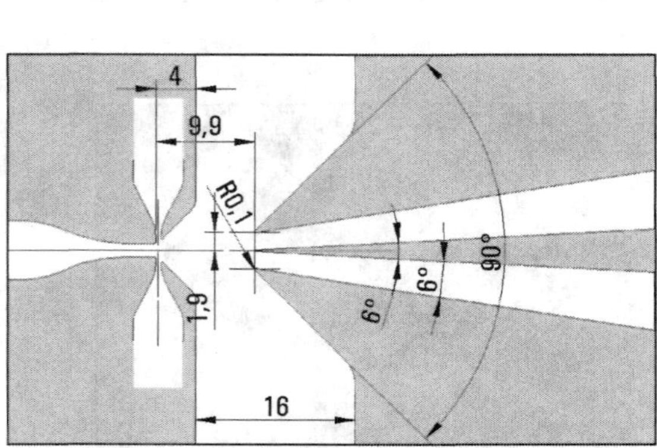

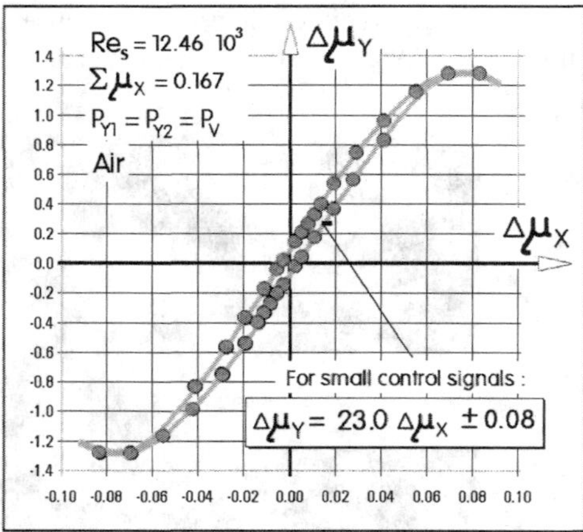

Figure 4.82 (Left) An example of a symmetric proportional amplifier as the next step of development using the same geometry of the nozzles as shown above in Figures 4.64 and 4.70.

Figure 4.83 (Right) Flow transfer characteristic for the differential signals in the valve shown in Figure 4.82. Despite any attempt at optimization this characteristic is reasonably linear (in its usable central part) and the flow gain as high as 23 is a good result.

horizontal axis and the difference $\Delta\mu_Y$ between the output flows on the vertical axis. The overall character seen in Figure 4.83 is indeed that of a proportionality, a linear dependence passing through the origin. The deviations at the both ends of the useful range are due to the jet being deflected too far, beyond the entrances of the collectors. This quite good linearity is no mean achievement considering the inescapable nonlinearity introduced by the jet velocity profile (Figures 3.53 and 3.54) and the complexity of the deflection mechanism (compare with Figure 4.81). It should be noted that this favorable behavior was obtained (as indicated in Figure 4.83) with the basic control flows only $\mu_{X1} = \mu_{X2} = 0.0835$, much less than the value 0.27 in Figure 4.81. Despite no optimization of the geometry, the slope (proportionality constant) evaluated in Figure 4.83 shows a good flow gain, 23, sufficient for applications like amplification of a signal from a sensor (note that it may be easily increased to 530 by using two valves in a two-stage cascade, corresponding to Figure 4.77).

The characteristic in Figure 4.83 is, however, not perfect. There is still a recognizable effect of hysteresis, no doubt due to actions of remaining Coanda effect. It is instructive to examine the improvements achievable by eliminating this residual drawback. Essentially, the improvement was attempted by widening the vent and increasing the vent wall inclinations so as to suppress the tendency of the jet to attach to these walls. In particular, the wall a (Figure 4.84) in the improved geometry is at inclination angle 80°, increased from the original 45°. To make it possible, the control nozzles had to be made asymmetric, while retaining their original area contraction ratio. Also the slope of the wall b, which was seen in Figure 4.74 to facilitate formation of the "vacuum bubble" by turning back the part of the flow not entering the collector, was changed so as to suppress this tendency (compare the angle 77° with the earlier 90°). The resultant transfer characteristic in Figure 4.85 shows now a total elimination of the hysteretic behavior.

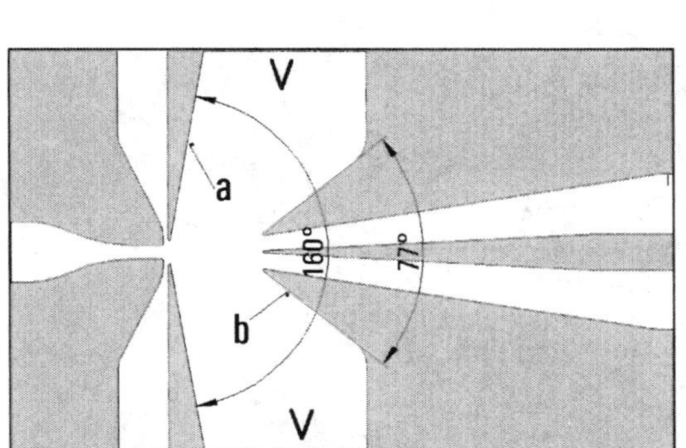

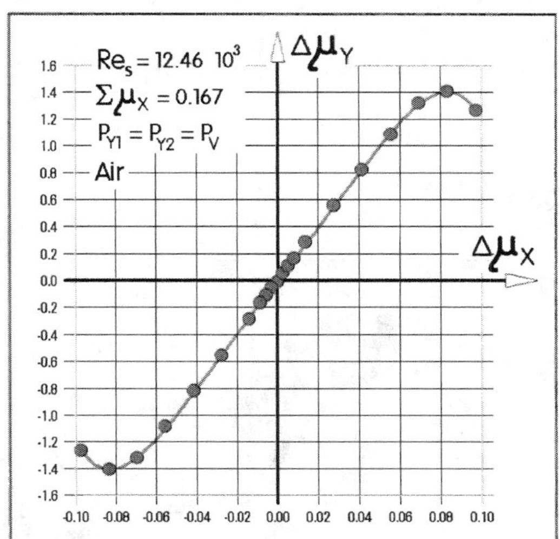

Figure 4.84 (Left) The improved version of the amplifier from Figure 4.82. The inclination of the walls defining the vent entrances were changed to further suppress the Coanda attachment of the jet to the wall a and the return of a part of the jet not entering the collector and turned away by the wall b.

Figure 4.85 (Right) The nearly perfect flow transfer characteristic of the improved valve version from Figure 4.84.

4.3.4 Laminar proportional amplifiers

An important development step in amplification of fluidic signals, made by Manion, Mon, and Drzewiecki in the United States, was the introduction of the *laminar amplifier,* in contrast to operation with turbulent jets in all the valves discussed above. Turbulent fluctuation in fluid flow is a disadvantage when working with tiny fluidic signals requiring great amplification, especially by an amplifier cascade, because the turbulent noise is inevitably also amplified. With the laminar proportional amplifiers (LPA), it is possible to convert very small input effects into organized large-amplitude fluid flows, even changes as small as those caused by thermal gas expansion when the gas at the amplifier input is irradiated by light brought into the fluidic system by fiber optics. Such extreme sensitivity is very useful mainly in processing fluid flow variations as input signals from various sensors.

The basic operating principle in LPA, the deflection of the jet, is the same as in the turbulent amplifiers. The difficult part of the amplifier design are measures needed to keep laminar character of the flows. This means protecting the jet from all disturbances that tend to cause early transition. This is easier if Reynolds numbers are low, preferably below $Re \sim 1\ 500$. It is important to select small depths of the cavities (transition into turbulence depends upon Reynolds number evaluated from the depth h). In particular, laminar amplifiers are characterized by careful design of the vents. These are typically separated by protruding partitions into several parts, seen Figure 4.86 (though the individual outlets are usually mutually connected further downstream – for the design from Figure 4.87 the connection is according to Figure 4.88). A characteristic feature is also a partition separating the interaction region (the presence of which would be dangerous in a turbulent amplifier as it would cause the Coanda-effect diversion of the jet). The supply nozzle is long, to dampen the disturbances caused by the 90° turn in the inlet. These amplifiers operate well in Reynolds numbers of the order of hundreds, typical for present-day microfluidics [5]. Characteristically, the gain is usually found to decrease (Figure 4.89) with decreasing Re. There is a limit below which the amplification effect ceases to be of useful magnitude, though there are microfluidic applications where a need arising to divert the flow at any rate, even with by control flows more powerful than the deflected flow. Typically, the operating frequency range of an LPA cascade (especially when the amplifiers are small) reaches

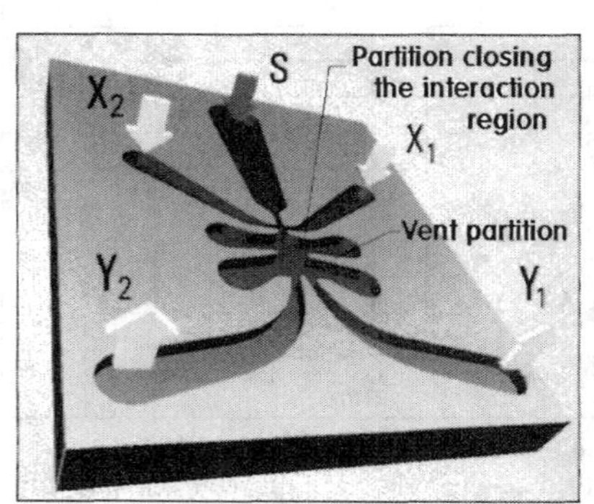

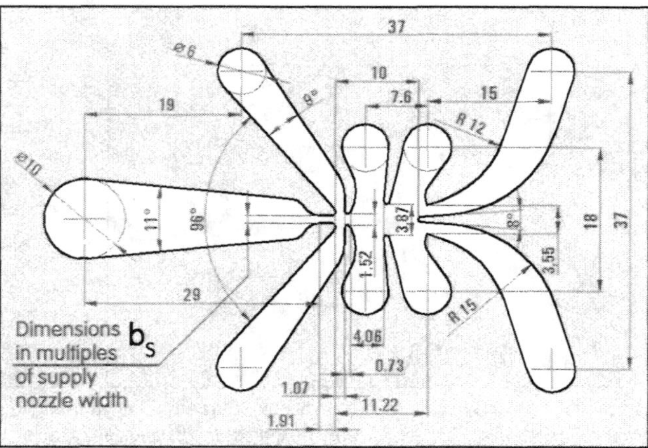

Figure 4.86 (Left) Typical laminar proportional amplifier (LPA) is characterized by vents divided by partitions onto several parts each with its separate outlet.

Figure 4.87 (Right) Geometry of LPA developed for applications in MEMS according to Athavale et al. (*After*: [5].)

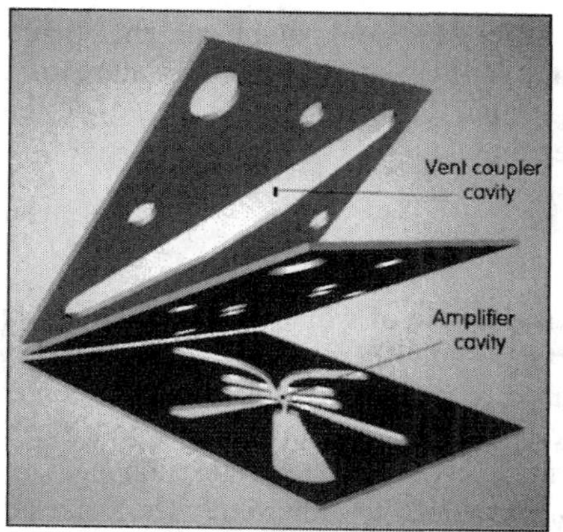

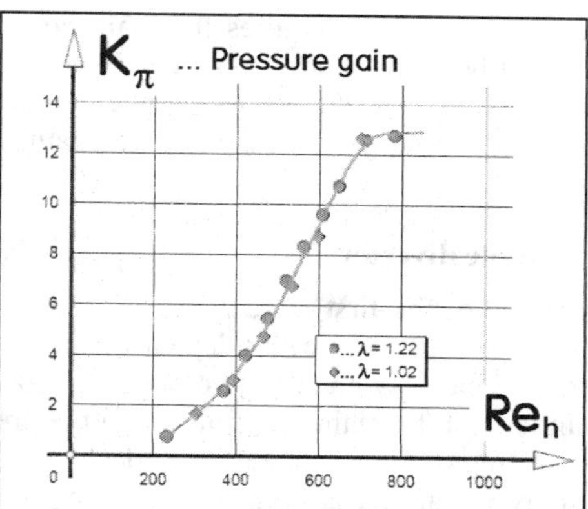

Figure 4.88 (Left) The individual vent exits in the laminar amplifier are actually connected together, in another parallel plate, through the common coupler cavity.

Figure 4.89 (Right) Amplification in all amplifiers decreases with decreasing Reynolds number due to the increasing effect of the friction. The decrease is here demonstrated using data from [5] for amplifiers of $b_S = 124\ \mu m$ supply nozzle width and two different aspect ratios (*After:* [5]).

up to ultrasonic range. It is commonly wide enough to make possible an operation such as amplification (an example of application: Kosher synagogues, where electronics must not be used). The range is generally wider than needed for most industrial or biomedical flow control task applications.

4.4 SWITCHING VALVES BASED ON THE COANDA EFFECT

The other large family of fluidic jet-type diverter valves differs from the proportional variant by retaining and making use of the Coanda effect. Like all jet-type devices, these valves are usually applied as flow diverters. They are extremely useful, applied in a considerable percentage of current fluidic circuits, though, again, their use needs to avoid the really low Re range of microfluidics. Typical for behavior of these valves is their operation with sudden jumps between different regimes and also their hysteretic response to the input. This makes them impractical for a continuous operation. On the other hand, this makes them useful for switching between various paths with the advantage that the jet can remain in the deflected state without any control action, while in a proportional valve the jet may be kept deflected only by a constant control flow and hence power spent on the input side. In the Coanda effect valves, the fluid flows are typically controlled by brief control pulses applied only at a time of switching the flow.

There are two basic variants. Both are *active,* which means the main flow into the output is supplied from the supply nozzle, controlled by much smaller flow(s) admitted into control nozzle(s); though as the Reynolds numbers tend to go down in microfluidics, even large control flows have to be accepted. The much-less-often-used passive load-switched monostable version was already discussed in Section 4.1.2. The jet-deflection controlled monostable version, with single control nozzle, is discussed in Section 4.4.4.

Both versions are used and known to exist in a wide range of sizes and layouts. Of course, the Coanda effect, on which the flow deflection is based, is much more effective if the jet is turbulent.

Roughly speaking, this requires the main nozzle exit flow Reynolds number higher than ~ 800 and preferably more. Such relatively low value may suffice because the curvature of the jet promotes earlier transition into turbulence. On the other hand, with specialized designs, the bistability was demonstrated also in purely laminar regimes.

4.4.1 Bistable diverter

Development of the first variant, the *bistable* one, an example of which is shown in Figure 4.90, originated from the symmetric layout of the proportional amplifiers discussed previously in this chapter. The basic symmetry (two control nozzles and two collectors) makes this version actually very similar and the main difference is in the presence of the attachment walls on both sides of the interaction region. They have to be designed and made with particular attention paid to equal opportunity for the jet issuing from the supply nozzle to attach to either one. The valves may be provided with vents (Figure 4.91), which makes them less sensitive to changes in the loading dissipance and indeed applicable even with total blockage of the output terminals. However, more important at present tend to be the unvented versions, because of the growing occurrence of applications, such as in microchemistry, where it is not desirable to let the different transported fluids to escape into the atmosphere or mix indiscriminately in the common space into which the vents lead.

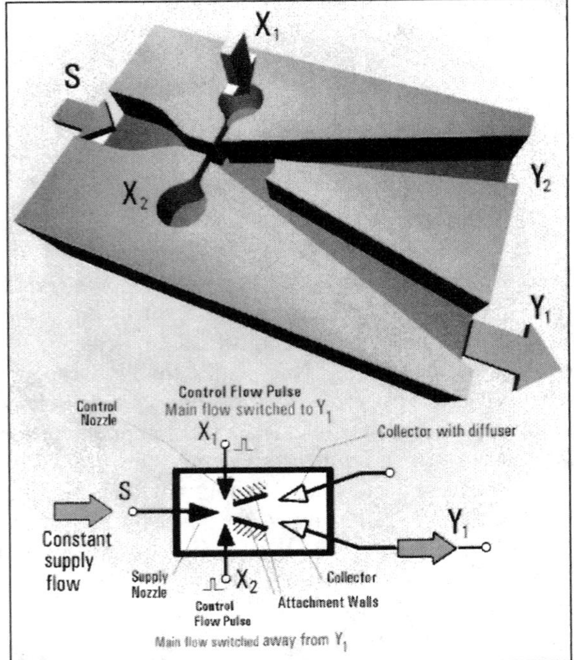

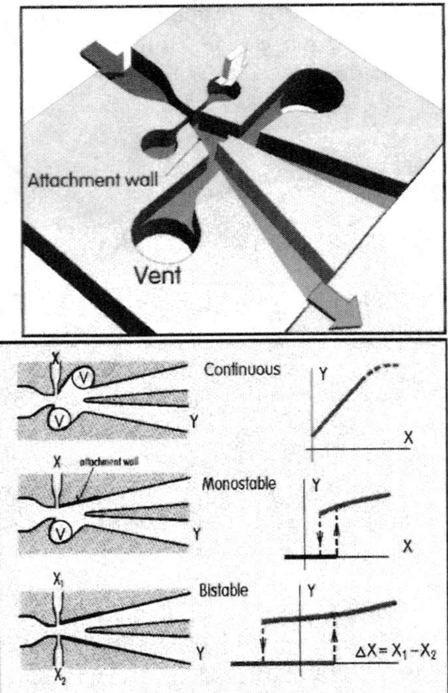

Figure 4.90 (Left) An example of an unvented bistable diverter valve. The schematic representation emphasizes the attachment walls keeping the jet deflected in absence of the control flow.

Figure 4.91 (Right, Top) An example of a vented bitable diverter. The inlets into the vent channels are downstream from the trailing edge of the attachment walls.

Figure 4.92 (Right, Bottom) Simplified representations of continuous, monostable, and bistable flow control valve layouts and their schematic transfer characteristics.

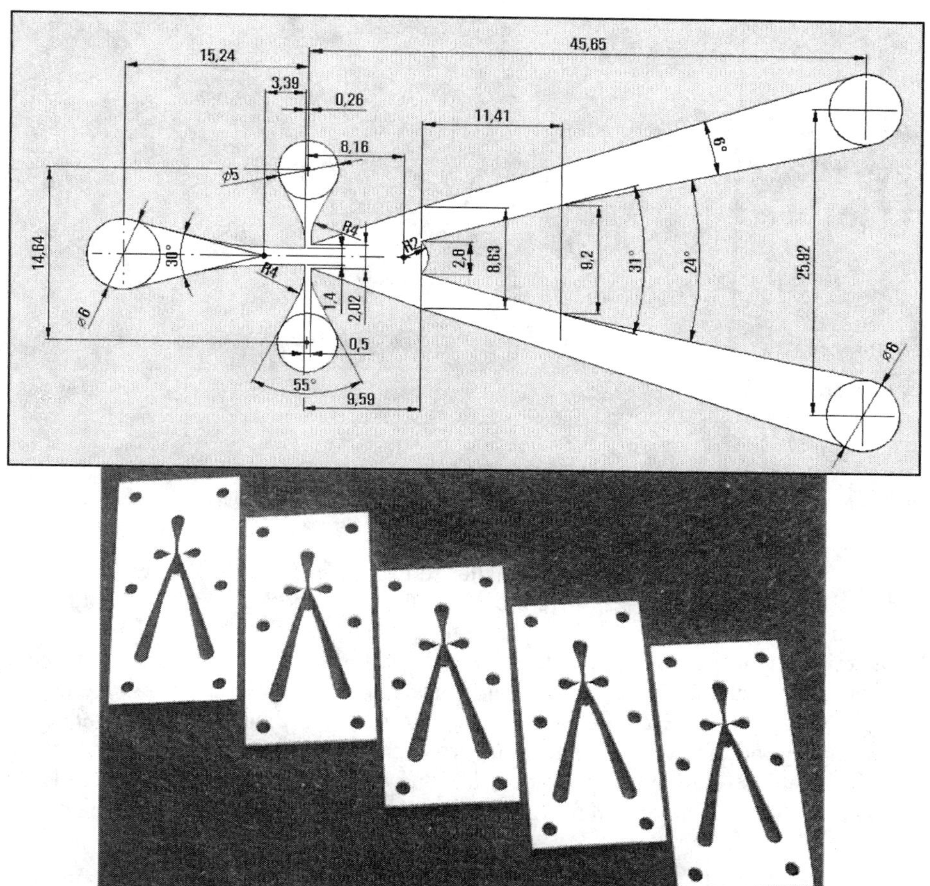

Figure 4.93 (Top) Geometry of the model bistable valve used as an example for investigating typical properties of this valve type. The dimensions (in millimeters) are those of the scaled-up model Figure 4.96.
Figure 4.94 (Bottom) A set of valves etched from both sides in thin foils. The foils may be stacked to get a larger aspect ratio, which in general improves valve performance. (*From:* [6]. © 2006 Elsevier. Reprinted with permission.)

Characteristic for the bistable version is the presence of two mutually opposed control nozzles and transfer characteristics corresponding to the bottom of Figure 4.92. The control flow pulse may be admitted only into one control nozzle, the one at the side towards which the main jet is deflected. The pulse moves the jet to the opposite side. On the other hand, these valves also may be connected into a cascade where the downstream valve receives the control flow simultaneously into both control nozzles (one of these flows being the spillover flow in the upstream valve). The switching is then dependent on the difference between the two control flows.

The flow-switching process evaluated for the example valve of Figures 4.93 to 4.96 is presented in the form of the flow transfer characteristic in Figure 4.97. In this case, the control action was not a pulse, but gradually increased steady control flow. The switching took place when the relative magnitude of the control flow was slightly above 7 % . The corresponding value of the flow gain is $K_{\mu} = 14$. If desired, this value may be increased (at the price of decreasing the pressure gain) by a choice of more narrow control nozzles.

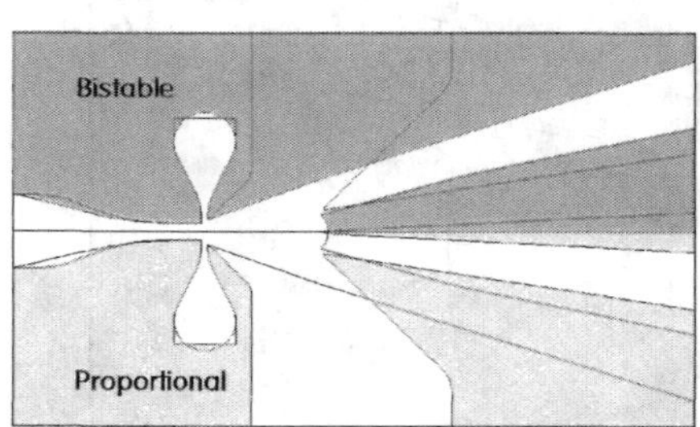

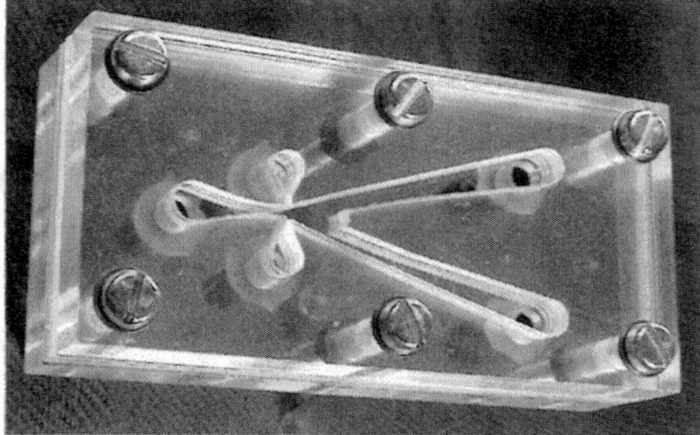

Figure 4.95 (Left) Comparison of the original version of the proportional amplifier (Figure 4.82), bottom half, and the bistable valve in Figures 4.93 and 4.94, top half of the picture. The core of the flow interaction region is identical, collectors are wider apart because of the presence of the cusped splitter between them.

Figure 4.96 (Right) The model valve for laboratory tests, consisting of the stack of thin transparent PMMA, bottom plate with inlets and outlets, and a transparent top cover plate. (*From:* [6]. © 2006 Elsevier. Reprinted with permission.)

Also helpful for higher gain is choosing a smaller setback of the attachment walls (in Figure 4.93 this width is 22% of the main nozzle width b_S), but this may endanger stability of the attachment and cause a tendency toward easy load-switching.

Characteristic features of these valves are discontinuities and hysteresis in the transfer characteristic. As the main jet is switched from the state A (Figure 4.98) to the opposite attachment wall, the relative output flowrate μ_{Y2} plotted on the vertical coordinate in Figure 4.97 is suddenly changed from the positive value (flow out from the terminal Y2) to the negative value, corresponding to the jet-pumping entrainment into the main jet. The presence of the jet-pumping was actually already observable from the magnitude $\mu_{Y2} > 1$ at the beginning of the curve, at

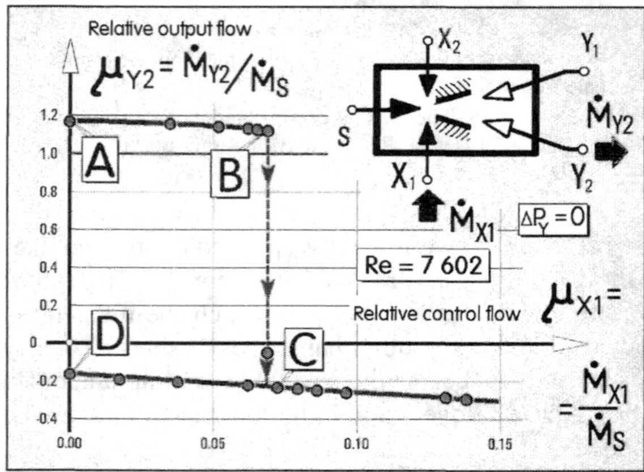

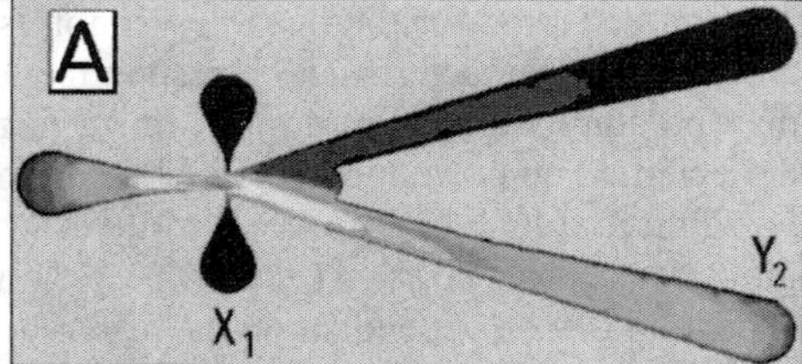

Figure 4.97 (Left) Transfer characteristic of the valve from Figure 4.93 shows the switching response to the control signal admitted to the control terminal X1 when the jet is attached to the wall on the same side. Once the jet is switched, the conditions do not return to the initial state when the control signal is decreased.

Figure 4.98 (Right) Grayscale reproduction of the original color-coded distribution of absolute velocity in the state A of Figure 4.97 in the midplane (at $h/2 = 2.5$ mm from the bottom of the cavities) in the example of the valve from Figure 4.93. (*From:* [6]. © 2006 Elsevier. Reprinted with permission.)

zero control flow. The character leads to the different return trajectory in Figure 4.97, followed when the input control flow is decreased. It does not return to the initial position, of course, because once the main jet is switched to the opposite attachment wall, the control signal in the input terminal X1 ceases to have any influence on the valve.

4.4.2 Internal stabilizing feedback

For many applications, sensitivity to changes in the load—and, in particular the extreme of load-switching phenomenon, discussed in Section 4.1.2—is undesirable. In particular, the measures meant to increase gain by decreasing the control flow magnitude required for the switching, such as the small setback of the attachment walls, may lead to small stability of the attachment. Especially in applications where the load may vary unpredictably there is a danger of the valve responding by unwanted switching. To avoid this, the deflected position may be stabilized by providing the feedback loops shown in Figure 4.99. In fact, by adjusting the feedback loops, the bistability may be obtained even without the attachment walls. Such feedback-loop bistability may be useful especially at low Reynolds numbers, where the Coanda effect ceases to be effective, being based on entrainment into the jet, which is much less efficient in laminar flows. Feedbacks, especially if they introduce some delay and phase shifting, must be applied with care as they may easily cause self-excited oscillation. Some bistable valves are built with two (or several) control nozzles on each side so that one of them may be used for the external feedback. They may also perform the logical operation **OR** between two alternative incoming switching signals.

The loop pipes of Figure 4.99 are inconvenient. The same stabilizing effect may be obtained in a more elegant manner by providing the splitter between the two collector entrances with cusps and a hollow turning-vane shaped wall between them (this explains the wider separation between the two collector inlets in the comparison in Figure 4.95). This geometry generates the standing vortex in the interaction cavity, as shown in Figure 4.100. The vortex, in effect, creates one of the two feedback loops shown in the bottom part of Figure 4.99. In the case of Figure 4.100 it is the

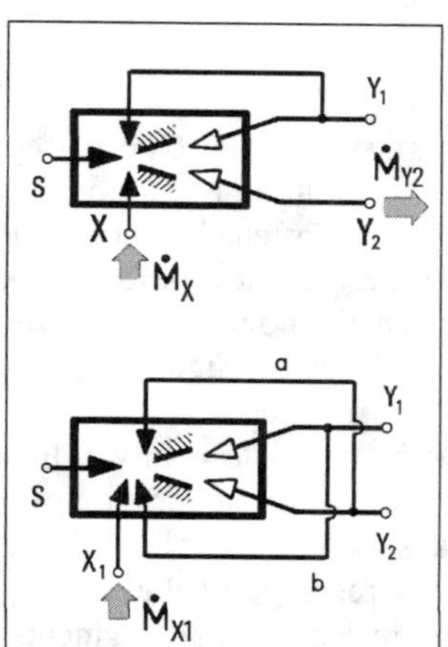

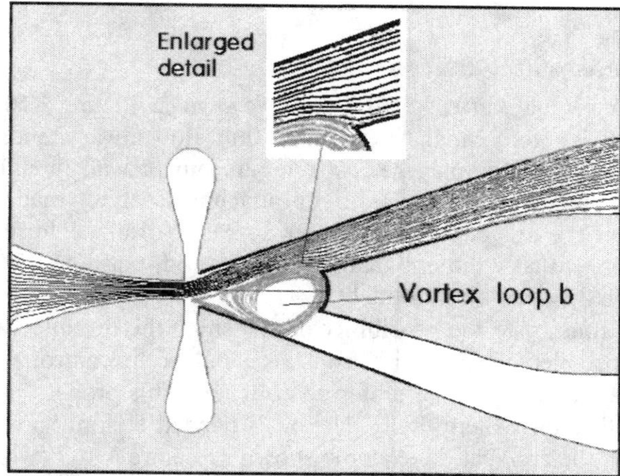

Figure 4.99 (Left) Stabilization in support of a weak wall attachment by external feedback loops. Top: monostability obtained by the single loop always giving preference to the output Y2. Bottom: positive feedback enhancing bistability.

Figure 4.100 (Right) The large standing vortex created inside the valve. Apart from the feedback action, it influences the pressure recovery by causing a sudden increase in flow cross section at its trailing side.

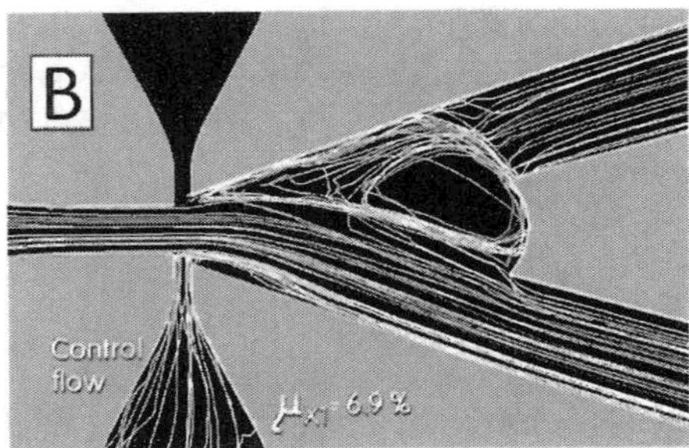

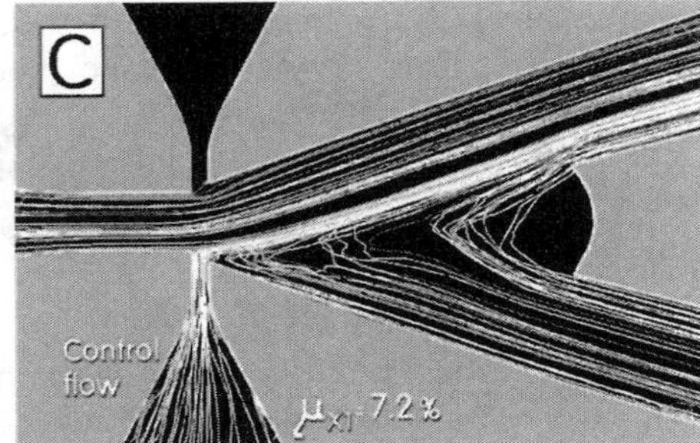

Figure 4.101 (Left) Flow pathlines in the state B of Figure 4.97. The vortex generated by the cusped splitter nose leads to an internal positive feedback loop effect.

Figure 4.102 (Right) Pathlines in the state C of Figure 4.97, when applied control flow overcame the internal feedback so that the jet is switched to the opposite attachment wall.

loop "b." Its presence becomes important when an external effect tends to switch the jet. It may be an increase in the dissipance of the connected downstream load or an applied control signal. The latter case is presented in Figure 4.101. It shows the internal flowfield for the conditions of the main jet directed to the bottom collector leading to the terminal Y2 with rather large control flow applied to the bottom control nozzle X1; this is the state B in Figure 4.97. A part of the flow is caught by the lower cusp and turned back towards the upper control nozzle X2 where it acts against any separation of the jet from its lower attachment wall. If this separation tendency increases, for example, due to an increased loading, the amount of the flowrate turned back increases and so does its deflecting effect returning the flowfield into its original regime. Only when the combined effects of the control flow and loading (note that the transfer characteristic in Figure 4.97 as well as the pathlines in Figure 4.101 were obtained for zero load, conditions are somewhat different with loading) overcome the combined influences of the Coanda-effect pressure forces and the internal feedback loop, then the flow switches from the state B into C (Figure 4.102).

4.4.3 Monostable diverters

The monostable variant of the valve, shown in Figure 4.103, is asymmetric. One of the output terminals is preferential; the supplied fluid flow always tends to leave through it in the absence of control flow. There may be just one attachment wall though there is often the second auxiliary wall at the opposite side with its attachment effect made weaker: it may be set at a larger divergence angle or with a larger setback (distance form the main nozzle exit). One of the collectors (usually the one that is not preferred, but note the case shown in Figure 4.104) may be substituted by just a vent outlet.

In contrast to the possibility of switching the bistable valve by short flow pulses, the control of monostable valves requires constant input of the control flow for the whole duration of the flow deflection away from the preferred collector. This property makes the monostable variant suitable for applications where the deflection is needed for only a short percentage of the total operation time. In the application example shown in Figure 4.104 (which presents a small segments of four valves from the very large number actually used) it is characteristic that only a single valve that is in the OPEN state needs the presence of the control flow. In all remaining CLOSED valves the sample flow (not really closed but diverted) is sent to the vent by the Coanda effect, not needing any control action.

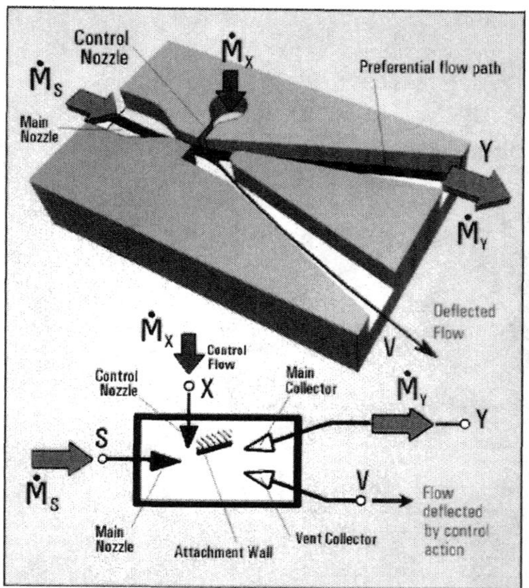

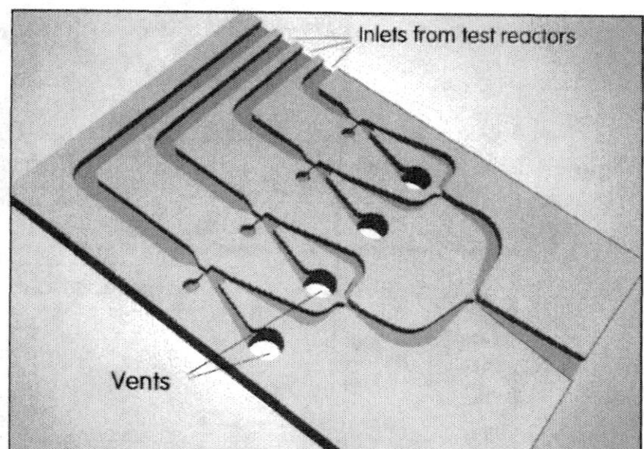

Figure 4.103 (Left) An example and schematic representation of the monostable diverter valve. The presence of the control nozzle makes it different from the passive valve load-switched valve in Figure 4.21.

Figure 4.104 (Right) An example of use of monostable switching valves: fluidic sampling unit consisting of an array of the monostable diverters. A single sample from reactors is chosen and delivered to a composition analyzer while all remaining sample flows are dumped into vents. Switching, compared with blocking the unneeded sample flows, does not disturb the reactor operating conditions.

A practical example of a recently tested monostable valve geometry is presented in Figure 4.105 with an example of numerical flowfield solution in Figure 4.106 and experimental data obtained on a model in Figures 4.107 and 4.108. The plotted data there show the loading characteristic with zero control flow (compare with Figures 4.20 and 4.22). This case is rather typical for such characteristics, as they are found not only for the monostable valves but for the bistable valves as well. It demonstrates the Eulerian similarity (may be compared with Figures 4.7 and 4.8, in spite of the negative values there due to the orientation convention).

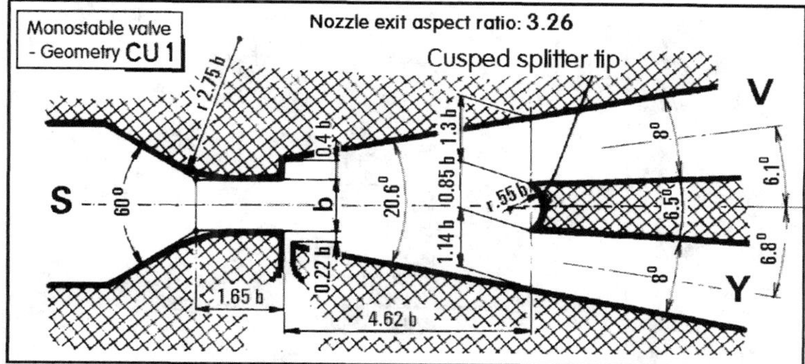

Figure 4.105 An example of monostable valve geometry recently investigated in the form of a large, scaled-up model. The small, 12.9° divergence of the collector axes was chosen in an attempt at a high performance.

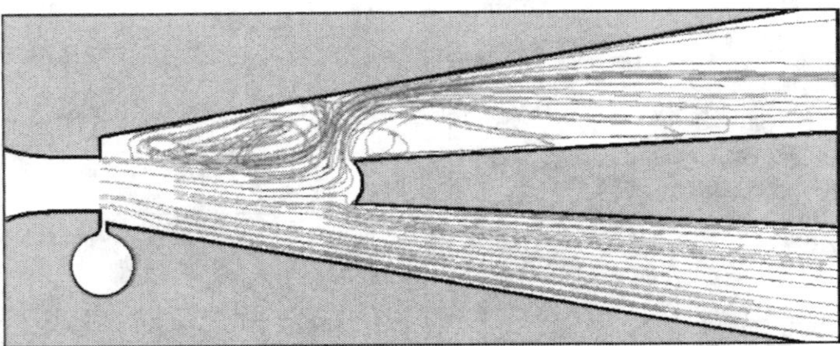

Figure 4.106 An example of computed pathlines in the valve from Figure 4.105 showing the effect of the cusped splitter.

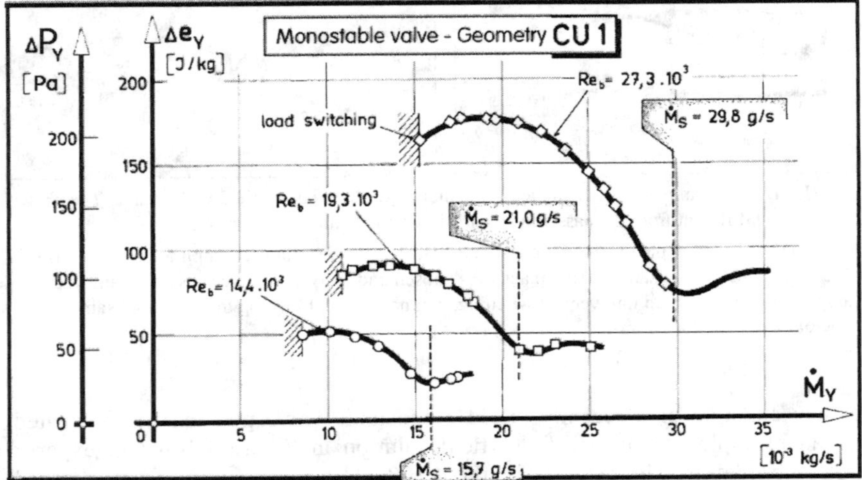

Figure 4.107 Experimental loading curves of the monostable valve from Figure 4.105 obtained with the scaled-up model (and therefore rather large Reynolds numbers) at three different supply flowrates.

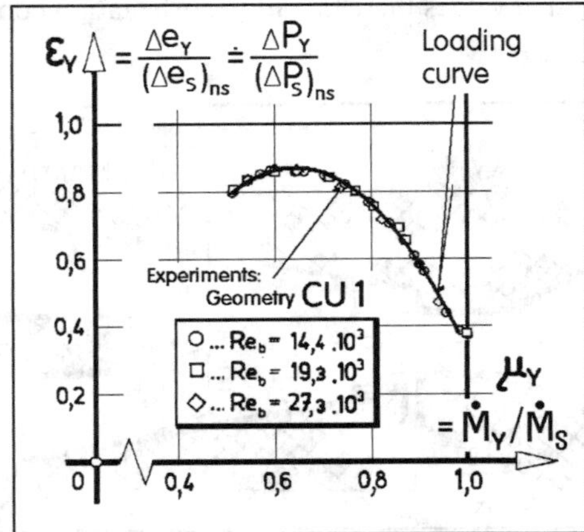

Figure 4.108 The demonstration of the Eulerian similarity: all three results from previous Figure 4.107 collapse into the single universal curve in the relative coordinates.

4.4.4 Pressure recovery

The control action in the interaction cavity needs accelerating the fluid so that its inertial effects prevent it from leaving the valve through outlets other than the desired collector. If this collector is loaded by the connected device, such as by the reactor in Figure 4.109, which makes it more difficult to fluid to get through compared with the empty parallel exit path, the acceleration has to be quite significant. It is achieved in the main nozzle at the price of the decrease in pressure which, however, has to be recovered by reconversion from the jet kinetic energy. The diffuser in the collector is there for just this purpose. The recovery is unfortunately never complete. The success is measured by the magnitude of the pressure recovery factor π_Y introduced in Figure 4.110. It varies with the loading; to give an indication of the success in this aspect of the valve design, it is usual to indicate the value at the no-spillover regime. In the example shown in Figures 4.105 and 4.108, the value is 0.38, a not very impressive though perhaps typical figure (this author's more successful bistable valves exhibited a value as high as 0.7 [1]).

Somewhat surprisingly, recent investigations show a more complex picture of the mechanism actually taking place. The actual distributions of pressure along the flowpath plotted in Figure 4.111 exhibit first a rise from the local minimum A followed by a decrease to the second local minimum B. The local extremes are associated with the presence of the large vortex set up for producing the internal stabilizing feedback loop.

The most important effect caused by the vortex is the fact that the pressure recovery is by far not due to the effects taking place inside the output collector diffuser. Most of the pressure rise takes place before the fluid enters the diffuser, at $X_1/s < 1$ in the rather typical Figure 4.111. More than 3/5 of the output pressure difference ΔP_Y between the valve terminals obtained by the pressure recovery mechanism is seen to take place inside the interaction cavity. The diffuser adds only the remaining ~3/5 of the energy conversion pressure rise. Detailed study of the flowfield,

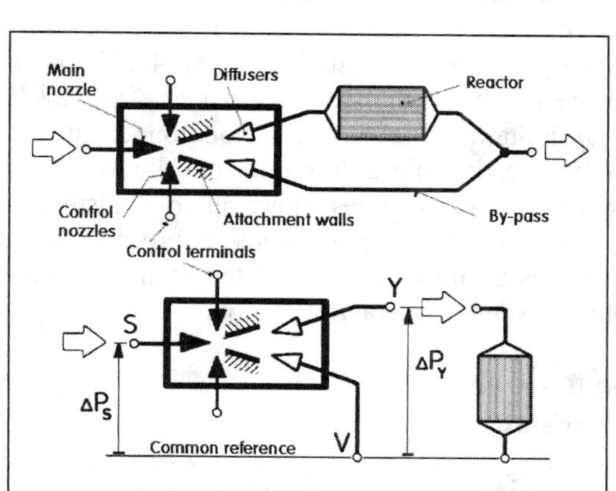

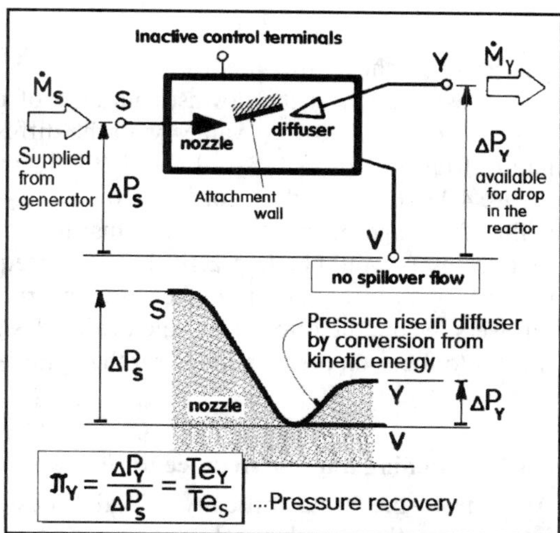

Figure 4.109 (Left) Schematic representations of the fluidic control, in the bypass mode, of the flow of premixed reactants in a through-flow (short residence time) chemical reactor. Top: the actual mutual position of the devices, with the reactor in one of two parallel paths. Bottom: topologically equivalent connection of a three-terminal device and a two-terminal device with common reference, see Figure 4.1.

Figure 4.110 (Right) Simplified schematic representation of the valve from Figure 4.109 leaving out components irrelevant for zero control flow, fully OPEN no-spillover state. Also shown schematically is the basic idea about the changes of pressure along the flowpath from S to Y: first pressure drop in the nozzle and then recovery in the diffuser.

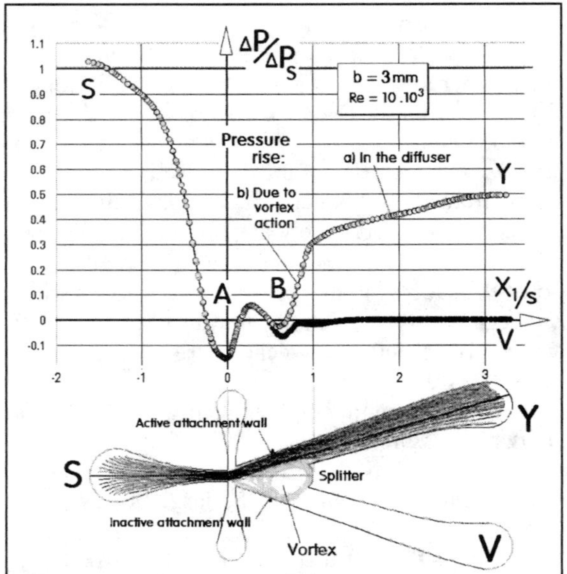

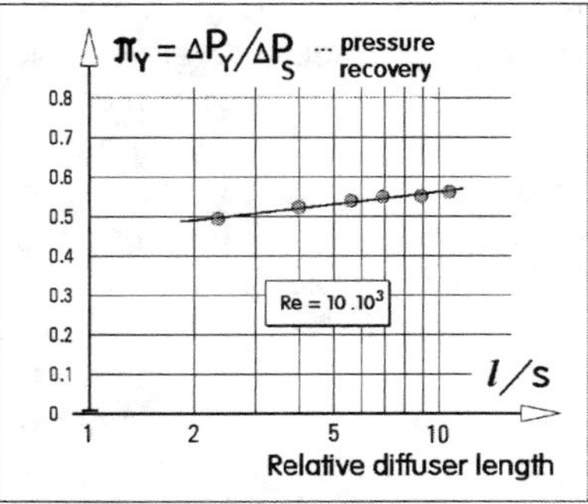

Figure 4.111 (Left) Distribution of pressure along the projection of the flowpath into the horizontal direction. Explanation for the two minima A and B and the spectacular pressure rise before the entry into the collector is provided by the existence of the large standing vortex. It makes possible a strong diffuser effect by the increase in cross section of the main flow without separation, which would take place in a diffuser with such rapid divergence.

Figure 4.112 (Right) Pressure recovery values in no-spillover state for the valve family Figure 4.114, differing in relative diffuser length. The improvement with diffuser length is discernible, but less than could be expected.

Figure 4.100 shows this prediffuser pressure rise to be due to the separation-less increase in the flow cross section at the downstream side of the vortex. This is also documented by the small slope of the growth of π_Y values with the diffuser length increase, plotted in Figure 4.112 for the family from Figures 4.113 and 4.114 (the lengths are related to the splitter distance $s = 7.24$ b, Figure 4.115). There does not seem to be a physical limitation (the growth is monotonous, at least within the range of investigated lengths; this may be questionable at low Re), but the extremely long diffusers would be impractical. The consequences for the valve design are obvious: instead of trying in vain to obtain the higher pressure recovery by lengthening and shape improving of the diffusers, where the effort of fluidic device designers has been so far concentrated, it may be more fruitful to focus on the conditions for the dominant vortex in the interaction region.

4.4.5 Matching and importance of the no-spillover state

Apart from the basic difficulty of valve design, setting up a control action based on delicate interplay of fluid-mechanical phenomena, it is also important to consider the constraints arising from the requirement of proper matching with other components in the circuit. Essentially, the task is to find an optimum compromise between the large and small valve size.

A striking feature of successful no-moving-part valves is often their large size in relation, to its load, for example the chemical reactor (Figure 4.109). Far from being the expected small axiliary device, the valve often dominates the whole assembly. The large cross sections in the valve may be chosen for small hydraulic loss, but this is contradicted by the need to have kinetic energy of

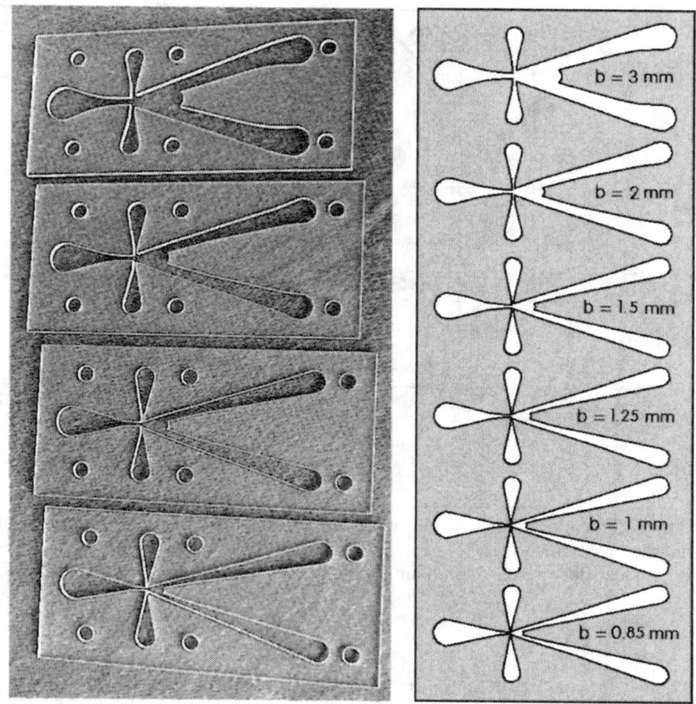

Figure 4.113 (Left) Photographs of the valve plates made for experimental verification study by laser cutting in polymethylmethacrylate.

Figure 4.114 (Right) Cavity shapes of the investigated valve model family. While the core geometry is scaled in proportion to the nozzle width b, the external size of the plates and location of terminals is invariant. As a result, the valves with small b have larger relative length of their diffusers.

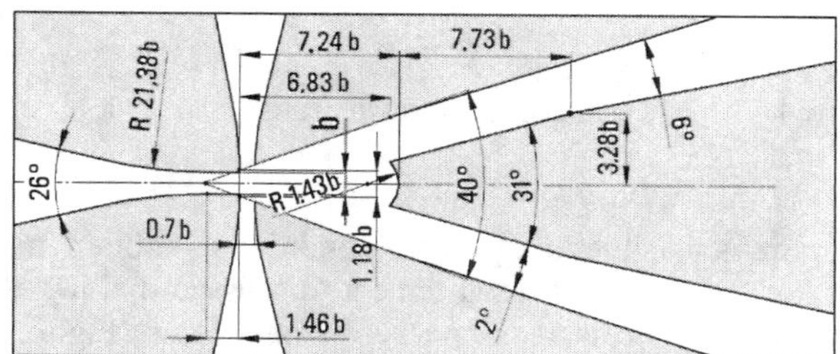

Figure 4.115 Geometry of the core part of the family of model fluidic valves.

the jet large enough for a sufficient control action (which depend on the jet energy). If, on the other hand, the valve is small (with the correspondingly high jet velocity in the small nozzle), a large proportion of the kinetic energy is lost because diffusers tend to be lossy devices. This is particularly unpleasant in situations where the available energy is limited and has to be handled with care. Also, the pressure conditions in the high-velocity valve may topple the pressure balance and produce undesirable reverse jet-pumping flows. Early experiments (e.g., [7]) suggested that optimum matching is obtained with equal minimum cross sectional areas in both the valve and the load. Obviously, this cannot be more than a crude rule of thumb; the cross-sectional area alone cannot properly characterize the differently shaped minimum cross section parts.

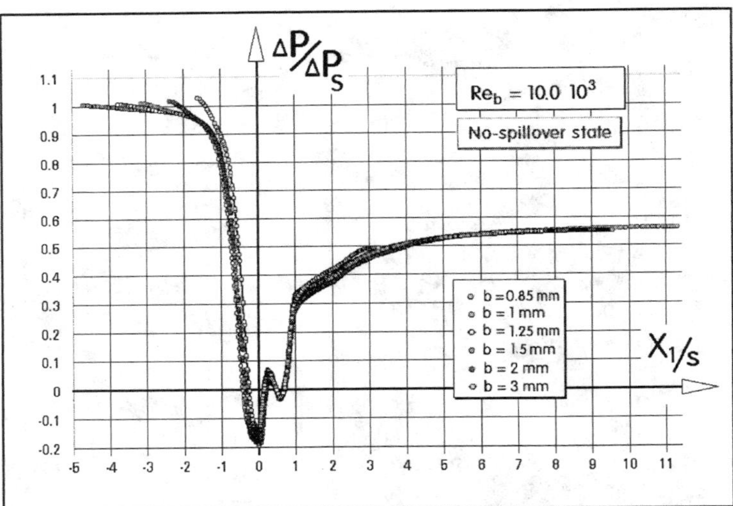

Figure 4.116 Superimposed nondimensional pressure distributions evaluated for all six valves of the family from Figure 4.109. All exhibit the two minima A and B as well as the pressure rise before entering the diffuser, as seen in Figure 4.111. Pressure increase in the diffuser takes place mainly downstream from the entrance.

Power transfer criterion of performance

This more exact matching condition operates with the quadratic model of device behavior, usually plausible as long as the Reynolds number is not extraordinarily low. As shown in Figure 4.117, the load is characterized by its dissipance Q_L while the characteristic of the valve is characterized by Q_S. The ratio of the two quantities is the loading factor $q = Q_L / Q_S$. The relative magnitude α of the hydraulic power (product of pressure drop and volume flowrate) transferred from the valve to the load is:

$$\alpha = \frac{q}{2}\left(\frac{3}{1+q}\right)^{3/2} \tag{4.5}$$

This is plotted in Figure 4.118. The maximum power transfer, $\alpha = 1$, is achieved for the loading factor value

$$q = 2 \tag{4.6}$$

Unfortunately, it is not always feasible to adjust the conditions to obtain this value. Apart from the possible conflict with the no-spillover requirement discussed below, there is the limitation associated with the no-spillover pressure recovery π_Y parameter (Figure 4.117), which is also uniquely related to the loading factor q. A diagrammatic presentation of this dependence is shown in Figure 4.119. This shows that the optimum loading condition (4.6) is obtained with pressure recovery

$$\pi_Y = 0.666 \tag{4.7}$$

In the example in Figure 4.112 the family of valves from Figure 4.114, this would require for the optimum loading a monstrous diffuser, $l = 76.2$ s long.

No-spillover criterion

There is yet another factor to be considered. Sending the fluid to an unwanted output path is generally undesirable. If it is spilled over inside the valve into the bypass, the conditions would correspond to the schematic representation in Figure 4.120. This is what usually happens with a valve that is very large in relation to the size of its load. Especially when handling special fluids

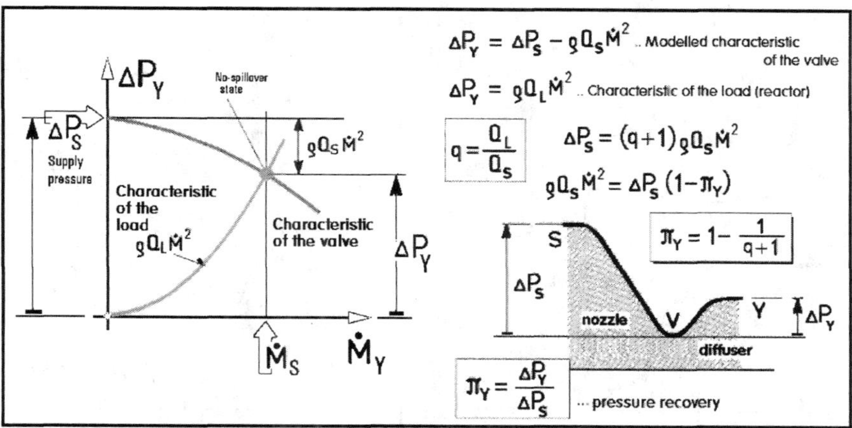

Figure 4.117 The typical loading characteristic of the valve and the characteristic of the load (such as the chemical reactor, Figure 4.109) both approximated by simple quadratic dependences. These are used to derive the relation (4.5) between loading factor q (the ratio of the reactor dissipance Q_L to the output dissipance Q_S of the valve) and the parameter a of power transfer into the load, Figure 4.118. This simple consideration of pressure changes shows the importance of the recovery parameter π_Y, the measure of the performance of the diffuser in the main exit of the valve.

such as reactants in chemistry applications, which should not get into wrong places in the circuit, the necessity of avoiding this regime in the basic state without acting control signal is of obvious paramount importance.

The no-spillover condition may be also not met for the opposite reason: some fluid being ingested back into the valve through vent terminal V, as shown in Figure 4.121. The reverse flow is due to the entrainment (jet-pumping) action of the jet in the valve. The effect usually arises if the valve is smaller than the properly matched valve and jet velocity in the small cross-section is high. Admittedly, the recirculation may sometimes be an advantage; there may be indeed situations where the pressure drop aspects are less important and passing the fluid once more through the reactor may be useful. This is, however, unusual. Figure 4.121 shows that a considerable proportion of the pressure recovery obtained in the diffuser is in such case wasted on generating the return flow into the vent V. This wastage of the painstakingly achieved pressure recovery is generally hardly acceptable.

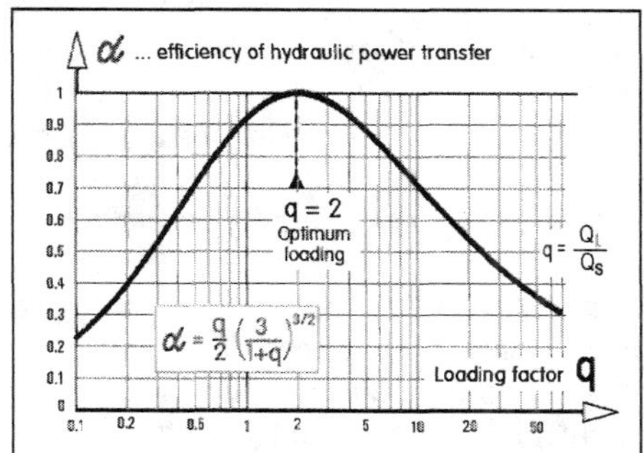

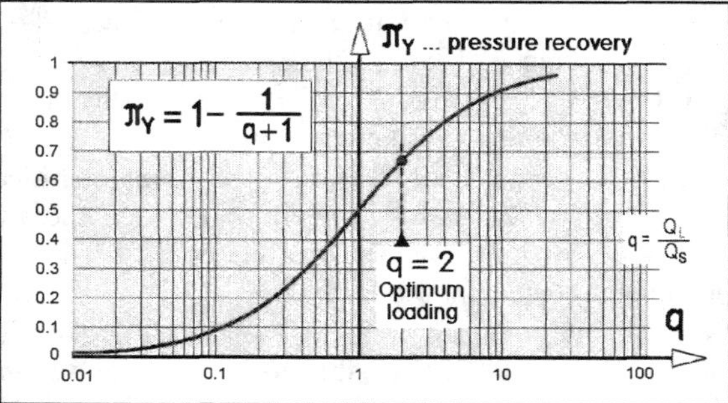

Figure 4.118 (Left) The usual criterion for optimum loading of the valve. The condition of maximum hydraulic power transfer is met at the loading factor value q = 2.

Figure 4.119 (Right) Mutual dependence between and the pressure recovery parameter (Figure 4.117) and the loading factor q, the ratio of the load dissipance Q_L to the output dissipance Q_S of the valve.

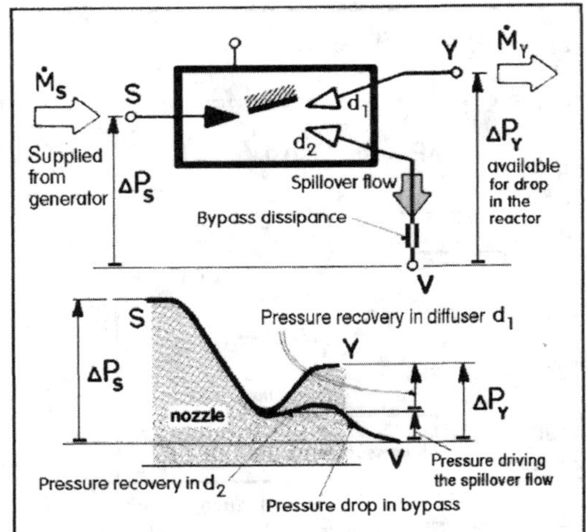

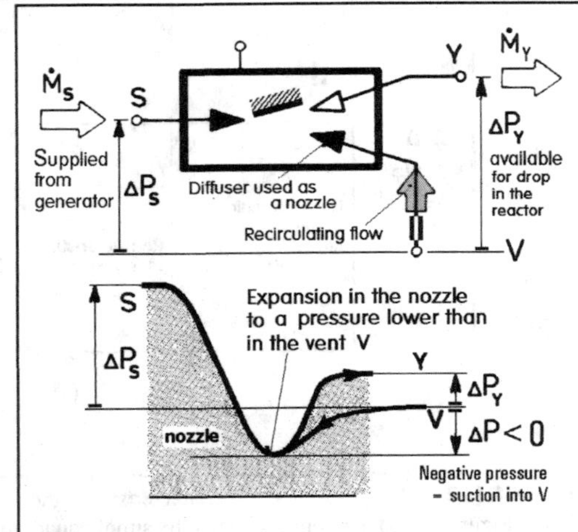

Figure 4.120 (Left) If the no-spillover condition is not met due to the valve being too large for its reactor load, some fluid is wasted by being spilled over into the bypass.

Figure 4.121 (Right) The other mismatch: the no-spillover requirement is not met due to the valve being too small for the particular load. The jet-pumping effect of the jet in the valve recirculates some fluid.

Obviously, a reasonable criterion for circuit design is the requirement of zero spillover flow. Applying it requires knowledge of the pressure drops, obtained either experimentally or from numerical flowfield solutions. In the first step, the value of the load pressure drop ΔP is evaluated at the required flowrate. Then the output pressure differences ΔP_Y between the valve terminals in the no-spillover state are evaluated for a range of valve sizes. This may be rather laborious and may be replaced by an evaluation made only for a single size and then computing the pressure differences ΔP_Y as being inversely proportional to the square of the minimum cross section of the valve (that is assuming Euler number invariance). In the third and last step, the proper valve size is identified as that for which the pressure valve and reactor drop values are equal: $\Delta P_Y = \Delta P$.

This procedure may be demonstrated on loading the family of valves shown above in Figure 4.113, all of them having the same external size and identical relative geometry of the central core parts (Figure 4.115), but different absolute sizes of this core—apart from the depth of cavities, which was the same. The core size is defined by the main nozzle exit width b, varying from b = 0.85 mm to 3 mm. Because of the identical external dimensions of the plates in which the valves were made, which did not scale with the core, the smaller members of the family have longer diffusers in their collectors. The properties were investigated at nozzle exit Reynolds number Re = = $10.0 \cdot 10^3$. The valves were loaded by devices shown in Figure 3.19, basically a nozzle of different size, but again keeping the same relative dimensions, as shown in Figure 3.20, with the downstream chamber of the same size. Some chemical reactors, stirred by the inlet jet, actually do correspond to such an arrangement of a nozzle and a chamber.

Figure 4.122 shows the consequences of the matching criteria from the perspective of the dependence between the minimum cross-sectional areas in the two devices. Because of the assumed equality of the depths, the values actually plotted in the diagram are the nozzle widths. The diagram makes it possible to choose the suitable element from one family for some given other device, for example, a reactor from the family in Figure 3.19 with the proper inlet nozzle width b_L to a given valve from the family in Figure 4.114, specified by its main nozzle width b. The dependence is determined by the heavy line for the no-spillover condition. The lower than desirable pressure recovery in the valves results in a power transfer deficiency penalty

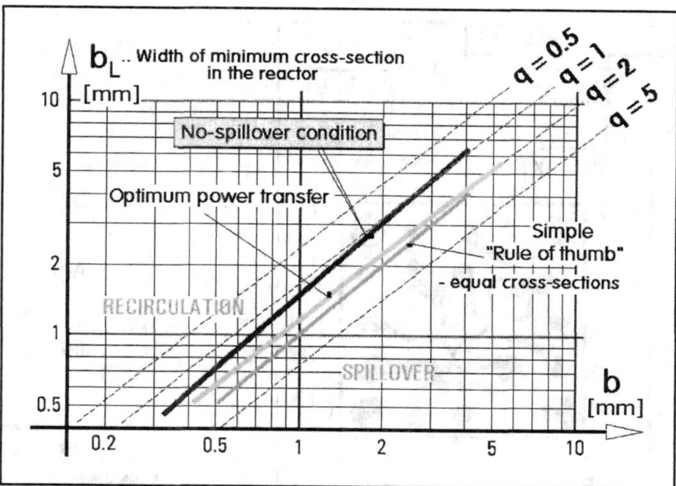

Figure 4.122 A typical diagram for choosing a suitable valve and reactor size (specified by their minimum cross-section widths, b for the valve and b_L for the load). The diagram is valid, of course, only for the investigated two families of the devices shown in Figures 4.113 and 3.19. Spillover avoidance results in q lower than the power transfer optimum q = 2.

(inaccessibility of q = 2). This is, fortunately, small and for most purposes inconsequential. Interestingly, even the old approximate rule of equal cross sections leads to a matching acceptable from the transfer efficiency point of view, though it would result in spilling over of the flow into the bypass.

4.5 MULTIELEMENT VALVES AND MODULES

Very few tasks are carried out by a single valve alone. More common are more complex fluidic circuits containing several valves as well as other devices. Some of the devices operate so often together in similar circuits that it is economical to develop modules consisting of cooperating pairs, especially since the devices need to be mutually matched anyway. An interesting development trend is an *integration* of the devices in the module, which then share some parts or components. In fluidics, with the devices usually having a high-velocity interaction region with a nozzle upstream and a collector with a diffuser downstream, it is quite reasonable to integrate the neighboring devices to avoid unnecessary conversions: deceleration in a diffuser followed after a very short distance by again an acceleration in a nozzle.

Of a large number of existing variants, it is useful to mention the following three examples.

4.5.1 Amplified logical operations

Even though the task of performing logical operations, especially of a complicated sort, is better left to electronics whenever possible, there are several, mainly simple, examples where the operation may be usefully carried out in the course of doing other things in the fluidic circuit. The advantage is avoiding the otherwise necessary signal conversions and the price is often just a small adaptation of the channel carrying the fluid flow. There are some useful techniques, taken over from what was earlier developed for the fluidic logic circuits, that may be useful. For example, let

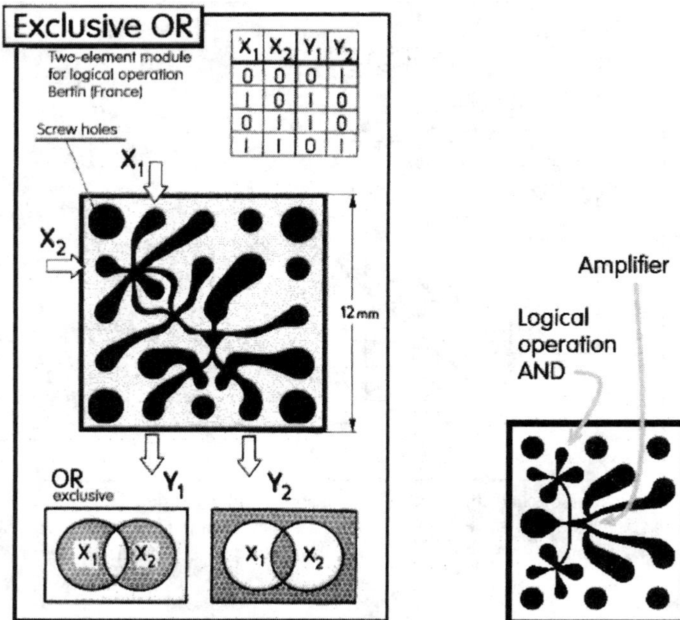

Figure 4.123 (Left) An example of an early microfluidic module containing the passive logical element and a bistable switching valve. Originally developed for fluidic logic operations, similar two-element modules are useful for a number of other tasks.

Figure 4.124 (Right) Another example of a module: two passive AND elements in the inlets of a bistable diverter. Note the absence of exit diffusers in the AND devices and of the classical contraction in the diverter control nozzles.

us assume the sample selection valves in the example shown in Figure 4.104 are to be brought into the OPEN state in a repeated sequence by an electric signal. The expensive way of providing each valve with its electro/fluidic (E/F) transducer may be replaced by much more elegant solution: using bistable valves so that they, together with several additional logical elements, form a binary counter with a single input and therefore just a single input electro/fluidic transducer (such as a solenoid valve). The proven modules like those shown in Figures 4.123 and 4.124 may be suitable building blocks for this and similar tasks. Worth noting in Figure 4.123 is the shape of the interaction cavity of the bistable valve. There are no Coanda attachment walls in the shape discussed in Section 4.4. Instead, the mechanism of keeping the jet deflected is based on the low pressure in a sideways cutout (similar to what will be shown in Section 4.7.4 and in Figure 4.163).

4.5.2 Bistable vortex valves

Bistability, in the sense of controllability by only short pulses with no control flow between them, brings so many advantages for circuit design that it became desirable to look for a similar property in valves capable of turning the flow down. The simple vortex valve, if its ratio of diameters D/d is chosen so small that its flow transfer characteristic (Figure 4.49) gets the unstable positive slope, is also operated in two stable regimes, but this is the monostability, with the need to apply the input power for the whole duration of the second regime. The solution exhibiting the desired qualities was found according to Figure 4.125 in the module containing both the Coanda-effect bistable diverter and the vortex valve. Apart from the bistability, the advantageous result is also the very low control pressure, as opposed to the standard vortex valves (in Section 4.2.3), for

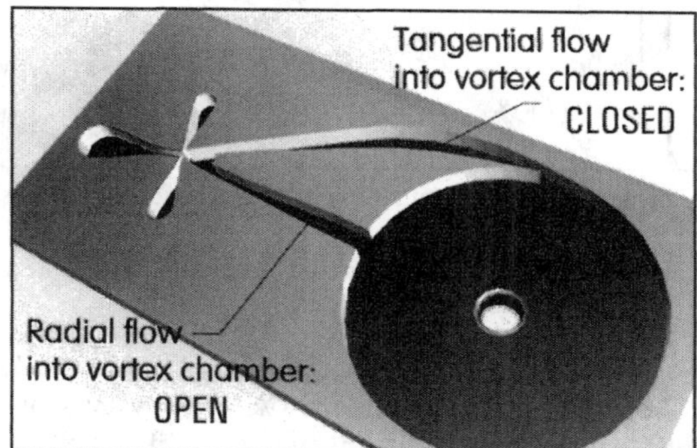

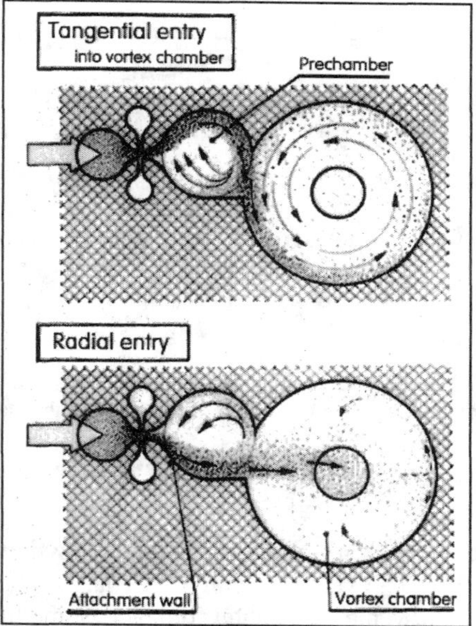

Figure 4.125 (Left) A two-element module combining the bistable diverter and the vortex valve. The advantage of the bistability is short switching pulses suffice for the control. Also, the control pressure needed is quite low.

Figure 4.126 (Right) The integrated version of the module from Figure 4.125: it is made more compact by removal of the collectors.

which is characteristic the necessity to have the control flow source at a pressure considerably higher than the pressure of the controlled flow at the entrance into the vortex chamber. The devices forming the module may be integrated as is shown, for example, in Figure 4.126 by the absence of the diverter collectors and the inlet nozzles of the vortex valve. The resultant remarkable combination has several uses, so far mainly, however, in power fluidics—though its potential for microfluidics is obvious. Further development toward the higher turn-down ratios

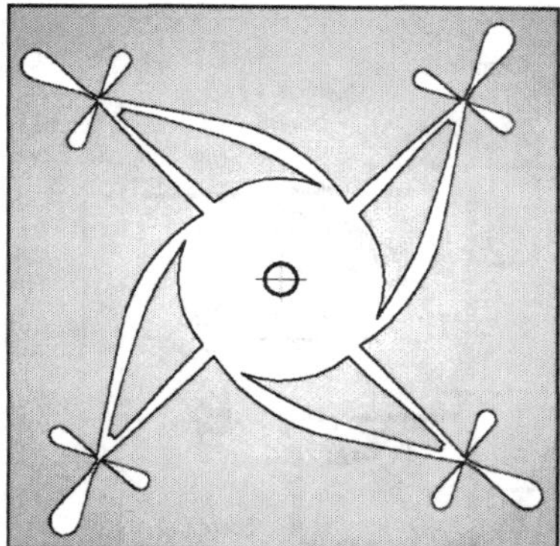

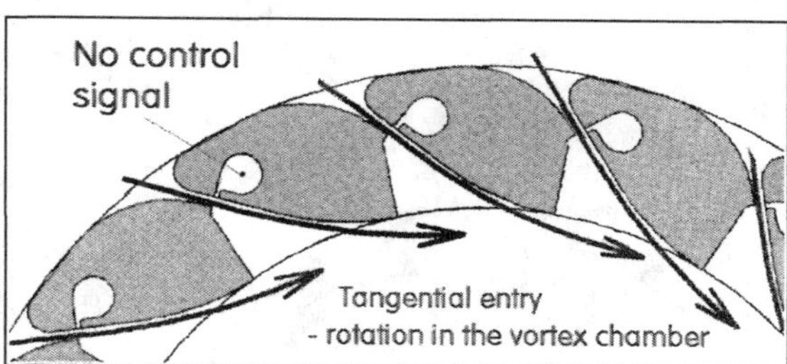

Figure 4.127 (Left) To obtain really good turn-down ratio requires symmetry of the flowfield inside the vortex chamber, not obtainable with the single inlets of Figures 4.125 and 4.126.

Figure 4.128 (Right) With a very large number of the switched valves on the vortex chamber outer boundary—and with the removal of the splitter wedges—the module of Figure 4.127 is turned in effect into a stator wheel with cascade of blades that are controllable by a fluidic signal. This is a monostable version in the CLOSED state.

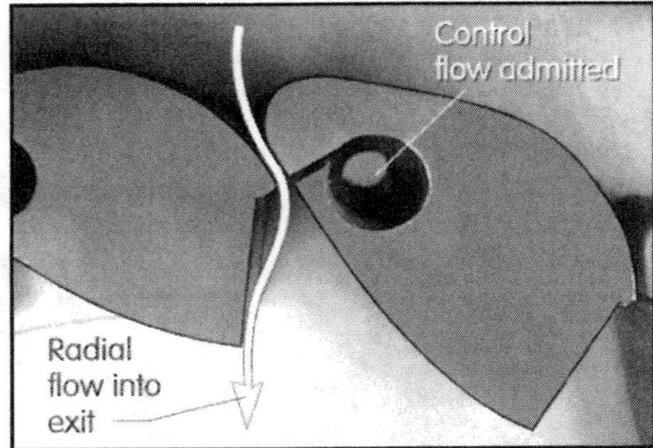

Figure 4.129 (Left) A view of the model of the monostable turn-down valve with the peripheral cascade of elements from Figure 4.128.

Figure 4.130 (Right) A close-up of the peripheral element in the OPEN state, caused by the action of the control flow.

requires measures leading to a better symmetry of the flow in the vortex chamber (Figures 4.127 to 4.130). The versions with a very large number (16 in Figure 4.129) of the switching valves on the periphery may be quite compact, something one cannot say about the versions like the one in Figure 4.127. Of course, the bistable variant of the switching valves in Figure 4.127 is equally well possible. The monostable version has simpler control circuitry, but, of course, needs the constant control input to maintain one of the states (the OPEN state in the case shown here).

4.5.3 Valves with "guard" flows

Yet another example of an integrated module is the valve developed for sampling units selecting one sample at a time and delivering it into a composition analyzer. The path towards the analyzer is provided with an array of parallel valves which may be, in dependence on the admission of a fluidic control signal either in the OPEN or CLOSED state—a symbolic description, because what the valves actually do with the samples not needed at a particular moment is diverting them into the

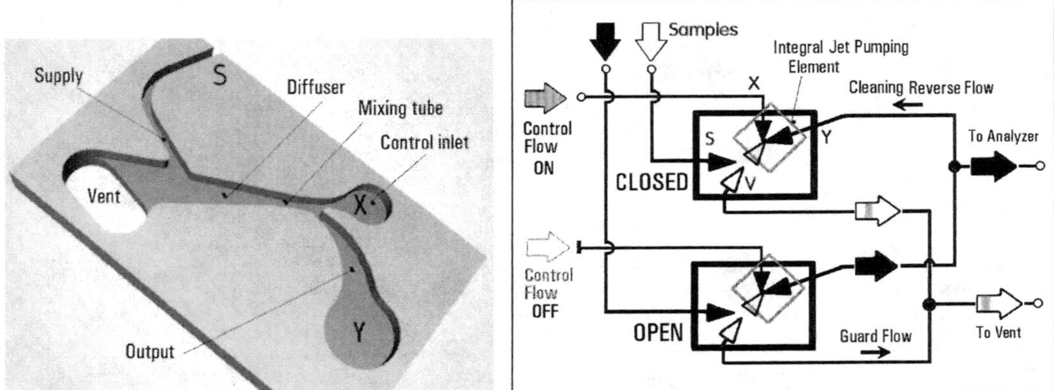

Figure 4.131 (Left) The pressure driven very low Re valve for sampling units with cleaning of the cavities.

Figure 4.132 (Right) Schematic representation of a pair of the valves from Figure 4.131, one of them in the OPEN state (sample to analyzer), the other in the CLOSED state. Note the indicated auxiliary flows for securing sample purity.

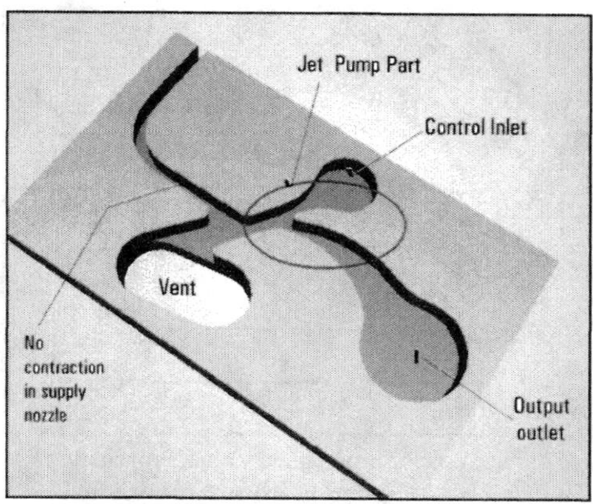

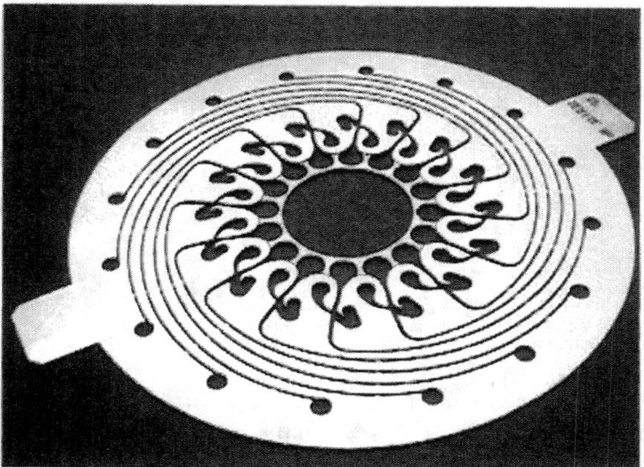

Figure 4.133 (Left) The version of the valve with the integral jet pump that was actually used. The cleaning reverse flow needed is rather small (it actually decreases the amount of the sample available for the analyzer) so that the jet pump part could be simplified.

Figure 4.134 (Right) Photograph of the tested sampling unit with 16 valves.

vent outlet. Because of the particular nature of the chemical reaction tests, there was an exceptionally strict emphasis on purity of the analyzed sample and avoidance of any mutual contamination between the individual samples. To be absolutely certain, an unusual demand was specified to generate two sorts of auxiliary flows. One of them, the "guard" flow (Figure 4.132) is a small part of the sample flow directed into the vent in the OPEN state (while the remaining large percentage flows into the output). This is secured by control of the pressure conditions in the vent and secures no entrainment into the main flow of the uncontrolled mixture of dumped samples from the common vent space. The other demand was to generate a return flow from the output Y in the CLOSED state, removing from the downstream channels all remains of the samples previously flowing to the analyzer. To secure this, the valve, originally a jet-deflection diverter, was integrated with a jet pump driven by an exceptionally strong control flow. To make the jet-pumping effective, the control flow has to be turbulent. This has led to an exceptional situation of there being a slow, very low Re laminar main flow, controlled by two orders of magnitude stronger turbulent control flow.

4.6 CAPILLARY VALVES

A special part of this chapter is devoted to flow control valves, the operation of which is based on the surface tension forces acting on fluid. In fact, these valves are currently not used very frequently. Their special treatment here is due to two reasons. First, their operation is completely different from the other valves discussed in this chapter, which are based on inertial forces. Second, this is the type of phenomena that are recognized as being of increasing importance in the future. As the typical size of microfluidic devices show the constant trend to a decrease, the importance of surface phenomena in general is increasing because of the increasing surface-to-volume ratio (Figure 1.17). Surface tension is one of those phenomena usually completely neglected in large-scale fluidics that are becoming more interesting and more often used, with this trend no doubt likely to continue in future.

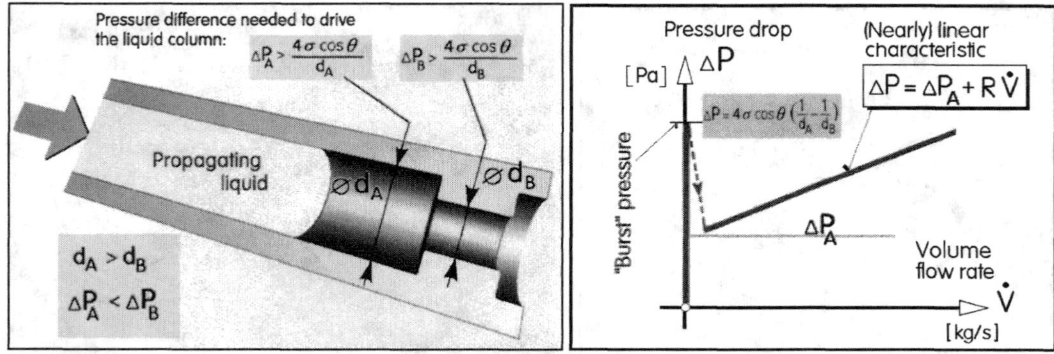

Figure 4.135 (Left) The basic principle of the passive valves based on the surface tension phenomena. Propagating liquid column is stopped at the entrance into (if the channel material is hydrophobic; Figure 4.126) or at the exit from (Figure 4.125, if the material is hydrophilic) the local constriction. The motion can continue only when the driving pressure rises to the level, causing the blocking meniscus to burst.

Figure 4.136 (Right) Pressure-flow characteristic of the passive capillary valve. The busting phenomenon is of a hysteretic character: return to the initial state requires removing the liquid from the channel.

A group of utilizable flow control principles using the capillarity has been there already for some time, not to speak about thermally or electrically influenced capillary phenomena which do no belong into the present chapter on purely fluidic interactions. The capillarity phenomena take place on the interface, in the present context mainly between air and a liquid, The common denominator of these principles is the necessity to work in the microfluidic system with two phases, a liquid and a gas. Usually, it is the liquid which propagates throughout the channels of system and displaces the gas which was there before. The details of the propagation is different for hydrophobic or hydrophilic material of the walls, but both have in common the change in the propagation occurring whenever the meniscus at the front of the liquid column comes to a local constriction, as shown in Figure 4.135. In each case, at different locations, as was already shown

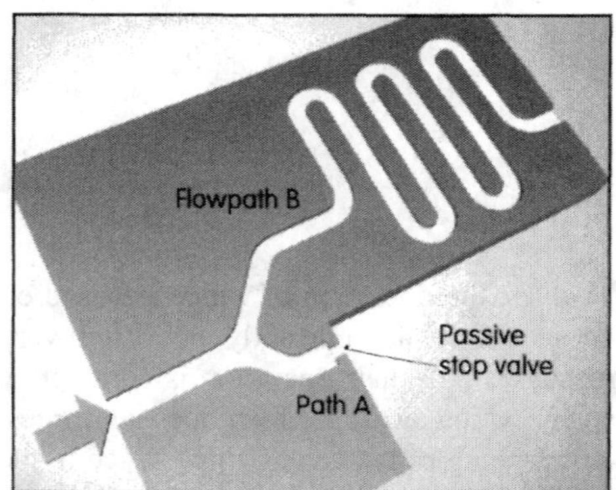

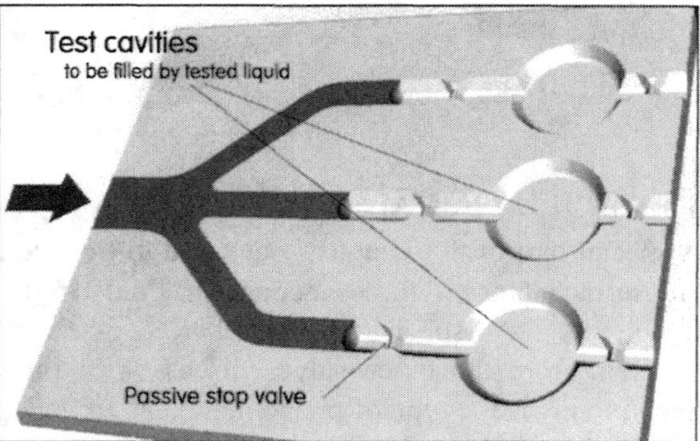

Figure 4.137 (Left) An example of the meniscus-bursting phenomenon used to switch the flow between the two branches. The pressure drop across the stop valve rises with the gradual increase of the friction length in the meander.

Figure 4.138 (Right) Use of the passive capillary stop valves at the entrance into the test chambers ensures simultaneous entry into all chambers even if the propagating liquid column lengths are different.

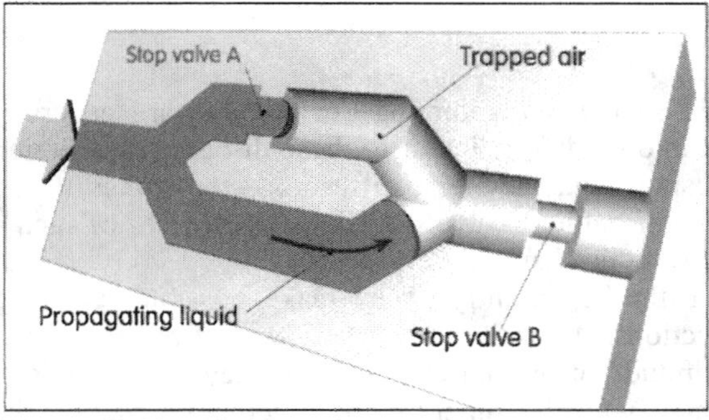

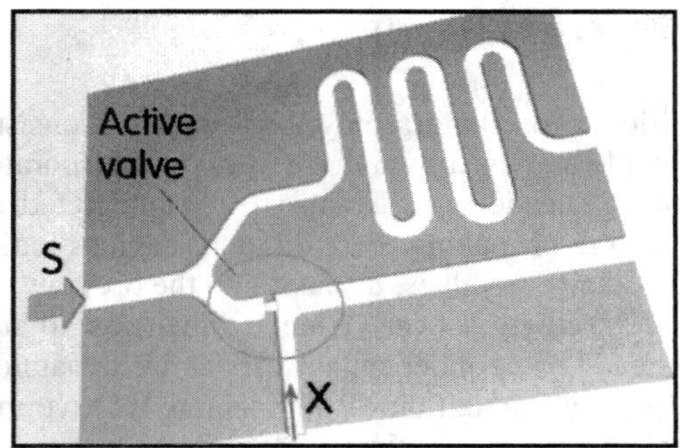

Figure 4.139 (Left) A frequent task in the two-phase liquid/gas microfluidic systems is formation of air bubbles, which may be done using a stop valve in the bypass as shown in this example.

Figure 4.140 (Right) The real beginning of interesting operations that can be done with two-phase liquid/gas: the active valves in which a large main flow S is controlled by the much weaker control flow X, which in this example, similar to that in Figure 4.137, opens the valve before the normal bursting pressure level is reached.

in Figures 2.125 and 2.126, the propagation stops there. What happens next depends on the driving-pressure difference. If it grows and reaches the "burst" pressure (Figure 4.136), then the motion of the liquid column can continue. The description is figurative, the meniscus actually does not rupture, but starts moving. There is, nevertheless, a certain analogy with a "fuze" valve with rupturing of a diaphragm – with the advantage of the capillary "diaphragm" being automatically renewable when the liquid column retreats. Figure 4.137 shows an example of a flow bifurcation based on this mechanism. Initially, the flow of the incoming liquid into the gas-filled system is stopped at the constriction in the path A, but continues through the parallel flowpath B. As liquid propagates there through the meandering channel, the friction it has overcome in path B gradually increases. When the "burst" pressure level is reached in A, it is the propagation in path B that stops. In another application example (Figure 4.138), this stopping at a constriction and waiting for the further pressure increase is used for injecting the liquid into the test chambers simultaneously, despite possible differences in the incoming flows.

A task encountered very often in liquid/gas two-phase flows is generation of an air bubble and its propagation through the channels. Figure 4.139 presents a solution, again based on the assumption of the driving pressure continuously growing as the liquid is pushed into the channels. Such growth is almost invariably the case as a consequence of the increasing liquid column lengths and the associated friction. In this example, the liquid flow is again stopped at the constriction A, but continues through the parallel flowpath, trapping some gas downstream from A. Then the liquid is stopped at the smaller constriction B. Before this becomes open, the meniscus blocking the larger constriction A opens and the gas bubble is thus pushed forward.

The opposite operation of removal of a gas bubble is very simple: just a small orifice made in the side of the channel. This forms a stop valve for liquid (the more difficult to pass, the smaller the size), but presents no real obstacle for the gas.

As already mentioned in Chapter 1 (Figure 1.27), the capillary valves also exist in the *active* version, with the more powerful main flow from the supply source controlled by a control flow. The control action in Figure 4.140 is based on the removal of the constriction blocking meniscus by the control liquid that gets on the downstream side.

4.7 OSCILLATORS

Fluidic oscillators may also actually consist of several collaborating devices, but are generally made and used with these devices incorporated into modules, similar to those discussed in Section 4.5. The devices are also often integrated (losing, e.g., their exit diffusers)—sometimes to the degree of making the individual devices indistinguishable. Very often, the basic devices of such a module are valves, discussed in the previous part of this chapter.

Oscillators bring a new dimension to microfluidic circuitry. They make possible replacing steady, time-independent effects by dynamic actions. Also, they provide a way of bypassing the essential limitation of diminishing relative magnitude of inertial forces at low Reynolds numbers: even though the time-mean effects remain small, with the superimposed oscillatory motion the associated inertial effects—characterized by the Stokes number, (1.4)—may be quite large. Nature has known this for a long time: the smaller the bird and insect (so that their smaller Re would make it difficult for them to fly), the higher is their frequency of their wingbeat (Figure 4.141) Superimposed oscillation, easily produced by a simple, inexpensive, and reliable fluidic oscillator, can improve transport phenomena. In Figure 4.142 there is an example of using oscillation to increase heat transfer by impinging hot air jets.

Oscillators are supplied with steady fluid flow. The self-excited oscillation they produce are a consequence of an inherent hydrodynamic instability. Essential for this concept—sometimes very obvious, in other cases not so easy to find—is the feedback action: the change in the output flow is associated with return flow of a small percentage of the fluid into a location where it can act against the cause that generated the output effect. This necessity of the powerful action of a

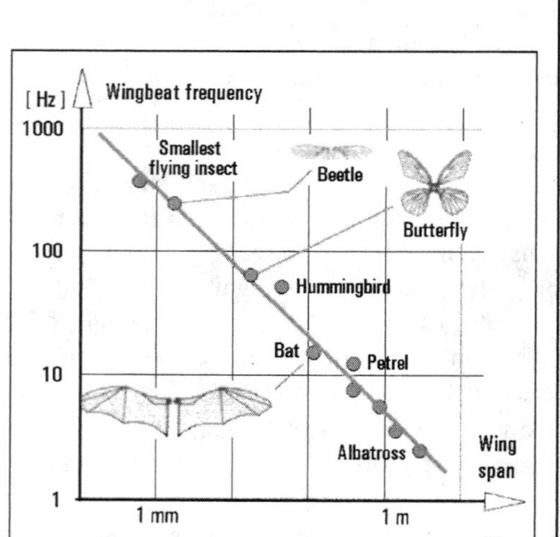

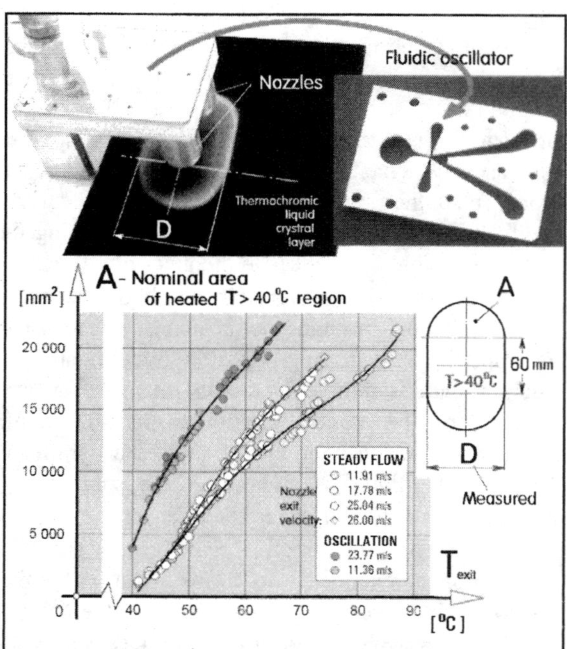

Figure 4.141 (Left) Increasing oscillation frequency with decreasing size as developed in nature. Note that the general trend near to frequency^2 ~ 1/span, so that the Stokes number (1.4) actually increases rather fast with diminishing size.

Figure 4.142 (Right) Typical example of heat transfer improvement by oscillation generated by fluidic oscillator: increased heated area (measured by liquid crystals) under a pair of impinging nozzles.

relatively small amount of flowing fluid, of course, may be found in the signal amplifying properties of typical fluidic valves.

It is useful to classify the principles of the oscillators by dividing them into four groups:

(1) Twin valve oscillators with phase-shifted mutual blockage. This is a rather rarely used principle.

(2) An external feedback loop added to a single valve (or several valves forming an amplifying cascade). This is the very obvious and also very common operating principle, mostly used to generate an oscillatory or pulsatile fluid flow in a connected load.

(3) Internal feedback oscillator, sometimes using a geometry reminding a fluidic valve, sometimes with rather remote from it and perhaps retaining only topological similarity. Also used to generate an output fluid flow.

(4) Oscillator, usually using the internal feedback, with immediate fluido/electric conversion (usually by an inbuilt E/F transducer). This version is very popular for use as a flowmeter. It is not a part of a fluidic circuit but in effect a peripheral device of an electronic circuit.

4.7.1 The twin valve flip-flop

An example of this version is shown schematically in Figure 4.143. The valves A and B are of the vortex type, as described in Section 4.2.3. They are placed in two parallel branches between the common fluid source S and the common exit. In series with each amplifier is a fixed restrictor R. The control nozzles of each valve are connected to the opposite parallel branch, between the valve and its series restrictor. The drop of specific energy across the valve A acts as the control specific energy difference for the other valve B and vice versa. Figure 4.144 presents the graphic solution for the specific energy (vertical coordinate) and mass flowrate (horizontal coordinate) for one of the two parallel branches, it is assumed both branches behave identically. The total specific energy drop, assumed to be constant, is the sum of two parts: (1) the drop Δe_R across the restrictor R, and (2) the drop Δe_V across the vortex valve. The division into these parts is determined by the intersection points of the characteristics. There are two curves for the valve: The specific energy drop Δe_{VCL} across the valve in the CLOSED state is significantly higher than the specific energy drop Δe_{VO} across the valve in the OPEN state.

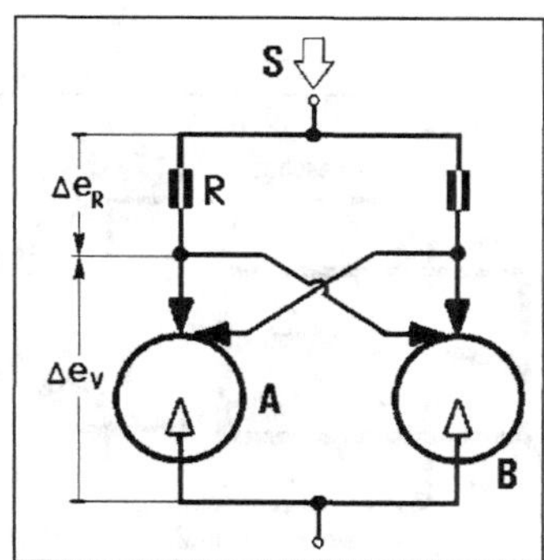

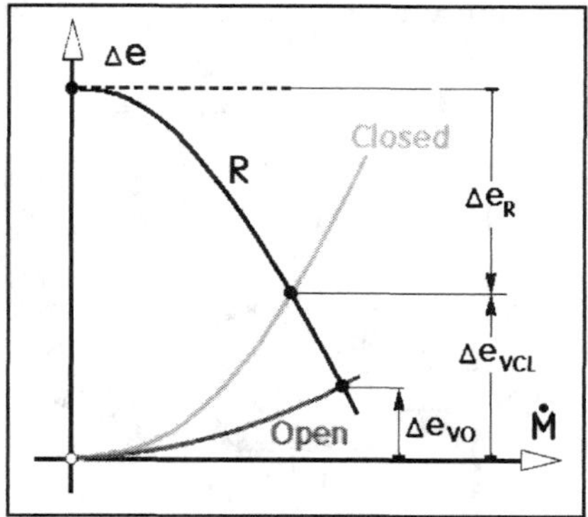

Figure 4.143 (Left) Oscillator using the principle of two mutually blocking valves (in this case the vortex valves).

Figure 4.144 (Right) Specific energy drops across the devices as well as the mass flowrates in the two parallel paths with vortex valve and restrictor R in Figure 4.143 are determined by the intersections of the characteristics. Note the difference in the vortex valve characteristic in the OPEN and CLOSED states.

Let us begin with a situation where the valve A is in its OPEN state while the other valve B is in the CLOSED state. The specific energy difference Δe_{VCL} across the valve B is acting as the control specific energy difference for the valve A. As was mentioned in Section 4.2.3, vortex valves require, for the control action being effective, a control specific energy difference higher than the supply difference fed into the radial inlet. This is fulfilled and the control signal has the tendency to generate rotation in the vortex chamber and close the valve A. On the other hand, the control drop Δe_{VO} available for controlling the valve B is smaller than the acting drop Δe_{VCL} between its radial supply nozzle and its outlet. Therefore, the valve B cannot remain in its CLOSED state. As a result of these control pressure conditions, the circuit switches so that the valve A enters the CLOSED and the other valve B changes its state to OPEN. Then, however, the energetic drops across the valves change and also changed are the control signals. After this change, the valves again cannot remain in the states they have just assumed. They switch back to their former states, and this is periodically repeated. It is necessary to ensure there is sufficient time delay in the control so that the valves can actually get into the other regime before the tendency to switch them back prevails. Usually, sufficient time delay is provided by the inertia of fluid in the interconnecting pipes and also helpful is the sluggishness of vortex amplifiers, which need a certain time before the rotation effect in the vortex chamber becomes effective.

4.7.2 Jet-type valve with feedback loops

Very common and in fact typical for oscillators in fluidics is the use of a single jet-type diverter valve with negative feedback, as presented schematically in Figure 4.145. The amplifier properties of fluidic valve are necessary: the weak input flow has to be able to control a much stronger output flow from which it is derived. The more common oscillators use two feedback loops (Figure 4.146), and because of the common "two-sidedness" (the symmetry of the usual valve design may be thought of as there being two valves in parallel), this concept is actually in principle related to the mutual blockage of the two valves in the flip-flop circuit of Section 4.7.1. Since the very dawn of modern fluidics (Figures 4.147 and 4.148) it continues to be popular.

The feedback—in contrast to the stabilizing feedback action discussed in Section 4.4.3—has to be adjusted so that it has a destabilizing influence, creating states that may seem to be impossible from the static point of view. Let us assume that the jet issuing from the main nozzle supplied by S is attached to the upper attachment wall a leading it to the output Y_1. A part of this flow, however, is returned by the feedback loop 1 to the control inlet X_2. Outflow from this nozzle

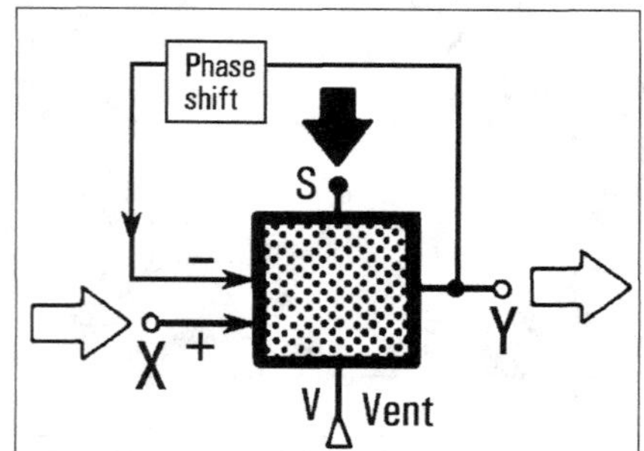

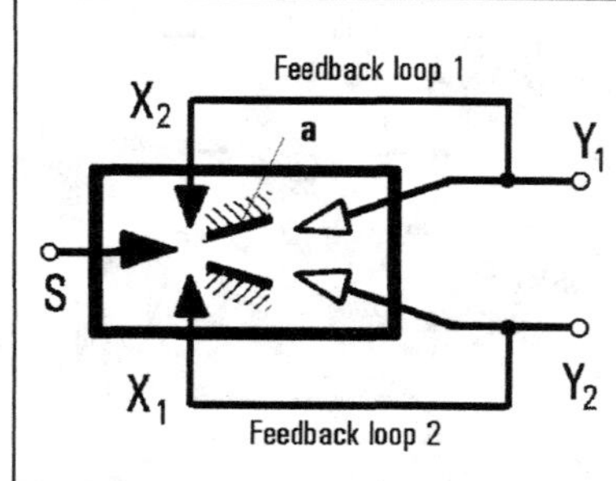

Figure 4.145 (Left) Basic principle of the negative feedback: input signal X is amplified and fed back to the input of the amplifier in a manner opposing the original input effect.

Figure 4.146 (Right) The oscillator with two feedback loops, typical for the standard symmetric layouts of fluidic bistable valves.

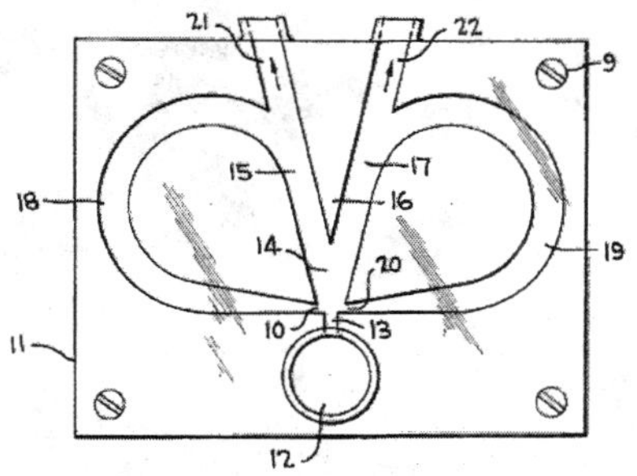

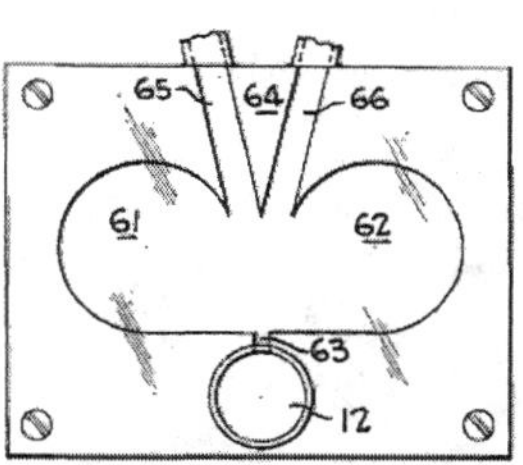

Figure 4.147 (Left) An illustration from the patent filed in 1962 by one of the "founding fathers"of modern fluidics: R.W. Warren [8]. The bistable diverter valve with the two output-input feedback loop.
Figure 4.148 (Right) Another illustration from Warren's patent already shows the oscillator with missing "islands." Seemingly nearly identical, with its absence of attachment walls it is actually based on a significantly different mechanism.

makes the attachment to the upper attachment wall a "impossible"; it switches the jet to the other side of the interaction cavity. But this deflection of the jet is impossible as well, because it means the jet is leaving through the output Y_2 and due to the presence of the feedback loop 2 a flow from the control inlet X_1, switching the jet upwards. Indeed, the jet does not remain in any of the two possible deflected states and oscillates between them. As in the case of the flip-flop, it is essential to introduce some delay into the feedback loops so that the jet can actually get temporarily into one of the deflected states before the tendency to switch it back prevails. Warren (Figure 4.147) found that inertia of the fluid repeatedly accelerated in the feedback loop channels sufficed for this delay action.

In some applications, it may be useful to generate asymmetric oscillation, for example, to produce short pulses separated by longer intervals, and the presence of the two loops makes this possible by making the feedback paths unequal.

If the required oscillation frequency is very low, relying solely on fluid inertia may lead to very long feedback channels. This is not a fundamental problem, experience has shown that the feedback signal is not dissipated even at channel lengths of the order of tens of meters (see Figure 4.154), but such lengths are likely to be impractical. It is then useful to increase the delay by the RC member in the loops. The fluid can get to the control nozzle only after it has filled an accumulation chamber. The filling may be made slower by placing a restrictor R upstream from the chamber, as shown in Figure 4.149 (and Figure 4.150, showing the use of gas compression

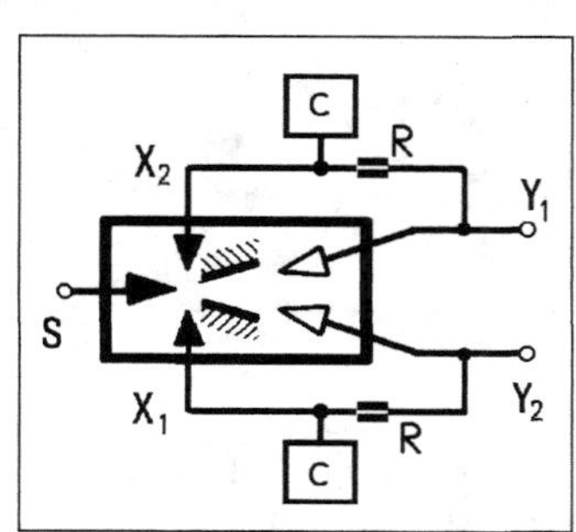

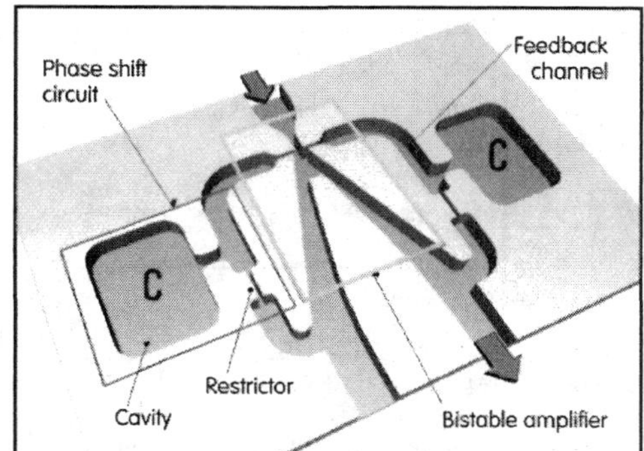

Figure 4.149 (Left) and Figure 4.150 (Right) The usual way of introducing the phase shift to the feedback signal transfer is delaying the control nozzle flow by the time required to fill the accumulation vessel C through the restrictor R.

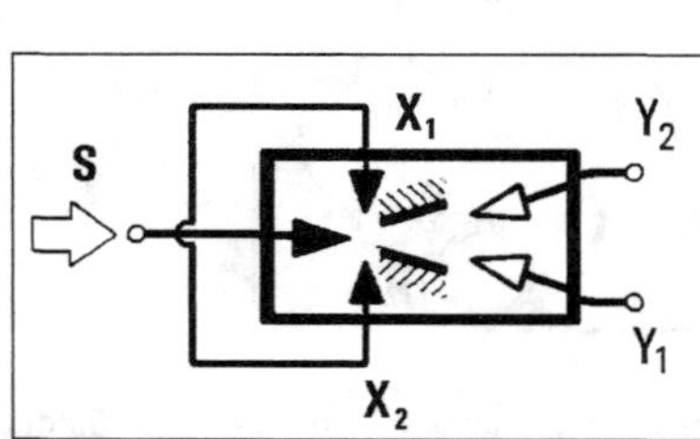

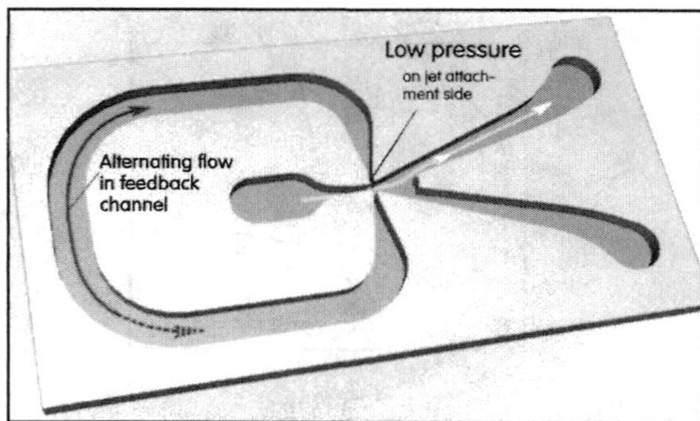

Figure 4.151 (Left) Schematic representation of the oscillator consisting of an amplifier valve and a single feedback loop connecting both control nozzles.

Figure 4.152 (Right) An practical example of the oscillator from Figure 4.151: a high-frequency version with rather exceptionally short feedback loop (compare with the lengths in Figure 4.154).

accumulation). The restrictor has to be chosen with care, considering its dissipance with that of the control nozzle. If the restrictor cross section is much smaller, the feedback flow may be reduced to a degree that makes it ineffective. With liquid as the working fluid, to get sufficient accumulation properties in reasonably sized chamber may require using some of the more complicated mechanism discussed in Section 2.8.

The two-loops feedback, however, is not the only possibility. The alternative is the simpler single loop connection according to Figure 4.151. While the twin-loop version can use a proportional amplifier to generate harmonic (or rather hear-harmonic) oscillation, the single-loop version uses almost always the bitable valve, generating the rectangular wavetrain, as shown in Figure 4.153 (with the typical superimposed chaotic high-frequency turbulence).

The single-loop feedback mechanism is perhaps less obvious. It uses the fact that the Coanda attachment is a pressure effect; the jet is kept deflected by the low pressure at the attachment wall. The low pressure is also in the control nozzle on the attachment side. Figure 4.152 captures the

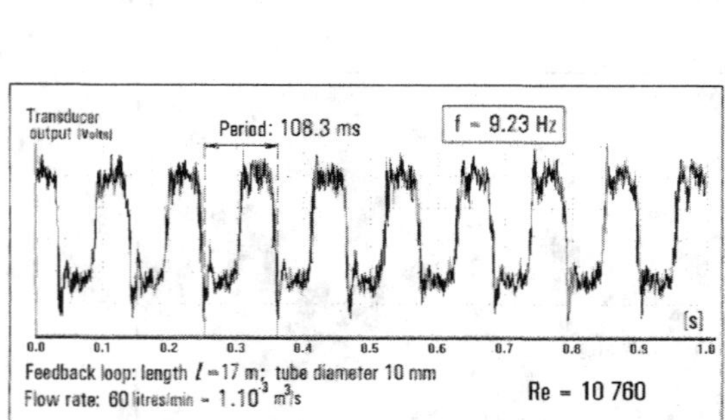

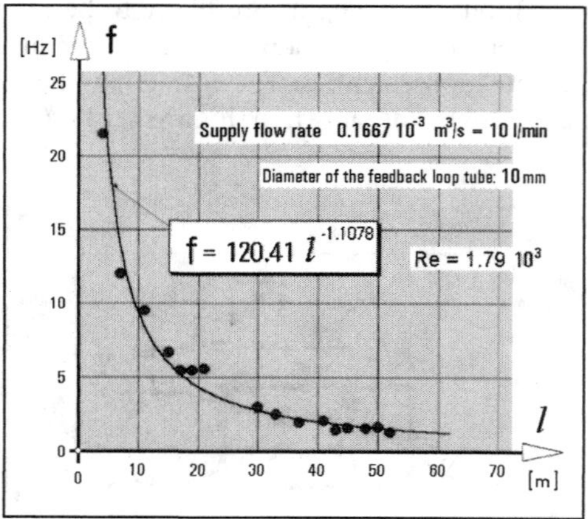

Figure 4.153 (Left) Typical oscilloscope trace of the anemometer signal from the output of the amplifier valve model (the one from Figure 4.93 with very long 10 mm diameter loop connected according to Figure 4.160. (*After:* [6].)

Figure 4.154 (Right) Results of experimental investigation of the influence of the feedback loop length on the oscillator with the amplifier valve from Figure 4.93 and the 10 mm diameter loop. Oscillation frequency f is practically inversely proportional to the feedback loop length *l*. (*After:* [6].)

moment with the jet diverted to Y_2 and low pressure in X_1. The loop connects it with X_2 where at this moment the pressure is higher as the Coanda effect does not act there. This pressure difference between the ends of the loop generates the flow from X_2 to X_1. This results in a control flow issuing from the control nozzle X_1, deflecting the jet to the opposite attachment wall. This generates the low pressure at the other end of the loop, the flow in which reverses its direction, so that the process continues by permanent oscillation. The necessary phase delay of the feedback action is due to the inertia of the fluid in the loop, which has to reverse its flow direction twice per period.

4.7.3 Feedback loop mechanisms

To provide again an specific example, investigations were performed with the bistable amplifier valve model described in Section 4.4.1 (geometry in Figure 4.96 and photograph in Figure 4.93). It became an oscillator of the single feedback type, by being provided with the loop of 10 mm internal diameter Tygon tube of various lengths, connecting the control terminals X_1 and X_2. Experiments performed with air found it oscillating reliably across a wide range of supply flowrates. Figure 4.153 shows an example of a typical regular oscilloscope trace of hot-wire anemometer signal in the output terminal. Were there not the superposed chaotic turbulence, the trace would be a succession of identical rectangles. Figure 4.154 is a plot of the frequency of the generated switching oscillation as a function of the feedback tube length. These measurements were made at a constant air supply flowrate. The length *l* of the feedback loop is important because in this oscillator with just the plain tube the feedback action depends mainly on the phase delay caused by inertia of the fluid in the tube. Since the jet switching in the valve is a much faster process, having insignificant influence, the frequency is determined by the delay inversely proportional to loop inertance J that (Figure 2.85) is proportional to channel length. The

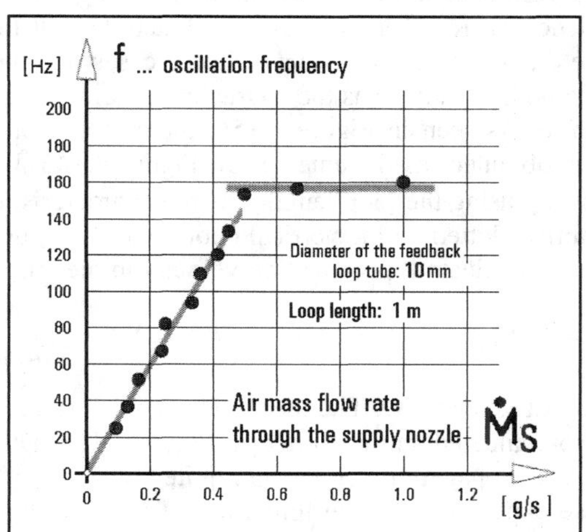

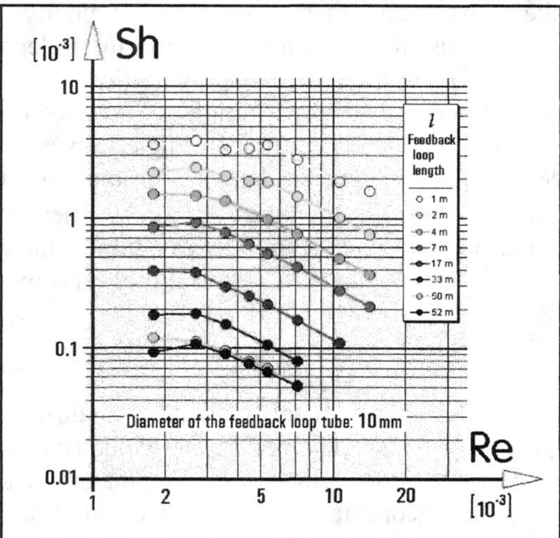

Figure 4.155 (Left) Frequency of generated oscillation of the oscillator build with the bistable valve from Figure 4.93 is plotted here (for a particular loop length) as a function of the supplied air mass flowrate.

Figure 4.156 (Right) Standard Strouhal number, (1.5), plotted as a function of the main nozzle exit Reynolds number. As the velocity (and hence Re) reaches a certain limit value, Strouhal number ceases to be constant. The value of the critical Re at which this change takes place is similar to the value of laminar flow transition into turbulence in the round pipe, but here this is a mere coincidence. The real reason behind the change is the propagation speed limit.

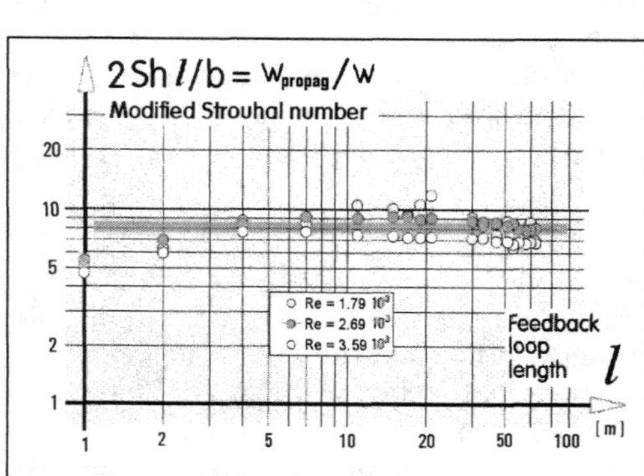

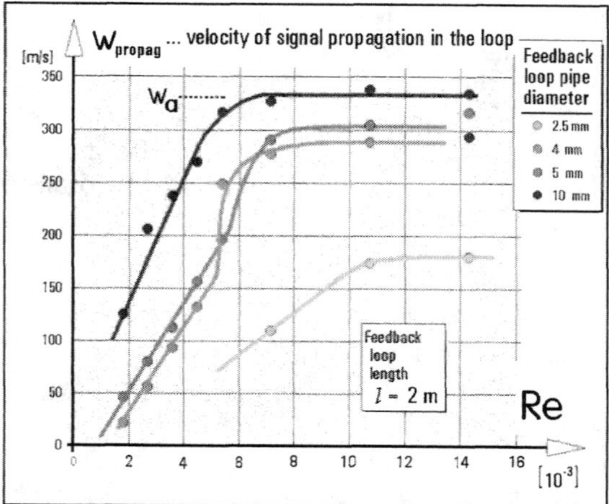

Figure 4.157 (Left) The nearly constant Strouhal number evaluated at low Reynolds numbers for this particular case using the more appropriate characteristic dimension, the loop length l rather than the nozzle exit width b.

Figure 4.158 (Right) Propagation velocity evaluated (approximately; neglecting the switching time in the valve) according to (4.8).

practically inverse proportionality $f \sim 1/l$ in Figure 4.154 (deviation no doubt due to scatter and small errors) comes, therefore, as no surprise.

The experiments revealed two important facts. First, in Figure 4.155 there is the measured dependence on the other variable influencing the frequency, the supplied flowrate. This time the loop length was constant and the air flowrate was varied. It is characteristic for most no-moving-part aerodynamic oscillations that the frequency increases in proportion to the flowrate, or, alternatively expressed, to nozzle exit velocity. This results in constant Strouhal number Sh, (1.5), which is usually plotted as showing the independence on Reynolds number (an example will be seen in Figure 4.161; the small variations there are due to secondary effects). In the cases like the present one, the usual definition of Sh based on the nozzle width b as the characteristic dimension does not lead to the desirable single universal curve. As seen in Figure 4.156, there is a strong dependence on the loop length. The universality is obtained, as documented in Figure 4.145, by multiplying Sh by the ratio l/b, or, in other words, by using the loop length l as the characteristic dimension. An interesting interpretation of the quantity plotted on the vertical coordinate in Figure 4.157 should be noted. It is the ratio of (approximately evaluated) propagation velocity in the loop,

$$w_{propag} = \frac{2l}{\Delta t} = 2lf \tag{4.8}$$

to the (also approximately calculated) main-nozzle exit velocity w. The mean value of this ratio is quite large, ~ 8. The nozzle exit velocity was also evaluated and is plotted in Figure 4.158. Due to the value of the velocity ratio, the magnitudes of the propagation velocity are quite high.

The second fact of importance found in these experiments is the limitation of the constant Sh regime. In Figure 4.155 the linear growth of frequency with the supplied flow is found only at small flowrates, or small Re, though in this case the phenomenon is not a Reynolds number effect. The reason for the deviation from the linearity is the propagation velocity in the loop reaching the speed of sound: the acoustic propagation velocity w_a (Figures 1.43 and 1.44). This limit is apparent in Figure 4.158. Even the decrease of the limit with diminishing tube diameter is in agreement with known facts (Figure 1.47).

The conclusion from this exercise is as follows: there are two different regimes of the signal propagation in the feedback loops of fluidic oscillators. Both are capable of producing the oscillation. One of them is the hydraulic regime with essentially (there may be small deviations)

$$\mathrm{Sh} = \mathrm{const} \tag{4.9}$$

The other is the regime with propagating pressure waves, for which

$$w_{propag} = w_a \tag{4.10}$$

The situation found in the discussed case of both mechanisms being present and actually with transition between them is rather exceptional.

4.7.4 The internal feedback

The distinguishing feature of the third and fourth oscillator groups listed at the beginning of Section 4.7 is the absence of visible feedback channels, at least visible at first sight. In reality the hydrodynamic instability due to a feedback mechanism is always present. It may be even distinguished whether is belongs to the mechanism described by (4.9) or (4.10).

In some of these oscillators, it is possible to follow the development line leading from the principles with the external loop. This may be even verified by existence of devices representing an intermediate development stage. This is the case of Figure 4.159 in which there are distinguishable transitional steps leading from the twin-loop external feedback oscillator to the target type device Figure 4.160. In fact the boldest step, the one to the "islandless" version B in Figure 4.159, was made already at the very beginning of fluidic oscillator history (Figure 4.148). The target oscillator does not possess a fluidic output channel as it was developed for immediate

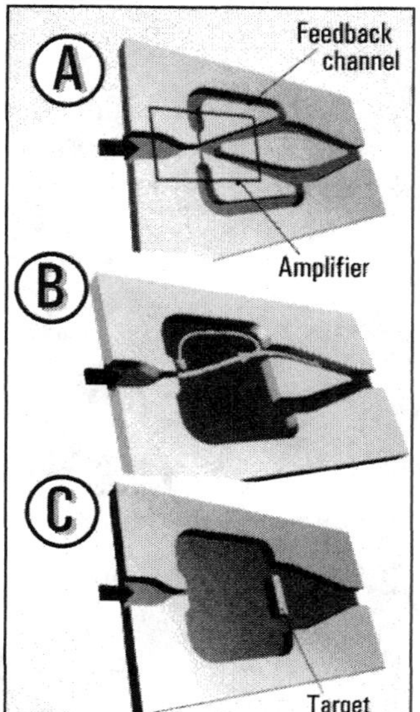

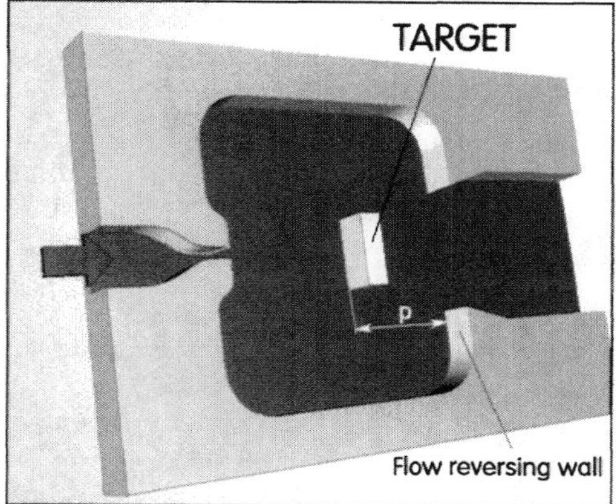

Figure 4.159 (Left) Development leading from the standard oscillator version with two feedback loops A through the "islandless" version B of Figure 4.148 (no Coanda-effect jet attachment) to the target type oscillator C.

Figure 4.160 (Right) A typical target version of the oscillator. The oscillation mechanism is somewhat different with the target body at the distance p in front of the inward turned walls of the cavity.

generation of an electric signal by an in-built transducer. The constancy of Sh means there is a direct proportionality between the nozzle velocity (and hence fluid flowrate) and the number of oscillation periods per time interval. The latter is easily measured, transferred (no effect of resistance changes in the transition lines), and converted to digital signal by electronic counting techniques.

To ensure suitability for flowmetering and uses like mixing (the colliding jets oscillator described elsewhere in this book: Figures 3.123 and 4.164) there are several design factors that are necessary to consider:

(a) A factor of high importance, especially for use in microfluidics, is the lowest Reynolds number Re_{min} at which the device still keeps oscillating. This defines the lower end of the operation range.

(b) To ensure high flowmetering accuracy and similarly in their use as mixers, an important criterion is designing the fluidic oscillators to pass the smallest possible fluid volume per period.

(c) Also important is the Euler number Eu, the dimensionless magnitude of pressure loss. It has to be small, being the usual limiting factor at the upper end of the range.

A useful dimensionless parameter, high value of which ensures high performance in all three respects (a, b, and c above), is the Markland number

$$Mk = \frac{1}{Re_{min}\sqrt{\lambda\sqrt{Eu}}} \tag{4.11}$$

A list of the values for several flowmetering oscillators is the following table:

Hermann & Tesař 1991	$Mk = 0,35 .10^{-3}$
Kalsi, Markland et al. 1988	$Mk = 2,70 .10^{-3}$
Boucher & Mayharoglu 1988	$Mk = 6,45 .10^{-3}$
Kawano et al. 1986	$Mk = 7,10 .10^{-3}$

It includes an attempt by the present author with his diploma student, showing the typical values obtained without proper optimization. A useful fact to remember is the importance of small friction on the top and bottom flat plates, ensured by high aspect ratio $\lambda = h / b$.

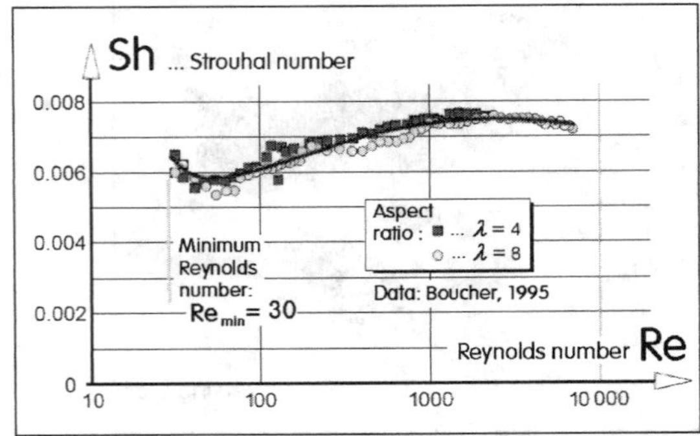

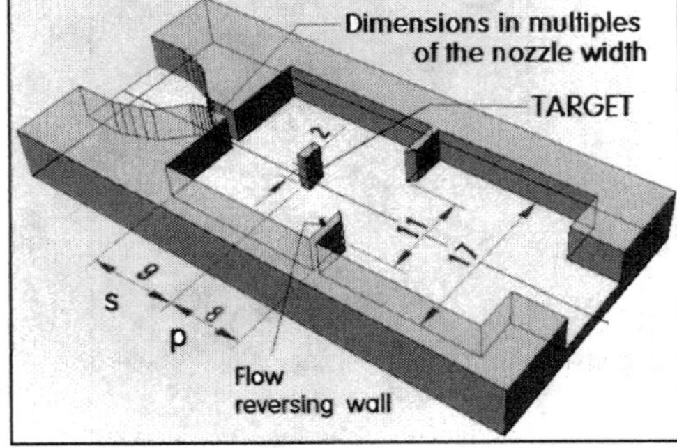

Figure 4.161 (Left) Results of frequency measurements with the flowmeter oscillator shown in Figure 4.162. Strouhal number is reasonably constant and the oscillation is present down to very low Reynolds numbers. (After [9].)

Figure 4.162 (Right) The favorable results were obtained with this particular (actually quite typical) example of the target type oscillator thanks to its large aspect ratio λ.

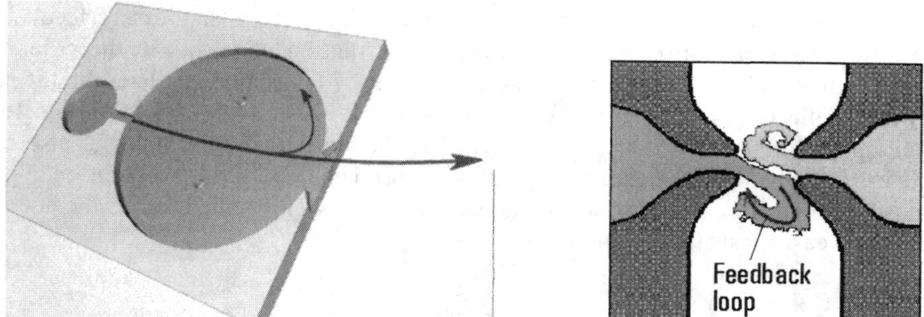

Figure 4.163 (Left) Water jet oscillator developed for cleaning surfaces by the issuing transversally oscillating free water jet (*After:* Nakayama, [10]). An interesting feature is the jet deflection by "attachment to an edge," in fact pressure force deflection to the low pressure recession, which is protected by the jet flow past the edge from ingress of higher pressure outer air. A similar bistability mechanism is used in the amplifier in Figure 4.168.

Figure 4.164 (Right) Flow visualization of the colliding jets oscillator from Figures 3.125 and 3.126 with the picture processed by an extreme reduction of the color palette (so-called posterization) indicates that even there the oscillation is due to actions of internal feedback loops.

Characteristically, the high performance of the Kawano's oscillator was obtained with extremely high aspect ratio $\lambda = 23.7$.

An important feature of the target oscillator (Figures 4.162 and 4.160), is the flow reversing walls leading to the feedback flows. As a matter of fact, the oscillation can be present even without them, but the generation mechanism will be different: it will be the vortex shedding flow past a blunt body, producing the periodic the wake, such as the Kármán vortex street in the flow past a cylinder. There the feedback, producing the alternative shedding of the vortices on one and then

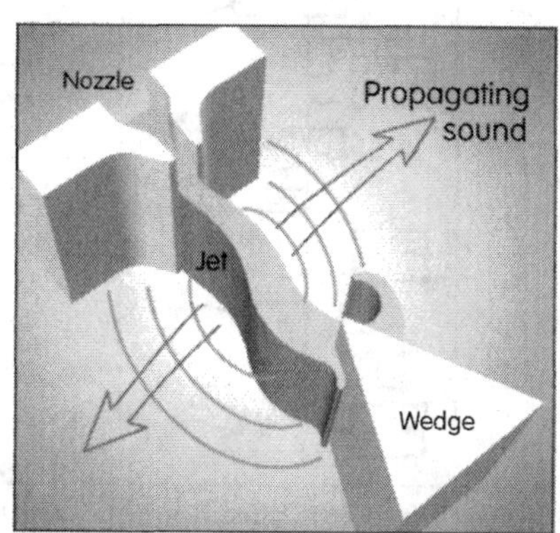

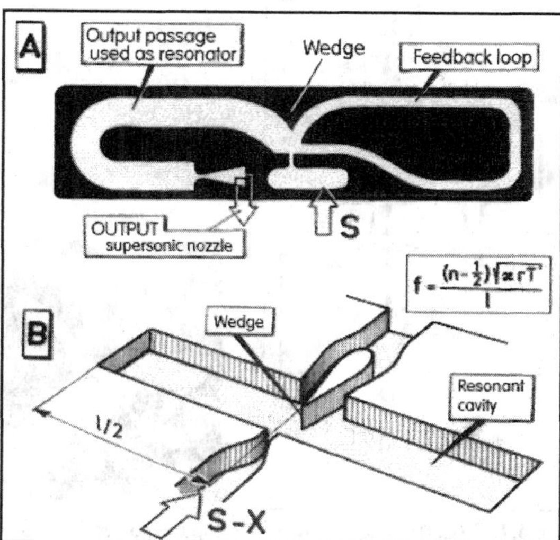

Figure 4.165 (Left) The edgetone phenomenon: a jet generated by a slit nozzle impinging on a sharp-edged obstacle undergoes transversal oscillation because of the feedback pressure action from the vortices cut off by the edge on alternate sides of the edge. The mechanism has been used since time immemorial in musical instruments of the flute family.

Figure 4.166 (Right) Fluidic edgetone type oscillators with oscillating frequency stabilized by the adjacent resonant cavities.

the other side of the body, is of the pressure-transmitted kind. A similar pressure feedback effect transmitted by pressure action causes the transversal oscillation of the jet in the case shown in Figure 4.165. When used in fluidic oscillators, the frequency is usually stabilized by the returning pressure waves reflected back from the end of an adjacent resonant cavity (Figure 4.166). The dependence on the flowrate is lost here, nevertheless this type of the oscillator is also useful. It may serve, in particular, as a temperature sensor. This use is based on the temperature dependence of the acoustic propagation velocity w_a (Figures 1.41 and 1.42). Again, there is the advantage of the output signal easily transferred and converted to digital form.

4.7.5 Frequency dividers

In general, it is inconvenient and space consuming to use long feedback loops or loops with large accumulation chambers needed to generate low-frequency oscillation. On the other hand, high-frequency oscillation at ultrasonic frequencies up to 35 kHz with special devices and about a decimal order of magnitude less in nonoptimized designs may be produced rather easily. There are, however, applications requiring very low frequencies. It is useful to know there are circuits capable of decreasing a generated frequency by a constant integer number factor. As an example,

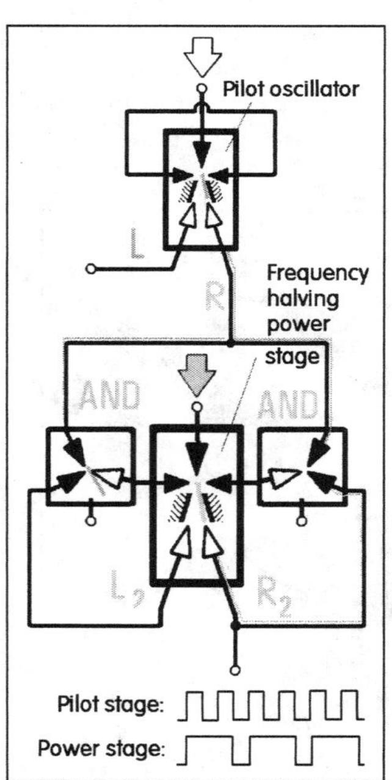

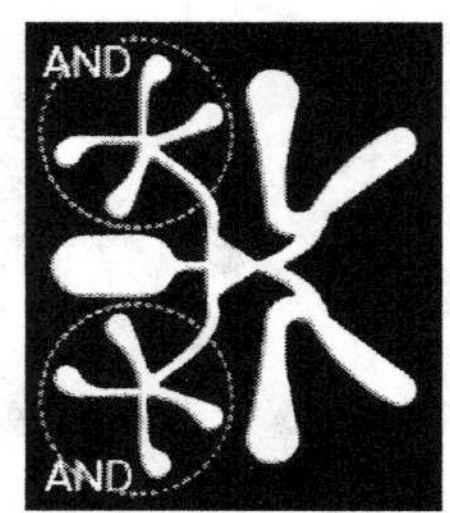

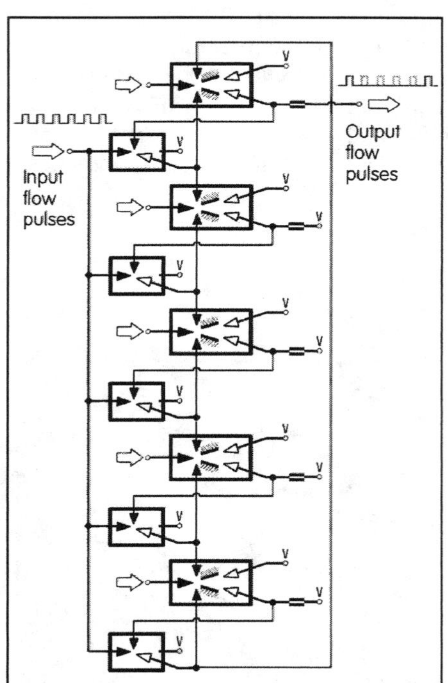

Figure 4.167 (Left) Pilot oscillator driving the power stage through frequency halving connection. The circuit with two passive AND devices uses essentially a binary digital circuit technique for frequency reduction.

Figure 4.168 (Center) One of the "fluid logic" devices retaining their usefulness are the AND jet-type devices shown here at both inputs of the bistable amplifier: the output signal is available only if both input signals are present.

Figure 4.169 (Right) Circuit diagram of a frequency divider: with this five-stage ring counter, only each fifth input pulse is passed to the output.

the frequency of oscillation generated by the pilot oscillator may be halved, with simultaneous power increase, by the circuit of Figure 4.167. The technique used is essentially a digital one, a leftover from the time when fluidics was developed for signal-processing digital systems. The circuits, actually quite simple, which divide the frequency are the ring counters. Their main part is an array of bistable amplifiers connected so that only one of them is in an activated state at any time. At each incoming input pulse from the pilot oscillator the activation is shifted to a neighbor in the ring. An output pulse is produced only when the activation progresses full circle to the initial position. A necessary device cooperating with each of the bistable diverters is a vented jet-type "logical AND," with two nozzles and a single collector (Figure 4.168). It delivers the exit flow only if the first and the second input flows are present simultaneously. In the example shown in Figure 4.167, only one of the return conduits in the power stage, resembling feedback loops, brings a signal into the AND device (to the right-hand one in this figure). This means only one pulse from the generator progresses to the power stage. Flow pulses occur in its right-hand side output R_2 only at each second pulse coming from the right-hand side output R of the generator.

The principle is evolved further in the fully fledged ring counter shown in Figure 4.169. The input pulse can progress only through the AND device immediately below the currently excited bistable amplifier. It is then branched into two lines, one extinguishing the excited valve while the other excites its nearest lower neighbor.

4.8 FLUIDIC RECTIFIERS

A rectifier in principle carries out a task inverse to that of an oscillator. It is supplied with alternating flow and converts it into the one-directional motion. Many microfluidic systems are driven by an inbuilt miniature reciprocating pump with rectifier circuits used to produce the steady output flow, usually with some remaining pulsation, which is smoothed out by accumulating cavities and RC modules naturally present or sometimes built for this purpose. Most microfluidic pumps described in the available literature (e.g., [11]), are in principle a scaled-down version of the classical large-scale reciprocating pumps, with mechanical suction and delivery valves. The development trend towards the pure fluidics is not felt very strongly in these cases where, anyway, the displacement motion generating the alternating flow is usually produced by a mechanical actuator. Nevertheless, the advantages of the no-moving-part versions are recognized and lead to the so called "valve-less" purely fluidic pumps. The name "valve-less" is obviously a misnomer. Not only are the valves are there, but they are actually the most interesting parts of these pumps, presenting challenging opportunities for development. What is absent are the movable and deformable components of the mechanical valves.

4.8.1 The Grätz bridge circuit

The usual present-day pumps in microfluidics are of the simplest configuration corresponding to Figure 4.170. The actuator, represented symbolically by the traditional crankshaft-connecting rod-piston mechanism, generates the alternating flow rectified by two fluidic diodes (Section 3.5). During the first half of the cycle, the liquid is sucked into the displacement cavity through the upstream or suction diode. No fluid is coming from the output, because return flow is prevented by

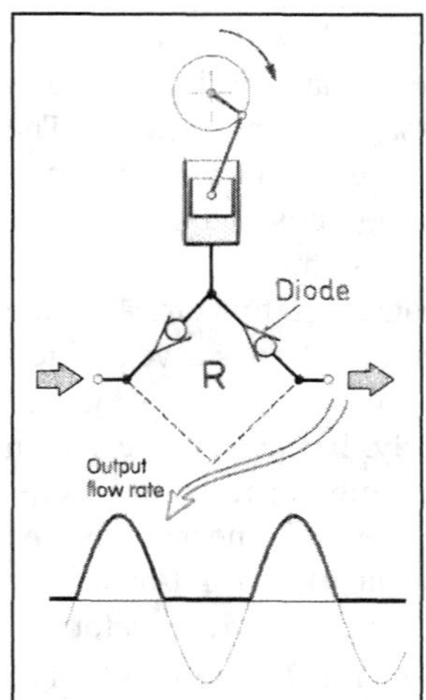

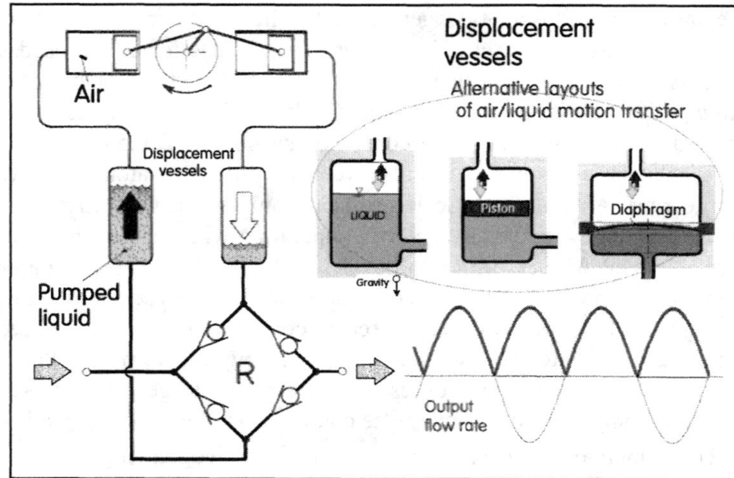

Figure 4.170 (Left) Schematic representation of a single-action reciprocating pump. The alternating flow, shown symbolically as generated by a piston mechanism, is rectified by two fluidic diodes - the upstream suction valve and the downstream delivery valve.

Figure 4.171 (Right) Schematic representation of the double-acting two-phase reciprocating pump. Two alternating flows, mutually shifted by π rad phase shift, are rectified by four fluidic diodes connected into the Grätz full-wave bridge. Also shown is one of the advantages of energy transfer by the alternating flow: the possibility of using different fluids in different parts of the fluidic system.

the upstream or delivery diode, which is oriented so that in this phase of the cycle it is in its CLOSED state. Its state changes into OPEN in the subsequent half of the cycle when it allows the liquid, displaced from the cylinder, to flow to the output terminal. Obviously, the output flow is intermittent, available only during one half of each operating period. The output flow pulsation is an obvious disadvantage.

In applications where the pulsation can cause problems and also in those aiming at a higher efficiency it is advisable to use the full-wave (Figure 4.171) Grätz bridge rectifier (Figure 4.172). It needs to be supplied by two-phase alternating input flow, carried in two channels. Generating it is usually no essential problem since the to-and-fro motion of the front side of the mechanical actuator (symbolized in the drawing by the piston) means availability of a motion of the same amplitude with the opposite phase on its reverse side. The second phase may be also used differently, according to Figure 4.173, to increase the generated output pressure. This is welcome in situations where the pump has to drive a high-dissipance load.

In the "valveless" pumps the mechanical valves with closure by a movable (or deformable) component are replaced by the purely fluidic diodes. Generally speaking, the phenomena in no-moving-part diodes are based on dynamic effects and this may lead to a decreased rectification efficiency in very small-scale microfluidics. Also, these diodes never achieve a complete stopping of the flow and are therefore inherently less efficient than their mechanofluidic counterparts. Fortunately, in common layouts, a central single pump supplies the whole system with sometimes a large number of devices. This leads to increased total delivered flowrate and hence larger pump Reynolds number even if Re in individual devices themselves are small. Generally, characteristic Re for pumps are larger than 1 000, which is a safe value for the use of dynamic effects. Another aspect for the relation between the inertial and friction forces is the fact that the former ones

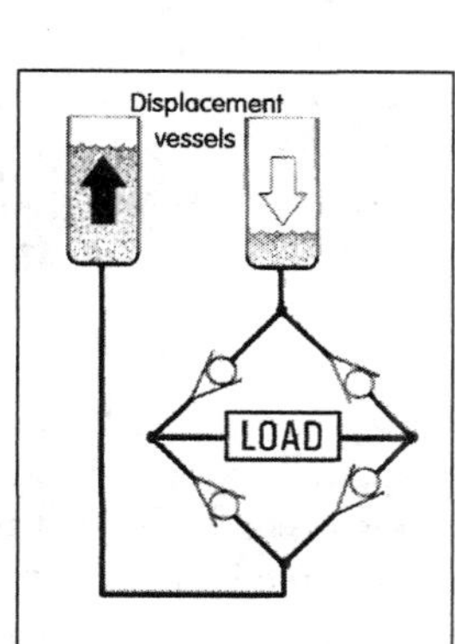

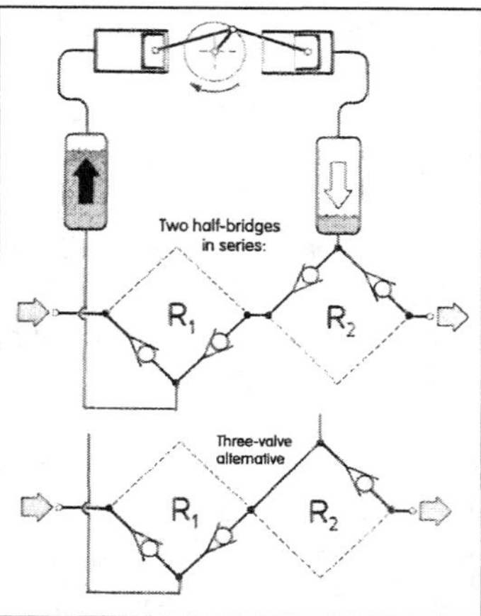

Figure 4.172 (Left) The reason why the Grätz circuit is called a "bridge": the load with the one-way fluid flow is bridging the two parallel lines of diodes.

Figure 4.173 (Right) The double acting, two-phase driving does not necessarily mean use of the Grätz full-wave bridge. As shown here, the two phases may be used in half-bridges connected in series. The pulsation of the output flow is not suppressed, but the rectifier generates a higher output pressure.

increase their relative importance with increasing operating frequency of the alternating input flow. The frequency is characteristically chosen higher as the size of the pump decreases (compare with Figure 4.141). Rather than the Reynolds number, the proper criterion in these cases is the Stokes number (1.4).

Of the two-terminal fluidic diodes, the highest rectifying effect offer vortex diodes (Section 3.5.2). Their use in microfluidic pumps is limited by two factors. First, considering the structure of the full-wave bridge even in the schematic representation of Figure 4.174, it is obvious how difficult it is to accommodate the exits leading at right angles from the vortex chamber into the

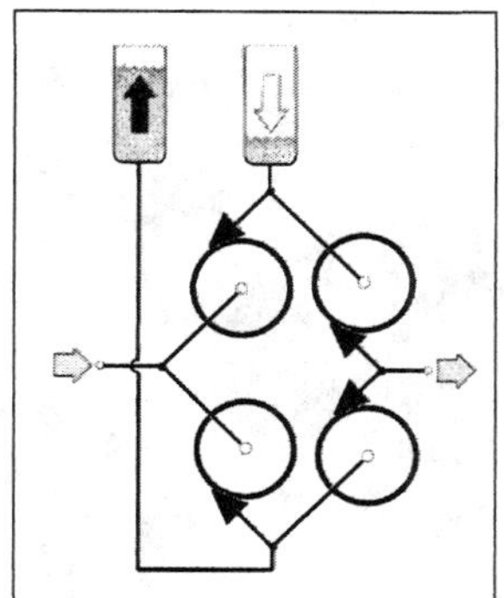

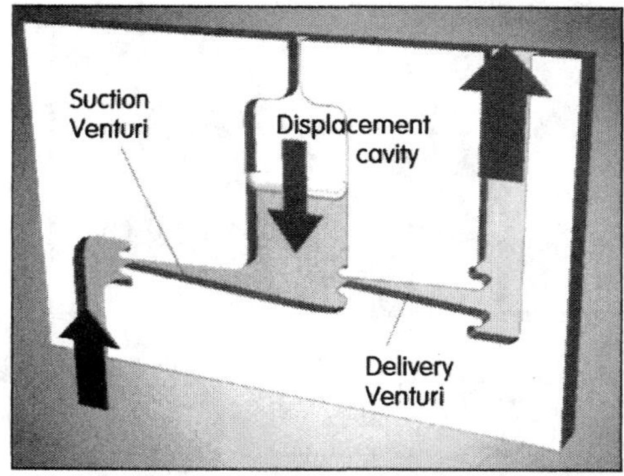

Figure 4.174 (Left) Schematic representation of rectification by vortex diodes – in the double-acting two-phase Grätz full-wave bridge. The disadvantages are the slow start-up of diodes limiting pumping frequency and the necessity of complex spatial layout.

Figure 4.175 (Right) Single-action pump using the rectification effect of the Venturi (convergent-divergent) diodes is currently the most common type of the "valveless" pumps used in microfluidics. Shown here as driven by alternating air pressure, it is commonly driven by various electro/fluidic (E/F) transducer actuators.

typical planar configuration fabricated on a substrate plate. Even more problems would occur in the multiport inlets required in the high-performance vortex diode versions. The second factor is the rather poor frequency range. It takes a relatively long time to start up the rotation after the nominal onset of the reverse CLOSED state. This counters the modern trend towards the operation at a high frequency.

More suitable for the planar layout are the Tesla diode (Figure 3.83), and the various versions of the labyrinth (Section 3.5.1) and Venturi (Figure 4.175) diodes. Their diodity $\mathscr{D}$, (3.1) is, however, rarely higher than 1.2 to 1.6. An improved version, the Tesser valve, is described in [12]. To improve the rectification effect, these diodes are usually designed in multistage arrangements. A better performance and, because of the short and simple flowpath, also a high frequency range are characteristic of the Venturi diode (Figure 3.91), consisting of a nozzle and a diffuser placed in series. Under various other names, it is currently the most popular among the no-moving-part rectifiers. An application of a two-Venturi rectifier circuit in a flow control actuator for generation of the hybrid synthetic jets [13] is an example of a use of the rectifier circuits other than in simple pumps.

4.8.2 Jet-type rectifiers

Use of the rectification principle based on properties of a fluid jet is not new. They were mostly used in devices of little importance, such as aquarium ventilators, coolers, or humidifiers, where the dominant requirement was low cost. Sometimes the decisive factor for their use was the fact that inertia of moving parts in mechano-fluidic diodes prevents them from operation at 100 Hz,

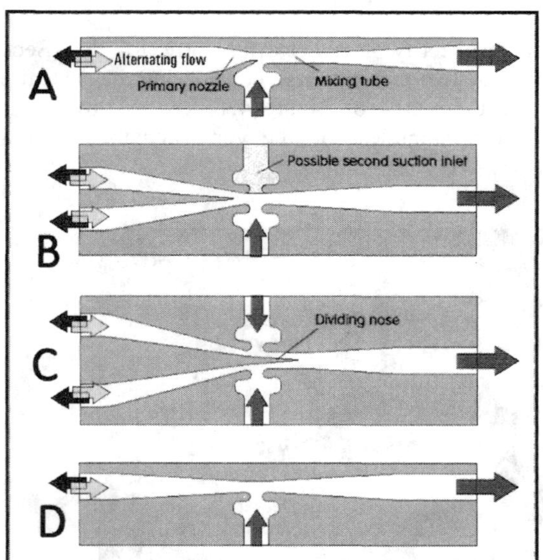

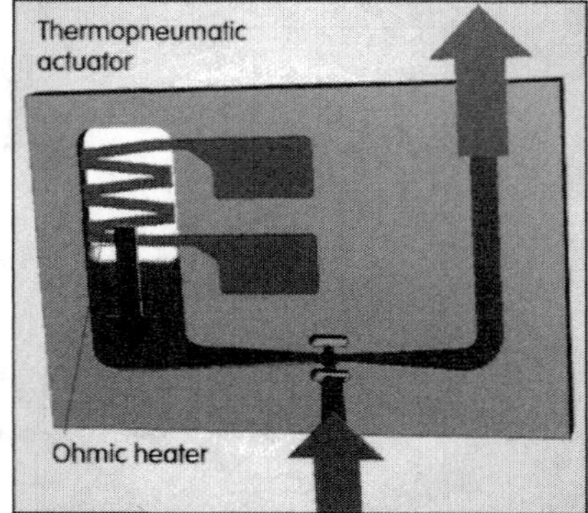

Figure 4.176 (Left) Survey of the linear jet-type diodes. A – The standard jet pump may exhibit a reasonable rectification effect. B – The Walkden two-phase full-wave rectifier was developed from a jet pump with two parallel "primary nozzles," actually better performing when shaped as diffusers. C – Tippetts' analysis of the long-nosed symmetric Walkden rectifier found it actually consists of 3-terminal devices. D – The 3-terminal Venturi.

Figure 4.177 (Right) A simple single-action pump with the 3-terminal Venturi rectifier. Less known than the version from Figure 4.152, it may exhibit a higher rectification efficiency.

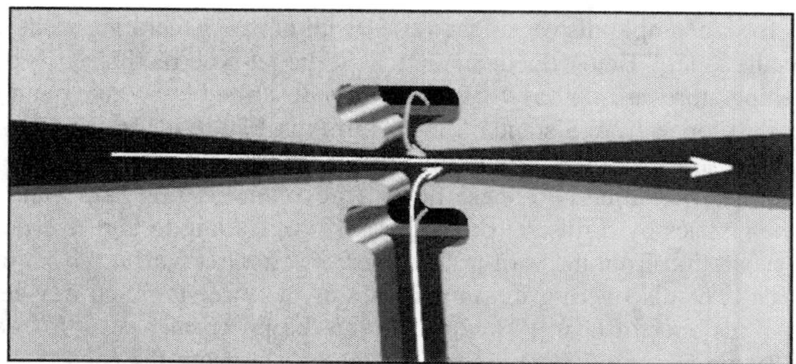

Figure 4.178 Operation of the three-terminal Venturi rectifier (also known as RFD). Fluid from the pump inlet suction channel enters the rectifier due to the jet-pumping effect of entrainment into the jet, which is generated (oriented differently) in both halves of the operating cycle.

the frequency of a solenoid actuator driven by AC mains (also the wear of contacting moving components would be too rapid at frequencies of this order). Probably the first reference in literature is from 1957, when Dauphinne described in a short note the use of the "loudspeaker wind" for simple closed-circuit cooling. The loudspeaker was provided with a front plate with a

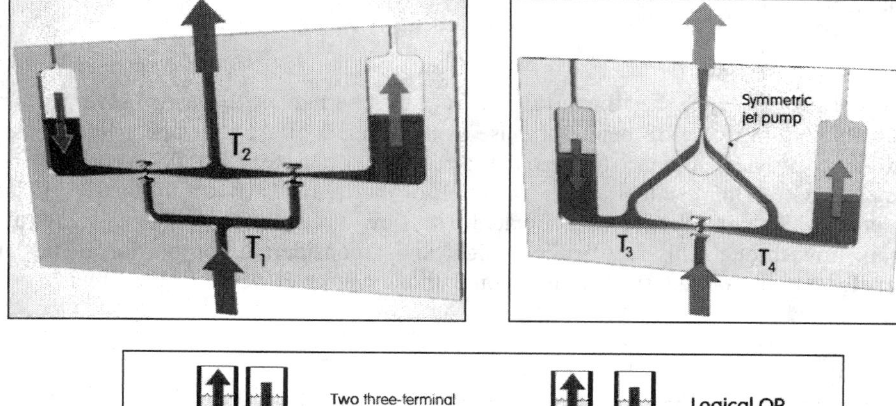

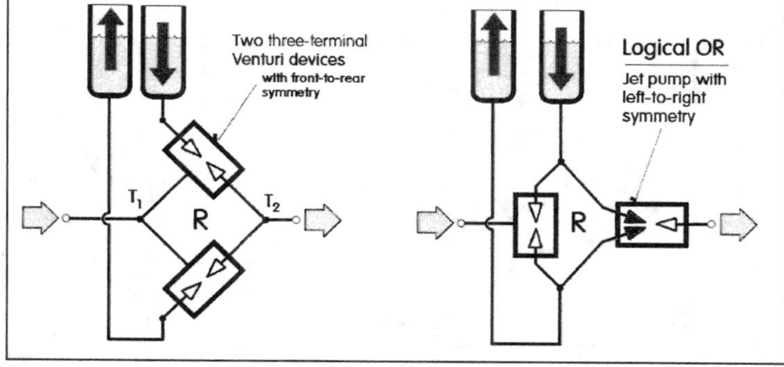

Figure 4.179 (Left, top) The double-action full-wave pump with two three-terminal Venturi rectifiers. Basically, two single-action pumps of Figure 4.177 meet at the two Tees T_1 and T_2.

Figure 4.180 (Right, top) The Tippetts' version of the double-action pump with the single three-terminal Venturi rectifier and a symmetric jet pump in the opposite vertex of the Grätz bridge.

Figure 4.181 (Bottom) Schematic circuit diagrams of the two pumps from Figures 4.179 (at left) and 4.180 (at right).

central orifice to form a simple displacement vessel with nozzle generating what would nowadays be called the "synthetic jet." Better documented use of the jet-type rectifierêin a pump for moving molten salts at temperatures as high as 1 500 K —was described by Walken and coworkers [14]. His rectifier was developed from a standard jet pump A in Figure 4.176. To obtain the full-wave rectification, Walkden provided his jet pump with two primary nozzles in parallel, each supplied with one phase of the alternating two-phase flow. An even more important improvement was his replacement of the nozzles by diffusers, B in Figure 4.176, leading to higher efficiency due to the absence of flow separation from the wall in the reverse flow direction. Analysis of the Walkden's rectifiers led Tippetts to discovering the importance of the three-terminal device shown as D in Figure 4.176 as a member of C in Figure 4.176 and the proposal to use this device alone in a reciprocating jet pump (Figure 4.177) [16]. The return flow from the load during a part of the cycle was acknowledged as inevitable and actually made use of by arranging the rectifier device (Figure 4.178) so that the jets are formed in both flow directions and the pumped fluid is sucked in from the input inlet due to the entrainment effect into the jet.

Though also using the same two diffusers as the design of Figure 4.175, the pumps according to Figure 4.177 are more effective. Tippetts himself went on later and extended this principle to the full-wave two-phase operation. He did not follow the more obvious additive layout of Figure 4.179, presented also schematically in the left half of Figure 4.181. Instead, he placed the rectifier devices into the other vertices of the Grätz bridge, where the single direction of the output flow has made it possible to replace the upstream three-terminal device by the symmetric jet-pump OR logical element.

4.8.3 Traveling wave pump

Essentially also a jet-type rectification, this is an alternative featuring several important improvements [17, 18]. One of them is the use of the Coanda-effect attachment to a curved wall. This wall is positioned so that it turns the direction of the forward flow coming from the displacement cavity. The resultant angular position of the channels ($\beta < \pi$; in fact $\beta < \pi/2$; Figure 4.183) decreases the hydraulic losses associated with flow direction change. The general direction of the main flow through this rectifier is straight and a considerable proportion of the pumped liquid actually continues in the straight direction without ever entering the displacement cavity.

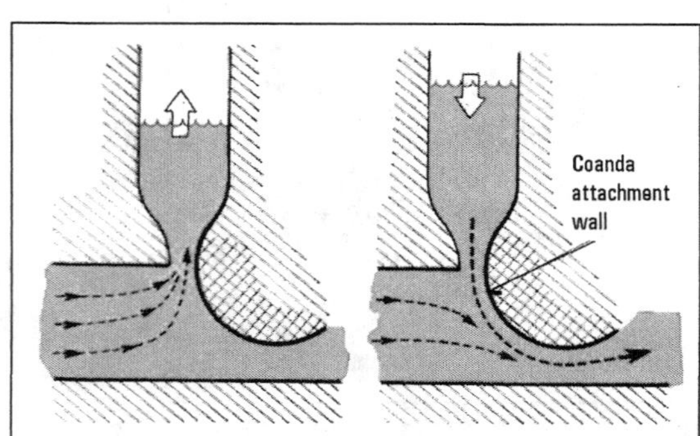

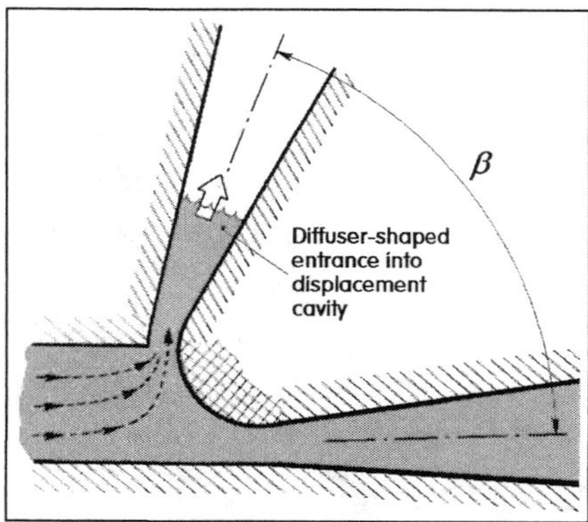

Figure 4.182 (Left) Operating principle of the jet-type rectifier with the jet turning by the Coanda-effect attachment to a curved wall. Original version with nozzle-shaped entrance into the displacement cavity.

Figure 4.183 (Right) Improved efficiency of the suction stroke achieved with the diffuser-shaped entrance and its inclination.

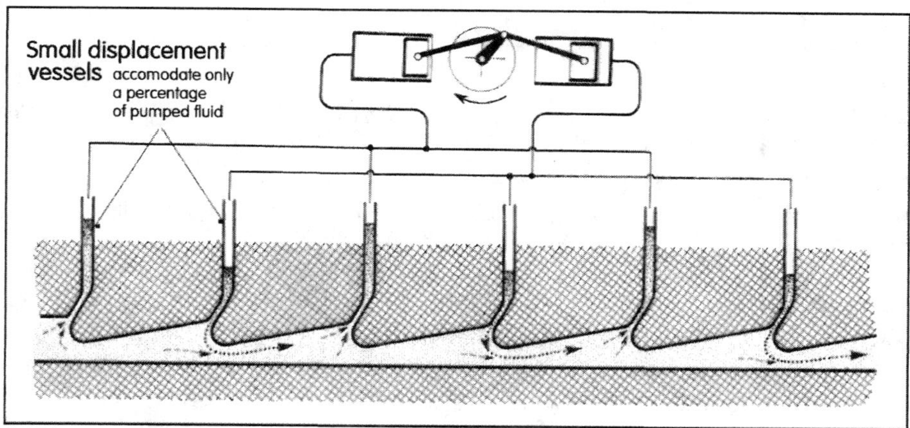

Figure 4.184 Principle of the traveling-wave two-phase jet pump with Coanda-effect deflection of jets issuing from the displacement cavities.

Traveling-wave pumps are relatively recent development. The tested version uses the principle from Figure 4.184, with a number of displacement cavities arranged along the channel length. The entrances into the cavities are placed at a streamwise pitch which corresponds to the distance traveled from one displacement cavity to the following one by the attached Coanda jet during one-half of the period. As a result of this timing, the outflows from the cavities form a traveling wave passing through the channel. With proper adjustment of operating frequency it is possible to enter the resonant regime in which the efficiency reached very high values. Figure 4.185 presents a version in which the nozzles driven in the second phase are positioned on the opposite side of the channel than nozzles operated in the first phase. The direct electric driving employs the volume changes of an electrostrictive gel (Figures 5.62, 5.63, and 5.64).

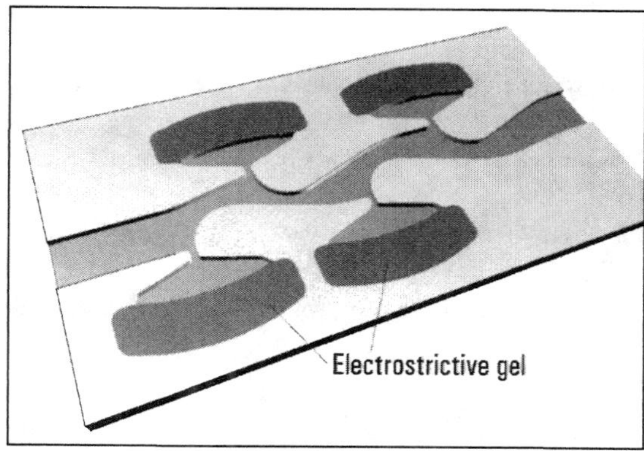

Figure 4.185 The basic plate of a recent traveling-wave micropump with equal phase nozzles positioned on one side of the channel. Displacement cavities operate with newly developed gel changing its volume when exposed to the electric field between electrodes.

References

[1] Tesař, V., "A Mosaic of Experiences and Results from Development of High-Performance Bistable Flow-Control Elements," *Proc. of Conf. Process Control by Power Fluidics*, Sheffield, 1975.

[2] Tesař, V., "Microfluidic Turn-Down Valve," *Journal of Visualisation*, Vol. 5, No. 3, 2002, p. 301.

[3] Wormley, D. N., and H. H. Richardson, "A Design Basis for Vortex-Type Fluid Amplifiers Operating in the Incompressible Flow Regime," *Trans. ASME, Journal of Basic Engng.*, Vol. 98, 1972, p. 82.

[4] Tesař, V., "'Fluid Plug' Microfluidic Valve for Low Reynolds Number Fluid Flow Selector Units," *Journal of Visualization*, Vol. 6, No. 1, 2003, p. 77.

[5] Athavale, M.M., et al., "Modeling 3-D Fluid Flow for a MEMS Laminar Proportional Amplifier," *Proc. of MSM 98, Intern. Conf. on Modeling and Simulation of Microsystems*, Santa Clara, CA, April 1998, p. 522.

[6] Tesař, V., C.-H. Hung, and W. B. Zimmerman, "No-Moving-Part Hybrid-Synthetic Jet Actuator", *Sensors and Actuators A—Physical*, Vol. 125, 2006, p. 159.

[7] Tesař,V., "Bistable Turn-Down Power Amplifier—Matching the Two Dominant Cross Sections," *Proc. of the 9th Internat. Fluidics 'Jablonna' Conf.*, Paper A-5, Jablonna, Poland, September 1982.

[8] Warren, R.W., "Negative Feedback Oscillator," U.S. Patent No. 3158166, filed August 7, 1962.

[9] Boucher, R. F., "Low Reynolds Number Fluidic Flowmetering," *Journ. of Physics E—Scientific Instrum.*, Vol. 21, 1988, p. 977.

[10] Nakayama, Y., et al., "Characteristics of Vortex Chamber Oscillation Device," *Bulletin of JSME*, Vol. 29, 1986, p. 3313

[11] Laser, D. J., and J. G. Santiago, "A Review of Micropumps," *Journal of Micromechanics and Microengineering*, Vol. 14, 2004, p. R35.

[12] Forster, F. K., and B. E. Williams, "Parametric Design of Fixed-Geometry Microvalves—The Tesser Valve," *Proc. of IMECE 2002*, ASME, New Orleans, LA, November 2002.

[13] Trávníček, Z., A. I. Fedorchenko, and A.-B. Wang, "Enhancement of Synthetic Jets by Means of an Integrated Valve-Less Pump, Part I: Design of the Actuator," *Sensors and Actuators A – Physical*, Vol. 120, 2005, p. 232.

[14] Stutely J. R., and A. J. Walkden, "Improvements in or Relating to Pumps," U.K. Patent specification No. 1.132,442, October 1968.

[15] Tesař, V., "Fluidic Jet-Type Rectifier: Experimental Study of Generated Output Pressure," *Journal of Fluid Control /Fluidics Quarterly*, Vol. 14, Issue 4, November 1983.

[16] Tippetts, J. R., and J. Swithenbank, "Fluidic Flow Control Devices and Pumping Systems," U.S. Patent No. 4,021,146, filed October 30, 1974.

[17] Tesař, V., "Fluidic Pump Driven by Alternating Air Flow," *Pneu-Hidro' 81 - IV. Colloquium on Pneumatics and Hydraulics*, Györ, Hungary, September 1981

[18] Tesař, V., "Introductory Notes on Microfluidics," Chapter 13 in *Microfluidics: History, Theory, and Applications*, Springer, New York, 2006

Chapter 5

Conversion Devices

This chapter discusses devices that may be collectively described as peripheral. They provide inputs into and mediate outputs from the assumed central pure-fluidic devices and circuits. In the reality of present-day fluidics, these so-called peripheral sensors, transducers, and actuators often occupy central positions. Fluidic devices almost always cooperate with devices of other types—electronic, mechanical, optical, and others—and there are many applications in which the devices performing the conversion are the most important ones in the system. Indeed, a brief survey of the programs of several recent conferences on microfluidics has shown that something like 80% of current development efforts in this field is actually devoted to the progress in these so-called "peripheral" devices.

The distinguishing feature of these devices is their operating, besides with the fluid-carried energy, also with other forms of energy (mechanical, electric, and so on), and performing energy conversions into or from a form other than the fluidic one. A list of the most important conversions between fluidic and other energy forms is presented in Figure 5.1. This list mirrors the division of this chapter into its individual parts and is, of course, incomplete because of the very wide variety of possible exceptions. The classification system used in this list is inevitably somewhat subjective. For example, strictly speaking, optical energy is carried by electromagnetic radiation as well as is thermal radiation (which is a part of the thermal action), so there may be theoretically substantiated reasons for treating them together under the heading of electrical energy. Nevertheless, the techniques for handling light, heat, and radio-frequency signals (the latter used, for example, in nuclear magnetic resonance sensors included here into the last special group) are sufficiently different, from the practical point of view, to justify a separate treatment. Another debatable aspect is the absence in Figure 5.1 of a separate bulleted point for chemical changes. It was decided to simply include the chemical composition of the fluid among its other properties, such as its velocity and pressure.

It should be said that the devices with conversions between fluidic and other forms of energy represent a vast subject. There are many conversion mechanisms that are known and may be—and, indeed, are—used. A list of devices using these mechanisms would be correspondingly vast.

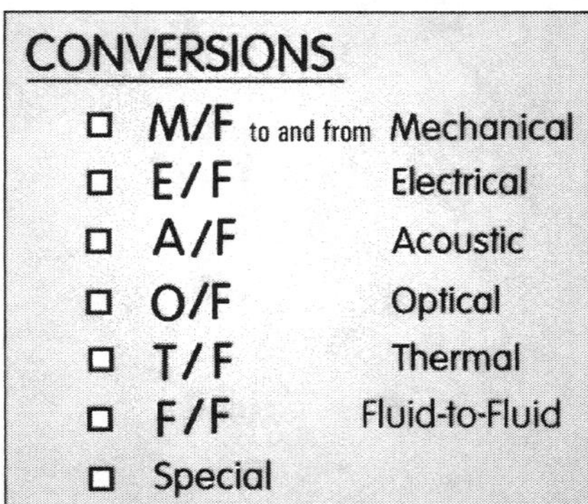

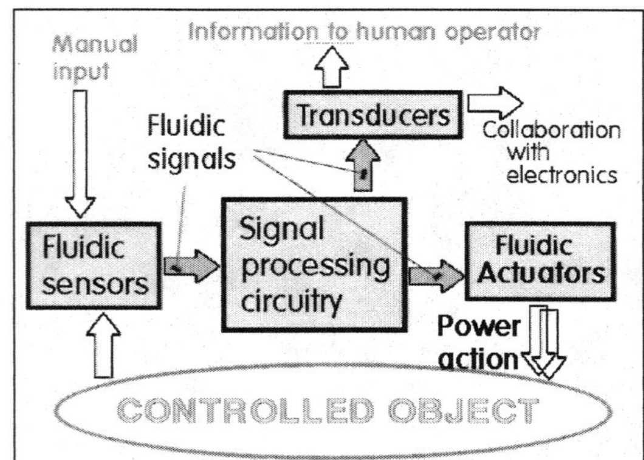

Figure 5.1 (Left) Basic list of the forms of energy, the conversion of which into the fluidic form—and the reverse transformation—are the task of the conversion devices discussed in this chapter. Sensing the chemical composition of fluid is included here among the sensing of other fluid properties.

Figure 5.2 (Right) A block diagram of a fluidic controller. In present-day microfluidics, the use of similar information-processing fluidics is not common—the signals are more conveniently processed in the electronic forms—but they may find useful applications in situations (perhaps in implants, where the signal processing is rather simple, already involves the handling of fluid flows, and perhaps where supplying the driving electric power is not easy). This figure presents the mutual relations between the types of conversion devices: sensors, transducers, and actuators.

A detailed discussion would itself require not a single chapter, but a book (or several extensive books). The brief descriptions in this chapter simply aim at providing a general survey. In line with the character of this book, this chapter concentrates on discussing the processes taking place on the fluidic side of the conversions.

5.1 CLASSIFICATION AND BASIC CONCEPTS

5.1.1 Signals

It is useful to distinguish between conversions at a (relatively) high power level, performed to make energy available in different forms, and low-power transformations, the objective of which is obtaining and converting *signals*. Although the signals—and information in general—cannot be handled without an accompanying energetic transformation, the power levels involved in their conversions and transmission are typically much lower, because most of the handled energy is, after all, lost by dissipation. In principle, the lower are the signal processing energy levels, the smaller and more effective are the involved devices and circuits. There is, however a limit: the quality of a low power level signal may be endangered by noise. In the extreme, a weak signal may be completely lost in the noise produced by the operation of the devices. Noise is a problem with fluidic signals (carried by flowing fluid) particularly if the flow is turbulent. Even if the conditions are chosen so as to keep the flow laminar, there is still a possibility of noise generated by vortices produced by flow separation at sharp edges or obstacles in the flow. Considering this, it is preferable to associate the signal with pressure energy rather than with the kinetic energy of fluid flows. This is one of the reasons behind the frequent use of diffusers in fluidic circuits and devices (and also the use of nozzles to perform the conversion back into kinetic energy, when the

	Conversion TO fluidics	Conversion FROM fluidics
Converting POWER	PUMPS VALVES	Fluidic ACTUATORS
Converting SIGNALS	Fluidic SENSORS	TRANSDUCERS

Figure 5.3 "Window panes" showing the four types of peripheral devices. The classification is based on the dual discrimination of power level and conversion direction.

kinetic action is really needed). In general, it may be said that the fluidic transmission of signals is typically less effective than their transfer in electric or optical forms, which are also much faster. Fluidics is therefore very often used together with electronic signal processing, and this fact, after all, is also one of the reasons for the need of signal converting transducers.

A schematic representation, using only building blocks, of a self-standing, autonomous fluidic control systems is presented in Figure 5.2. To provide a specific example, this may be thought of as a representation of an implanted microfluidic unit, the task of which is to correct a malfunction due to some inborn disease, perhaps an improper operation of a secretory gland. The system obtains information from the surrounding human or animal body about the level of the substance synthesized by the gland, processes this information, and if it finds the level insufficient, stimulates the gland.

Essentially there are three types of peripheral conversion devices in such a system, apart from the pumps and valves, as shown in Figure 5.3:

(1) ***Sensor*** devices are at the beginning of the signal transfer route. In a traditional fluidic control system, they are expected to provide the central fluidic circuits with information about the part of the outer world that is of immediate interest: the controlled object or process. They detect changes of their particular sensed variable and generate a fluidic signal, in the form of fluid flow, containing information about the change.

(2) At the opposite end of the fluidic control system, the signal processing chain involves peripheral devices of another sort, the ***actuators,*** which use the processed signals for an action on the outer reality, beyond the boundaries of the actual fluidic system. Very often, this is also associated with a conversion to another form of energy in the cases in which performing the desirable action is easier or better done by other means than by an action of a fluid.

(3) If the conversion of the fluidic signal into another form is done for a different reason than for an action on the controlled object, the output converting devices are called ***transducers***. With the general preference for signals in electronic form, transducers are very often of the F/E type, generating an electric output. In the above-mentioned case of the implant, a transducer would be useful for providing information about the proper condition of the implanted system and about the values of variables such as the remaining stored volume of the dispensed liquid. The advantage of the electronic form is the capability to respond to radio-frequency interrogation through skin (and perhaps tissue). In the case of the implant, the electric part may be passive, or excited by the interrogation signal, thus circumventing the need for batteries.

In the implant example, because of its inaccessibility, it will probably be missing one special family of sensors and transducers that is often found in other control systems: those intended for communication with human operator. The typical members of this family are indicators and displays on the one hand, and adjustment turning knobs and switches, on the other one. The

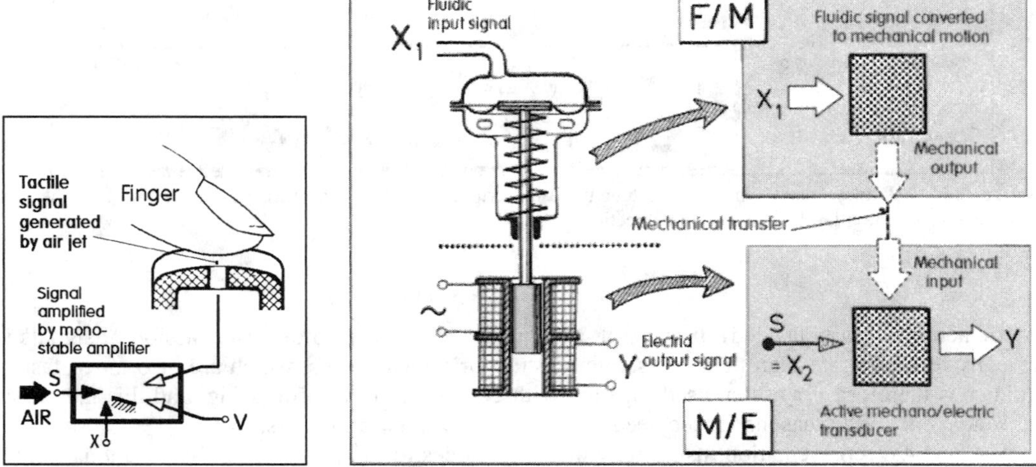

Figure 5.4 (Left) An example of a fluidic transducer for generating a tactile signal—an air jet—to communicate information to a human operator. Such a transducer may be extremely inexpensive (in fact, consisting of just a hole in the device body) and suitable for extreme miniaturization.

Figure 5.5 (Right) A schematic representation of a conversion chain—fluidic to electric by way of mechanical motion—with a very classic example of the devices. In the absence of a direct F/E conversion (apart from using special fluids), using transducers with a mechanical intermediate transformation is the standard way of performing this most important of conversions, despite the inherent low frequency range, which is limited by the inertia of moving parts.

general trend toward the decrease in dimensions, especially in microfluidics, makes these classical modes of communicating with the operator increasingly difficult to incorporate. Figure 5.4 presents an example of a fluidic device used for this communication purpose, which employs a tiny air jet to generate a tactile sensation. This example demonstrates the achievability of an exceptionally small volume and very low cost, advantages perhaps limited by the necessity of using air as the working fluid and by the (relatively) large air consumption. In general, the development toward smaller sizes as well as the trend of the systems being more autonomous (not needing human supervision) leads to communication devices being less important. They are mainly replaced by radio-frequency, infrared, or ultrasound transmitter communication, perhaps using a handheld controller.

5.1.2 Conversion chains

Many devices mentioned in this chapter do not perform a conversion to or from fluidics directly. Instead, the solution – as shown schematically in Figure 5.5, may involve an intermediate transformation to a signal of a different sort from either the input or output form. In particular, the important E/F conversions, plagued by the practical absence of any direct method of generating electric current by fluid flow, very often use mechanical inter-conversion. This is not an entirely good approach. Mechanical motions limit the frequency range due to the moving part inertia. They also cause undesirable sensitivity to acceleration (moving the fluidic system, especially if leading to impacts, can generate false signals), and make the manufacturing more expensive. Alternate solutions are therefore sought, using, for example, intermediate conversions to thermal effects. Thermal processes are generally known as slow, but in microdevices with extremely short heat transfer distances, the speed of temperature changes may be surprisingly fast.

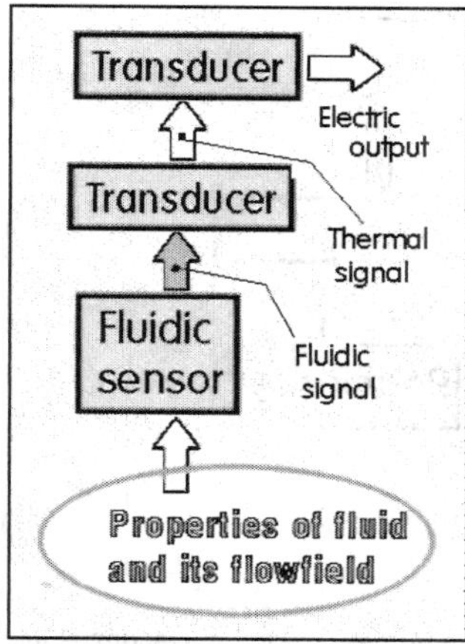

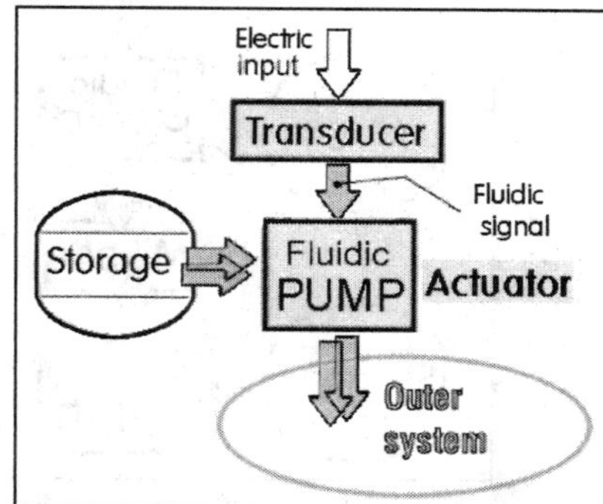

Figure 5.6 (Left) An example – in schematic representation – of a fluidic system for sensing the state (pressure, flow rate, or composition) of a fluid. The signal is immediately converted to an electric output (by way of an interconversion using thermal effects, which is necessitated by the absence of a direct F/E conversion). Such "incomplete" fluidic systems are effectively just special kinds of electronic sensors.

Figure 5.7 (Right) Another example of an incomplete fluidic system, which is in effect a special electric actuator operating with fluid. This may be a schematic representation of an electronically controlled implanted drug dispenser or perhaps the active part of an inkjet printer.

An example of a typical use of an intermediate thermal conversion may be fluidic flowmetering, as discussed in Section 4.7.3. In the oscillators with internal feedback and a single outlet, no fluidic output terminal is available that could be used to read the oscillator frequency (proportional to the measured fluid flow rate). An oscillation confined in the internal cavities, for example, those shown in Figure 4.162, is usually sensed by a heated electric conductor placed into a suitable location in the cavity, where it is, in each oscillation period, cooled by a flow reaching the location a periodic manner. Because of the temperature dependence of electric resistance, the cooling is easily detected and converted to an electric digital signal. A schematic representation of this F/T/E chain conversion is given in Figure 5.6.

There are many other similar examples of the use of the thermal effect and also of uses of other intermediate conversions – for example, the optical signals (Figure 5.1) are almost always converted to an electric output. Sometimes the internal members of the chains are considered so obvious that they are neglected in a discussion that mentions only overall effects. This makes the classification of conversions less clear cut. Because of this, some chain conversions are placed into sections of this chapter in which they actually only partly belong.

5.1.3 "Incomplete" fluidic systems

Systems operating with the complete set of fluidic sensors, transducers, and actuators covering the whole flowpath as shown in Figure 5.2, are actually very rare. Much more often, fluidics is used only in segments of systems.

The so-called "incomplete" systems without the actuator end of the path, as schematically represented in Figure 5.6, are actually quite widespread. There are many applications of microfluidics in which the fluid is used just to bring the information—either in the properties of the fluid or by carrying investigated particles or small objects—into the system, for example in an objective medical diagnosis by the analysis of bacterial DNA. The input part is usually fluidic, but

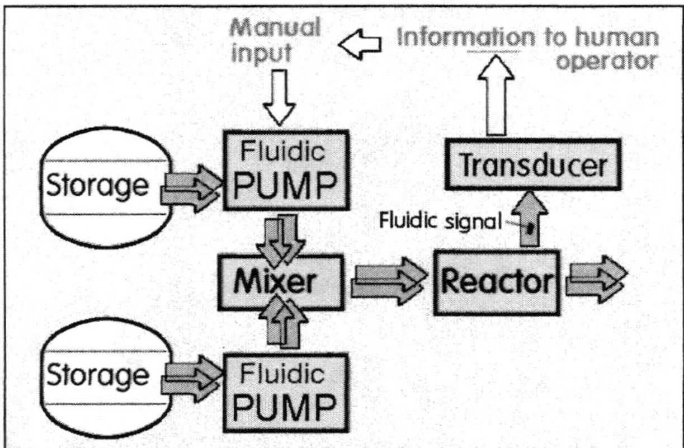

Figure 5.8 An example—again in a schematic representation—of a fluidic system not interacting at all with any outer system (apart from a human operator). It therefore needs neither sensors nor actuators. This may be the case of a microchemical system for combinatorial drug discovery.

the fluid, usually a body fluid, serves as just the carrier bringing bacteria to the analyzer. Similarly, the handling of the fluid carrying the samples of suspect substances in antiterrorist detection systems may be rather complicated, but the overall principle is not much different from the schematic representation in Figure 5.6. The handling of the fluid and the transfer of fluidic signals end at the location where the analyzer unit produces the output signal in an electric form.

There are also many examples of the opposite sort of "incompleteness," with no fluidic input segment of the processing. This case is schematically presented in Figure 5.7. The fluid handling, in effect, serves as just the actuator part of the electronics. As a particular example, this may represent the system for delivering anti-inflammatory agents to manage the negative effects of the presence of neuroprosthetic MEMS electrodes in living organisms.

Finally, there are frequent examples of fluidic systems lacking both the input or sensor part as well as the output or actuator part. Figure 5.8 presents schematically a typical example. As the very extreme are cases such as that of the accelerometer to be shown later in Figures 5.38 and 5.39, where fluid is actually confined to a single small cavity.

5.2 M / F CONVERSION TO AND FROM MECHANICAL MOTION

Until a relatively recent point in the history of fluidics, the inclusion of mechanical components was practically inevitable, and it was proper to describe all fluid handling systems as M/F, mechano/fluidic. Mechanical components may be made very robust, and themselves are generally reliable – the problem areas are the bearings or similar supports making their motions possible and the parts designed as flexible or deformable, such as springs. The bearings have the unpleasant tendency to wear and they also may seize. The flexible components tend to break, due to overstressing, for example, at accidental impacts or due to fatigue. The most unpleasant feature of mechanical components, however, is their inertia, which limits their range of operating frequency. The problem tends to become more serious as the size is diminished. Due to all of these reasons, there has been a general trend to eliminate mechanical motions in fluidics and especially in microfluidics. On the other hand, the drawbacks are not insurmountable, as documented, for example, by the centuries-old history of mechanical watches. There are several areas where fluidic and mechanical components will remain in successful coexistence. Some typical examples are shown here in four sections, organized according to the four "window panes" in Figure 5.3.

5.2.1 Mechanical pumps and valves

Even though the primary source of energy for driving fluidic systems is, in the majority of applications, electrical energy, the absence of a direct E/F transfer of power to simple basic fluids has led to mechanical fluid handling as the most natural solution. The use of the E/M/F power conversion chain means that the M/F conversion remains important and is likely to continue at least in the foreseeable future. Some methods of this power transfer copy the successful pattern of large-scale devices. Centrifugal pumps may be made in tiny sizes by etching, with the resultant stator and rotor (impeller) fluid-handling components corresponding to the low aspect ratio example shown in Figure 5.9. The more problematic aspects are the bearings and the electric motor, but even these have been demonstrated.

Generally speaking, however, such direct transfer of large-scale ideas, developed for a different range of Reynolds numbers, does not lead to an optimum solution. Much more promising are solutions making use of new materials, such as the shape memory alloys, to be discussed in the section on F/M conversions. Figure 5.10 presents an example of the reciprocating pump, which, instead of trying to scale down the connecting rod and crankshaft mechanism, has employed an electroactive material for the deformable diaphragm. The alternating flow generated by the reciprocating motions is to be rectified. In Figure 5.10, this is done by mechano-fluidic one-way valves, also with deformable diaphragms. Their rectification efficiency is much better than that obtainable with the pure fluidic rectifiers discussed in Chapter 4, but they suffer by wear of the contact surfaces, fatigue breakages, and limited operation frequency, which it would otherwise be preferable to select rather high to counter the small inertial effects at low time-mean Re values.

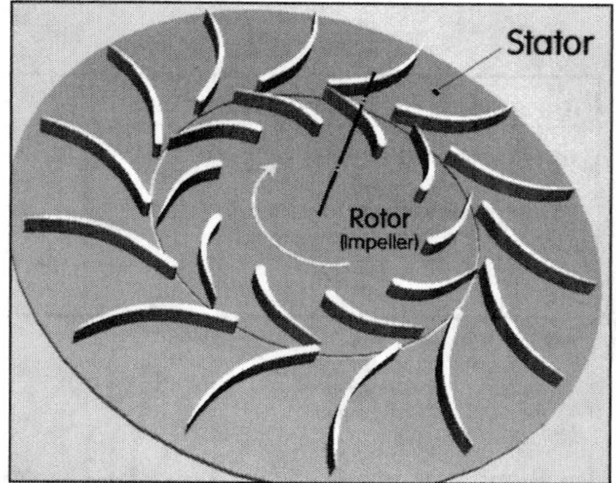

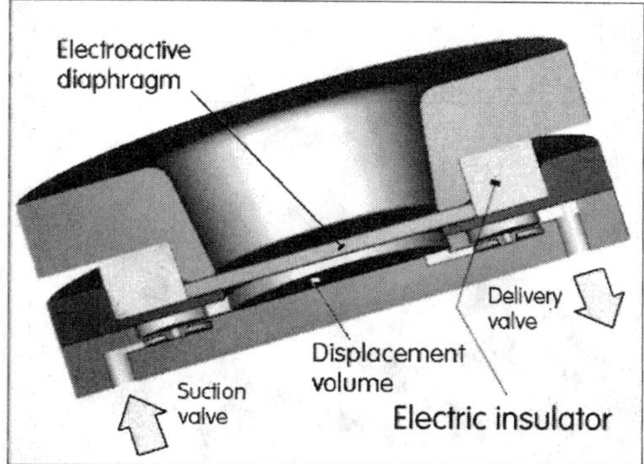

Figure 5.9 (Left) The stator and rotor (impeller) of a centrifugal pump, using the principle of large-scale pumps buts made in a small scale by etching. The efficiency, decreasing with diminishing Reynolds number, is generally poor, and the need for a rotational drive causes problems, but quite successful examples are known and in use.

Figure 5.10 (Right) An example of a microfluidic pump with fluid displacement by diaphragm deformations. It produces alternating flow, rectified by mechanical one-way valves.

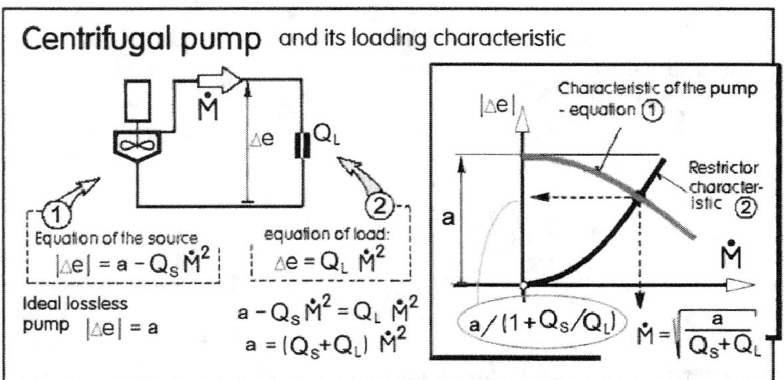

Figure 5.11 The basic relations for loading dynamic pumps with their lossless specific work a not dependent on flow rate, so that the characteristic would be a horizontal line. Internal losses taking place inside the pump cause the loading curve to be below the horizontal ideal by the amount dependent (usually quadratically) on the Q_S.

5.2.1.1 Two basic pump types

Each of the two pump examples shown in Figures 5.9 and 5.10 belongs to a different pump family, differing in their response to loading. The centrifugal pump in Figure 5.9 belongs in the dynamic pump family, together with jet pumps and also with the pump in Figure 4.185, with behavior dependent on the inertia of the jets. If there were no losses taking place inside such a pump, the loading characteristic would be a *horizontal* line at a constant vertical distance a [J/kg], which represents the magnitude of the specific work transferred to the fluid in the pump. All intersection points with the characteristic of any load that is connected to the pump's output terminal would exhibit the same pressure drop $\Delta P = a / v$ (neglecting compressibility, v [m^3/kg] is fluid specific volume) across the load. If internal losses are considered, the characteristic may usually be well approximated by ① in Figure 5.11. Some real centrifugal pumps may have slightly more complicated characteristics, dependent on the exit angle shape of their rotor vanes, which can

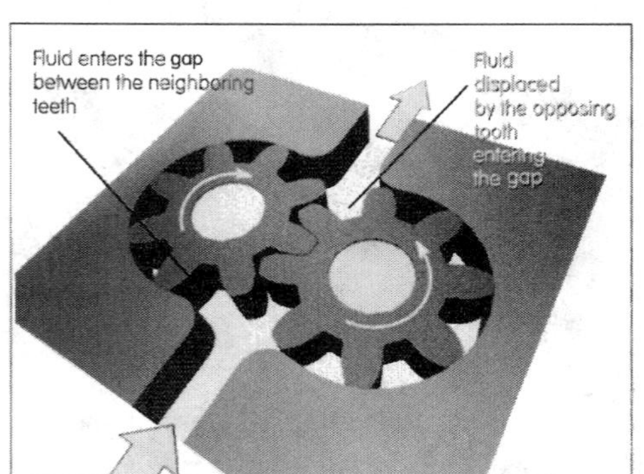

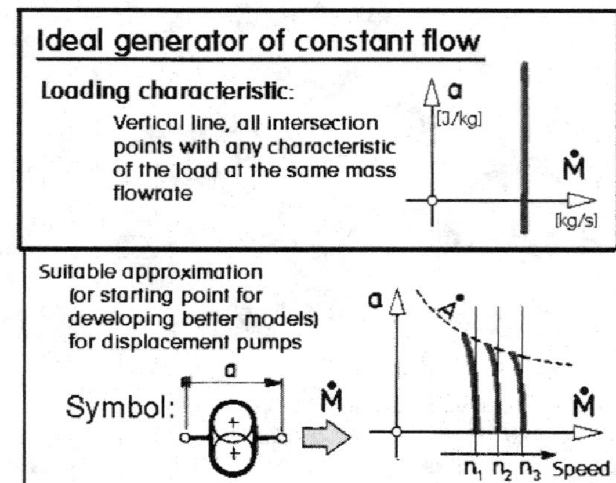

Figure 5.12 (Left) The gear pump: a typical displacement pump with continuous rather than reciprocating operation.
Figure 5.13 (Right) The basis of the displacement pump loading characteristics.

replace the horizontal basic lossless line by a line inclined either upwards or downwards. The upwards sloping chase leads to an inherent instability, a phenomenon not very important at the small sizes of interest here, because the usually large subtracted magnitude of the losses makes the resultant loading characteristic stable – at least in the vicinity of the usual operating point. Pulsations may arise if such a pump is improperly operated with a large dissipance load.

The other family of pumps are those of the volume displacement type. They may be subdivided into the reciprocating ones, with a principle corresponding to Figure 4.170 (the pump from Figure 5.10 belonging in this group), and continuous ones, an example of which is the gear pump in Figure 5.12. The pumps belonging in this family tend to deliver the same flowrate at each turn of the driving shaft. As a result, their ideal characteristics are *vertical* lines, each one for a particular shaft speed; see Figure 5.13. The deviations from this ideal are mainly due to inevitable leakages past the gears or imperfect rectification in the case of the no-moving-part rectifiers, discussed in Section 4.8.

5.2.1.2 Mechanical valves

The motion of the component that blocks the fluid flow in the mechanical valves may be rotational, like the one in a stopcock, but this is rarely used in the small-scale valves discussed here, where the motion is usually straight, along a line. There are the three different operating principles, listed in Figure 5.14. Of them, the most common in microfluidics is the case C, the closure motion made possible by the deformation of the blocking component or one of its parts.

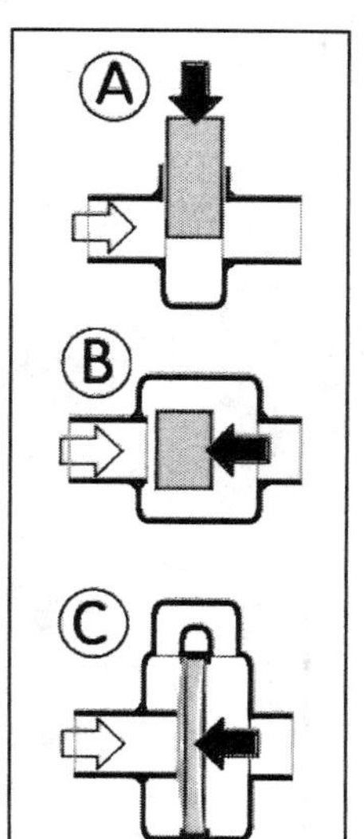

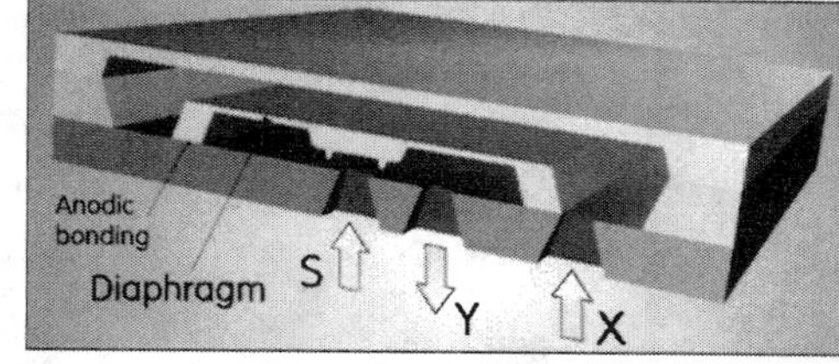

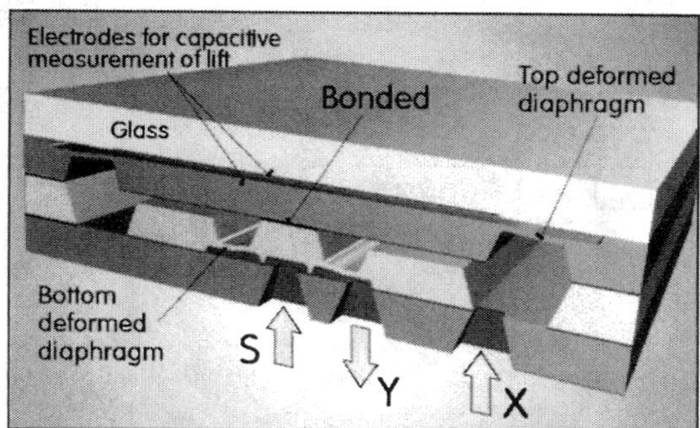

Figure 5.14 (Left) Principles of mechanical turn-down valves with linear (nonrotational) motion of the component, which decreases the available flowpath cross-section.

Figure 5.15 (Right top) An example of a mechano/fluidic (M/F) microvalve. This is a simpler version, open if the control signal X is absent.

Figure 5.16 (Right bottom) An example of a mechano/fluidic microvalve closed in the absence of the control signal X. The central boss of the top diaphragm is bonded to the boss of the bottom diaphragm. Both are lifted by the control signal pressure acting on both diaphragms, of which, however, the top one has a larger area.

The examples in Figures 5.15 and 5.16 show the F/M/T case of closure by a fluidic signal, acting on a diaphragm made by thinning the component by etching. There is usually a central boss left, which may carry the valve seat. This has a protruding small contact area ring allowed to deform slightly on contact with the opposite wall, thus preventing leakages (usually negligible in contrast with the leaky character of the pure fluidic versions described in Chapter 4).

5.2.2 Sensing position and motion by fluidics

5.2.2.1 Restrictor-type sensors

The direction of the M/F conversion in sensors is the same, toward the fluidic side, as for the pumps and valves described above. The difference is in the low power levels involved in signal generation. The distinction may be considered somewhat arbitrary. Indeed, the pump from Figure 5.9, as well as the valve from Figure 5.15, may be used to generate a signal containing information about the magnitude of the mechanical action driving them. Conversely, the signal-generating sensors, such as the variable restrictors listed as sensors in Figure 5.17, may be, at an increased size and power level, used as flow-control valves. Nevertheless the distinction exists: small power levels lead to different layouts because the designs are guided by different considerations. At the small sizes of the sensors, it is acceptable to lose the fluid with which they are supplied—the sensors typically operate with air, releasing it into atmosphere—in exchange for the advantages of the noncontact character of the sensing.

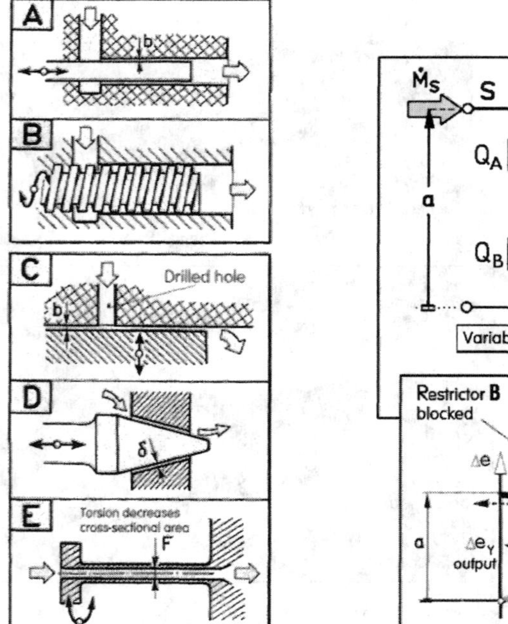

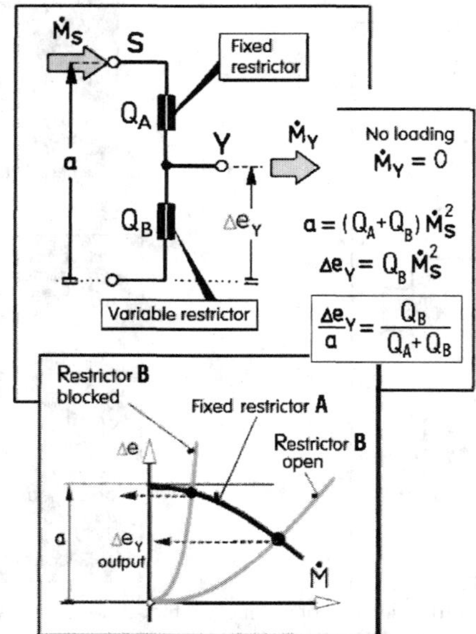

Figure 5.17 (Left) The basic principles of variable restrictors, with the mechanical change of dissipance. In A and B the change is by varying the channel length, in C, D, and E, by variation of the cross-sectional area.

Figure 5.18 (Right) The "pressure divider," a simple circuit producing an output pressure drop (or, in the accompanying diagram of characteristics, more generally a drop in the output specific energy) in the unloaded case, proportional to the ratio of dissipances. The mechanical sensors are usually based on the use of one from the variable restrictors of Figure 5.17 in the B position of this circuit.

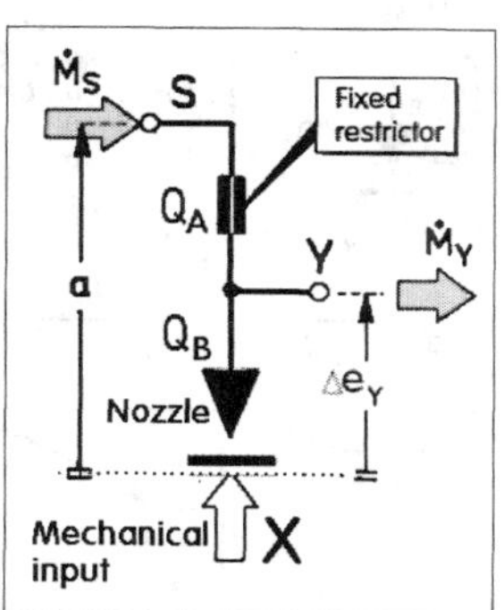

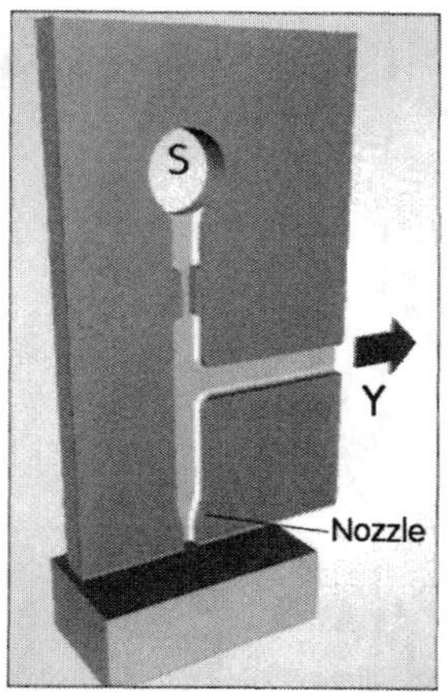

Figure 5.19 (Left) Schematic diagram of the most popular fluidic position sensor operating with air. In the B location of the circuit in Figure 5.19, a nozzle is used, with its exit blocked to various degrees by an object the position or movements of which are monitored.

Figure 5.20 (Right) An example of the appearance of the nozzle sensor from Figure 5.19. Its popularity is due to the extreme simplicity and low cost of fluidic sensing.

Typical mechanical position sensors are of the restrictor type. They are usually in the "pressure divider" circuit, shown in Figure 5.18. Of the principles in Figure 5.17, the most important is case C, the variation in the distance between the moving component and a hole through which air escapes into the atmosphere. This hole may be described as a nozzle (Figure 5.19), and this is often emphasized by its being given the characteristic nozzle shape, with an area contraction and an exit channel of a smaller cross section, see Figure 5.20. A nozzle sensor is usually used to detect and measure the position and motion of an external object. In that case, a position sensor may be justifiably described as a noncontact type, having no moving parts. These properties and its extreme simplicity make it very attractive. The cost is low – even the fixed restrictor need not be made as a manufactured element, since its role may be taken over by the inevitable losses in the channel bringing in the fluid. A factor in the popularity of this sensor is the ease with which it can be adapted to sensing various different variables – the water depth (Figure 5.23), rotation speed (Figure 5.25), operator's fingers (Figure 5.41), and many others. It even performs simple logical operations (Figure 5.24).

The only disadvantage is the lossy character: the operation is based on the dissipation of energy in the fixed restrictor (Figure 5.21) and, of course, the consumption of the fluid (usually air). The consumption is usually accepted because the flowpath cross-sections may be (and usually are, unless there are exceptional requirements for large detection distances) very small, resulting in a small absolute magnitude of the dissipation. Consumption ceases only in the extreme state of the detected body coming into contact with the nozzle exit. This may be avoided by using an auxiliary component, perhaps supported by a spring, blocking the outlet into the atmosphere and removed by the approaching detected body. This, however, is not welcome, since it eliminates the advantage of contactless operation.

Besides being dependent on the sensed position of the object, the output pressure signal may also show unwanted variations. It depends on the loading (Figure 5.22), but this usually changes in a predictable manner. More dangerous in large systems with a single supply source may be changes of the supply pressure, caused by processes occurring, perhaps, in distant parts of the system. Another disturbing effect is temperature, especially if at least one of the used restrictors is

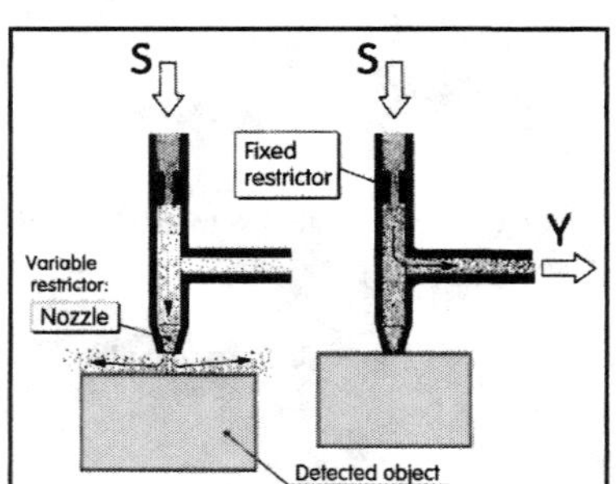

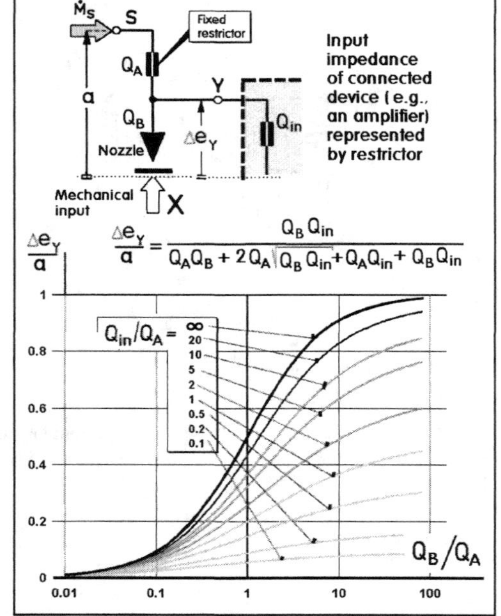

Figure 5.21 (Left) This is how the sensor from Figure 5.20 works: if the detected object is sufficiently far from the nozzle, the whole supplied energy is dissipated in the fixed restrictor. The output pressure in Y is low. It increases if blocking the nozzle exit stops the air flow and eliminates the dissipation mechanism.

Figure 5.22 (Right) The operation of the sensor in Figure 5.21 becomes complicated if the output Y is loaded by a connected device (e.g., a signal amplifier) extracting air from the sensor. The properties of the load may be represented by its input dissipance Q_{in}, leading to the expression for the relative output pressure displayed and presented graphically. Note that for infinite Q_{in} (no air extracted), the expression is equivalent to the simple version in Figure 5.18.

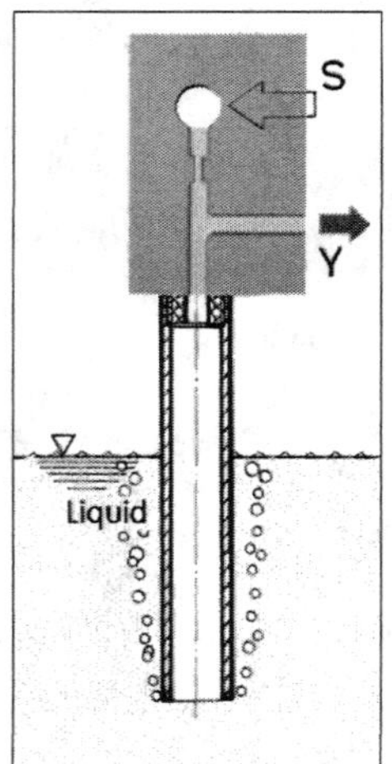

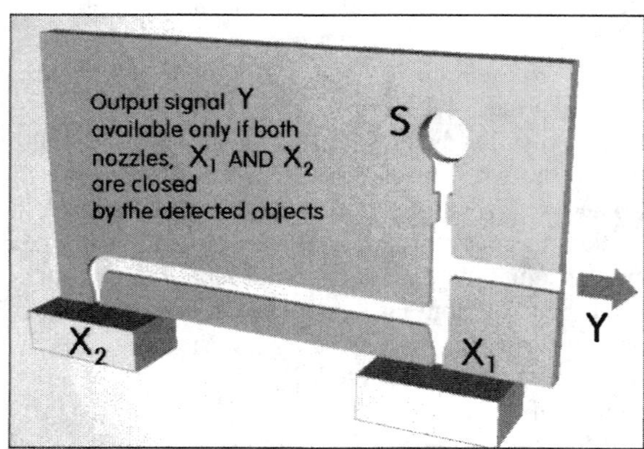

Figure 5.23 (Left) An example of the universality of the "pressure divider" sensors in Figure 5.18. They may be used to measure the height of the liquid level, if the nozzle of Figure 5.21 is replaced by a submerged tube.

Figure 5.24 (Right) Another example: with two nozzles in parallel the sensor can detect whether or not both objects are present simultaneously or if a single larger body is present and properly oriented so that its different parts block both nozzles.

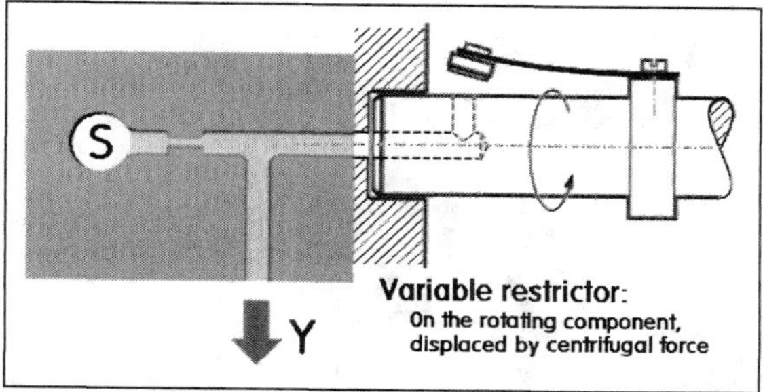

Figure 5.25 Yet another example of a use of the basic sensor body: here it measures the rotational speed of a shaft.

of the friction type (capillary or long channel, e.g., either case A or B in Figure 5.17). The output signal may also vary considerably with temperature (due to the temperature dependence of viscosity). Orifice-type restrictors are much less dependent on viscosity, but some temperature effects may also be felt. These effects may adversely influence accuracy. The low cost of these resistor sensors makes it possible to suppress unwanted influences by using the sensors in pairs, in the fluidic analog of the Wheatstone bridge (Figure 5.26). The disturbance effects act in both parallel pressure-divider branches. The output signal may be extracted as the difference between them – across the diagonal of the bridge (Figure 5.28).

If the bridge is balanced (Figure 5.27) by setting up an equal dissipance ratio in both parallel paths, the effects of temperature and supply-pressure changes are eliminated completely. The evidence of the balanced state is the zero difference read on the bridge diagonal. Operation in this balanced regime, however, requires a different layout of the system, usually with the fluidic proportional amplifier in the feedback loop.

Another use of the bridge circuit is the disbalancing of each of the two pressure-divider branches in opposing senses. The resultant effect is an increased sensitivity. If the sensors can be arranged so that the input action increases the dissipance of one variable resistor sensor while decreasing it in the resistor in the other branch, it is possible to generate much larger pressure

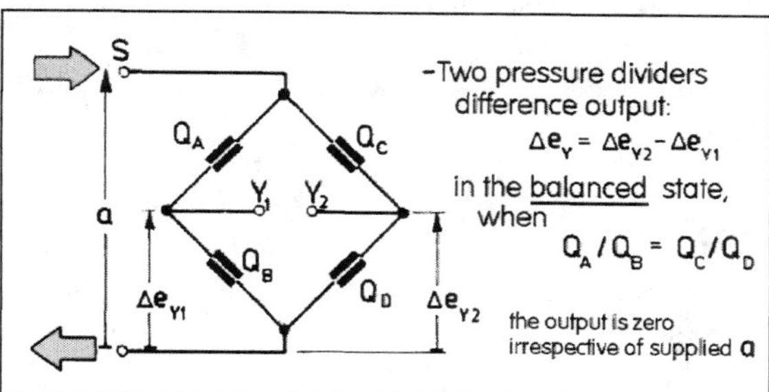

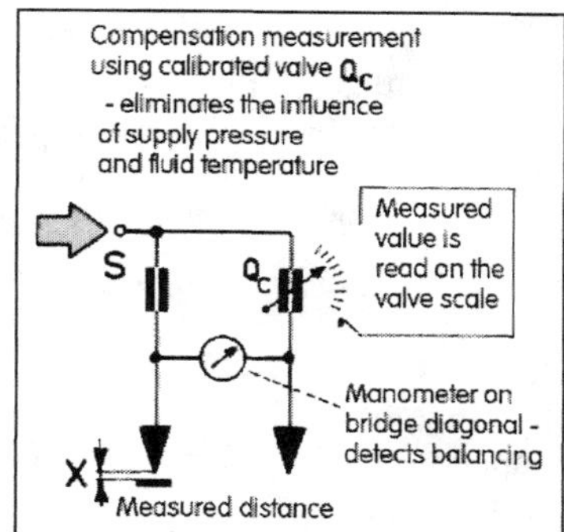

Figure 5.26 (Left) There are several advantages associated with using fluidic sensors of the pressure-divider type in the Wheatstone bridge, two dividers in parallel.

Figure 5.27 (Right) The principle of compensation-type sensing, with the variable restrictor in one branch and a balancing restrictor in the other branch. The most disturbing effects are eliminated, since they act in both branches. The arrangement shown here, with manual balancing, uses a manometer in the bridge diagonal for detecting the balanced state. The manual adjustment may be replaced by a servo-compensator analogous to an electric circuit in the constant-temperature anemometer bridge.

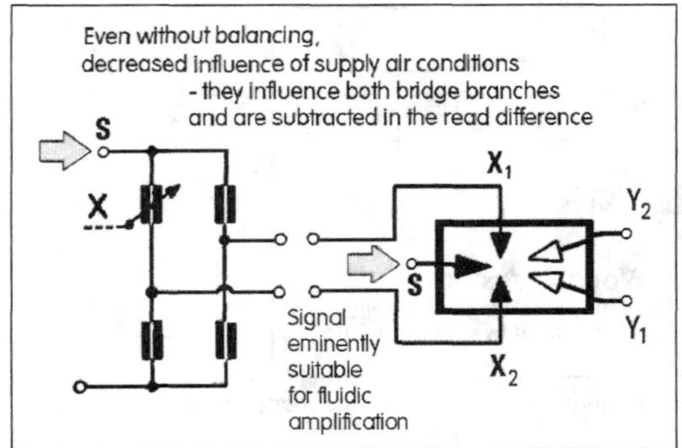

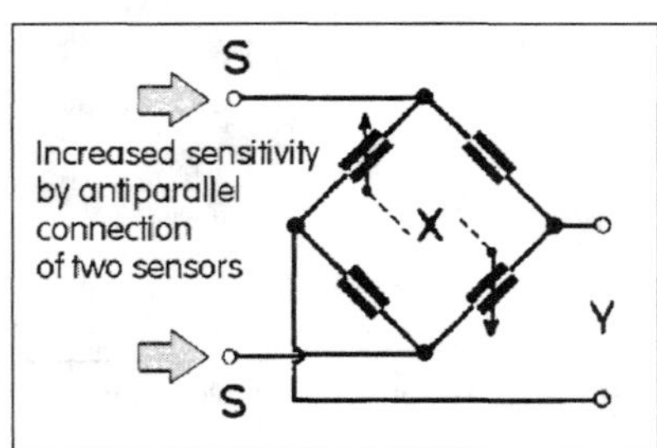

Figure 5.28 (Left) A more precise detection of the balanced state is possible using a fluidic amplifier. The differential character of the output signal makes particularly suitable for this purpose the symmetric proportional jet-deflection amplifiers, discussed in Section 4.3.3.

Figure 5.29 (Right) Connecting the sensors into the fluidic-restrictor bridge may also increase sensitivity by the opposition of two sensors in the connection that disbalances the bridge.

changes at the bridge diagonal. Alternatively, the changes may be the same, but the disbalancing of the bridge may be obtained by placing the variable resistors at different locations in the two pressure dividers (Figure 5.29).

5.2.2.2 Jet-type M/F transducers

In unsteady regimes there are some advantages to using sensors having dynamic properties similar to those of the amplifiers and other central devices of the system, preferably selecting the sensors so that they use the same operating principles. Suitable mechano/fluidic (M/F) transducers for cooperation with the jet-type no-moving-part amplifiers may be built using the operating principle of deflection or other modulation of a jet formed in a nozzle and captured in a collector. The historical course of the development was, actually, in the opposite direction. The M/F jet-type transducers with mechanical deflection of the nozzle, built for autopilots, preceded the jet-deflection amplifier by three decades.

The *guided jet* principles A and B in Figure 5.30 may also be built with deformable a jet guiding strip replacing the flap. Less used, though more sensitive, is case C, in which the guiding flap may suffer from a difficult-to-handle hysteresis. Principle A uses the Coanda attachment to the flap, though there may be problems with the attachment deteriorating by the ingress of additional fluid above and below the flap.

Relying purely on the Coanda attachment are the M/F transducers, the principles of which are shown in Figure 5.31, as well as case B in Figure 5.32. The latter operates in only a two-position manner (alternating attachment to one of the two translated walls). The spoiling of the monostable attachment by the protrusion of the spoiler from the wall (A in Figure 5.31) is suitable only for the binary operation mode. The seemingly simple screening of part of a jet, in case A in Figure 5.32, actually leads to a complex flowfield with the separated parts of the jet having a tendency to attach to various nearby surfaces.

One of the problems of the nozzle-type sensing in Figure 5.21 is the necessary growth of size and air consumption associated with an attempt to sense an object at a distance farther than about ~ 0.5 mm from the nozzle. The jet-type sensor based on A in Figure 5.31 can detect an object moving quite far from the nozzle, but it has the disadvantage of needing a collector on the other side of the object. Solutions sought in the jet's bouncing and turning back from the object failed.

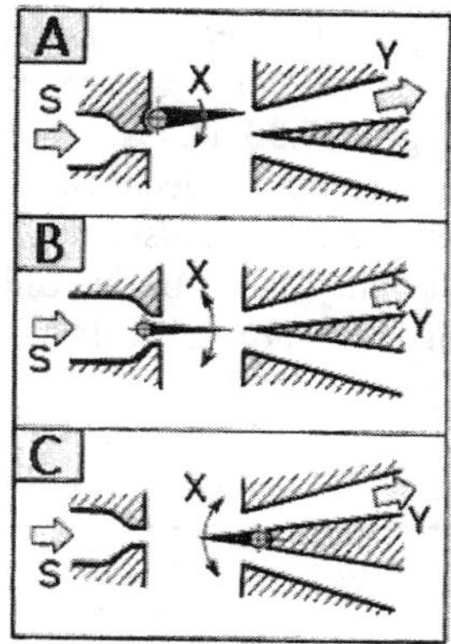

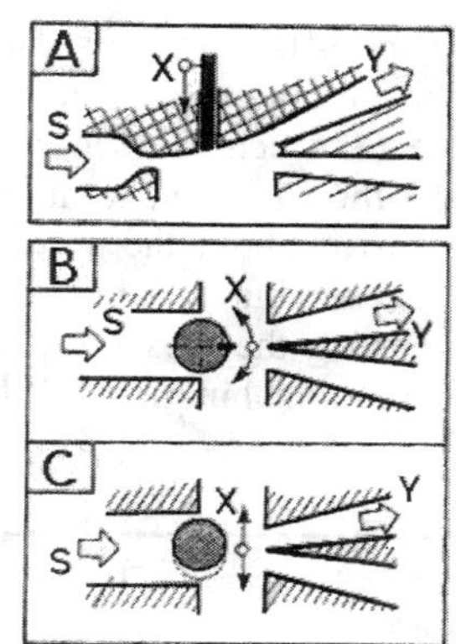

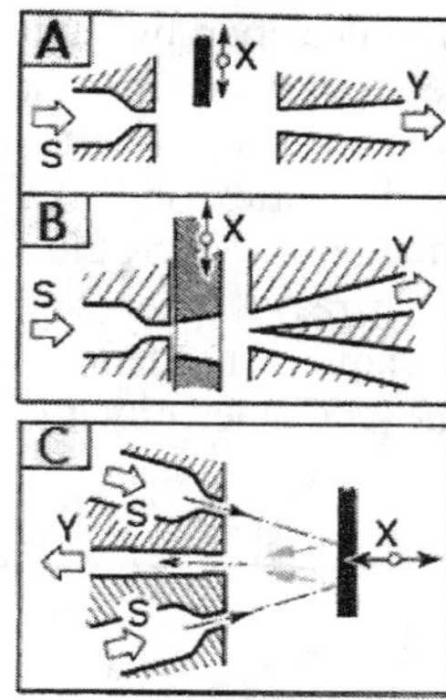

Figure 5.30 (Left) Jet-type mechano/fluidic (M/F) transducers with rotated flap. Especially in case B, the flap may be replaced by a deformable strip acted upon for example, electromagnetically.

Figure 5.31 (Center) Other examples of the jet-type mechano/fluidic transducers, here using the Coanda attachment to curved surfaces. In A the flow separates from the wall by the action of a mechanical spoiler, in B and C the attached jets collide with opposed curved wall-jets.

Figure 5.32 (Right) Jet-type M/F transducers: A and B are the principles suitable for binary, two-positional signals. Jets cannot be "reflected" when impinging at an angle on a wall, but it is possible to have the output signal collector on the same side of the wall, if there are two nozzles, as in C.

There is no "jet reflexion" – although it is possible to achieve it with a semicylindrical hollow cavity on the surface. The jet also "reflects" from a perforated plate (which reflects a part of the jet while allowing the other part to pass through [3]). If the surface cannot be adapted, the problem is solved by the layout according to C in Figure 5.32 or by its axisymmetric version, the annular impinging jet of Figure 5.33 with the vortex ring "bubble" between the sensor and the surface of the sensed object. This geometry recently became of particular interest for the problems associated with the antiterrorist warfare [2] application described in Chapter 6. The configuration of Figure 5.33 is used to collect trace amounts of explosives from the clothing of suspected terrorists,

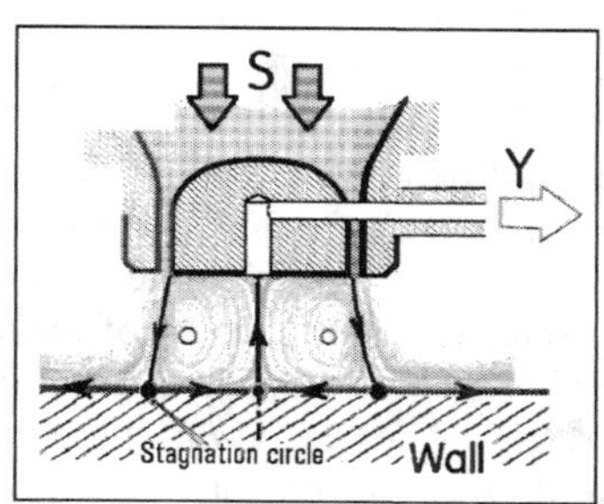

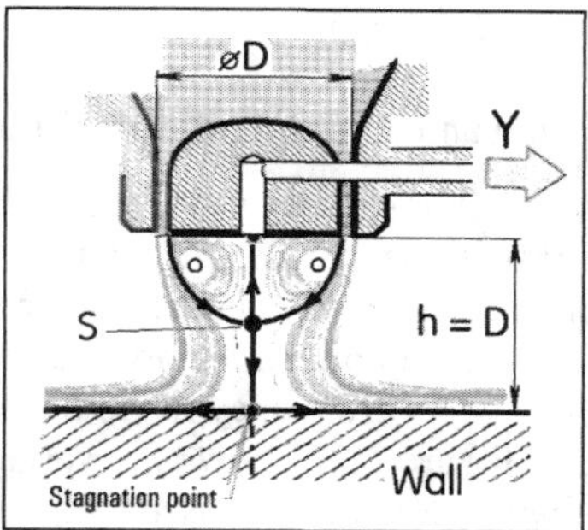

Figure 5.33 (Left) A sensor with annular impinging jets is used to detect the presence of an object at larger distances, at which the full cross-section nozzle from Figure 5.21 would require air flow rates that are too large. Also, detection is based on low pressure inside the annular vortex rather than pressure drop on a fixed upstream restrictor.

Figure 5.34 (Right) The problem with the annular nozzle sensor of Figure 5.33, when trying to use it for reaching to large distances, is the impossibility of keeping the stagnation circle on the wall closure when the distance h is larger than the diameter D. Instead, the vortex ring region is closed at the stagnation point S, and the sensor ceases to respond to the wall's presence.

analogous to detection by canine sniffing [4]. The impinging air jets are used to *strip* and *scrub* the surface, even generating the flutter motion of the cloth to release the samples. The unpleasant factor is the limiting largest distance to which the sensor in Figure 5.33 can reach. The low pressure in the vortex ring – caused by the entrainment into the jet – tends to decrease the size of the recirculation region by closing it way up above the wall (Figure 5.34). Large sensing distances require an increase in the sensor diameter, but this increases the air consumption, diluting the collected sample. An attempt at countering the closure is shown in Figure 5.35. The idea is to increase the jet diameter by giving it a rotational motion.

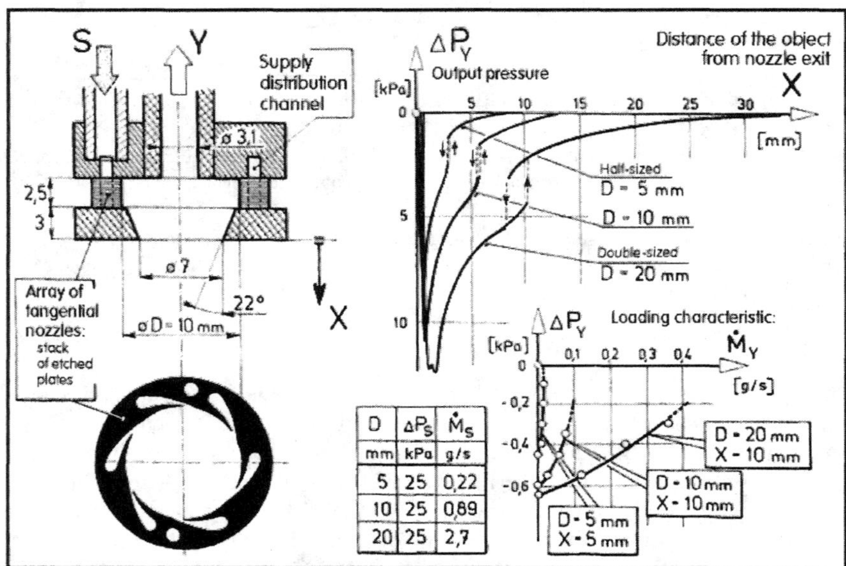

Figure 5.35 Jet-type sensor with the recirculation "bubble" size increased by centrifugal force action on the swirling jet. A low output pressure is detectable at the distances of the opposing wall ~ 1.8 times the exit diameter.

5.2.2.3 Sensors with an integral fluidic amplifier

A signal generated by the sensors usually has to be amplified before it becomes useful for further processing. It may be practical to perform the amplification by fluidics, especially when the final purpose of the signal is to handle fluids. The amplifier may then be integrated with the sensor. The amplification can produce collateral improvements, such as a shorter path of mechanical motions and, consequently, a wider frequency range or longer life associated with less deformation.

As an example, Figure 5.36 shows the variable resistor integrated with the vortex amplifier. In both variants shown, the main part is the same simple position sensor. In one case, it serves as force sensor, using the elasticity on the lid in the role of a calibrated spring. Pushing the lid toward the body decreases the radial flow into the vortex chamber, while the tangential inflow through the grooves is not affected. The rotational speed increases and the centrifugal action on the fluid in the vortex chamber reduces the flowrate much faster than would correspond to the small deflection of the lid. In the second version, integrated with an E/F transducer, the electromagnetic deformation of the diaphragm replaces the lid motion.

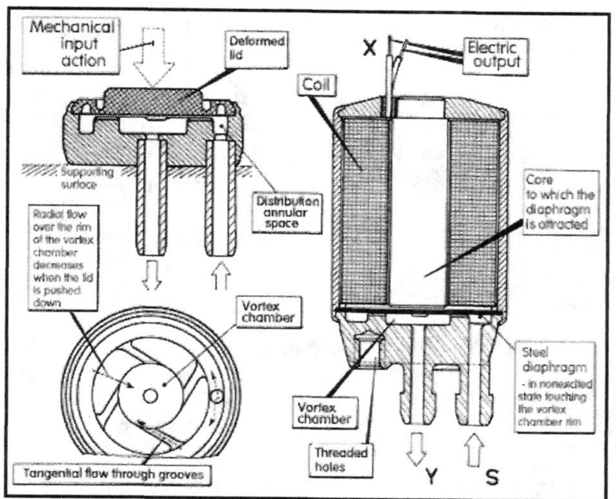

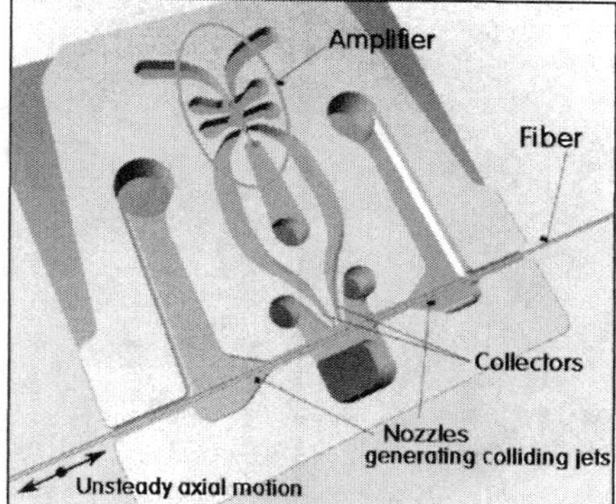

Figure 5.36 (Left) Integrating a mechanical motion sensor with a vortex amplifier. At left, the basic sensor with mechanical input deforming the elastic lid; at right, an electro/fluidic (E/F) transducer using the same principle in the role of mechanical intermediate transduction.

Figure 5.37 (Right) Fiber motion measuring sensor used to monitor weft insertion in shuttleless looms. There are two colliding jets, one of them coflowing and hence accelerated by the moving fiber, the opposite one contraflowing and hence decelerated. This moves the stagnation point where the jets collide and come to a stop, the position of which is detected as the difference in what is captured by the two collectors.

More typical are integral layouts in which the amplifier, usually of the jet-deflection type, is made by etching simultaneously with the sensor. In the example presented in Figure 5.37, the amplifier is made in the same body with the two opposing nozzles, in the axes of which it passes the measured thread or fiber. It would be very difficult indeed to measure unsteady axial fiber motion by other means. In a loom, the thread is accelerated and then decelerated up to 200 times per minute, to velocities over 100 m/s. Any attempts at mechanical sensing by components in contact with the thread would result in the weft tearing. Especially on very regular synthetic fibers, there are no features that could be used in an optical or electric sensor. Fluidics, using the aerodynamic friction forces acting between the thread and the two opposing air jets (one decelerated, the other accelerated by this force) was the sole way to solve this task.

5.2.2.4 Accelerometers

Acceleration and angular-rate-sensing MEMS devices are manufactured in large quantities, in fact they are currently representing the third largest MEMS sales volume (after inkjet printing heads and pressure sensors). Their use is mainly in automobiles for initiation of airbag crash systems and for chassis stabilization. Other uses are auxiliary functions in GPS navigation (measuring tilt with respect to gravity), military use in intelligent missiles, motion control of robots, and measuring vibration for machine condition monitoring. Earlier simple g-switches are nowadays replaced by proportionally measuring devices to make possible computer analysis of acceleration profile for distinguishing noncritical or noncrash events and eliminating false alarms. The typical response range required for airbag systems is 50 g with resolution < 0.1 g. Vehicle stabilization requirements are 2 g for range at 0.01 g resolution, while for navigation, it is 1 g, sensitivity $4 \cdot 10^{-6}$ g, and a dc response.

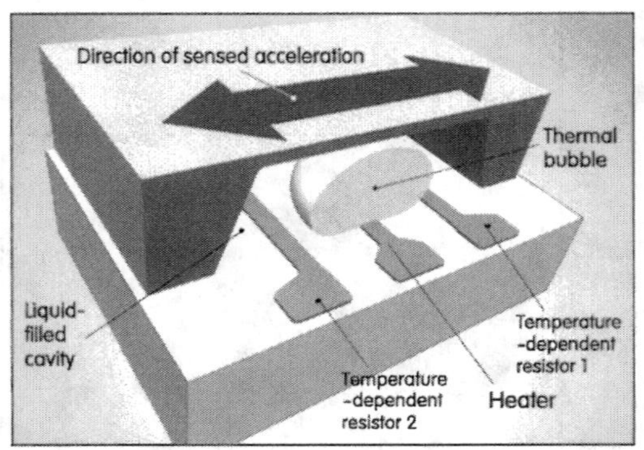

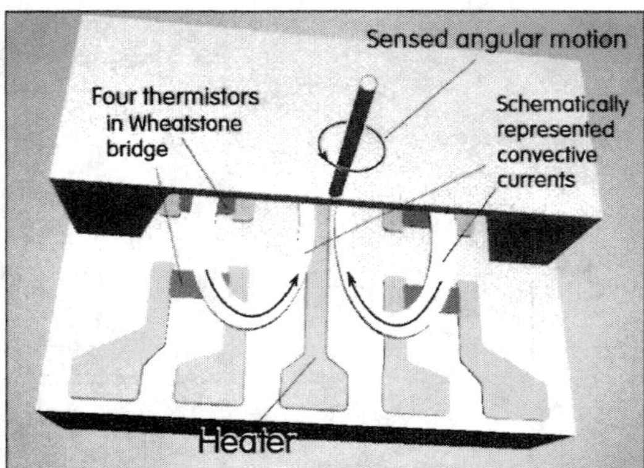

Figure 5.38 (Left) The principle of the bubble accelerometer with thermal readout of the bubble position. Earlier versions used a real vapor bubble. A faster response is obtained replacing it with a region of heated liquid surrounding the heater. It moves due to the specific volume difference.

Figure 5.39 (Right) Another version of a liquid-filled accelerometer, or, here, angular rate sensors. Its operation is based on thermal convective currents generated in the liquid. Their distortion by an angular motion is sensed by the array of symmetrically positioned temperature sensitive resistors.

While earlier accelerometers used a spring-supported movable mass, recent versions measure a thermal field in a tiny volume of liquid, making them a subject of interest for microfluidics. The original impulse for the change in the sensing principle was that of military application, such as for "intelligent" missiles with inertial-navigation-type control of trajectory. Movable mass does not survive the extremely high accelerations that occur when the missile is fired.

The bubble type fluidic accelerometers (Figure 5.38) originally used a gas or vapor bubble in liquid. This is replaced in more recent versions, for faster response, by a volume of heated liquid surrounding the heater strip. For a good response, the fluid must have a product of viscosity and thermal diffusivity as small as possible. A suitable example is methanol. There are now usually four (rather than the two shown in Figure 5.38) temperature-sensing resistor strips, arranged for two-axis sensing. The characteristic is linear, typically a 1°C response to a 1 g (= 9.81 m/s^2) acceleration. The convective currents of type angular rate sensor, as shown in Figure 5.39, typically respond with a 10 mV misbalance of the Wheatstone bridge to an angular rate 200 deg/s.

5.2.3 F/M actuators

The small intensity of fluid flows in microdevices does not provide much opportunity for fluidically driving mechanical motions at a higher power level. Much more often, the F/M conversion is completed as a part of an F/M/E chain in a device generating electric output effects, in a manner schematically represented in Figure 5.5.

A real power action is usually performed in large-scale fluidic devices. Despite the size, these devices may be a part of a microfluidic system, as in the case of clothing ruffling by air jets in security devices for screening for example airline passengers for samples of illegal substances (Figures 6.161 to 6.167). The samples are processed in microdevices, but the movement of the clothing surface should be rather vigorous to free the attached traces of substances, though it should not involve air flows that are too large, which would dilute the sample.

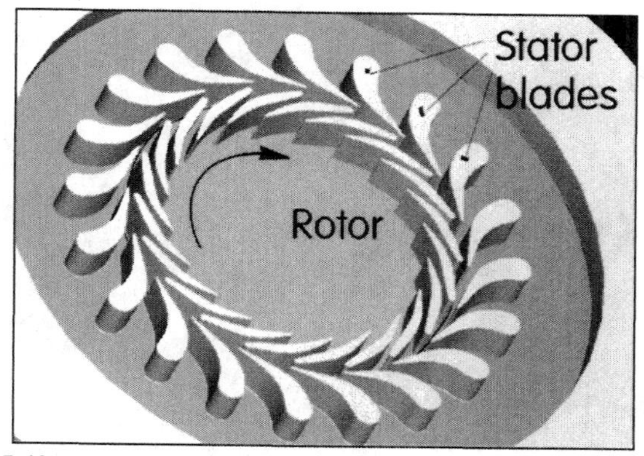

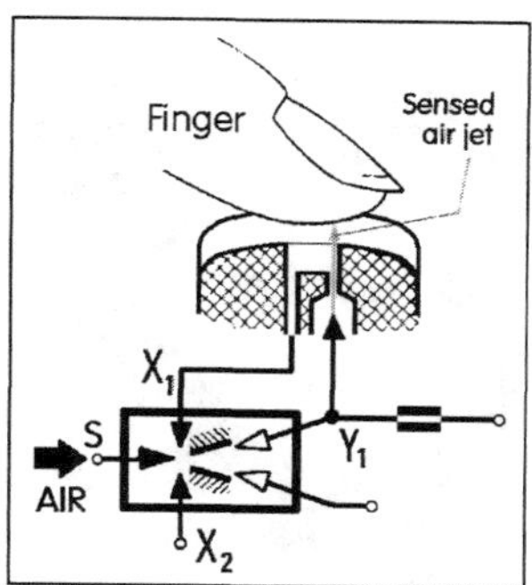

Figure 5.40 (Left) A radial turbine made by etching. To avoid low Reynolds number losses, rotational speeds are extremely high, typically above 200 000 rpm.

Figure 5.41 (Right) Finger-operated no-moving-part microswitch with tactile readout by an air jet. If the finger blocks the outflowing jet, the air enters the control nozzle of the bitable amplifier and switches it. The tactile signal disappears. Switching the flow back is done by a pulse to the other input, X_2.

A permanent activity exists in designing and developing microturbines, the active components of the example in Figure 5.40. The interest is fueled by a potentially much higher specific chemical energy, available to power portable equipment, in hydrocarbon fuels (typically 45 MJ/kg) or in hydrogen (119.9 MJ/kg), compared with the mere 0.5 MJ/kg stored in batteries. The basic problem of the combustion engine utilizing the energy is the low efficiency of turbomachinery at low Reynolds numbers. With a typical rotor diameter on the order of millimeters (a 10 mm rotor diameter axial turbine is described in [5] and the rotor of a radial turbine developed and tested at MIT is of 4 mm diameter), avoiding a too low Re calls for operation at extreme rotation speeds. The turbine from [5] delivered 28 W mechanical power at 160 000 rpm, and yet the efficiency was a mere 18%, with an energy density of 5-10 times lower than for large turbines.

Relatively large power, considering the conditions prevailing in microfluidics, uses pneumatic transducers for communicating with human operators by tactile signals. This is an interesting application. The output from the microfluidics is a small air jet sensed by a human operator on the control-stick grip (or perhaps keyboard or steering wheel). It is used to notify the operator about an occurrence of a less important event, without requiring that his or her attention be distracted from the visual observation of the main process. The example shown in Figure 5.41 is unusual because it combines in a single small no-moving-part body the output (tactile signal) from the terminal Y_1 of a bistable valve with an input action (control switch). By blocking the exit hole, the operator diverts the air flow to the control terminal X_1 of the amplifier, switching it to the other stable state.

A much more important power F/M action is moving hollow needles in medical applications. A pneumatic drive is smaller and lighter than electrical drives, either electromagnetic or other types (as they are listed in the next section in Figure 5.48). The current interest is not only in hypodermic needles, their typical microfluidic use is the automatic delivery of insulin for management of diabetes or delivery of an antimalarial drug (chloroquine phosphate). The needles made from Si by microfabrication may be so tiny that their penetration of human skin would be described as semi-invasive. For some purposes, such as monitoring a physiological state, these small needles penetrate only the stratum corneum, the outermost layer, causing virtually no harm to the living epidermis. An even more important impact from the use of extremely tiny hollow needles is expected from the injection of DNS, proteins, or pharmaceuticals into living cells. In

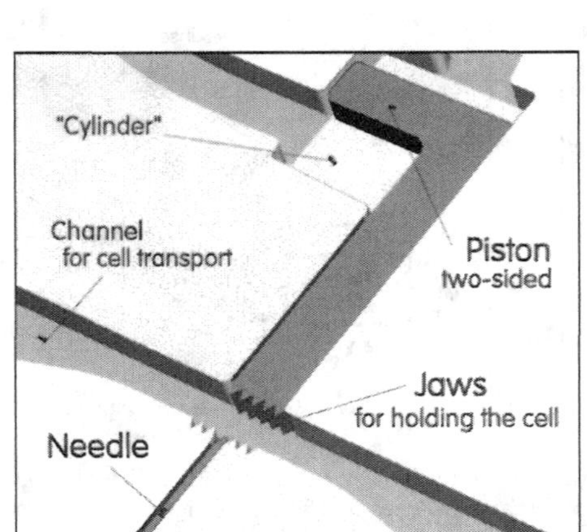

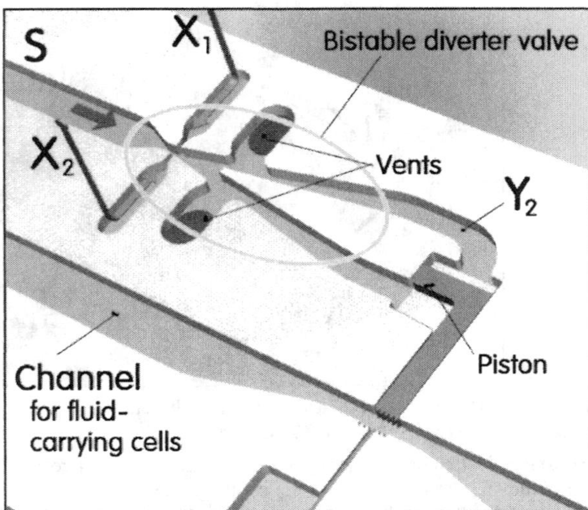

Figure 5.42 (Left) Piston-operated microjaws may be used to hold a living cell or bacterium in position while a hollow needle, moved in a similar manner from the other side of the channel, injects genetic material into it.

Figure 5.43 (Right) Fluidic driving of the movable jaw: the fluidic bistable diverter switches the fluid alternatively to the action side of the piston through the output terminal Y2 or through the other terminal to the other piston side to withdraw the jaw.

medicine, the manipulation of cells in a high-throughput manner—in large numbers operated in parallel—can change the course of a disease. Similarly, the transfer of plasmid DNA into bacterial cells forms the basis for genetic engineering.

An example of the fluid-operated handling of these tasks is presented in Figures 5.42 and 5.43. The microfluidic device punctures the cellular outer membrane and delivers genetic materials into the cell at a rate > 10 cells/s. The cells, suspended in a buffer liquid, are brought in by the tapering channel. During the manipulation, in the narrowest location, the cell is held by the jaws. Driving the movable jaw, similar to the motion of the needle from the opposite side, is another interesting example of the use of an F/M actuator. Despite the dangerous-looking teeth, the cell membrane is not harmed. Of course, the unhindered motion of the piston and the jaw requires a careful choice of materials or at least of surfaces. The size of the cells is on the order of 10 μm, and this is a scale at which it may be difficult to maintain a sufficiently high Reynolds number for the operation of the diverter valve shown in Figure 5.43.

5.2.4 F/M sensors and transducers

Generating a mechanical signal by fluidics is currently a very rare activity. If performed at all, it is almost always just a part of a more complex conversion chain. Examples are presented in Figures 5.44 and 5.45. Fluid drives a mechanical oscillation of a tiny cantilever. The oscillation is maintained by the feedback action of a tiny fluidic amplifier controlled by the fluid flow that is distributed to the right- and left-hand-side control inlets by momentary position of the edge on the cantilever near its end. The final output signal, however, is the oscillation frequency measured optically by a photodetector.

The oscillator detects the presence of material in the fluid supplied into the supply terminal S and deposited at the cantilever end. In particular, it may utilize the affinity of various biological

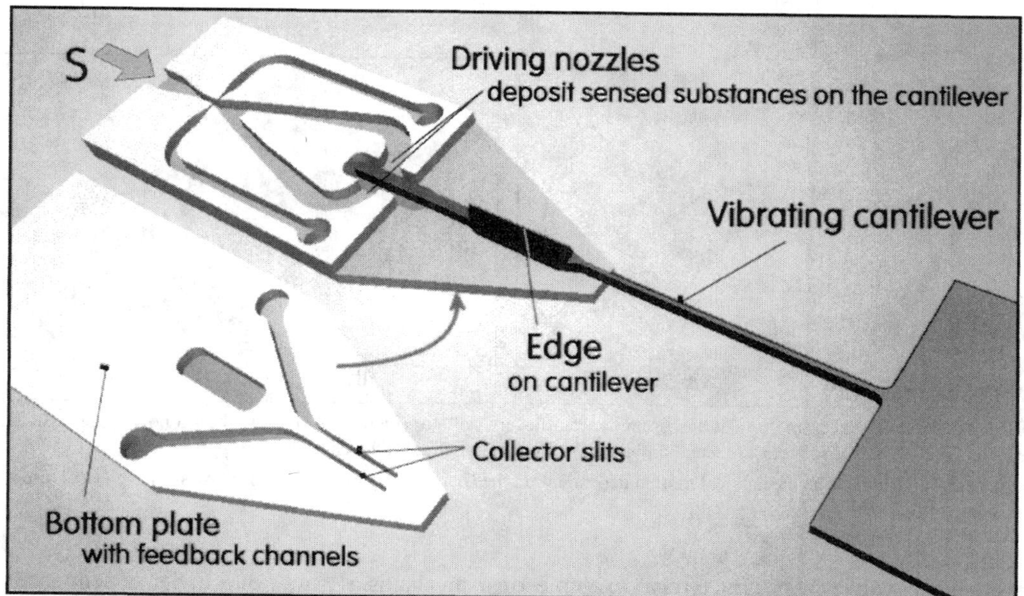

Figure 5.44 Fluidic oscillator with frequency determined by mechanical resonance properties of the cantilever. Exploded view, without the top part containing the slit nozzle (above the edge) for the control fluid, and with the bottom plate bringing the fluid into the control nozzles of the amplifier shown at left. The signal, in the form of decreased frequency, is caused by deposition of biological material at the end of the cantilever.

materials to the antibodies immobilized on the side walls of the cantilever end and exposed to the driving nozzles of the amplifier.

The selectivity of this detector is based on (1) the sensitivity of the resonant frequency to even minute changes in the vibrating mass, and (2) the recognition capability of the immune systems of organisms. In vertebrates, the immunity is due to antibody proteins produced as a defense against infection. There are specific antibodies forming aggregates with bacteria, viruses, or simply with foreign molecules. In a living organism, the aggregates are ingested by phagocytic cells. In the described oscillator, the aggregates are kept on the cantilever surface, changing its resonance properties. For the aggregates resisting washing, the antibodies have to be firmly anchored to the cantilever wall in a procedure called *functionalization*. The surface is silanized, then covered with a binding protein and finally covered with antibodies that are produced artificially (and available on commercial basis for immunoassays). In particular, a similar oscillator was used to detect spores of human pathogen fungi. Although the majority of fungal spores are smaller, some pathogenic ones are as large as 0.1 mm so that they can change the frequency very significantly. Subsequent changes in frequency, after canceling inflow of spores, can even detect their growth. This discriminates the harmless inactive ones.

Many cantilever detectors use electromagnetic driving (the cantilevers may also be as small as to be driven by the Brownian motion) so that the fully fluidic version from Figure 5.44 is not typical. Nevertheless, deposition on functionalized surfaces is important, in fact the key method for medical, forensic, and antiterrorist identification of organic materials.

Finally, Figure 5.46 presents yet another example of a mechanical response to a change in properties of the fluid. The response is a mechanical motion capable of turning down the flow

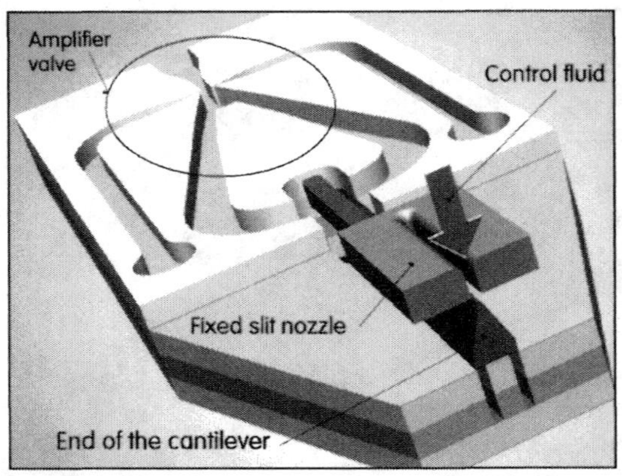

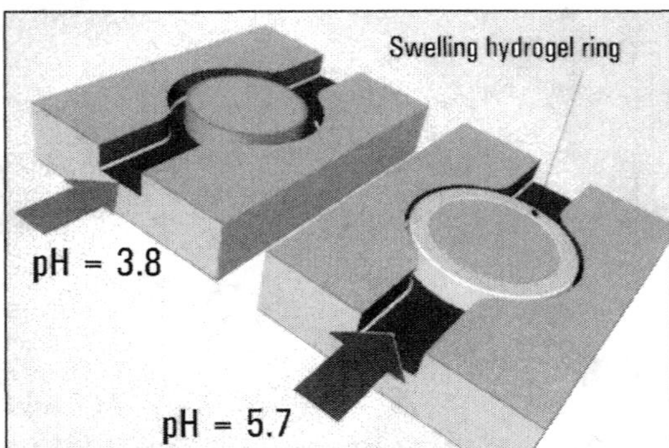

Figure 5.45 (Left) Section through the assembled cantilever oscillator (Figure 5.44) shows the slit nozzle from which the jet enters one of the slit collectors, depending on the position of the edge on the cantilever.

Figure 5.46 (Right) The mechanical motion dependent on the fluid properties: valves closed by swelling gel in response to change in fluid pH.

through the valve. This is a typical example of a mechanism that – due to its dependence on diffusion through the gel, would be hopelessly slow at a large scale but operate reasonably fast at the microscale.

5.3 E / F : CONVERSION TO AND FROM ELECTRIC EFFECTS

In view of the advantages of electronic information processing and the small degree of difficulty associated with combining microfluidic devices with standard integrated-circuits fabrication processes on the same chip, the electro/fluidic E/F and reverse F/E conversions (symbolic notation according to Figure 5.1) between fluidics and electronics are the most important among the transducer tasks. While fluidics handles the actual fluid flow, electronics processes the signals and information processing needed for effective handling. Unfortunately the direct transformation mechanisms for E/F and F/E are not known. The indirect paths, such as the F/M/E shown schematically in Figure 5.5, are therefore in general use. Another possibility is to use special fluids exhibiting nonneutral electric properties. Though this may seem to be rather impractical, the use of special fluids—in the form of polar liquids containing ions produced by the dissociation of dissolved salt molecules—is very widespread, actually so much so that it forms a large branch of fluidics competing with the pressure-driven fluidics discussed in this book.

The discussion of possibilities and typical examples of E/F and F/E conversions is here again divided into four parts, organized according to the four *window-pane* boxes in Figure 5.3.

5.3.1 Electric pumps and valves

5.3.1.1 Using ordinary fluids

E/F conversions at higher power levels are needed in the actuator parts of micropumps and microvalves. Typical designs base their operation principles on the E/M/F conversion chain (Figure 5.5 shows a classical large-scale version, with the variable core transformer on the electric side), with mechanical force action as the intermediate step.

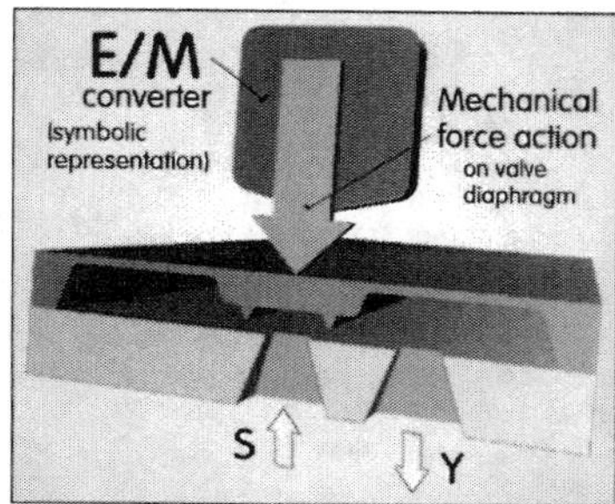

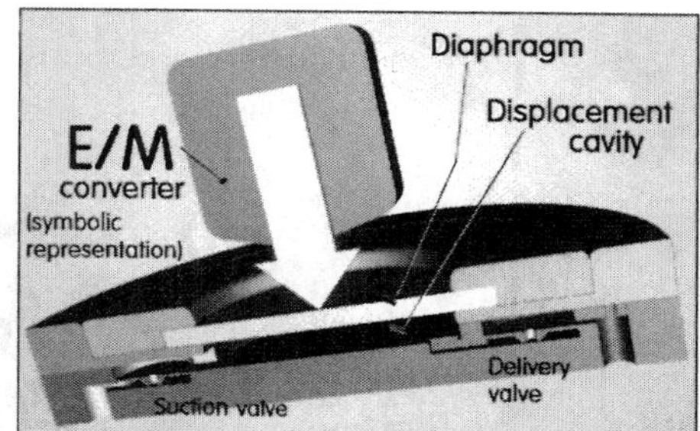

Figure 5.47 (Left) Representation of the standard electrically controlled valve with the E/M/F conversion chain. The interesting part from the fluidics point of view is the same as that in the fluidically controlled valve shown in Figure 5.15. The E/M conversion symbol represents one from the seven actuator principles shown in Figure 5.57.

Figure 5.48 (Right) Electrically driven mechano-fluidic pump, shown with a symbolic representation of the E/M conversion part. The basic component performing the M/F part of the conversion is the deformed diaphragm.

A discussion of the transducers needs a separate consideration of the E/M and M/F parts. Some aspects of the M/F conversion as used in mechanical pumps or valves, were already discussed in Section 5.2.1. In contemporary microfluidics, the component performing the mechanical action on a fluid is usually a deformable one, very often a diaphragm, which either occupies a space through which the fluid would otherwise flow (the protruding boss made on the diaphragm obstructs the flow passage in Figure 5.47) or displaces the fluid from a displacement cavity (Figure 5.48). Mechanical components limit the operation mainly due to their inertia. If a wide frequency range is important, the mechanical components may be reduced to just the very indispensable one, the diaphragm.

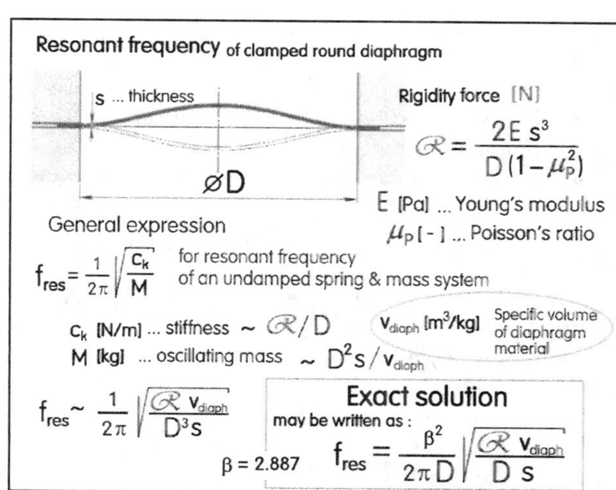

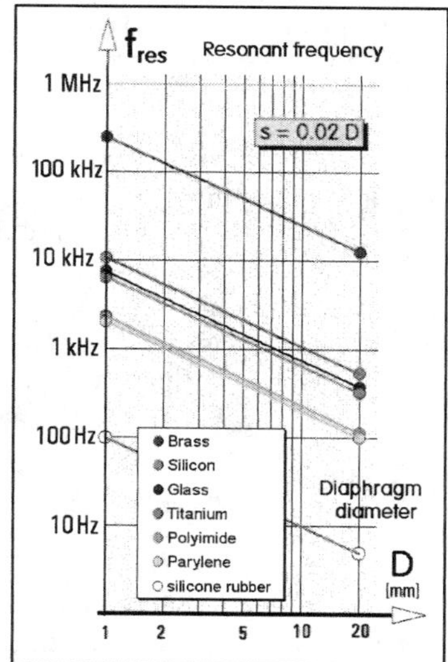

Figure 5.49 (Left) Formula for resonant frequency of circular diaphragms. It should be noted that this derivation neglects the interaction of the diaphragm with the fluid (pressure forces adding to stiffness and also the oscillating fluid mass adding to inertia). The assumption of rigid clamping (zero slope) may be not exactly met in real diaphragm designs.

Figure 5.50 (Right) Example computations of resonant frequency, using the formula from Figure 5.49 for different diaphragm materials, those actually used in published micropump designs. This may provide some idea about the typical range of resonant frequency. The same diaphragm thickness s, equal to 1/50 of the diameter D, is assumed.

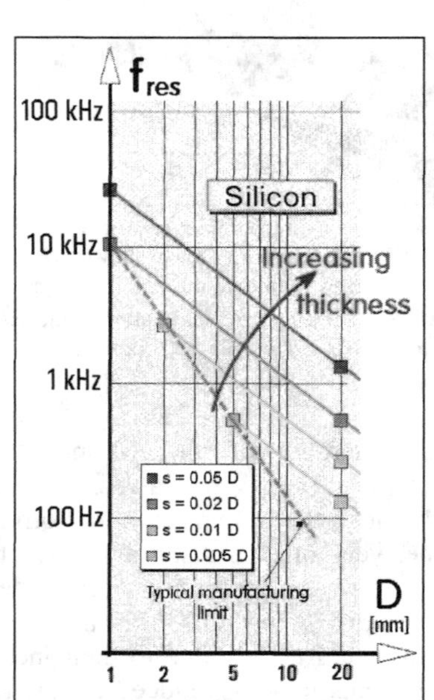

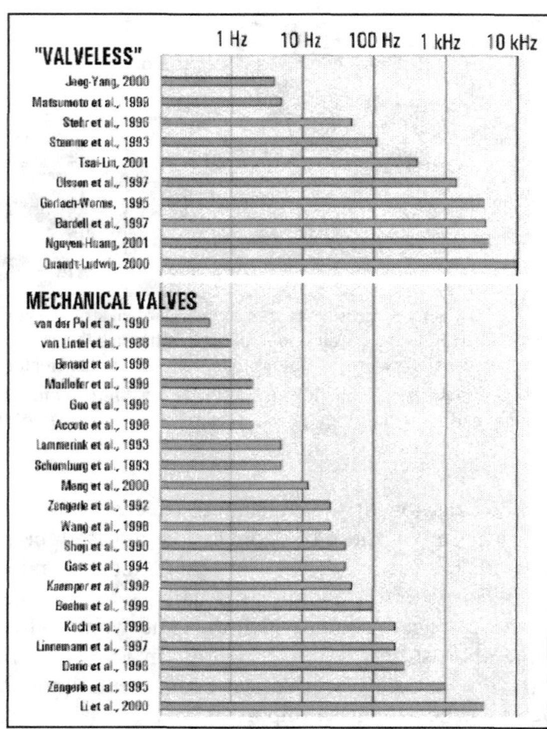

Figure 5.51 (Left) Another use of the formula from Figure 5.49 to evaluate diaphragm resonant frequency, this time assuming the same material, silicon, but different thicknesses.

Figure 5.52 (Right) An idea about actually chosen operation frequencies in diaphragm pumps may be provided by this list of values in the available literature.

A higher operating frequency is particularly advantageous in pumps. As a rule it results in a smaller pump and the presence of considerable dynamic effects in the fluid, which may improve the operating conditions of the rectifying pure fluidic components.

The essential factor limiting the choice of operating frequency is the resonant properties. Basic information about these properties of the diaphragms is presented in Figures 5.49 to 5.52. Note that the expression derived in Figure 5.49 is simplified. It neglects the interaction of the diaphragm with the fluid, assumes ideal clamping on the circumference, and also assumes the simplest diaphragm shape (circular and without the often-used bosses, as shown, for example, in Figure 5.47). In a flow control valve, the frequency range of the input electric signal has to be well below the resonant conditions for the input signal changes, being closely followed by the diaphragm motions. On the other hand, for operation in a periodic regime (in pumps), resonance may be a desirable phenomenon, decreasing the input power requirements.

Figure 5.53 presents some published data on the ratio of the operating frequency to the resonant frequency in micropumps. It shows a recognizable correlation between this ratio and the Stokes number, which is a measure (like the Reynolds number for steady flows) of the relative importance of dynamic effects in fluid flows in relation to viscous effects. A higher operating frequency obviously brings the advantage of viscous damping being less influential.

Indeed, although the time span in Figure 5.52 is too short, relative to typical delays in publication procedures, to reveal a clear chronology, it reveals a recognizable general trend in present-day microfluidics toward operation at a higher frequency, even though in actuators the

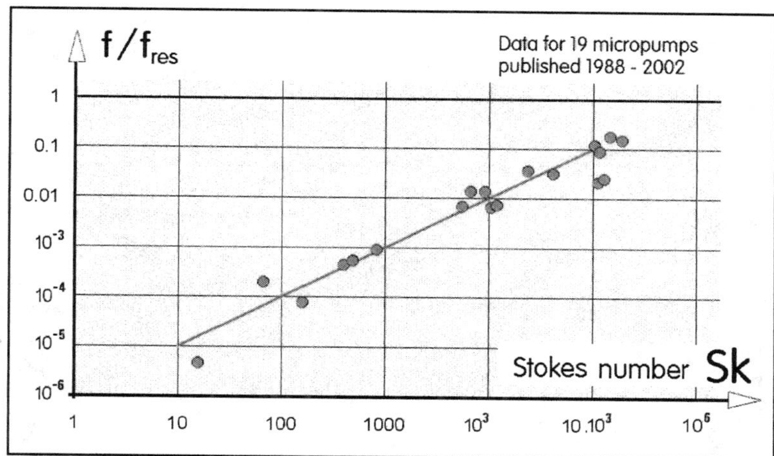

Figure 5.53 Published data on *E/M/F* micropumps shows a distinct dependence (first noted by Laser and Santiago [7]) between the ratio of the operating frequency f to the resonant frequency of the diaphragm (the value shown here was evaluated from published data on thickness and diameter using Figure 5.49) and the Stokes number, the similarity parameter for fluid periodic motion.

frequency range is generally less important than in signal processing. Achievable frequency is limited by inertial effects and a safe way to decrease them is to design the actuators with a smaller displacement cavity and smaller strokes. This dependence between size and frequency is recognizable, despite the large scatter in the data, from the statistical evidence in Figure 5.54.

The smaller size of the mechanical action components, however, is not always advantageous. Another statistical processing of the published data, for the pumps listed in Figure 5.55, is presented in Figure 5.56, in the form of the dependence of the effective stroke on the diaphragm size. The data shows that pumps having small diaphragms are characterized by much smaller effective stroke lengths than what would correspond to the same geometry of the deformation

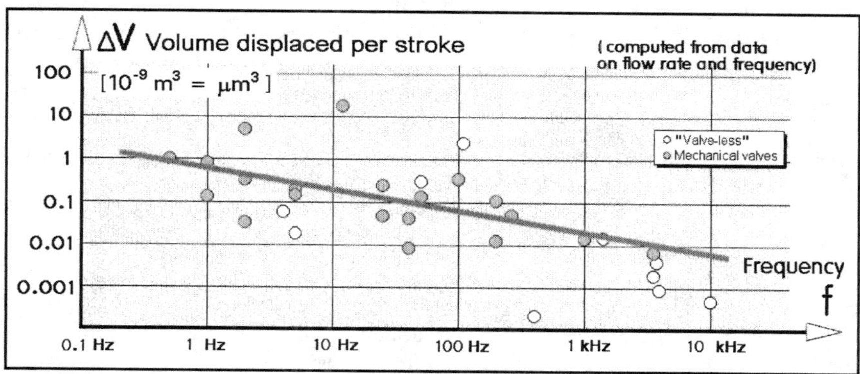

Figure 5.54 Available data on diaphragm micropump flow rates and operating frequency (corresponding to the list in Figure 5.52) were used to evaluate the volume of fluid delivered in the course of a single stroke. Somewhat surprisingly, considering the generally less effective purely fluidic rectification in the "valveless" pumps, there is no indication of their expected forming a distinct lower-performance group. More significant is the general trend of smaller displacement with increasing operating frequency.

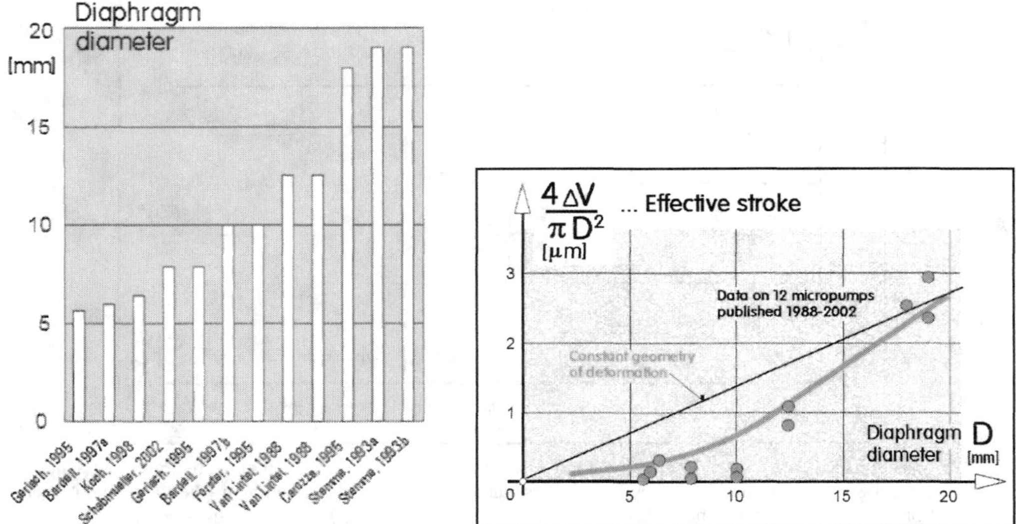

Figure 5.55 (Left) List of diaphragm diameters used in micropump designs described in literature.

Figure 5.56 (Right) The data for the pumps listed in Figure 5.55 were used to evaluate from the displaced volume per stroke (Figure 5.54) and the diaphragm area the effective magnitude of the stroke. There is a distinct trend of the stroke decreasing with decreasing diaphragms size. The effect is more pronounced than what could be expected for equal relative deformation.

(same ratio of diaphragm deflection to its diameter). One part of the explanation for this fact may be the loss of geometric similarity as the diaphragm size is decreased. First, the diaphragms are simply made relatively thicker to make the device more reliable; otherwise, it can be easily destroyed by a small accidental pressure pulse of a magnitude that, at a larger scale, would be completely safe. Second, it should be said that the notion of clamping used in the derivation in Figure 5.49 is just a convenient way of introducing a boundary condition on the diaphragm's circumference. No real clamping at these sizes would be reliable. The diaphragms are usually made by local thinning, with a transition to a a thicker outside material. At smaller sizes, the rounding radius of the transition is almost always relatively larger, again leading to larger effective diaphragm stiffness than the one expected on the basis of the formula from Figure 5.49. The other part of the explanation may be sought in the properties of the rectifiers. It must not be forgotten than the data in Figure 5.56 was derived from the pump's overall behavior. At the smaller size—and the correspondingly smaller Reynolds numbers—the rectifiers are generally less effective.

The second part of the discussion of the E/M/F transducer chains is consideration of the E/M conversion. This, of course, is not a specific problem of fluidics, and therefore it need not be discussed in detail here. What is needed for the present purpose is a survey of available possibilities, which are listed in Figure 5.57. Useful information may be also found, in particular, in Chapters 6 and 7 of [1]. An important circumstance is that, while some of these principles have been used for a long time and belong to the established methods, there has recently been rapid progress in the development of new electroactive materials, which can change the overall picture. A long-established method is use of the electromagnetic transducers. Another classical principle is the piezoelectric conversion, listed in Figure 5.57 under the more general name of electrostriction. Modern electrostrictive materials are currently the object of rapid development progress, and some of them, such as the electroactive polymers, are of a character so much

E/M actuators	Efficiency		Speed of response		Typical Voltage	Energy density [kJ/m^3]	
O Electromagnetic	high		fast	2 kHz	24 V	low	4
O Electrostatic	very high		fast	4 kHz	200 V	medium	200
O Electrostrictive	high	27 %	very fast	100 kHz	1 - 2 kV	medium	100
O Thermal expansion	very high		medium	5 Hz	12 V	high	2000
O Shape memory	low		slow	0.9 Hz	5 V	very high	9000
O Magnetostrictive	very high	67 %	fast	30 kHz	8 V	medium	70
O Electroactive polymer	potentially high		fast	1 kHz	10 V	very high	9000

Figure 5.57 The list of electric-to-mechanical actuator principles used in the *E/M/F* conversion chains in power transducers. Typical numerical values shown here are from published data for practical micro-actuators. They do not necessarily represent the limits obtainable with a particular conversion principle. Note that the "thermal expansion" in this table means an expansion of mechanical component (e.g., of a bimetallic strip) not expansion of the fluid (Figure 5.63).

different from the classical piezoelectric crystal devices that they deserve to be listed as separate items, though the basic idea of volume changes in the electric field is the same. The polymers respond to much lower voltages. The electrostatic principle has been known for a long time but has been little used because of small generated forces and short stroke lengths, while requiring inconveniently high driving voltages. This situation, however, may be less adverse in microfluidic applications, where the forces and strokes are generally small anyway, and the gaps between the electrodes may be small enough (the generated force is proportional to (voltage)^2/(gap)^2) for the necessary driving voltages to be within reasonable limits.

The limiting factor for the actuator speed (or frequency range in a periodic operation) being generally on the side of the mechanical components, the response speed aspect of the E/M part of the conversion is usually not a decisive factor. The electric action is generally very fast, even in those cases in which the effect involves thermal changes, because the thermal effects scale favorably as the device size is decreased. The only exception is the rather slow operation of shape-memory-alloy actuators. A useful survey of the typical frequency range of various E/M principles is presented in Figure 5.58. Basically, the speed of response is traded off for the weight of the devices. General dependence follows roughly the lines of constant specific power: higher frequency is available in those conversion principles that provide less work per unit of mass.

Though rather exceptional, there are cases in which the limiting factor for the speed is also on the electric side. The inductance of relatively large coils used in the *electromagnetic* converters may limit the speed with which the current can be varied; this may be the problem in actuators built in the form of solenoid valves. Solenoids with a large number of coil turns are used in applications requiring the generation of very large forces. The same principle of the electromagnetic E/M conversion, however, is available also for high-frequency operation in the voice-coil version. This can reach very high frequencies, in fact up to the general limit, which for all E/M may be said to be at about 100 kHz. What makes the electrodynamic principle less popular in microfluidic applications is not this aspect, but the problems and disadvantages associated with the magnetic circuit, which always tends to be big and heavy. On the other hand, quite popular at the microscale is the *electrostatic* principle. To generate a higher force, however, large electrode surfaces are needed. This has led, in microfluidics, to a version with interdigital electrode shapes and electrode arrays. The limiting factors in the use of the *electrostriction* in the piezoelectric crystals or ceramic version are very high driving voltage and the small relative deformation. The latter may be circumvented by use of a stack of piezoelements, with the

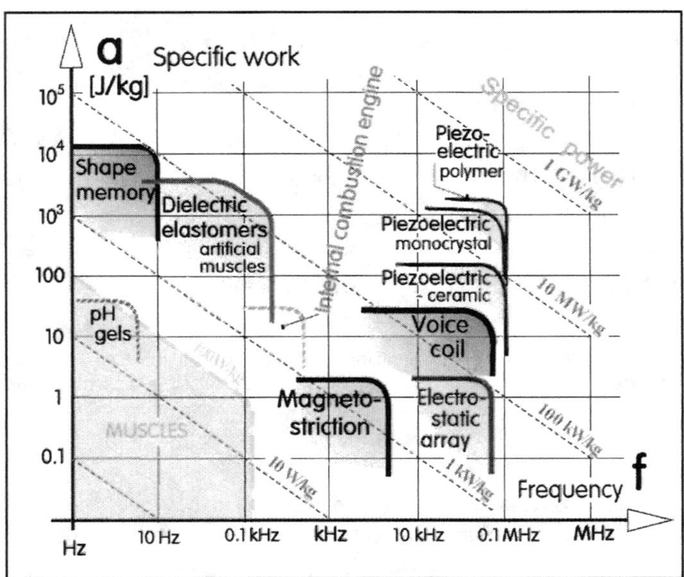

Figure 5.58 Approximate performance limits of E/M actuator principles. Generally speaking, high operating frequency means low specific work, the dependence roughly following the lines of constant specific power. The exceptions are modern electrostrictive materials: piezoelectric monocrystals and, in particular, polymers, which are capable of exceptional specific power performance. For comparison, the diagram includes the approximate boundary for muscles, swelling gels sensitive to changes in fluid pH (Figure 5.46) and data for an airplane model combustion engine.

with the resultant motion being the sum of the individual contributions. The mechanical thermal expansion effect in Figure 5.57 is actually a case of an E/T/M chain. Its typical layout uses a bimetallic element. Depending on the opportunities for heat transfer of the element, this conversion method may also be rather slow, though this is a less difficult problem in microdevices, with their generally larger surface-to-volume ratio. An unpleasant aspect plaguing most E/T/M chain principles is the removal of the heat for deactivation, which may require additional components and heat sinks. Other E/T/M chain devices are those using the *shape memory* elements. These are really very slow and also perform rather poorly from the conversion

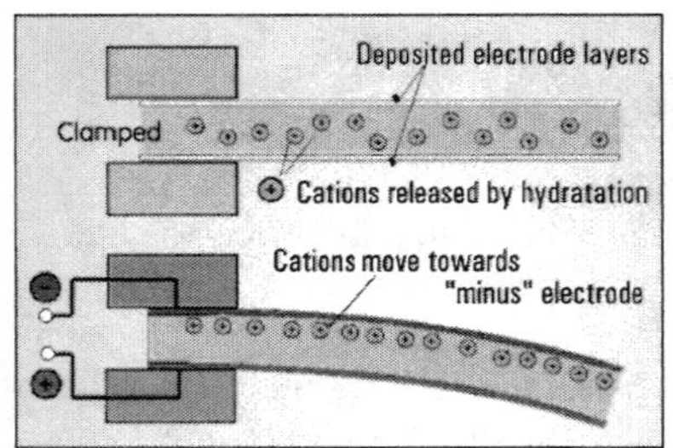

Figure 5.59 (Left) The principle of operation of an ionic electroactive polymer strip: the deformation is due to the migration of ions to one of the electrodes. The strip has to be covered with (or submersed in) water for the ions to be released.

Figure 5.60 (Right) A Nafion strip deflecting in response to an applied voltage.

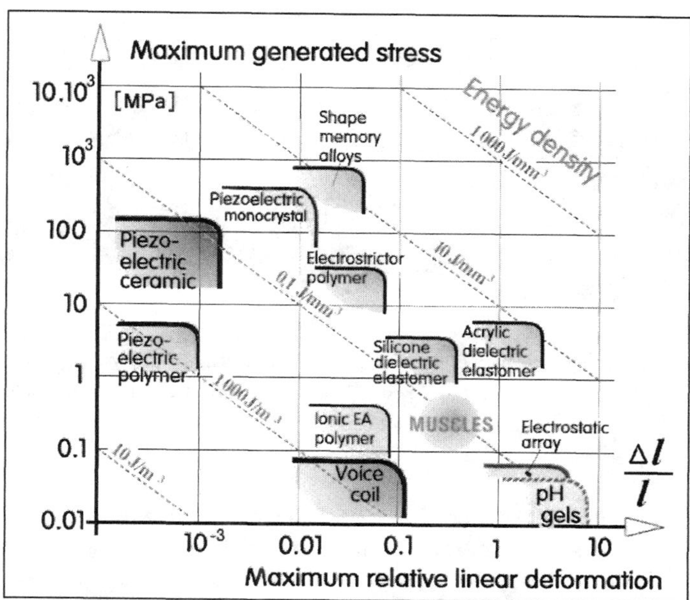

Figure 5.61 A survey of *E/M* actuator materials generating the stroke by elongation of the basic volume (with the voice coil, muscles and electrostatic arrays converted to equivalent deformations). The presentation is similar to that in Figure 5.58 but it plots other properties of importance. This different point of view can substantially change the relative merits of some principles: piezoelectric polymers, the winners in Figure 5.58, are among the worst performers from the energy density point of view.

efficiency point of view. On the other hand, the actuators based on this principle have the advantage of being very simple and usually can occupy minimal volume. The principle being relatively new, there has still been considerable progress in recent years in the development of better materials. Also relatively new and still in progress are the *magnetostrictive* actuators, already quite effective and small, apart from the complication caused by the magnetic circuit. Again, in operation they develop heat that must be removed, perhaps by the controlled fluid flow.

Some currently very promising actuator materials, under continuous development, are *electroactive polymers*. They are of two quite different sorts. Figure 5.59 shows the principle of generation of mechanical motion in *ionic* polymers and polymer-metal composites. A typical version consists of perfluorinated ionomer membrane (Nafion by DuPont), usually ~ 0.2 mm thin and plated on both faces with noble metal (platinum and/or gold) electrodes. When a cantilever strip of this trilayer composite is subjected to an applied electric potential of several volts (typically 3V), it bends (Figure 5.60) toward the anode, due to the movements of the electrically unbalanced ions. The behavior is rather complex; the initial bend with sustained voltage may be followed by relaxation and in periodic excitation the strain may fall significantly during initial cycles. A complicating factor is the necessity to keep the membrane or strip in a hydratation environment. This property is put to use in applications operating in water, such as autonomous underwater microvehicles. There are also, however, variants operating in a nonliquid medium, air. These very new materials [12] are electrically conducting polymers derived from aniline, thiophene, or most importantly pyrrole. Sometimes referred to as *artificial muscles*, their shape changes are also due to the motion of ions inserted into the polymer by electrochemical doping. The advantage is driving voltages as low as 1V. Ionic polymer actuators recently described in the literature include polyaniline actuators (which may also be driven chemically by doping and de-

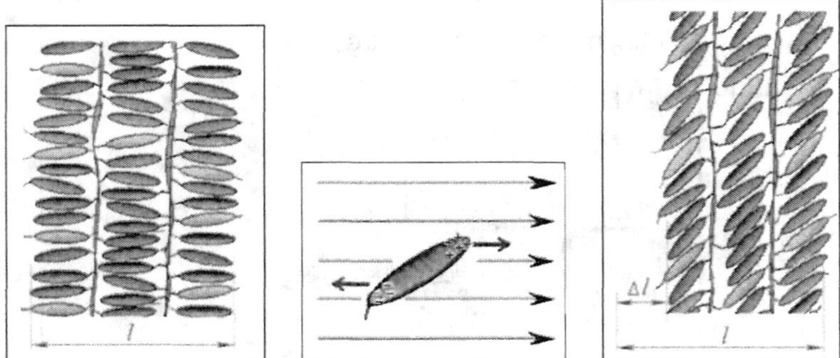

Figure 5.62 (Left) Schematic representation of the structure of electroactive liquid crystals.
Figure 5.63 (Center) Rotation of the element of the crystalline structure in an applied electric field.
Figure 5.64 (Right) The resultant length change of the electrostrictive gel.

doping) and carbon nanotubes. The mechanism for the swelling of the latter actuators is actually somewhat different (non-Faradaic electrochemical charging). They promise exceptional properties. Used in bundles, the nanotubes exhibit rather small actuation strains, on the order of only 0.2%, but with their high Young's modulus, approximately 640 GPa, the stress from a bundle may be as high as 1280 MPa.

The other sort of electroactive polymers is mainly those characterized as liquid crystal elastomers. They theoretically should be put into the group of direct E/F conversion, but their mechanical properties in most cases are actually akin to those of gels rather than fluids. They are in fact composite materials: the monodomain nematic liquid crystal elastomers containing a distributed network structure of conductive polymers (Figure 5.62). The actuation mechanism is a phase transition between its nematic and isotropic phases. As shown in Figures 5.63 and 5.64, this is associated with a rapid contraction due to the alignment of the liquid crystalline phase in the electric field. The reverse process is slower and may require effective cooling to return to the original length. The other promising actuator materials are electrostrictive poly(vinylidene fluoride-trifluorethylene) copolymer films. They exhibit a large deformation (~ 5%) and high elastic energy density (as shown in Figure 5.61) and seem to be an attractive possibility for high-frequency operation in nonresonance mode (up to what is effectively the general practical limit at ~ 100 kHz) and high-load applications.

Mainly due to the problems with achievable frequency range—now gradually removed by newly developed materials, especially elastomers—but also because of the possible unreliability due to sticking or breakages, there has been a distinct trend in microfluidics to replace the mechanical motion by the no-moving-part E/T/F conversion chain. The most popular mechanism is the thermal expansion of gas or vapor (Figure 5.65) in response to the passage of electric current in Ohmic heaters. Figure 5.66 lists the two versions as a continuation of the table in Figure 5.57. Also included are order-of-magnitude indications of typical performance. The advantage is the very large obtainable expansions, which are roughly an order of magnitude larger than with electrostriction and, at the same time, the very low required driving voltage. Some particularly imposing volume changes are achievable with the phase change, the boiling and evaporation of a liquid. Also, the expanding gas may act as the standard working fluid in other parts of the fluidic system, though a captive gas volume, separated by the gas-liquid interface as shown in Figure 5.65, is also used in systems operating with liquid. The thermal expansion of liquids is rarely used, since the expansion effect is small. This is evident, together with the large strokes and better

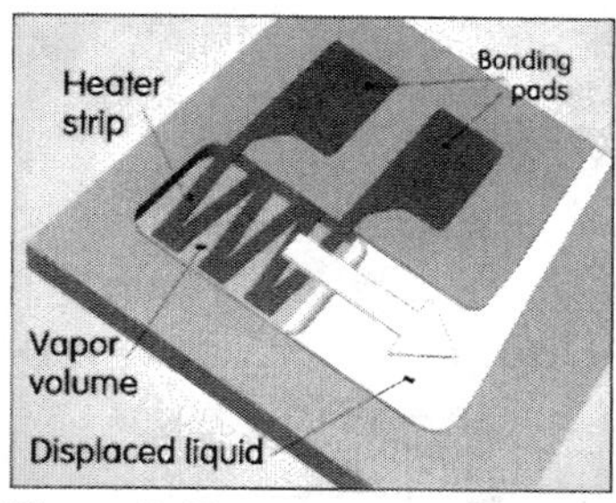

E/T/F actuators	Efficiency	Speed of response	Typical Voltage	Energy density [kJ/m³]
O Thermal expansion	very high	medium 400 Hz	12 V	medium 200
O Phase change	very high	slow 5 Hz	12 V	high 4000

Figure 5.65 (Left) Conversion E/T/F using the thermal expansion of gas heated by the electric current. It is a different principle than the thermal expansion of solid components listed in Figure 5.48.

Figure 5.66 (Right) Electric-to-fluidic actuator principles using the thermal inter-conversion (rather than the mechanical motion, for which the principles are listed in Figure 5.48). Again, the numerical values are data found in literature for some practical micro-actuator realizations.

energy density obtainable by the evaporation of a liquid, from the representative diagram for a typical fluid with phase changes in the left-hand part of Figure 1.42. Although only schematic, this figure also shows the small expansion in liquid for an isobaric thermal change. Of course, as with all thermal effects, it is necessary to ensure a similarly fast cooling for the return motion.

Operation with two phases—either air and water or liquid and its vapor—separated by mere surface tension interfaces may be not without difficulties. Once it is accepted, however, a number of interesting possibilities of the actuation become open, mainly based on influencing the surface tension. Since the surface tension forces scale with length, while inertial force scales with volume, that is $(\text{length})^3$, the surface effect may easily become dominant as the size of the devices is diminished. Inducing a motion by surface force requires creating a spatial difference by local control of the surface-tension magnitude. This is possible by chemical means (the release of surfactants), by Marangoni thermal effect, or by electrocapillary phenomena. The Marangoni effect (Figure 5.67) propels the liquid toward the warmer surface by the differences in the pressure governed by the Young-Laplace law (Figure 2.128). Figure 5.68 presents useful data on the surface tension changes with temperature for water.

A layout somewhat similar to that in Figure 5.67, with the sequential switching of electrodes rather than heaters, is used in the otherwise very elegant method of propelling liquid drops by electrowetting, changing the capillarity in the electric field. These were originally observed (by

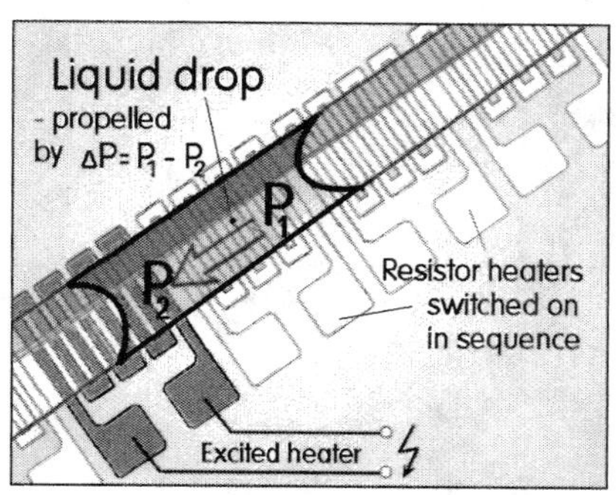

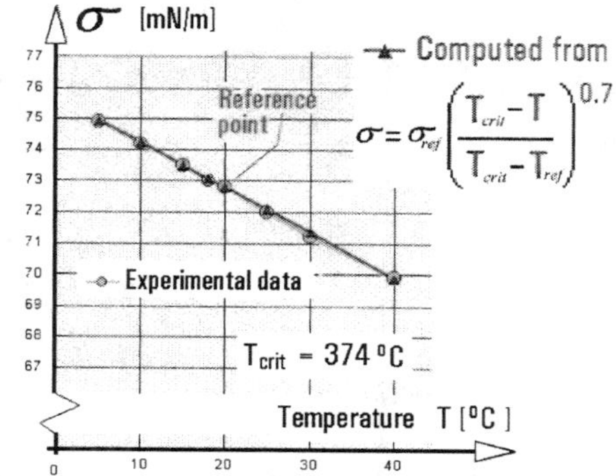

Figure 5.67 (Left) Motion of liquid drop by thermal effects due to sequentially switched microheaters. The acting force is due to the Marangoni effect: the thermal dependence of liquid surface tension.

Figure 5.68 (Right) Water surface tension σ dependence on temperature.

E/F special fluids	Efficiency	Speed of response	Typical Voltage	Energy density [kJ/m³]
○ Electrorheologic	low	medium < 100 Hz	high 2 kV	high
○ Magnetorheologic	low	medium	low 10 V	medium
○ Ferrofluidic	good 60%	medium 30 Hz	low 10 V	medium
○ Dielectric fluid EHD	low 4%	medium	very high 10 kV	high
○ Electro-osmotic	low 1%	medium	medium 100 V	very high
○ Electrostrictive	high 60%	fast	high 1 kV	very high 34 000

Figure 5.69 Direct E/F conversions are possible with special electroactive or magnetically active fluids.

Lippman in the nineteenth century; the equation he set up is still the basis of the mathematical description of the effect) in mercury immersed in electrolytes. The current embodiments make use of the electric double layer in an ionic solution, and this makes them belong in Section 5.3.1.2 of this survey. The control of capillarity was demonstrated to achieve not only the transport of liquid droplets, but also operations such as droplet merging and division. Considerable current research effort is, however, directed elsewhere: on the development of electrowetting displays, which use optical effects associated with presence or shape changes of liquid droplets.

5.3.1.2 Electric pumps and valves using special fluids

Working with special fluids is the price paid for simplicity associated with direct electric action on the fluid. There are, of course, applications in which this approach does not come into question at all, since fluid properties may be dictated by the performed task, as is the case, for example, if the fluid is needed as a reagent for a particular chemical reaction. Also, it may be noted in the listing of these principles in Figure 5.69 that none of the special liquids is gas; the principle is in practice limited to a special choice of liquids.

Some of the special liquids are practically suitable only for valve-type actuators: a typical case is the electrorheologic and magnetorheologic liquids, which react to the presence of an electric or magnetic field by increasing their viscosity (Figure 5.70). This is used for a turning-down effect, a decrease of the flowrate. On the other hand, the electro-osmotic and electrostrictive liquids may be applied in electrically driven pumps.

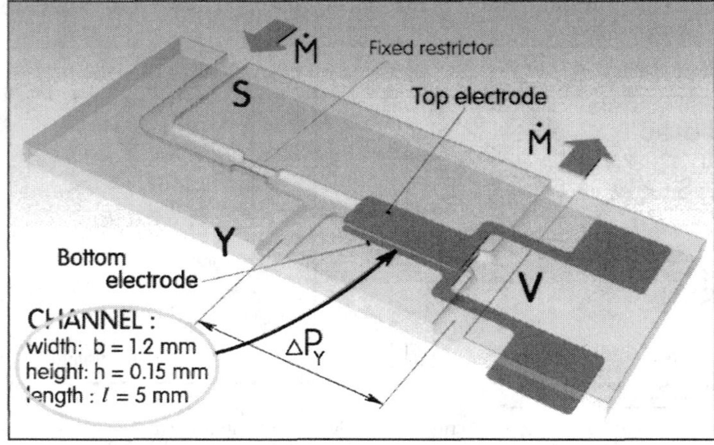

Figure 5.70 An example of using the electrorheological control of fluid flow by direct E/F conversion in a channel between two electrodes. The channel operates as the variable restrictor in the pressure divider circuit (Figure 5.18).

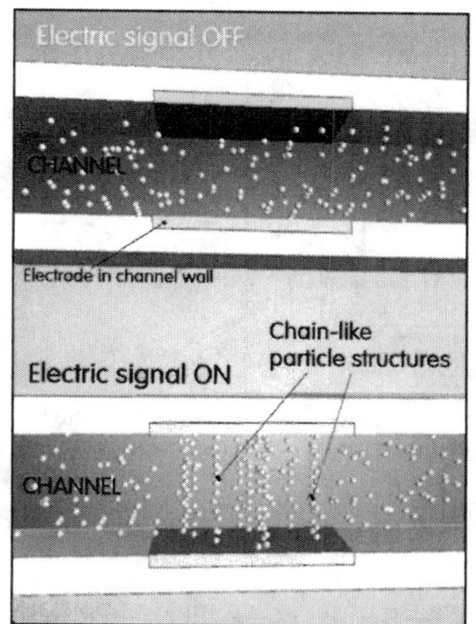

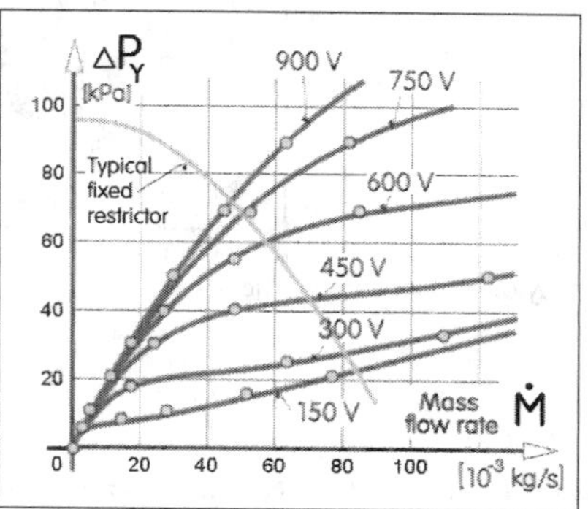

Figure 5.71 (Left) The reason why the viscosity of an electrorheological liquid increases in the applied electric field between the two electrodes is the behavior of the suspended particles. Distributed stochastically in the fluid (top) with the electric field switched off, they form fibrous structures when the voltage is switched on (bottom).

Figure 5.72 (Right) An example of the behavior of electrorheologic liquid crystals in a channel of the dimensions indicated in Figure 5.70, at different constant values of the voltage difference applied between the electrodes. For zero loading output flow, the intersections with the shown fixed restrictor characteristic determine the changes in the output pressure difference ΔP_Y with the voltage change. The data is similar to those in [8], where, however, the bottom electrode was not necessary, as the channels themselves were etched in a conductive material used in lieu of the bottom electrode.

In the extreme case, the increase in viscosity of an *electrorheologic* liquid may be by up to a factor of 10^5. The effect may then be described as stiffening or almost solidification. The change in the viscosity is reversible. Fluids exhibiting this property—and also the related *magnetorheologic* effect—are most commonly colloidal suspensions, as are also the *ferrofluids* (which are often related to magnetorheologic fluids to the degree of exhibiting also a considerable rheologic effect). The electrorheologic change in properties is also called the Winslow effect (after the first investigator, W. Winslow, in 1949). It is caused by the particles forming chainlike structures (Figure 5.71) parallel to the lines of the applied field. Typical current electrorheologic fluids contain particulate polymers of lithium, sodium, potassium, or caesium salts of methacrylic acid. There is, however, also the special category of electrorheologic liquid crystals. They are better suited to microfluidic applications, because they circumvent the problems (such as clogging) that may be caused by suspended particles. An example of experimental results that provides an illustration of achievable performance is in Figure 5.72, in the form of the pressure/flow characteristic at a different constant electric input signal.

Magnetorheologic fluids [9] are suspensions of ferromagnetic particles of about 1-5 μm in size, covered with nonmagnetic film to avoid agglomeration. The base liquids are special oils. Despite the high weight percentage of the particles, usually ~ 80%, without the magnetic field they behave like a conventional Newtonian liquid.

Ferrofluids contain magnetic particles, usually magnetite (Fe_3O_4) in carrier liquid: water, kerosene, or various oils. The particles are coated with long (~ 2 nm), chained, organic molecules of a surfactant, such as oleic acid. Typical is the submicron size of the particles so that they do not sediment, as they are kept in Brownian motion by the thermal agitation of liquid molecules. The flow control is achieved by magnetic force acting on the particles (Figure 5.73).

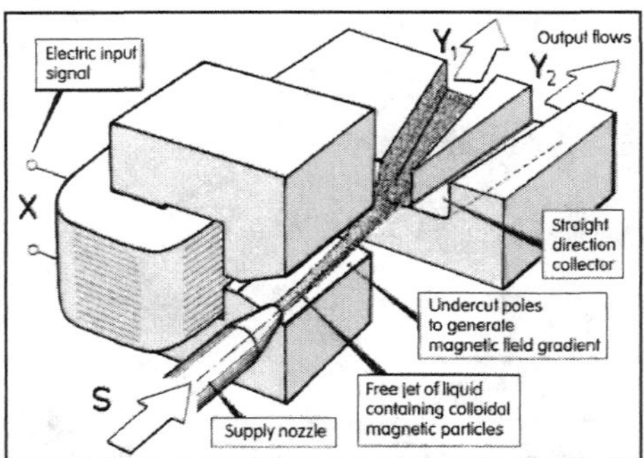

Figure 5.73 Principle of deflection control of a free ferrofluid jet. The deflection is due to attraction into locations with a higher intensity of magnetic field. The drawing is from one of the earliest patents [10] on the use of ferrofluids in fluidics.

Electrokinetic driving

This method deserves a more extensive discussion because of its exceptional importance. It is the basis of a rather large field of applications, especially in microchemical analysis. The electrokinetic mechanism operates by the electric forces acting on particles carried with the fluid. The particles, of course, must be chargeable (or at least polarizable). If they are large enough to be of comparable size with the microchannels, the effect is described as electrophoresis. The particles may be solid, but typical for electrophoretic devices is electric action on very large organic molecules, such as proteins. Though the carrier liquid, in principle, may remain stationary, the relatively large hydrodynamic drag of the particles causes the liquid to be moved along with them.

Electro-osmosis differs in using small-size particles, as small as commensurable with the size of the fluid molecules. The slip velocity between the particles and fluid is so small that the fluid effectively behaves as a single phase. This is the case with ions of dissolved salts, Figures 5.74 - 5.77. For dissolution, the liquid molecules have to be of a polar character. Although this is, considering the vast majority of nonpolar liquids, very exceptional, in practice this limitation is not very serious. Of course, this driving mechanism cannot be used with gas. On the other

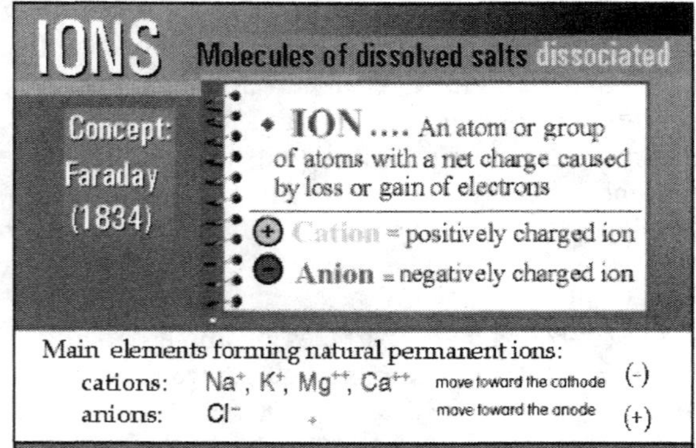

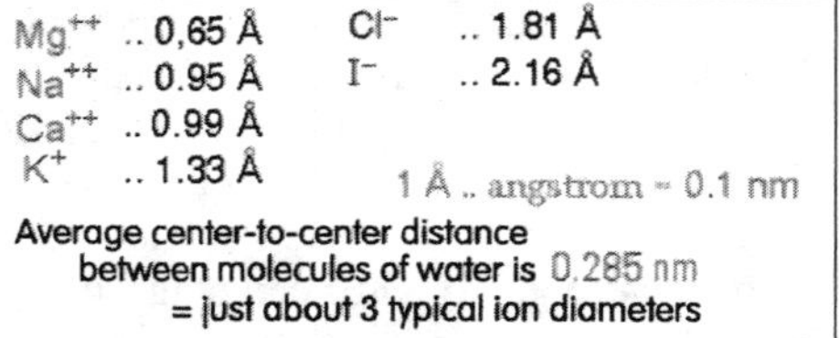

Figure 5.74 (Left) Currently the most popular electric-to-fluidic conversion methods is based on electric force action on free ions in the liquid, produced by the dissociation of molecules of dissolved salts.

Figure 5.75 (Right) Size of several important ions. They are much smaller than cells of living organisms and some actually can pass through a cell membrane. This transport is selective and essential for maintaining proper conditions in cell cytoplasm. The idea of a spherical shape is a simplification: X-ray diffraction studies of NaCl crystals show the distance between the ion centers to be 2.81Å, not 2.76Å (= 0.95 + 1.81).

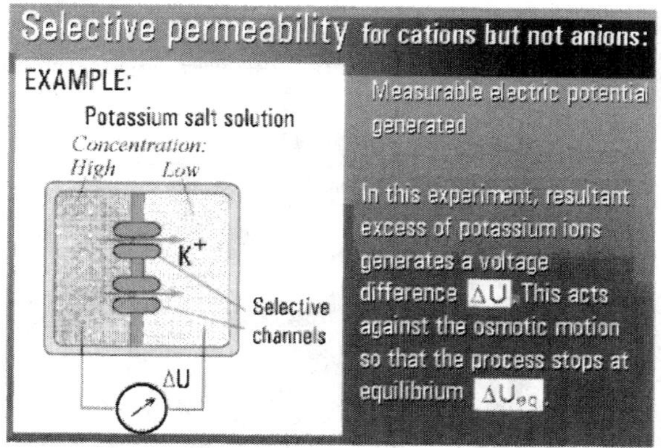

Ion concentration and Nernst potential
for typical mammalian cell

Ions	Cytoplasm [mM]	ΔU_{eq} [V]
Na^+	12	67
K^+	140	-95
Ca^{2+}	10^{-4}	131
Mg^{2+}	0.8	8
Cl^-	4	-90

Figure 5.76 (Left) Idealized example of voltage difference due to the spontaneous transport of ions (generated by dissolving a potassium salt) across a membrane possessing the selective permeability for potassium cations but not for salt anions.

Figure 5.77 (Right) Typical concentrations and voltages encountered in biological ion transport problems. Note the molality used as the measure of the concentration.

hand, water as an "exceptional" polar liquid is abundant and very often required to be handled in applications. Water with dissolved salts is the environment in which life took its origin, and handling salt ions is an important process taking place in living cells. Data in Figure 5.77 provide an idea about the numerical values. Cells posses an active transport system called a "sodium pump" which continually extrudes sodium ions Na^+ from cells. From the effects at the cellular level also follows an intensive interest in electro-osmotic driving in biological applications of microfluidics.

The concept of the dissociation of salt molecules dissolved in water is from M. Faraday (1834). He introduced the terms *cation* (ion moving towards the cathode when the solution is electrolyzed) for a positively charged ion and *anion* for a negatively charged ion, one moving toward the anode (Figure 5.74). The typical sizes of ions are shown in Figure 5.75. The thermal motion velocities in water are on the order of 100 m/s, but the motion of ions may be much slower, since they drag along a *hydration shell* of oriented water molecules. The size of the shell is on the order of the *Debye-Hückel length.* It is not constant, but varies with ion concentration: dissolved NaCl has a Debye length of 96 Å (~ 100 ion diameters) at 1 mM falling to 9.6 Å (~ 10 diameters) at 100 mM.

The name *electro-osmosis* suggests a close relation to the spontaneous passage of water through tiny pores, called ***osmosis.*** (This term, from the Greek word for "push", was introduced in 1854 by British chemist, T. Graham.) It was originally discovered in animal bladders, where the pores are inobservably small (several orders of magnitude smaller than the wavelength of visible light). It is of particular importance for biology, in particular for transport across semipermeable membranes, selectively blocking the passage of some particles.

If a membrane is selectively permeable for cations but not for anions, a measurable electric potential is generated due to an excess of ions with the same charge on one side. Such ionic channels are common in the plasma membranes of living cells, separating the internal cytoplasma from the external plasma. In experiments like that in Figure 5.76, the voltage difference due to the excess of potassium ions on the right-hand side of the membrane acts against the ion motion, so that the process comes finally to a halt in an equilibrium state. Although the voltage differences are small, their gradients—because of small cell-membrane thickness, typically about 3 nm—may be very high, perhaps more than 20 MV/m.

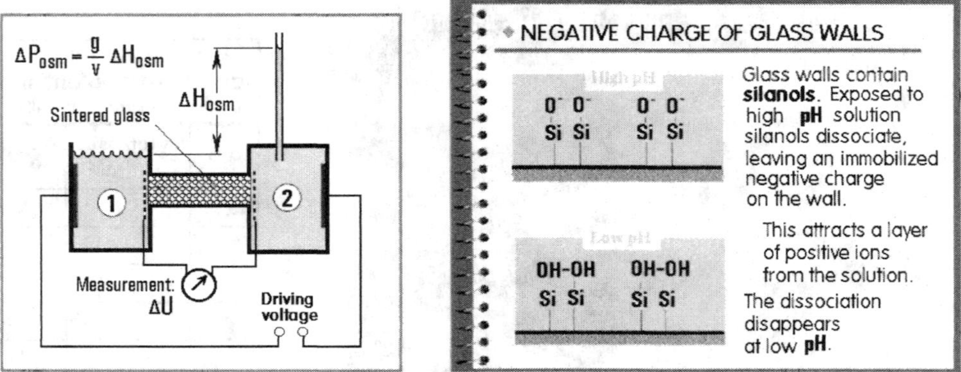

Figure 5.78 (Left) A typical large-scale experiment to determine osmotic pressure actually uses huge number of parallel irregular channels between the sintered glass particles. It is possible either to measure the voltage difference generated by osmotic transport or to generate the motion by applying the electric potential between the driving electrodes.

Figure 5.79 (Right) Spontaneous negative charging of glass walls used in typical electro-osmosis-driven fluid flows.

If a salt solution in vessel (2) in Figure 5.78 is separated from the pure water solvent in vessel (1) by passages that are selectively permeable to the solvent but not to the solute, the solution will become more dilute by absorbing the solvent through the passage. This way the system tends to reach osmotic equilibrium. This process can be stopped by increasing the pressure on the solution side by applying *osmotic pressure*, which provides an alternative measure of equilibrium properties. If, on the other hand, in the experiment in Figure 5.78, the pressure is kept constant, an electro-osmotic flow may be generated in the passages by applying the voltage between the electrodes.

Double layer

The electro-osmotic flow is possible if only one sort of ions (usually anions) is free to move, while the oppositely charged ones (cations) are immobilized. This situation is found surprisingly often. The immobilization is due to the channel walls becoming charged, so that they behave like a giant anion attracting the oppositely charged particles from the fluid. This way, they form a *double layer* of the wall charges plus the layer of immobilized cations. This happens, in particular, if the passages are made of glass, due to the dissociation of silanols (Figure 5.79) contained in the glass walls.

In fact, the double layer always occurs at any interface where dissimilar materials make contact. The electro-osmotic flow then can take place by the action of an electrostatic field on the anions inside the liquid, which remain free to move, causing the rest of the solution to move along with them. The term *double layer* (Figure 5.80) was introduced by Helmhotz in his early writings.

The distribution of ions in a fluid near a solid wall is more complicated than the simple Helmholtz layer model. Although attracted toward the wall, some cations may diffuse away from it, due to thermal agitation. They form the Gouy layer (Gouy and Chapman, 1910), in which their concentration decreases with increased distance from the wall. Some cations are found even in the bulk fluid solution far away from the wall. Those that are immobile (held by adsorption) form the high-concentration near-wall Stern (1924) layer. Examples of the cation distributions are in Figure 5.81. The corresponding charge density distributions are essential steps in computing electro-osmotic flows (Figures 5.82 and 5.83) in very narrow channels. The basic question is the Coulomb force interaction between an ion and the surrounding atmosphere of other ions. The fundamental

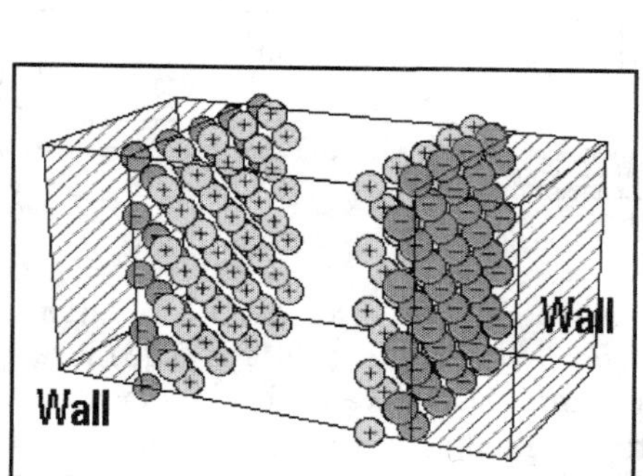

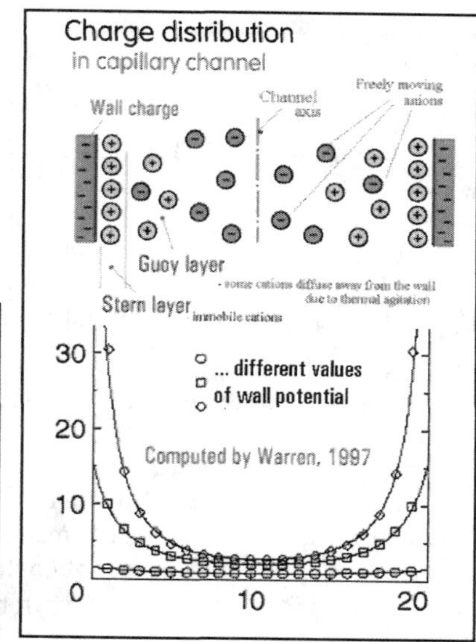

Figure 5.80 (Left) Schematic representation of the immobilized cations in a double layer on the walls of a glass channel.

Figure 5.81 (Right) Actual distribution of the cations in the channel's cross-section. Thermal agitation causes them to diffuse away from the walls.

paper containing the theory (for dilute electrolytes) was published in 1923 by Debye and Hückel. Their aim was actually quite different – an explanation for anomalies in freezing point depression and related phenomena. As shown in Figure 5.84, Debye and Hückel proposed that the charge density should be governed by the Boltzmann distribution law for the number n of ions per unit volume, essentially comparing the energy of the ionic interaction $ze\Phi$ (in Figure 5.84, for $z = 1$) with the disrupting thermal energy kT. The original version of the theory used the linearization of the equation valid for $ze\Phi < kT$. The equation for the cylindrical capillary (Figures 5.83 and 5.85) and the solution in Figure 5.88 reveal the importance of the *Debye-Hückel length scale* x_{DH} in

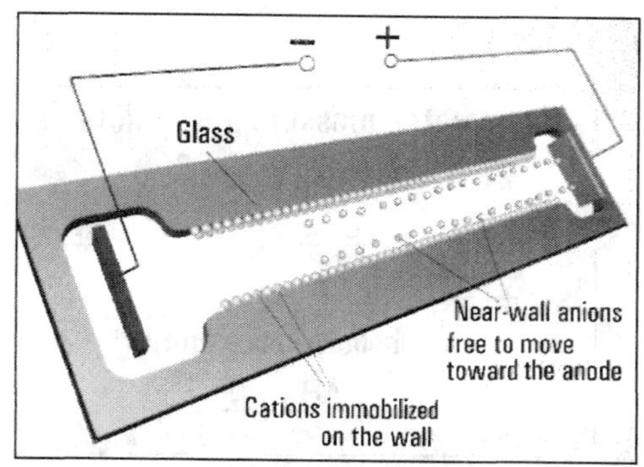

Electric field in the cylindrical capillary:

Equation for the electrostatic potential Φ
due to continuous distribution of charge density ρ

– Poisson equation $\nabla^2\Phi = -\rho/\varepsilon_0\varepsilon$

... ∇^2 is Laplace operator
ε_0 permittivity of empty space
ε permittivity of liquid (solvent)

In the cylindrical coordinates,
axially and tangentially uniform distribution:

$$\frac{1}{r^2}\frac{d}{dr}\left(r^2\frac{d\Phi}{dr}\right) = -\frac{4\pi}{\varepsilon_0\varepsilon_r}\rho$$

Figure 5.82 (Left) Liquid flow driven by electro-osmosis in a glass channel. Liquid is dragged along by the free anions moving to the anode.

Figure 5.83 (Right) Simple axisymmetric model of electro-osmotic channel flow. An equation describing the continuum spatial distribution of the electric charge inside the channel.

Debye and Hückel (1923):
Local numerical concentration n of ions governed by Boltzmann equation

$$n = n_b \exp(-ez\Phi/kT)$$

n_b .. bulk concentration
ez .. multiples of electron charge e

Hyperbolic sine solution: $\rho = -2ne \sinh(e\Phi/kT)$

$z = 1$ .. simple 1:1 ions

$e\Phi/kT$ small, so that $\sinh(e\Phi/kT) \approx e\Phi/kT$

Charge distribution equation

$$\frac{1}{r^2}\frac{d}{dr}\left(r^2\frac{d\Phi}{dr}\right) = \frac{1}{x_{DH}^2}\Phi$$

$$\Phi = B\, J_o(r/x_{DH}) \qquad B = \Phi_w / J_o(d/2x_{DH})$$

J_o .. zero-order modified Bessel function of the first kind
capillary wall at $r = d/2$ $\quad \Phi_w$ .. wall potential

$$\rho = -\frac{\varepsilon_o\varepsilon_r}{4\pi x_{DH}^2}\Phi = \frac{\varepsilon_o\varepsilon_r}{4\pi x_{DH}^2}\Phi_w\frac{J_o(r/x_{DH})}{J_o(d/2x_{DH})}$$

Figures 5.84 (Left) and 5.85 (Right) The axisymmetric model of electro-osmotic channel (capillary) flow: the approximate expression for the charge density due to the presence of the ions.

Figure 5.86. The electro-osmotic flowfield computation then uses the distribution of electric potential obtained from the Poisson-Boltzmann equation (Figures 5.84 and 5.85) inserting it into the usual Navier-Stokes equation. This, after the simplification valid for the developed flow in a circular geometry under the action of the uniform longitudinal pressure gradient $\Delta P / l$, leads to the parabolic velocity profile (Figure 5.87). With the additional term expressing the electrostatic force per unit volume in Figure 5.88, this basic axisymmetric case results in velocity profile solutions with the Bessel function J_o.

For electro-osmotic flows without any pressure difference between the ends of the capillary, the velocity profiles obtained by the solution of the equation, with only the second right-hand term in Figure 5.88, are shown in Figure 5.89 for various ratios of the capillary diameter to the Debye-Hückel length. In this context they may be thought of as the effective thickness of the double layer. Figure 5.89 shows there is a limiting value at about $d / x_{DH} = 20$. If the channel diameter is smaller, the boundary layers will overlap and the velocity profile will have a distinct maximum at the axis. In present-day microfluidics, the usual transverse dimensions (diameters) are much larger than this limit. The electro-osmotic flow then has a more or less plug-flow character, with a constant velocity across most of the cross-section, with the exception of the boundary layers at the

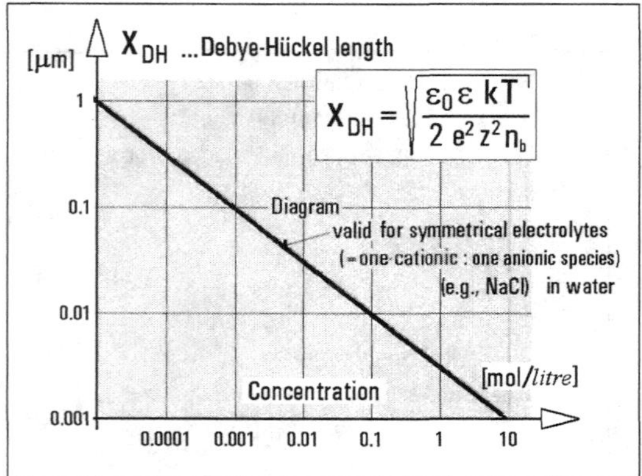

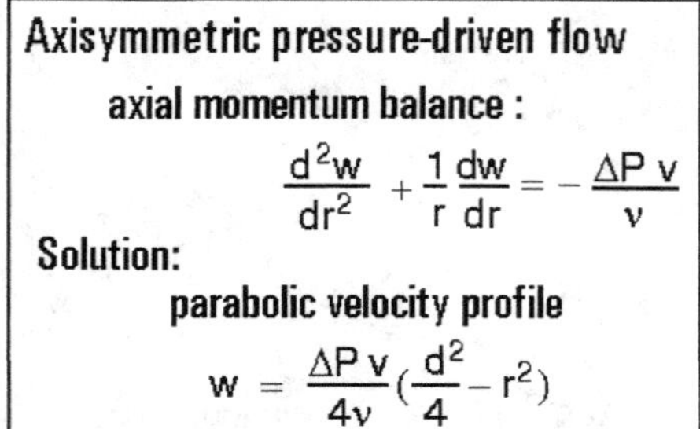

Figure 5.86 (Left) The characteristic length of the charge distribution in a channel. It may roughly be interpreted as the thickness of the double layer. It depends on the bulk ion concentration.

Figure 5.87 (Right) The standard equation of the pressure-driven flow in an axisymmetric channel and its solution, the velocity profile, is included here for comparison.

General equation for velocity distribution in an axisymmetric capillary

$$\frac{d^2w}{dr^2} + \frac{1}{r}\frac{dw}{dr} = -\frac{\Delta P\,v}{\nu} - \frac{\Delta U\,v}{l\,\nu}\rho$$

$\Delta U/l$ [V/m] ...Intensity of applied electrostatic field

$$w = \underbrace{\frac{\Delta P\,v}{4\nu}\left(\frac{d^2}{4} - r^2\right)}_{\text{pressure driving}} - \underbrace{K\frac{\Delta U}{l}\left[1 - \frac{J_o(r/x_{DH})}{J_o(d/2x_{DH})}\right]}_{\text{electro-osmotic driving}}$$

$$K = -\frac{\varepsilon_o\varepsilon_r\Phi_w v}{4\pi\nu}$$

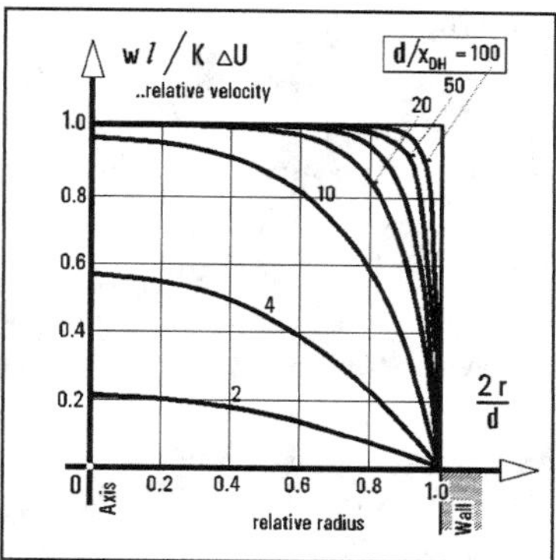

Figure 5.88 (Left) Equation of the velocity profile in the axisymmetric model case of a flow driven simultaneously by electro-osmotic as well as pressure force action.

Figure 5.89 (Right) Velocity profile solutions of the electro-osmotic (no pressure difference) flow computed for the axisymmetric model case.

wall. These are usually, for the short relative lengths in typical devices, quite thin, and yet, at least 10 times thicker than the double-layer thickness. Electrokinetic flows are in fact mostly used in applications where the much less "smearing" by shear, compared with the pressure-driven flows, is used to advantage. The liquid slugs in Figure 5.90 remain as distinct entities, even after passage through a relatively long channel. Of course, real flows in actual channels with noncircular sections are more complex than this model, especially if the blocked outlet at the end of the passage generates a pressure difference similar to that in Figure 5.78. The pressure term, analogous to the right-hand one in Figure 5.88, then becomes important and deforms the shape of the velocity profiles.

The generated pressure difference may then drive the liquid flow in the other, electrically neutral parts of the fluidic system, which is simply driven by the electro-osmotic pump. However, many applications, especially in the so-called lab-on-chip field, also use the electro-osmotic or electrophoretic effects for tasks other than mere pumping: the electric signals may control the flow. This uses the ease with which the fluid flow may be started, and its direction changed, by

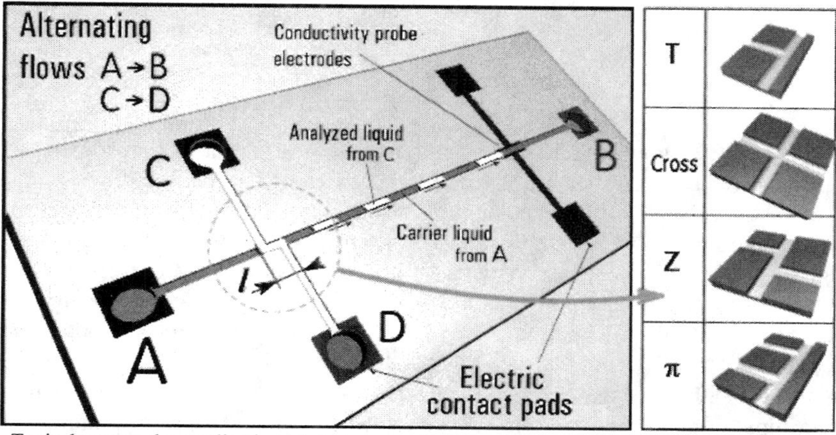

Figure 5.90 Typical present-day application of electro-osmotic flows in microfluidics: an electrically driven conductivity composition detector for sequential measurements of slugs of liquid from the inlet C. At right: alternative configurations of injection of the analyzed fluid. The one used in the detector at left is the Z version.

switching the electric voltage applied at different electrodes of a multi-electrode system. This way of distributed driving of fluid in system channels (rather than concentrated in a pump) is effective for very viscous biological fluids at extremely small Reynolds numbers. The electrophoresis (rather than electro-osmosis – but the distinction is rather tenuous) is especially useful for handling colloids. An example of application for detecting changes in the properties of the sampled fluid brought into inlet C is shown in Figure 5.90. A certain disadvantage is the necessity of working with quite high voltages – a typical value is 200 V. The sample is injected upstream in the Z-injector by switching the voltage between the electrode pairs AB and CD. Alternative injectors of the tested fluid are shown in the right-hand side of Figure 5.90. A simple T, one-sided injection introduces too much spatial and therefore concentration inhomogenity, and is out of use. Also, similarly as the Cross, a simple crossing of the channels at right angles, it delivers too small an amount of the analyte. That is why the Z at left in Figure 5.90 (and similar π, with C and D terminals on the same side of the main channel) are used most often. The amount of the tested analyte is defined by the common length l. There are, however, situations in which even the amount delivered by the plain Cross-type injector is too large. It is then possible to decrease it by the pinching effect shown in Figures 3.119 and 3.120.

The composition analysis tasks made possible in such a layout may be more sophisticated. Often the different electrophoretic mobility of the substances is used for analytical purposes. On the distance between the detector location and the injection port—the Z kink in the transversal channel from C to D—the individual components of a mixture move at different velocities, so that they arrive to probe at different times. This produces peaks on the record of the signal of the detector. Instead of conductivity, the biochemical applications often use fluorescence tagging and then optical sensing.

The last items in the list in Figure 5.69 are the electrohydrodynamic principle and the electrostrictive liquids. The former has been used in a dielectric fluid motor, useful in the situation in which the actuator's action has to have the character of rotational mechanical motion. The advantages are the absence of the usually bulky and heavy magnetic circuit and no need to switch the current. These features make the motor suitable for extremely small sizes. On the other hand, the typical efficiency is a mere 4% and the driving voltage is inconveniently high – for example 15 kV. The electrostrictive liquids are still under development. There is a promise of very high efficiency and extreme available energy density, but most liquids developed so far have inconveniently high viscosity, and the actuation requires rather high voltages.

5.3.2 E/F transducing

This part of the E/F conversion discussion concentrates on devices processing information carrying signals. Of course, most of the principles discussed in the previous section on power actuator conversions may be used for signal transducing as well. There is, after all, no clearly defined demarcation line. The difference is, nevertheless, usually distinguishable from the practical point of view. For a signal conversion, it is possible and often profitable to employ even principles exhibiting rather poor conversion efficiency. The control signals transferred often are so energetically weak that the dissipation of their power has no importance. Also the power-density or energy-density aspects are not particular decision factors.

There are, instead, other criteria that become important. One of them is - because of the often encountered input signal weakness, the sensitivity of the transducer in responding to small input changes. Because of the general trend of the continually increasing amount of processed information, the transducers often have to operate fast. This places at a disadvantage the earlier

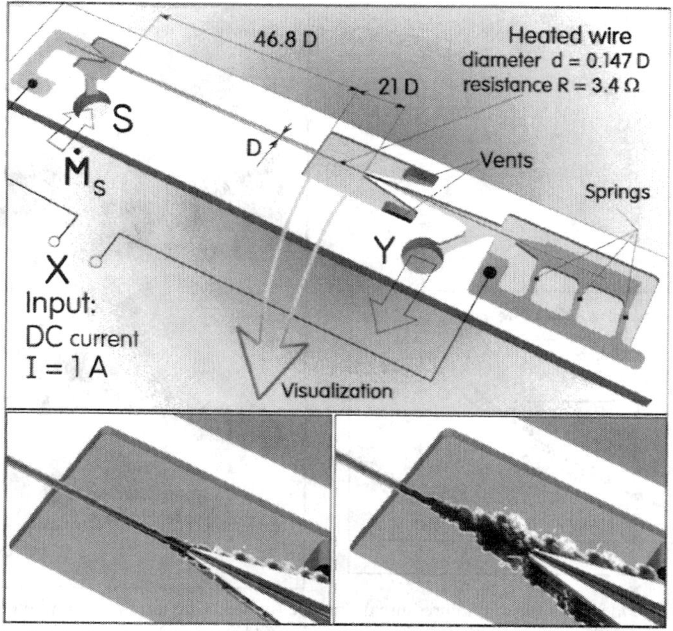

Figure 5.91 The author's laboratory model of a no-moving-part E/F converter. The electrical heating of air by the thin wire passing through the axis of the jet causes an earlier transition into turbulence. This results in a decrease in the output pressure measured in output terminal Y (Figure 5.92). The model was (perhaps unnecessarily) complicated by the spring component that holds the downstream end of the wire, eliminating any sag due to thermal expansion.

E/M/F conversion chains, burdened by the inertia of the moving components. If this is circumvented by elastic support of the moving components, with the support stiffness adjusted for high resonant frequency, then overcoming high stiffness calls for disproportionately high actuation forces and, hence, low sensitivity. In view of the general slowness of thermal effects, it may sound surprising that the E/M/F chains tend to be replaced by no-moving-part E/T/F conversion. This, of course, is due to the very short heat-transport distances and small volumes in microfluidic devices.

In the example presented in Figures 5.91 and 5.92, the input electric signal is used to heat a very thin wire that in turn heats the surrounding gas jet, increasing the air viscosity. The transition of the jet into turbulence takes place at a lower flowrate (Figure 5.92). Keeping the jet laminar on a long distance downstream from the nozzle requires a very long nozzle exit channel. Since the wire passes through the whole length of the channel, a considerable influence on the earlier transition may be due also to the air's thermal expansion and, hence, faster exit-flow velocity. The position of the collector at the downstream end of the wire and the supplied flowrate are adjusted so that the cold jet enters the collector just prior to its transition to turbulence (left visualization picture at the bottom of Figure 5.91) and then are kept constant. With the heating effect of the electric signal ON, the jet becomes turbulent and has to travel a longer distance to the collector in the turbulent state (right visualization picture at the bottom of Figure 5.91), which is characterized by more intensive mixing with the surrounding air. As a result, the jet reaches the collector with much of its momentum lost so that the output pressure drop ΔP_Y decreases. This decrease is used in the fluidic circuit, perhaps amplified by a fluidic amplifier. With the electric signal OFF, the return to the laminar state is fast, because the wire is effectively cooled by the jet flow.

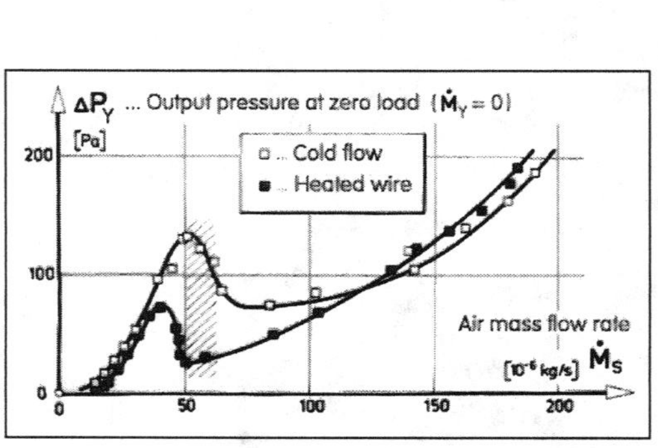

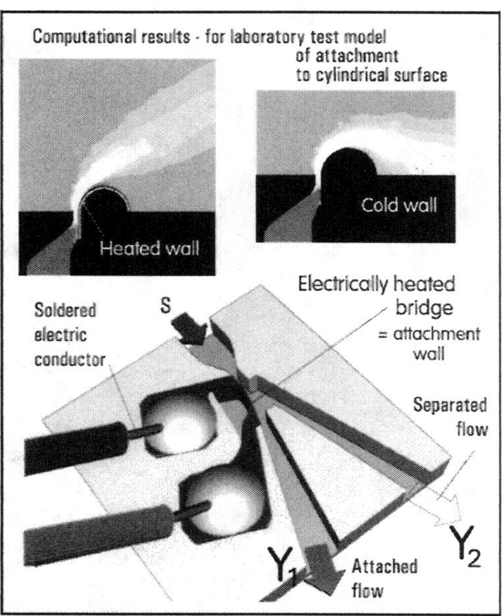

Figure 5.92 (Left) Dependence of the output pressure difference (between the terminal Y and the atmosphere, to which the vents are open, Figure 5.82) on the supplied air flow with cold and heated wire. The transducer is to be operated as binary (with two output signal states) with the constant flow rate adjusted to the shaded region.
Figure 5.93 (Right) E/T/F signal converter based on the separation of an air jet held in the cold state at the curved attachment wall by the Coanda effect, as soon as this wall is heated by the passing electric current.

Another use for signal conversion of the aerodynamic effects associated with gas heating by the electric input is in the E/T/F transducer shown in Figure 5.93. On the fluid side, it is a jet-type diverter valve with a monostable attachment of the air jet, in the cold flow state, to the slightly curved metal wall by the Coanda attachment effect, leading the jet to the output collector Y_1. Heating the wall, which is a thin but deep (spanning the whole depth of the valve cavities), high-resistance bridge between the two bond pads, causes the jet to separate from the attachment wall and switch into the other collector, leading to the output Y_2. Again, the return switching is made fast by the continuous cooling of the attachment wall by the air jet.

5.3.3 F/E power conversion

Considering the typically low power levels handled in microfluidics, it is obvious that extracting energy from fluid flow is not one of its typical jobs. The exception, combustion microengines (turbines), aimed at utilizing the much higher specific energy stored in fuels rather than in batteries, was already mentioned in Section 5.2.3 when discussing F/M actuators. The turbine is shown there in Figure 5.40. The small, microfluidic size is suitable for the proposed applications in portable electric and electronic appliances requiring electrical power at Watt levels. The electric conversion of the mechanical power supplied by the turbine is the task of microgenerators, described, for example, in [13]. Typical layouts have rotors of 6 – 8 mm in diameter made as a plane disc from rare earth elements, permanent magnets like $SmCo_5$ or NdFeB. The stator consists of a system of electroplated planar coils. The sources with a higher power level needed for military applications and yet portable—carried by a soldier—require high energy density. Requirements may be met by a numbering-up of the micro-devices. Another current approach is the use of small fuel cells, which, however, require even more development effort. They also provide opportunities for microfluidics, such as the obvious use of micropumps to move the fuel.

There are several emerging applications for independent, small electric power sources replacing chemical batteries, which are heavy, bulky, and need replacement usually in most inconvenient situations. Sometimes the replacement may be impossible due to inaccessibility, for example inside human or animal bodies. Independent power supplies would be also welcome for monitor devices that are planned to be distributed around in large numbers, such as those for structural monitoring of buildings and bridges, detecting of illicit substances (e.g., explosives, drugs) for antiterrorist and anticrime warfare, or environmental monitors distributed across the landscape. Sometimes intelligent autonomous sensor systems are needed in locations that are accessible only with difficulties, such as on rotating objects or in downhole drilling tools and equipment. Recent progress in VLSI has resulted in very small sensor power requirements, on the order of 10 μW to 100 μW, even in systems involving RF signal transfer.

The most promising answer in situations in which there is a moving fluid available is the generation of fluid oscillation by a fluidic oscillator, as described in section 4.7, and using it to drive an electrostrictive converter producing electrical energy. Recent development has led to converters suitable for this purpose, using thin-film electrostrictive materials, such as lead zirconate titanate $Pb(Zr, Ti)O_3$, on flexible substrate cantilevers designed to resonate at specific frequencies. A cantilever beam sized 170 μm x 260 μm forced to oscillate was recently demonstrated to deliver 1 μW of continuous electrical power at 2.4 V dc, with an energy density of 0.27 J/mm^3.

5.3.4 F/E signal transducers

Basically, the signal, usually for the purposes of monitoring the fluid flow, may be either derived from fluid pressure, or flowrate. It may be also derived from fluid chemical composition and other properties, but these cases will be discussed in Section 5.8. Typically, pressure evaluation uses mechanical action, such as the deformation of a diaphragm. The sensors are therefore of the F/M/E type. On the other hand, while mechanical actions dependent on the flowrate are also known to be used, the general trend is to prefer thermal effects in F/T/E-type sensors.

5.3.4.1 Pressure measurement and monitoring

Pressure is one of the most frequently measured quantities. Pressure sensors were actually among the first MEMS devices manufactured at an industrial scale, and their sales (after inkjet printer heads) still rank as the second-most important business in microfluidics – mainly in automobiles and, perhaps somewhat surprisingly, also in the watchmaking industry. Pressure is converted by employing its deforming action on a thin membrane. Monocrystal silicon membranes suffer much less from the creep, fatigue, and hysteresis problems that plagued earlier metal membranes. Their typical sizes in present-day microdevices are 0.8 mm x 0.8 mm in area and a thickness of 10 to 25 mm (dependent on the intended pressure range). This small size, high Young modulus, and lower density of Si result in a very high resonance frequency. The deformation may be sensed by the changes it produced in electric resistance, R, electric capacitance, C, or electric inductance, L. Generally, the advantages (and hence popularity) of these three principles decrease in this RCL order. Inductance is the most inconvenient of them, because it usually needs a large and heavy magnetic circuit; also the coils, the inductance of which is varied, are not particularly suitable for the typical planar layout of microfluidics. Capacitive sensing readout methods are more attractive.

The two capacitor electrodes are the deformed membrane and a conducting plate fixed at a small distance above it. The advantage is a small sensitivity to temperature changes. The necessary circuitry converting the capacitance changes to a signal in a usable form used to be considered a disadvantage, but with electronic circuitry routinely made on the same chip this ceases to be an important factor. Nevertheless, the most-often-used membrane deformation pickup mechanism is piezoresistance, a variation of electric resistance with mechanical stressing. The advantage is simplicity, linearity of response, and convenient character of the voltage output signal. The piezoresistors may be actually formed in the silicon membrane by a diffusion process, though sometimes it is preferable to make them on top of the membrane (and an insulation layer, usually SiO_2) by deposition. The complicating factor may be anisotropy of silicone electric properties, and a real problem is the temperature sensitivity and drift.

Piezoresistive transducer, with its characteristic dependence on an external supply of electric energy, is a demonstrative example of a conversion mechanism with excellent suitability of generating or converting signals, which would be useless for the electric power generation, as discussed in Section 5.3.3.

The usual fabrication of a monocrystal silicon chip with the membrane uses the n-type substrate oriented so that the plane with (100) Miller indices (Figure 2.13) is the upper suface. The typical size of the chip may be 3 mm x 3 mm and the membrane itself 0.8 mm x 0.8 mm. The membrane thickness is usually between 10 μm and 25 μm, depending, of course, on the intended pressure range. Although the theoretical fracture stress of silicon is much higher (Figure 2.1) the maximum recommended operating stresses for the (100) orientation are at about 7 MPa. The etching producing the thin diaphragm is done from below, which leaves the upper plane surface conveniently available to make there the piezoresistors, the metal conductive strips with the bond pads, and possibly also the signal-processing electronics (or at least the temperature-compensation circuits). Very thin diaphragms are not advisable, giving rise to nonlinearity due to the complicated geometry of the deformation and also to errors due to the increased influence of imperfect material homogeneity. For sensing very small pressure differences, the diaphragm may be bossed to concentrate the stresses near its edges. Diaphragms are usually rectangular (though round shapes are not unknown), with one of the diaphragm edges oriented in line with the (110) direction of the chip. The electric resistance of mechanically stressed silicone is determined by four parameters:

	Resistivity [Ω mm]	π_{11}	π_{12} [10^{-12} Pa^{-1}]	π_{44}
p-Si	0.78	66	-11	1381
n-Si	1.17	-1022	534	-136

The piezoresistors made by diffusion are of the p-type and are oriented so as to use the very large value of the coefficient π_{44}. The expression for the resistance change of a piezoresistor positioned so that it is parallel with this (110) edge is given in Figure 5.94. It is placed near the diaphragm edge, where the stresses reach maximum values (the precise spot depending on how far the boundary conditions approximate the ideal clamping assumption, for which the maximum is at the very edge). To suppress the unpleasant temperature influence, there are almost always several piezoresistors connected into the Wheatstone bridge. For increasing the sensitivity, the bridge is very often of the antiparallel version, with the two piezoresistors in each branch reacting to the diaphragm stressing in opposite manner. This is easily achieved by utilizing the fact that, for the

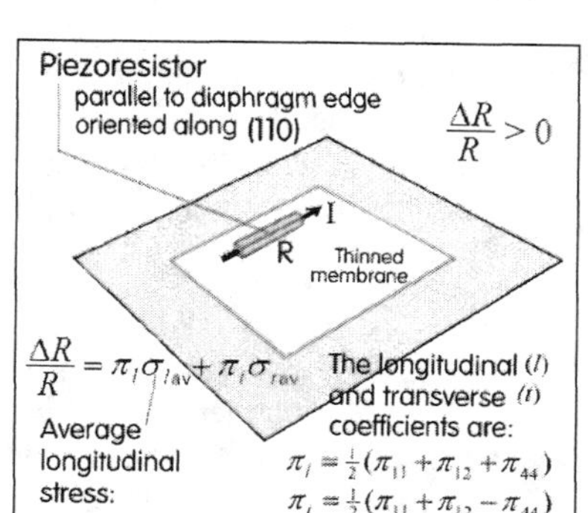

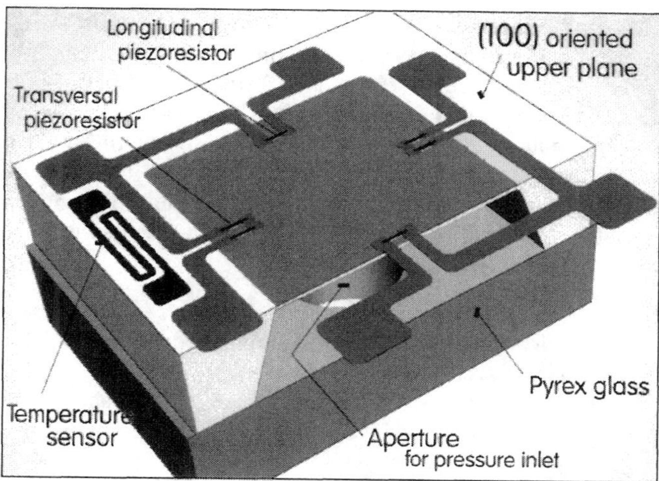

Figure 5.94 (Left) Dependence of the relative change in electric resistance on the acting stress is complicated by silicon anisotropy. The expressions given here are valid for the resistor parallel to the (110) Miller index direction.
Figure 5.95 (Right) An example of a typical monolithic pressure sensor with silicon diaphragm thinned by etching and its deformation sensed by the Wheatstone bridge of the piezoresistors. The front part of the sensor is shown cut off to reveal the cavity under the diaphragm. The piezoresistors are connected by deposited metal film conductors.

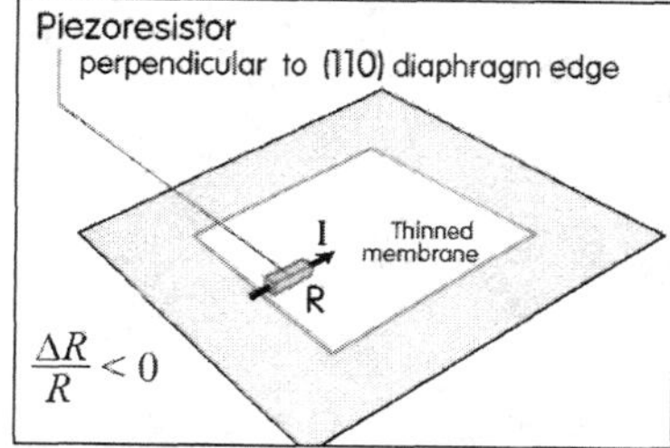

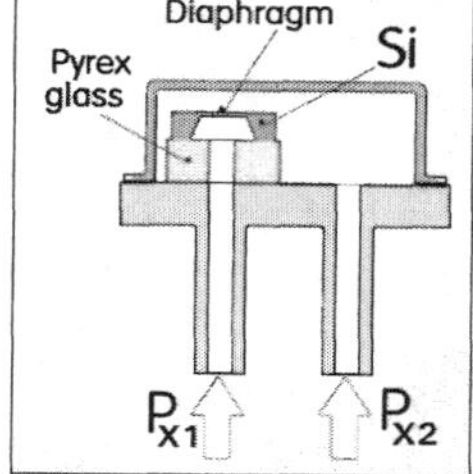

Figure 5.96 (Left) Electric resistance R of a piezoresistor oriented equally as in Figure 5.94 but located at the transversal edge of the diaphragm changes in the opposite sense (it decreases when the diaphragm is bulged upwards).
Figure 5.97 (Right) Characteristic packaging of the sensor from Figure 5.95 when used to measure a pressure difference $\Delta P = P_{X1} - P_{X2}$.

second piezoresistor with the same parallel orientation but placed at the other, perpendicular diaphragm edge (Figure 5.96), the coefficients π_l and π_t will multiply stresses that in effect exchange their meanings (and magnitudes). With the p-type resistor, the different sign of the large coefficient π_{44} causes the coefficients π_l and π_t to be of opposite signs. As shown in Figures 5.94 and 5.96, while the resistance increases ($\Delta R > 0$) in the longitudinal resistor in Figure 5.95, it decreases ($\Delta R < 0$) in the transverse resistor. In Figure 5.95 the transverse piezoresistor is divided into two parallel parts, so that it is shorter and thus located inside the high stress part of the diaphragm while keeping the same absolute resistance.

The silicon chip is anodically bonded to the Pyrex glass support, which also closes the cavity under the membrane. The cavity may be evacuated for absolute pressure measurement. If the measured quantity is a differential pressure, a hole is drilled there in the glass connecting the cavity with one of the pressure taps (Figure 5.97).

Slightly different MAP (manifold absolute pressure) sensor with a proprietary X layout of the piezoresistors is manufactured in large quantities (over 300 million pressure sensors shipped so far, the business well over the \$ 540 million estimated in 2006), with integrated on-chip circuitry, and with two (originally three) operational amplifiers and temperature compensation. Apart from engine control, a considerable market exists for the tire pressure monitoring system required by safety standard to be fitted to all vehicles in the United States by November 2006. Another large-scale use is also the altimeter/barometer module for some wristwatches. These can read altitude variations as small as 1 m, which produces a pressure change ~ 12 Pa (in fact, the resolution limit of the used sensor is stated to be 3 μbar, corresponding to a mere 30 mm altitude change). Sensor power consumption is a mere 1.3 μW.

5.3.4.2 Flowrate measurement and monitoring

This is also a very frequently occurring task. The classical approach used the pressure drop across, for example, a fluidic resistor (as will be shown in Figure 5.141), supplementing it by a F/M/E transducer which performed the conversion to the requested electric form of the output signal using again, for example, the principle schematically shown in Figure 5.5. Of course, the mechanical motion shown there would be highly impractical. The necessity of oiling, the contact surface supporting the moving stem, and the danger of diaphragm rupture or spring breakage would be totally incompatible with the ideas of no-moving-part fluidics. Modern versions limit the mechanical motion, as shown in the example of a modern version made by microfabrication technology in Figure 5.98, where there is just the small bending of the cantilever.

The component corresponding to the orifice plate, generating the pressure difference increasing with increasing measured flow rate (though not simply proportional to it, because of the inherent nonlinearity of the underlying Bernoulli equation), is here made deformable. The deformation is picked up and converted into the electric signal by using the piezoresistors as discussed in Section 5.3.4.1.

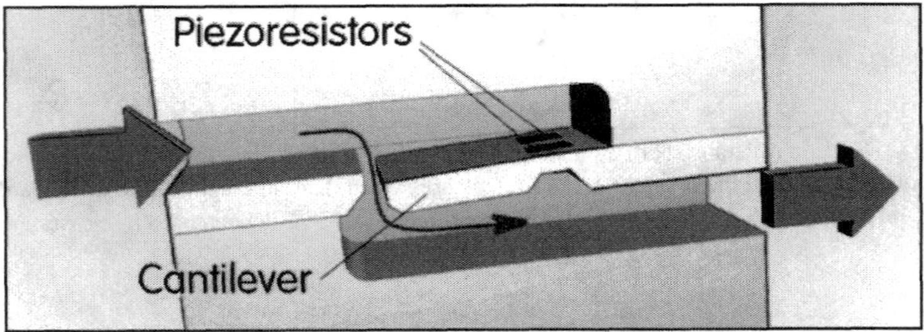

Figure 5.98 Flowrate sensing of the F/M/E type, here with a deformable cantilever rather than a movable component. The deformation is measured by piezoresistors made by diffusion in the most stressed part of the cantilever.

In Figure 5.99 there is another approach, again based on sensing the pressure drop on an orifice that is a part of the device. The output signal is the difference between the readings of the two pressure sensors. It is, of course, much more complex, but the cost may be kept down by using the manufacturing methodology developed for the large-scale production of pressure sensors, identical to the type shown in Figure 5.95. The accumulated experience obtained with suppressing various disturbing factors, such as the effect of temperature, and their compensation, is also important. A typical example of the device from Figure 5.99 may be mentioned, which has been produced for the full-scale measured flow 5 ml/h = $1.39 \cdot 10^{-9}$ m^3/s.

As was stated already in the introduction to this section, currently more popular are the devices using a thermal intermediate conversion. This is because making a deformable component in the

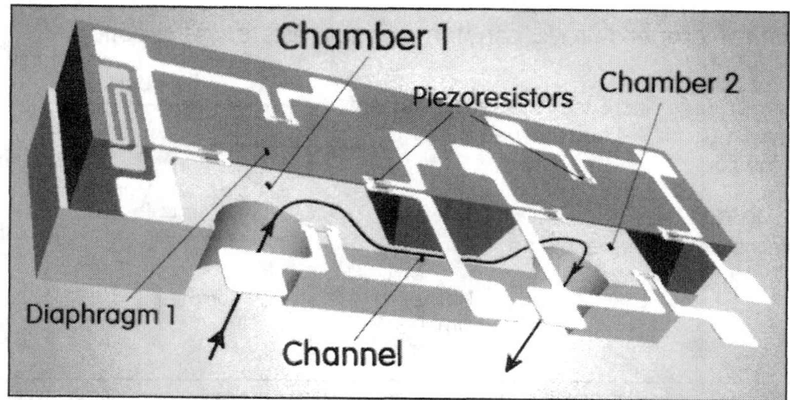

Figure 5.99 Manufactured as a flowrate sensor are two pressure sensors with piezoresistive pick up membrane (as shown in Figure 5.96) made in a common chip together with a precise channel between their chambers.

device is always more difficult than staying within the solid state realm, and besides, however small, there is always the possibility of the deformable part being broken, for example, by an inadvertently applied flow that is too large.

Thermal effects are simply produced by the Joule heating in electric resistors. The very similar (and simultaneously made) resistors then also form the simplest pickup transducers of the thermal effects into the electric signal. This conversion is based on the temperature dependence of electric resistance, a property common to most materials. Usually, for example, in metals, the resistance increases with temperature. Carbon and certain semiconducting oxide materials, called

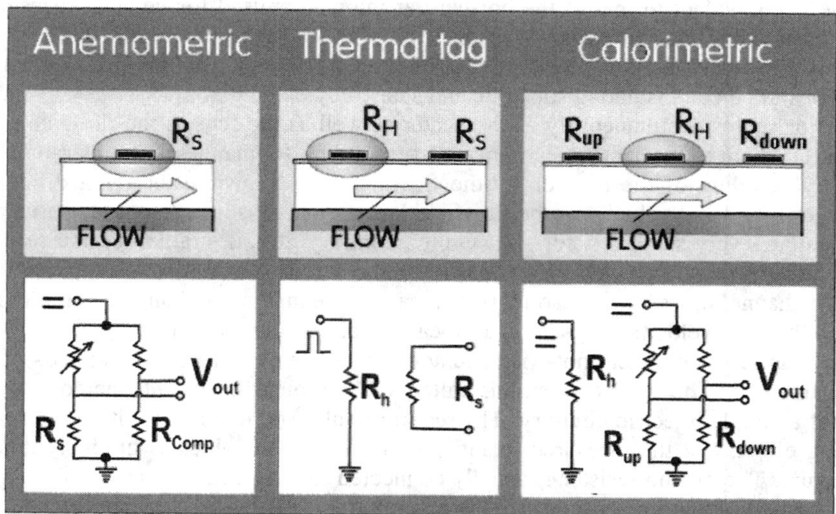

Figure 5.100 The three basic principles and electric circuits of the three most important methods of F/T/E measuring fluid flow rate. The measured quantity is actually mean velocity over the exposed part of the resistor, to read the flowrate the device has to be calibrated, usually assuming constant shape of the velocity distribution in the channel.

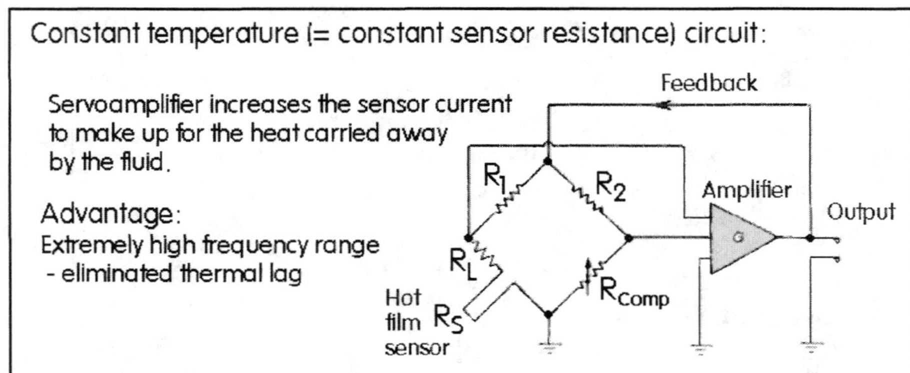

Figure 5.101 Keeping the temperature of the pickup resistor (usually of the deposited metal film type) constant by the feedback from a servoamplifier is the standard way of increasing the frequency range of thermal transducers up to several kHz.

NTC (negative temperature coefficient), exhibit a decrease with rising temperature. They are used in sintered mixtures as the *thermistors*, often exhibiting an order-of-magnitude larger temperature coefficient (common values are higher than 6% per K) than metals (less than 1% per K). The advantage of metals, which are usually used in the form of vacuum-deposited thin films, is the better stability of properties – platinum (3.93 % per K) thermometers are the usual calibration standards. They can reproducibly resolve 0.01 K and the amounts needed are so small that the high price is irrelevant.

There are three basic versions of the thermoelectric flowmeters in Figure 5.100, which also presents the basic forms of the typical electric circuits.

The simplest version, with just a single resistor in the flow, may be called *thermoanemometric.* Its principle is equivalent to that of the hot-wire anemometers used for velocity measurements in fluid mechanics. However, the resistor sensor for fluidic flow measurement is hardly ever made in the classical form with the extremely thin wire. Instead, it is usually in the more robust form of a thin film deposited on an insulating substrate and shaped by etching.

The resistor serves simultaneously as the heater as well as the sensor: the flowrate is evaluated from measured changes in electric resistance corresponding to changes in temperature caused by the convective cooling of the resistor by the flowing fluid. Flowrate metering differs from the measurement of the local velocity of the flowfield because there is no attempt at spatial resolution, which necessitates very small resistor size. On the contrary, it is desirable that the sensor is large and spans the whole channel width. It would be even better if it would cover as much as possible of the whole channel cross section, so that the measured temperature changes would represent the effects of all the fluid velocities present in all locations across the section.

Similarly, as in the classical hot-wire anemometry, there are two basic methods of operating the sensing resistor. The simpler method, called CCA (constant current anemometry), has the advantage of a simpler electric circuitry. The resistor is allowed to decrease its temperature by the fluid cooling effect, and the measured quantity is its resistance, which is uniquely dependent on the temperature. The pickup resistor is usually connected to the electric Wheatstone bridge (Figure 5.100). The actual output of the sensor is the magnitude of the bridge disbalancement. The other, more sophisticated method is called CTA (constant temperature anemometry). Its principle is shown in Figure 5.101. Again, the sensing resistor (R_S in Figures 5.100 and 5.101) is connected (together with the resistors R_1, R_2, and R_{Comp}, as well as R_L which is needed if the sensor resistance cannot be small enough) in the bridge, the output signal corresponding to its disbalancing. This, however, is here amplified by the servoamplifier, the output of which supplies the bridge. The negative feedback connection results in balancing the bridge. The actual output of the anemometric sensor is the magnitude of the amplifier output current required for the balancing. As a result of this negative-feedback principle, the sensing resistor is kept at a constant resistance and hence constant temperature.

The constant temperature property essentially increases the frequency range of the sensor, as it is no more limited by thermal lags. In classical anemometry, this is used for extending the anemometer response towards the high frequency signal component of the smallest, energy-dissipating turbulent eddies. In fluidics this component is usually an unwelcome noise, outside the range of interest. Nevertheless, the extended frequency range made possible by CTA is useful. Sometimes it may be useful to have information about the noise, however unwanted it is. As it is not necessary to suppress the thermal lag by making the probe very small, CTA also makes possible a less delicate and less easily destructed sensing resistor.

The next one of the three basic operating principles (Figure 5.100) of thermoresistive flow sensing uses two electric resistors exposed to the fluid, as shown in Figure 5.102. It is the *thermal tagging* principle. By a current pulse passing through the upstream resistor, fluid in its immediate vicinity is heated. Despite thermal diffusion, tending to gradually eliminate the thermal tag, at the usually rather small downstream distances l its presence is recognizable in the output signal of the downstream pickup restrictor. Measuring "time-of-flight" Δt (which is easy), the flowrate is computed as proportional to $l / \Delta t$. This measurement method becomes more accurate at low Reynolds numbers at which diffusion due to turbulence ceases to be effective. It is therefore of particular importance for microfluidics, where the timing of the electronic circuit can be easily made on the same chip. The advantage is the easy generation of a digital output signal. The interval Δt is measured by counting the number of pulses generated by a constant-frequency clock oscillator before the arrival of the peak to the sensor R_S. Of course, evaluating the flowrate actually requires computing the reciprocal value, not a problem in digital signal processing.

Finally, the last of the methods in Figure 5.100 uses three resistors exposed to the flow. It is called *calorimetric* sensing. One of the resistors serves as a permanently heated thermal source. The other two monitor the temperature distribution in the fluid. Due to the flow, the temperature distribution is deformed. Both symmetric and asymmetric versions are used. In the symmetric version in Figure 5.103, the upstream pickup resistor becomes cooler, while the temperature of the downstream one increases. The resistors are connected in a bridge (Figure 5.100). Experience

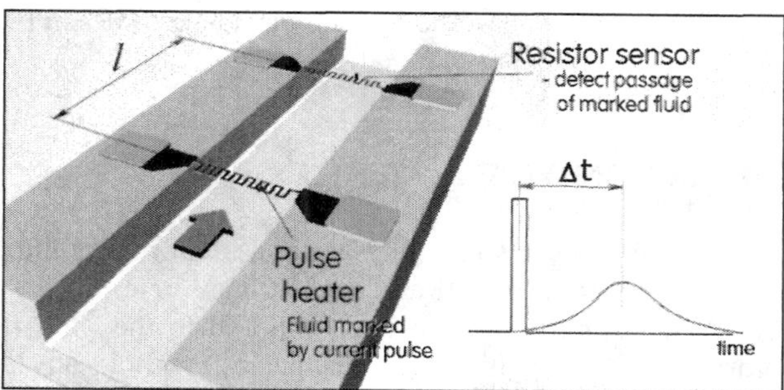

Figure 5.102 F/E fluid flow sensor with two resistors exposed to the flow. The measured flowrate is proportional to the ratio of the resistor distance l to the measured "time-of-flight", Δt.

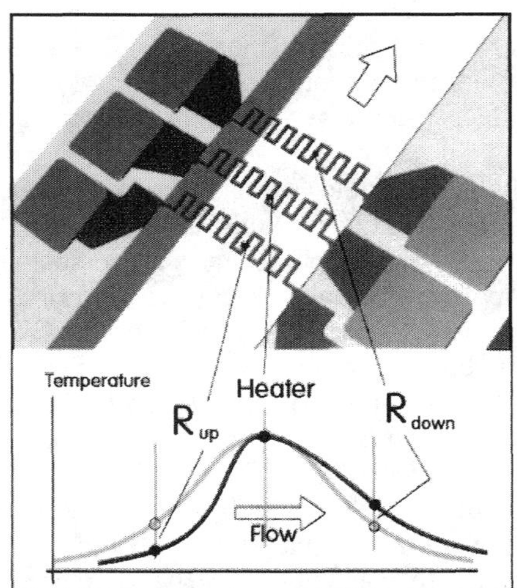

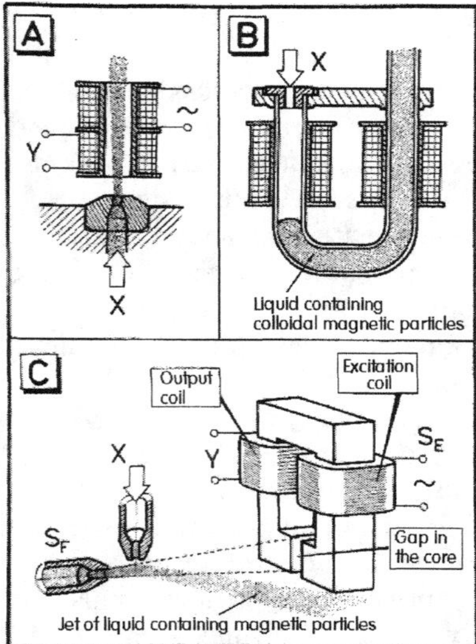

Figure 5.103 (Left) The calorimetric method measures the flow rate by considering the difference in heat propagation to the upstream and downstream temperature sensors from the centrally located heater. The characteristic features are three resistors exposed to the flow.

Figure 5.104 (Right) The desirable direct F/E sensing of fluid presence or flow is unfortunately possible only with fluids having special properties; these are old patent drawings for the F/E conversions with ferrofluids.

has shown that with otherwise identical resistor elements, this last method can be most sensitive, delivering an order-of-magnitude higher voltage difference on the bridge diagonal then the anemometric method. This is an advantage for the often encountered very small flowrates in microfluidics.

Finally, the fluidic oscillators with internal feedback, already discussed in Section 4.7.3, should be mentioned in this context, since many of them are developed for the purpose of electric flow measurement. Their advantage is a simple and robust electronics (just counting oscillation periods per unit of time), resistant to errors and imperfections. Their eminent suitability for the generation of digital signals makes them a very important alternative.

As in the E/F conversions, a direct action is also possible in this F/E case, but it requires special electrically active fluids, as shown in the historic patent examples in Figure 5.104.

5.4 A / F : COLLABORATION WITH ACOUSTICS

In fluidics it is easy to generate and use sonic (and in the case of microfluidics mainly ultrasonic) effects. Both fluidic sound generators as well as detectors may be extremely simple. In principle the generator is a whistle consisting of just an edge exposed to the fluid flow. It can be made, usually with a resonator cavity for stabilizing the frequency, without any additional fabrication step, together with other components of the fluidic system, and it costs virtually nothing. Despite this, the general trend is to generate as well as measure and employ the acoustic phenomena using electronic components, such as speakers and microphones, as offered by manufacturers of electronic accessories. Indeed, it is the easy availability of cheap mass-produced devices that may be the decisive reason behind this situation. There are, however, also technical advantages of using electronics: the frequency of the sound generated by a whistle varies inconveniently with the supplied air pressure, while an electronic oscillator keeps the frequency of a driven speaker sound constant. This is certainly convenient in a laboratory experiment, in which the cost of the electronic instruments is immaterial. The situation may change if the devices are to be mass deployed. First, even the smallest cost difference may be very important for devices that are made in large numbers. Second, many envisaged deployments require the microfluidic devices to be autonomous, or usable in locations where the electric mains will be not available.

As a result of the present situation, many contemporary fluidic devices discussed in the literature as employing an acoustic effect are a part of an E/A/F conversion chain. The most popular way of setting up the E/A part of this conversion is the use of electrostrictive components, usually in the form of a plate covered on both sides with deposited metal electrodes.

5.4.1 Acoustically driven pumps and separators

The devices belonging in this category are usually electrically driven. They differ from the pumps discussed in Section 5.3 by employing the acoustic (mostly ultrasonic) action of the fluid. Nevertheless, the distinction is (as usual) rather tenuous and due to the general trend towards the higher operating frequencies it is not uncommon for the otherwise more conventional E/M/F pumps to operate at frequencies where they generate audible sound; if not operating in the ultrasonic range. They may perhaps even use the same piezoelectric driver as is used in Figure 5.105 in an example of an acoustic pump. A practical difference is in the acoustic pumps operating at frequencies so high that the associated alternating movements of the fluid are too small to produce any rectifying effect in a fluidic diode. The fluid is forced to change its motion direction so vigorously that its volume expands and contracts in each cycle rather than moving any

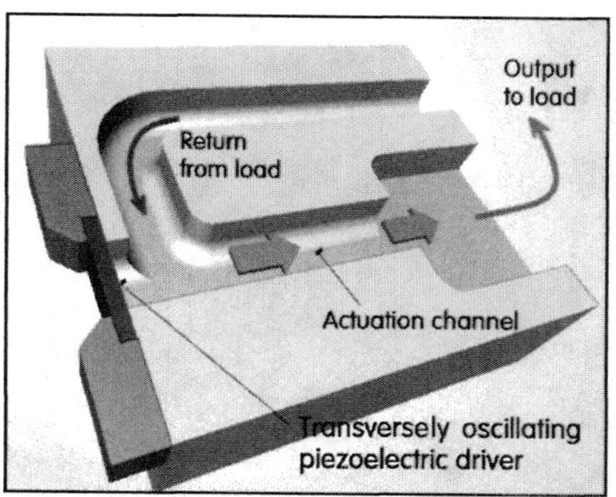

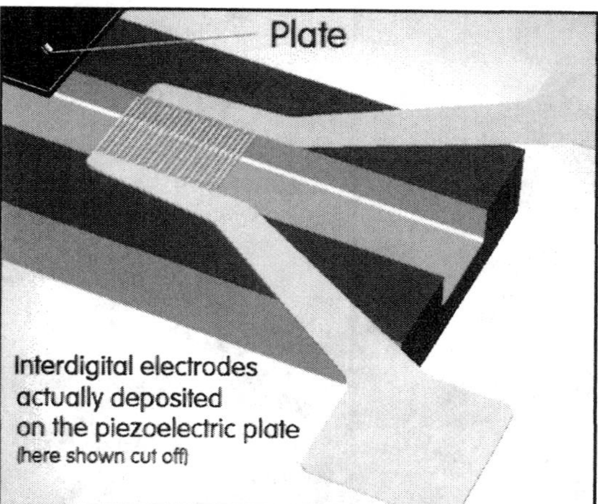

Figure 5.105 (Left) An example of the utrasonic pump using the acoustic streaming ("quartz wind") rectification effect in the actuation channel exposed to radio frequency (200–500 Hz) acoustic waves propagating from the driver. The wavelength on the order of μm is so small that the waves do not significantly diverge.

Figure 5.106 (Right) Detail of the system of interdigitated electrodes. They are used to generate and also pick up deformation waves propagating on the surface of the plate of the piezoelectric material on which the electrodes are deposited. The plate is removed here to reveal the electrodes on the bottom side.

significant streamwise distance. As a result, the acoustic pumps do not use the diodes. The rectification, as is shown in Figure 5.105, takes place inside a simple straight channel. The effect is made possible by the inherent nonlinearity of the governing equations. The same basic nonlinearity is actually used in the jet-type rectifiers, already discussed in Chapter 4 (Section 4.8.2). As a matter of fact, the basic layout of the pump in Figure 5.105 is identical to the low-frequency jet-type rectifier presented in Figure 4.178. The slightly more complicated shaping, with the diffusers, is used there to improve the performance in alternating flows, which at the lower frequencies have considerable amplitudes, comparable to or larger than the channel diameter. On the other hand, in the high f case in Figure 5.105, the alternating motion of the fluid has extremely small amplitudes relative to the channel size. This brings into account some additional mechanisms in the fluid, in particular the compression waves propagating downstream at very high propagation speeds, especially in liquids.

The pump in Figure 5.105 was actually designed to operate at frequencies as high as the radio frequency range. The corresponding wavelength is so small, on the order of μm, that the fluid behaves as if it were stationary, with rays of acoustic waves propagating in it. At these wavelengths, diffraction effects are also are negligible and the propagation from the driver indeed has a character of a practically not diverging acoustic "ray." If desired, the waves may even be guided down a curved channel, thanks to the total internal reflection at the interface (a critical angle for a water-glass boundary is 38°). The theory of oscillating channel flows is complicated by the complexities of the boundary layer, especially if the flow is turbulent, but even in the laminar case more common in microfluidics there may be complicated mutually phase-shifted layers near the walls. At the RF range, however, the boundary layers are also very thin and not important. The absorption and damping of the waves may be characterized by the absorption length. This is usually proportional to $1/f^2$ and may be rather short in the RF range. At 50 MHz this length is ~ 8.3 mm in water at the intensity ~ 0.1 mW/mm^2 typical for the pumps of the type in

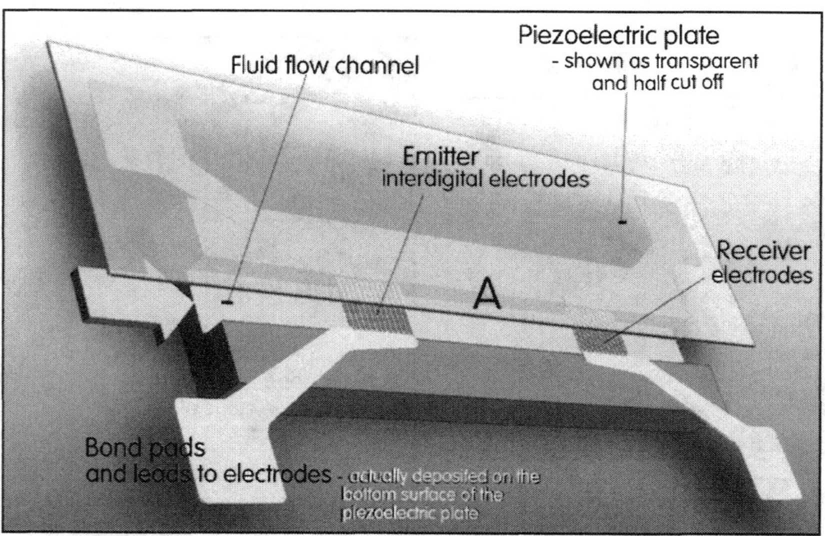

Figure 5.107 Device making use of the important interaction between fluid in the channel and the surface waves (SAW) propagating on the interface with piezoelectric material from the emitter to the receiver electrodes. In a SAW pump, the waves generate fluid flow (and the receiver is not necessary). In other devices, the change in the propagating conditions between the emitter and receiver can be an exceptionally sensitive detector of changes in fluid properties.

Figure 5.105. At higher intensity levels the absorption may increase due to heating and cavitation phenomena.

Also operating with high frequency ultrasonic phenomena is another class of devices – the pumps using waves produced on surface of solids. These are known as SAW (surface acoustic waves) devices. They are a type of waves first discussed by Rayleigh and studied in association with earthquakes. The SAW devices are manufactured in large quantities for applications not involving fluids, mostly for RF signal processing (filtering) in mobile communications. The waves propagate so that surface elements perform a retrograde elliptical motion with both in-plane and out-of-plane components. Both components become negligible at a depth larger than the wavelength. Depending on operating frequency, wavelengths λ are on the order of μm, while amplitudes are as small as only ~ 1 nm (the high waves shown in Figure 5.108 are an exaggeration). Operating frequencies may reach ~ 10 GHz, and typical values in the pumps for microfluidics are ~ 400 MHz. In microdevices, these waves are generated in piezoelectric materials, and as a result the mechanical waves are actually accompanied by electric waves, though because of the relatively weak electrostriction in typically used materials the electric fields are relatively weak. Most of the energy (> 95%) propagating in a SAW is in the mechanical component. The waves – and the accompanying electric field, travel at the speed of sound in the substrate. This is typically about 3 000 m/s.

The surface waves are generated by a system of electrodes deposited on the same side of the plate made of electrostrictive material, the side on which the waves are to propagate. Usually, they are deposited in the typical interdigital form (Figure 5.106). The material is usually aluminium. For pumping liquids, the corrosion-resistance aspects may dictate using gold (on a chromium or titanium adhesion auxiliary layer). The alternating compression and expansion action which produces the waves is schematically presented in Figure 5.109. Obviously, the electrodes have to be very small and separated by distances commensurable with the wavelength.

If the propagating wave encounters along its path a liquid (for example, a liquid drop on the plate's surface) the mechanical wave is radiated into the liquid, causing an internal streaming inside the drop. In a lager volume of liquid the radiated component causes a considerable pumping effect, moving the liquid along the wall in the direction of the wave propagation. The action on

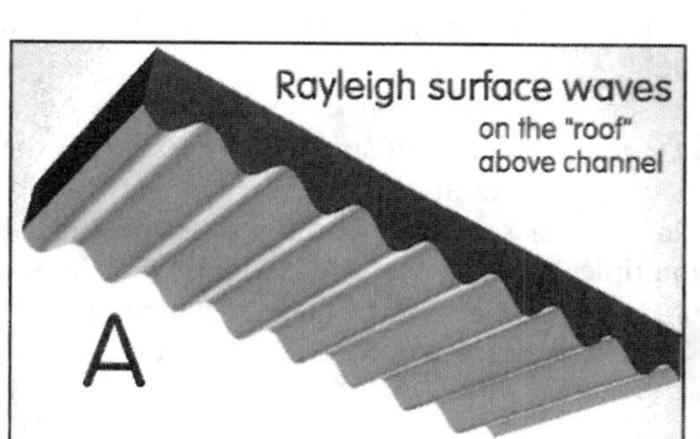

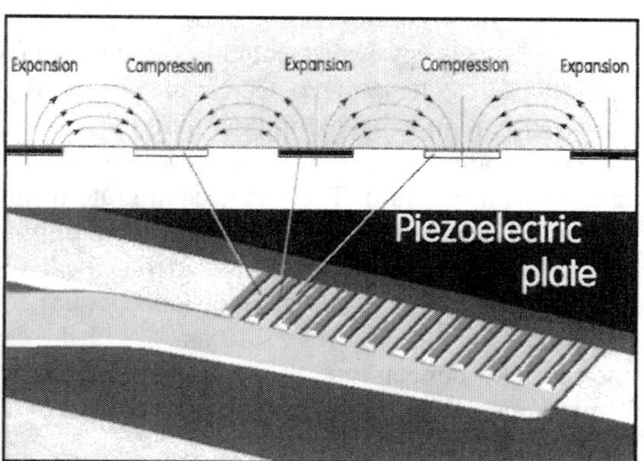

Figure 5.108 (Left) Very much enlarged detail of the waves propagating on the surface of the piezoelectric plate (in the location A, Figure 5.107) from the emitter interdigital electrodes to the receiver electrodes. Because of the piezoelectric character of the plate, the mechanical wave motion is accompanied by a propagating electric field. Typical actual wavelengths λ are on order of μm and amplitudes of only ~ 1 nm.

Figure 5.109 (Right) Generation of the surface waves by the compression and expansion stresses in the piezoelectric plate between the two interdigital emitter electrode systems.

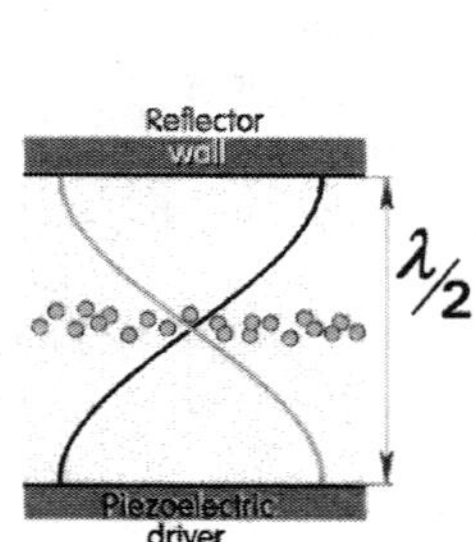

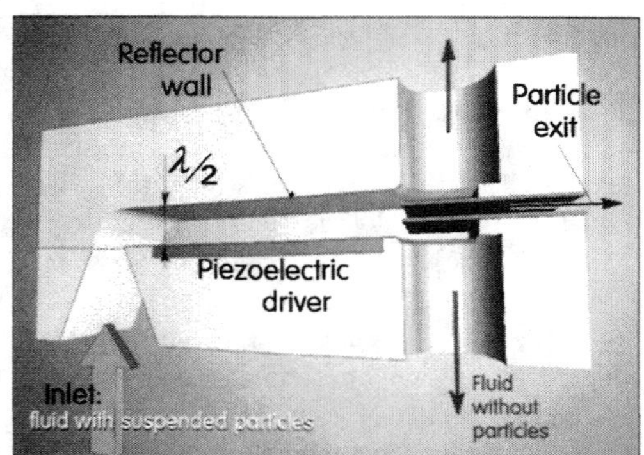

Figure 5.110 (Left) Ultrasonic standing waves generate forces on particles within the fluid, moving the particles toward the nodes. In the megahertz frequency range the wavelengths may be comparable with channel width. If the width (the distance between the driver and the reflector) is one-half of the wavelength, the particles concentrate on the channel axis.
Figure 5.111 (Right) Ultrasonic concentrator and separator of particles with piezoelectric drive and a reflecting opposite wall at the half-wavelength distance.

liquid takes place mainly in a thin layer near the surface. The generated velocity profile therefore has the character of a wall-jet, similar to the one shown in Figure 3.62 (of course, without the driving nozzle shown there). In the example in Figure 5.107, apart from the emitter system of interdigital electrodes, there is a similar receiver system. This is not necessary in a pump (where it may then be necessary to provide an absorber eliminating reflection of the waves). Monitoring the propagation of the waves by the receiver electrodes, however, is of importance for another class of the devices. Anything influencing the mechanical properties of the transmission in region A — such as a change in liquid properties or in the particular deposition of particles carried with the liquid on the wave-propagation surface—results in a significant change in the received signal. The layout from Figure 5.107 can therefore serve as a sensitive detector of bacteria, viruses, proteins, and other substances immobilized on the surface in the same manner that was discussed in association with the resonant frequency detector in Figure 5.44. In this detector application it is useful to employ a different type of wave, a horizontally polarized one. Without the out-of-plane motion, less energy is dissipated into the liquid (which carries the detected objects). It may be also appropriate to chose piezoelectric material with a dielectric constant approaching that of water; a sufficiently high value exhibits lithium tantalate $LiTaO_3$.

Yet another acoustic actuator is important; the separator based on concentrating particles carried with the fluid. The particles migrate toward the pressure node of standing waves inside the channel, because this is the location of minimum acoustic potential energy. For this application, the acoustic driver is positioned in one wall of the channel so that it emits waves toward the reflecting opposite wall, at a distance equal to multiples of half the wavelength [17]. In the example shown in Figures 5.110 and 5.111 the channel width is one-half of the wavelength, so that the particles concentrate on the channel axis from where they are removed. The overall properties remind one of those of a centrifugal separator.

5.4.2 A/F signal conversion

A simple and elegant solution to this particular transducing task was found in the sensitivity of laminar jets to disturbances, including those caused by acoustic waves. The phenomenon was first described in the 1858 *Philosophical Magazine* by an American professor named Le Conte, who noted at an evening musical party that the flames in lamps exhibited pulsations synchronous with

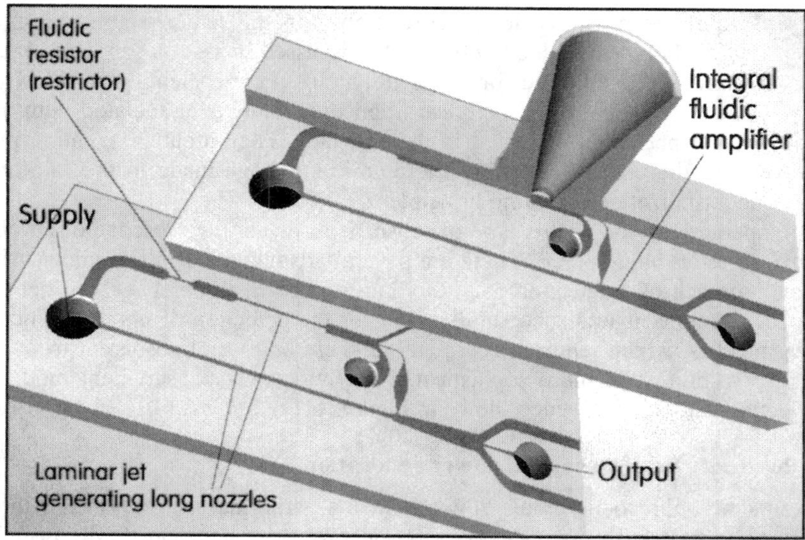

Figure 5.112 A fluidic ultrasonic sensor with an integral fluidic signal amplifier. The sensor is based on the high sensitivity of a laminar jet at conditions near the transition to turbulence (as in Figures 5.91 and 5.92) to any disturbance, including in this case the acoustic ones. The integral signal amplifier uses the same turbulization effect, this time caused by the sensor output flow.

some audible tones. Irish physicist John Tyndall, with whom the effect is often associated, developed the observation into the 18-page paper, "Sensitive Naked Flames", and used the specially adjusted flames to perform acoustic measurements. It was soon discovered that the combustion process is irrelevant; it just helps the observation. Of importance is just the adjustment of the flow on the verge of transition into turbulence.

An example of the sensor presented in Figure 5.112 was developed for sensing ultrasound waves at a frequency slightly above 50 kHz. They are concentrated by the conical horn on the

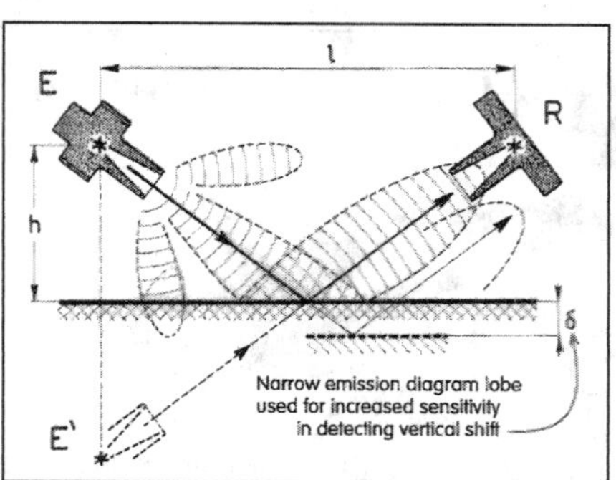

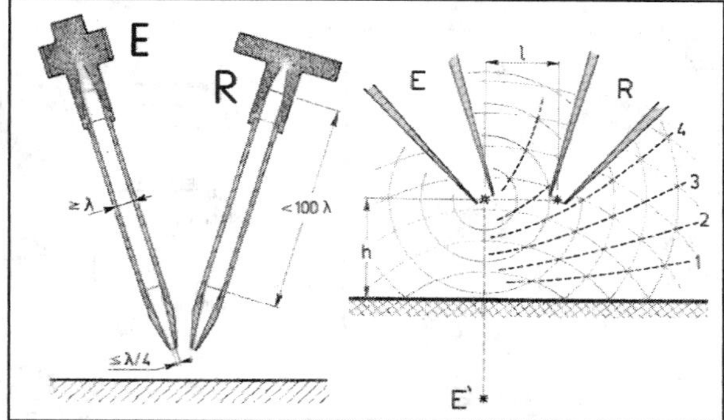

Figure 5.113 (Left) With the conical horns, both the emitter and receiver directional diagrams are quite narrow (the acoustic energy is concentrated into a slender beam with insignificant side lobes). This not only increases the detection distance, but together with the reflectance of most surfaces it makes possible a very sensitive detection of object position.
Figure 5.114 (Right) With waveguide extensions, the same emitter and receiver bodies may be used for very precise reflecting surface positions using the interference.

laminar air jet issuing from a very long, narrow nozzle—the length stabilizing the flow and producing a nearly parabolic exit velocity profile. As discussed in association with Figure 5.95, the supply conditions have to be adjusted for the jet Reynolds number being just below the value at which the jet spontaneously undergoes the transition to turbulence associated with a substantial decrease of the output pressure, recovered in the collector. The output from the primary sensing stage is directed into the control nozzle of the amplifying stage made in the same body and is based on the same turbulization operating principle.

The basic operating mode of this type of sensor is to detect the presence of an object placed between it and an ultrasound source. There are several advantages of ultrasonic sensing over the more common approach of a detection by photodiode detecting an object placed between it and a light source: for example, it was successfully used for the detection of photographic films in the manufacturing process, which required complete darkness and was transparent to infrared. There are interesting possibilities of more sophisticated detection and measurement modes, using, for example, reflection and/or interference shown in Figures 5.113 and 5.114.

5.4.3 F/A fluidically driven acoustic power generation

Generating sound and ultrasound by air flow has been a traditional method; in fact the brass wind musical instruments belong in this category. Neglecting here mechano-fluidic principles with a moving component—the reed—the pure fluidic generation principles (Figure 5.115) are mostly based on the alternating separation of vortices on an edge exposed to a jet flow (Figure 4.165). The separation can actually take place across a considerably wide range of frequencies, and the particular oscillation frequency is almost always stabilized by locking to a resonant frequency of a cavity resonator.

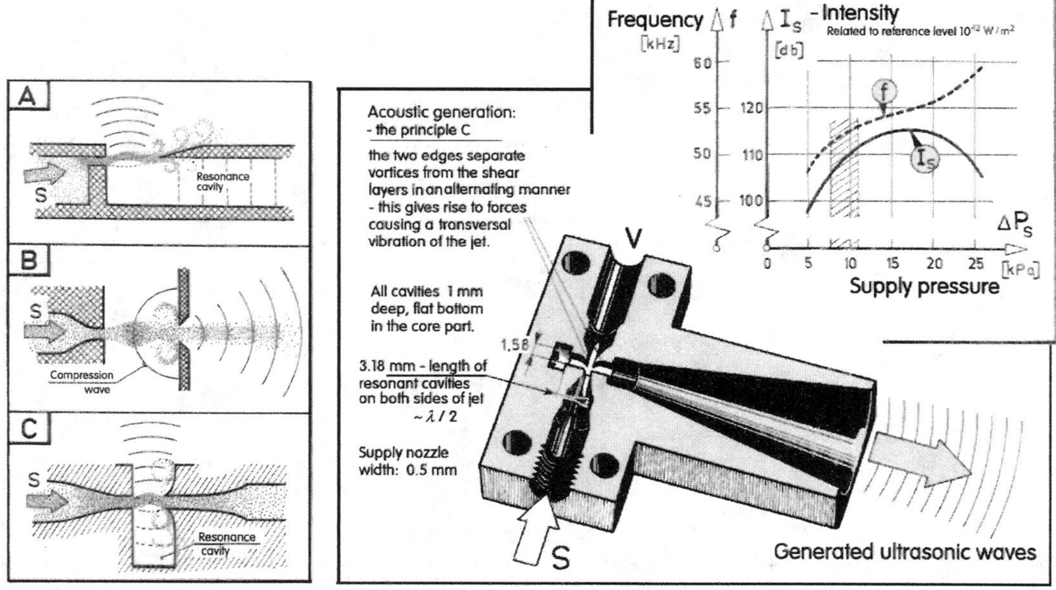

Figure 5.115 (Left) Typical F/A conversions: the simply made fluidically driven acoustic generators use the principles of the organ pipes (A), the "nightingale whistle" (B), or (C), a two-edged whistle with a sideways oriented resonance cavity. Note also Figure 4.165 and Section 4.7.3.

Figure 5.116 (Right) Successful fluidic generator of ultrasonic waves based on the principle C of Figure 5.115.

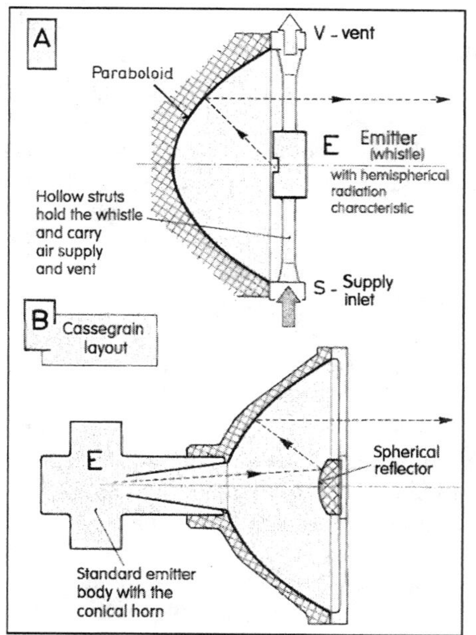

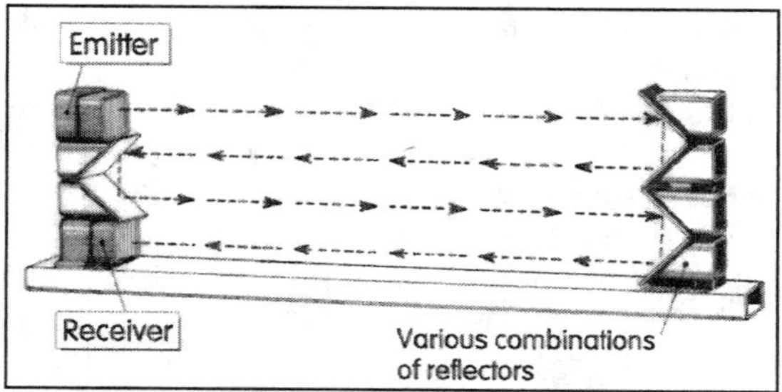

Figure 5.117 (Left) The sensitivity as well as direction characteristics of ultrasonic emitters may be significantly improved by the addition of a parabolic reflector.

Figure 5.118 (Right) With the exit aperture increased by the parabolic dish, the distance covered by the ultrasonic protective fence may be long enough to make possible the folding of the path by reflectors to make the fence more dense.

Whenever available space permits, it is useful to provide the generator, such as the one shown in Figure 5.116, with the parabolic reflector (Figure 5.117), and the same applies for the receiver (Figure 5.118). The magnitude of the obtained improvement depends on the area ratio of the reflector aperture and the entrance window in the basic device.

5.4.4 Fluidic and acoustic signal processing

It may be useful to briefly mention here two examples of the very classical fluidic amplification of acoustic signals. In the first one, fluidic signal processing entered a direct competition with electronics for voice transmission. The circumstances that made it successful were exceptional: the restrictions of orthodox Judaism do not allow electronic sound amplification, because they prohibit, on the Sabbath, the performing of any labor, which is defined as the creation of anything new. The electronic amplification of a rabbi's voice actually involves the creation of several new states – the motion of a diaphragm that was previously at rest and the creation an electric signal that did not exist before. Fluidic processing is accepted, as long as it restores the original sound to its original level. At a 50 ft distance, laminar proportional amplifiers (Figures 4.86 and 4.87) provide essentially 0 dB sound level gain at about 50 dB acoustic power gain. The electrically driven compressor supplying 10 psi air, not affected or modulated by the voice, is permissible as long as it is switched on before the Sabbath. The system was developed by Dr. Drzewicki, one of the pioneers of fluidics in the United States.

Perhaps more in line with current trends in microfluidics is another application developed by the same author: the detection and quantification of insect larvae inside grain, which would visually pass as flawless. In particular, fluidic sensors were used to detect rice weevil (*Sitophus oryzae*) larvae's chewing noise at distances measured in hundreds of millimeters. In principle, autonomous microfluidic detectors may be made to be distributed inside the volume of the stored grain.

Most uses of acoustic devices that collaborate with fluidics and are under current development aim at the detection of bacteria and biological warfare agents using the immunoassay techniques of binding the detected objects so that they change the ultrasonic wave transmission properties.

5.5 O / F : COLLABORATION WITH OPTICAL DEVICES

Photonics has recently developed into a very viable branch of MEMS, with devices of a similar size as the electronic or fluidic ones. It handles light (photons) usually in planar structures made on the surface of a plane substrate plate. The key elements, analogous to fluidic channels, are optical waveguides; they keep the light inside due to the total internal reflection on their walls. Waveguides not only carry light, but they can also perform a number of interesting operations with it, such as coupling, splitting, merging, switching, and multiplexing. Devices requiring a more complex structure then can perform more complex operations, such as amplification and polarization. The planar structures very often involve O/E and E/O conversions in light-emitting diodes or photodetectors. The optical transmission of a signal between two electronic circuits (made in optocoupler devices) is a reliable way of ensuring their electrical separation. Conversion chains involving acoustics are very common: a typical E/A/O device is a frequency shifter with a piezoelectric transducer generating surface acoustic waves, which behaves (the Bragg effect) as a grating interacting with the propagating light.

Interactions with fluid flows are numerous. Nevertheless, they are somewhat outside the focus of interest in contemporary integrated optics or photonics, whose main area is communication systems with high data-transfer rates. The main advantage is the high frequency of light, ~ 200 THz. This makes possible the sharing of the same waveguide by a large number of communication channels. One technique is known as WDM (wavelength division multiplexing). Using up to 32 wavelengths, commercial WDM systems transmit up to 400 Gbit/s. Another technique, time domain multiplexing, operates with extremely short pulses on the order of picoseconds.

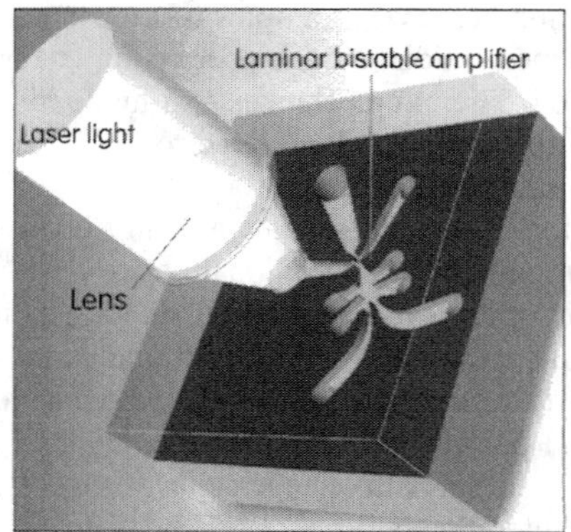

Figure 5.119 Some laminar bistable amplifiers are sensitive enough to be switched by the flow pulse produced by a laser heating a small chamber. With transparent fluids, it suffices to heat the chamber walls (covered with black, light absorbing paints).

Historically, the optical effects used to be rather weak to produce any mechanical effect capable of influencing fluid flow. The availability of compact and yet powerful laser light sources have changed even this aspect of collaboration with optics.

5.5.1 O/F : driving a fluid flow by light

This is not a common conversion. A known successful design uses the high sensitivity of the latest fluidic amplifiers in an O/T/F conversion chain involving thermal effects. In particular, a bistable version of the laminar diverter amplifier, which differs from the proportional version shown in Figure 4.86 by the presence of the attachment walls downstream from the interaction region of the main nozzle and the two control nozzles, may be switched by flow pulses so tiny that they may be generated by laser irradiation of a small cavity. Figure 5.119 represents the demonstration with air flow. Because of air transparency, the heat is actually absorbed by the chamber walls, dyed black.

The irradiation can also change the mechanical properties of monomer fluids by polymerization, but using this effect is complicated by the irreversibility of the change. The literature also mentions the effect of photostriction, but this also tends to be limited to materials not of direct interest in fluidics.

5.5.2 F/O : optical power controlled by a fluid

The inverse effect, F/O conversion at a significant power level, is also not very common. A very simple solution is using a ray from an external light source, and directing it by a mirror rotated by the force action of the flowing fluid. Another obvious possibility is using the *chemiluminescence*, the release of photons that have received their energy from a chemical reaction without involving heat. This technique is quite popular at low power levels as a detection method in analytical chemistry (as it needs no external light source). If thermal effects are also involved, then what the chemical reaction generates is a flame, a very intense light source, but the associated heat makes it less convenient to handle. A particularly intense light may be generated by chemical lasers, with the optical pumping provided by a chemical reaction, usually in a gas, again a process usually run at high power levels, rather than being suitable for microfluidics purposes. What can be done

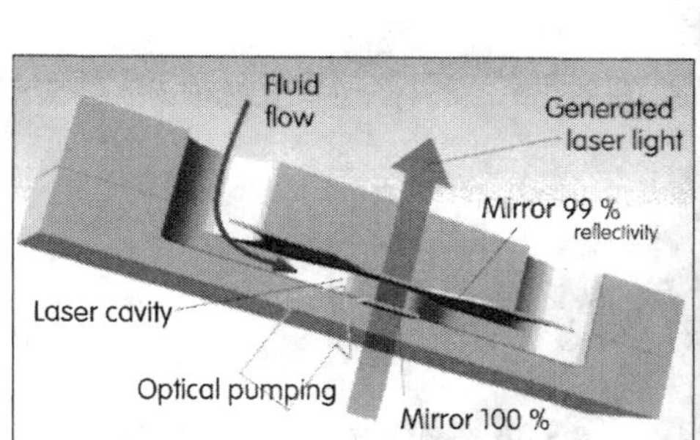

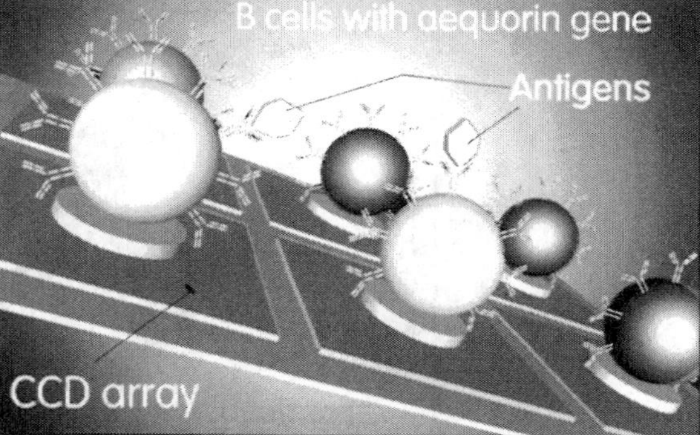

Figure 5.120 (Left) A low-power fluidic laser. In this typical version, it is an optically pumped dye laser. The active medium is a dye (e.g., *Rhodamine 6G*) dissolved in ethanol.

Figure 5.121 (Right) A sensor employing genetically modified living cells to generate light in response to a change in the composition of a fluid: the cell culture liquid. The change here is a presence of suitable antigen that triggers biochemical reactions that cause the aequorin in the cells to glow. The cells are attached to transparent sites on top of the ccd array, which convert the light to an electric signal.

relatively easily at a microscale is the fluid-flow-controlled generation of laser light using dyes (Figure 5.120). This is merely a light amplifier, with the optical power input as the optical pumping. In this particular example the necessary pumping power has to be rather high: 34 $\mu W/mm^2$ at wavelength λ = 570 nm. The liquid laser operates essentially in the same manner as solid-state dye lasers, with the advantage of avoiding problems with bleaching of the immobilized dye molecules.

Yet another possibility is to employ bioluminescence. If the aim is just to generate light, it is possible to use the enzyme luciferase from the firefly (*Photinus pyralis*). More interesting are the uses of captive living cells for specific detection purposes. The idea uses the extreme sensitivity and selectivity of white blood cells in detecting pathogens. The particular cells mentioned in the literature (e.g., [14]) are the B lymphocytes. Their surface is covered with antibodies that detect particular antigens by binding to them (Figure 5.121). For use as detectors, the cells have to be adapted by the addition of aequivorin, the protein that causes a species of jellyfish to glow. Binding the antigens triggers biochemical reactions that amplify the single binding event and release many Ca^{++} ions. These induce the aequivorin to emit photons, and the cell emits light.

5.5.3 O/F : optically generated change in a fluid at the signal level

What a relatively weak optical signal can do is generate bubbles in liquid. Figure 5.122 presents an example of a device in which the generated bubble fills the whole cross section of a small channel, and its presence or absence causes switching of light brought into and out from the device by optical fibers. The switching is based on total internal reflection of light on a flat surface exposed on the other, outer side to the alternative wetting by the liquid flowing in the channel. The liquid is chosen so as to have an (almost) equal refraction index, so that if the channel is filled by the liquid, the reflection surface virtually disappears. The light coming from A progresses unobstructed to the output, C. When the liquid motion brings the bubble into the gap between the two reflection surfaces, the light from A is reflected to the output, D. The device in Figure 5.122 actually contains two simultaneously switched but mutually independent switches—one supplied by light

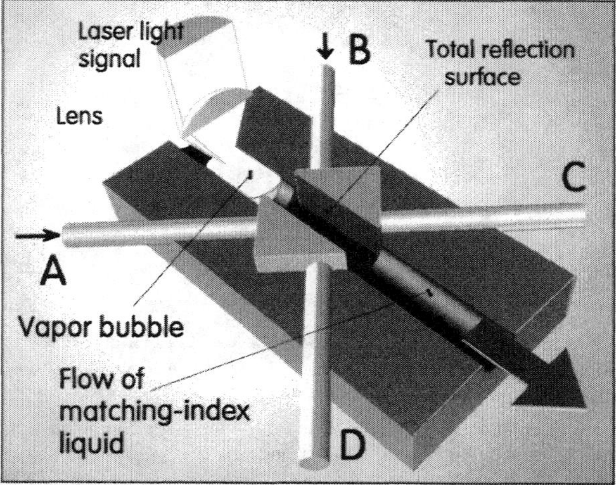

Figure 5.122 Generation of bubbles as signal carriers in liquid using laser irradiation. In this particular example, the bubble is used immediately downstream to switch the normal light passage (A-C and B-D) into the switched A-D and B-C.

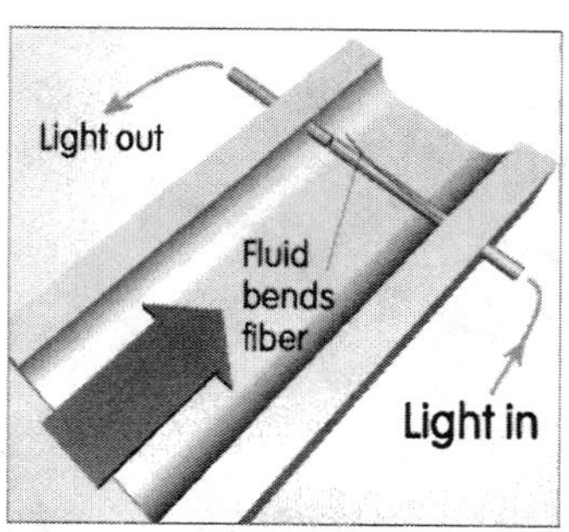

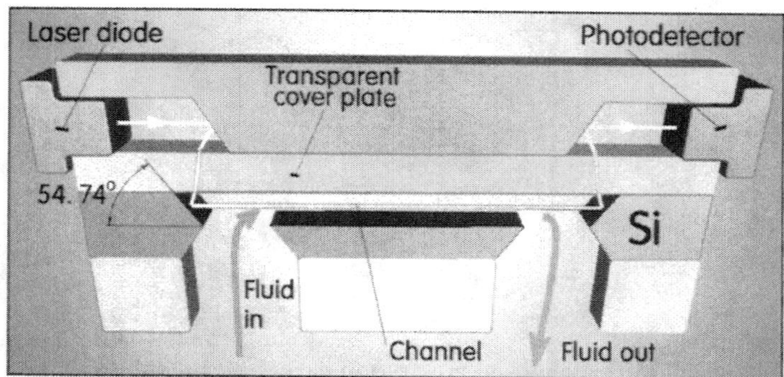

Figure 5.123 (Left) A simple optical measurement of the fluid flowrate by two optical fibers with their ends opposite one another. The hydrodynamic drag force deflects the fibers (especially the longer one), and this decreases the amount of light passing through.

Figure 5.124 (Right) Absorbance sensor detecting the composition changes of the fluid associated with its optical transmission properties. The light comes into the fluid flow channel from above, through the transparent cover plate, and is reflected by the plane with Miller indices (111), produced when the channel is manufactured in silicon by the anisotropic etching (Figure 2.14).

from A and the other supplied from B. In contrast with the impossibility of the mutual crossing of two fluid streams, there is no interaction between the two light bundles when they pass through one another, if the space between the reflection planes is filled by the liquid.

5.5.4 F/O: optical signal generated in response to fluid flow or properties

A very simply built detector of fluid flow is shown in Figure 5.123. It is particularly effective if the fluid is a transparent liquid, so that when it fills the gap between the ends of the two optical fibers protruding from the channel walls, the light passes through both fibers unobstructed. This is no more the case than when the fibers cease to be aligned when deflected by a hydrodynamic force. The decreased output light intensity is a very effective measure of the flowrate, capable of following even fast changes.

Increasingly often there are demands to detect and measure variations in fluid properties resulting from chemical reactions or from the separation of fluid mixtures into their individual components, for example, by the chromatographic method or due to their electrophoretic mobility (compare with the discussion in association with Figure 5.90). Quite often, the fluids to be distinguished possess different optical absorbances or optical densities. The detection is rather simple, only requiring means for passage of an optical beam through a cavity, a microcuvette, filled by the fluid. The sensitivity of the detection increases with the length of the optical path inside the cuvette, and as long as the fluid is not too opaque, it is usual to arrange the path axially in a long straight channel. The optical access may use the transparence of the cover plate and the mirror surfaces made in the anisotropic etching process (Figure 5.124). A more simple possibility is embedding two optical fibers into the walls at the end of the channel, as shown in Figure 5.125. Similar sensors, some not needing the supplied light source at all, some requiring UV irradiation of the fluid, use fluorescence and luminescence. This is particularly popular in investigations of proteins, which contain three aromatic amino-acid residues (tryptophan, tyrosine, and phenylalanine) capable of contributing to their intrinsic fluorescence, and biological objects in general.

An efficient optical response to ultrasmall quantities of biological and chemical agents is obtained in interferometric devices, making them very attractive for healthcare, bioengineering,

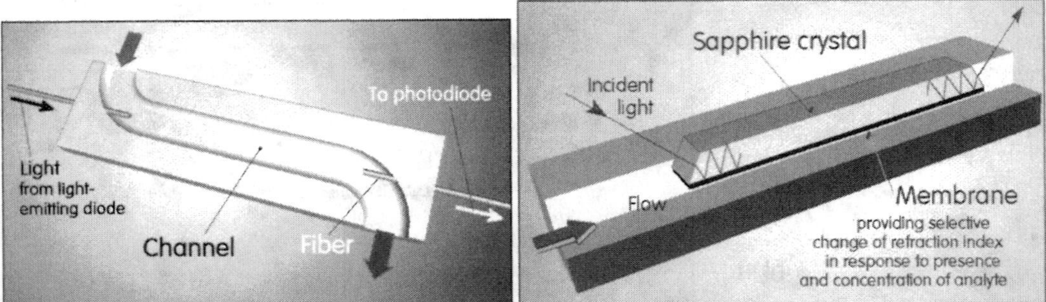

Figure 5.125 (Left) A simpler version of the optical absorbance microcuvette. A light inlet as well as an outlet is through optical fibers embedded into the channel walls. To make small absorbance changes more observable, the light is arranged to pass axially through a long leg of a meander.

Figure 5.126 (Right) Generation of an optical output signal by multiple reflection interference in the crystal forming the top cover of a fluidic channel. The channel top cover surface is formed by a chemically responsive membrane changing its refraction index in the presence of the detected agent.

and detection of traces of prohibited or dangerous substances for defense purposes. Their operation is usually based on the variations in the total internal reflection in optical components, caused by the changes in the refractive index of the surrounding fluid or due to the formation of adsorption layers on the surface. In the design shown in Figure 5.126, based on the Fabry-Pérot interferometer configuration, the analyte layer is immobilized in the membrane covering one of the reflecting surfaces and produces there changes in the optical properties. Among other detection uses, a pH sensor was also built in a similar layout, using the swelling and shrinking of polymer microspheres (polyvinylbenzyl chloride cross-linked with divinylbenzene and derivatized as dicarboxylate) dispersed in a hydrogel membrane. Thanks to the simplicity and low cost of chemical membranes, it is feasible to use an array of them in parallel to solve the problems associated with their often-low selectivity.

Other types of interferometers, the Michelson interferometer (perhaps not instantly recognized in its micro-scale guise in Figure 5.127) and the Mach-Zehnder interferometer presented in Figure 5.128, are also used increasingly often for these purposes. They may be described as typical devices of integrated optics, in particular with the Y-junctions in the rectangular waveguides for splitting the optical power into the two branches, and the subsequent

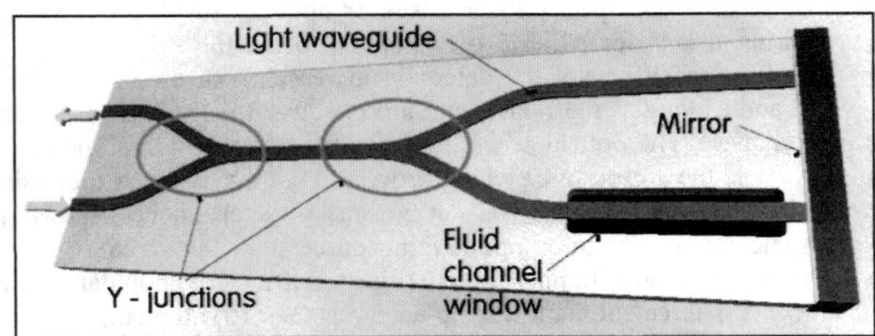

Figure 5.127 MEMS version of the well-known Michelson interferometer with light propagating in rectangular cross-section optical waveguides. In the window the waveguide is exposed to fluid and responds to variations of its refraction index or to the formation of an adsorbed layer on the surface.

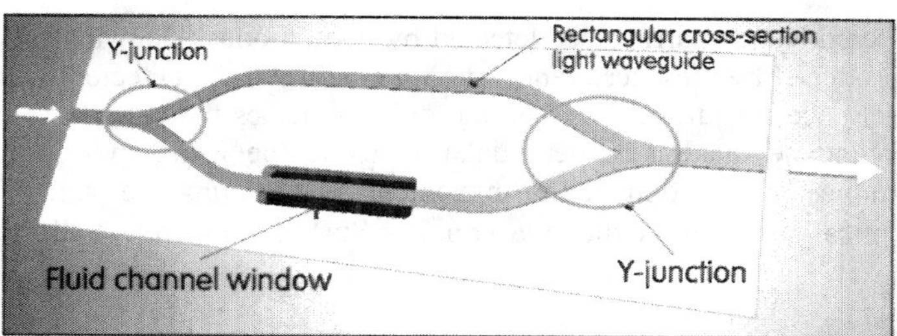

Figure 5.128 A currently very popular device for converting changes in fluid properties into an optical signal is another interferometer, the Mach-Zehnder type.

merging. The collaboration with fluidics is made possible by the windows in which the waveguides, carrying the optical signal, are surrounded by the fluid. They serve there as the sensing elements, reacting to the changes in the optical condition on the interface.

Another principle seen as very promising is the *thermal lens*, the radial gradient of the refraction index produced in fluid by the laser light when the fluid contains molecules or particles absorbing the incident power. In Figure 5.129 it defocuses the light on its path to the detector. This makes it possible to evaluate the concentration and character of the absorbing substances.

Quite often it is not the fluid itself but the suspended particles that are of interest. The motion of the particles may be easily detected by sensing the scattered light. This is the well-known LDA technique, which the recent progress in integrated optics made it possible to miniaturize to very small scales. Figure 5.130 shows a complete system fabricated by NTT Labs, Japan, with dimensions smaller than 1 mm. The fluid flow is illuminated by the pair of coherent laser beams,

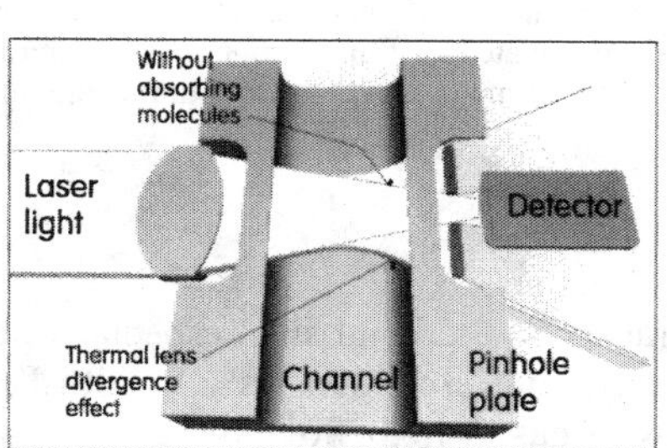

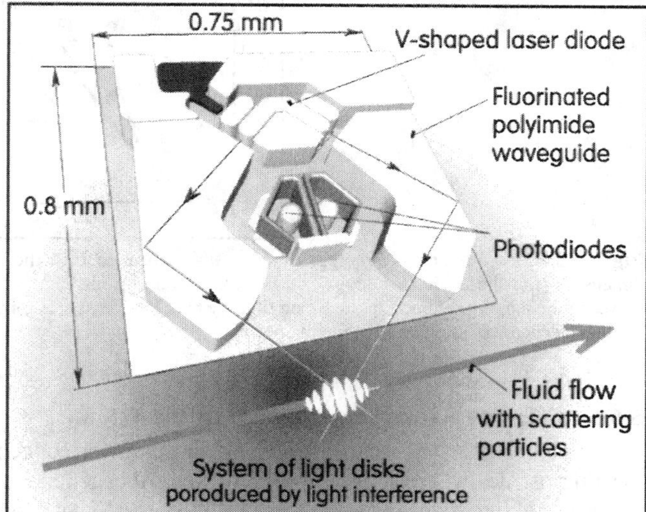

Figure 5.129 (Left) Thermal lens sensor: light-absorbing molecules present in the inspected fluid absorb some of the incident light. This locally increases the temperature of the fluid and changes its refractive index. Usually, the radial distribution of the index produces the same effect as if there were a concave lens, diverting the light rays from the detector.

Figure 5.130 (Right) Particle motion sensor based on the principle of LDA (- laser Doppler anemometer) that was recently built in submillimeter size. The photodiodes react to the flashes of light produced when light-scattering particles pass through the periodic system of bright disks produced, in the intersection of the two coherent beams, by interference.

and the light scattered by the particles is detected by photodiodes. Knowing the distance between the bright disks produced by the interference, the time between the detected flashes, and therefore the time the particle needs to travel between the disks indicates the particle velocity, without any calibration. A homodyne quadrature demodulation technique using two photodiodes makes it possible to determine the direction of the movement by detecting the sign of the signal phase difference. The optical power of the diode laser at $\lambda = 850$ nm is 4 mW, with measured velocities up to 0.4 mm/s.

5.6 T / F : THERMAL EFFECTS

Temperature changes are associated with several important conversion principles already discussed in previous sections of this chapter. As intermediate steps in the transformation chain, they are used in accelerometers in Figures 5.38 and 5.39, in the Marangoni-effect control of surface tension, many of the E/F conversions (e.g., Figure 5.57), and the F/E sensing of fluid flows. As already mentioned, while at large scales thermal changes are notorious for being very slow (it takes a long time for an accumulated heat in a fluid to be removed by heat transfer), in microscale the conditions can be quite different. As shown in the schematic Figure 5.131, heat transfer may there become the dominating effect with the consequence of the changes being very fast. This makes the use of the thermal intermediate step for example, in the E/T/F chain, a reasonable proposition even if it were not for the fact that the heating (mostly by the Joule effect) may be done in a very simple and compact manner.

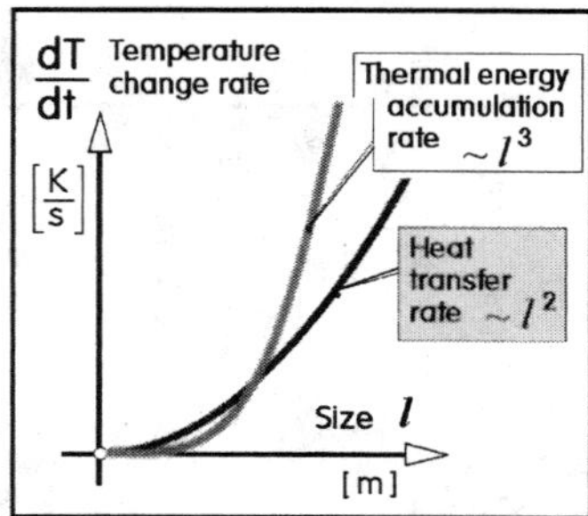

Figure 5.131 Schematic representation of the dependence of the speed of thermal changes on the size (linear dimension l) of the device. In large-scale device, the removal or addition of accumulated heat, proportional to the third power of the length (volume), takes a long time by heat transfer, depending on the second power (~ area), but the ratio of the two mechanisms is reversed at the microscale.

5.6.1 F/T: thermal power produced by fluid flow

These processes usually call for a really large power concentration on the input side, especially at the microscale, where it is not easy to keep the generated heat from escaping. As a result, the devices making such a conversion possible usually operate with gas at high flow velocities, in contradiction to the usual conditions in microfluidics, in which it is generally desirable to avoid the dissipation of fluid kinetic energy by conversion into heat. Of interest in the present context are principles operating without any moving components in the conversion device. The temperature growth would contradict the second law of thermodynamics: the devices, despite being seemingly not very complex, actually employ some form of cyclic process that is similar to that of a com-

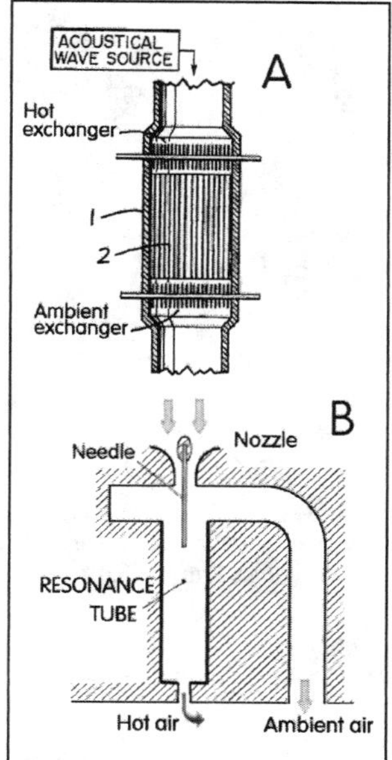

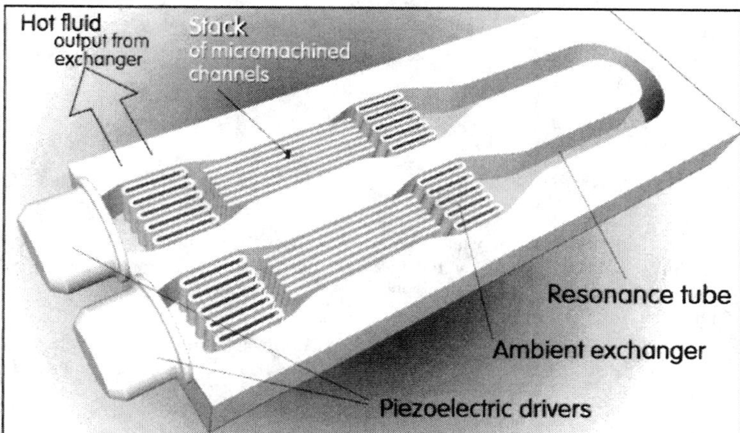

Figure 5.132 (Left) Two thermoacoustic devices that can convert the kinetic energy of fluid flow into temperature changes, so that they can be used for heating or cooling. A: use of traveling acoustic waves (picture from the original illustration in Ceperley's patent [14]). B: the Hartmann-Sprenger tube uses standing waves in a resonance tube opposite of the nozzle.

Figure 5.133 (Right) A microfluidic version of the thermoacoustic heater, which may also be used (taking the air from the other exchanger) for cooling. Microfabricated stacks of channels in which the conversion actually takes place are actually used even in large-scale versions.

bustion machine. The periodic processes usually involve acoustic waves and heat transfer to and from nearby walls. There are essentially two alternatives (Figure 5.132) both involving resonance in a cavity (so that, to a certain degree, there may be seen some acoustic analogy with lasers). The standing-wave version driven by steady flow is older; the effect was discovered by Hartmann in 1922 in Pitot probes in supersonic flow and investigated by Sprenger in 1954. The closed end opposite the nozzle is heated up to 2.6 times the ambient temperature, for example, enough to initiate an ammunition charge. Though usually operated in the supersonic jet regime, especially the versions with the rod (called needle) protruding from the nozzle were operated even at subsonic velocities. The traveling-wave device (Figure 5.133), driven by an alternating flow, has been a subject of intense investigation since the Ceperley's realization in 1979 that this is actually a pistonless Stirling engine. Gas compressed in the first part of the wave cycle reaches a high temperature while moved to the heated part of the channels, heating the walls there. Then in the next part of the cycle, it expands while moved to the cold end of the stack. Its temperature there is below that of this end of the channel walls, removing the heat from them. Cooling by $\Delta T = 20°C$ below ambient temperature is reported with He or Ar. The performance with air is usually worse.

5.6.2 T/F : driving a fluid flow by heat

Just like the Stirling engine, the device from Figure 5.133 can be also operated in an inverse role. Driven by supplying heat to the hot-fluid exchanger, it produces acoustic waves.

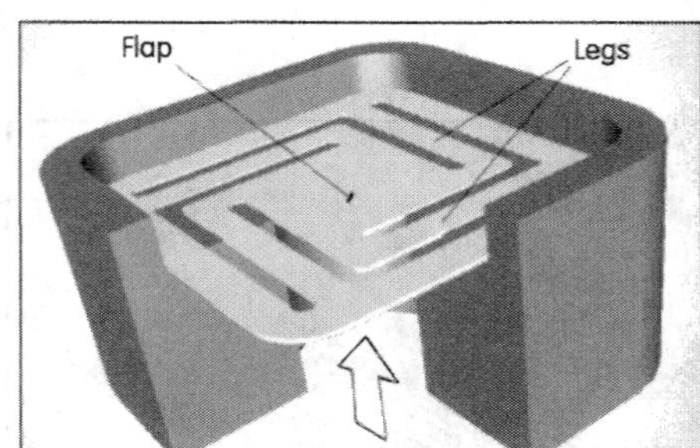

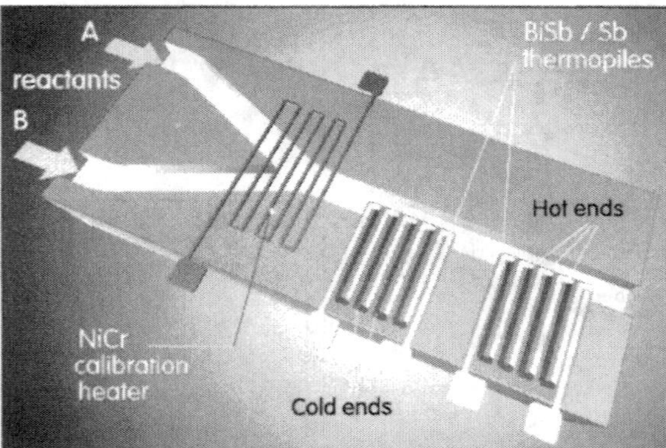

Figure 5.134 (Left) Thermostat valve using the Nickel-Titanium (NiTi) shape-memory alloy. As fluid temperature rises from 69.3°C to 86.8°C, the alloy undergoes a reversible austenitic transformation, as a result of which the flap closes and the flowrate decreases by as much as 90%, depending on the fluid pressure. The size of the square flap is 0.6 mm, the width of the legs, 50 μm, and the thickness is 12 μm.

Figure 5.135 (Right) An example of a fluidic calorimeter. An exothermic chemical reaction is monitored by measuring the streamwise increase of the temperature in the channel. The chip contains a heater for calibration by heating a nonreacting fluid.

5.6.3 Generating and using a temperature-dependent signal in a fluid

Heat not being a conserved quantity, signals in the thermal form are impractical, especially in fluids and in the microscale, where they are difficult to keep at the value they attained. Most fluidic signal conversions involving thermal effects are therefore of the temperature of the fluid at the input side.

5.6.3.1 Mechanical output

The T/M converters usually employ some form of thermal expansion of the mechanical component. Very often, this is a part of a T/M/F conversion chain, the overall effect being the use of the temperature signal for some control action on the fluid. An example is shown in Figure 5.134. The task to solve in this case is to adjust the flowrate of a heat-carrying fluid so as to maintain a constant temperature. The device is a mechano/fluidic valve, and the point of interest is the use of the shape-memory alloy to obtain a much longer stroke of the moved component (the central square-shaped flap) than in the relatively narrow range of temperatures that would be possible by mere thermal expansion. The flap is moved by contraction due to the martensitic change (or elongation due to the martensitic transformation, which takes place during cooling from 44.3°C to 25.6°C) of the four legs that move the flap. The legs are, of course, also deformed by the acting pressure difference: in the example reported in [16] the valve at increasing temperature reduced the flowrate by up to 90% at 13.8 kPa and only by 60% at 34.5 kPa. Arrays consisting of a large number of these devices were used in [16] to control flow in large-scale conduits.

5.6.3.2 Electric output

The most common conversion is into an electric signal using various thermometer principles—thermistors (sensitive but nonlinear), resistance sensors (small sensitivity), diodes, and transistors have been used—the most popular being, however, the thermocouples. The generated voltage is rather low, of the order of μV, but it may be easily increased by use of thermopiles, consisting of a large number of thermopairs, as shown in Figure 5.135. A design using repeated parts is typical

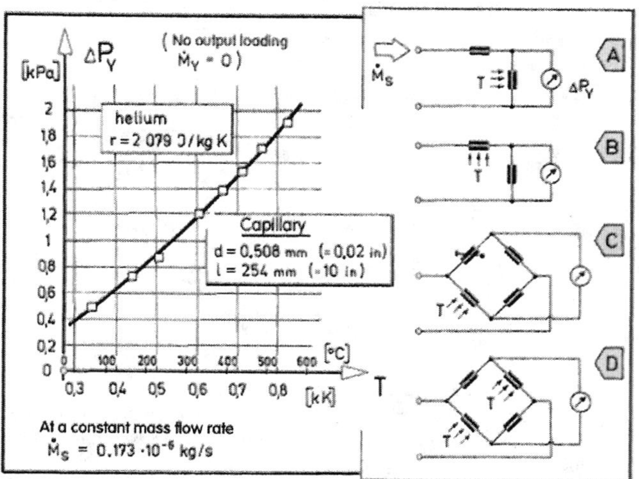

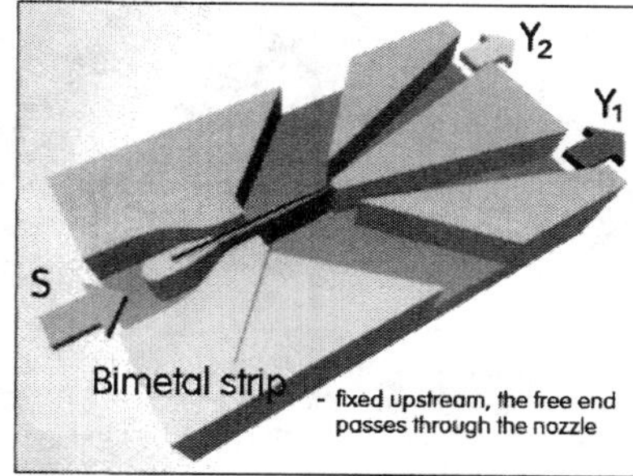

Figure 5.136 (Left) An extremely simple fluidic temperature sensor, with fluidic output signal in the form of pressure difference, based on viscosity variations with temperature in a capillary. The four examples at right show alternative connections of the capillary into the fluidic circuits.

Figure 5.137 (Right) Temperature sensor with fluidic output signal. Deformation of the bimetallic strip diverts the fluid jet, changing the ratio of its capture in the two collectors.

for microfluidics anyway, the devices themselves are commonly fabricated by repeating a pattern many times on the etched wafer and the repetition of the thermocouple detail is just an extension of the principle. The microfabrication (by deposition, requiring extremely small amounts of deposited materials) also makes it possible to use rather exotic modern thermocouple materials with extraordinarily high Seebeck-effect factors. If the output variable is to be the temperature, it is necessary to provide a suitable location, preferably with a known temperature for the cold ends of the thermocouples (the hot ends being exposed to the fluid in the channel) or to use an electronic compensation circuit with an amplifier drift compensating for the temperature change. When the interest is in the generated or transferred thermal power rather than the temperature, the measured difference for the calorimetric output may be easily obtained by positioning the cold and hot ends at known distances across a component with a known thermal conductivity, using the generally good linearity of most thermocouples.

Less well known is the use of the Peltier effect (Jean Peltier, 1834), the inverse of the Seebeck theromoelectric effect, for heating or cooling of the fluid (Figure 5.138).

5.6.3.3 Fluidic output

It is simple and easy to measure the pressure drop across the microfluidic channel, making use of the temperature dependence of viscosity. If the measured temperature is that of a fluid outside the fluidic circuit, such as that inside a chemical reactor, the friction resistor has a form of a capillary inserted into the location of interest (Figure 3.136). It need not necessarily be the classical capillary tube, perhaps a better solution is a long channel, similar (apart from having only a single inlet) to the one in Figures 3.117 or 3.118. Such use of a separate component is promoted by the fact that in order to obtain a large sensitivity, the channel has to be very long, and the fluidic circuits evaluating the pressure difference can be positioned away from the sometimes high sensed temperatures, though this is not a very important factor, fluidic circuits being resistant to even very high temperature environments. Figure 5.136 presents an example of the achievable sensitivity and

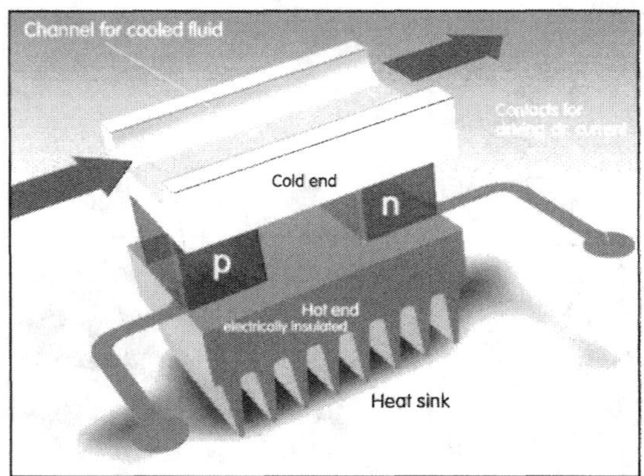

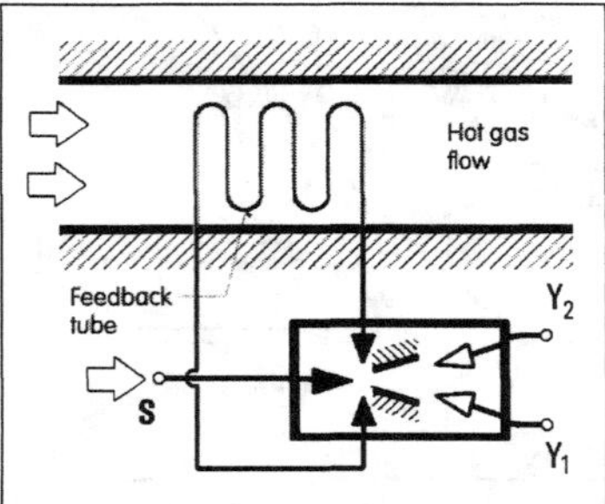

Figure 5.138 (Left) Peltier cooler for microfluidics: temperature difference between the ends of a thermocouple generated by the supplied electric current. The Si channel for the fluid (which may be longer when handling a faster floiwing fluid) represents the cold end, connecting p- and n-type Bismuth Telluride (Bi_2Te_3) semiconductor cubes.

Figure 5.139 (Right) Schematic representation of monitoring temperature by the changes it causes to the signal transfer in the feedback loop channel and therefore to the frequency of generated oscillation.

ty and the usual alternative connections. The sensing capillary is usually used together with a fixed, preferably a separation-type (Figure 3.74) restrictor, to form the "pressure divider." Its role is analogous to that of the variable restrictor for position sensing in Figure 5.18. In case A in Figure 5.136 the output is directly the pressure difference across the capillary. It is also possible to arrange the output signal as derived from the pressure drop across the series-connected fixed restrictor (case B in Figure 5.136) the pressure drop across which actually decreases with the increasing temperature in the sensing capillary, thus generating the opposite sign of the output signal change. If it is difficult to secure a stabilized fluid supply, it is useful to use the resistor bridge connection (Figure 5.27), either compensated (case C in Figure 5.136) or with increased sensitivity to temperature and suppressed sensitivity to variations of supply pressure, according to case D. An example of a sensor for sensing the fluid temperature and producing a fluidic output signal is shown also in Figure 5.137. Its body is essentially that of a jet-deflection proportional amplifier (Figure 4.76), with the deflection by control flows replaced by the *guided-jet* attachment to the bimetallic strip. As in other similar cases using jet deflection, there is the advantage of a similar fluidic impedance to that of the amplifiers in the fluidic circuit, making mutual impedance matching easier.

Also in the temperature-sensing task it may be advantageous to generate frequency-coded signal, using a fluidic oscillator. The resultant frequency-coded output signal has the obvious advantage of being independent of (nonconstant) dissipative losses. The signal is also easily converted to a digital form. An example is shown in Figure 5.139. This sensor is essentially the oscillator from Figure 4.152. The effect of temperature on the flow in the feedback loop is two-fold. There is some effect on the tube flow of viscosity change as in Figure 5.137, but the main influence is due to the change of the acoustic propagation velocity. The oscillator has to be designed therefore to operate in the regime of the signal transfer in the acoustic propagation mode. Contrary to the usual design of fluidic oscillators, here it is necessary to suppress the usual constant Strouhal number Sh property (even though the flowrate through the temperature sensor is usually kept constant by a fluidic regulator) and use the operation locked-in to the acoustic transmission in the feedback loop (Figure 4.158). As a result, the sensor oscillators usually involve a resonant cavity. Sensitivity dictates operating at a high frequency so that the feedback loops are often replaced by an internal feedback effect. The two oscillators of this type in Figure 4.166 were actually developed for temperature sensing—one of them for measuring the surface heating of a hypersonic airplane and the other for measuring the turbine inlet temperature in a jet engine. In this case, the sensed temperature causes a change in the resonant frequency due to the temperature dependence of the acoustic propagation velocity in the cavity.

5.7 F / F : FLUIDIC INPUT AS WELL AS OUTPUT

This conversion is not so illogical as it may seem to be at first sight. Some fluidic signals are simply not in the form we need them to be for further processing, such as when it is in the form of fluid flowrate, while the needed value is, for example, the pressure difference. Sometimes, of course, the signal is too weak, but the amplification task need not be discussed here, as it was already the subject of discussion in Chapter 4. What should be mentioned, however, is the nontrivial task of changing the character of the fluid somewhere en route, for example, converting a gas flow into a liquid flow while retaining the variations of the flowrate magnitude. Also included in this section are the cases in which the extracted output signal has to contain information about the chemical composition of the input fluid. The design task may even involve the separation of individual components of a mixture.

5.7.1 Pressure signal derived from measured flow

Fluidic signals may ideally contain the information they transfer either in the magnitude of a flowrate or the magnitude of a pressure drop. Real signals are somewhere between these two extreme cases, but usually tend to be near to one of the two. If the signal acts on an input diaphragm of a mechano-fluidic amplifier, the decisive value is the pressure magnitude, at least in steady states, the available flowrate being important in the transitory regimes. On the other hand,

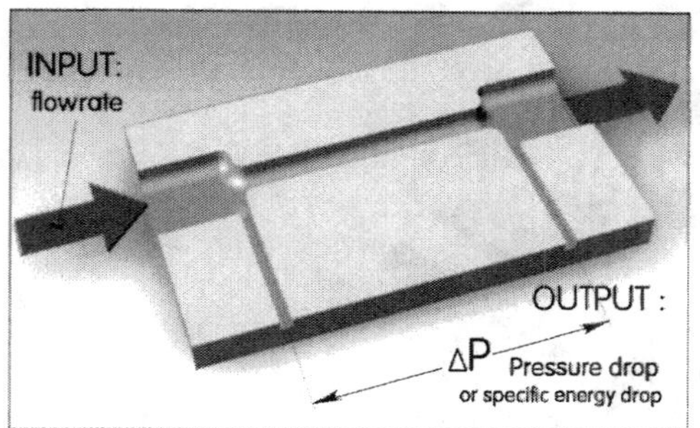

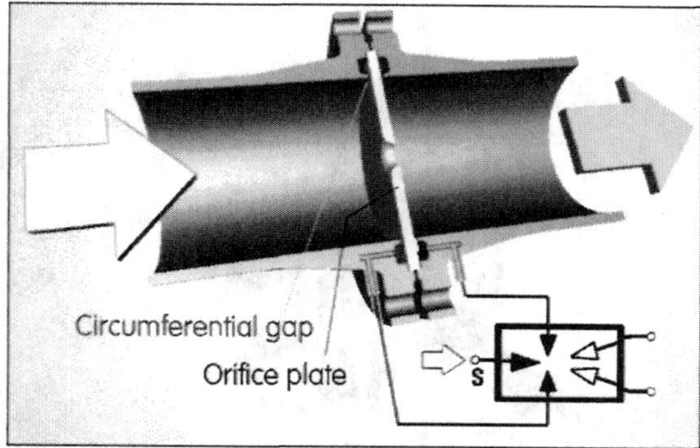

Figure 5.140 (Left) Conversion of fluid flowrate as the input quantity into a pressure drop using a fluidic resistor (restrictor), which must be provided with the two pressure tap channels, and, most importantly, has to exhibit known characteristics, which are usually obtained by calibration measurements.

Figure 5.141 (Right) Orifice flowmeter: use of the principle from Figure 5.140 to measure a large flowrate in a large cross-section of a pipe. Again, the complex interplay of hydraulic loss mechanisms with the changing ratio of their influence with changing Reynolds number necessitates laboratory calibration of this device. There are standardized orifice flowmeter geometries for which various authors have already made the measurements. In this particular case, the measured flowrate, despite the large pipe size, is small, calling for a very small aperture in the orifice plate and an amplification of the output signal.

hand, if acting in the control nozzle of a pure fluidic jet-type amplifier, the important parameter is likely to be the flowrate, depending on the cross-section of the control nozzle: if its exit is small, the generation of the control flow may need a high input pressure. Obviously, it is possible to convert between the two aspects of the signal by the choice of the dissipance of the flow device.

Similarly, using a device with a known dissipance (or preferably whole characteristics, since dissipance is rarely constant enough to be a reliable value for precise measurements in a large flowrate range) it is also possible to convert the flowrate as the input variable into the pressure drop output, as shown in Figure 5.140. Figure 5.141 shows the extension of the idea for large-scale flow measurements. It also gives an idea of how the signal of the orifice-type flowmeter may be amplified. Of course, with the nonzero signal flowrate needed to control the used jet-type diverter proportional amplifier, this amplification provides an example of the signal that is not of the pure pressure character.

5.7.2 Regenerative circuits

Even among the F/F conversions there are those handling higher power levels, not only just signals. An important example are the heat exchangers, for transfer of heat from one fluid to another. The microfluidic exchangers are exceptionally effective, due to the scaling law of Figure 5.131, and there are a number of applications of microfluidics based on using this advantage. The more usual devices are built on the principle according to Figure 5.142, but the less known regenerative exchangers of Figure 5.143 may be more effective. Their generally better thermodynamic performance is paid for by the more complex circuits necessary to handle the switching of the flows. This is handled by a class of fluidic circuits that deserve to be mentioned here, because, apart from their use in heat transfer they can perform several other interesting and useful F/F conversion tasks.

The essential component of the regenerative circuits, associated with flow reversals, is a module called the fluidic *alternator*. It may be built from several fluidic devices, such as the vortex amplifier (Figure 4.41) and in addition needs to be controlled by what is in the schematic

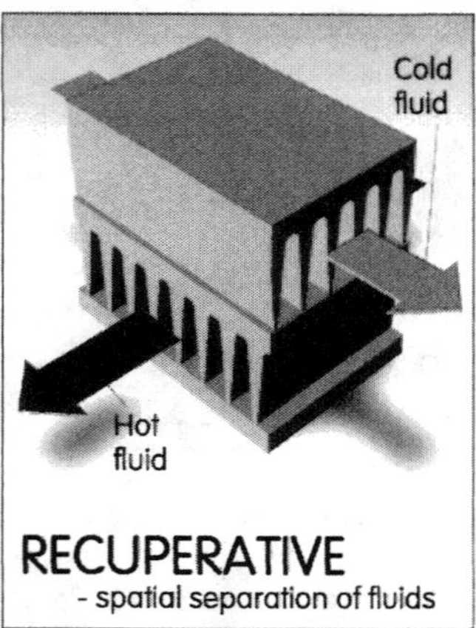

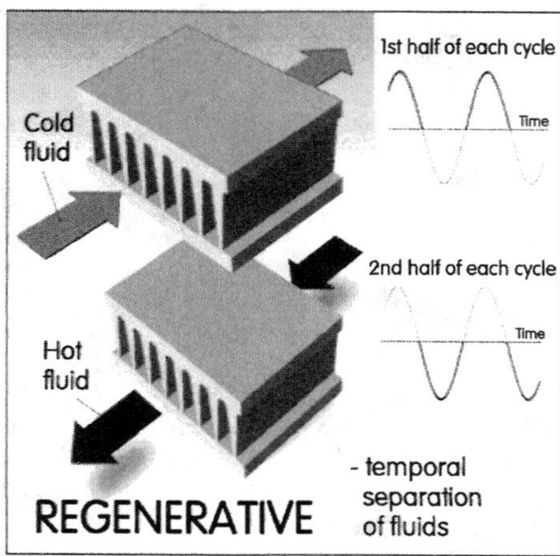

Figure 5.142 (Left) Recuperative heat exchanger: the two fluids each pass through their own cavities, separated by solid walls through which the heat is transferred by conduction.

Figure 5.143 (Right) Regenerative heat exchangers: both fluids use the same system of cavities, each occupying it at a different part of the operating cycle.

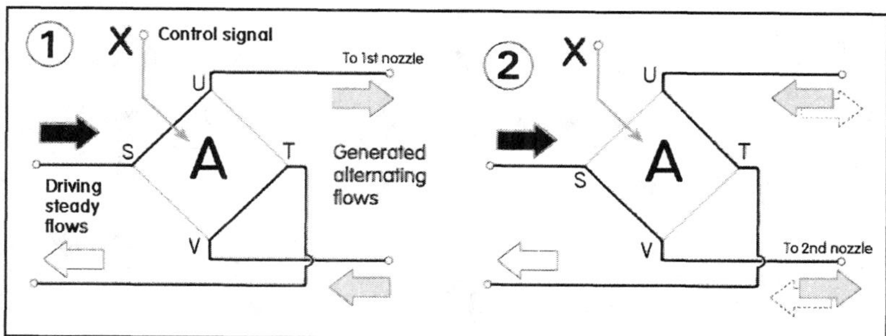

Figure 5.144 Principle of a fluidic alternator: this schematic representation of the circuit presents the flowpaths in the first half (1 – at left) and then in the second half (2 – at right) of the operating cycle. Topologically, the layout resembles the Grätz rectification bridge (Figure 4.174), the difference being the necessity of the components in the bridge arms (or vertices) being controllable.

representation of alternator operation in Figure 5.144, presented as an external control signal source. This illustration shows the two-phase case, generating two alternating output flows mutually phase-shifted by 180 deg. In principle, just a single-phase version is possible, but in most applications this would cause the problem of what to do with, for example, the hot fluid during the period when the regenerating exchanger is being filled by the heated cold water. On the other hand, there are the theoretical possibilities of three-phase and multiphase alternators, rarely if ever made because of the unnecessary ensuing complexity. In fact, even the two-phase case in Figure 5.144 may seem to result in quite complex circuitry. Indeed, the circuit in Figure 5.146 (a direct counterpart to the rectifier Grätz bridge with diodes in Figure 5.145) is far from simple (considering especially the fact that the vortex amplifier exits are perpendicular to the vortex chamber plane, leading to a three-dimensional system). This impression is misleading; there are designs in which the alternator consists of just a single standard device (see Figure 5.153), even

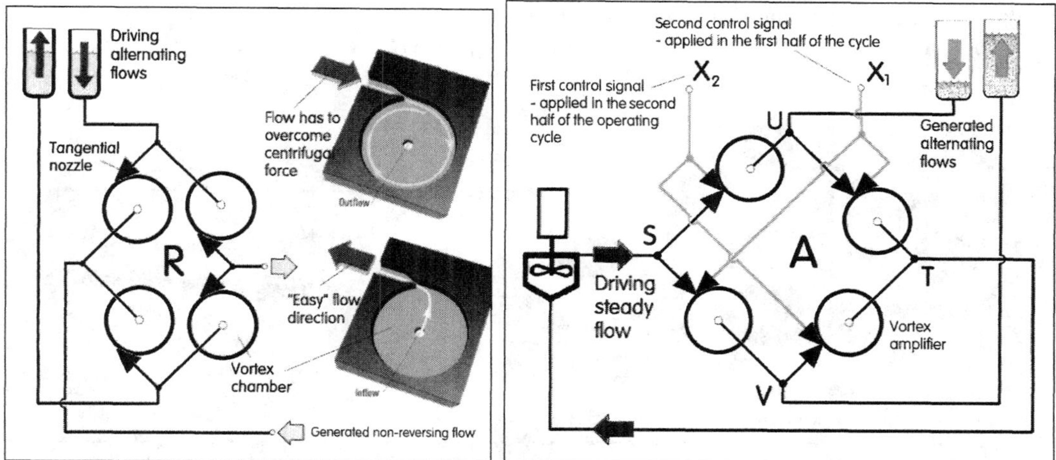

Figure 5.145 (Left) An example of the Grätz rectification bridge with vortex diodes in the bridge arms. The rectification is needed to get a steady flow of heated fluid from the regenerative exchangers (Figure 5.143).

Figure 5.146 (Right) Schematic representation of the fluidic alternator corresponding to the version with the vortex-type turning down devices in Figure 5.145. The obvious difference is the need to use vortex triodes (vortex amplifiers controlled by the input signals X_1 and X_2) in place of the diodes.

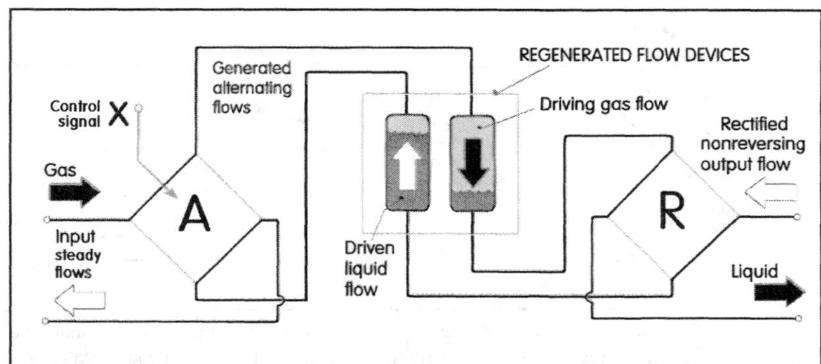

Figure 5.147 Schematic representation of a complete fluidic two-phase regenerative circuit. Actually, such a complex circuitry with both alternator (A) and rectifier (R) bridges is very rare, as it is usually possible to simplify it considerably with an integrated version of some components.

including the feedback loops, as shown in Figures 4.145 and 4.151, which convert it into a self-excited oscillator, eliminating the need for an external control signal source.

Mentioning the fluidic rectifiers in this context is appropriate, not only because of the topological analogy (compare Figures 5.145 and 5.146), but because the rectifier at the output type is an important part of the complete regenerative circuit, Figure 5.147. Unless the alternating output flow is acceptable, the rectifier is needed to convert the flow into a nonreturning and, with some capacitive low-pass filtering (Figure 2.101), even steady flow.

Apart from the flowpath switching to generate and handle the input and output flows for the pair of the regenerative heat exchangers (Figure 5.148), the regenerative circuits offer many other interesting application possibilities. The accompanying figures present some examples of the alternative twin devices replacing the exchangers in the central part of Figure 5.147: F/F conversions between different fluids or fluids in different thermal or pressure states.

One possibility, presented in Figure 5.149, is to use the regenerative circuit at the interface between fluidic circuits operating with vastly different pressure levels. The conversion devices in

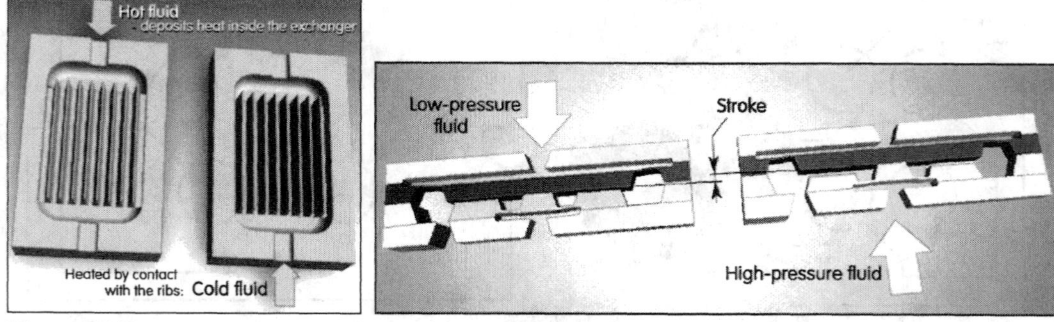

Figure 5.148 (Left) Example of regenerated devices: a pair of heat exchangers in which the hot fluid transfers heat to the ribs past which it flows during the first part of the cycle. Then in the next half-cycle, it is replaced by incoming cold fluid, which is heated by removing the heat from the ribs.

Figure 5.149 (Right) Another example of a regenerated device: the pressure exchanger. The same force (transferred by the contact with the top of the diaphragm bosses) represents high pressure acting on a small area (bottom) and low pressure (top) acting on a large area (dark diaphragm).

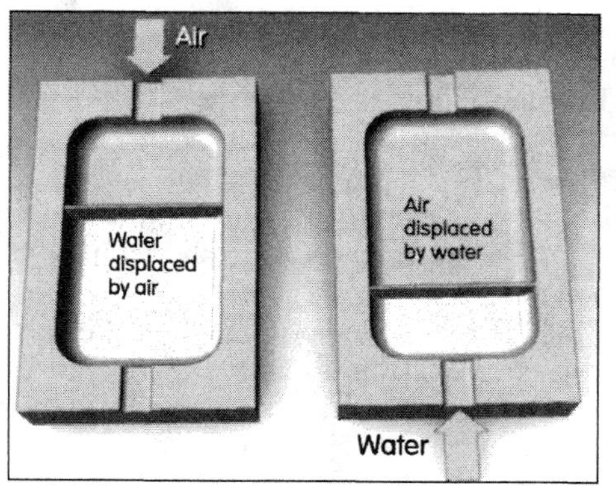

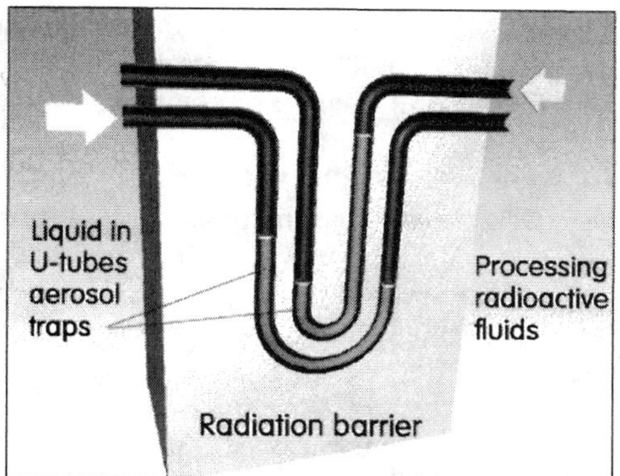

Figure 5.150 (Left) Yet another example of a pair of regenerated devices in the circuit of Figure 5.147. In this case the regeneration is used for F/F conversion from gas (air) to liquid (water).

Figure 5.151 (Right) Another example of regenerated devices: U-tubes are used in transferring energy by alternating air into a space that must be kept totally isolated from the outer world. The U-tubes are positioned inside the barrier. The liquid in them stops any (rare but conceivable) passage of aerosols.

this case use a moving part, with a large area exposed to the low-pressure fluid and a small area on the other side (bottom in Figure 5.149) to the high-pressure fluid. A thin slice of the device image shows (in Figure 5.149) the moving (here, actually deformable) parts in their two extreme positions. Each device functions as a transformer: the high-pressure side operates with small flow amplitudes and the large-area low-pressure side with large flows so that the product of the maximum flowrate and pressure is constant, there being, of course, no power amplification.

F/F conversions between different fluids are presented in Figures 5.150 and 5.151. The resultant overall effect may be that of a gas-driven liquid pump. The questions of the interface between the fluids (piston, diaphragm, and so on) are analogous to those discussed in the context of the accumulative capacitance devices (Figures 2.98, 2.99, and 2.100).

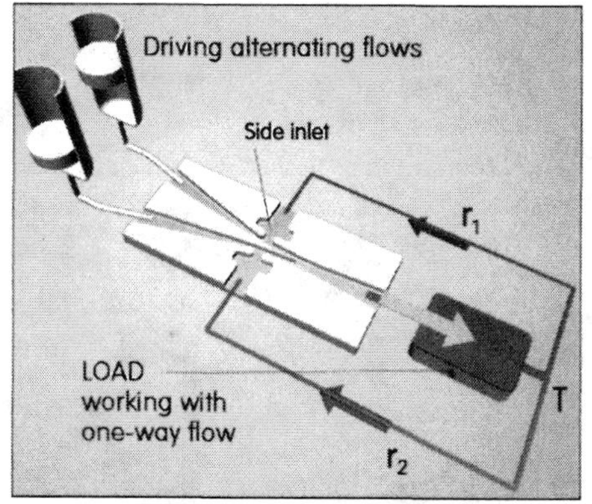

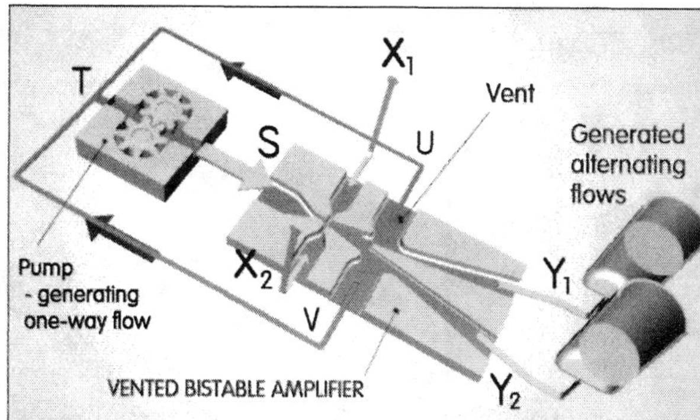

Figure 5.152 (Left) Planar Walkden-type airflow rectifier (Figure 4.176): the basic character of the Grätz rectification bridge (Figure 5.145) may be difficult to recognize, as the bridge arms r_1, r_2, and the vertex T do not exist as manufactured components – their role is taken over by the atmosphere.

Figure 5.153 (Right) The alternator may actually be quite simple, consisting just of a single standard fluidic diverter valve (Figure 4.90, here in the vented version, Figure 4.91). Apart from the necessary control inlets, this is obviously a direct counterpart of the rectifier from Figure 5.153, including the effective absence of some parts of the theoretical bridge in Figure 5.144 (compare the vertices S, T, U, and V).

Figures 5.152 and 5.153 show that using integral jet-type rectification as well as alternator devices may result in the regeneration circuits being actually quite simple, despite the impression that may follow at first sight from Figures 5.144, 5.146, and 5.147.

5.7.3 Generating a signal carrying information about fluid composition

This is a task of particular importance for chemical and biological microfluidics. Full composition analysis is, of course, usually performed in devices (gas or liquid chromatographs or infrared and mass spectrometers) producing an electric output signal. Typical fluidic principles make possible the building of simple passive sensors that are best suited to sense composition changes in a binary mixture, as they are reflected in the resultant change in the physical properties of the fluid, such as its viscosity.

A typical viscosimetric pure fluidic device is the gas concentration sensor shown in Figure 5.154. It was developed to detect CO_2 concentration in a respiratory apparatus. The idea is to measure the pressure drop across a friction-type (capillary) resistor of the type discussed in connection with Figure 3.75. It makes use of what is otherwise a disadvantage of friction-type resistors: their dependence on the fluid viscosity of the pressure drop, indeed an unpleasant property, causing temperature dependence of the pressure drop. In the present case, the viscosity does not vary with temperature (which is kept constant) but with the measured percentage of carbon dioxide contents. To make pressure drop changes possible when connected in a fluidic circuit supplied, as usual, with the more or less constant supply pressure, the variable restrictor is used in series with the "fixed", nearly viscosity-independent orifice type restrictor, the circuit corresponding to the sensor circuit explained in Figure 5.18.

Of course, the temperature dependence remains, and if the circuit cannot be kept at a constant temperature, this tendency has to be suppressed by using the same devices in the parallel arm of the fluidic Wheatstone bridge (Figure 5.26) supplied with the reference gas – clean air. The complete bridge is planar, made by etching, and the generated output pressure difference is amplified by the fluidic proportional amplifier according to Figure 5.28.

A similar idea, differing in the use of the vortex diode (Figure 3.79) as the viscosity-dependent component, is presented in Figure 5.155. One of the advantages is larger cross sections throughout

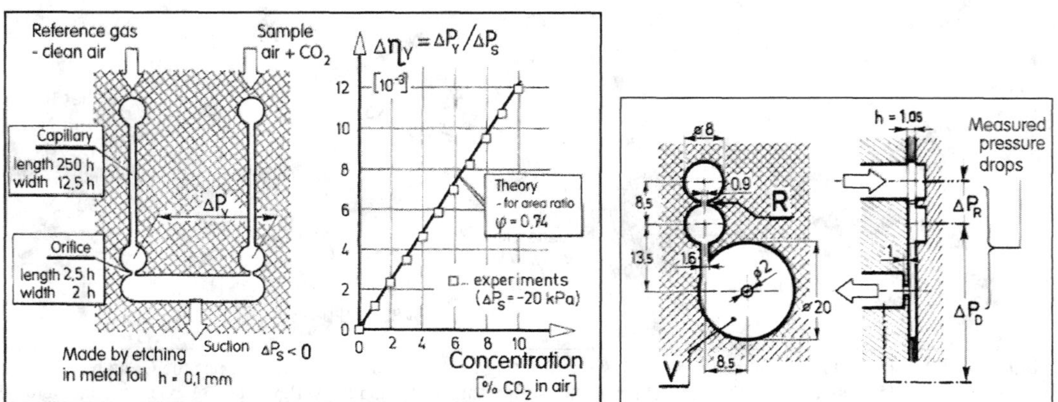

Figure 5.154 (Left) Simple fluidic chemical composition sensor: the measured quantity is a percentage of the carbon dioxide CO_2 in breathed air in a closed-loop respiration apparatus.

Figure 5.155 (Right) *F*luidic composition sensor for detection changes in hydrogen-air binary mixture. Vortex valve V is series-connected with orifice-type resistor R. The fluidic output signal is the ratio of the pressure drops across the two components.

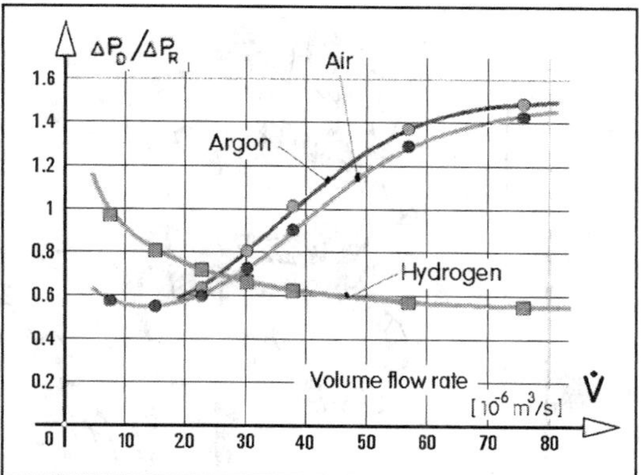

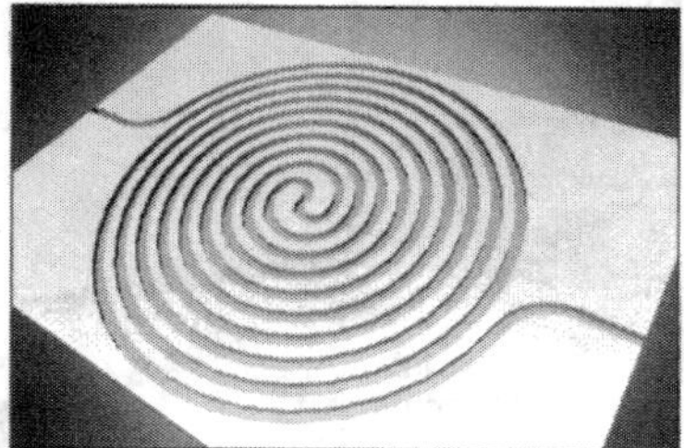

Figure 5.156 (Left) The ratio of the two pressure drops measured in the simple sensor from Figure 5.155 obtained experimentally as a function of the volume flowrate for air, argon, and hydrogen.

Figure 5.157 (Right) An example of a typical chromatographic "column" as used in microfabricated chromatographs –a long channel in which the components of a fluid mixture separate due to their different laminar channel flow properties.

the whole fluid flowpath—with less danger of clogging by carried particles—and yet greater sensitivity. The diode exhibits a nonmonotonous variation of Euler number with increasing Reynolds number. Figure 5.156 shows the ratio of the two measured pressure drops, across the diode (V) and the orifice (R), that demonstrates the resultant large, in fact, qualitative changes in the behavior between air and hydrogen, due to their different resultant Reynolds numbers. Otherwise, for some other binary gas mixtures, the Reynolds numbers are not sufficiently dissimilar. The diagram shows a very small, in practice hardly distinguishable change for air and argon (the latter was used to simulate uranium hexafluoride, used in nuclear fuel reprocessing).

In mixtures containing more than two fluid components, the sensing is inevitably more complex. The usual approach to composition analysis is a cyclic, rather than a continuous operation. In each cycle, the components in the sample are separated using their differences in some mobility mechanisms. Some of the used methods are outside the present context of F/F conversion devices, for example, the electrophoretic separation already discussed in association with Figure 5.90. A method belonging here is another type, the chromatographic separation. This is defined as distributing the substances between two phases, a mobile phase and a stationary phase. The mobile phase is either a gas or a liquid, which gives rise to two basic types of *chromatography*: gas chromatography (GC) and liquid chromatography (LC). The stationary phase is either a liquid or a solid. The typical microfluidic chromatographic devices are of the type that use the differences in the sample components propagation through a very long channel (Figure 5.157). The substances leave the channel end in the inverse order of their propagation speed and their amounts may be then measured by a sensor sensing the change in properties. This sensor, however, usually generates an electric signal, processed and recorded by electronic means, with the character of individual components evaluated from their order in the record. Some of the suitable sensors were discussed in Section 5.3.4. Some chromatographic sensors also use an optical intermediate method for F/O/E chain conversion, as discussed in Section 5.5.4.

5.7.4 Fluidic power amplifiers

An interesting case of the fluidic/fluidic conversion from low power to high power levels is offered by the use of the scaled-up fluidic valves discussed in Chapter 4. Figures 5.158, 5.159, and

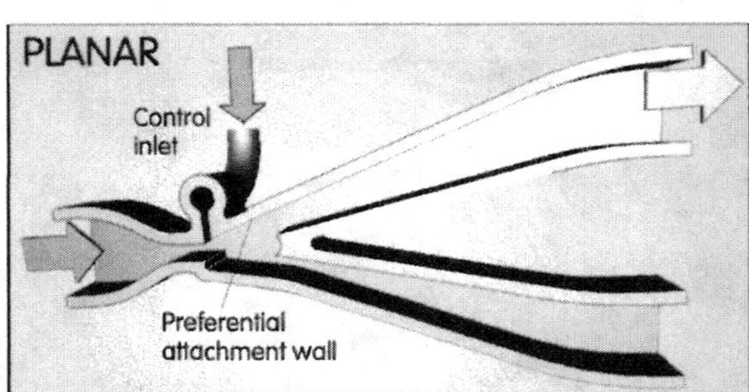

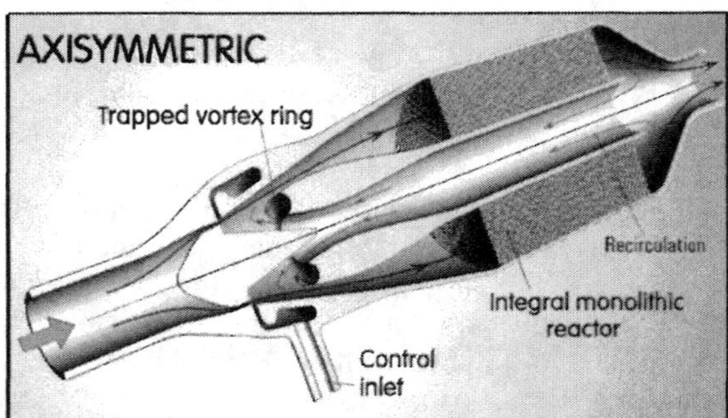

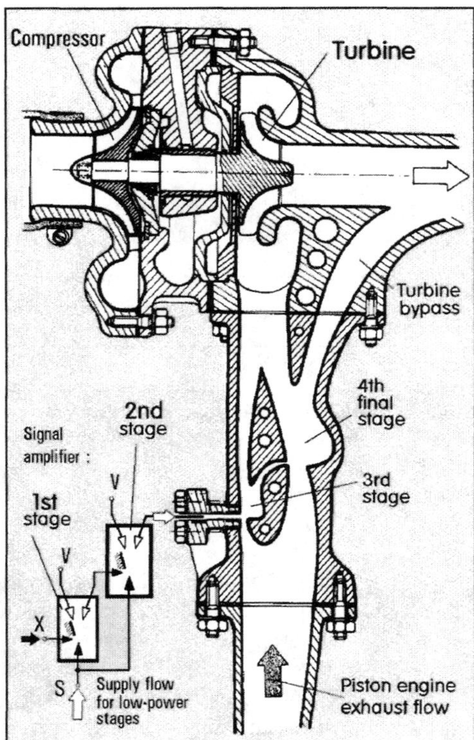

Figure 5.158 (Left, Top) A very large-scale fluidic monostable amplifier as an example of the final, power stage of an amplification cascade. Compared with the more usual mechano-fluidic valves for the control of large flows, this pure fluidic solution is cheaper, much more reliable, and resistant to heat and vibration. To design it with a small hydraulic loss, however, may be not easy; very much depends on successful design of the diffusers.

Figure 5.159 (Left, Bottom) Spatially very compact axisymmetric version of a fluidic power amplifier, operating otherwise in the same way as the planar version from Figure 5.158, the diverter control valve is here an integral part of a monolithic reactor, built into its entrance. Control input switches the main flow into the central bypass pipe.

Figure 5.160 (Right) An example of an amplification cascade handling in its last stages a really high-power gas flow: a no-moving-part alternative to the waste gate of a supercharger.

5.160 present three examples; by chance, the available illustrations all depict a monostable diverter valve, which is a rather exceptional operating mode, used less often than the bistable or proportional valves. In any case, the cost (especially if manufactured in a single casting operation) may be much lower than the price of a conventional valve with moving components (especially eliminating the expensive assembly). Figure 5.159 shows how the valve may be built in a single body with the device in which the flow is to be controlled (here a catalytic reactor). Figure 5.160 provides an example of an application (control of exhaust gas bypassing the turbine) utilizing the resistance to high temperatures (in operation, the amplifier body is glowing red hot).

5.8 SPECIAL CASES

Finally, at the end of the present Chapter 5 we discuss several interesting conversion devices that do not conveniently fit into any of the previous categories of sensors and transducers and yet deserve being at least briefly mentioned.

5.8.1 Sensing based on nuclear magnetic resonance

This method is based on detecting the spin of subatomic particles, electrons, protons, and neutrons in the nuclei of atoms. In some atoms the particle spins in the nucleus are all mutually compensated, so that no overall spin can be detected. Nuclei with a noncompensated spin have two possible orientations and therefore different energy in a magnetic field. There are slightly more nuclei whose magnetic moment due to spin agrees with the external field and are in the state with the lower energy. These nuclei may be excited to the higher level by a radio frequency pulse. As they subsequently relax, they release radiation, which is used to set up an NMR spectrum. In complex organic molecules, some atoms generate relaxational radiation in different ways when in different locations in the molecule (they are differently covered by the screen of electrons). Carbon and oxygen, unfortunately, are not detectable because the spins in their nuclei are compensated. It is, however, possible to identify, for example, ethanol (CH_3–CH_2–OH) in the fluid channel, because the different radiation of H atoms in the methyl group CH_3 and in the methylene group CH_2 form two characteristic peaks in the NMR spectrum.

The microfluidic device serving for NMR analysis usually looks like the example in Figure 5.161. The essential part is the coil for detecting relaxational radiation. In particular, alcohol could be reliably detected using such a device. The other necessary components are problematic; the source of the radio frequency pulse and of the magnetic field (which is required to possess a high degree of spatial homogeneity) tend to be inconveniently large.

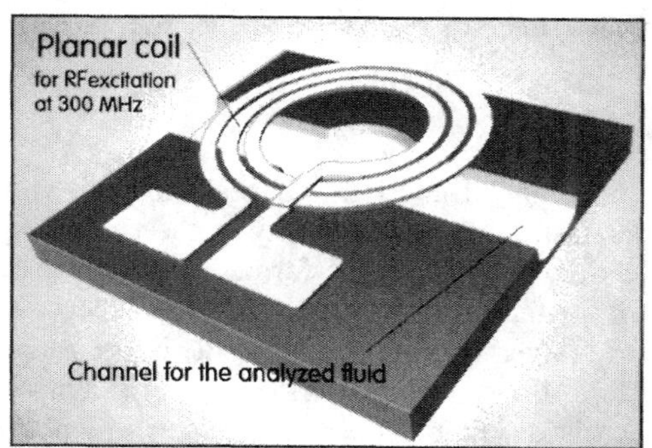

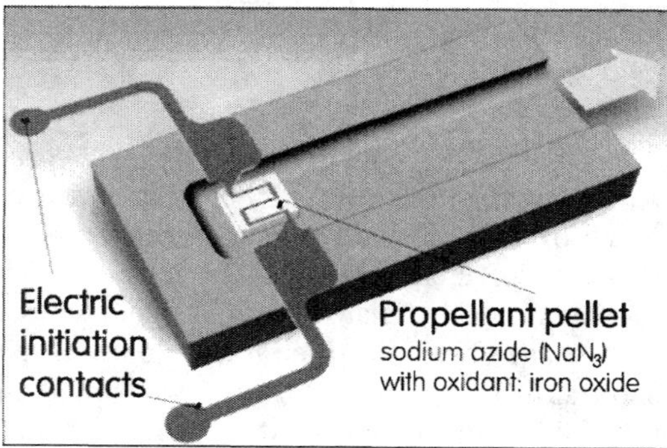

Figure 5.161 (Left) Fluid composition analyzer based on the nuclear magnetic resonance.
Figure 5.162 (Right) Micropyrotechnics: generator of one-time nitrogen flow pulse.

5.8.2 Micropyrotechnics

It may be useful to generate a one-time, large gas flowrate pulse by inciting an explosion of a small amount of explosive material (Figure 5.162). This is particularly of value for security systems that are expected to operate only once in their lifetime. An example may be car airbags, which actually use the same principle. The materials actually used belong into the category of propellants, such as those used in ammunition or rockets. Because they are manufactured in large quantities, these materials are rather cheap. Using the detonation phenomenon, the one associated with the propagation of an explosive shockwave at the speed of sound in a particular material – is not advisable. The substances, such as lead styphnate, in contrast to the propellants are unstable and difficult to control. Some typical propellants are glycidyl azide polymer and ammonium perchorate. Their energy density, typically 10^7–10^8 J/m^3, is actually lower than that of hydrocarbon fuels, but the energy may be released at a high power level in a short time. It is common to generate pressure of ~ 20 MPa within less than 3 ms. Pellets as small as 600 μm in diameter are used, capable of generating up to ~ 150 mW power in the form of almost pure nitrogen gas at a temperature of ~ 1 300 K, instantly cooling as it undergoes the adiabatic expansion. The high concentrated energy was used for purposes such as the transdermal injection of drugs and the sudden opening of a microvalve (rupturing of a membrane).

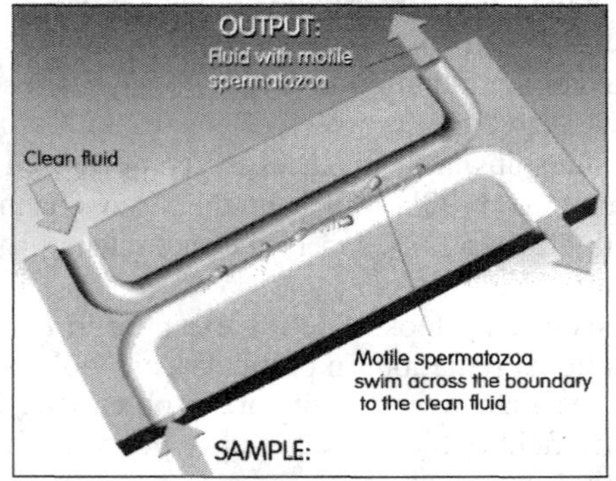

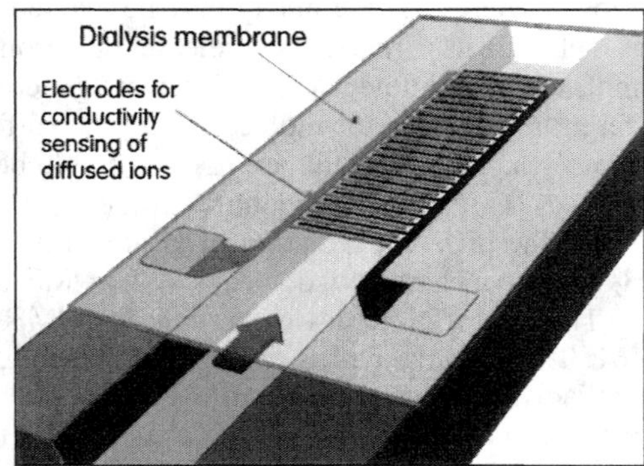

Figure 5.163 (Left) Microfluidic selection of healthy motile spermatozoa using their capability of swimming, at a preselected speed, across the thin laminar diffusion mixing layer.

Figure 5.164 (Right) Fluid composition sensor based on a conductivity readout of transport across a dialysis membrane.

5.8.3 Motility of micro-organisms

An exceptional microfluidic device is based on the capability of many microorganisms to swim. Some of them, in particular spermatozoa, swim rather fast, and this capability is discriminating criterion for their good, healthy condition. This selection is regularly performed for diagnostic purposes, and the relatively simple fluidic devices built for this task are one of the very successful ideas. The device in which it takes place (Figure 5.163) is very simple. Use is made of the rather short diffusion distance between two parallel laminar fluid flows (Figures 3.113 to 3.115). One of the fluids carries the tested organisms, and the other is a suitable clean fluid into which the organisms themselves move across the thin diffusion layer. Only healthy organisms with higher motility than the limit (adjusted by the flowrates and the length of the contact zone) are available in the output fluid.

5.8.4 Devices based on properties of special membranes

In the example shown in Figure 5.164, there are also two parallel fluid flows with one of them being a clean fluid. The electroconductive probe (with interdigital electrodes) senses in it the presence and amount of ions transported there by diffusion across a dialysis membrane from the other fluid. The mechanism of the transport simulates the transport of substances in and out of living cells through the cell's outer membranes; the dialysis membrane in this fluidic device is composed of regenerated cellulose.

The conduction method may be also used to sense the composition of multicomponent mixtures. The electrodes are used in parallel and are adjusted to sense only a certain substance by either placing them behind a selective membrane exposed to the fluid or by embedding the thin electrodes in a membrane containing the chemically sensitive material changing its electric properties (Figure 5.165).

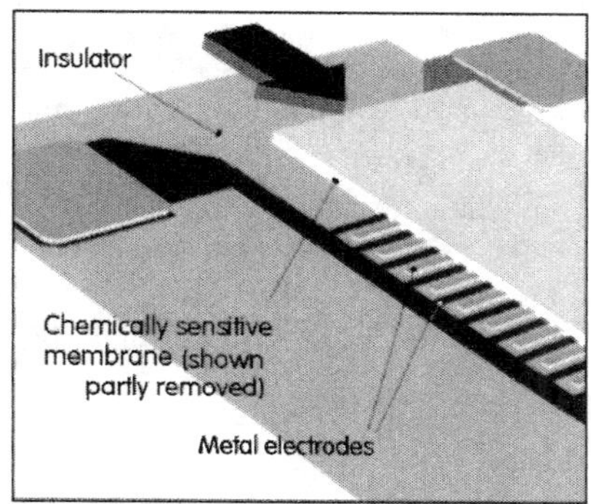

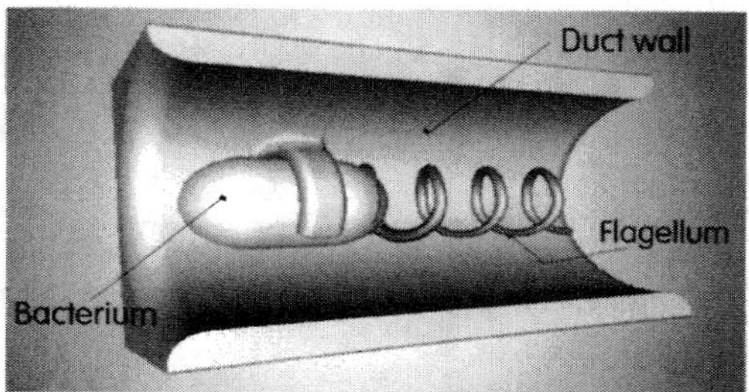

Figure 5.165 (Left) An example of an active membrane with embedded interdigitated electrode sensor for detecting chemical and biological species.

Figure 5.166 (Right) A detail of a pump in which the fluid flow in the duct is produced by flagella motion of captive bacteria.

This approach is also applicable in analyzing the practically nonconductive gas mixtures. Some organic polymers with substituted metal molecules change their conductivity with remarkable selectivity as well as sensitivity when exposed to a certain gas: for example, a monolayer of Cu-substituted phthalocyanine on 50 interdigitated Au electrodes 25 μm wide was demonstrated to react by changing the electric conductivity to the presence of NO_2 at concentrations as low as 0.5 ppm.

The array of electrodes with AC excitation may be used to react to changes of capacitance rather than the resistance. For example, a polyphenylacetylene (PPA) layer exhibits sensitivity to CO and polyetherurethane is sensitive to toluene. These *chemicapacitors* have been also applied in sensing humidity in air; devices measuring 5% to 95% relative humidity are available commercially.

An important role is played by semipermeable membranes, which are capable of detecting *biological* materials in medical and biological applications. They are often of a lipid nature and very thin, in the range of 4.5 ± 1.5 nm (i.e., thinner than the walls of living cells). Again, the sought-after property is the capability of extracting the analyte selectively so as to discriminate between interfering species. The small thickness helps in getting a fast response, but it makes the membrane vulnerable. Often, the membranes are therefore incorporated in relatively thick polymeric bulk membranes. Plastified PVC membranes, consisting of ~ 33 wt% of PVC matrix and ~ 64% plasticizer are typical. The electrical properties are determined by ~ 1–5 wt% ion selective ionophore and the actual active substance. Other high-polymer membrane matrices in use are made of silicone or polypropylene. When used as biosensors, the essential problem of membranes is their biocompatibility—not only they must be nontoxic, but they must also not trigger the coagulation processes (this may be suppressed by anticoagulants in the fluid, such as heparin, but this may cause other problems). The biological recognition additives are usually of enzymatic character.

One successful application of the membrane technique is the measurement of nitrates, the major contaminant of drinking water causing significant risks to human health. The detector uses anion-permeable membrane for selective (much faster than other anions) diffusion of nitrate ions and their detection, using the reduction on silver electrodes.

5.8.5 Fluidic pumps employing captive bacteria

The trick of how to harness bacteria to perform useful work is in fastening them to the walls of a microfluidic channel leaving their flagella, which they normally use to swim, free to move and pump the liquid through the channel (Figure 5.166). The fastening is made possible by the tendency of some, especially the pathogenic, bacteria to contact the tissue they are about to attack. Typical bacterial flagella are ~ 12 μm long and impart to the surrounding liquid a motion with a velocity of ~ 25 μm/s. Raising the temperature or adding glucose makes the bacteria pump faster.

References

[1] Nguyen, N.-T. and S. T. Wereley, *Fundamentals and Applications of Microfluidics*, 2nd ed., Artech House, Norwood, MA, 2006.

[2] Settles, G., "Fluid Mechanics and Homeland Security," *Annual Review of Fluid Mechanics*, Vol. 38, 2006, p. 87.

[3] Tesař, V., "The 'Reflection' of a Fluid Jet from a (Perforated) Wall—A Flow Visualisation Study," *FLOW VISUALISATION V.*, R. Reznicek (ed.), Hemisphere, New York, 1990.

[4] Syrotuck, W., "*Scent and Scenting Dog*," American Publ., Rome, NY, 1972.

[5] Peirs, J., D. Reynaerts, and F. Verplasten, "A Microturbine for Electric Power Generation," *Sensors and Actuators A*, Vol. 113, 2004, p. 86.

[6] Metz, P., G. Alici, and G. M. Spinks, "A Finite Element Model for Bending Behaviour of Conducting Polymer Electromechanical Actuators," *Sensors and Actuators A: Physical.*, Vol. 130, 2006, p. 1.

[7] Laser, D.J., and J. G. Santiago, "A Review of Micropumps," *Journal of Micromechanics and Microengineering*, Vol. 14, 2004, p. R35.

[8] [8] De Volder, M., et al., "The Use of Liquid Crystals as Electrorheological Fluids in Microsystems: Model and Measurements," *Journal of Micromechanics and Microengineering*, Vol. 16, 2006, p. 612.

[9] Zipser, L., L. Richter, and U. Lange, "Magnetorheologic Fluids for Actuators," *Sensors and Actuators A: Physical*, Vol. 92, 2001, p. 318.

[10] Tesař, V., "Způsob a zařízení k elektrickému řízení průtoku tekutiny" (Ways and Means for Electric Control of Fluid Flow – in Czech), Czechoslovak Patent No. 134391, October 1966.

[11] Yamahata, C., et al., "Plastic Micropump with Ferrofluidic Actuation," *Journal of Microelectromechanical Systems*, Vol. 14, No. 1, 2005.

[12] Pelrine, R., et al., "Dielectric Elastomer Artificial Muscle Actuators: Toward Biomimetic Motion," *Proceedings of SPIE*, Vol. 4695, 2002, p. 126.

[13] Rasisigel, H., O. Cugat, and J. Delamare, "Permanent Magnet Planar Micro-Generators," *Sensors and Actuators A*, Vol. 130-131, 2006, p. 438.

[14] Primmermann C.A., "Detection of Biological Agents," *Lincoln Laboratory Journal*, No. 1, Vol. 12, 2000, p. 3.

[15] Ceperley, P. H. "Travelling Wave Heat Engine," U.S. Patent 4,114,380, March 1977.

[16] Shin, D. D., et al., "Thin Film NiTi Microthermostat Array," *Sensors and Actuators A: Physical*, Vol. 130-131, 2006, p. 37.

[17] Harris, N. R., et al., "A Silicon Microfluidic Ultrasonic Separator," *Sensors and Actuators B: Chemical*, Vol. 95, 2005, p. 425.

[18] Rife, J. C., et al., "Miniature Valveless Ultrasonic Pumps and Mixers," *Sensors and Actuators A: Physical*, Vol. 86, 2000, p. 135.

[19] Lin, C.-F., et al, "Microfluidic pH-Sensing Cips Integrated with Pneumatic Fluid-Control Devices," *Biosensors and Bioelectronics*, 2006.

[20] Stone, B. M., and A. J. de Mello, "Life, the Universe and Microfluidics," *Lab on a Chip*, Vol. 2, 2002, p. 58N.

[21] Gardeniers, J. G. E., and A. van den Berg, "Lab-on-a-Chip Systems for Biomedical and Environmental Monitoring," *Analytical and Biaoanalytical Chemistry*, Vol. 378, 2004, p. 1700.

[22] Morgensen, K. B., H. Klank, and J. P. Kutter, "Recent Developments in Detection for Microfluidic Systems," *Electrophoresis*, Vol. 25, 2004, p. 3498.

Chapter 6

Applications

The objective of this chapter is to present ideas about practical uses of the fluidic devices discussed in the previous parts of this book. It should be emphasized that at the time of this writing the applications have been developing at a rapid pace. There are some, among those discussed here, that may soon prove to be impractical and become obsolete. Others almost certainly will emerge and become important, but so far are considered unrealistic. Despite this danger of the discussion soon becoming outdated, it is certainly useful, and it may even be inspiring to see how wide a spectrum of uses has been already suggested and tested.

As mentioned in several locations in this book, there is a characteristic trend of current development leading to progressively smaller sized devices. Some of the currently successful applications reveal the early date of their origin by using devices that are larger than the present fabrication technology would allow them to be made. Indeed, in many cases discussed here, the definition of microfluidics is met only with considerable stretching of the rules, and a better characterization of the devices would be to describe them as meso-fluidic rather than microfluidic. Nevertheless, it is considered a sign of current development in a technology field that is still far from being mature.

Perhaps an even more characteristic and immediately apparent sign of the early stages of the current development is the fact that most really successful businesses concentrate on rather simple cases. In fact, the financially most rewarding applications actually use just a single fluidic device rather than fluidic circuits or systems. A typical case is inkjet printer heads, with fluidics involved only on the very short path of the fluid—ink—moving from the supply vessel to an exit nozzle. Another case may be the successful use of the piezoresistor pressure sensors in the altimeters and barometers built into watches, a typical example of an "incomplete" system, from the fluidics point of view (as discussed in Section 5.1.3). In spite of the simplicity of the fluidics in these cases, their widespread use fulfills a very important role in paving the way and gaining experience with the manufacture and utilization of the small devices.

In line with the general trend to entrust signal and information processing to electronic circuits (perhaps made on the same chip), in some applications promising really substantial breakthroughs, fluidics is left to perform only rather mundane roles. Its job may be just to take care of bringing the investigated fluids to a detector array, while the essential task of producing meaningful conclusions from the generated signals is done by electronic data processing. This is the case with the various artificial "noses" used in increasing numbers, for example for objective food-flavor monitoring. Nevertheless, this unglamorous job does not make the presence of fluidics less important and deserving of attention in this survey; without it the nose would not work.

In general, the current state of development provides a very optimistic perspective. Some applications among the proposed or already developed ones indicate that microfluidics, in cooperation with electronics, optics, acoustics, and other MEMS, can indeed substantially change our world. They are likely to change the classic automobile by replacing its reciprocating engine (a primitive and wasteful machine, in spite of all its current sophistication). Microfluidics and MEMS devices are expected to change the character of health care by making possible continuous health monitoring of individuals and exact diagnoses of illnesses in the earliest stages. Microfluidics is also foreseen to be one of the main defensive weapon in the antiterrorist war. In the end, by making possible communication with the central nervous systems of animals, microfluidics brings not only the hope for a new relationship with our fellow living creatures, but also a possible danger of misuses of the sorts not deemed possible earlier.

6.1 SIMPLE SOLUTIONS

At the beginning of this chapter it may be useful to discuss several application examples characterized by simplicity due either to the use of a minimum number of devices or to following well-proven circuit principles of large-scale devices. The simplicity reflects the current early state of development. There is hardly any doubt that more sophisticated microfluidic circuits will follow, as they are likely to be demanded by the increasing complexity of the solved tasks. Nevertheless, the so-far financially most successful uses of microfluidics are the uncomplicated ones. This is far from being meant as a derogatory statement. Even the simplest cases are very useful, as they help us to accumulate experience with their characteristic features, in particular with small sizes and the microfabrication methods used to produce them. In some of the applications—like inkjet printing—it is the small size of the generated ink droplets and the consequent high resolution of the printing that made the application viable. Elsewhere, small size not only leads to performance improvement, but also can bring forward uses that until recently were considered to be unrealistic dreams.

6.1.1 Controlled injection of liquid droplets

The technology of generating individual droplets gave rise to the so-called inkjet printers (no real jet is, however, formed in most present-day devices) sold by the tens of millions each year. These printers completely relegated the earlier dot-matrix printers into oblivion. The main advance was in the increase in resolution – the droplets are small, between 30 μm and 100 μm in size.

There are two basic printing methods, the older continuous droplet formation and the drop-on-demand principle. The continuous method depends on the electrostatic deflection of the droplets, which are formed by driving the ink by high-pressure pump through a micronozzle; in this case there really is a jet formed that, however, breaks up into a spray of droplets immediately downstream from the nozzle due to the Kelvin-Helmhotz instability of the jet surface. Indeed, William Thomson (the first baron of Kelvin) himself patented the idea of the continuous method's use for printing in 1867. The first practical printer, however, was put on the market by Siemens in

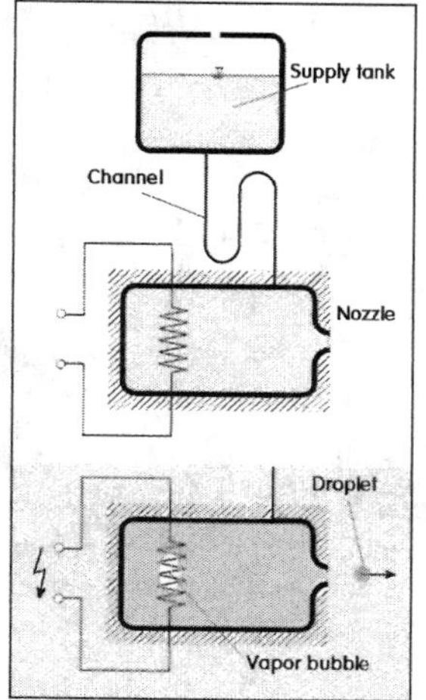

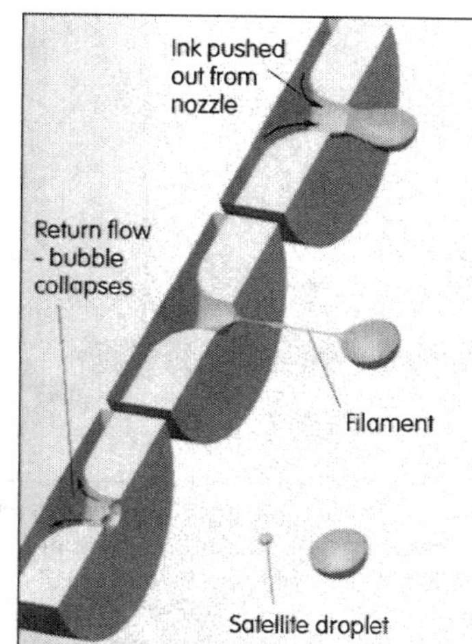

Figure 6.1 (Left) Schematic representation of the drop-on-demand inkjet printhead principle with droplet ejection by expanding vapor bubble, corresponding to the E/F transducer from Figure 5.65. The bottom (shaded) part of the picture shows the droplet formation in response to the input electric pulse.

Figure 6.2 (Right) Formation of the liquid droplet by the liquid expansion pulse. The growth phase (top) ends when the extruded drop forms a neck. This is stretched (center) until it becomes a thin filament, known already to Lord Rayleigh who photographed the phenomenon (calling it ligament) in 1891. When the droplet is finally detached (bottom), the filament, especially if it contains a local diameter increase, coils up into one or more smaller satellite droplets.

1951. The method is still used for code-marking of products in industry (the "best-before" dates on cans is the characteristic example). The breaking-up of the jet is made more regular by piezoelectrically introduced oscillation. To make the droplets deflectable, they pass through a charging electrode, but not all of them; interspersed neutral droplets were found useful in suppressing droplets' mutual repulsion. Undeflected and neutral droplets hit a collector that returns them to the tank. The advantage is high achievable droplet velocity ~ 50 m/s so that they can fly quite a long distance, making possible printing on round and uneven objects. Since the nozzle is in constant use, it does not have a tendency to clog. This makes possible the use of inks based on volatile solvents. Not only do they dry fast, but before that they partly dissolve the top coating layer on the impingement surface so that they are not easily removed.

Printers sold as computer accessories operate with the drop-on-demand method using ejection by a liquid pulse. The repetition frequency of droplet formation is up to 5 kHz. The typical nozzle diameter is 60 μm– 80 μm, the droplet size being close to the nozzle size. Most printers on the consumer market produce pulses by the electric input signal, generating a vapor bubble, (hence the trade-name *Bubble-jet* used by Canon), as shown in Figure 6.1. Note the channel separating the pressure chamber from the ink supply manifold. Its inertance (Figure 2.84) prevents losing the E/F transducer's power by moving uselessly liquid back into the supply. The ink is usually water-based, with glycol and dyes or pigments. The usual ejection velocity is from 7 m/s to 10 m/s. The heater is made by a deposited Ni layer, with typical resistance of 20 – 40 Ω. In most designs, the heater is located at the other side of the pressure chamber from the nozzle. Less common is the "side shooter" location, with nozzle in one of the side walls.

The other method of generating a pulse uses a piezoelectric E/F actuator, usually of the ceramic type (see Figure 5. 58) and usually based on lead zirconium titanate (Figure 6.3). This is more expensive, but it allows a wider choice of inks with a higher quality in the resultant prints; it is therefore typical for industrial and commercial printers—the exception being the Epson

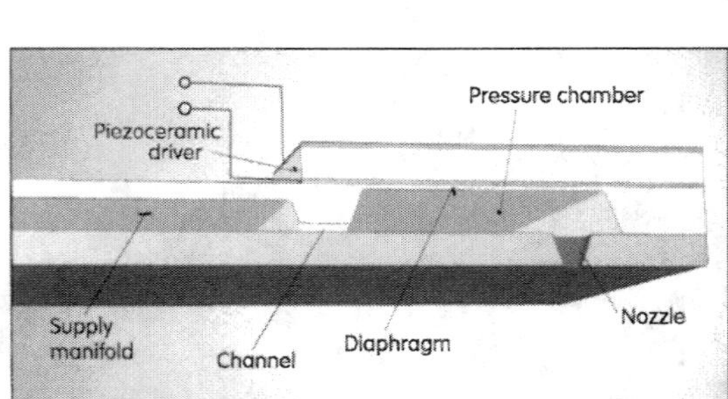

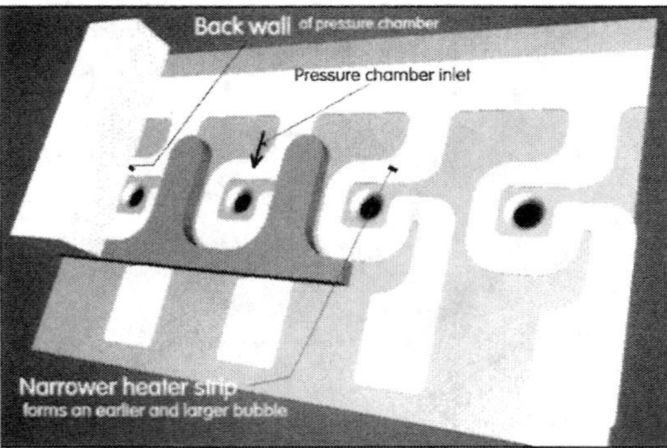

Figure 6.3 (Left) Apart from the "bubble-jet" F/E conversion producing the flow pulse by vapor bubble, some manufacturers also produce printheads for the ink droplet printing with the pulse generated by a piezoelectric driver, as shown here. This is more expensive and therefore usually made more permanent than the version in Figure 6.1 (with disposable heater foils).

Figure 6.4 (Right) The "tail-cutter" layout, originally proposed by Theng, Kim, and Ho (University of California), aims at suppressing satellite droplet formation. The heater strip is arranged into a loop around the nozzle exit. It forms a bubble ring collar, the growth of which pinches off the droplet.

consumer-level printers that also use this principle. The head is not replaced every time the ink runs out. The consumable costs are lower, and the head itself can be more precise than a disposable version.

The current development challenge is suppressing the generation of satellite droplets that accompany the main droplet (Figure 6.2) smearing and blurring the produced image. One approach is the "tail cutter" idea (Figure 6.4). Note the narrower heater strip at the inlet into the pressure chamber. The bubble there is formed earlier and blocks the way back to the supply manifold. Later on, the bubble loop is formed around the exit hole, terminating bubble growth before it can progress to formation of the neck, from which develop filaments and, later, satellite droplets.

The technology of droplet formation, having matured in mass-produced printing cartridges, tends to apply its advantages in other fields than printing on paper. Characteristic is the use of the so-called "functional inks," exhibiting such properties as electrical conductivity. Electrical circuits are made by inkjetting conductive strips on polymeric substrates and organic materials. Successful applications include displays with the inkjetted transparent electrodes, replacing the current laborious sputter-coating by tin-doped indium oxide machined to form conductive paths by photolithography (one of the factors making inkjetting an interesting possibility is the increasing price of indium, not needed in this method). Another use is manufacturing solar cells, with all the components deposited in a single inkjet machine, rather than in a number of sequentially used different operations. Conductive ink containing silver nanoparticles (~20 nm diameter) were even used to manufacture thin-film transistors by depositing the electrodes on doped n-type silicon wafer with 0.2 μm thick SiO_2 insulation layer. In another attempt, the active transistor material was a semiconducting polymer, again deposited between the ink-jet printed silver electrodes.

Interesting applications are also found in the medical field, a typical example being tests for pregnancy or diabetes using enzymes deposited by printing and covered by a protective layer, eliminating airborne contamination. The test then consists of wetting the strip with body liquids (saliva, urine). Alternatively, still in the development stage, are inkjetted electronic circuits on the same strip, supplied by the electrochemical potential between the wetted electrodes. These can perform even complex evaluation procedures.

Recently, it was demonstrated that it is possible to deposit living cells by the inkjet printer. They remain alive not only while passing through the nozzle, but also when reaching the substrate. This is one of the ways towards *tissue engineering*, building organic materials consisting of live cells within "scaffolds" of biocompatible fibers. Nutrients later jetted into the structure allow the cells' DNA to control the process of cell multiplication and joining.

6.1.2 Cooling garment and the problem of portable power units

The recent shift of the centers of military operations to countries with hot climatic conditions (Afghanistan and Iraq) highlighted the question whether—and how far—can modern technology reduce the thermal stresses to which military personnel are exposed. The reduction of fatigue and an increase in the duration of missions may be particularly significant, if the personnel are required to operate in protective suits – from the bulletproof vest and helmet, to the nuclear, biological, and chemical protective clothing, up to full explosive ordnance disposal suits, inconvenient to wear even in mild climatic conditions. Then there are, of course, other cases in which hazardous duty personnel must wear protective clothing: firefighters, workers exposed to heat or to aggresive chemical materials, police wearing body armor, chemical spill cleanup teams, and others. Many of them find enduring the garment almost as tough as doing the work.

There are other reasons why a garment provided with a cooling system is beneficial, for example for therapeutic purposes. A typical case is multiple sclerosis, the disruption of myelin in the nervous system. Studies proved that a decrease in temperature improves the signals' transmission across damaged nerves. In this case, the solution may be easier because the protective suit does not necessarily have to be fully autonomous. A similar situation is found in flyers' cooling suits and for astronauts where, at least to a certain degree, the garment may be connected to an external cooling unit. In these cases effective solutions are known and in use.

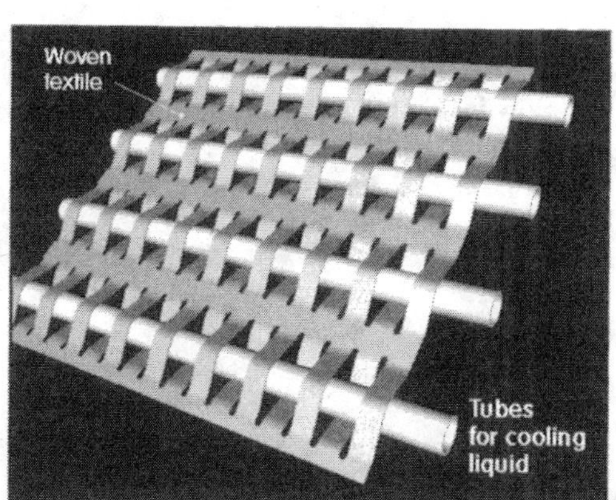

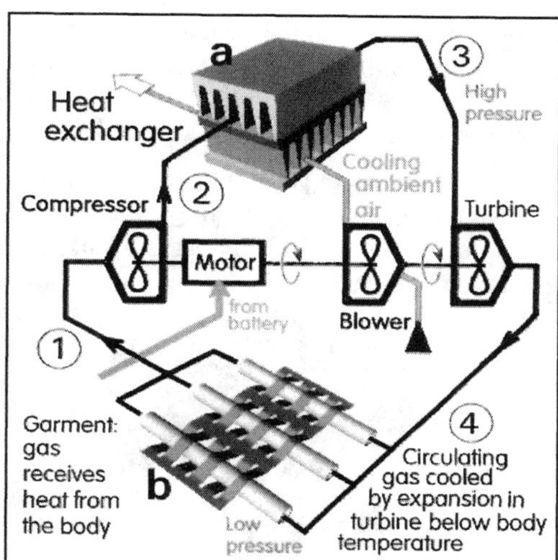

Figure 6.5 (Left) The cooling garments remove heat from the body by low-temperature fluid (in a nonautonomous system preferably a liquid, such as a glycol mixture) flowing through polyvinylchloride tubing distributed in the inner textile layers.

Figure 6.6 (Right) Schematic layout of a cooling system operating with gas—air—circulating between the garment tubes and a heat exchanger in which the heat is rejected into the ambient air. For the proper direction of the heat flux, the air has to be compressed from **1** to **2**, to increase its temperature above ambient temperature and then it has to be cooled by expanding in the turbine between **3** and **4**, as shown in the schematic diagram in Figure 6.8.

The problem with autonomous cooling for unrestricted movement across terrain is much more difficult. There are no real troubles with the garment itself. It may use the proven layout with the integrated polyvinyl chloride tubes (Figure 6.5), as used in glycol circulation cooling for astronauts and flyers. Because of higher expected physical activity levels, it is necessary to adjust the proper humidity transmission properties to avoid sweat condensation on the tubes.

The core of the problem is the sheer weight of the system and of the available stored energy needed to drive the machinery. It is estimated that at least 350 W cooling power is necessary for the effect to be felt. At the same time, ~ 5 kg is the maximum mass of the system considered acceptable for an 8-hour operation. The available energy storage in batteries makes this target beyond reach if energy were used in a direct manner. The energy density of even the most effective, lithium-ion batteries is 137.5 W/kg/h, so that the batteries alone would weight 2.54 kg for each hour of the mission. A promising solution uses liquid hydrocarbon fuels, with their ~130 times higher energy-storage density, and also the heat-pump approaches, in which the generated cooling power is larger than the energetic input into the system (which only covers the difference between the thermal input and output). The question is how to design a light and at the same time very effective cooling system. Attempts using standard technology almost uniformly failed. Typically, the mass of a conventional 350 W capacity 8-hour cooling suit with fuel using 1999 technology was estimated in [1] to be ~ 10 kg, twice the value considered acceptable. In fact even the effort required to carry a 5 kg cooling system – in addition to the other equipment the personnel has to carry – almost eliminates the benefits obtained by the cooling.

However, a substantial proportion of this weight represents heat exchangers and mass transfer units. Due to the basic square/cube law of Figure 1.17, these may be made much lighter using the microfluidic technology. The cooling power per unit volume of conventional cooling systems is about $20 \cdot 10^{-6}$ W/mm^2, while tests with microfabricated devices indicated $1.25 \cdot 10^{-3}$ W/mm^2, an improvement by a factor of more than 60 [1]. This improvement seems to place the cooling-garment idea within the reach of what is technically possible. Even so, this application of microfluidics still awaits its full commercial success.

6.1.2.1 Reverse Brayton cycle

The conceptually simplest version of the heat-pump cooling circuit is presented in Figure 6.6. This version uses air as the circulating coolant, including the flow inside the garment tubing. To remove heat from the wearer's body surface, the air in the tubes has to be at least 7–12°C cooler, not much more, however, especially if the whole body cannot be covered with the cooling: the

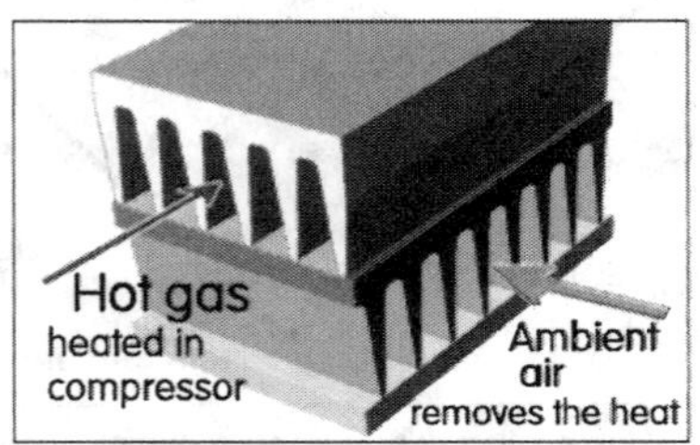

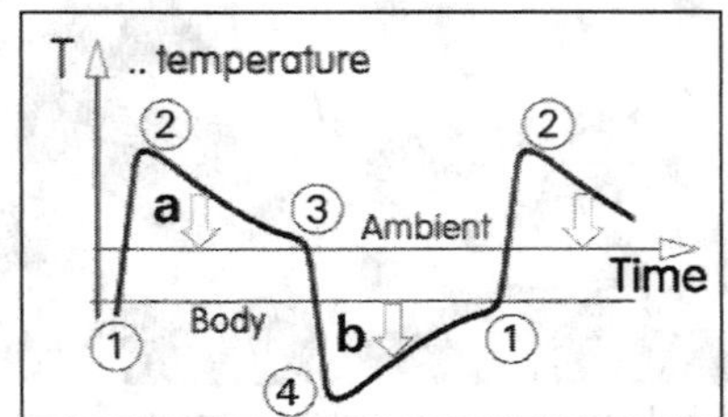

Figure 6.7 (Left) A detail of the microfluidic heat exchanger, extremely compact and light, which can make the circuit from Figure 6.6 a realistic proposal, though considerable development is necessary to decrease the inconvenient weight of the machinery aggregate (compressor, turbine, blower, and the electric motor to make up for the difference between the small power obtained from the turbine and the power demands of the compressor and blower).

Figure 6.8 (Right) Schematic representation of the temperature changes of circulating air in the inverse Brayton cycle presented in Figure 6.6 (named after George Brayton (1892), an American engineer who developed the cycle in 1872).

difference relative to the uncooled parts would cause serious discomfort. The decrease in the air temperature is achieved by an expansion prior to entry into the garment. This expansion takes place in the turbine. It is a process that happens so fast that there is no real opportunity for a significant heat transfer. Air performs work (by displacing the downstream air with which it is in contact) at the price of losing its internal thermal energy. The decrease of energy means a decrease of temperature to which the internal energy is in direct proportion.

As the air absorbs the heat in the garment, its temperature rises and the cooling effect decreases (Figure 6.8). When the temperature reaches the state marked as **1** in Figures 6.6 and 6.8, the air must be removed from the garment tubes. It passes through the compressor (Figure 6.6). Again, the compression process is nearly adiabatic, with only negligible heat transfer. As a result, the pressure increase is associated with a rise in temperature. It reaches the compressor exit (point **2** in Figure 6.8) at a higher temperature level than the ambient temperature, however high the latter may be in the hot climate. This temperature difference makes possible the transfer of heat into the ambient air. This essential step is made in the heat exchanger, which is usually of the recuperating type (Figure 6.7). To make the heat-removal part of the apparatus more compact, the ambient air is usually blown past the heat-exchanging surfaces by a blower (Figure 6.6). The temperature of the circulating air decreases, and when it reaches point **3** at the exchanger exit, the cool but still pressurized air is again left to expand and cool in the turbine, leaving it in state **4**, thus closing the cycle. Of course, the power delivered by the turbine does not suffice for driving the compressor as well as the blower (the rotors of both are shown in Figure 6.6 arranged on the common shaft). The power deficiency is made up by a mechanical drive. In Figure 6.6 this is shown as done by an electric motor. This may be a suitable solution only for short durations. As shown in Figure 6.11, the only suitable solution for the specified 8-hour operation at present is to use a combustion engine; in [2] it was a small reciprocating engine. Even this, with the corresponding tank holding the necessary fuel, was found in this design exercise (as will be shown later in Figure 6.11) to be too heavy for the 8-hour goal, not mentioning the obvious disadvantages (such as noise, vibration, general unreliability) of the engine for the proposed military use.

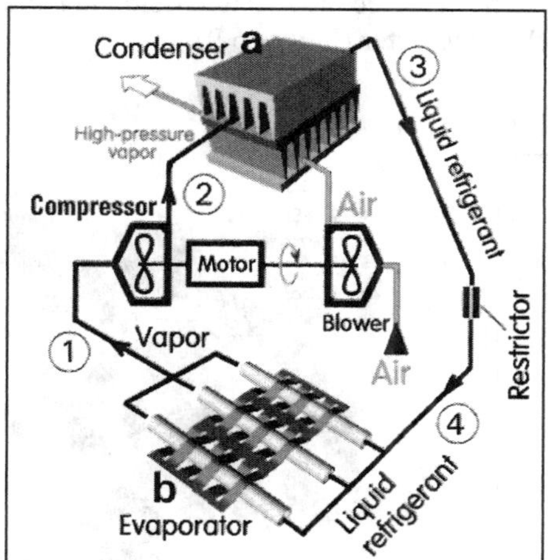

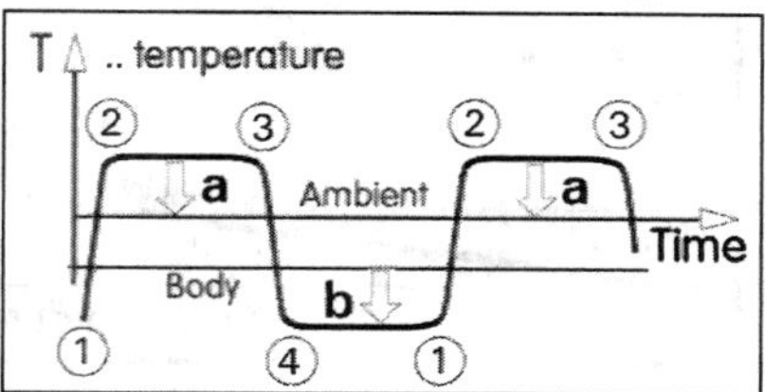

Figure 6.9 (Left) The Rankine evaporation cooling cycle is thermodynamically more effective (since the refrigerant temperature does not drop during evaporation and condensation) and also the heat-transfer coefficients are higher than for circulating gas. The power obtainable from expanding liquid in a turbine is negligible, so that the turbine is usually replaced by a cheaper, smaller, and lighter restrictor (such as an expansion valve).

Figure 6.10 (Right) Analogous to Figure 6.8, schematic representation of the temperature changes of the circulating refrigerant in the inverse Rankine cycle, Figure 6.9, shows the favorable consequences of the lower degree of freedom during the phase changes: the temperatures in the evaporator and condenser remain constant during the heat transfer.

6.1.2.2 Reverse Rankine cycle cooling

The Brayton gas cycle has the advantage of using a cheap and light coolant, but its basic concept is not very efficient from the thermodynamic point of view. This is reflected in the fact that usual household refrigerators operate on a somewhat different principle. It is the cycle with phase changes of the circulating refrigerant (Figures 6.9 and 6.10). Since during both evaporation and condensation, as shown in Figure 1.42, thee degree of freedom of fluid states is reduced, with the result of temperature and pressure being mutually dependent. Until all the fluid undergoes the change, temperature cannot decrease without a simultaneous decrease of the pressure, which, however, remains essentially constant (apart from small and insignificant hydraulic pressure loss). The advantages of the constant-temperature heat input as well as rejection may be expressed theoretically as the better approximation to the ideal Carnot cycle. In the present context it may be sufficiently explained by the constant temperature differences across the heat exchanging walls in the temperature history of the recirculating refrigerant in Figure 6.10 – and therefore a more effective heat transfer across the whole of the exchanger. This higher efficiency of the cycle is accompanied by the beneficial higher heat transfer coefficient of the (condensing or evaporating) liquid. Refrigerants with suitable range of phase change temperatures are available and cheap (mass produced). Due to reasons which will be apparent in the next section, those considered for the cooling garments are either water or ammonia. In the first case the expansion to the state 4 before entry into the suit has to be to very low pressure so that the water evaporates at the temperature as low as that of the human body. This may be helped by filling the evaporator with a neutral gas so that water properties correspond to the low partial pressure. On the other hand, ammonia is gas at room temperature (boiling point at -33°C) and the system has to be pressurized so that ammonia becomes in state 4 a liquid evaporating when the tubes are in contact with the body. In the Rankine cycle (named after W. Rankine, 1820–1872, who studied it in connection with steam engines – the cooling was invented in 1846 by F. Carre) the pressure decrease from the state **3** to **4** takes place in a liquid phase, with a very small volume change (Figure 1.42). The corresponding expansion work performed in a turbine would be too small to justify the expenses and complexity and most of the considered systems are reconciled to dissipating the pressure difference, using for the expansion just a simple flow restrictor (Figure 6.9).

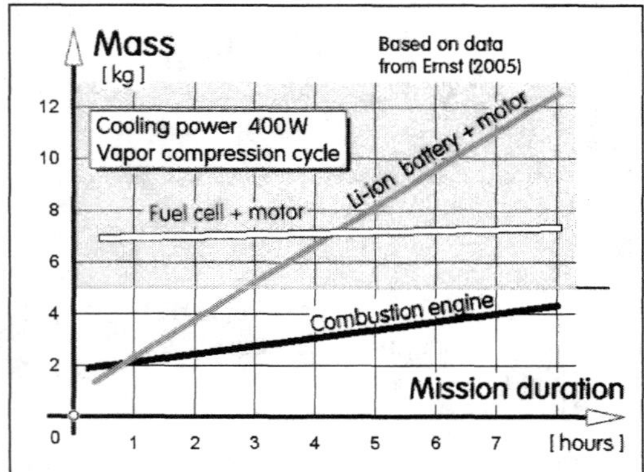

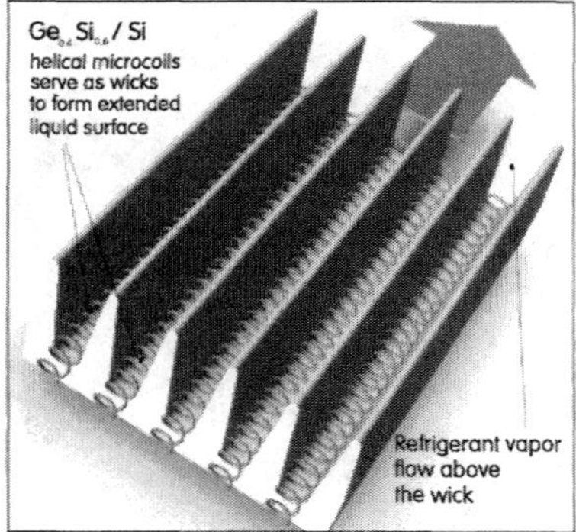

Figure 6.11 (Left) The basic reason why, so far, none of many known attempts to make a wearable cooling garment has been really successful: all known alternative compressor drives are too heavy. (*After:* [2]).

Figure 6.12 (Right) An example of detail of the absorber. In the channels, the refrigerant vapor comes into contact with the carrier liquid into which it is absorbed. For intensification of this process the surface of the liquid exposed to the vapor has to be as large as possible, here it is increased by the liquid stretched by capillarity on the wicklike helical structures.

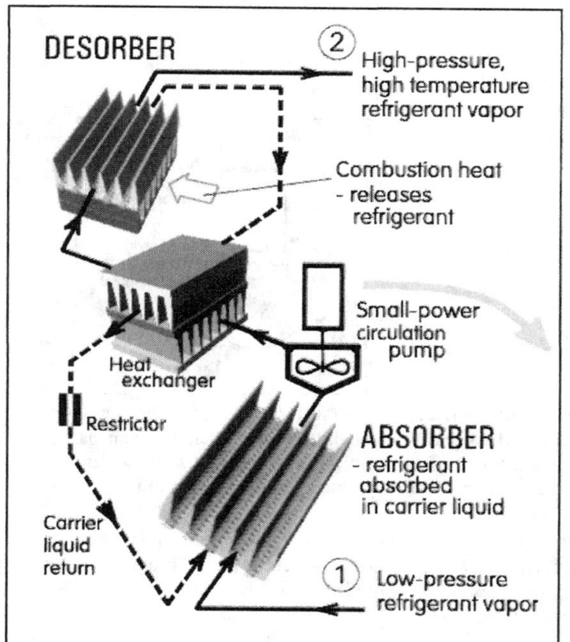

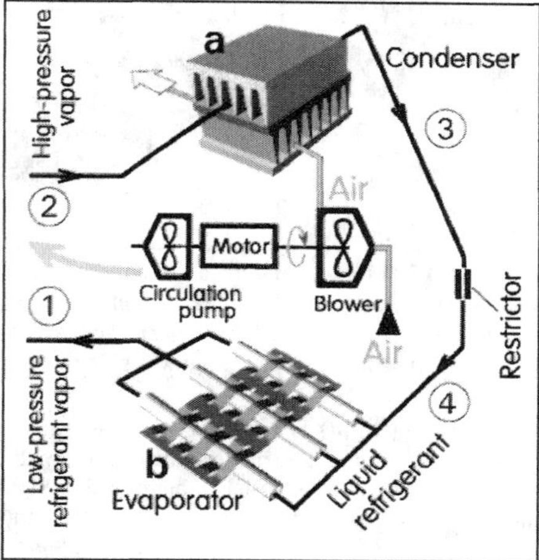

Figure 6.13 (Left) In the absorption cooling system, a subcycle with a circulating carrier liquid substitutes the compressor of the Rankine cycle in Figure 6.9. The absorption system needs more heat and mass transfer devices, but these, in contrast to the compressor, may be small and light, profiting from the increase of efficiency from the microfabrication.

Figure 6.14 (Right) The main (refrigerant) circuit of a cooling cycle with absorption remains exactly the same as in the Rankine evaporation cooling cycle (Figure 6.9).

Again, certain mechanical work (the difference between the thermal input in the garment and the thermal output in the condenser) is needed to drive the compressor. Although functioning demonstration models were successfully built (e.g., [2]), operating according to the schematic representation in Figure 6.9, the experience has been generally frustrating – again, mainly because all of the available machinery (Figure 6.11) for driving the compressor is too heavy.

6.1.2.3 Absorption cooling

Much more promising than the cooling with rotating machinery is absorption cooling (Figures 6.13 to 6.15). It was invented in 1922 by B. van Platen and C. Munters, who were at the time students in Sweden. In principle, the compressor (Figure 6.16) is replaced by the action of a liquid subcircuit that transports the refrigerant vapor to the high-pressure part of the cycle by utilizing its absorption in a carrier liquid. Practically only two pairs of refrigerant and carrier liquid are used: either water as the refrigerant and lithium bromide as the carrier and absorbent, or ammonia as the refrigerant and water as the carrier.

In the original Platen and Munters stationary absorption refrigerator, the carrier liquid with the absorbed refrigerant is recirculated by a bubble pump, in principle a version of airlift pumping (liquid column rising due to the lower effective density obtained by introduced vapor bubbles). This is impractical for the cooling garment application because there would be no steady direction of gravity. The garment system (as shown in Figures 6.13 and 6.14) contains a mechanical circulation pump. This pump's energy input to the liquid is actually insignificant. The primary

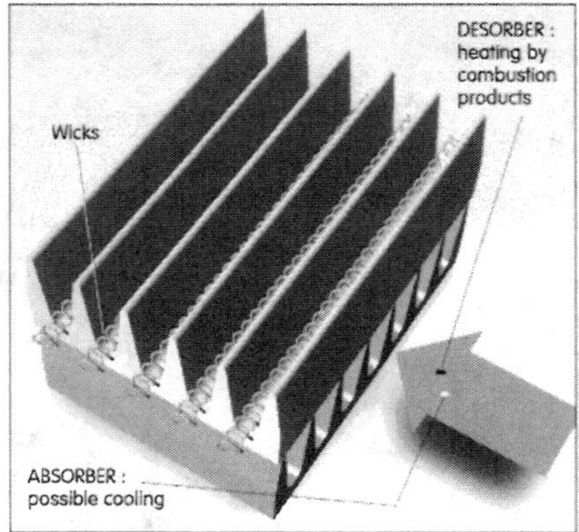

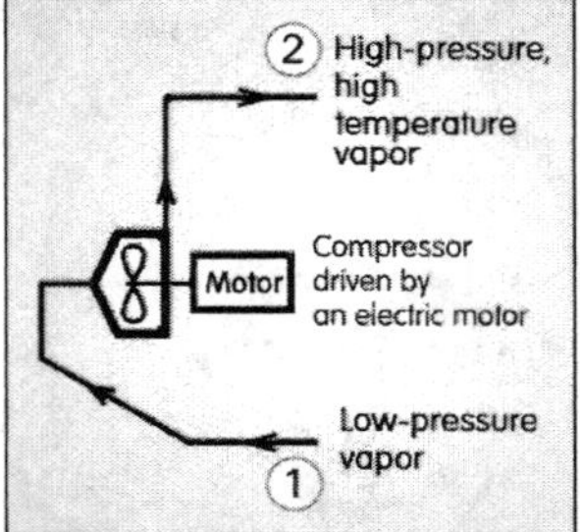

Figure 6.15 (Left) A small detail of an example from the microfabricated mass-transfer devices, the absorber and desorber (which differ only in heat brought either in or out by fluid passing through the other system of channels, the device also acting as a heat exchanger). This layout again uses the thin helical objects to act as wicks for the carrier liquid at the bottom of microfabricated grooves. In the desorber, the driving energy for the processes is input in the form of combustion heat.

Figure 6.16 (Right) The compressor – essential component of the compressor cycles (Figures 6.6 and 6.9), here shown schematically – cannot be made significantly smaller and lighter. This is the main obstacle to further progress. Its role may be taken over by the absorption subcircuit (Figure 6.13), which not only can profit from the increased transfer rates in microfabricated components but also operates without moving parts (apart from the small circulation pump).

energy input into the cycle is combustion heat transferred in the desorber, which may be made by micromachining according to the idea presented in Figure 6.12. Combustion in a simple burner is, of course, a much lighter and less annoying driving method than using the machines mentioned in Figure 6.11 and is probably the best route to follow in designing a garment. The blower for the condenser need not be used (it is, after all, also absent in domestic refrigerators) and the only remaining moving-part unit, the circulation pump, may be replaced according to Figure 6.17 by the no-moving-part fluidic pump consisting of the self-excited alternator from Figure 5.153 and the rectifier built as shown in Figure 5.152. This proposal has not yet been tested. Another suggestion (shown in Figure 6.15) is the possible layout of microfabricated mass transfer devices (absorber and desorber) with the wick structures at the bottom of the channels. The absorbing liquid is spread on the wick, offering for the mass transfer a very large surface.

6.1.2.4 Integral power solutions

The proposal according to the combined Figures 6.12 and 6.15 seems to be the most promising solution to the cooling garment problem. Nevertheless, some of the recently published ideas suggest other solutions may be also attractive, especially because they consider the whole question from a wider angle. In particular, these approaches view cooling as an integral part of the general problem of the needs of military personnel. A modern soldier in the field consumes a considerable amount of electric power to drive all those items of electronic equipment gradually becoming necessary for survival. After all, people in their everyday civil lives, in increasing numbers cannot imagine their existence without a cellphone and a personal digital diary if not a laptop computer. A soldier out in the field needs a more powerful and sophisticated communication system, an infrared night-vision equipment, a global positioning unit, a torch, a laser sighting device, and perhaps electronically stabilized binoculars – apart from the equipment needed by specialists. The electric supply for most of them is batteries of different sorts chosen independently by designers of these individual equipment items. Since he cannot jeopardize his survival by any of these batteries

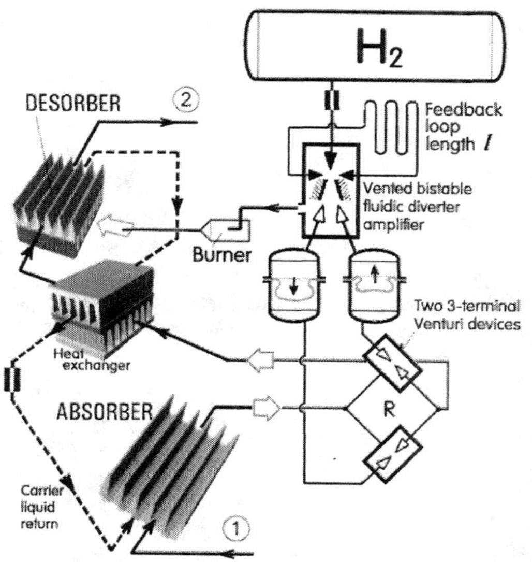

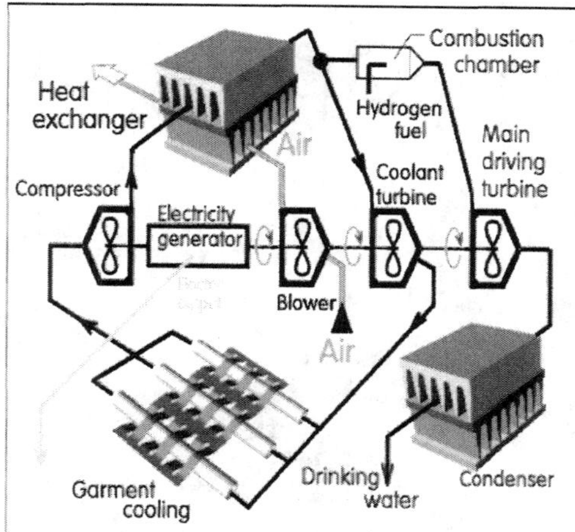

Figure 6.17 (Left) Using a fluidic no-moving-part circulation pump in the absorption subcircuit is a way toward a cooling system characterized by a total absence of any mechanical drives. The system is mainly driven by the chemical energy released by combustion in the burner. The proposed fluidic pump handles much smaller amounts of power so that it may be driven by the expansion of gaseous fuel before it enters the burner.

Figure 6.18 (Right) The other promising solution to the autonomous wearable cooling garment problem is the use of an air cycle driven by a combustion turbine. While less thermodynamically effective than the evaporator cycle, there is an advantage of generating electricity useful, for example, for a computer or other electric appliances. Also useful may be the production of drinking water.

running out, he carries spares. A special forces soldier is reported to lug along 12 kg of batteries on a three-day mission. It is a reasonable idea to adapt the driving machinery for the cooling battledress to incorporate a more powerful electricity generator and a small gas turbine driven by hydrogen (or even by a hydrocarbon fuel) as a common supply source for all of these items. In that case, the proposals revived the Brayton cycle idea. In fact, condensing the water vapor in the turbine exit may produce drinking water, another item the soldier is otherwise expected to carry in a hostile hot-climate terrain. This reasoning may lead to the system presented schematically in Figure 6.18.

The other alternatives, electricity generation by fuel cells (discussed in Section 6.2.2) or by a thermoelectric generator (see later Figure 6.24) cannot currently compete with the combustion-driven gas turbine, in the overall efficiency and weight aspects.

Of course, the turbine and compressor—the basis of the proposed machinery—also cannot remain at their current development stages, which were inherited from large-scale devices. A qualitative change is needed, and microfluidic technology is likely to provide help in this direction, too. Not impeded by the inertia of reciprocating components, no fundamental obstacle limits turbomachinery development toward extremely high rotation speeds. Typical considered values are on the order of millions of rpm. This results, on the one hand, in sufficiently high Reynolds numbers to avoid large viscous losses. On the other hand, this may result in sizes so small that it is reasonable to consider manufacture by the planar methods typical for microfluidics. The turbine thus made was already shown in Figure 1.8 (a typical power level for a microturbine etched in silicon is 50 W obtained with a 4 mm rotor diameter). The compressor essentially

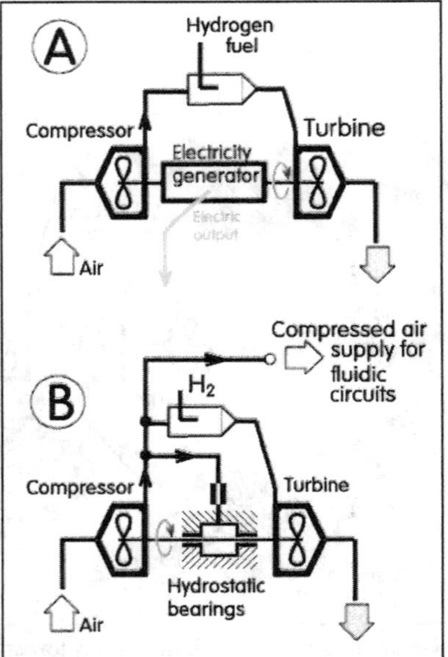

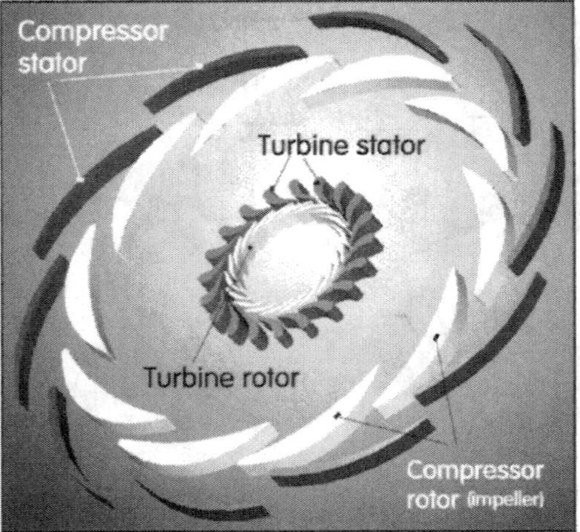

Figure 6.19 (Left) Circuit diagram representations of: A – a hydrogen fueled combustion-microturbine electricity generator and B – a compressed air generator.

Figure 6.20 (Right) Blade systems of the compressor-turbine aggregate of the compressed air generator (Figure 6.19B) with both systems formed by microfabrication techniques on the same side of the common rotor plate. The larger size of the compressor reflects the situation in which only a small percentage of the compressed air progresses to the turbine.

follows the idea of the centrifugal pump shown in Figure 5.9. Figure 6.19 presents cases in which it is the portable power unit that is of primary interest, for climatic conditions not requiring a cooling garment (and perhaps requiring an electrically heated garment). There are applications in which the primary target may be the generation of compressed air, for example, for use in fluidic systems. This leads to a configuration without the electricity generator (though most foreseen uses are likely to need both outputs). A typical required power level is around 500 W. In the schematic representation in Figures 6.18 and 6.19, the two main parts are shown arranged back-to-back on the common shaft, as is the customary design in aeroengines. Experience has shown, however,

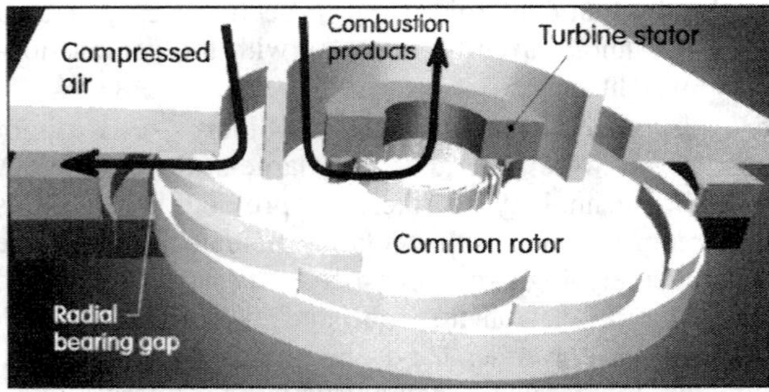

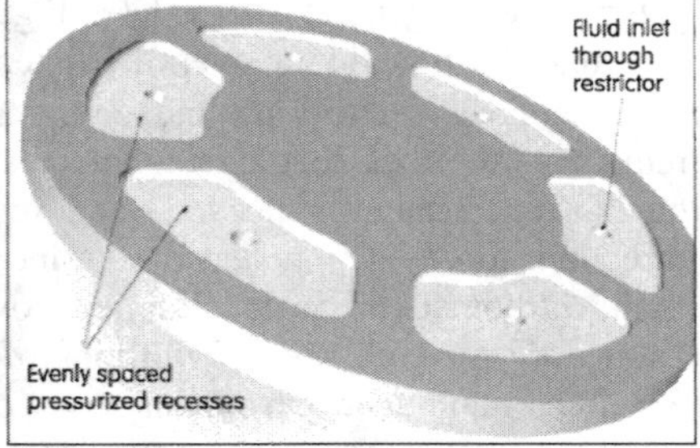

Figure 6.21 (Left) Layout of the aggregate from Figures 6.19B and 6.20 with indicated paths of air and combustion gas.

Figure 6.22 (Right) An example of the "hydrostatic" bearing plate supplied with compressed air to support the rotor from the bottom, unbladed side.

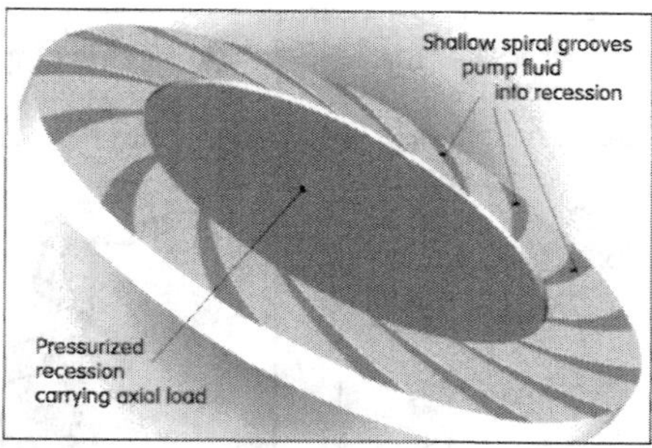

Figure 6.23 Another alternative of the axial (thrust) bearing for supporting the rotor from Figure 6.22. A pressurized air cushion is formed under the rotor by moving the air in the spiral grooves radially inwards. The air is impelled by viscous effects on the bottom side of the rotor.

that more suitable for microfabrication is the layout shown in Figures 6.20 and 6.21. Not only does the one-sided etching make both blade systems in a single operation, but it ensures their concentricity, which is essential for the assumed high speeds. Since no microscale ball bearings can be made to oppose the high pressure forces at the typical 10^6 rpm range, the rotor has to be supported by pressurized gas bearings (Figures 6.22 and 6.23), the design of which [4] is one of the major challenges, especially considering the requirements of stability and the suppression of the tendency toward precession movement when the whole portable assembly is moved. The supporting gas layer has to be divided into a number of separately supplied pads, each forming a "pressure divider" (Figure 5.18) with the pad in the role of the variable restrictor: if the rotor tilts, increasing the gap above one of the pads, the gas pressure decreases there and increases in the pad on the opposite side where the rotor gets nearer to its support. Of course, the pressure-supply system complicates the design. This may be evaded by using the passive bearings with spiral grooves (Figure 6.23), a very promising alternative at the small scales. The problem, perhaps requiring a combination with an active externally pressurized layout, is avoidance of dry friction contact when the turbomachine starts running. The bearing retaining radial position of the rotor is on the rotor's circumference (Figure 6.21); its operating conditions are particularly severe [4] and may also require an active type of support with several separate pads.

The other component that cannot simply be scaled down from the large-scale gas turbine is the combustion chamber. Due to the square/cube law, the heat loss to the walls requires a different design principle, preferably with the escaping heat used to warm the fresh fuel/air mixture. This is seen in the "swiss roll" layout in Figure 6.25, where, however, the escaping heat is used only in the plane of the double spiral (and left to leave in directions perpendicular to this plane). At any rate, an efficient combustion reaction requires using a suitable catalyst on the combustion-channel walls. A suitable material able to withstand high combustion temperatures is alumina (Al_2O_3).

6.1.2.5 Why not thermoelectric generators?

The heat escaping from the burner may be used to generate electric output directly, with no moving parts, using thermoelectricity (Figure 6.25). It is obvious that if this were an efficient method of using combustion heat, turbomachinery could be left as superfluous. An experience with a device essentially similar to the one in Figure 6.25 was described in [5]. There were two channels of 800 μm^2 cross-section area, 0.1 m long, each on one side of a sintered alumina plate of 12.5 mm x 12.5 mm outer size. A platinum catalyst was deposited on the channel walls by injecting hydrogen hexachlorplatine and then heating it in hydrogen. The used bismuth-telluride thermoelectric modules limited the hot-end temperature to 100–115°C. A catalyst permitted

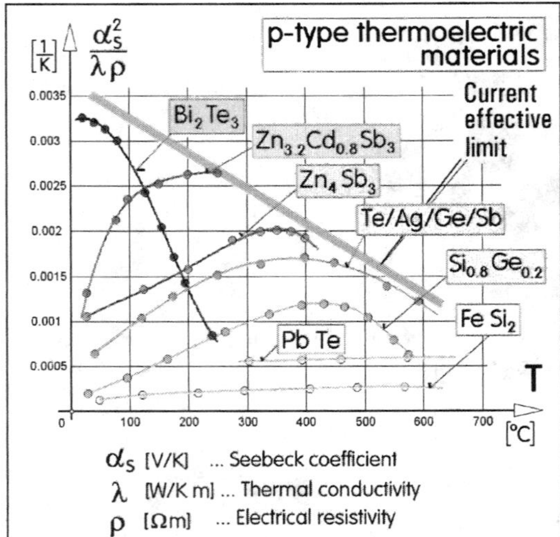

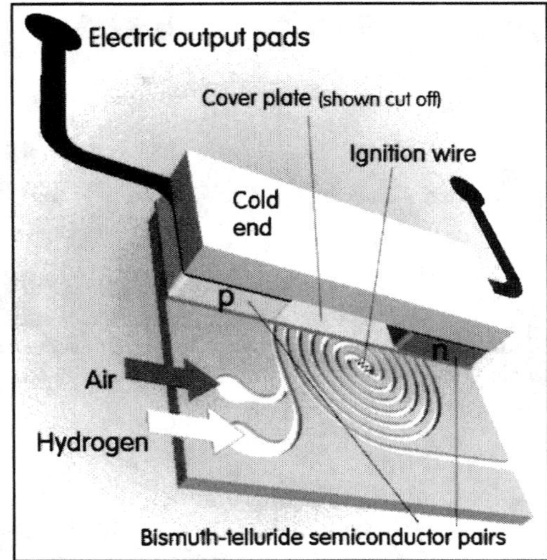

Figure 6.24 (Left) Figure-of-merit dependence on temperature for several modern thermoelectric materials. Currently only bismuth telluride (Bi_2Te_3) is commerically available. Its properties unfortunately deteriorate quickly as the temperature increases. This is a field of intense research activity, and new materials are reported almost every month.

Figure 6.25 (Right) A catalytic mircrombustor and its use in a thermoelectric generator. To suppress the loss of heat, the combustor channel is arranged into "swiss roll" double spiral. The supplied air and hydrogen mixture is thus heated from the combustion products. Direct electricity generation is attractive because of its absence of moving components, but the overall efficiency is generally poor, usually less than 1%.

sustained burning at this required low temperature. Instead of the single p- and n-type pair shown in Figure 6.25, there were 31 thermoelement pairs (series connected electrically and in parallel thermally) covering 38% of the outer surface. The maximum output power varied from approximately 30–55 mW for chemical input power 6.8 - 9.1 W – which means the conversion efficiency was between 0.44 and 0.57%. This, of course, is too low for any practical application. There is a general hope that recently discovered materials, not yet available commercially – such as the "holey" Skudderudite crystal structure alloys, rare-earth filled – may in the near future bring higher figures of merit (Figure 6.24) at higher temperatures.

6.1.3 Filling a vessel or keeping a constant liquid level

This next discussed example represents a really simple solution, typical for fluidics, in this case, rather, mesofluidics, since it requires a size large enough for a Reynolds number of at least 800 (or more whenever possible). It may be used to automatize filling liquid samples into a large number of vials to an exact surface-level height, determined by the location of the inlet into the sensing pipe. The control inlet (X_2) of the bistable valve (Figure 4.90) is partly blocked by a restrictor. As a result, the liquid jet leaving the supply nozzle always attaches to the preferred attachment wall a_1 (Figure 6.26), if the sensing pipe is open to the atmosphere and admits more air being sucked in by the jet-pumping effect through X_1 than through X_2. The valve switches and the jet is attached to the other attachment wall when the air inflow in X_1 is stopped by the rising liquid level, and the flow through X_2 then becomes the dominant control effect. The used valve is of a diverter, not a turning-down type, and hence it is necessary to provide for the outlet Y_2 a dump outlet or a return

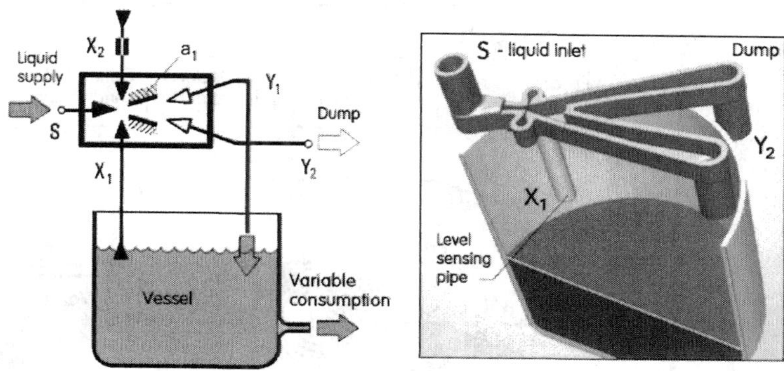

Figure 6.26 (Left) Schematic diagram of a simple circuit keeping a constant liquid level in the vessel by a bistable diverter valve. **(Right)** This is how the liquid-level-keeping device might look. As shown here, the fluidic diverter valve is, of course, a meso-scale rather than a microfluidic device.

to the upstream supply tank. Alternatively, the device may be used to regulate the level height – in a two-position manner and hence with some small pulsation of the controlled variable – in a vessel with unpredictable variable outflow or inflow.

6.1.4 Chromatographs

In a long channel, called for historical reasons a column (Figure 5.157), individual components of an investigated multicomponent analyte move at different speeds. The recorded time dependence of the passage of the individual components by the column exit exhibits characteristic peaks, the configuration of which, against a background of data from previous tests, makes it possible to obtain information about the analyte's composition. The difference in propagation speeds in traditional large-scale chromatography is mainly due to different solubility in two substances, one of them immobile, fixed to the column walls, and the other (usually a neutral fluid, either gas, in gas chromatography, or liquid) flowing through the column. A component that is quite soluble in the stationary solid-phase substance will take longer to travel than a component that is not very soluble in the stationary phase but soluble in the mobile phase. A crude measure of how fast a component travels in the gas chromatograph column is its boiling-point temperature. While some compounds can be separated by either technique, gas chromatography is usually used for the separation of volatile materials. Liquid chromatography is used mainly for the separation of involatile liquids and solids, such as in the identification of peptides, polypeptides, proteins, and other large biopolymers that are of importance in biotechnology.

Recently, especially in microfluidic devices for liquid chromatography, the separation may be based simply on a purely hydrodynamic mechanism. The velocity distribution in a column cross section is used to discriminate between differently sized molecules. Large organic molecules are transported quickly, since they are not so much held in the near-wall, slower layers.

The simplicity characteristic of the devices discussed in this section is, in the case of chromatography, in the direct transfer of conventional, macroscale separation techniques to microfluidics-size format. Otherwise, of course, chromatographs, and especially their electronic control, are rather complex, especially if they incorporate various means necessary for achieving the often-demanded highest precision in identifying very small, trace amounts of substances. A typical example of such measures is the "oven" keeping the chromatographic column—and very often also the detector—at a precise constant temperature.

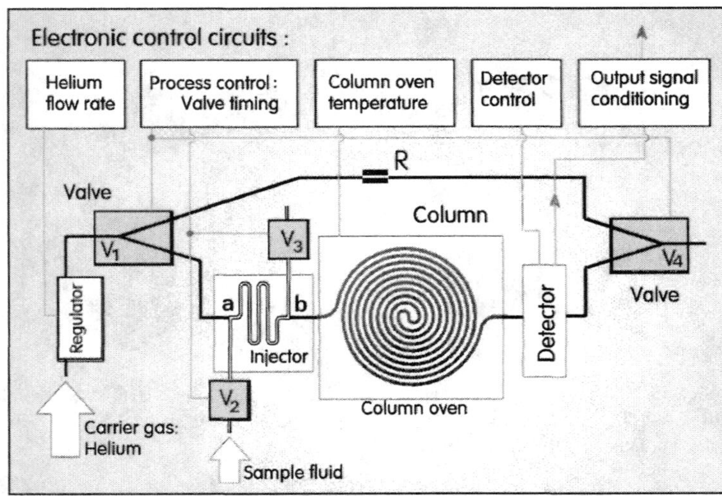

Figure 6.27 Schematic representation of a microscale chromatograph – an invaluable instrument for analyzing mixtures, especially of organic compounds. An inseparable part of modern chromatographs is a number of electronic controllers, preferably made on the same chip as the fluidic components.

A typical principle of chromatograph operation may be followed in Figure 6.27 for one of the most usual cases: a gas chromatograph with helium as the carrier gas. The analysis operation is not continuous; it is performed for a small sample volume injected into the column before the beginning of the test. During the filling of the injector, helium is diverted by valves V_1 and V_4 into the bypass. The restrictor R (which may be of a friction type, in the form of a long, convoluted channel) serves for adjustment of the pressure conditions in the system, so that it does not change (at least not substantially) when the helium flow goes through the bypass. At this stage, valves V_2 and V_3 are open so that the sample enters into the channel between them. The amount of the injected analyte is determined by the length the channel between the locations a and b. If the required amount is small (because large volumes lead to broadening of the peaks in the output record, resulting in loss of component resolution), this length may be reduced to the Z shape injector shown in Figure 5.90, provided the detector is sufficiently sensitive. Sometimes even the volume inside a plain crossing of the channels may be too large for the given purpose, and the amounts are decreased using the "focusing" effect, as shown in Figures 3.119 and 3.120.

The actual test run begins by switching all the valves to the other position. The bypass is closed, and the analyte is injected into the column. Its components then arrive at the detector at different times. The spectrum of alternative detectors is very wide. Apart from some very specific sensing methods, use may be made of many of the principles discussed in the Chapter 5. Of them, because of the need to record the time history of the sensed changes, the choice is almost always on principle with the electric output signal. The electroconductive method may be typical (as in Figure 5.90, perhaps with larger electrodes according to Figure 5.70), also perhaps with the use of selective membranes (Figures 5.164 and 5.165), optical (Figures 5.124, 5.125, 5.126, 5.127, 5.128, and 5.129), nuclear magnetic resonance (Figure 5.161), and if the attachment of biological substances to an antibody is reversible, then the acoustic (Figure 5.107) or vibrating cantilever (Figure 5.44, with electronic pickup and perhaps also with an electric driving coil rather than the pure fluidic layout shown in this figure).

The microfabrication provides the opportunity for making chromatography portable and robust enough for use in the field. It already has quite a long history. As long ago as 1975, S. Terry [6]

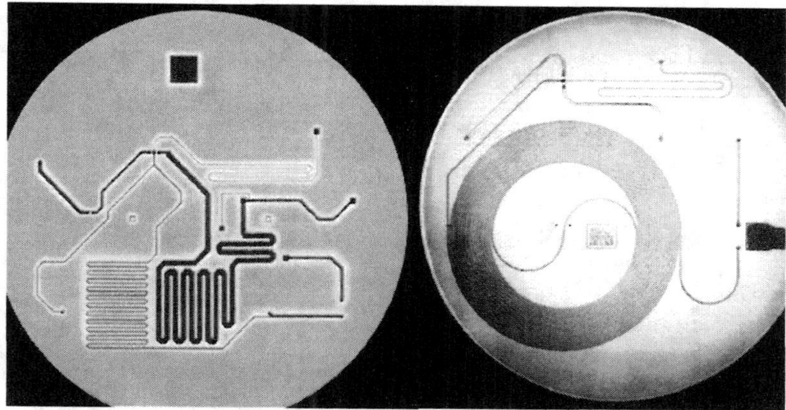

Figure 6.28 Terry's original microfabricated gas chromatograph from 1975. Considering the complexity of the task, it is astonishing that something so difficult to make was actually the harbinger of all present-day microfluidics. (*After:* [6]).

S. Terry [6] reported the first example of a microfabricated gas chromatograph, pictured in Figure 6.28, incorporating a 1.5 m long column channel. It was actually one of the earliest microfluidic devices in the full sense of the term. The thermal conductivity detector that was used was separately fabricated and mounted on the substrate. Other designs came much later. The first microfabricated (on a substrate 5 mm x 5 mm) liquid chromatograph was made by A. Manz in Hitachi laboratories in Japan in 1990. Later efforts focused on improvement of separation power, which in Terry's device was poor compared with large-scale chromatographs, due to the inhomogeneous layer of the stationary phase. An important development direction in liquid microchromatographs is packed columns: relatively large (typical cross section 300 μm x 300 μm, compared with widths of 10–30 μm in the "empty" columns) and filled with porous material, usually in the form of beads covered with a solid-phase layer and kept in place by frit retainer plugs at the column end. This, however, causes manufacturing complications. A solution is to bring beads made in situ by the controlled polymerization of a low-viscosity liquid monomer, which is easily introduced. Polymer beads attach to walls, making retainers unnecessary. Another variant uses micrometer-sized organized structures inside the column, which are made by micromachining narrow gaps between and around the structures.

An unpleasant feature of the very small cross sections are the resultant very high necessary driving pressure differences, leading to mechanical sealing problems and pressure-dependent retention effects or to acceptance of very slow flows and long duration of the analysis. An interesting solution is the use of shear stress in a Couette-type flow for the generation of the mobile phase flow, which became a popular operating principle despite the necessity to accommodate moving parts.

Special stationary phase substances are commercially available for specific tests. A common procedure is analyzing gasolines, which are multicomponent mixtures of very similar compounds. The substance available for the columns consists of bonded dimethylsiloxane, retaining the solutes approximately in the order of their increasing boiling points.

In many instances, a pretreatment of the samples is necessary, and microfluidics is often found useful in devices for this purpose. Many involatile substances such as amino acids, anabolic steroids (in antidoping tests of athletes), and high-molecular-weight fatty acids can be derivatized to form volatile substances that can be separated by gas chromatographs. In tests for the presence of aromatic hydrocarbons—toxic and carcinogenic—in food and drinking water, the concen-

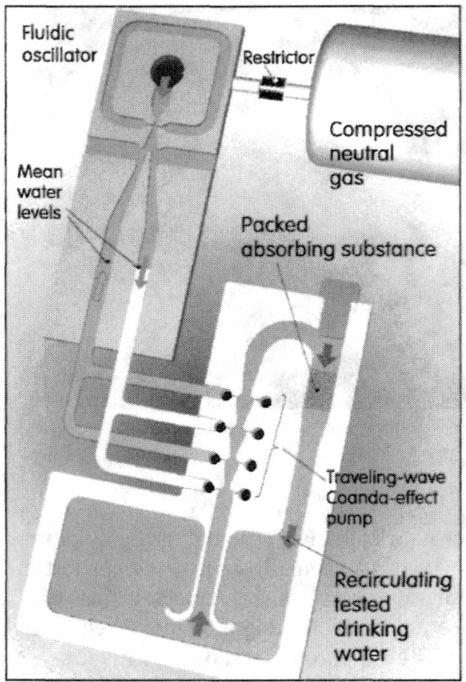

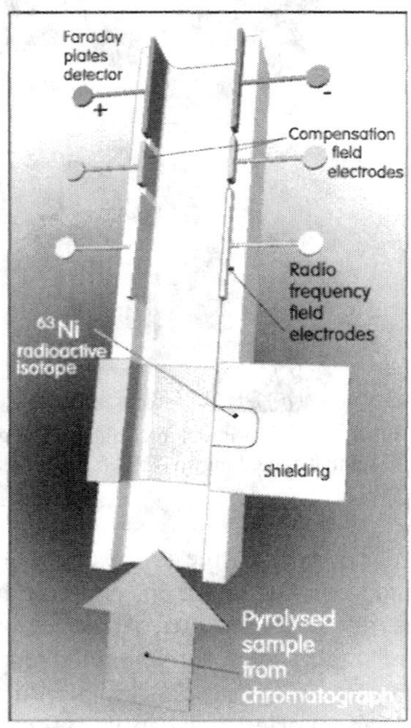

Figure 6.29 (Left) Fluidic preconcentrator: traces of organic material in water are, prior to chromatographic investigation, concentrated in charcoal by recirculating the water for a long time through a packed charcoal bed. Ordinary moving-part pumps are out of the question, because they cannot be completely free of organic material (lubricating oil, particles from worn seals), while the pictured combined alternator from Figure 5.153 and a symmetric version of the rectifier from Figure 4.185. Made by etching from stainless steel, it may be annealed to destroy all organics.

Figure 6.30 (Right) A differential ion mobility spectrometer is an alternative, better suited to miniaturization, to the mass spectrometer usually used for additional processing of the sample in chromatographs..

trations are very small (measured in ppb). The sensitivity is routinely increased by sample preconcentration: recirculating it through a packed absorbing substance (Figure 6.29), from which it is later released thermally. Pretreatment is also used to identify bacteria by their volatile fatty acids. A common procedure detects, for example, the *clostridium* genus of bacteria, which are particularly dangerous as they are very resistant to heat. They are known as the causes of some of the most deadly diseases in man, such as tetanus (*clostridium tetani*), botulism (*clostridium botulinum*), and gas gangrene, the killer of wounded soldiers in the trench warfare of WWI. Another preconcentration is used for detecting drugs—or mostly drug metabolites—in blood by liquid chromatography. It involves the recirculation of the sample through a selective adsorbent, from which the metabolites are extracted with a small amount of a solvent. A typical example is the determination of tetrahydrocannabinol carboxylic acid, found in the urine of subjects that have recently smoked marijuana. It is extracted by circulating it through packed beads containing octyldecyldimethyl chains.

Gas chromatograph analysis is very often combined with the following measurements on a mass spectrometer to provide the additional information required for unambiguous compound identification for each chromatographic record peak. The fragment ions of each gas chromatograph elutant provide what amounts to a "fingerprint" in the mass-spectrometer-generated pattern and this is particularly useful for the identification of the *Bacillus* species.

Unfortunately, mass spectrometers tend to be very expensive (usually more than $ 50 000), large, and heavy. An inconvenient feature is the need to evacuate the active region. Figure 6.30 shows the principle of a recently developed (Charles Stark Draper Laboratory) alternative capable of being microfabricated and operating at atmospheric pressure. Solid and liquid samples are fragmented by pyrolysis, and the resultant gaseous sample, after chromatograph separation, is ionized by a radioactive source. The motion of the ions is influenced by two electric fields: an asymmetric-waveform radio-frequency field alternating between high and low strength and a DC compensation voltage. At each setting of the latter, only particular ion species are allowed to pass through and collide with the detector electrode. The continuous variation of the compensation voltage produces a record allowing the identification of ions by comparison with compound libraries. This was found particularly useful for detecting extremely small (ppt level) concentrations of nerve gas and blister agent chemical weapons with a hand-held device.

6.1.5 Keeping a (nearly) constant flowrate

In this example, a simple single fluidic valve of the vortex type, performs a task that would normally be expected to require an electronic control system, consisting of a flowrate sensor, the actual control circuit, and an E/F power actuator. The fluidic controller was actually used by this author to keep a constant flowrate of material that was rather difficult to handle: molten metal. Handling it would require a very special, heat-resistant versions of the sensor and the actuator, which cannot avoid contact with the handled liquid.

The simple solution offered by fluidics is shown in Figures 6.31, 6.32, 6.33, and 6.34. The flowrate is maintained by the vortex valve operated in the transition between the low-dissipance radial-flow "OPEN" regime (zero flow through the tangential nozzle) and the high-dissipance "CLOSED" regime. The characteristics of both regimes are shown in Figure 6.32, together with the

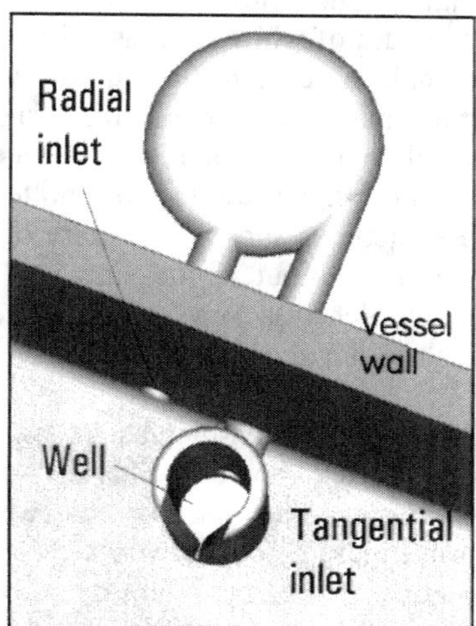

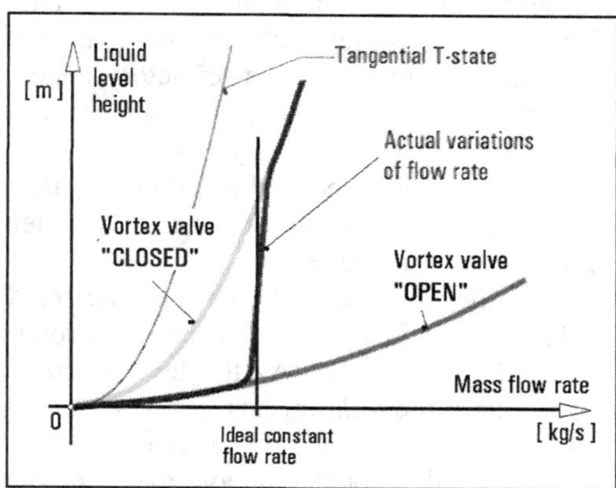

Figure 6.31 (Left) The fluidic device for keeping a (nearly) constant output flow, a vortex valve at the exit from a vessel in which the liquid level varies in an unpredictable manner. The radial inlet (no rotation in the vortex chamber) is simply open into the vessel. The tangential inlet becomes active only when the liquid reaches sufficient height. It flows over the walls of the well (which actually has a contoured notch in the wall that admits a smaller amount of radial flow before the lever reaches the well rim).

Figure 6.32 (Right) The principle of the flowrate control: as the liquid level in the vessel rises, the vortex valve gradually undergoes a transition from the OPEN regime—with radial flow—into the CLOSED regime, with a significant influence of the tangential flow.

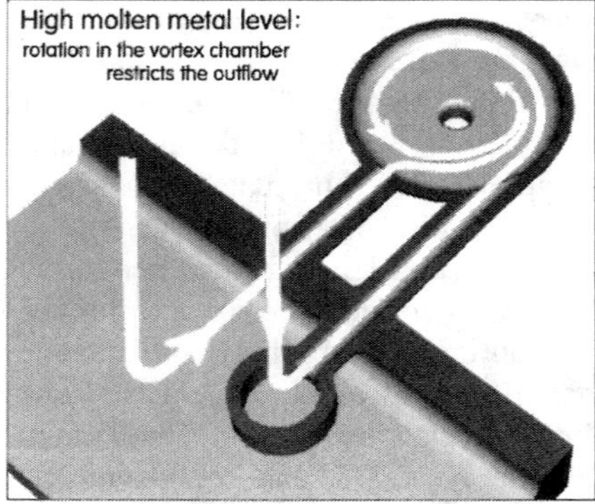

Figure 6.33 (Left) When the liquid level in the vessel is initially is high, liquid enters the vortex chamber also by tangential flow (over the rim of the tangential inlet well). The rotation in the chamber makes the liquid outflow difficult, due to the necessity to overcome the centrifugal force, which is quite high because the rotational speed increases to high values as the liquid gets nearer to the central exit.

Figure 6.34 (Right) Radial, easy flow OPEN regime in the vortex valve when the level in the vessel is low.

transitional behavior followed by the regulator. The flowrate is not kept exactly constant, but the deviations are small enough so this regulation is sufficient in most applications. The control range is limited by the inaccessibility of the full tangential state T (Figure 4.43), because reaching it typically requires a pressure in the tangential inlet higher than that in the radial inlet. In the present case, the pressure values are almost equal. The flowrate division into the vortex chamber may be, to a certain degree, adjusted by the choice of area contractions in the nozzles.

In principle, the vortex valve amplifies and inverts the signal provided by the variable restrictor in the tangential inlet. For proper variation of the tangential branch flow with the liquid level height in the vessel (Figure 6.31) use is made of a weir, the width of which increases in a vertical direction, so that this flowrate increases very fast with a small increase in the liquid level. The interesting application in maintaining a constant flowrate of molten metal from a vessel, independent of the metal level height inside, used the typical robustness of fluidic devices. The valve was, of course, made from a refractory material. A control signal transfer by molten-metal flow is certainly unusual, indicating the often unusual and sometimes surprising nature of fluidic control methods. The shaped, varying-width weir is made as a cutout in the side wall of the circular well, which is located at the entrance of the tangential inlet. It is easy, of course, to apply the same principle to controlling the flows of other difficult-to-handle liquids, that may be hot, corrosive, toxic, or radioactive.

Figure 6.33 shows the condition inside the regulator when the molten-metal level is initially high. The large flow through the weir causes a rotation in the vortex chamber. This decreases the total flow leaving the vessel. As the level decreases, so does the flow through the weir. The intensity of rotation then diminishes. In the end, as shown in Figure 6.34, the weir's flow ceases. Metal flows fast through the vortex chamber in its "OPEN" regime (the regime R in Figure 4.43).

The quality of the control action may be increased substantially with the addition of a proportionally operating fluidic amplifier, as shown schematically in Figure 6.35. The amplification actually increases the gain of the feedback loop, leading to an overall characteristic better approaching the ideal vertical line. Also the range over which the regulator operates is widened, since the jet-deflection amplifier can divert all flow into the tangential inlet, so that the problem with attainability of the full tangential state T of the vortex valve (Figure 4.43) is removed. On the other hand, the layout ceases to be really simple, losing the advantages typical of fluidic solutions with their robustness and low cost.

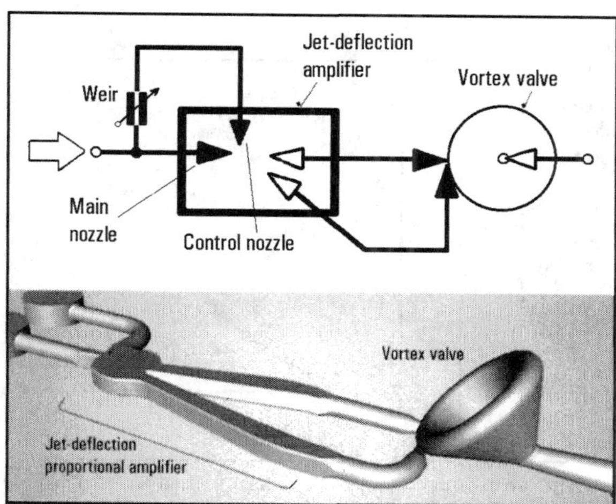

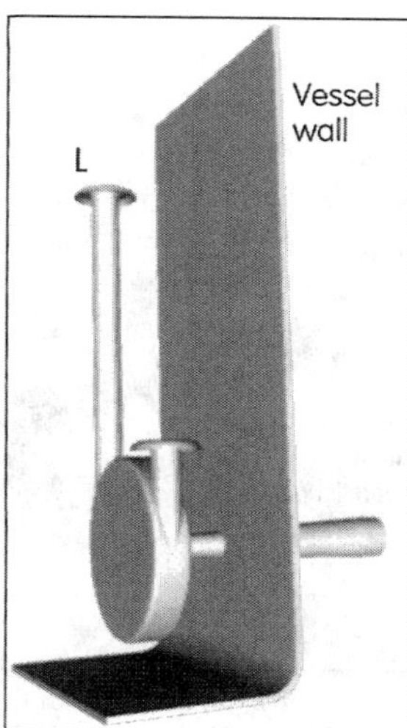

Figure 6.35 (Left) The flowrate-keeping performance could be improved by including a fluidic amplifier. The price of the increased complexity and overall hydraulic loss was considered excessive, but a unique fact was the demonstration of signal transfer and amplification in a molten-metal flow.

Figure 6.36 (Right) Simple regulator with the turn-up valve at the exit of vessel. This keeps the liquid level at the height of the entrance L into the vertical pipe. An increase above this level destroys the vortex in the chamber and makes the outflow easier.

The variant shown in Figure 6.36 also improves performance by making attainable the tangential "CLOSED" regime (T in Figure 4.43). This time, the vortex valve is of the turnup type, with two tangential inlets (Figure 4.44). It was successfully used to work with the difficult-to-handle suspension of solid particles flowing into the vessel at an unpredictably varied rate. The task was to maintain the suspension surface in the vessel at the level of the top inlet L of the sensing pipe, which is connected to one tangential nozzle. As long as the surface is below this level, the vortex valve remains in its high-dissipance, tangential "CLOSED" regime T. The regulator waits for the supply flow to increase the liquid level. When the liquid rises so high that it enters the vortex chamber through L, the vortex valve is turned up into its "OPEN" regime, facilitating a fast fluid outflow.

6.1.6 Simple pressure regulator

An analogous layout with the vortex valve (Figure 6.37), may be used to perform the more-often-demanded task: keeping a constant pressure level. It is suitable for doing it with gas as the working fluid. The valve again operates in the transition regime but in the opposite manner than in Figure 6.32: as the flowrate increases, the valve regime changes from the "CLOSED" to the "OPEN" state, following—as far as possible—the horizontal line (Figure 6.38). With gas, the weir's sensing principle is out of question. The varying ratio of the tangential and radial inlet flows is here a consequence of difference in the characteristics of a nozzle and of a friction-type restrictor. The laminar-flow narrow channels with characteristics near to linear are in the radial inlet. The pressure loss on them dominates at small flow rates. The vortex is in the "CLOSED" regime – although it does not attain the fully tangential T state, Figure 6.38. As the flow increases, the quadratic loss in the tangential nozzle becomes dominant. The flow in the valve then gradually approaches the "OPEN" regime. Mantaining the pressure is usually far from ideal, yet it is certainly better than having no pressure control at all.

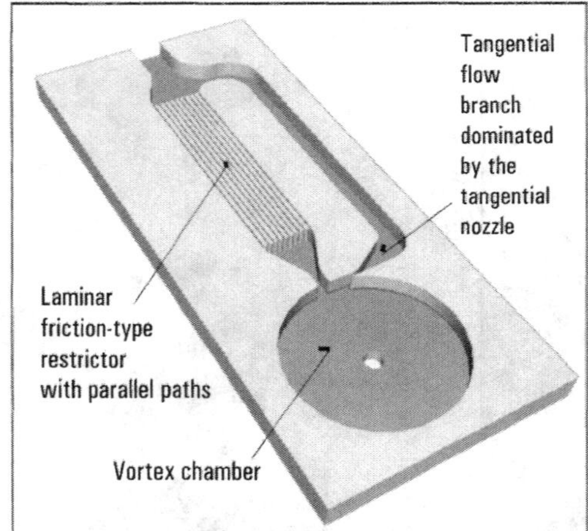

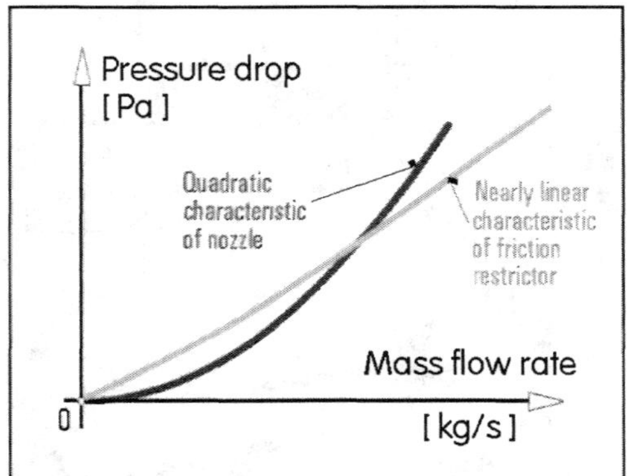

Figure 6.37 (Left) The basic idea of the no-moving-part pressure regulator with laminar friction resistance in the radial flow branch. **(Right)** Characteristics of the two flow branches. The steeper slope of the tangential branch causes the vortex valve to be turned up when the total flowrate through the circuit increases.

An example of the use of this idea in a macrofluidic control—but certainly one well known to those working in the field of microfluidics—is shown in Figure 6.39. The glove box is used to manipulate etchants and other fluids presenting a health hazard, which must not be left free to enter the laboratory atmosphere. The blower (or central suction system) used as the vacuum source removing contamination and neutralizing it by filtering or chemical treatment has to be dimensioned considering the worst case of the glove becoming ruptured (or perhaps even completely torn off). To fulfill this requirement, the vacuum source must be quite powerful, and if it were left to act on the glove box directly, the internal pressure there would be very low, stressing the box walls too much (and also stressing the gloves or, at any rate, making finger motion inconvenient). As a result, a pressure regulator must be used to maintaining, perhaps not very precisely, the pressure inside the box at a certain level. The regulator costs money and can never

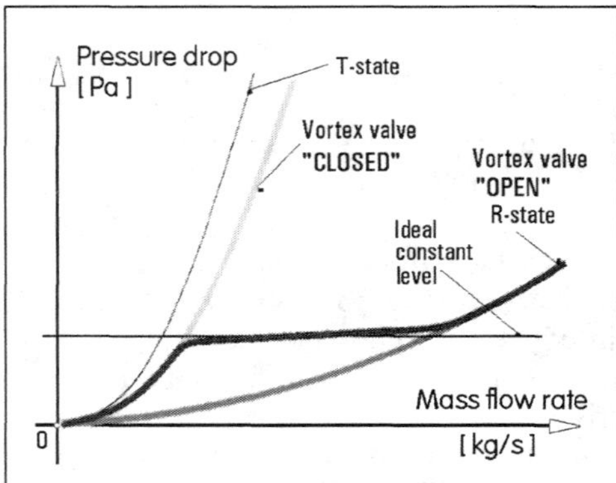

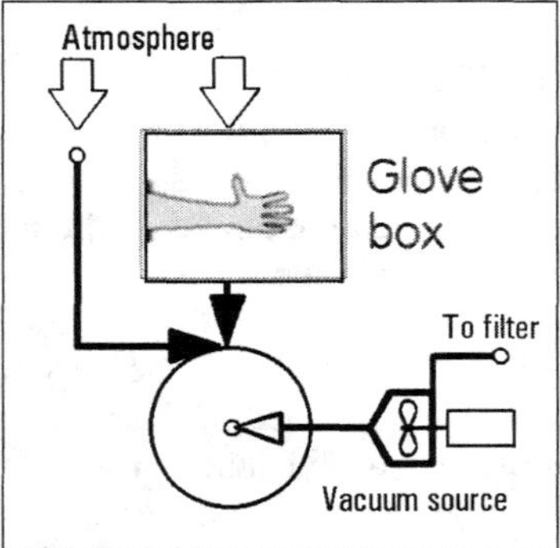

Figure 6.38 (Left) The (nearly) constant pressure is maintained during the transition of the flow in the vortex chamber from the "CLOSED" state into the "OPEN" state.

Figure 6.39 (Right) Large-scale fluidics – but of direct practical interest in microfluidics: pressure drop regulator for keeping a constant pressure in the glove box.

be absolutely reliable. As shown in Figure 6.39, the pressure regulation task is solved simply and with nearly perfect reliability by the vortex valve regulator. In principle this is noting more than just an empty cylindrical space with the two inlets. If the glove is intact, air is sucked from the atmosphere through the tangential inlet. This causes the air in the valve to rotate, generating a large pressure drop on the valve – with only a small remaining pressure drop stressing the glove box. If, however, the glove is torn, the radial inflow suppresses the rotation. The vortex valve becomes "OPEN," admitting a large airflow through the hole in the glove.

The same principle may be used in other similar pressure control tasks, even in microdevices, provided the characteristic Reynolds number is above the critical value (~ 750) (Figure 4.48).

6.2 TAKING PART IN REVOLUTIONARY CHANGES IN CARS

The ubiquitous automobile, perhaps the most characteristic symbol of our civilization, is expected to undergo fundamental changes in the foreseeable future. These may considerably change its very character.

The first change is generally described as giving to cars a certain degree of intelligence – the capability to perform some tasks so far completed by the driver. An example of the features expected to become common is automatic parking, already available in some models. There are under discussion plans to make changes so profound as to convert the car from an object handled by the driver at his free will into a component of a centrally governed transportation system, at least on highways (where, after all, the driver is not left many free choices anyway). This should avoid congestion, which is now becoming an everyday experience and, most important, increase safety. In Europe alone, more than 40 000 people die every year in car accidents caused by driver errors. In view of this fact, perhaps even the extreme of centralized control and certainly the more modest case of "intelligent" cars taking over some tasks from the driver should be considered acceptable. Accidents would be avoided or mitigated by sensing the nature and significance of possible danger, and also taking into account the driver's physical and mental state. In particular, the "intelligence" should decrease the incidence of accidents caused by wrong maneuvers, lack of anticipation of other drivers' movements, missed road signs, and technical condition of the cars. Obviously, as the first step toward making the correct decisions, the control system requires more information than it currently has. Cars must have many more sensors, and fluidics is expected to be the basis of many of the sensors to be incorporated.

The other really revolutionary change that seems to be inevitable is the replacement of the present combustion engine. This change is necessitated by environmental issues but also by the expected lack of availability of the present fuel. Among the considered replacements, the most likely is the use of fuel cells. Microfluidics is expected to take part in this development also.

6.2.1 Sensors for intelligent cars

The most important contribution to accident mitigation system development is, naturally, expected to come from electronic data processing and laser sensors obtaining the main brunt of the data about the outside world. Nevertheless, the contribution of sensors and devices in some way associated with fluid flows—and therefore falling under the heading of fluidics—is certainly not negligible. This is mainly due to those many fluids encountered and used in a car. Figure 6.40 lists and locates the most important fluids, the presence (or, in some cases, the absence) and state

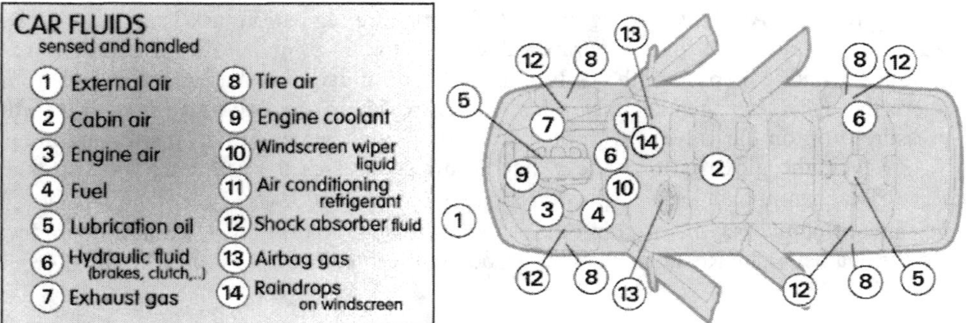

Figure 6.40 List of fluids which a car control system has to handle – or at least take into consideration – and the locations in a car. The handling requires an information about the fluids presence and state, and changes of the state during the handling. There are also many more sensors of nonfluidic sort and the fluidic signal is mostly immediately converted to the electric one. Nevertheless, the very number of the fluids gives an idea about the importance of the fluidic sensor principles.

(temperature, pressure) of which have to be monitored by sensors in a modern intelligent car. The close control of car's operation requires more data about it, for example under the heading of *external air* it is now necessary to also measure the external airflow velocity vector (the side-wind intensity). Also as an example, the number of probes measuring the oxygen content in exhaust gas has recently risen to three as the minimum (and it has to be complemented by measured temperatures). The design of these probes, of the solid electrolyte (yttrium-stabilized zirconium oxide - ZrO_2) type, now follows the planar patterns of microfluidics (Figure 6.41) in place of the well-known former thimble shape – sometimes even applying the same manufacturing techniques.

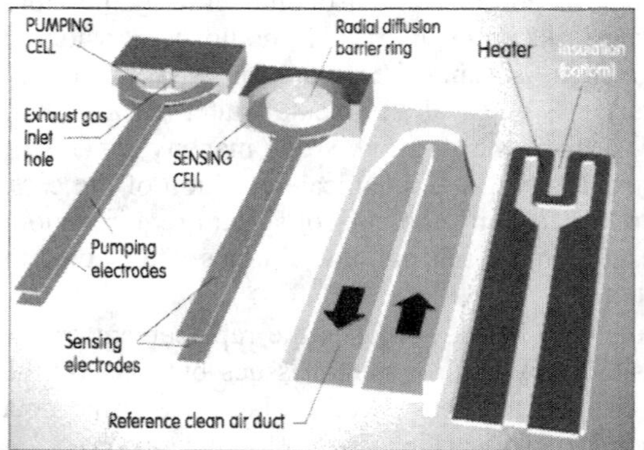

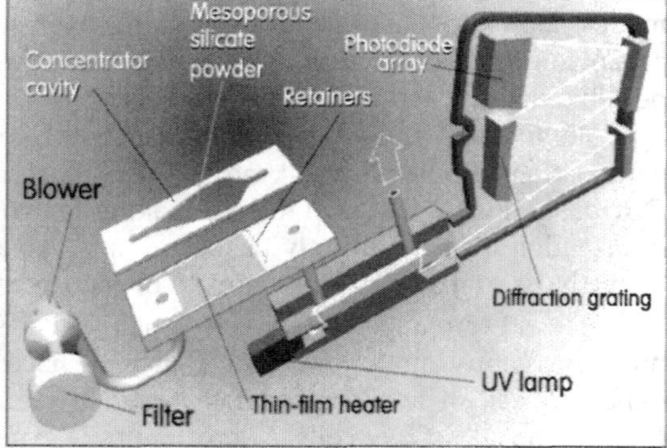

Figure 6.41 (Left) Exhaust gas oxygen sensors are manufactured at a rate of nearly 100 million sensors per year. This is a modern planar amperometric version for a very wide air/fuel ratio range, obtained by a dual-cell layout with a diffusion-limited pumping cell and Nernst-type sensing cell (with clean air reference), both operated in a closed-loop electronic circuit.

Figure 6.42 (Right) Sensor for measuring ppb levels of aromatic volatile organic compound concentration in cabin air—in this case not a sensor for car operation, but as a check of " new car odor" caused by internal trim materials. This is mostly enjoyed by customers but sometimes not only considered unpleasant, but actually dangerous. The aromatic compounds even at these low concentrations are mutagenic and carcinogenic.

Some designs need the clean air reference and this may even lead to the presence of real microfluidic channels.

It is not possible to name here all the cases of the use of fluidic and microfluidic principles, sometimes used in a role that has no apparent connection with fluids, as is the case of the accelerometers (Figures 5.38 and 5.39), which were earlier used just to initiate the airbag operation, but now are of decisive importance for stabilizing the motion of the car and controlling its cornering behavior. Just to give a characteristic example of the present trends, Figure 6.42 presents an example of a sensor incorporating fluidic components. The sensor is used to check the interior smell of the car (Renault car company actually uses for this purpose an "artificial nose," to be discussed later in the present chapter) mainly in order to detect airborne aromatic volatile organic compounds released from paints, rubber, and foam materials in upholstery. Benzene, toluene, xylene, styrene, and ethylbenzene are toxic and mutagenic or carcinogenic even at a concentration of a few ppb. They are also released into the atmosphere in automobile exhaust gas: in Japan, this source of benzene alone was responsible for an estimated 16 000 tons per year in 1997. The level has been decreasing, and these compounds are also being gradually replaced in paints, inks, adhesives, and solvents.

A sensor similar to the one shown in Figure 6.42 uses the characteristic peak of these compounds in the UV absorption spectrum at wavelengths of 200–300 nm. The small photodiode spectrometer used is not sensitive enough, and the sensitivity of detection (10 ppb at signal-to-noise ratio > 3) is obtained by the fluidic concentration in adsorbent - silicate powder - packed in the channel of the concentration cell (made in a Pyrex plate with dimensions 10 mm x 30 mm). Air passes through it for 50 minutes. After that time the detected compounds are released by heating the cell to 250°C and passing them to the detection cell.

6.2.2 Replacing the combustion engine

Cars are known to be worse than mass transportation from the efficiency and emissions points of view. Individual transport, however, offers the advantage of freedom of movement and flexibility, advantages so attractive that the use of cars in Europe has increased from 65% in 1960 to present 80% of the total transport (persons x kilometers), with the railways' share decreasing to a mere 8% and coach transport also to 8%. In the United States, there are now 752 motor vehicles for every 1 000 people. This trend is unlikely to reverse, even under possible social and legislative pressures that may come. Worldwide, use of cars is likely to become even more ubiquitous, with an expected substantial increase in car ownership in countries such as China and India.

Since the time of their inception—electric battery and steam-driven cars being just short-lived episodes—automobiles were inseparable from the reciprocating internal combustion engine, a quite inefficient machine based on a primitive basic operating principle. Efficiency slightly higher than the 50% claimed for exceptional conditions is practically out of question in practical use, where 30%–40% are the top values. On a highway, ~ 17 kW is the lowest required power, which means that at the optimum operating regime of ~ 100 km/h speed, a small highly supercharged ignition engine (with specific consumption ~ 210 g/kWh) is not likely to consume less than 1 g of fuel per second at its best. Fuel is becoming less available at a frightening rate. Existing sources are located in politically unstable regions, and the price exhibited a practically linear rise from $20 per barrel in 2002 to $75 per barrel at the time of this writing. After all, world supplies of fossil fuels are limited.

In addition, the reciprocating engine is a tremendous source of pollution. In France, for example, the incidence of cancer increased by 35% between 1980 and 2005 (the increase is even higher, by 50%, for lung cancer), and this has been mostly attributed to the pollution to which cars are significant contributors. A study [9] uses data obtained from car trip from Paris to Lyon and re-evaluates it in terms of medical data on health damage due to dose response. Converted to monetary terms (using as unit financial cost of year of life lost YOLL unit), the health damage value per kilometer is only slightly lower than the cost of the fuel. In other comparisons, the years of life lost due to pollution were found to be on the same order as those lost due to traffic accidents. This was recognized, and since 1990 governments imposed legislation limiting the emission of particulate matter (soot), CO, and unburned hydrocarbons, caused by imperfect combustion, as well as of NOx, produced chemical reaction of atmosphere components at high temperature. The most stringent limits are those imposed by State of California, which also introduced the categories LEV, ULEV, SULEV (low, ultra-low, and super-ultra-low emission vehicles), to be met in certain years by a percentage of manufacturer's fleet. No limits are set for CO_2, which, after all, cannot be evaded in an engine burning hydrocarbon as the fuel. Some CO_2 is decomposed by natural processes, in particular by absorption in sea water (more then one-half of the generated amount, but this is already changing the basic chemistry of the ocean). The 1860 concentration of 290 ppm in the atmosphere has increased to the present 380 ppm, and this is suggested as being responsible for the climate-changing greenhouse effect. Some states require cars to display labels with score values; this reflects the fact that higher fuel efficiency means less generated CO_2.

6.2.2.1 Fuel cells

Fuel cells—with an electric motor driving the wheels—offer an alternative promising substantially higher efficiency and also, in effect, zero emissions. Their development is far from settled, but their introduction may be accelerated by social and legislative pressures. Perhaps not surprisingly, microfluidics, or at least some of its ideas, may play a considerable role also in this development which also emphasizes processes on a small scale and micromanufacturing techniques using planar substrates. The promised (though not yet reached) high fuel efficiency is due to the absence of wasteful conversion through a thermal, desorganized form of energy (this means no Carnot cycle limitation). The processes taking place may be roughly described as a reversal of water decomposition by electric current: hydrogen and oxygen are combined, and the electric current is generated. This form of output means that car wheels are to be driven by electric motors. These tend to be heavy, but may easily be incorporated one per each wheel and may use already existing effective semiconductor control systems. There being no combustion, the conversion of chemical energy directly into electric energy may take place at low temperatures, though an increased temperature improves the reaction kinetics. Keeping the temperature low requires the use of efficient catalysts, the development of which is currently one of the critical progress steps. Platinum is too expensive, its availability is limited, and may not, after all, be the best choice.

Fuel cells are by no means a new idea. They were invented in 1839 by a Welsh justice of the High Court and amateur scientist Sir William Grove (1811–1896). His device was a mere scientific demonstration, but functional cells were also built a long time ago, by Mond and Langer in 1889. Nearly forgotten for half of a century, they became an object of renewed interest in the 1950s, when they were used to drive an Allis Chalmers tractor in 1959, and then employed in NASA space missions (Gemini in 1963 and Apollo in 1968).

In principle, the fuel cell (Figure 6.43) is based upon dissociation of hydrogen atoms into positive ions (H^+) and electrons (e^-) on the anode side of an electrolyte, usually a solid electrolytic membrane. This permits the protons (H^+) to travel across toward the cathode. Selective permeability is essential: the electrons are also attracted by the cathode but cannot get through and are forced to travel through the external circuit. Oxygen (usually the O_2 contained in air) is pumped to the cathode, where it reacts with both H^+ and e^- to form water, the harmless emis-

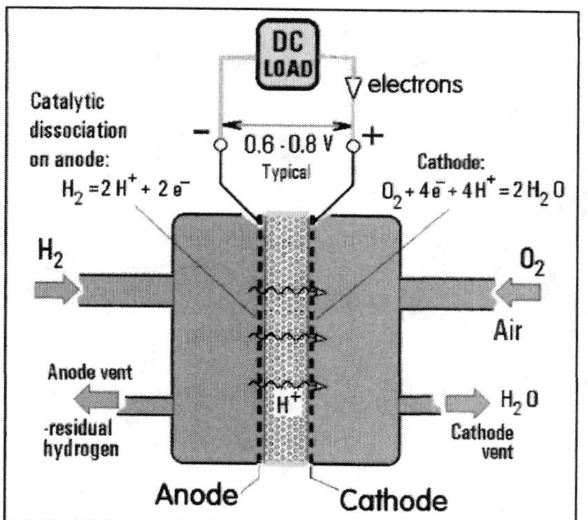

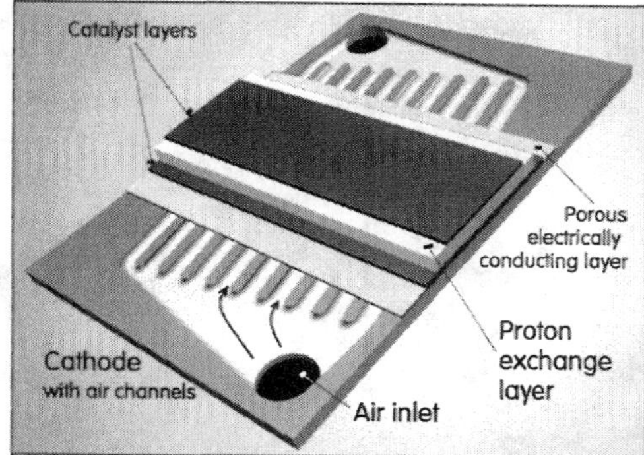

Figure 6.43 (Left) The principle of fuel cells may be described as the inverse of water decomposition by electric current. Hydrogen, as the fuel, dissociates on the anode side and the H^+ ions – actually protons – pass through the electrolyte to the cathode where they combine with air oxygen to form water. The current of the excess electrons from the anode to the cathode drives the electric load.

Figure 6.44 (Right) An example of the layers that form the solid electrolyte fuel cell. The cathode-side outer plate is micromachined in a manner typical for microfluidics to form a system of channels for air inflow and water (plus excess air) outflow, here arranged in a parallel layout. The anode-side outer plate is not shown.

sion product. The produced voltage, less than 1V per cell, is low, and the cells are usually stacked and generally used as a battery, except that they do not need recharging. They can run for an unlimited time, as long as there is a fuel supply. In contrast with an internal combustion engine, the efficiency is not limited by the usual Carnot cycle thermodynamic considerations. Theoretically, the processes are reversible, and the efficiency would reach 100%. In practice there are dissipations due to Joule heating, diffusion, and electrode surface effects. These increase with increased output current. Also the delivery of air to the cell requires a compressor consuming some power. As a result, at a full load, the practical efficiency levels are between 40% and 65%.

Table 6.1 lists types of fuel cells currently under development. They differ mainly in the character of the electrolyte. The early alkaline-type cells were used in NASA space missions. They tend to be too costly. Molten carbonate cells are efficient and can use carbon-based fuels. Their higher operating temperatures require some time to reach. The slow start limits the applicability for vehicles. This is also the main problem with the solid oxide cells, usually based on the yttrium-stabilized zirconium oxide (ZrO_2). The sensing cell in the sensor shown in Figure 6.41 is actually a fuel cell of this type. Hot exhaust gases carry away heat; this may be utilized in a Brayton or Rankine cycle (Figures 6.6 and 6.9), but this makes the system larger (the size of low-temperature heat exchangers may even be found prohibitive), heavier and more expensive.

The most promising therefore is the low-temperature fuel cell with a solid polymer membrane. A suitable membrane material is Nafion manufactured by DuPont, the electroactive polymer already mentioned in association with Figures 5.59 and 5.60. The ion freedom of motion requires that it be kept in a hydratation environment, but this fortunately is always available (Figure 6.43), due to the water production on the cathode side. In fact, the handling of the generated water may require a water management system, such as a gas separator using small holes in a hydrophobic plate through which the excess air can leave while water is left behind. On the other hand, the removal of excessive water is proposed, using channels in a hydrophilic plate.

	Polymer electrolyte	Alkaline	Phosphoric acid	Molten carbonate	Solid oxide
Electrolyte	Polymer ion exchange membrane	Potassium hydrated solution	Phosphoric acid	Carbonate	Yttrium-stabilized zirconium oxide
Operation temperature	80 °C or less	120 °C	200 °C	600-700 °C	1 000 °C
Efficiency	40% - 45%	40%	40% - 45%	45% - 60%	50%-65%
Fuel	hydrogen, methanol (LNG)	hydrogen	hydrogen (LNG, methanol)	hydrogen, CO (coal gasified gas, LNG, methanol)	hydrogen, CO (coal gasified gas, LNG, methanol)
Features	Compact, rapid start	Special uses: spaceships, submarines	Suitable for static applications	High generation efficiency, slow start	High efficiency, slow start

Table 6.1 The most important types of fuel cells. Only the solid electrolyte (polymeric or oxide) versions are currently considered for the replacement of combustion engines in cars.

A typical configuration of the layers comprising a cell is shown in Figures 6.44 and 6.45. Apart from the proton exchange membrane, there are on its both sides the catalyst layers and porous, electrically conductive diffusion layers. The latter—though some designs evade using them—are there for several good reasons. They distribute the fuel and oxidant on the membrane surface, remove water from the membrane on the cathode side, and conduct the electrons. An often used material is carbon paper, sometimes with the conductivity increased by filling it with electrically conducting powder.

Of basic importance are the favorable consequences of the scaling factor. Fuel cells operate more efficiently if they are made small, the required output power obtained by numbering-up rather than scaling up. The smaller the cell, the more the dominant surface effects improve the electrochemical reaction conditions. The smaller size, as usual, requires higher pressure for driving the fuel and oxidant, but this is beneficial for the cell's operation, improving the pressure-driven diffusion of fuel and oxidant to the membrane. Of course, the high pressures need not be perfectly

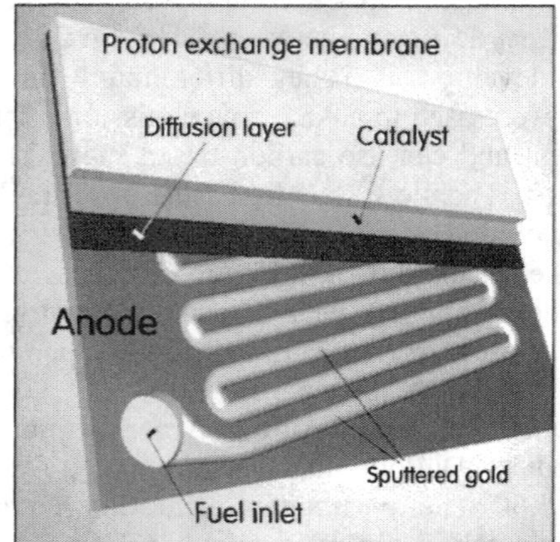

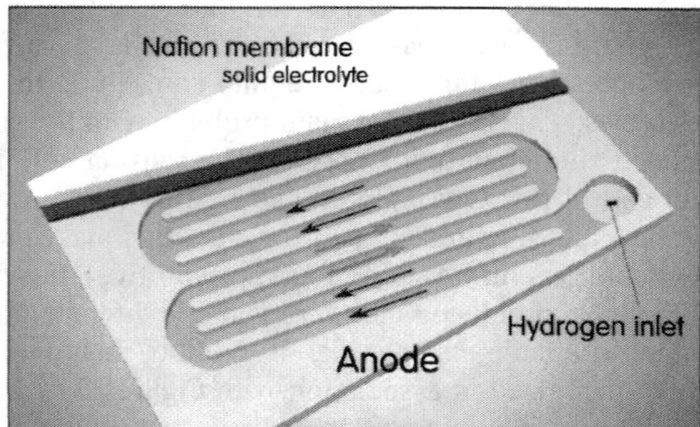

Figure 6.45 (Left) Another layout of the distribution channels – here for the fuel on the anode-side outer plate. They are made in a way that is characteristic of the planar layout of microfluidic devices. The electrically non-conducting plate material needs the formation of the current pickup, here by a deposited thin precious-metal layer. The fuel is high-pressure hydrogen, and this has led to the series connection of the channels (compare with the parallel connection in Figure 6.44).

Figure 6.46 (Right) Yet another possibility for the channel layout: pairs of parallel channels are connected in series.

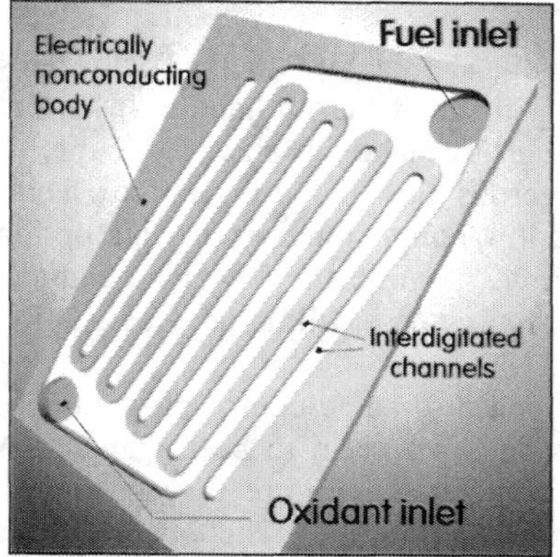

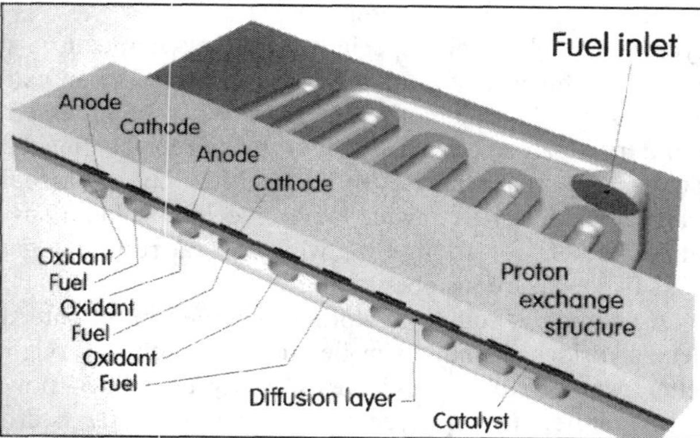

Figure 6.47 (Left) One-sided layout of fuel cells with both fuel and oxidant channel networks located on the same side of the membrane. The reaction takes place between the pairs of interdigitated channels. There is an obvious manufacturing simplicity, but the channel flows are not too effective – in particular, in this simplistic case, there is no easy way out for the generated water and the unused components of the atmospheric air. More sophisticated channel layouts exist.

Figure 6.48 (Right) A section through the assembled one-sided cell design. Proton motion is parallel to the membrane plane.

balanced on both sides of the membrane everywhere; and this is why the membrane and the system of other layers need supporting. The anode as well as cathode cavities are divided into a system of channel recesses surrounded by the supporting remaining parts of the original flat faces of the outer plates. It is the proper design of these channels where the experience of a microfluidics expert is employed. There are two basic layouts: the two-sided (Figures 6.43, 6.45, and 6.46) and the one-sided. Perhaps less efficient, but more compact and easier to make (Figures 6.47, 6.48, and 6.49), the dead-ended channels are actually used only in inexpensive, small versions for portable devices, not for cars. Designing the channels is actually quite a challenging task, requiring to take into account the gradual depletion of the fuel and oxidant along the channel length, water management on the cathode side, and the generated temperature fields.

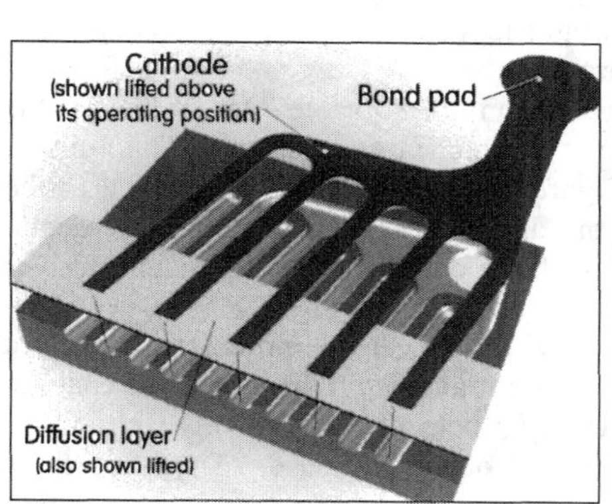

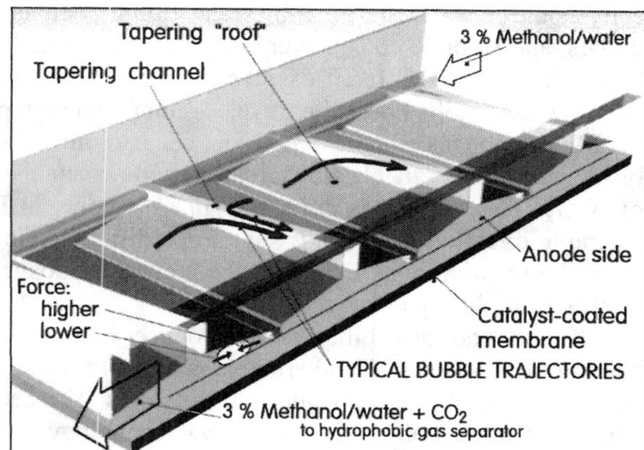

Figure 6.49 (Left) Not only channels, but also the current pickup electrodes (here the cathode) have to be designed interdigitated shapes in the one-sided fuel cell layout.

Figure 6.50 (Right) The passive driving of CO_2 out from the anode side of the direct methanol fuel cell: it uses the force imbalance on a bubble in a wedge-shaped cavity or a tapered channel. The resultant force difference drives the bubble towards the side with a greater height between the "floor" and the "roof" above it.

6.2.2.2 Liquid fuels and reforming

Hydrogen as the fuel, especially in a low-temperature fuel cell, should lead to the ideal zero-emission vehicle. Unfortunately, it is known to be an extremely difficult fuel to handle. If kept at low pressure, its volume is too large. When compressed enough to make the size reasonable, it would need an excessively heavy tank, representing a dangerous bomb in the case of a car accident. Hydrogen may even be liquified, but the associated low temperatures make this option very impractical; even with expensive and voluminous thermal lagging, the low temperatures cannot be kept for a long time while the car rests in a garage. In any case, filling at a fuel station looks like a hazardous operation.

A solution is sought in storing hydrogen in the interstitial space of some compounds, such as metal hydrides. It may even be stored in metals, though this does not seem to be promising. A better perspective is provided by newly developed polymeric materials. Before they become a reality, preference is given to the idea of using hydrocarbon liquid fuels. The huge logistic advantage is the already existing vast and dense net of petroleum stations.

The first possibility, already proven by including novel catalysts, is based on taking the hydrogen directly from methanol in the *direct methanol fuel cell.* Methanol is handled like gasoline, though it is more toxic and more chemically aggressive. The equations of the electrochemical processes included in Figure 6.43 are replaced by the process on the anode side:

$$CH_3OH + H_2O \rightarrow CO_2 + 6\,H^+ + 6\,e^-$$

and on the cathode side:

$$1.5\,O_2 + 6\,H^+ + 6\,e^- \rightarrow 3\,H_2O$$

Even though the basic layout, as shown in Figure 6.43 and the arrangement of the layers from Figure 6.44, remains the same, the physics of the cells operated with liquid fuel is substantially influenced by the fact that the Schmidt number (ratio of kinematic viscosity to the diffusion) of liquids is about three orders of magnitude higher than in gases. At the same Reynolds number, therefore, the laminar diffusion processes in the liquid fuel are $\sim 10^3$ times slower. They are so slow, in fact, that there were recently quite successful attempts at layouts with a liquid oxidizer and the reaction taking place at the liquid-liquid interface.

It is possible to produce methanol from crude oil and also from coal. However, the necessary conversion of production facilities and filling stations to supply methanol would cost billions of dollars, and its price could never beat the present price of gasoline. As opposed to the emissionless hydrogen-fuel-cell-powered car, the methanol version would leave a trail of CO_2, though at a rate reduced by 58% per mile traveled, compared with present cars.

Most problems with methanol fuel cells are minor and seem to be solvable. One of them is the blockage of the available flow area of anode channels by the attached gradually growing bubbles of CO_2. The solution is quite elegant. As shown in Figure 6.50, the bubbles are moved in the desirable direction by the sideways force difference in the direction of increasing channel depth. Once the growing bubble touches the inclined opposite wall, the continuing depth increase drives it away from the anode cavity.

The second possibility is to produce hydrogen on board a car by processing standard automobile gasoline. This alternative is very attractive because it avoids all the expensive changes of the fuel distribution and sale infrastructures. The necessary processes are relatively well known and in principle follow the *reforming* technology as carried out in refineries. There are three alternative solutions:

(1) *Steam reforming.* As in the classical refinery process, the primary fuel is mixed with steam, which, in the presence of suitable catalysts, decomposes the gasoline into a mixture of H_2, CO, and CO_2. This is a highly developed method. The process is endothermic and requires

running an auxiliary combustor, which is also needed for the steam-generation vaporizer. Systems of this type were demonstrated to produce hydrogen from diesel fuel as well as the aviation kerosene JP-8. They are further complicated by the necessity to convert CO, which poisons the catalyst on the anode side (the typical polymer membrane cell catalyst can tolerate only 10–20 ppm CO).

(2) *Partial oxidation.* Some components of the fuel react with oxygen, liberating H_2. The advantage of this reaction is its being exothermic. It can be run after a brief start with combustion heating, so that there is no delay in starting the car. On the other hand, the conversion rates are typically only 70% to 80%.

(3) *Autothermal reforming* combines the partial oxidation process with steam reforming, balancing the thermal effects. This is a field that has seen much recent research.

Current fuel-cell test cars use steam reforming, since this method is most developed. As shown in Figure 6.51 and in its disposition in the test car in Figure 6.52, this fuel processing is a multistep process requiring a quite complex layout. A percentage of the fuel is burned in the combustor, which generates the heat sustaining the reaction, transferred to the process reactors by the heat exchangers. Together with the water vaporizer, the combustor exhaust gas is also the source of the steam. This is mixed with vaporized fuel and led to the conversion reactor. The CO cleanup is a two-step process involving a water-gas shift reactor and selective oxidation.

There is, nevertheless, a profuse abundance of problems yet to be settled, such as the presence of organic sulphur compounds in current gasoline. Catalysts for fuel cells demand total sulphur of less than 1 ppm. Although desulphurization processes are known, further complication of the

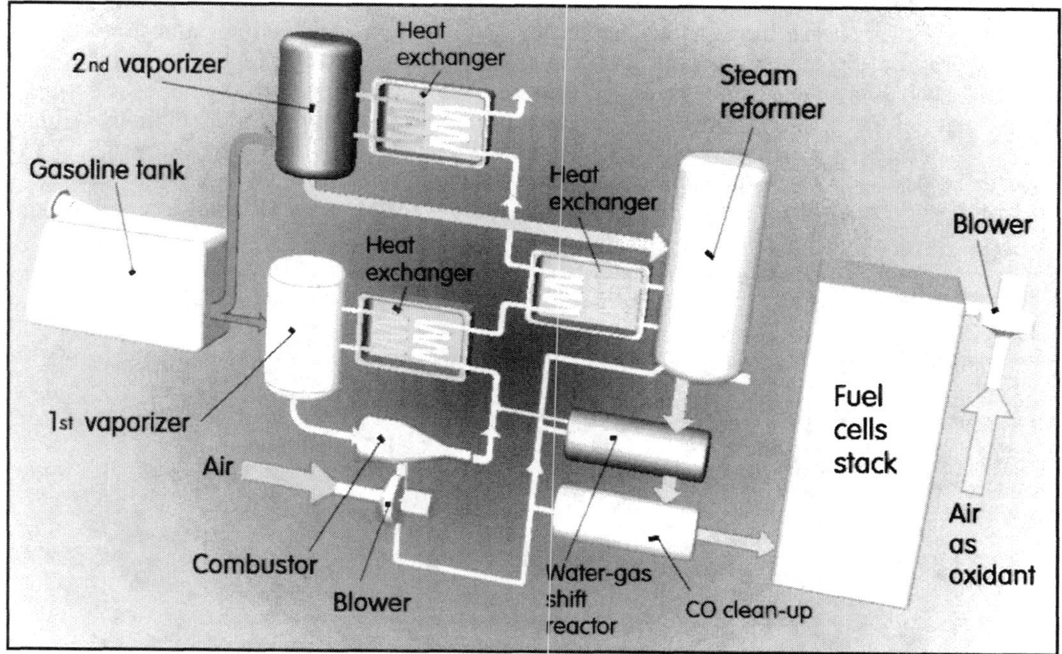

Figure 6.51 The basic components of an on-board facility for producing hydrogen from gasoline by the steam-reforming process. The actual system is more complex: pumps, control valves, and some flowpaths (e.g., the anode and cathode gas from the stack brought into the combustor to be burned there) are not shown so that the picture is less cluttered.

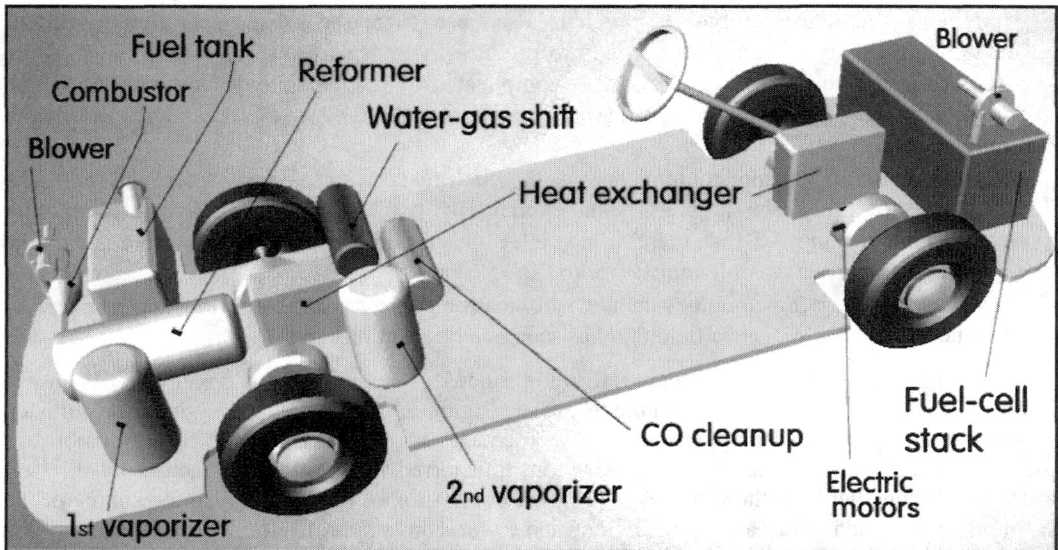

Figure 6.52 With the classic technology, the on-board reforming fuel-processing system producing hydrogen for the fuel cells leaves no space in the car for baggage (and, indeed, very little for the passengers). Applying microfluidics technology can decrease the volume between 10 and 60 times.

system does not seem to be a reasonable proposition. A better solution is demanding removal of sulphur in gasoline refineries.

Most test cars so far use conventional technology based on fixed-bed reactors. While useful for feasibility demonstration, the typical size of the components in Figure 6.52 shows that this is far from suitable for practical use. However, the size of the reactors and heat exchangers is exactly the field in which a considerable progress is expected from the application of microfluidics. Various estimates predict between 10 and 65 times smaller resultant size. Very much also depends on development of suitable catalysts. The cleanup reactors may be replaced by the use of the hydrogen permeable membrane, another field requiring closer study. In traditional reforming

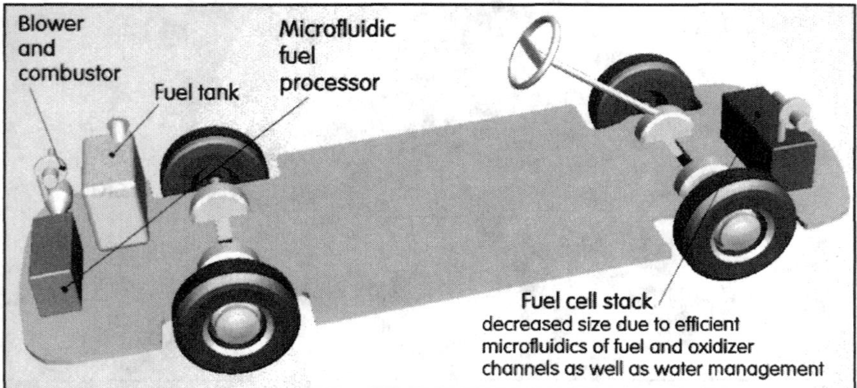

Figure 6.53 Improvement made possible by applying microfluidics: a substantial decrease in the volume of heat exchangers, vaporizers, and reactors allows the integration of the whole fuel reforming into a single body. Also the fuel cells themselves are smaller and lighter.

technology contact times are on the order of seconds. In microfluidic devices, contact times of milliseconds suffice for proper operation. This is reflected in the lower temperature required to obtain the > 99% conversion. The expected resultant decrease in the sizes of the processing devices is seen in comparison of Figures 6.52 and 6.53.

6.3 DISCOVERING NEW MATERIALS AND DRUGS

The preceding section, in indicating how the introduction of a new operation principle depends on improvements obtainable by catalysts and polymeric materials, provides an example emphasizing the need of concentrated effort in materials research. This is also a field in which microfluidics is capable of bringing a substantial change in current practice. The basic obstacle is the tenuous and indeed sometimes unpredictable mutual dependence between the chemical composition of a material and its practical, useful properties – mechanical, optical, electrical, and sometimes even aspects such as fragrance or taste. In view of this near-impossibility, the only guidance being previous experience with related materials, the development, or more properly, discovery of better new materials has to be made empirically. The new materials are individually synthesized and subsequently tested in the laboratory. Depending on the complexity of the compound preparation and the laboriousness of its testing, an experienced investigator may be expected to perform between 200 and 1000 studies of a new compound material per year. Sometimes the numbers are much smaller, and the whole process is slow and expensive. As an extreme, with exceptional requirements placed on the testing, the development of a safe and effective single pharmaceutical product is currently estimated to take, on average, 12 years, with the risk-adjusted cost of $ 500 million per single drug (i.e., discounting the average of ~ 10^3 discarded unsuccessful attempts made during this period). The simple A + B = C chemical reaction implied in the examples in Figures 6.54 and 6.55 are an exception. The number of input variables is usually large, and the necessary research effort increases rapidly with the fast-increasing number of possible permutations. There are some 10^{200} possible molecules with relative molecular masses less than 850. Of them, ~ 10^{40} organic compounds are expected to possess pharmaceutical properties. Testing them in the traditional way is a hopeless task. It became obvious quite soon that the discovery process can be dramatically changed by automation, involving rapid parallel preparation of the tested compounds. This field of research is called combinatorial chemistry, and it is becoming of vast importance. Microfluidics can provide the instrumental tools in this area also – mainly in the two directions of this development: in preparing a large number of tested materials and in taking and delivering samples for the test.

6.3.1. Combinatorial tests

The combinatorial tests consist of producing, under identical conditions, a large number of reaction products. Due to (usually small) variations in the mixing ratio of input reagents or the absence of some reagents, the compounds produced in a test run slightly differ. The properties of the produced compounds are evaluated, and the results are stored in the form of combinatorial libraries. Information extracted from the libraries is used for the identification of paths leading to a certain desirable optimum, such as the optimally behaving compound from the point of view of a performance criterion. The search for the optimum is based on combinatorial strategies that originated in pharmaceutical research. They often use advanced mathematical methods, such as genetic algorithms – function optimization based on the genetic principles of natural selection and

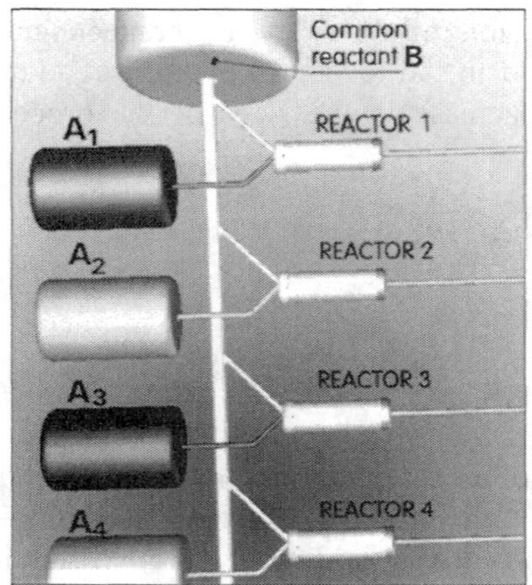

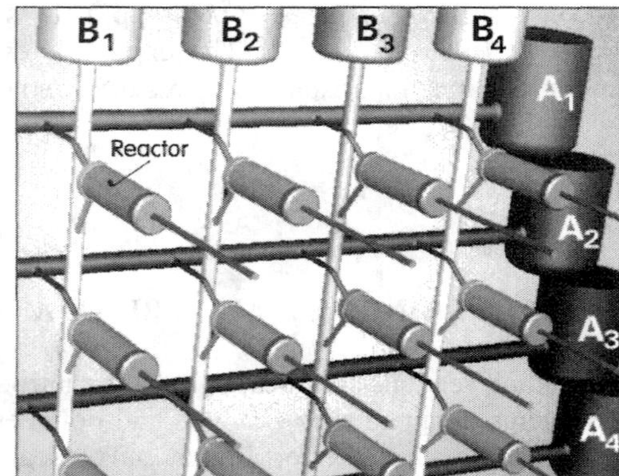

Figure 6.54 (Left) An example of a one-parameter test of chemical reactions involving two reactants, A and B, the latter common for the whole test run, while the reactants A_1, A_2, A_3, and so on are different. The picture does not show the flow control valves in the reactant inlets.

Figure 6.55 (Right) An example (again not showing the flow control valves) of a two-parameter test. The chemical nonpremixed reactions between two reactants, A and B, under the same temperature and pressure conditions, are arranged so that there are all combinations of different reactants A_1, A_2, A_3 and so on, with different reactants B_1, B_2, B_3, and so on, in reactor output terminals.

survival of the fittest [16], which work well on mixed (qualitative *and* quantitative variables) optimization problems and are robust against locking to local extremes. The search for optimum is often complicated by requirements to meet several, sometimes contradictory optimality criteria. There are traditionally at least two criteria – yield and selectivity – in the case of catalysts, and five criteria known as *ADMET* properties (adsorption, distribution, metabolism, elimination, and toxicology) applied to medicinal drug candidates.

The reliability, usefulness, and general quality of the information that may be extracted from a combinatorial library is very much dependent on the number of entries (for example, the number of the tested catalyst) and on the accuracy of maintaining identical test conditions. The safest way of securing this sameness of conditions is to perform the tests at the same time in parallel. The performance is judged by composition analysis of the reaction products. Since the amount of the product needed is small – and decreasing with the continuing progress in testing and analysis methods – microreactors offer the best solution. Because of their small size, they may be easily accommodated for example in a single metal block, assuring the constant reaction temperature. The actual layout of the reactors varies widely with the character of the reaction. Many reactions that produce durable compounds as the reaction products are made on the surface of a solid support. The products then may be retained on the support plates, which may be subsequently stored as a combinatorial library [10] in material form. This is perhaps not so convenient as storing just the virtual library of resultant data in a computer memory, but certainly has the unsurpassed advantage of the library being available at a later time for repeated analyses or an analysis by a different method, perhaps evaluating a different property of the compounds than was the property of interest in the original investigation. The open surface supports may actually be exposed to high pressure and even high temperature in an oven during the run. Nevertheless, this is usually the primary choice for reactions of organic reagents under atmospheric conditions.

On the other side of the spectrum of typical high throughput tests is the discovery of catalysts. Traditionally, this has always been an empirical process, and the high-throughput method offered

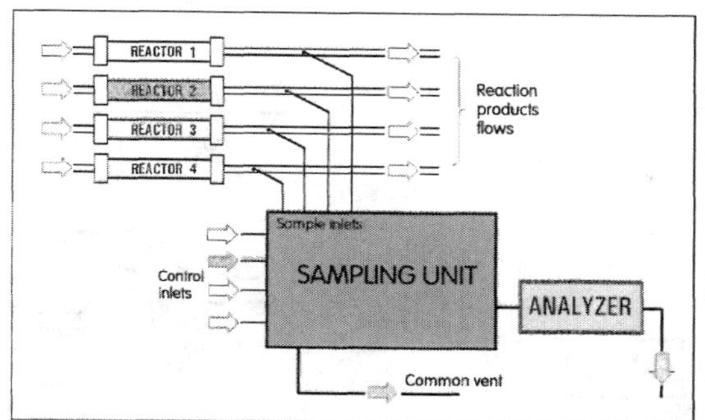

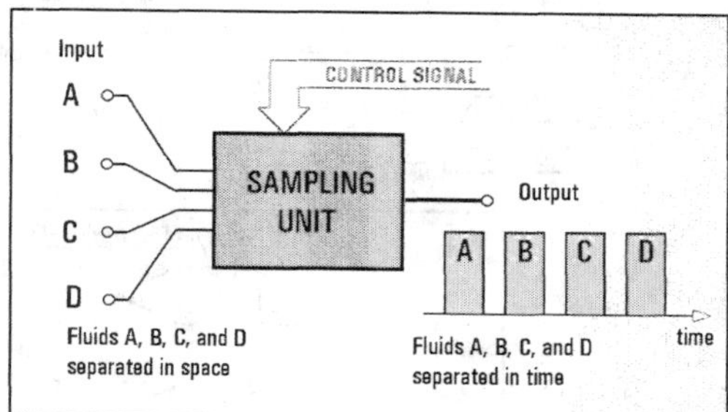

Figure 6.56 (Left) In the combinatorial testing it is usual to use a single analysis instrument for sequential, one-by-one evaluation of reaction products properties. This is made possible by the sampling unit.

Figure 6.57 (Right) The task carried out by the sampling unit is in principle a spatio-temporal conversion: the samples A, B, C, and D, from different locations in space are brought into the analyzer at different time instants.

by microfluidics is a substantial improvement over the time-consuming, one-by-one, individual tests. In this case, the test reactions, such as those of importance for fuel reforming in cars, are usually performed at both high temperature and high pressure levels. It is usual to use stainless steel reactors. Typical catalysis is heterogeneous, using solid catalysts immobilized in coating on walls (either the walls of the reactor or more effectively the walls of packed beads filling the reactor cavity). In different reactors, run in parallel, there are slightly different catalysts but the supplied reagents and reaction conditions are the same for all of them. The evaluation of the catalyst performance is done by composition analysis of the reaction products, usually already during the test run.

The tests and analyses used for the evaluation of the reaction products may be completed offline, when the reactions are finished. With organic samples on the tray, the test may consist of the laser irradiation of the whole tray and the detection of individual sample properties using their luminescence. The library then consists of a photograph of the tray with various light intensities in the well locations. For the microfluidic approach, it is typical to perform the test online using samples brought in channels from the reactors, while the reaction is still in progress. There is also a trend to use very small instruments, commensurable with the small microreactors, and some of the earliest gas chromatographs were developed in miniature size for this purposes. Due to the small analyzed sample amounts, there is often a requirement of exceptional sensitivity of the test instruments. This makes them rather expensive. Because of the price, there is almost always—provided the reaction is not too fast—just a single instrument performing the tests on all of the samples one-by-one, even with a very large number of the microreactors. After all, using a single analyzer instrument brings the advantage of performing all tests under the same conditions and with the same instrument setting. If there are two or more instruments used, they more often than not are used to test different aspects of the samples. The *ADMET* properties of drugs traditionally require expensive and time-consuming tests on live animals. Increasingly often, tests in microfluidic devices using entrapped living human cells are substitutes for tests on live animals.

6.3.2 Sampling

The use of a single analysis instrument for testing the success of the discovery process places a particular emphasis on the transport of the sample to the analysis. There is still a strong tendency, even in highly advanced systems, simply to simulate the activity of a human operator. A typical example of this approach is storing liquid or dissolved solid reaction products in vials on a tray and transferring them to the analyzer by a dedicated robot that with its arm opens the vials one by

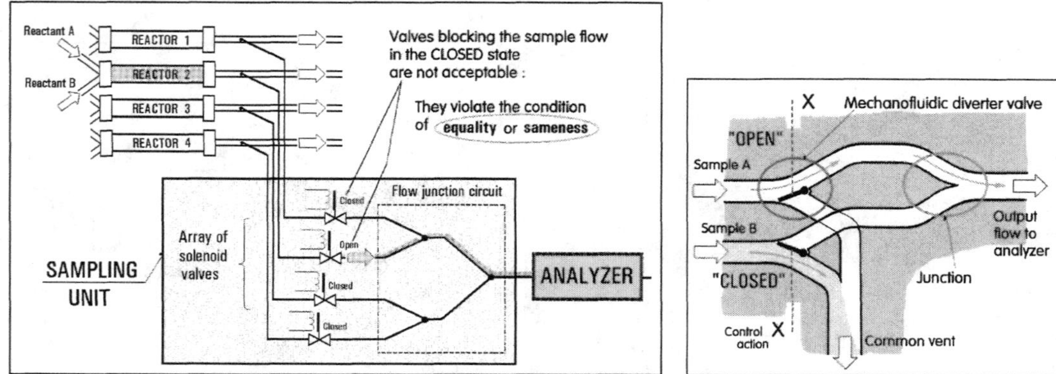

Figure 6.58 (Left) This idea of a sampling unit formed by an array of solenoid ON-OFF flow-blockage valves is wrong: opening the valve would change the pressure conditions inside the reactor.

Figure 6.59 (Right) With diverter valves like the one shown here the pressure conditions may be kept constant, provided the pressure drop across the vent outlet is adjusted to be the same as the one across the analyzer. The descriptions OPEN and CLOSED are related to the branch leading to the analyzer.

one by unscrewing their lids and then transfers, perhaps with another arm, the sample from each vial (e.g., using a pipette) to the analyzer. Even in tests performed in a more modern way, in through-flow reactors supplied with fluidic reactant-handling system the traditional approach with mechanical motion, based on the idea of the robotic arm, still survives. The sample transfer to the analyzer is still often made by a sampling unit with a moving probe [11]. In this case, the reactors are of the open-exit type, with the reaction products issuing into the common space where they encounter the probe and its traversing gear. The probe is connected to the analyzer by a flexible tube and is moved from one reactor exit to another by a mechanical positioning servomechanism.

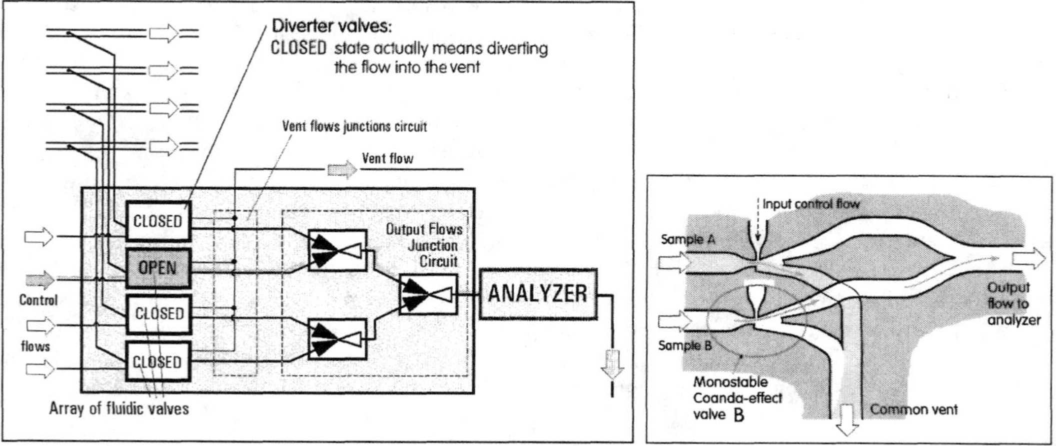

Figure 6.60 (Left) Schematic representation of a sampling unit built from diverter valves. In the CLOSED state, the sample flows are dumped into the common vent. The junction circuit contains the mutual enhancing/inhibiting devices, according to Figures 2.81 and 2.82. Their operation is based on dynamic action, which means that no proper functioning can be expected at very low Reynolds numbers.

Figure 6.61 (Right) Schematic representation of a simple (showing only two valves) pure fluidics sampling unit. The diverter valves used are the monostable variant (Figure 4.103) of the valves using the Coanda effect.

The progressive idea of a fluidic sampling unit has been slow in finding acceptance. As shown in Figures 6.56 and 6.57, what the sampling unit performs is the conversion of the spatial distribution of the samples into their distribution in time. The fluidic solution, less expensive, faster, and less prone to a failure, is characterized by having the reactor outputs directly connected with the analyzer by means of a system of channels or cavities (Figure 6.56). The *sampling unit* is positioned between the reactors and the analyzer. In its fluidic version, sometimes called a fluidic *selector or multiplexer*, this unit is essentially an array of valves which open the way leading to the analyzer one passage at a time. There is a large number of possible sampling unit variants. The fluid flows not selected at a particular instant of time may be halted (a turning-down action is not really acceptable in the present context, Figure 6.58), or preferably dumped to a vent outlet (diverting action in Figures 6.59 and 6.60). The durations of the individual time steps (Figure 6.57) need not be the same, though they most often are. The sequence of the samples at the output is mostly of the simple repetitive periodic character, though in special situations the sequence may be varied according to some program. There are features common to all sampling unit designs:

(1) The unit has several input terminals and a single output (an exception being aggregate designs with two or more outputs, serving two or more analyzers).
(2) The operation is controlled by an external signal (though the signal generator may be integrated into the unit, even to the degree of sharing some of its components).
(3) There is the essential requirement of elimination of any possibility of cross-contamination between the samples.

The sampling units, though this is not always recognized, actually consist of two essential constituent parts: apart from the array of the valves, there is the less obvious but also important flow junction circuit (Figure 6.60). In more conventional sampling units with mechanical valves, the junction circuit may be rather trivial, since the mechanical closure ensures there is no communication between the cavities containing different samples. All valve outlets may simply be connected together. The more elegant, pure fluidic, no-moving-part solution is schematically presented in Figure 6.61 with a particular valve type in Figure 6.62 chosen from several possible

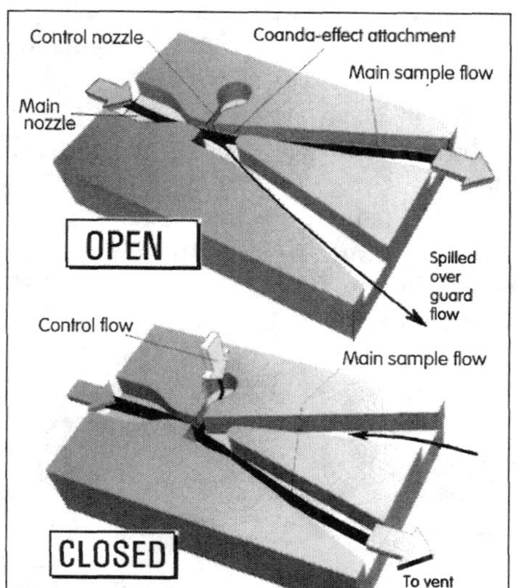

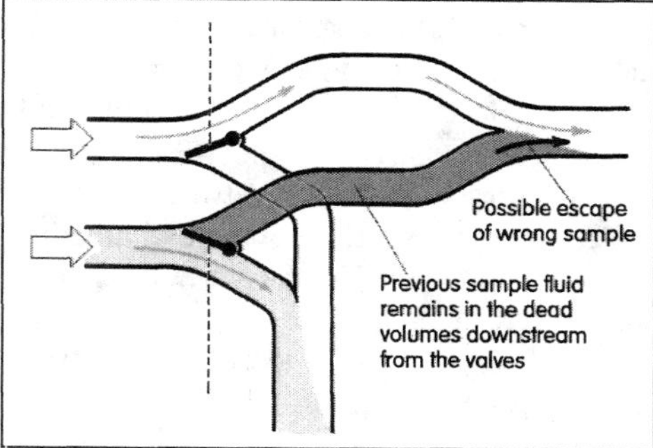

Figure 6.62 (Left) The controlled monostable switching valve that can be, as in Figure 6.61, the basic building block of a pure fluidic, no-moving-part sampling unit. Note the indicated small protective and cleaning flows.

Figure 6.63 (Right) One of the problems with the mechanical valves in the sampling unit is caused by the "dead" downstream volumes. The fluid trapped there differs from the actually tested sample, and the problem is the possibility of its escape into the downstream flow endangering the results of the often extremely sensitive composition analysis.

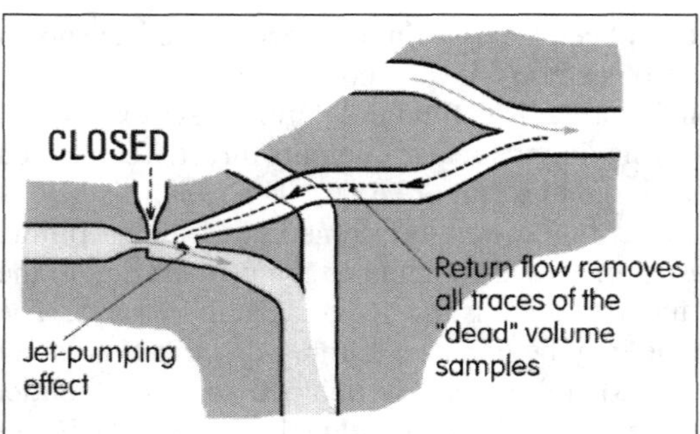

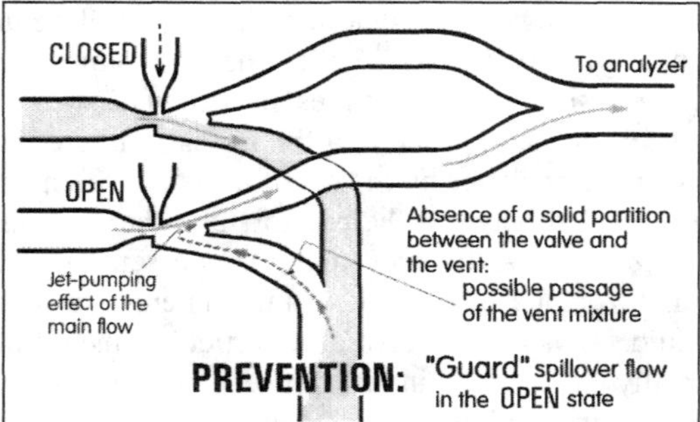

Figure 6.64 (Left) It is an advantage of the fluidic diverters that, with proper adjustment of pressure conditions in the circuit, it is possible to employ the jet-pumping property of the main jet for removal of the fluid from the "dead volume."

Figure 6.65 (Right) With the no-moving-part valves in the sampling unit, there is no solid-wall separation between the sample flow into the analyzer and the incontrollable mixture of fluids in the common vent. Although the danger of a return flow from the vent is not high, in the interest of absolute sample purity it is advisable not to lose sight of this eventuality and arrange in the OPEN state for a small "guard" overspilling flow (see Figure 6.62) protecting the sample.

valve types. The fluidic version is certainly much better suited than the solenoid valves to the often high temperatures that they are necessary in the research of catalysts.

The fluidic valves, however, are more sensitive to the proper distributions of pressure in the fluidic circuit. Using the flow interaction devices on the junction side may be instrumental for the proper flows and pressures adjustment. The junction circuit may then resemble Figure 2.80. The fluidic elements there are also no-moving-part devices (sharing the advantages of easy manufacture and extreme reliability under adverse conditions). Their task is to generate a specified mutual interaction of the flows that meet in them. The effects are dependent upon dynamic effects in flowing fluids, which—similarly as in the valves—limit the use of these elements to higher Reynolds numbers, on the order of at least 10^2. There are two basic cases of these flow interaction elements. Both consist of two nozzles and one collector and differ only in the angle at which the nozzle exits meet. If the angle is small, in the mutual flow enhancing flow junctions (Figure 2.82), the jet generated in one nozzle induces the flow in the other inlet by jet-pumping action. The opposite effect is achieved if the angle at which the nozzles meet is near to or equal to π in the mutual flow inhibiting devices (Figure 2.81). The flow admitted into one of the nozzles tends to generate a flow of the opposite sign in the other inlet. This element is usefully connected to the output terminals of two adjacent diverter valves. The sample flow from the valve in its **OPEN** state generates the desirable backward flow through the output collector of the other valve, which is in the **CLOSED** state. Of course, specific requirements of pressure and flow balancing in the junction circuit may call for devices in between, with intermediate values of the angle.

As a rule, the pure fluidic valves in the sampling units are controlled by fluidic control signal flows carried in channels. The valves may be individually switched, in which case the unit needs as many control signal input channels as sample flow channels. Another possibility (provided, as usual, that the switching is always in an invariant sequence) is building in the sampling unit of a fluidic ring counter (Figure 4.215). The sampling unit is then controlled by pulses carried into it through a single control channel. This does not influence the control of the sample flow switching valves; its inputs, instead of being connected to the external control, are each connected to one stage of the counter.

Quite often the combinatorial experiments are performed at a high level of precision, requiring a high level of sample purity. One of the possible causes of cross-contamination are the remnants

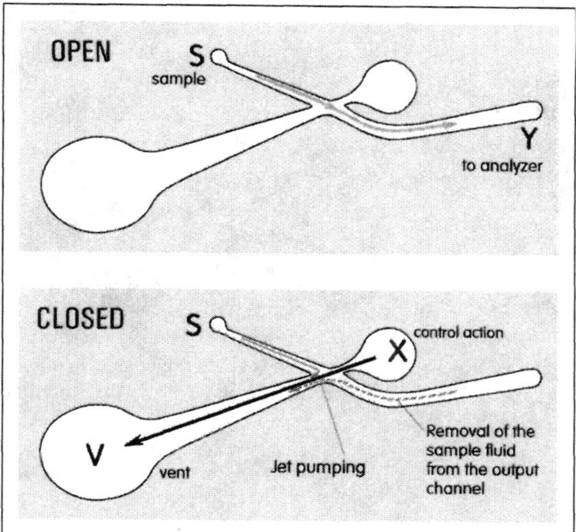

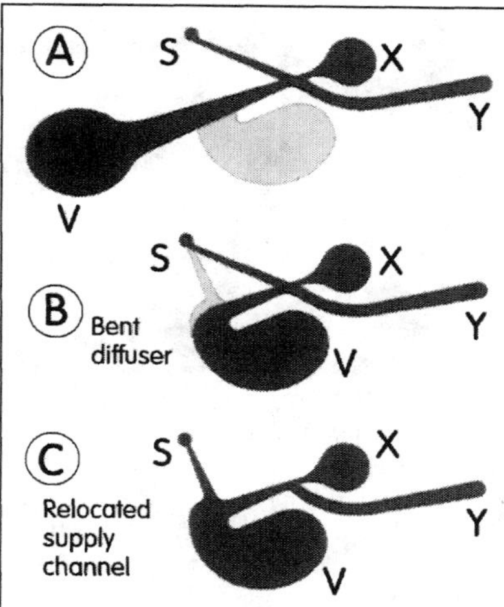

Figure 6.66 (Left) One of the initial geometric variants of the low-Reynolds-number valve developed for the sampling units. Instead of the insertion of a solid partition, the CLOSED state, stopping the flow of the sample to the output Y is accomplished by the powerful control jet from the control terminal X. The jet-pumping entrainment action of this control flow also performs the secondary task of generating a return flow from the downstream cavities, as shown in Figure 6.64.

Figure 6.67 (Right) The development of original valve geometry A corresponding to Figure 6.66 progressed through stage B with the bent diffuser and finally arrived at the shape C.

of a sample trapped in the "dead" volumes downstream from the valves (Figure 6.63). This is difficult or impossible to avoid with the mechanical flow-switching valves. A significant advantage of sampling units with fluidic valves is the possibility of the "dead" volume fluid pumped back into the CLOSED neighbor valve. This is done by the jet-pumping effect of the entrainment into the main jet (Figure 6.64), which is directed into the vent in the CLOSED state of the valve.

6.3.3 "Guard" flows

With the fluidic valves, there is another possible cause for cross-contamination. In principle, there is always a possibility that the jet-pumping action of the jet of supplied sample will suck into the valve some uncontrollable mixture of fluids from the vent (Figure 6.65). After all, the active cavities of the valves are not separated by any solid inserted partition, as is the case in the mechanical valves. The fluidic valves may actually by intentionally designed to be able to generate a strong jet-pumping effect to clear the "dead" zones in the CLOSED regime. At any time, all but one sample flow and all the control flows coming from the CLOSED valves meet in the common vent cavities and freely mix there before leaving the sampling unit. To make sure this uncertain composition mixture is prevented from coming into contact with the sample in the only OPEN valve calls for adjustment of the conditions in the fluidic circuit. What is needed in the CLOSED state is a spillover effect, the very opposite; instead of the pumping occurs the spillover effect. The small auxiliary protective spillover flow [14, 15] is called *guard flow*. It means sacrificing a small amount from the sample. To keep the resultant total sample loss acceptable, these cleaning flows must be very small. To generate the *guard flow* in the OPEN valve and the *cleaning reverse flow* in the CLOSED state of the same valve is not easy, due to their mutually contradictory natures.

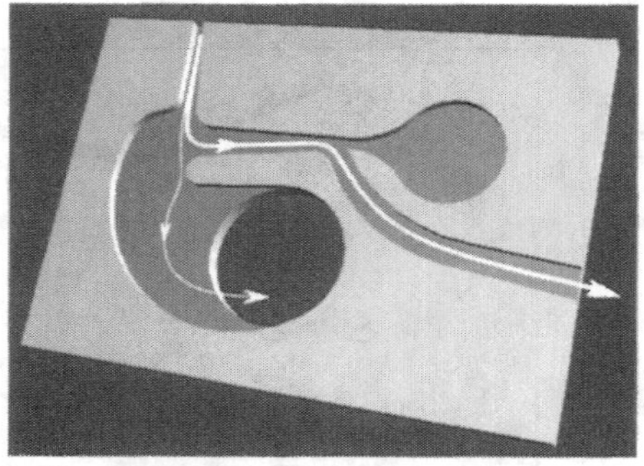

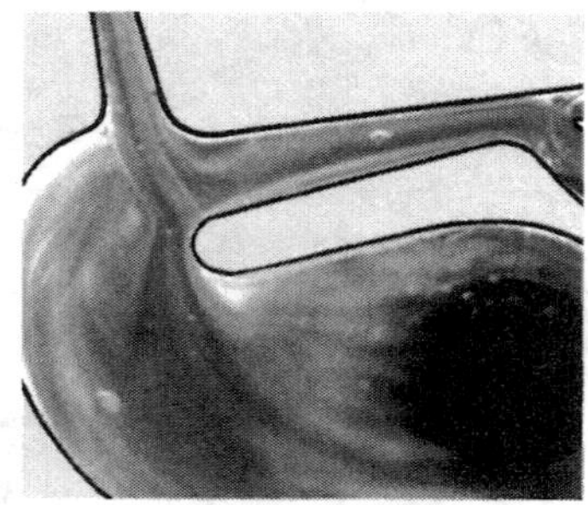

Figure 6.68 (Left) The sample fluid flows in the final version of the valve in the OPEN state: the relocated supply channel mouth (changed from shape B to C in Figure 6.67) produced the desired "guard" flow, a small sacrificial flow into the vent preventing the vent mixture from coming back into the valve cavities.
Figure 6.69 (Right) The flow in the valve from Figure 6.68 was visualized by adding dye to water in a scaled-up model to validate the idea of the "guard" flow in the OPEN state.

6.3.4 Low Reynolds numbers: jet pumping by control flow

At very low Re, the sampling unit discussed so far cannot operate as described above. The Coanda-effect attachment to a wall, employed in the valves in Figure 6.61, usually ceases to exist below about Re = 800. At even lower Re relying on the inertia of a jet of sample fluid accelerated in the nozzle is futile. Instead, it is necessary to drive the flow through the unit by a pressure difference applied between the vent (V) and the output terminal (Y). This pressure difference ΔP_{YV} is acting simultaneously in all the valves. It acts also in those valves which are in the CLOSED state and has to be opposed by the powerful control jet. The low Re flow from the supply inlet S cannot be expected to generate by jet-pumping the return cleaning flow in the output channel.

The solution to this problem was found in using another available fluid flow for the generation—the control flow which fortunately is there in the CLOSED state and has to be powerful

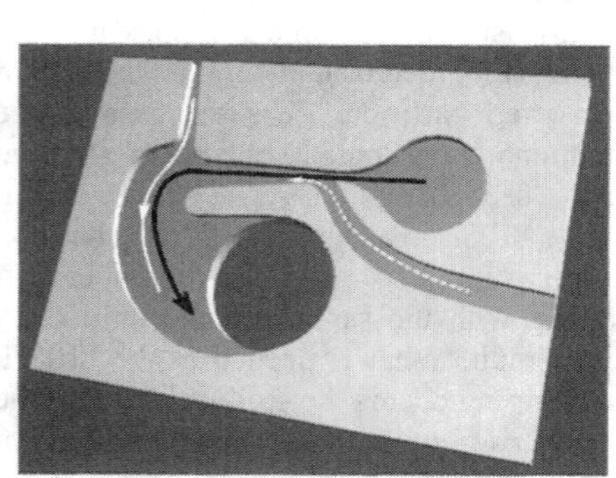

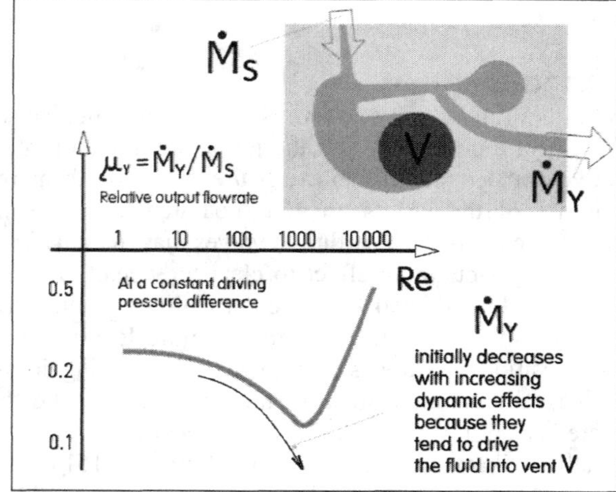

Figure 6.70 (Left) The control (black) and sample (white) fluid flow in the final version of the valve in the CLOSED state.
Figure 6.71 (Right) The relocation of the supply channel (change from B to C in Figure 6.67) resulted in an unusual character, already shown in Figure 3.165, of the transition into the subdynamic regime, presented (in a manner similar to Figures 3.159 and 3.162) in terms of the dependence of the relative output flowrate on the Reynolds number Re.

anyway. An example of the valve designed for the low-Re regimes is presented in Figures 6.66 to 6.71. In principle, this valve corresponds to the layout already discussed in association with Figure 4.131. As was the case in that figure, the valve incorporated a classic large-Re jet pump (Figure 6.66) driven by the control flow. The exit from the jet pump flows into the vent. Typical for low-Re microfluidic devices is the missing cross-sectional area contraction in the supply (main) nozzle, as there is no point in trying to accelerate the sample flow there. Originally because of spatial constraints, Dr. Tippetts, who redesigned this valve, found it necessary to bend the diffuser of the jet pump – B in Figure 6.67. This bent shape was later found useful for the required generation of the small *guard* flow heading into the vent V. This was achieved by directing the supply channel coming from terminal S so that it was inclined toward the final, curved part of the jet-pump diffuser (Figures 6.68 and 6.69). This changed location of the supply channel, of course, makes the flowpath from S to Y in the OPEN state more difficult and certainly loses whatever jet inertia might be still available in the low-Re regime. This is, however, unimportant in the fully pressure-driven subdynamic regime.

Figure 1.5 shows an early example of the sampling unit with these valves [14], as it was tested in a research facility for combinatorial development of catalysts for the Fischer-Tropfsch reaction (hydrogenation of carbon monoxide to ethanol). The valves are not very small—the supply nozzle channel width is 340μm—but they operate at the very low Re = 32 because of the small velocity and viscosity (is as high as 40.10^{-6} m^2/s) of the hot *syngas*, containing a substantial proportion of hydrogen. The valve is controlled by cold nitrogen control flow. In terms of standard fluidic amplifiers, the valve flow gain is hopeless: to get the jet-pumping effect required, the control flowrate had to be about 40 times the sample mass flowrate supplied into the valve. The Reynolds number of the control jet is around Re = 1 000, just enough to get a perceptible entrainment effect.

6.3.5 Biological tests

Testing the properties of developed drugs is traditionally done on animals as the test objects. Despite the justification—prevention of loss of the human life by the developed pharmaceutical—there are serious ethical objections to this approach. An alternative of increasing importance, *in vitro* tests on cultured cells, is another example in which it is possible to employ the techniques developed in microfluidics. Of course, this approach is reasonable only in the case of pharmaceuticals that do not affect whole organs (there is actually also a possibility of growing *in vitro* some organs starting from stem cells, but this is also a controversial subject, particularly with embryonic stem cells).

In the tests on cells, they are kept in chambers supplied with nutritional cell culture medium and provided with a drain removing metabolic products. There are two alternatives. The direct approach uses individual cells. The problem is they are typically a decimal order or two smaller than the typical channel size of present-day microfluidics, and this causes manufacturing problems. The other approach handles neutral organic microspheres instead, with the cells forming a surface coating. This makes it possible to operate in much larger and easier-to-handle scales. Things are made easier by the cells' tending to adhere to the walls, either the walls of the test chambers or those of the microspheres, by peptide adhesion ligands. On the other hand, the coating tends to be inhomogeneous: living cells split and duplicate and, on the other hand, may also die. The propagation is often limited by the contact inhibition phenomenon. It is not possible to expect a formation of test objects consisting of cells and having reproducible physical properties. The other problem is that, in contrast with physical censors for evaluating combinatorial tests, the biological tests result in irreversible changes. After each measurement, the cells become unusable for further experimentation and usually have to be replaced by fresh cells for the next measurements, with drugs having the slightly different compositions. Figure 6.72

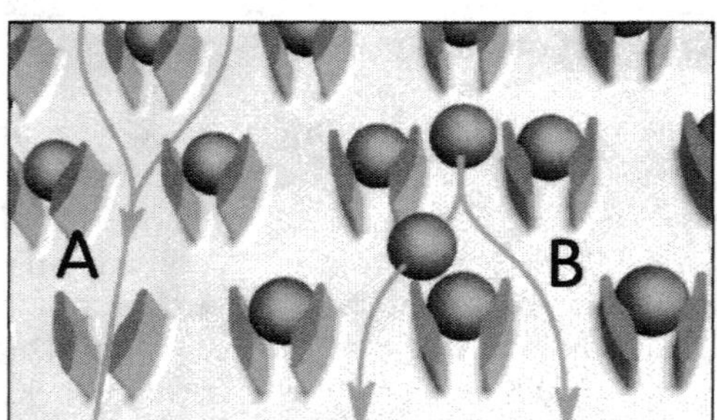

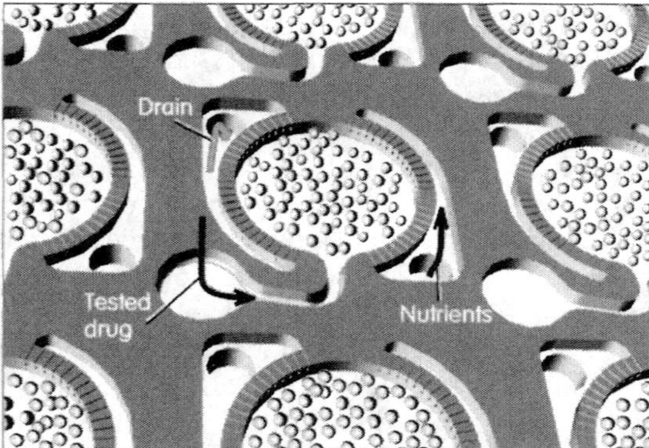

Figure 6.72 (Left) Automatic positioning of individual living cells by flowing culture medium into locations corresponding to a ccd array that is used to monitor the optical properties of the cells. A: fluid flow centered toward an empty position. B: streamlines (and a cell carried with the fluid) avoid an occupied position.

Figure 6.73 (Right) Two-dimensional testing of a drug by cells living in an array of circular chambers. The drug solution, together with the metabolic products, leaves into the drain exit.

presents an example of a biological sensor with an optical readout. For reproducibility of the readings, the cells (or cell-covered microspheres) are held in an organized pattern by the regularly spaced protrusions. Each location defined this way is positioned under the particular location of the ccd array above. Of course, placing cells in these locations in preparation for the next measurement could be extremely laborious. Fortunately, it was found to be possible to use the "self-assembly" property of the cell-carrying fluid flowing past the protrusions, to which the cells position themselves. After the testing of a drug, the cells are removed by reversing the flow direction (upwards in Figure 6.72).

Figure 6.73 presents an example of a biological test layout without the flow stitching by a sampling unit. Instead, the chambers are arranged in a 2D pattern corresponding to the 2D pattern of the reactors (Figure 6.55). The cells are permanently exposed to the tested drug and, at the same time, are permanently monitored by a ccd sensor array, each sensor located above one of the circular chambers in which the cells are kept.

6.4 ARTIFICIAL NOSE AND TONGUE

Recent interesting development in the field of systems imitating the human senses of smell and taste was also made practically possible by microdevices. (Earlier attempts did exist, but have not led to usable results.) It may be surprising how late these systems evolved compared with those for the senses of hearing and vision, which have been developed to quite high levels since the beginnings of electronics. One reason is perhaps a lower importance of smell and taste in present-day society – though their importance cannot be described as negligible, since they retain the roles of providing sensual pleasure (perfumes, cuisine) and warning of danger (spoiled food). A factor may be the difficulties in developing the "artificial nose" and "tongue." Not only did their development require a high level of sensor development, but it also needed very sophisticated data processing. At any rate, even the basic operating principles of the olfactory and gustatory senses are still not completely understood. In humans, the two sensations are related. Individuals with a temporarily disabled sense of their smell have their taste discrimination capability decreased to a mere ~ 20% of what they can sense with their smell cooperating.

In principle, both smell and taste respond to the presence of molecules of certain species. Taste responds to molecules dissolved in water (and ions resulting from dissociation). Smell detects molecules carried in the air, though the actual sensing also involves a dissolution step in mucus. For being detectable by olfactory organs, the odorants have to be volatile (it is impossible to detect metals). Besides this necessity of high vapor pressure, they must also be soluble in water and in fat (*lipophilicity.)* Interestingly, the detection mechanism can only handle molecules that are not very large: there is no known odorant with a relative molecular mass greater than 294.

Our present-day practical applications requiring a fine discrimination of smells are still heavily dependent on using trained dogs. Their replacement by engineering systems would bring obvious advantages, but for several reasons it is not yet possible. Technical solutions can currently exhibit comparable sensitivity as presented in Figure 6.75, but what is not yet fully solved is a unique correspondence between the recorded signals and the characterization of the odorants. Initial attempts at the artificial nose used only a single sensor. In some cases a sensor, such as the UV spectrometer in the example presented in Figure 6.42, could provide enough information in the absorption spectrum about the nature of the detected odorant, but more typically the idea was to perform a much simpler task: to determine the presence and perhaps the concentration of only one type of analyte. This may suffice for a particular specialized task. On the other hand, a characteristic of the present-day layouts is the use of an array of sensors, the signals of which are evaluated in a complicated way involving pattern recognition and comparisons with stored library of odorants.

Microfluidics comes into this process on the inlet side, where either the water or the air with the detected molecules is transported to the sensor elements. The basic task is to distribute the

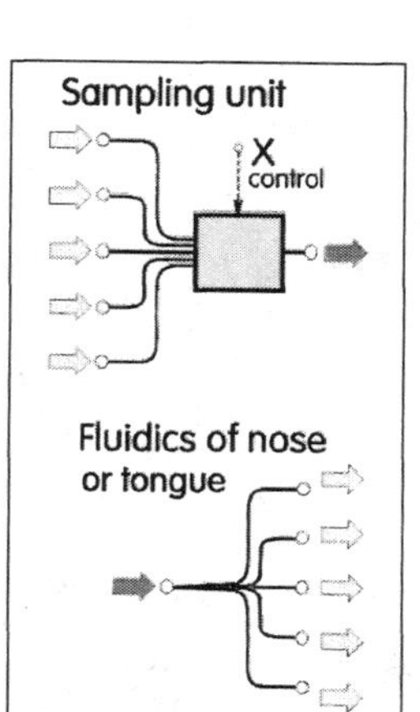

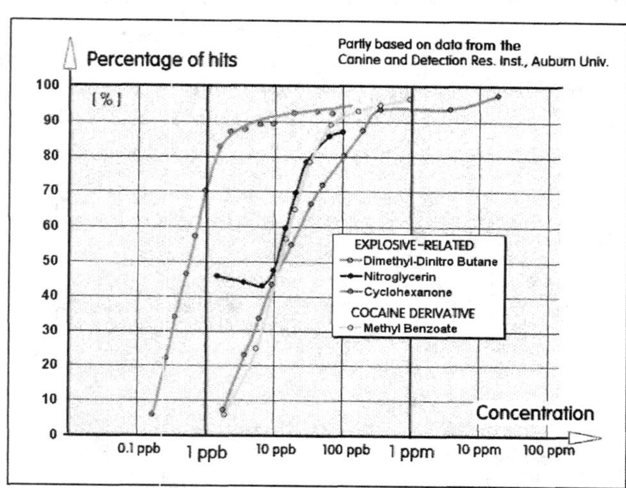

Figure 6.74 (Left) The task of fluidics in artificial nose or tongue systems is in principle the opposite of its task in the sampling units discussed in the previous Section 6.3.1.2, where fluids from many inlets are directed to a single outlet. Here there is a single inlet, and the fluid has to be delivered to a large number of sensors. The task is simpler as there is no need to impose the controlled operation in a time-dependent sequence.

Figure 6.75 (Right) Olfactory sensitivity of a typical dog, *Canis lupus var. familiaris.* The limit is very roughly at about 10 ppb, which is (compare with Figure 6.42) also the sensitivity limit of typical modern hand-held detectors. The problem is, however, the subsequent processing of the signals. The dog identifies a given complex of odors, handles it as a single conceptual whole, and recognizes it among other similar complexes. This is very difficult to do with instrument signals.

analytes to the sensor array. It is interesting to note that in this respect the basic configurations (Figure 6.74) are the inverse of the basic problem with the sampling units discussed in the previous sections. Studies of sniffing dogs, however, revealed that they do not just simply inhale. The curious shape of a dog's nose, with the lateral slit, is not accidental. In the exhaling phase, the generated jet is arranged so that it brings toward the nose the outer air with odor traces. Especially when the dog sniffs a possibly contaminated surface (Figure 6.76), the fluidics of the sniffing is quite sophisticated, and fluidic "noses" can learn quite a lot from this behavior. In applications in which sensitivity is of prime concern and no particular emphasis is placed on the speed of response, fluidics can improve performance with the addition of a preconcentrator stage, as is shown in Figures 6.29 and 6.42.

As the names *artificial nose* or *artificial tongue* indicate, the designs are guided by the idea of imitating the principles of the mammalian olfactory and gustatory systems, though the principles of the sensings are not suitable for direct imitation. Instead, most designs use the standard methods described in Chapter 5: mostly the surface acoustic waves (Figures 5.106 to 5.109), conducting organic polymer sensors, thin oxide layers (Figure 5.165), oscillating cantilever microbalance (Figures 5.44 and 5.45), or optical sensors (Figures 5.121 and 5.124 to 5.128).

The term "nose" in this context is not correct. The animal (and human) sensors are not located inside the nose but further along the breathing passages, in olfactory bulbs separated from the brain by only a thin cribriform bone plate. In humans, the bulbs are at the very top of the nasal cavity, just below the level of the eyes. There are two bulbs, one on each side, and they are quite small. Their sensitive epithelium surface covers only ~ 2.5 cm^2 in total. It contains approximately $50 \cdot 10^6$ receptor neurons plus their supporting cells. The receptor neurons are bipolar. Axons projected from the basal pole pass through the pores in the cribriform bone, in bundles of 10–100. On the brain side of the bone, the signals are transferred to mitral cells, which relay them to the olfactory cortex in the brain. On the other side, apical poles with the receptor neurons extend to the epithelial surface. There each neuron expands 8–20 cilia located in mucous layer. Cilia, 30–200 μm long, are the locations where the odorant molecules are detected. The odoriferous substance must first dissolve in the mucus. A constant secretion of the mucus (by the Bowman glands) washes the cilia, preventing them from sensing the same smell continuously. The molecules detected by the sense of taste arrive already dissolved, as ions produced by dissociation.

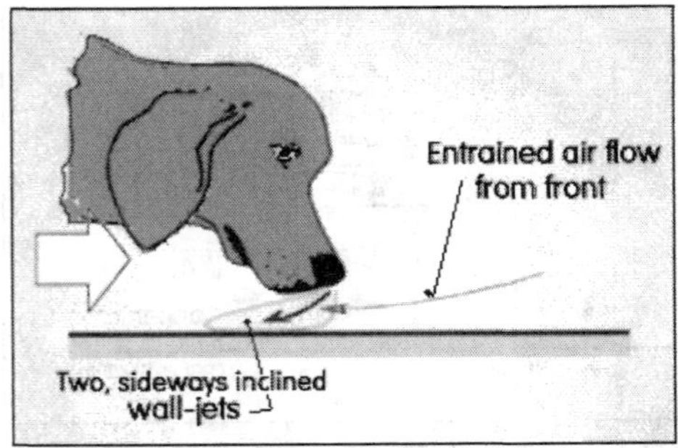

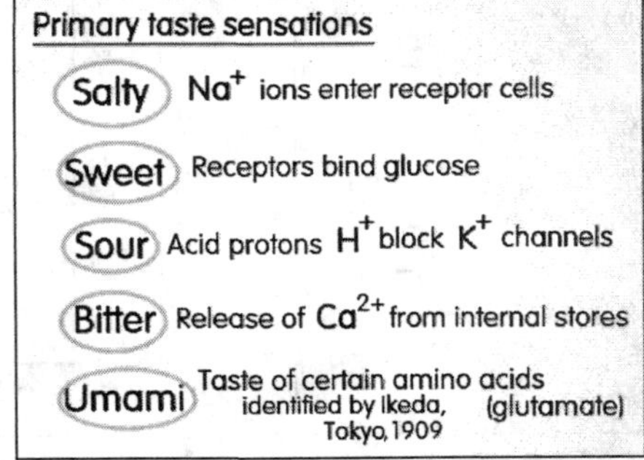

Figure 6.76 (Left) A sniffing dog not only inhales air carrying odors. In the exhalation phase, which is also very important, airflow patterns bring odors from the surrounding atmosphere. The flow is also arranged so as to release the odor molecules from the solid surfaces to which they adhere.

Figure 6.77 (Right) Until 1909, western civilization knew only four primary tastes. There is a fifth that cannot be composed from the four other ones. In principle, all other existing tastes may be decomposed into these primaries, but people cannot distinguish the individual contributions. Moreover, no sensors are available that would respond specifically to only a single primary taste: the sensitivities of existing sensors overlap.

They are detected by taste receptor cells, clustered in taste buds on the surface of the tongue. Classic textbooks used to show different taste areas (sweet, sour, and so on) on the tongue, but this is wrong. The receptors for the different sensations are distributed in the buds evenly. A single taste bud contains 50–100 taste cells. Each bud has a pore enabling the ions to reach the receptor cells inside. Similarly as all the colors seen by an eye consist of a combination of the three primary colors, all taste sensations may be represented as a combination of the five primary tastes in Figure 6.77. Somewhat surprisingly, in each of the five tastes shown there, the basic recognition mechanisms are different. Designing an "artificial tongue" is complicated by the fact that no known fluidic sensor can respond selectively to only one primary taste. The responses of real sensors overlap and the decomposition has to be completed at the signal-processing stage.

The basic problem with the "primary odors" is similar. Initial studies approached the problem with studies of subjects with *specific anosmia* defects in olfactory perception. This revealed six primary odors of the human sense of smell, observed as insensitivity to *isovaleric acid, l-pyrroline, trimethylamine, isobutyraldehyde*, 5-*androst*-16.*en*-3-*one*, and *-pentadecalactone*. The corresponding primary odors were: *sweaty, spermous, fishy, malty, ruinous, and musky*. This decomposition appears to be well adapted to perceiving food and body odors, thereby conveying nutritional information and pheromonal signals in animals and man. Later, around 1977, however, evidence from the known varieties of anosmia suggested that the total number of human primary odors is 10. This was used for some time and was found quite useful for identifying, for example, food fragrances (Figure 6.78). It lead to the layout of the basic sensor component of an artificial nose, as in Figure 6.80. The first problem arose from the practical impossibility of synthesizing

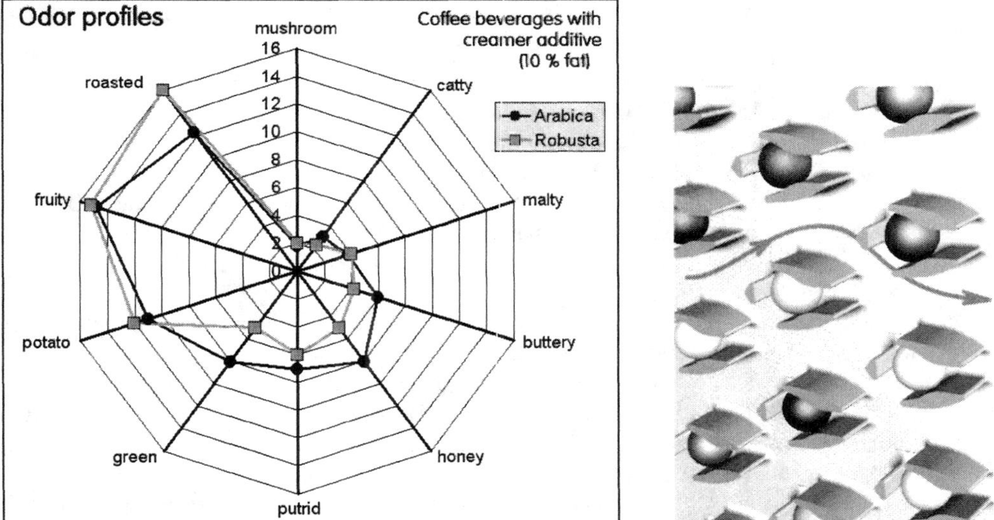

Figure 6.78 (Left) The situation with primary odors is even more complicated than that with primary tastes. For some time it was believed there are 10 basic smell sensations. Indeed, some food fragrances (e.g. for coffees shown here) can be relatively well characterized by using decomposition into these 10 primaries. Even highly trained individuals cannot identify all ten of them in a complex smell with absolute certainty. To make things worse, recent research results indicate there are many more primary odors, at least 32 and probably as many as 50. (After [49].)

Figure 6.79 (Right) Converting smells to colors: microobjects are available, each responding to presence of particular molecules in the fluid by changing their color. One type of them are cells generically modified to glow when antigens bind to their binding sites, Figure 5.121. Another possibility is colloidal polymer microspheres with antibody-derivatized surface coatings. In contrast to Figure 6.52, here the upstream "pin" holds the objects permanently in their locations.

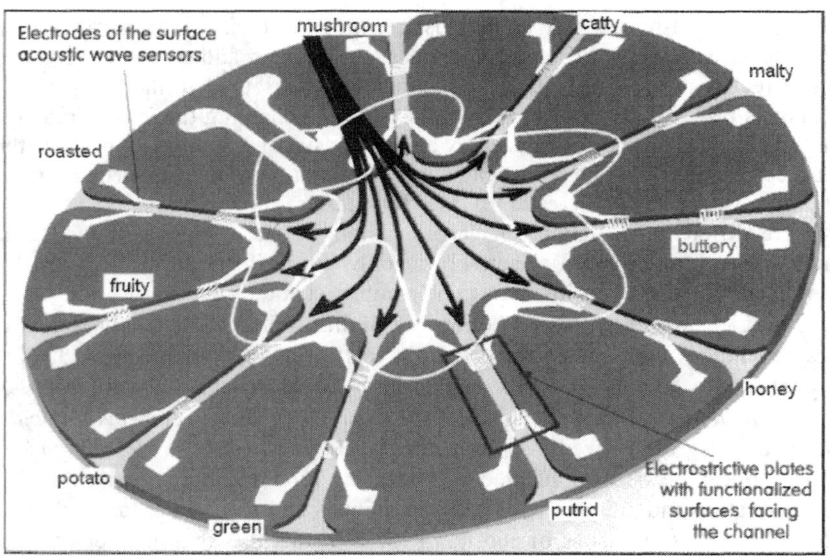

Figure 6.80 An array of olfactory sensors based on the surface acoustic wave principle (Figure 5.107). Ideally, a unique identification is possible if each sensor responds to only one of the 10 primary odors (Figure 6.78). Unfortunately, available practical sensors do not possess this selectivity of response. Existing sensors respond to odors in a mutually overlapping manner.

sensitive substances selective enough to respond to only one of the 10 primaries. More significantly perhaps, recent studies have progressively increased the number of what are considered the primary odors. Some scientist consider as the final blow to this concept the results of genetic studies of the receptor neurons. Analysis of their DNA can distinguish all the types of these neurons. Surprisingly, the number of the types is large: there are about ~ 1 000 odorant receptor genes. Actually, only about 350 odorant receptor genes are functional in humans, and the remaining 560 are pseudogenes. The receptor genes comprise nearly 2% of the 50 000 or so genes of the human genome. On the other hand, mice were recently found to have 1 296 receptor genes of which 20% are pseudogenes. Humans have obviously lost nearly two-thirds of their receptor genes compared to mice (and other mammals).

The nonavailability of sensors for the primary sensations—and in the smell case the nonexistence of a reasonable set of primary odors—has led to designs with large numbers of nonselective sensors. A typical example is the use of microspheres with color-responding coating (Figures 6.79 and 6.81). The task of describing a particular smell or taste and identifying it among a large number of similar ones is left to the pattern-recognition software part of signal processing. The use of colloidal polymer microspheres with molecular and polymeric reagents binding with particular analyte species for the detection is actually far from a new idea. They are in common use in clinical agglutination tests. Microspheres are commercially available with functional coatings in variants capable of identifying more than 60 analytes and routinely used for the detection of infectious diseases, narcotics, and early pregnancy. The microspheres covered with antibodies become linked by an antigen so that they agglutinate and become unable to pass through a filter. The microspheres are colored by distinguishing dyes so that their presence is detected by coloration changes after the removal of the antigen-carrying solution. In the artificial tongue detector, pictured in Figure 6.81, the microspheres are immobilized in cavities at the equidistant endpoint of the branches from Figure 2.77. The cavities are located under the pixels of

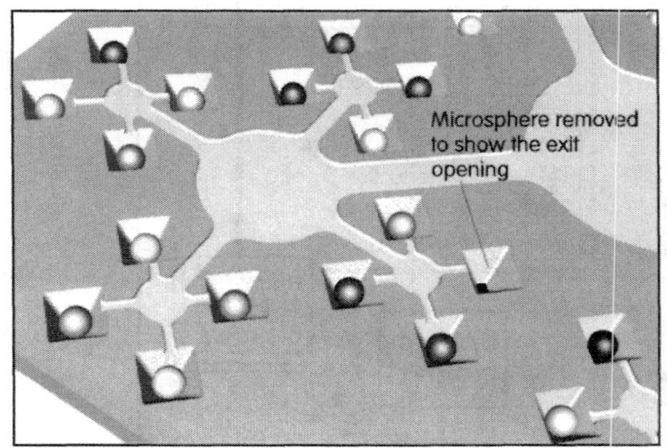

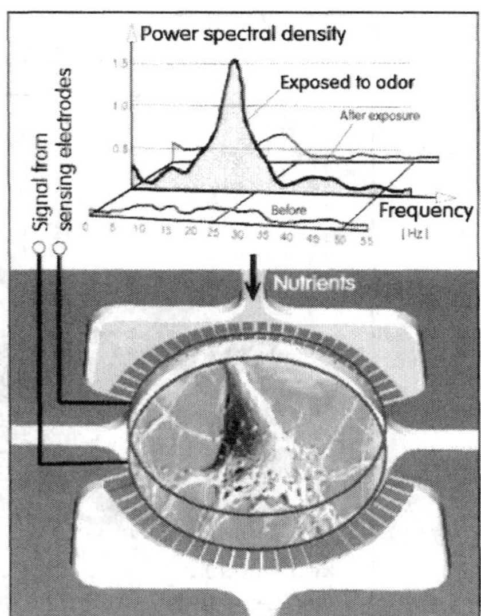

Figure 6.81 (Left) Example of "artificial tongue" sensor using a large array of functionalized microspheres: small polystyrene-polyethylene-glycol composite balls respond by changes in color (or fluorescence) to the detected molecules that are covalently bound to the termination sites. The changes are read by charge-coupled devices (CCD), and to ensure their proper correspondence, the microspheres, exposed to fluid flow, are individually immobilized in etched cavities.

Figure 6.82 (Right) Bionic odorant sensor: receptor neuron cell taken from a rat and living in a chamber formed on an n-type silicon. An electrolyte insulator SiO_2 layer 30 nm thin is formed on the surface by oxidization at 1 000 °C. The cell is regularly fed with fetal calf serum. The output in the presence of the deodorant is periodic (24 Hz, as shown in the top part of the picture) extracellular voltage sensed by the electrodes.

charge-coupled devices (CCDs of the type used in camcorders). These can detect grayscales of one part in 256 000 (three decimal orders better than the human eye) and may be interrogated at speeds on the order 10^5 pps (pixels per second). In the "tongue" application, the coating of *o*-cresophtalein attached to the spheres was particularly effective. This can actually serve simultaneously as the binding site for the detected dissolved molecules as well as the color responding compound. The reversible response is a characteristic purple color in the presence of Ca^{++} at high pH (due to strong absorbance at 550 nm). The color disappears in an acid environment.

Particular attention has been recently paid to bionic odorant detection sensors. Typically, the sensor is formed by a culture of living receptor neurons on a semiconductor chip. Their response is a change in extracellular potential, conveniently in the form of an electric signal. The investigated cultures involve cells from insect antennae and human embryonic cells. Figure 6.82 shows a cell chamber; the sensor array may use a large 2D system of such chambers. A particularly successful approach was found in the use of olfactory bulb cells from 5- to 7-days-old young rats. The advantages are high sensitivity, rapid response, and excellent selectivity. The disadvantages are associated with the biological character of the cultures, which involve more difficult handling and sometimes imperfect reproducibility. Although the typical lifetime of the receptor cells is limited to only ~40 days, this may be circumvented if the cells have the opportunity to propagate by division. (It may be necessary, however, to arrange for the removal of dead cells.) The cells have to be fed by a culture medium; this may be needed only once every 1–2 days. The response to an odorant's presence is a very distinct periodic electric signal at the frequency of 24 Hz, as shown in the spectrum of the output signals at the top of Figure 6.82.

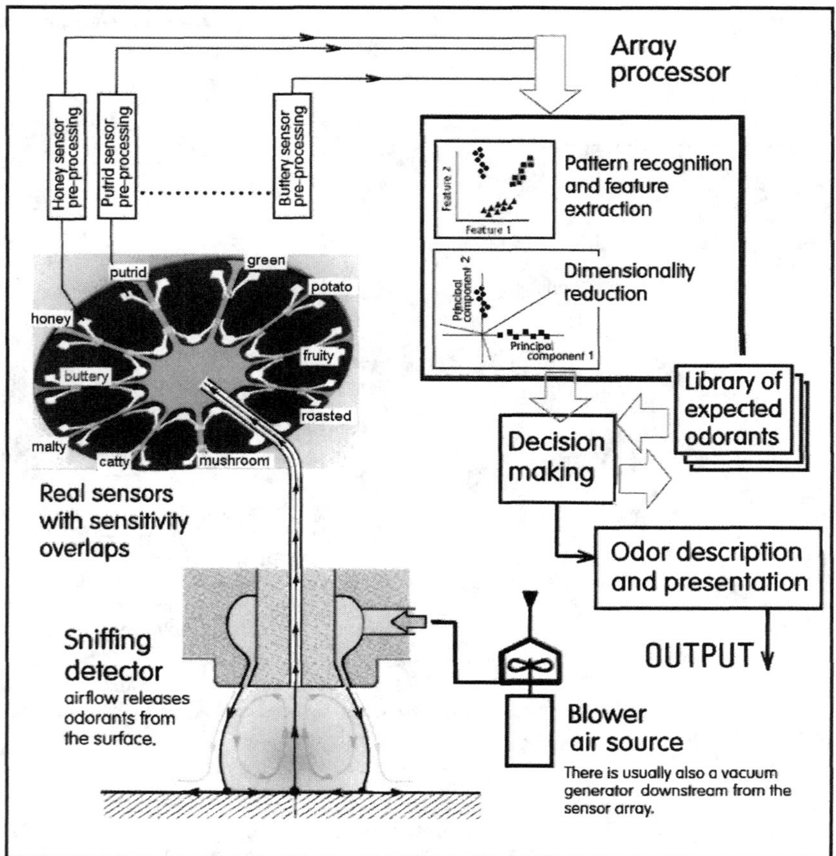

Figure 6.83 Components of a present-day artificial nose – in the clockwise direction: detector, system of sensors, signal processing, fitting to a pattern, and decision about the output based on comparison with a stored database. Fluidics, although essential, ensure just transportation of the odor molecules to the sensors. The complexity of signal processing is due to the nonavailability of both a system of primary odors and corresponding nonoverlapping sensors each responding only to its primary odor. The fluidics of the system may be more complicated if requirements of high sensitivity demand the use of preconcentration, similar to what is shown in Figures 6.29 and 6.42.

A schematic representation of the overall layout and signal processing in a typical artificial nose without odorant preconcentration and with an array of nonselective surface acoustic wave sensors is shown in Figure 6.83. The fluidic detector in this case uses an annular air jet impinging on the investigated surface for releasing the odorant molecules. The processing of the sensor signals involves comparisons with the stored library of signal pattern features. The ten-dimensional space of odors is for most practical uses reduced to a lower-dimensional problem by applying the procedure of the identification of principal components: the mathematics of this processing is described in [17]. An important part of the process is the teaching feedback loop: the library is updated using successful identifications.

Among recent applications of the artificial noses several interesting cases may be mentioned:

(1) Identification of spilled chemicals (for the U.S. Coast Guard);

(2) Quality classification of stored grain;

(3) Wastewater treatment monitoring;

(4) Checking interior smell of new cars (Figure 6.42);
(5) Detection and diagnosis of pulmonary infections (e.g., TB or pneumonia);
(6) Diagnosis of ulcers by breath tests;
(7) Food industry uses listed in Figure 6.84: in particular the identification of source and quality of coffee and the monitoring of the roasting process;
(8) Rancidity measurements for olive oil (due to accumulation of short-chain aldehydes);
(9) Freshness of fish.

6.5 FOOD – AND WASTE

One of the ways in which microfluidics can change our world is concerned with the most elementary of human needs: providing food. Consumed food may be divided into three general groups: animal origin, plant origin, and artificially produced. The first two methods are traditional, generally a heritage from the era when humans were a part of nature, and in order to stay alive they did not have any other choice than obtaining proteins from other living organisms. Increasingly, these sources are considered to be somewhat unethical, and the initially disapproved artificial synthesis of food to be an acceptable idea, one for which microfluidics can offer a number of opportunities. At any rate, the need to feed the permanently increasing number of inhabitants, now in the billions, has gradually led to food production in a factory-like style. The life of poor creatures such as hens in the egg factories, can hardly be described as dignified. Food habits and choices are among the most conservative and everybody prefers to have food from the traditional farming sources. Nevertheless, the social and financial pressures for changes are strong, and it is useful to consider the ways in which microfluidics can offer the means not only for a more efficient processing of food, but also for changes in the very principles of food production.

The other side of the food consumption coin is, of course, the reality of increasing amounts of produced wastes. The traditional ways of disposing of them, such as releasing them into rivers or discarding them in landfills are also near to their limits, even with investments in some processing, such as the present wastewater cleaning (imperfect and slow) or incineration (releasing dioxins).

6.5.1 Quality monitoring

Food quality is not only an esthetic or pleasure requirement. In the United States alone as many as 9 000 food-related deaths and close to 33 million cases of food-induced illnesses occur annually. The estimated cost of the resultant litigation and productivity losses to food industries caused by just seven specific pathogens is $35 billion annually. Traditional laboratory testing methods are not only expensive, requiring costly instrumentation and highly skilled personnel, but they are too slow: including transportation time for the samples, it takes several days before the results are known. By that time the food is usually sold and consumed. Handheld microfluidic devices can return results on the spot in a matter of minutes at most, if not seconds. They can also change the very principle of the monitoring: instead of checking the quality of the final products, the emphasis may be shifted to the points of entry for raw products. Bacteria or antibiotic residues may be detected, for example, in milk while still on the farm, before money is invested into its processing and before it is irreversibly mixed with other milk, making the traceability of culprits difficult and expensive.

FOOD INDUSTRY
applications of artificial nose

- ○ Raw substances quality
- ○ Storage monitoring
- ○ Cooking process monitoring
- ○ Fermentation monitoring
- ○ Maturation and ripening: wine, cheese, and so on
- ○ Quality of products
- ○ Packaging and shelf life

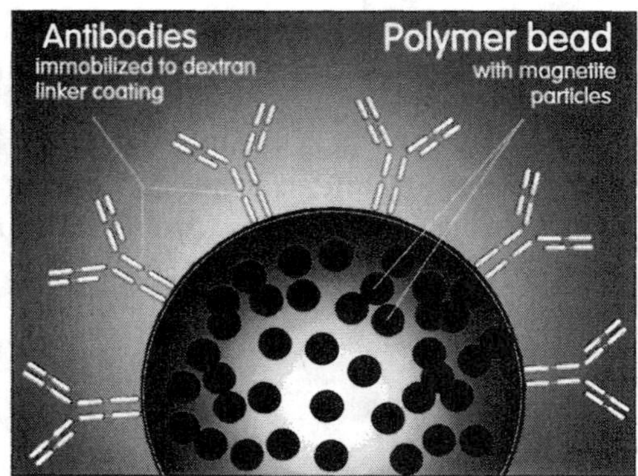

Figure 6.84 (Left) Gradually becoming important tools in food quality monitoring are the various "artificial noses" (Figure 6.83), replacing subjective evaluation by human experts.

Figure 6.85 (Right) Immunosensors typically use immobilized antibodies—obtained from laboratory animals (rabbits) exposed to a particular detected contaminant—for changing physical properties in a detectable manner. In this case, the binding to the antibodies changes the properties of beads used in the detector of antibiotic residues in milk, shown in Figure 6.86. The magnetic particles are used to move the beads by an external action: the shifting of a permanent magnet.

An important factor is also the permanently increasing number of contaminants to be monitored (Figure 6.86). Apart from the more-or-less traditional bacteria, *Salmonella* and *Listeria,* the food-illness outbreaks in 1982 underlined the danger caused by *Escherichia coli* 0157:H7. Other toxins included in the list are pesticides, toxigenic fungi, and mycotoxins. Until the hormone scandal of 1980, anabolic steroids were used for growth promotion in farm animals. They are now banned (in Europe since 1988), and this requires additional tests to ensure that the ban is respected. Later additions to the list are the β-agonists, which were originally therapeutic drugs, but in larger doses also reduce animal body fat and promote muscle growth. In Italy in 1994, their presence in rumpsteaks at a mere 0.5 ppm caused the food poisoning of 16 persons. Also an additional recent factor to be considered is the danger of bioagents used in terrorist attacks.

Detection using immunosensors:

Food contamintion:

- ○ Microbiological — Pathogens (*Salmonella, Lysteria*, and so on)
- ○ Chemical — • Antibiotics • Pesticides • Mycotoxins
- ○ Fungal — (*Aspergillus*, ..)

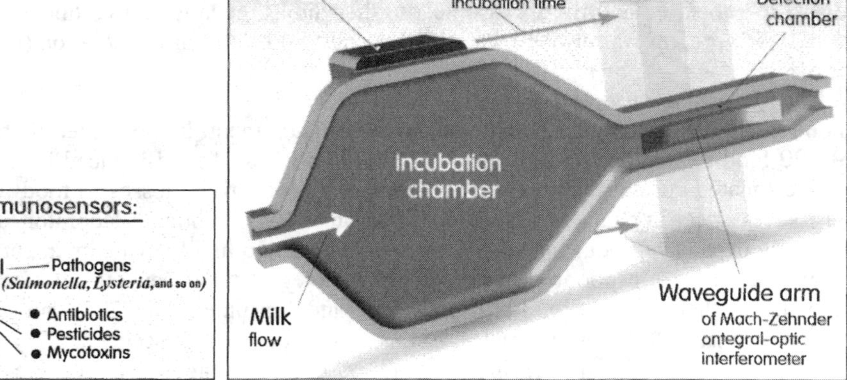

Figure 6.86 (Left) The main areas of microfluidic testing of food quality by means of the specific binding properties of antibodies. These may be attached to the beads (Figure 6.85), but also to electrodes exposed to the tested fluid.

Figure 6.87 (Right) Microfluidic device for detecting a specific contaminant in milk. The detection is made with the immunosensor beads of Figure 6.85. The readout uses an optical transducer shown in Figure 5.128.

In the example presented in Figures 6.85 and 6.87, microfluidics is applied to the detection of β-lactams, sulfonamides, tetracyclines, and chloramphenicol in milk. The screening is done at the farm before the milk is loaded in the collecting truck. The detection is based on the use of antibodies (Figure 6.85), biologically derived binding agents. For each of the antibiotic families, the device contains different antibodies immobilized on beads. Using the immunity mechanism, the detected antigens in milk bind to the antibodies and change their optical properties. The beads are synthesized with the contents of hematite particles so that they can be held magnetically in the incubation chamber, in spite of the flow of the surrounding milk, by a permanent magnet. After the incubation period, the magnet is shifted and moves the beads to the detection chamber. This is smaller, which results in an amplification of the sensing effect: a change in the optical properties of the fluid that surrounds the monomode optical waveguide passing through the chamber. It forms an arm an interferometer (Figure 5.128). The light propagation in the waveguide depends on total internal reflection. A certain small fraction of the light actually travels through the outer fluid (evanescent field), so that the change in the fluid optical properties caused by the presence of the antigens suffices for a substantial effect detected at the output of the interferometer.

An alternative are antibodies immobilized to electrodes (usually a thin layer of gold) connected to an electric circuit measuring the conductivity or electric capacitance of the fluid between the electrodes. Again, the presence of the antigens changes the properties (in this case, the electric properties) enough for a considerable output signal.

6.5.2 Food processing and meals preparation

Preparing dishes is not an area in which microfluidics has much opportunity. Its main features, handling fluids and functioning at a small size, tend to produce the proverbial "gray goo" rather than attractive, traditional meals. This is one of the most conservative areas. People prefer being served meals with consistency, shape, and visual properties resembling what their mothers served to them in their childhoods. Despite this difficult position, applications for microfluidics were found, and this section discusses two examples. One of them is mainly oriented toward preparation of beverages, an easier to change field because the final results do not directly reflect the

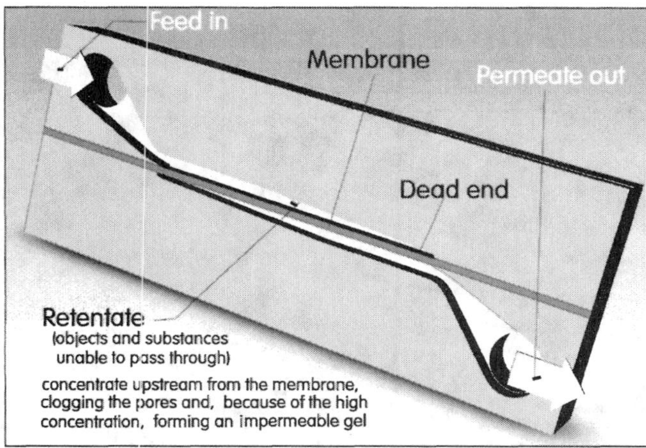

Figure 6.88 (Left) A detail, under large magnification, of a filter plate with regular, microfabricated, hexagonal pores. Pore sizes as small as ~4 μm may be made by standard methods; special techniques (laser interference pattern) makes realizable (less regular) pores as small as 65 nm, placing the process among ultrafiltration, Figure 6.93.

Figure 6.89 (Right) The "in-plane" version of the basic dead-end filtratrion. The fundamental problem is the gradual loss of effectiveness with time, due to pore clogging and accumulation of the retentate.

techniques used in their preparation. The other application somewhat paradoxically does not aim at a new character of the meals, but quite the opposite, a return to traditional meal consistency and character.

6.5.2.1 Microfiltration techniques

Filtration as a step in food production is performed mostly to remove unwanted components. The oldest well-known example is sieving – the removal of foreign particles, with the discrimination based on the particle size. Microfabrication processes developed for microfluidics, such as etching, may be quite easily applied to producing sieves with very small openings (Figure 6.88). Membranes with much smaller pore sizes, though not so regular, are available. They are small enough to prevent the passage of objects as small as bacteria. These membranes are usually polymeric but are also made from ceramics, metal oxides, or graphite. With very small pores, comparable to the dimensions of molecules and ions, the selectivity with which the membrane restricts transport may also be strongly affected by electric forces (Figure 5.76). A typical example the ion-exchange membranes introduced in the 1950s for purifying water. This was a particularly successful beginning that started gradually widening the range of various membrane uses in food processing. A membrane has to be thin for the transport of the permeate across it to not be very slow. Also the driving-pressure differences are then lower. Thicknesses as small as 100 μm are no exception. Even so, the pressures may be quite high and the mechanical stressing of the membrane considerable. The membrane cannot span a large transverse distance without support. This is why layout in Figures 6.89 (with improvements according to Figures 6.91 and 6.92) is usually replaced by the one in Figure 6.90. The features of microfluidics, the numbering-up principle and channel micromanufacturing, then find an application there quite naturally.

Some of the uses of membrane technology are listed in Figures 6.93 and 6.94. Of particular importance historically was the cold sterilization of beverages, accomplished simply by removing bacteria by filtration. It was a great success; compared with traditional thermal processing, it sufficed with substantially lower energy demands and avoided thermal damage to tastes and fragrances, in particular for fruit juices. In fact, modern nanofiltration membranes are capable of capturing much smaller components. In the case of juice, the captured objects may be so small

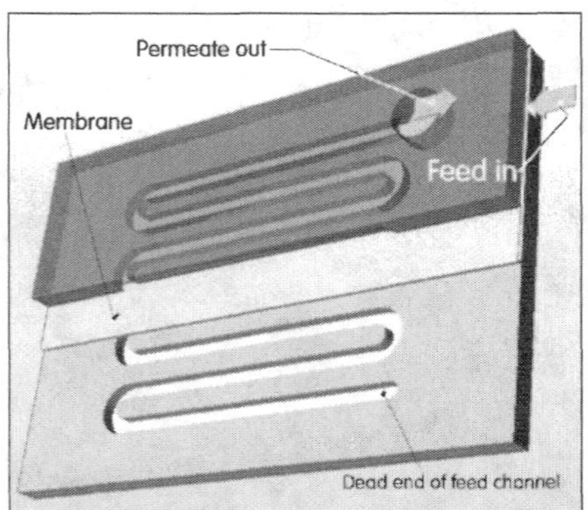

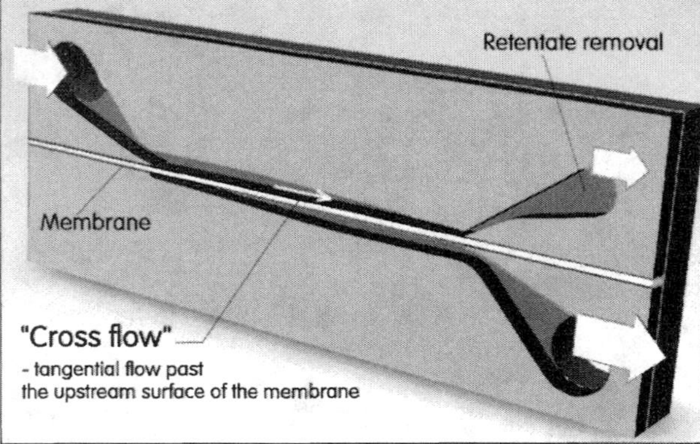

Figure 6.90 (Left) Typical microfilter of the "out-of-plane" dead-end variant. The meandering channel has one feed inlet and no way out except across the membrane into a similar meandering channel in the upper plate.

Figure 6.91 (Right) A considerable increase in the useful filtration time compared with the similar dead-end filter, (Figure 6.89) may be obtained by the "cross-flow" layout, providing an outflow of the concentrated substances not able to pass through the membrane. In some applications the loss of filtered fluid through the outflow may be a drawback.

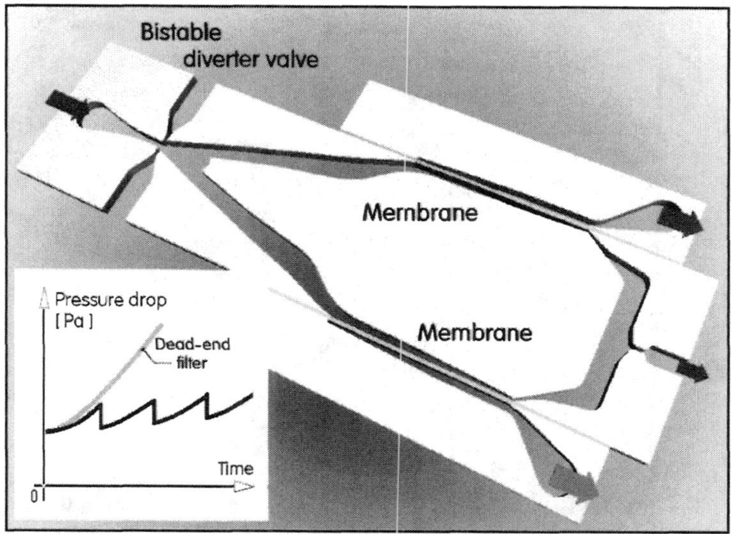

Figure 6.92 A fluidic diverter valve with a jet-pumping return flow in the OFF outlet removes a substantial proportion of the membrane pore blockage at each switching. This offers an opportunity to keep the pressure drop low and operate the membrane far beyond its lifetime in the usual dead-end filter layout.

that the permeate can actually lose color, fragrance, and taste compounds. This procedure is actually utilized in collecting these compounds for use as dried taste and odor extracts. The drying is also made possible by membranes, by allowing only pure water to pass through. This drying step is also used in producing dried eggs, even though, in that case, the final dry product needs a heating stage, the use of the membrane water-removal unit for thickening and concentrating the egg protein substantially decreases energetic demands and increases throughput.

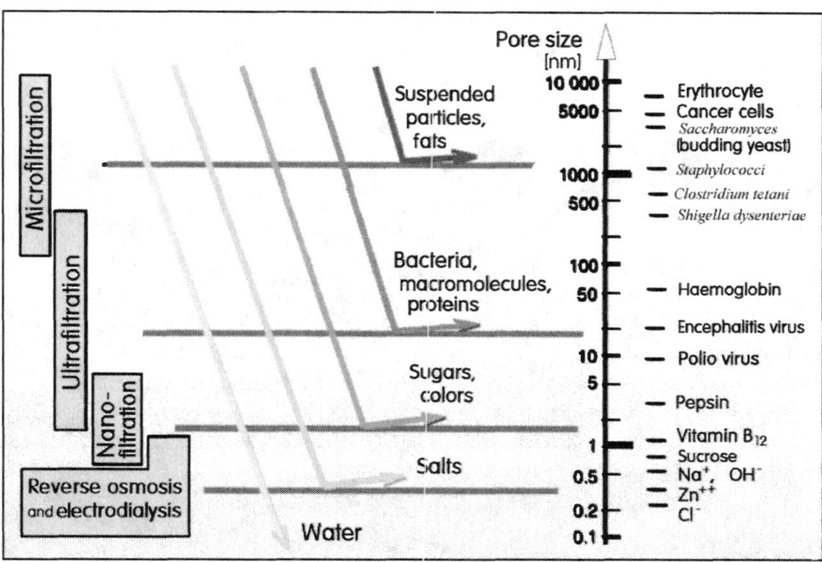

Figure 6.93 Typical examples of the uses of fine filtering techniques in food processing.

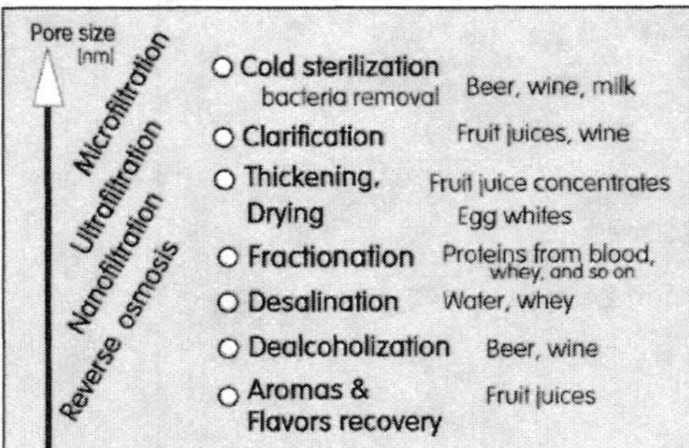

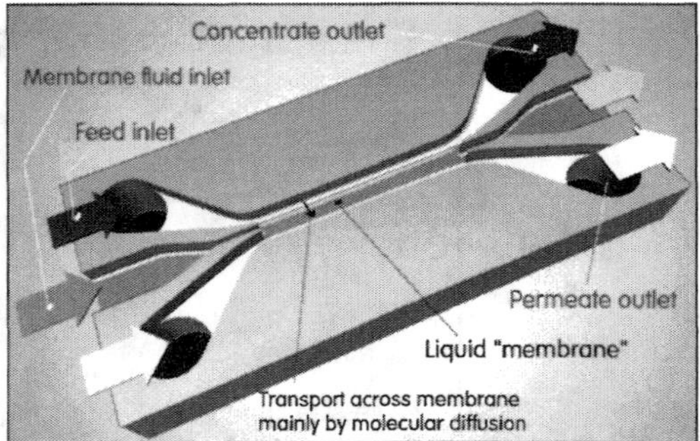

Figure 6.94 (Left) Fine filtering techniques in food processing, though typically performed in large-scale devices, may be applied with microfluidics devices to on-the-spot and personal use.

Figure 6.95 (Right) Perhaps surprisingly, in some filtering tasks the role of the membrane may be taken over by a thin laminar-flow layer of a third liquid, separating the feed and permeate liquids.

One of the basic problems, which initially seemed to be inevitable, was the limited lifespan of a membrane in operation due to the accumulation of the retentate – the constituents of the fluid prevented by the membrane from passing through. Apart from simple clogging (blockage of the pores by the immobilized particles), the flow through the membrane was also inhibited by the formation of a large upstream concentration of the retentate. The retentate objects may actually be very small (their size limited by the upstream filtration stages). The local concentration then produced a gel "cake" upstream from the membrane. A partial remedy was found in replacing the traditional dead-end that had no escape for the retentate (Figures 6.89 and 6.90), by the "cross-flow" design, according to Figure 6.91. Fluidics, however, brought an even better solution, according to Figure 6.92. It uses an upstream, simple-to-manufacture fluidic diverter valve. Switching it causes an alternating flow reversal (due to the jet-pumping action of the main jet in the valve; see Figures 3.152 and 4.97). The control terminals of the valve may be connected by a feedback loop, to form an oscillator (Figure 4.152) with automatic permanent switching. Of course, any membrane clogging is completely avoided in the recently introduced "liquid membrane" filtering (Figure 6.95). The processes involved in it are more complex, and the special

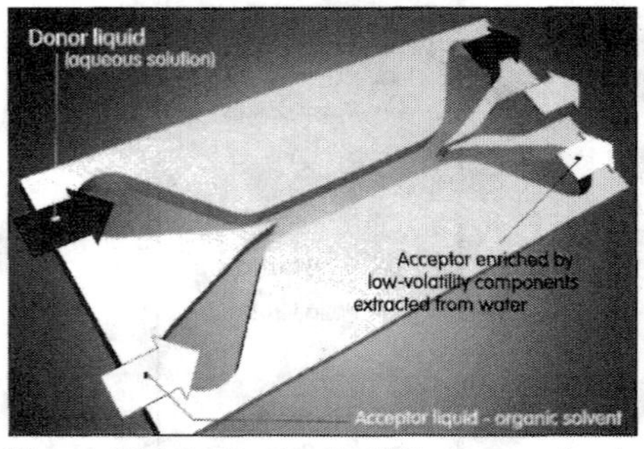

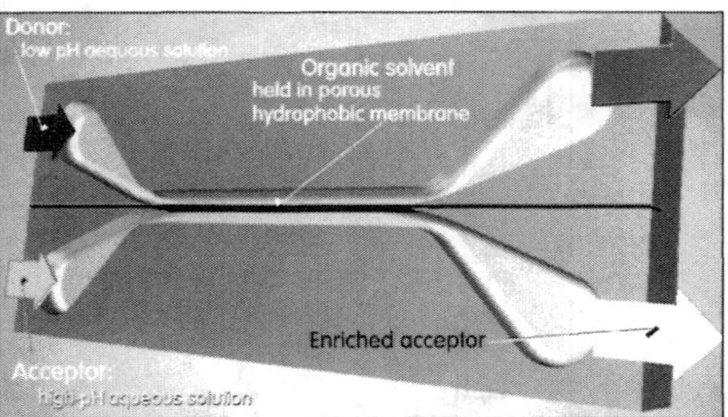

Figure 6.96 (Left) The liquid-liquid extraction: low-volatility substances dissolved in the water-based donor liquid cross the laminar interface between the two immiscible liquids and move to the organic solvent acceptor liquid. This method is used for the extraction of fatty acids in the edible oil industry, but may also find uses in general microchemistry.

Figure 6.97 (Right) The liquid-liquid extraction may also use a microporous hydrophobic membrane separating the two liquids. In this case both the donor and the acceptor may be water-based, differing in their pH value.

conditions of its operation cannot replace (and does not aim at replacing) the traditional sieving action. Nevertheless, this opens a way toward other techniques that also have considerable applicability in food processing. One of them is the liquid-liquid extraction, a diffusive exchange of a solutant between two immiscible liquids, in a device such as that according to Figure 6.96. It uses the low-Re laminar flow. The extraction takes place by transferring the extracted substance across the aqueous/organic interface into the organic flow. Other layouts, with the organic solvent kept in the pores of a hydrophobic membrane, are also used (Figure 6.97). This is an area in which further progress is ongoing, mainly made possible by the discoveries of new membrane materials (in turn made possible by the combinatorial microfluidic discovery and testing procedures).

6.5.2.2 Baking and roasting by hybrid synthetic jets

In spite of the decline in popularity of home cooking, a recent survey in seven European countries found on average 6 GJ of electric energy consumed yearly per dwelling for cooking and oven operation. A relatively new addition to the more traditional appliances for this purpose, the microwave oven, is attractive due to its fast operation, by far beating the traditional methods of convective and radiative heat transfer to the food. Its popularity reflects the general necessity to save time, even at a cost of having to be content with the less enjoyable quality of some processed food. The fast action of the microwave oven is obtained by the magnetron radiation transferring energy to the whole food volume. This, however, means that it fails to form a surface crust, essential for the proper perception and enjoyment of eating food. The products are often described as *soggy* or *doughy*. The crust formed by traditional heat transfer methods also keeps moisture inside the food. This is valuable for proper food quality and is of importance from the commercial point of view. In rotisseries and other point-of-sale processes, heated food products are usually sold by weight, and water escaping from crustless microwaved food results inevitably in loss of product weight to the point of significantly influencing the financial side of the thermal processing.

Solutions combining microwave processing with convective heat transfer by hot air are known but so far have failed to be a success. The heating of the outer surface of food by hot air is generally much slower than the microwave action on the rest of the food. Increasing the power-

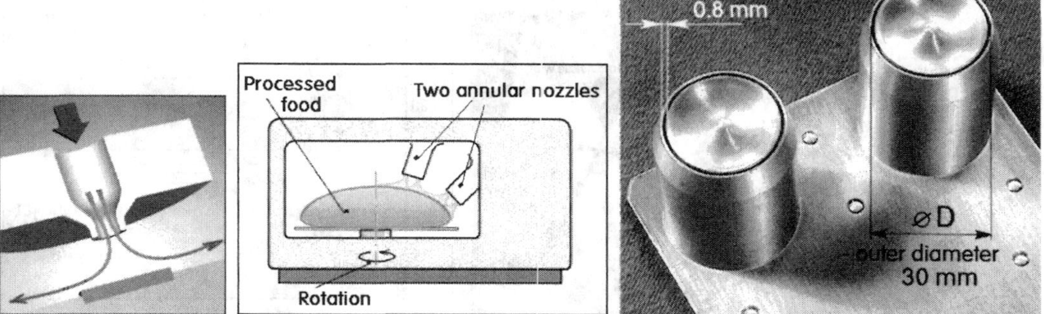

Figure 6.98 (Left) The principle of impingement heating by hot air, capable of the highest intensity of heat power transfer between fluid (air) and solid surface (processed food).

Figure 6.99 (Center) The idea of using impinging hot-air jets for the formation of a desirable crust on the surface of food baked or roasted in a microwave oven. The two heat-transfer methods are mutually not very compatible.

Figure 6.100 (Right) The two annular nozzles used in the experiments (see also Figure 4.142) with impinging-jets heat-transfer enhancing by flow oscillation generated by a fluidic oscillator. The results obtained with this nozzle pair are presented in Figures 6.105, 6.106, and 6.107.

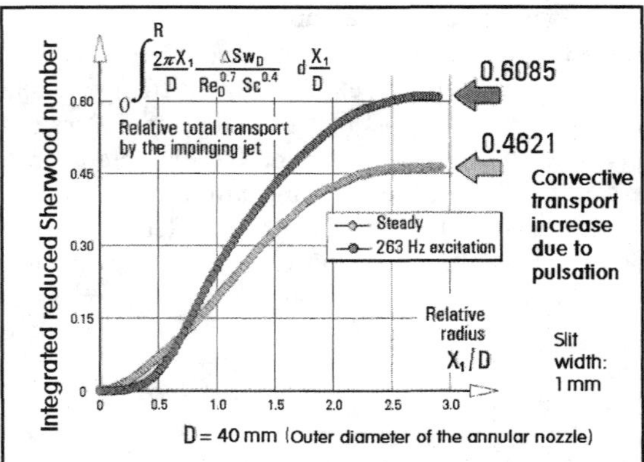

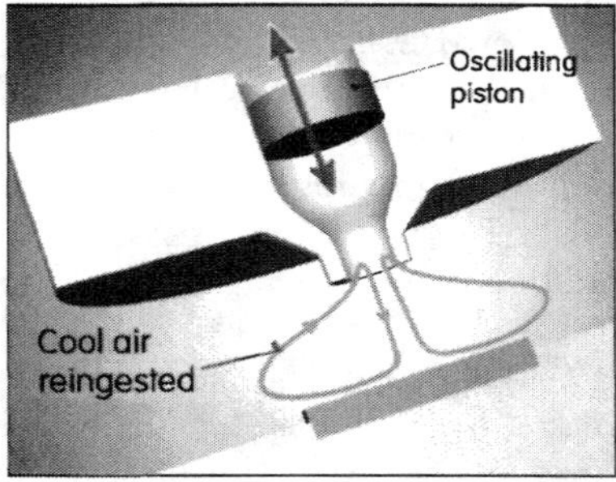

Figure 6.101 (Left) Experimental demonstration of convective transfer improvement in an impinging jet by superimposed pulsation: dimensionless magnitude of mass transfer under a single annular nozzle (of different, larger diameter than that in Figure 6.100), integrated over the impingement surface radially from the nozzle axis.

Figure 6.102 (Right) The idea of the synthetic jet—alternating inflow and outflow from a nozzle—in its basic form is unsuitable for impingement heating. In spite of its great potential for high heat/mass convective transfer due to the high level of flow pulsation, the absence of new hot air makes sustained operation impossible.

transfer rate the usual way, by increasing the air temperature, results in thermal degradation, and could lead to burning of the surface (besides the increased energy consumption). An idea, also already tested, of bringing the hot air to the food surface by impinging jets (Figures 6.98 and 6.99), helped in increasing the heat power transfer density, but, unfortunately was concentrated in only a few locations, at and near to the stagnation points directly under the nozzles.

The basic problem that causes convective heat transfer to be much slower than microwaving, is the existence of a very thin layer of nearly stagnant air, held on the surface by viscosity. The impinging flow exhibits an exceptionally high transfer rate, because in this case the air accelerated in the nozzle and oriented perpendicularly reaches very far toward the surface. The insulating

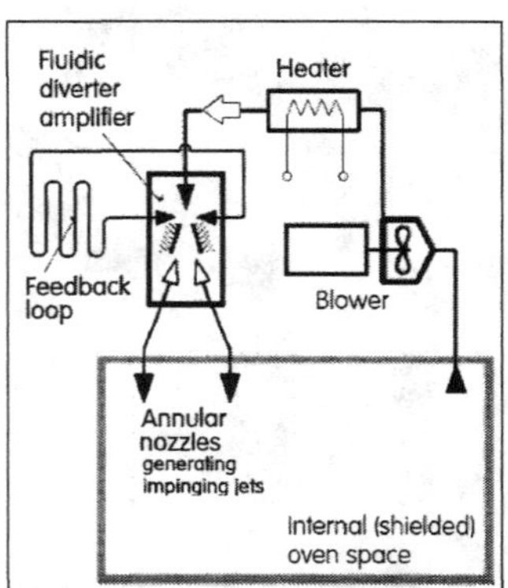

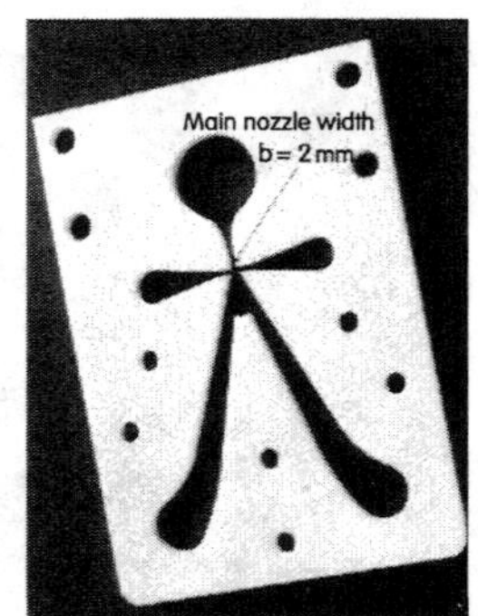

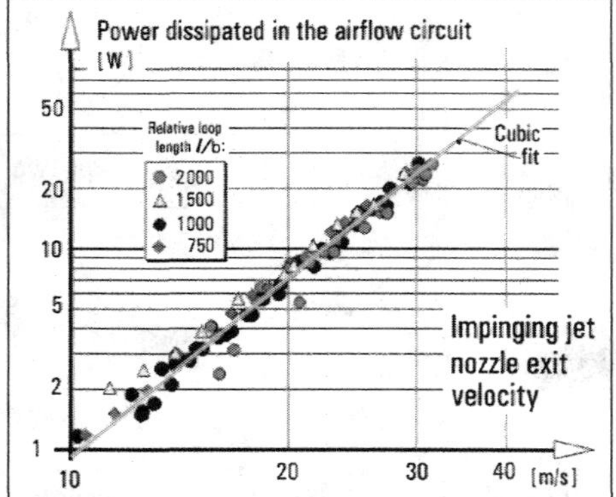

Figure 6.103 (Left) Circuit diagram of the fluidics added to the microwave oven. The main part is the oscillator (Figures 4.151 and 4.152) generating the hybrid-synthetic jet [20] for heating by impingement flow. The return flow into the OFF-state nozzle is due to the jet-pumping suction into the main nozzle of the diverter valve.

Figure 6.104 (Center) The diverter valve made by laser cutting in a Teflon plate.

Figure 6.105 (Right) The power needed to drive air through the system shown in Figure 6.103. It is practically negligible considering the heating powers (typically 1.7 kW) handled in microwave ovens.

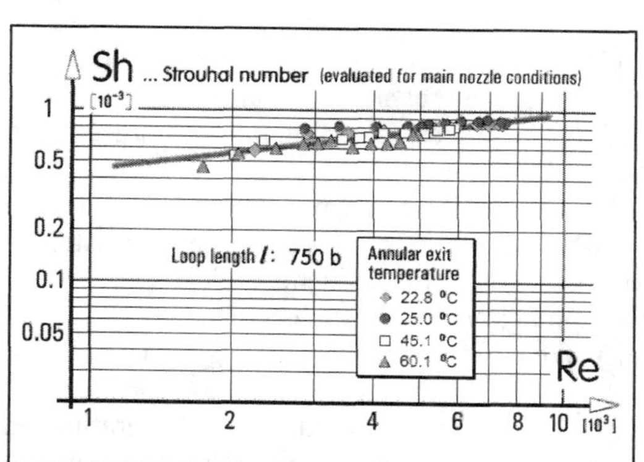

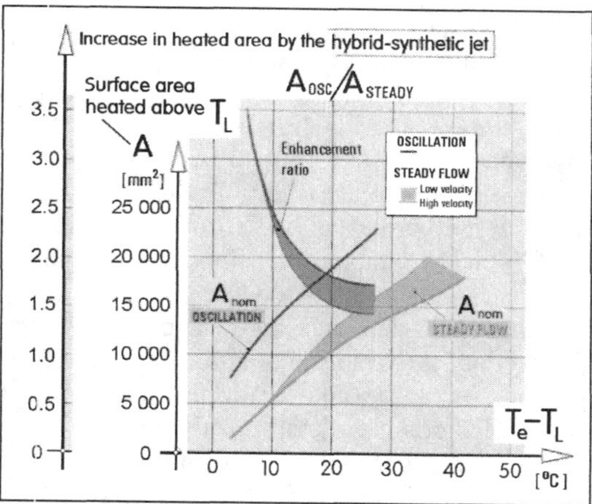

Figure 6.106 (Left) The frequency of the oscillator actually varies with air temperature, but when expressed in terms of the dimensionless Strouhal number dependence on Reynolds number (both evaluated for the velocity and viscosity in the main nozzle (Figure 6.104)), experimental data are fitted by a single line.

Figure 6.107 (Right) The intensification of the impinging jet heat transfer expressed in terms of the increase (more than three times) in area A bounded the isotherm in Figure 4.142. The spreading of the heat transfer over a larger area is welcome in the oven application, as it leads to a smaller number of nozzles needed and to better uniformity of the crust formed on the processed food.

stagnant layer thickness may only be on the order of 0.01 mm. Conduction in air is so inefficient, however, that the thermal resistance of even so thin a layer has a decisive influence on the overall power transfer. One possible way to suppress this influence is by pulsating the nozzle flow (Figures 6.100 and 6.101). It is not a straightforward method. It requires an adjustment of conditions (such as the pulsation frequency), otherwise the pulsation energy may be spent on the formation of vortical structures [21] that do not influence the heat transfer. The effect in Figure 6.101 was achieved with an annular jet, which is particularly susceptible. It is also better than the simple circular jet, because the thermal effect is spread on a larger surface area, decreasing the number of needed nozzles, ideally, to only two (Figure 6.99).

The zero-time-mean-flow synthetic jet (Figure 6.102) should be particularly effective in destroying the stagnant layer. Unfortunately, the effect is only short-lived [22]. A solution was found in the hybrid-synthetic impinging jets, with a superimposed small time-mean-flow component. These jets are usually generated in a different way, as described in [20], but a very attractive possibility is the use of a fluidic no-moving-part oscillator (Figure 6.103). This may be very easy and simple to make (Figures 4.151 and 6.104). The meso-fluidic oscillator of Figure 6.104 with the nozzles shown in Figure 6.100 was used in investigations with heat transfer to a simply planar surface (Figure 4.142). The results were encouraging. The aerodynamic power dissipated in the circuit (Figure 6.103) is small, incomparably smaller than the heating power levels handled in the oven (Figure 6.105). Operation at any temperature is governed by the universal dependence presented in Figure 6.106.

The improvement over the same (already quite effective) annular nozzle with a steady flow, without the fluidic oscillator, was evaluated in terms of the size of the impingement surface area A surrounded by an isotherm T_L, plotted in Figure 6.107 as a function of the nozzle exit temperature (T_e). The increase in this area is a useful criterion of performance, more so because in this application it is desirable to distribute the heating effect over as large as possible surface of the processed food. The improvement obtained with the simple no-moving-part fluidic oscillator, expressed by the ratio A_{OSC}/A_{STEADY} is quite impressive (Figure 6.107).

6.5.2.3 Producing food

The potential of microfluidics in providing food involves food production, in addition to just processing what is already an existing food. There are two stages of production. At the first, easier stage, the inputs are edible components. They are changed mechanically—sometimes with some involvement of chemistry—to obtain a new food. In the more interesting and sometimes somewhat controversial second stage, microfluidics applies biological processes analogous to those used in agriculture to grow and harvest organisms to be processed into food.

A typical example of the first stage is the generation of an oil/water emulsion, in its basic case equivalent to milk. Foodstuffs related to this emulsion have a more dense consistency: mayonnaise, salad dressings, sandwich fillings, and custards. This may be a first step toward production of various creamlike substances by whipping and also to icecreams and similar products made by cooling or freezing as the final production step. Also belonging here is the whole plethora of various spreads.

Emulsion components, essentially oils and water, are mutually immiscible. Generating an emulsion, sometimes described popularly as mixing, in principle means distributing the oil into the water in the form of tiny droplets. The total surface energy of the emulsion is high, consisting of the sum of surface energy over the vast number of the tiny droplets. The generation of an emulsion therefore involves the input of considerable energy; in the classic procedure, this is the mechanical energy for driving a stirrer. The common denominator of commercial problems with emulsion-type foodstuffs is their limited shelf life. They are metastable, not stable, and tend to decrease their total energy. As time progresses, the oil droplets tend to aggregate, attaining a more stable lower-energy state. This, unfortunately, means a change in consistency that does not correspond with customer expectations. The product ceases to be saleable. Manufacturers apply precautions in an attempt to postpone the phenomenon. One possibility is cooling or freezing. Another is the addition of emulgators. The latter is somewhat problematic; some emulgators are

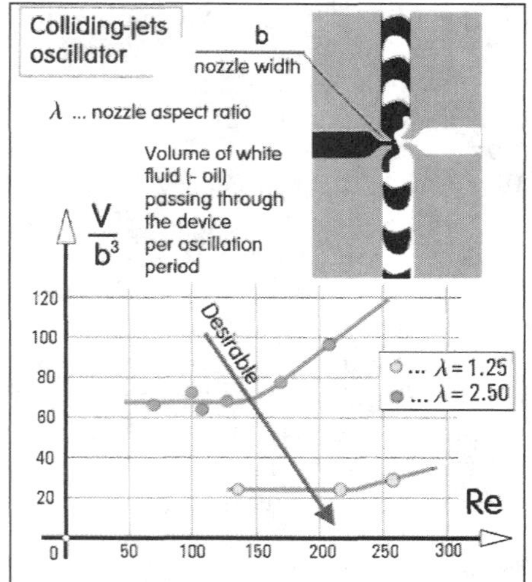

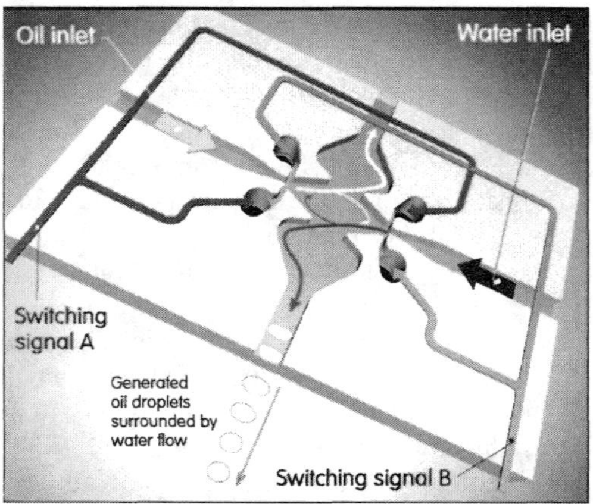

Figure 6.108 (Left) Results of experiments with microfluidics production of foodstuffs having the character of an emulsion: mixing oil and water. The tests with the colliding-jet oscillating mixer of Figure 3.126 revealed this Reynolds number dependence of the relative magnitude of the oil droplets. A small droplet size means emulsions with longer shelf lives. However, a reduction in droplet size may lead to strong viscous damping at low Re. The hydrodynamic instability leading to self-excited oscillation may cease to exist.

Figure 6.109 (Right) An externally controlled colliding-jets mixer is more complex, but the forced switching eliminates the problems with viscous damping at low Re.

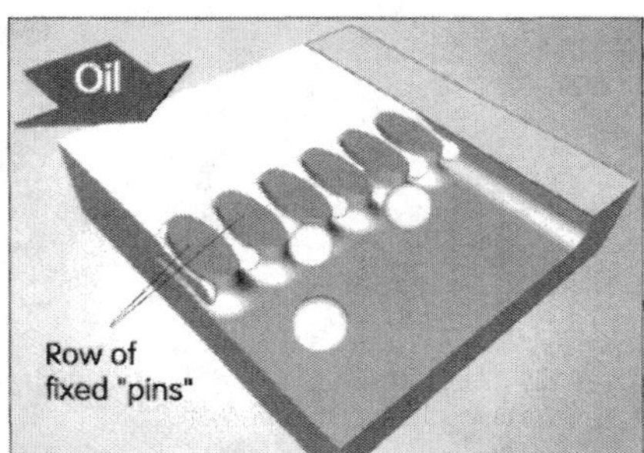

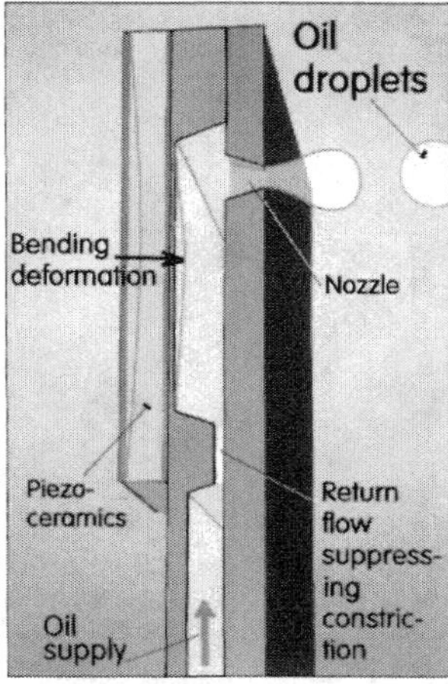

Figure 6.110 (Left) Producing monodisperse emulsions in a microfluidic device by percolation. The oil is pressure-driven through the small gaps in the row of tiny "pins." The problem is very high driving-pressure differences.

Figure 6.111 (Right) Producing a monodisperse emulsion based on the inkjet principle. Oil droplets are displaced through the nozzle by the periodic bending of the piezoceramic driver. The necessity of supplying the driving electric power may seem to be a disadvantage; actually the total power demands may be lower than in the percolator of Figure 6.110 (the advantage of which is the simplicity of the principle). The inkjetting can compete if it can use the mass-produced devices for which there exists a vast market in computer printing.

dangerous to health and others, not suspected earlier, were recently also added to the list of dangerous substances. At any rate, these steps represent additional expenditure.

What microfluidics can offer is producing the emulsion instantly, at the point of sale. The idea is to provide microfluidic machines operated by the customer to make the emulsion just when it is to be sold. No aggregate preventing additives are needed, and, as a bonus, the customer gets

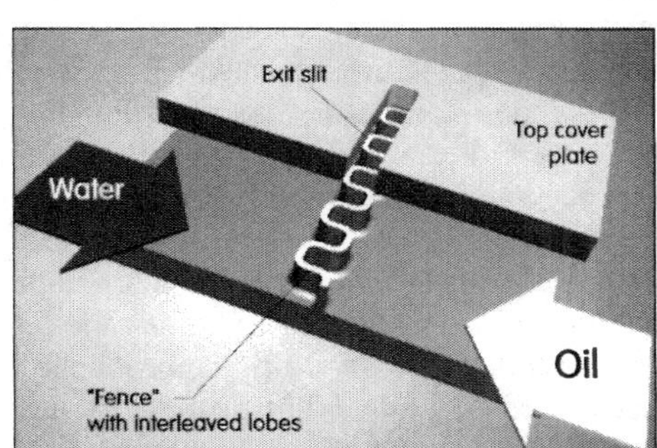

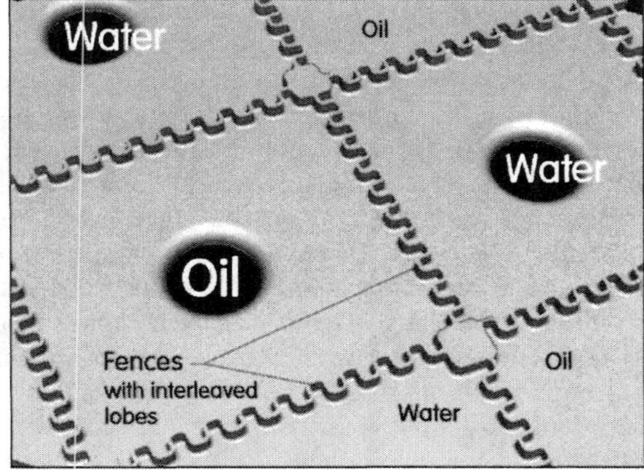

Figure 6.112 (Left) The mutually independent inlet flows of water and oil in a device with interleaved lobes has the advantage of freedom in adjusting the concentration within a wide range. Somewhat complicating is its three-dimensional character: the produced emulsion leaves at right angles to both water and oil inlets.

Figure 6.113 (Right) A practical layout of a high-throughput emulsion generator for distributed production at the points of sale. The throughput is obtained by the numbering-up principle, but this does not necessitate a large number of separate devices. The fences, as shown in Figure 6.112, form the boundaries of regions distributed in a plane, with fluid inlets in the centers of the regions. The cover plate with the exit slots above each fence is not shown here.

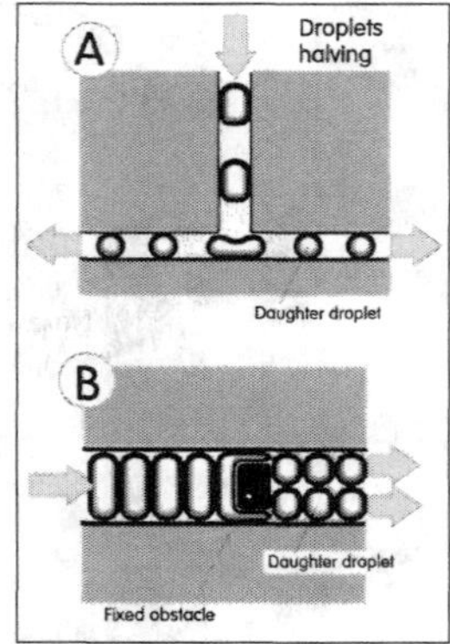

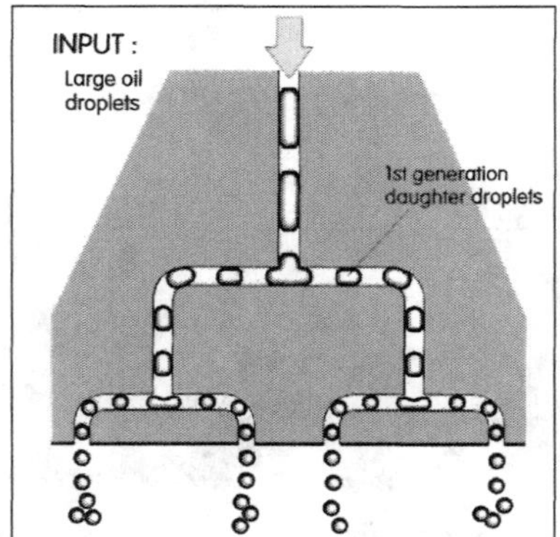

Figure 6.114 (Left) Producing smaller droplets from large ones. A: droplet splitting in a T junction, by deformation action of the inlet flow, and B: by an obstacle in the channel, which forces the large inlet bubbles to bifurcate.

Figure 6.115 (Right) A repeated application of principle A in Figure 6.114 may be used to obtain one-quarter of the original droplet volume.

a choice of taste and/or color or consistency alternatives. There are essentially four different methods used by microfluidic emulsion generators: oscillators (Figures 6.108 and 6.109) concentrating the needed driving energy in the region where two opposing accelerated jets collide, percolators (Figure 6.110), oil droplet injectors (Figure 6.111), and the layouts with a confluence of interleaved lobes (Figures 6.112 and 6.113). Even if the idea of point-of-sale machines is not adopted, the replacement of stirring by the microfluidic emulsion generators brings another advantage: the produced emulsion is of an almost perfectly monodisperse character. As long as the design avoids the formation of satellite droplets, the droplets are of uniform size as dictated by the size of the oil exit holes in the microdevice. The taste perception, or smoothness, of monodisperse products is recognizably more pleasant, and this gives them a considerable marketing advantage. If there is a limitation on how small the droplets can be made, such as the Reynolds number limit of sustained oscillation in the principle from Figure 6.108, there is a way to decrease the droplets by halving according to Figures 6.114 and 6.115.

As far as the second stage of food production is concerned, fewer problems arise with growing vegetable food. Of course, growing traditional large-scale fruit and vegetables is a slow process; it may be replaced by the much more intensive microfluidics growth of lower plants such as algae. These are better suited for harvesting in machinery-type facilities. The idea was first followed in association with astronautics, where algae were intended to provide food for astronauts on long interplanetary missions. The main remaining problem is processing the harvested products so as to obtain meals of acceptable shapes and consistencies. A solution is usually sought in using algae growth as a mere first step in a more complex food chain, the algae providing food for higher plants or animals such as shrimps, better corresponding to the traditional notions of human food.

Less known and less developed are attempts at producing animal-origin products in microfluidic machinery. Microfluidics is actually already used in animal fertilization. A more direct approach foreseen, however, to be important for feeding the future human population is the in vitro growing of animal parts to be harvested as meat. While growing plant cells meets with no objection (it is only the product that people in general do not find attractive in its as-harvested form), the idea of producing meat by growing muscle cells, though quite analogous, is considered by many as rather repulsive.

The idea is certainly not new. It has been discussed for 75 years [18]. Later it was studied also as a means for providing food for astronauts [19]. Recent progress in tissue engineering (from

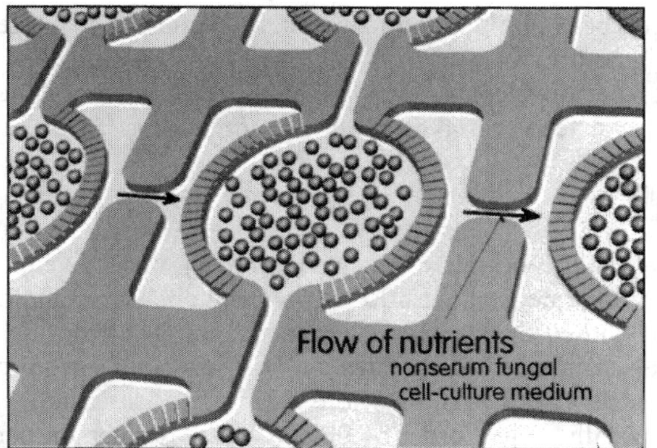

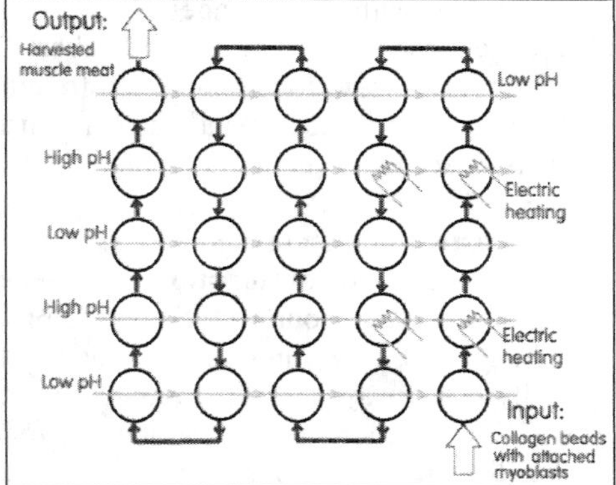

Figure 6.116 (Left) Producing meat for food by growing animal muscles in a fluidic device. The muscle cells grow on the surface of collagen microspheres, which are gradually moved through growth chambers, supplied with nutrients by a medium obtained from fungi.

Figure 6.117 (Right) Schematic representation of progress of the microspheres or beads with the meat cells growing on them through the chambers of the microfluidic device from Figure 6.116. The variations in pH and temperature cause the beads to swell and thus provides the physical exercise necessary for the differentiation of the myoblasts into proper skeletal muscle cells.

the point of view of which it is actually a rather simple task) made the idea technologically realizable. The meat to be produced is skeletal muscles that grow either from embryonal myoblasts by differentiation or from postnatal satellite cells. There are two directions in the current research. One is the scaffolding technique, using some form of support structures for the produced cells. The other one is more difficult to realize: self-organization growth. In principle, the latter is considered capable (in principle at least) of simulating actual animal growth, including the vessels for the nourishment fluid and perhaps even bones so that the harvested product may be turned into real steaks. The fluid-carrying vessels are needed in this case because cells become necrotic when further than 0.5 mm away from a nutrient supply. The distributed fluid is, of course, not blood but a fluid from the family of the cell-culture media. The standard ones are mostly based on bovine fetal serum, not a suitable choice for mass production purposes. In [19], the authors succeeded in using much more suitable medium made from maitake mushrooms.

Unfortunately, progress has been slow, mainly because most of the effort was wasted by researchers pursing on the idea of the producing highly structured meats and on attempts to obtain large-size products by scaling up whatever successful small-scale experimental results there were. Much more promising even here is to follow the basic idea of numbering up (rather than scaling up) that is typical for microfluidics. Figures 6.116 and 6.117 present an example of such production where the scaffolding are microspheres from collagen, an edible material. Instead of steaks, the product is boneless meat quite suitable for hamburgers and sausages. The boundary for the growth is the Hayflick limit, a number of doublings the cells can undergo. It is given by various authors at about 75. The corresponding number of cells, 2^{75} is, however, so large that this does not represent any real limitation. A more serious limit is the contact inhibition (cell proliferation arrested at a certain density that can be controlled by signal-regulating proteins, a technique evolved from studies aimed at accelerating the healing of epithelial wounds) and the layer thickness nourishment limit, 0.5 mm.

The starting point usually considered is myoblasts. Their differentiation into muscle myotubes requires some physical exercises at the initial stages. Studies suggest repetitive stretching by ~10% and relaxation, six times per hour. In the design shown in Figure 6.116, this is accomplished with size changes of the collagen beads by pH and temperature cycling as they are moved stepwise into progressive chambers (Figure 6.117).

If direct human consumption is found to be unwelcome, then, as in the case of the algae growth, the harvested meat may be used to feed carnivorous lower organisms in more sophisticated food chains.

6.5.3 Waste liquidation

The reverse of food consumption is the generation of food-related (as well as other) waste. Even in the least-developed countries, the waste production rate is about 300 kg per capita per year, and it is as high as ~900 kg in countries such as the United States. The rate is increasing fast, from 3% annually in Norway to a 4.5% increase each year in the United States. Fast growth is also in solid waste, dominated by paper, cardboard, food scraps, and plastics. More steady regionally and in time is the production of water-borne waste, usually on the order of 90 000 liters per person per year. Of this typically 85% is domestic waste, while the rest comes from manufacturing and trade activities. The world is gradually cluttered with garbage, which becomes a serious problem. The traditional ways of disposing of waste—landfills for the solids and biodegradation in rivers for the liquids—are near the end of their capabilities. Most of the solid waste is combustible, but incineration is not a perfect solution ([23] presents statistics of a soft-tissue cancer increase by 13.0% in the proximity of incineration plants). A solution may be the conversion to liquid fuel, by nonoxidation pyrolysis and a subsequent Fischer-Tropsch process. In this, fluidics helps indirectly—in the discovery of catalysts, on the properties of which the economy of the process is critically dependent. Fluidics may be employed directly in the hydrogenation of carbon monoxide using microreactors, which are more efficient than the large facilities considered so far.

The liquidation of water-borne effluents is done by breaking down the major pollutants—rganic matter, nitrates, and phosphates—using aerobic biological decomposition. The process is slow, mainly because the microorganisms performing the decomposition die due to lack of oxygen. It is common to bring it to them by aeration, the continuous supply of air bubbles. This is rather expensive. Aeration costs represent more than 80% of total energetic consumption of a typical municipal installation [24], and yet the oxygen transfer is not very effective. The authoritative monograph [24] says that the "... majority of current [wastewater treatment] plants suffer from oxygen deficiency." The reason is the small surface of the large bubbles and their fast motion towards the surface. The rising speed increases fast with size (Figure 6.118).

In an attempt to make the bubbles small, aerator designers choose very small-size exit apertures, even using – at the price of the resultant high pressure drop - air percolation through sintered porous materials with submillimeter pores. Unfortunately, this does not help. The result is the agonizing experience of the bubble size being practically the same no matter how small are the pores (Figure 6.119). Also, however large is the designed percolator area, the bubble formation concentrates into only a few locations. The only outcomes are the high pressure loss and the susceptibility of the small pores to clogging by various debris and particles carried with the air. The concentrated bubble formation at a single exit is sometimes also due to clogging, but a more fundamental reason is instability, as presented in Figures 2.140 and 2.141. A remedy has been sought in moving the aerators, placing them on rotating arms, for example. The shear stress on the aerator surface limits bubble growth, and there is the additional benefit of stirring. The cost of the driving motor, gearbox, struts keeping it above the surface, bearings, and rotating seals (which tend to become worn and need maintenance), and, of course, the consumed driving power make it expensive and far from the ideal solution.

The need to generate aeration bubbles smaller than the current size, between 8 and 10 mm is obvious: the ratio of their total surface to total volume (Figure 1.17) would increase and they would stay longer in the water, providing more opportunity to mass transfer. The basic idea of the

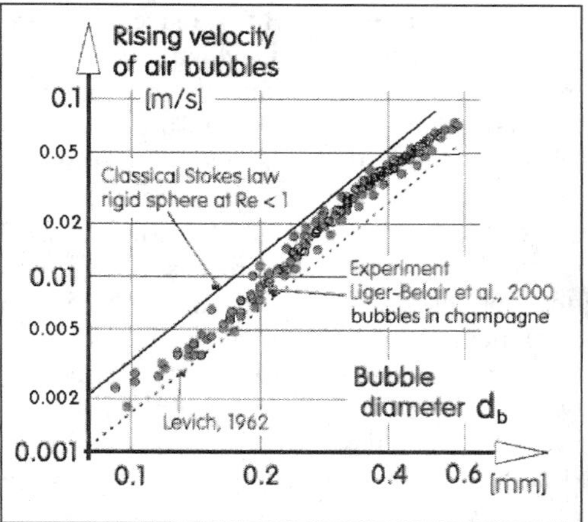

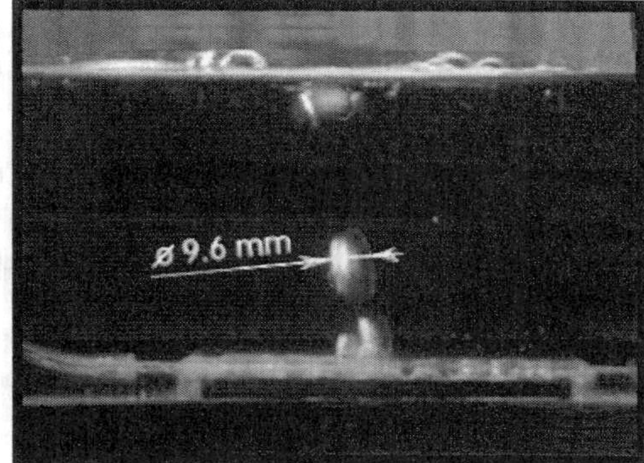

Figure 6.118 (Left) The high rising velocity of large bubbles in liquids (here including the interesting case of bubbles in champagne) is one of the reasons why air from large bubbles cannot diffuse into the surrounding liquid. The velocity increases practically as the second power of the bubble size.

Figure 6.119 (Right) The futility of attempts to produce smaller bubbles by decreasing the size of the exit aperture. In this example, steady airflow is supplied into a row of 0.6-mm-diameter apertures. Because of the instability of Figures 2.140 and 2.141, instead of the expected row of small bubbles a single large bubble forms, of a size more than a decimal order of magnitude larger than the aperture. The motion unsharpness testifies to the large rising velocity (Figure 6.118).

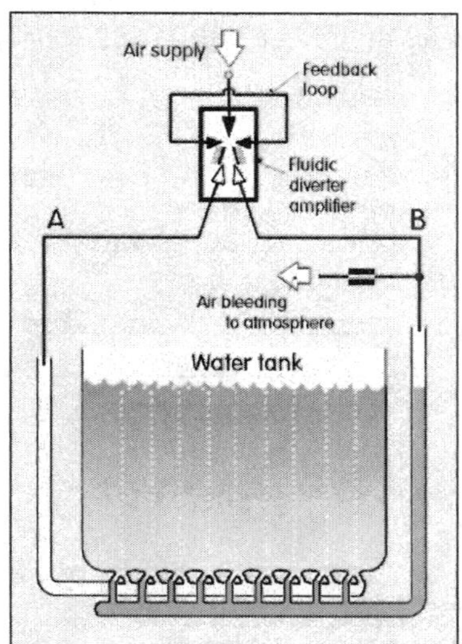

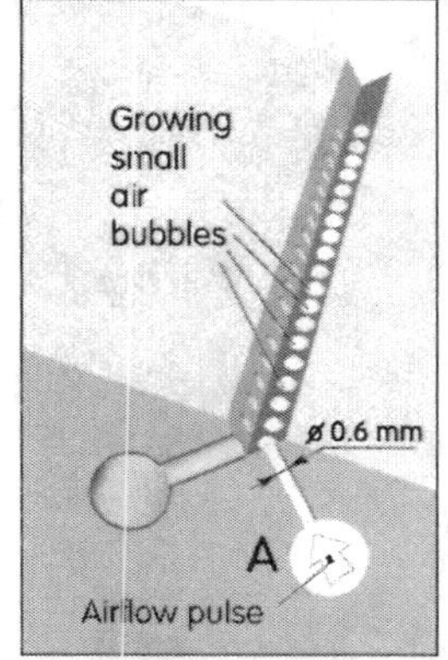

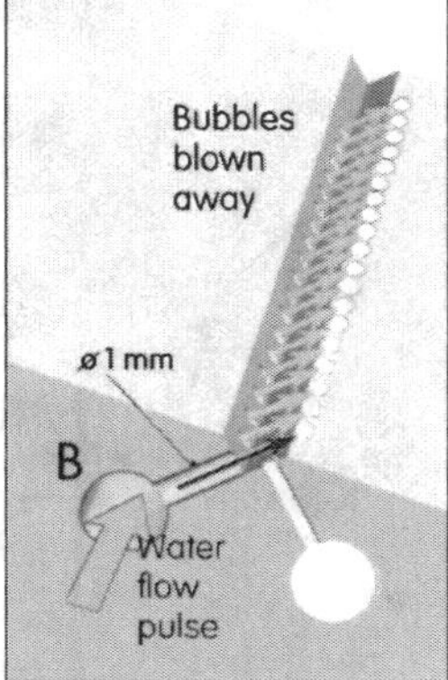

Figure 6.120 (Left) Schematic representation of fluidic aeration based on the bubble "knocking off" principle of Figures 6.121 and 6.122, due to oscillating the airflow. Each end of the oscillation half-period interrupts the (still stable, Figure 2.141) growth of bubbles. In the second half of the period the bubbles are separated from the exit aperture by an inflow of the liquid. The oscillator used in the tests was the single-feedback-loop version (Figures 4.151 and 4.152).

Figure 6.121 (Center) The tested layout of the aerator. The liquid and air apertures are arranged in mutually inclined rows. In the first half of the oscillation period, shown here, the bubbles are smaller than the instability limit, so that they grow equally.

Figure 6.122 (Right) The second half of the oscillation period: the bubbles stop growing and are displaced from the air aperture exits by perpendicular water jets.

application of fluidics is to operate the aerator in a periodic regime. Bubble size is governed by the duration of the oscillation period and not by the surface tension, which varies due to the presence of contaminants in the wastewater. Growth is terminated before the bubbles reach the hemispherical stability limit (Figure 2.141), while they all still grow. This ensures the full use of all exits. An attractive feature is the simplicity of the fluidic oscillator, the addition of which does not cause any significant increase in complexity and price. There is no need for supplying electricity. The tests were made with the single-feedback-loop oscillator, as shown in Figures 4.151 and 4.152. To make the detachment of the bubbles from the aerator exits more reliable, the availability of the other exit from the diverter-type oscillator was used for forced bubble separation. The mechanism is similar to the perpendicular collision of the main airflow and the flow from the control nozzles in the oscillator amplifier. As shown in Figures 6.120 to 6.122, the aerator surface was provided with grooves into which the water and air nozzle pairs issue at right angles. The bubbles grow attached to the air nozzles, and their separation at the end of the oscillation half-period is ensured during the next half of the period by the water flow pulse from the water nozzles (Figure 6.122).

The size of the bubbles produced by the novel method was demonstrated (Figure 6.123) to be smaller by a decimal order of magnitude than the smallest size produced by present aerators. The air exit aperture size is no more the decisive factor and may be larger than the current practice. This makes the aerators easier to manufacture and less prone to clogging and obliteration. The synchronized input pulses in the exits can produce a welcome stirring effect near the tank bottom; higher above, the water is stirred by the airlift pumping effect (decrease of effective density of the water/air mixture).

In continuation studies of bubble formation, other mechanisms were found that may be used to generate small bubbles by oscillating the air-supply flow. The principle is presented in Figure 6.124. The aerator uses just the row of the 0.6 mm air nozzles from Figure 6.121, all connected to

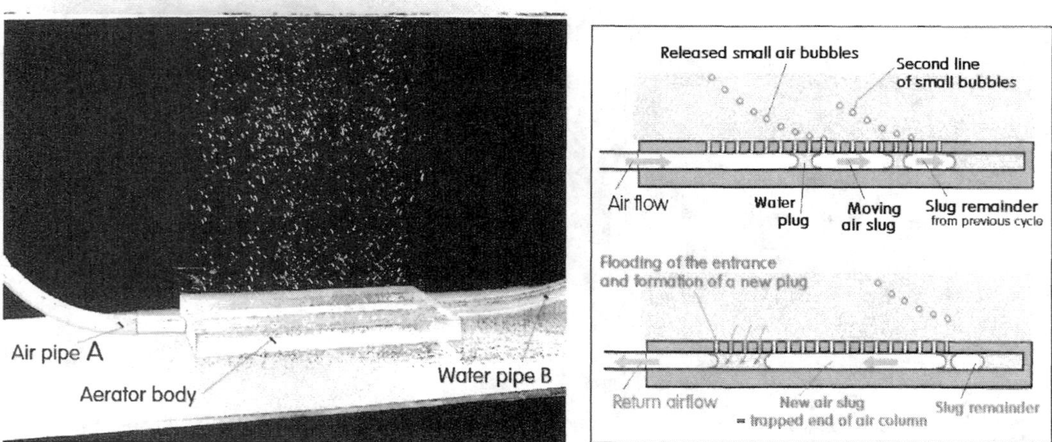

Figure 6.123 (Left) Photograph of the small bubbles generated in fish-tank tests using the principle from Figures 6.121 to 6.122. The original > 9 mm diameter large bubbles (Figure 6.119) produced with a steady airflow in the same aerator are seen here replaced by bubbles of less than 1 mm diameter.

Figure 6.124 (Right) Another mechanism to generate small bubbles. Air is pulsed into and out from the long horizontal passage with small exit holes on top. In the *out* part of the cycle, the air column is trapped by liquid entering through these holes and forming a plug. Upon moving *in*, the trapped air enters the small holes, where it is kept by surface tension. When the plug comes, it pushes the small air volumes from the holes as a row of small bubbles.

the horizontal 3 mm diameter manifold passage. The passage is filled by a succession of air slugs and liquid plugs between them. When the oscillator switches the flow into the passage, the air slugs and water plugs move to the right. Air from each slug enters the exit holes, but stops when air reaches the exit, in the same way as in the surface-tension microfluidics stop valve shown in Figure 4.135. The contents of a hole is pushed out into the water above when the progressing liquid plug reaches the hole. As the plug moves, the row of gradually released small air bubbles is well visible (Figure 6.125). When the oscillator switches the airflow direction in the inlet, the slugs and plugs are pulled out from the horizontal passage. This generates a low pressure there – not low enough for the liquid overcoming the stop-valve effect of surface tension anywhere but in the lowest pressure region at the left-hand side. The liquid comes in there and forms a new liquid plug. When the inlet airflow reverses again, this plug pushes in front of itself the new air slug formed by the trapped air in the passage. As the liquid plugs and air slugs form small bubbles, they are consumed and their column length in the passage gradually diminishes. They also tend to move towards the closed, right-hand side of the passage, where their size as well as the amplitude of their movement decreases. They are, however, continuously replaced by the newly formed long columns at the left-hand side. At first sight, this may not seem to be a promising foundation for an industrial process, but experience gained in experimental investigations indicates that it is actually a robust and very reliable mechanism.

Finally, it should be said that wastewater—mainly thanks to its content of organic matter—may be actually a source of electric energy in the *microbial fuel cell* (Figure 6.126).

This is not a new idea. Michael Potter, professor of botany at the University of Durham, United Kingdom, demonstrated the generation of voltage and current from *E. coli* in 1912. Again, it was practically forgotten until NASA, in the 1960s, generated an interest in producing electricity from organic waste on space flights. In the anode compartment, fuel is oxidized by microorganisms, generating electrons and protons. Electrons are transferred to the cathode through the electric circuit, while protons pass through the membrane. On the cathode side they combine

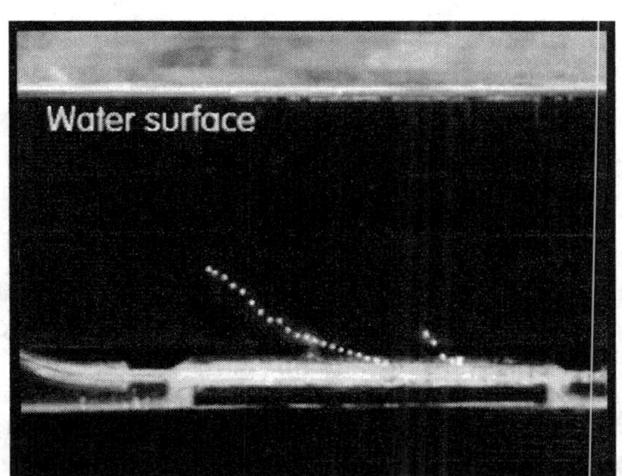

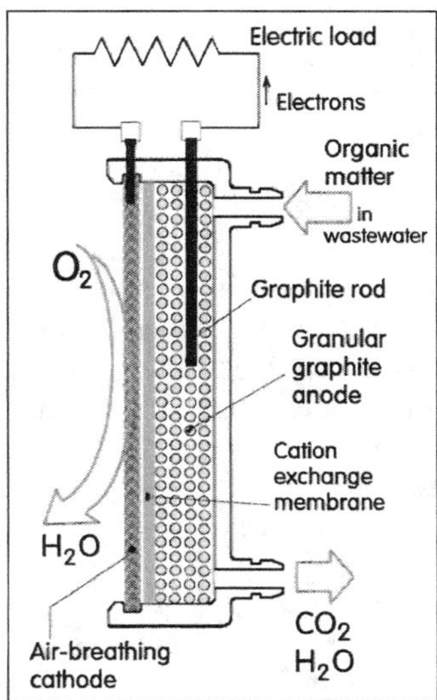

Figure 6.125 (Left) The row of small bubbles pushed out from the small holes by the liquid plug (Figure 6.124). The devices shown here and the conditions are the same as with the steady flow in Figure 6.119.

Figure 6.126 (Right) An example of a microbial fuel cell. Bacteria consuming organic matter from domestic or hospital wastewater convert it to CO_2-generating electrons and hydrogen cations (protons). The latter pass through the exchange membrane to the oxidation side. Electrons get there through the external electric circuit and combine there to water.

with the supplied oxygen to make water. Very much depends on finding suitable bacteria. Three fuel cells types are known. In the first type, redox mediator substances are needed on the anodic, bacterial side to penetrate bacterial cells and transport electrons. Other fuel cells oxidize fermentation products on electrocatalytic surfaces. The third type of cells uses metal-reducing bacteria, with initial interest being in the fylla *Geobacteraceae* or *Shewanellaceae*. A recent step forward brought bacterium *Rhodoferax ferrireducens* discovered at the bottom of Oyster Bay, Virginia. It reduces iron while converting glucose to CO_2 and transferring electrons to graphite anode. It can grow at 4–30°C (important compared with the earlier-known thermophilic and hyperthermophilic Fe^{3+} reducing organisms) under anaerobic conditions, needs no (toxic) mediator compound, and oxidizes glucose at 80% electron efficiency (*Clostridium* strains can show only 0.04%). In other fuel cells that use immobilized enzymes, glucose is oxidized to gluconic acid, generating only two electrons per reaction, whereas *R. ferrireducens* oxidizes glucose completely to CO_2, exhibiting a remarkable long-term stability. The current density of 31 mA/m^2 has been produced over a period of more than 600 hours.

Microbial fuel cells effectively process simple sugars (glucose), starch, and cellulose in industrial effluents. Experiments with domestic wastewater were successful, but the output was not constant, due to uncertain composition and concentration. An interesting application is a fuel cell processing the fermented bodies of garden slugs as an on-board energy source for an autonomous slug-catching robot. It sees the slugs at night as infrared sources. (Slugs are pests, farmers in United Kingdom alone spend about £ 20 million per annum on their eradication.) Some anaerobic microbial fuel cells, when supplied electricity from outside, can be used to convert biomass into H_2, producing it much more effectively than simple electrolysis. They may be an important factor in expected future hydrogen energetics.

6.6 IDENTIFICATION OF PERSONS

Human society depends in surprisingly large measure on the reliable recognition of persons in situations such as the authentication of those allowed to enter a building or to perform a particular task. Very much depends on personal recognition and passwords, but people also have to carry a continuously increasing number of various keys as their means for mechanical authentication. At an even faster rate, various swipe cards and "smart" cards are now used. Cards and keys are relatively easy to forget. They usually tend to be forgotten when most needed. A better solution, gradually taking over, is biometric identity recognition. The (nonexhaustive) list of considered or developed possibilities includes physiological and behavioral characteristics such as fingerprints, face shapes, voices, footprints, walking features (carpet sensors), human leukocyte antigens, hand shapes, palm prints, iris patterns, and at the highest level, DNA analyses. The common feature is difficult reproduction with an intent to deceive. Some biometric features, such as fingerprints, have a long tradition in forensic investigations for which this principle is eminently suitable. This use has led to vast databanks that are already in existence, the utilization of which may be an advantage. However, for use with identity verification, fingerprints have some disadvantages. The features to be read and recognized are very small, and this causes difficulties associated with extracting important small aspects (named *minutiae*). Because of the need to read tiny features, the sensor cannot be very robust. Also, there is a strong psychological barrier to fingerprint collection that is associated with an infringement on privacy. The use of iris patterns is considered exceptionally reliable, but the required readout devices are very expensive. They, of course, do not offer much opportunity for the application of fluidics. The natural domain of fluidics is methods that process body fluids. These, however, are generally rather inconvenient in the role of a key. Even though some body fluids (saliva or sweat) may be collected noninvasively, their use is impractical for purposes of repeated authentication on entering a room or accessing a computer.

6.6.1 Identification based on hand silhouette geometry

Hands as the keys to open access are convenient to use and available readily at all times (cannot be forgotten at home). Measuring hand geometry (Figure 6.127), can accumulate a large number of data values, from at least 10 to the usual practical value near 100. Hands are not small, so the sensor is not complicated by a need to detect very small features. Collecting a large number of read values is desirable, since the accuracy of reading and difficulty of overcoming the system are directly proportional to the number of the elements in the data vector associated with a particular person. In the design from Figure 6.127, the values (distances from the contour of the largest expected hand) are read pneumatically. It is inexpensive, hygienic (self-cleaning), and fast. The data vector is obtained simply by a sequential reading of the pressure values measured at each nozzle of the row positioned on the circumference of the silhouette.

Before the actual reading of the data values, it is useful (not necessary in principle, but for simplifying the data processing) to ensure that the hand is in a standard position. According to Figure 6.128 this is done by selecting several locations on the reader device against which the hand has to be pressed (they have to be positioned so that even the smallest hand touches them all). The nozzles in these locations (Figure 6.128) form a separate system; all of them have to be in their CLOSED state for the actual measurement to begin. The output pressure signals in the system of the remaining nozzles are between the two extreme values corresponding to the two situations presented in Figure 6.130. The output pressure values are read one by one and stored.

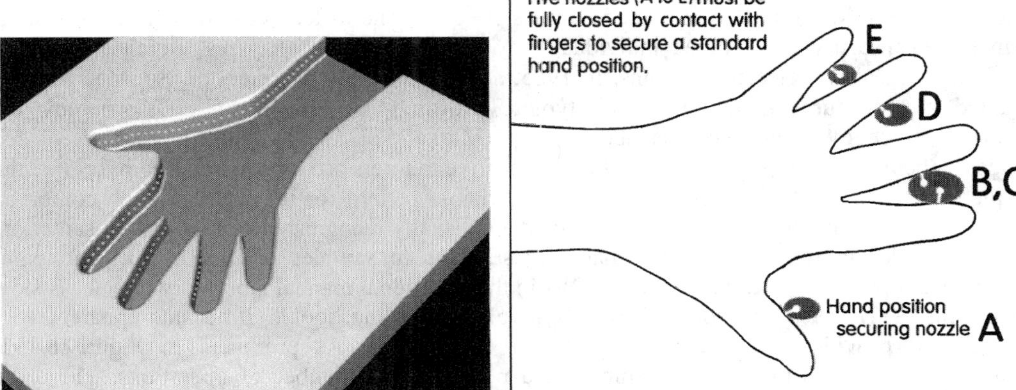

Figure 6.127 (Left) Fluidic sensing of a hand's silhouette by a row of air nozzles is quite convenient and fast. The recognition is based on a vector of output-pressure values collected by sequential interrogation in individual nozzle sensors. A typical number of the nozzles is 96, a number giving a reasonably high improbability of chance coincidence.

Figure 6.128 (Right) There are actually two systems of nozzles. One system with a relatively small number (here five) of nozzles is used to ensure a standard position of the hand the pressure values collecting can begin.

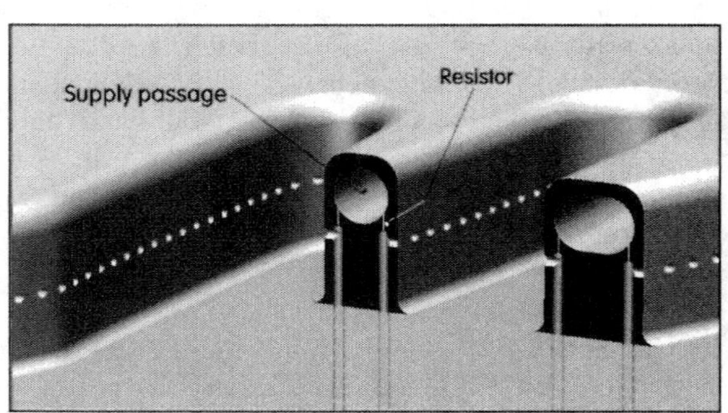

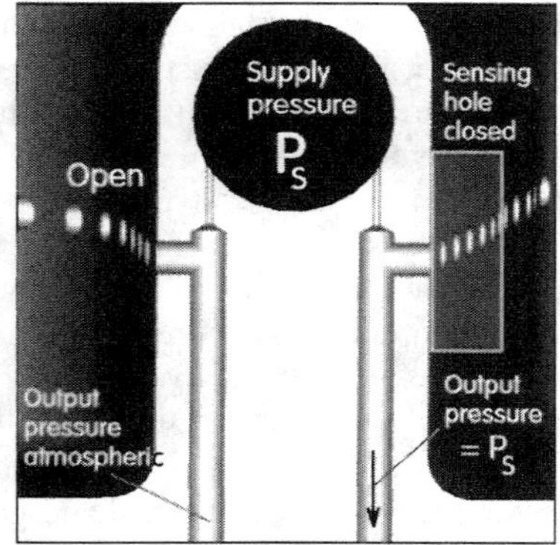

Figure 6.129 (Left) The air supplied to the sensing nozzles is delivered by common supply passages. Each nozzle is connected to the passage via a small cross-section restrictor. The supply pressure is quite low, so that sensing is not inconvenient. Air blowing out from the nozzles removes dirt and contamination, making the data collection hygienic and free from stoppage due to clogging.

Figure 6.130 (Right) The basic principle of the pressure reading in the sensing nozzle corresponds to the M/F conversion described in Figures 5.18, 5.19, and 5.20. This picture shows the two extreme cases: the nozzle fully open (finger shorter than in the maximum-sized hand cavity in the sensor, at left) when the small air flow through the resistor escapes into the atmosphere and the fully closed position (finger or other part of the hand completely obstructing the flow, at right).

The resultant vector is then compared with the stored-data vectors obtained earlier for the persons authorized to enter.

6.6.2 Fluidic devices for DNA analysis

In principle, an identification based on DNA is the most exact method, because it handles genes, the hereditary units controlling the functions of living organisms. The method is, at its current stage, not suitable for use as an entrance key. Instead, it is used mainly for legal identification. In DNA the entire genome, the genetic contents of the organism, is encoded. DNA is a polymer of the well-known double-helix shape; if fully stretched, it is about two meters long. Of course, the majority of the DNA from two unrelated humans is equal. The differences are in relatively short fragments. These may be handled individually by using restriction enzymes, which are able to slice off fragments at selected locations. In 1985, at the University of Leicester, Sir Alec Jeffreys invented a procedure (known as DNA typing, profiling, or “fingerprinting”) whereby the fragments move by electrophoresis and line up. Markers of known lengths are then placed alongside them in electrophoretic gel. When radioactive markers are administered to hybridize the sample, radiographs may be formed to provide a visible pattern for the DNA of each compared individual. This technique, resulting in an immediate visually recognizable, graphical presentation, became popular in forensic science (matching suspects to samples of blood, hair, and other evidence) and paternity testing. Handling DNA in the traditional manual laboratory manner is slow and expensive. Because the procedure mostly involves handling liquids, it became apparent very soon that it provides an excellent opportunity for fluidics. As presented in Figure 6.131, identification involves moving the samples and performing a number of operations. There are many preliminary steps that are particularly well suited for being performed in a fluidic device. At the time of this writing, there were very few complete, integrated, fluidic circuits performing all of the steps in Figure 6.131. Most reported devices performed only several steps at most. Also the detection and evaluation steps are far from standardized, and in the literature there is a continuous stream of descriptions of various methods.

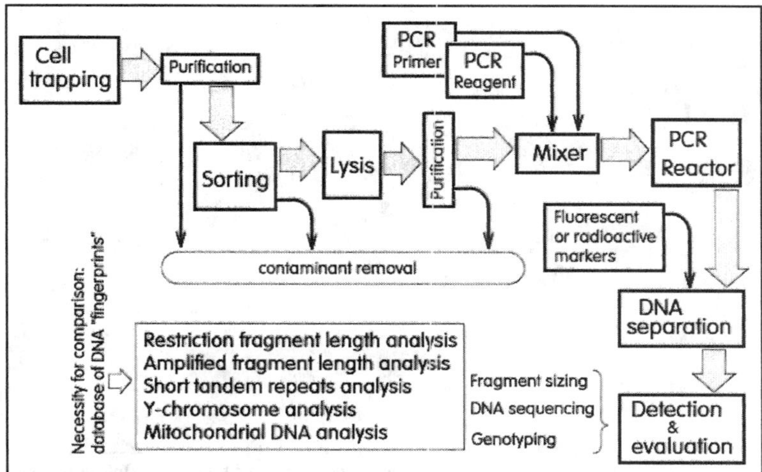

Figure 6.131 Schematic representation of the steps completed in the course of DNA analysis. The wide gray arrows indicate the flowpath of the processed sample, which may be arranged as a chain of fluidic devices – ideally in a single integral fluidic circuit. Some of the devices, for example, the mixer, are standard and were discussed in previous chapters.

Figure 6.131 lists several used alternatives. The radioactivity used for the readings is mostly replaced by fluorescence. Unfortunately, DNA itself does not exhibit fluorescence, and much depends on progress in development of suitable markers that attach themselves to DNA strands at the proper locations.

A typical example of the preliminary steps indicated in Figure 6.131 is the *sorting* of the cells, which means selecting those that are suitable for the analysis (for example, white blood cells) and discarding the rest (the nucleusless red cells). Several techniques have been performed using special microchips, usually more or less similar to the illustration in Figure 6.132. The diverter shown there uses for the diverting action a dynamic effect of flows from control nozzles. More common is, however, manipulation by an electric field switched between electrodes, using the principle of the electroosmotic flow from Figures 5.82 and 5.90. The electroosmotic flow control is more suitable for very low Re but suffers from a number of drawbacks, such as buffer incompatibilities and frequent changes of voltage setting due to ion depletion, evaporation, and pressure changes in the circuit. The detection of cells is often based on optical properties including, for example, the fluorescence of activated attached antibodies.

Another essential preprocessing step is *lysis*, or destruction to release nucleic acids. The known methods include mechanical crushing (adaptation of the layout from Figures 5.42 and 5.43), the chemical action of enzymes or surfactants disrupting the cell membrane, ultrasonic disintegration, thermal lysis (effective but mostly impractical because the involved temperature levels cause denaturing), and electroporation, the formation of openings in the cell membrane by the action of a pulsed electric field. Two examples of the electroporation devices are presented in Figure 6.133. By decreasing the distance between the electrodes (it may be as small as 30 μm, which is about three times the size of a typical mammalian cell), the applied voltage may be as low as 10V. It may be adjusted to utilize the difference in the transmembrane potentials for the outer cell membrane and the membrane of mitochondria, which have their own DNA. It may be desirable to leave this DNA undisturbed (but it may be useful for some tests that study genes inherited in maternal lines, which are of importance for forensic or historic sciences handling highly degraded samples: there are many copies of mitochondrial DNA in a cell, while there are only 1.2 copies of nuclear DNA).

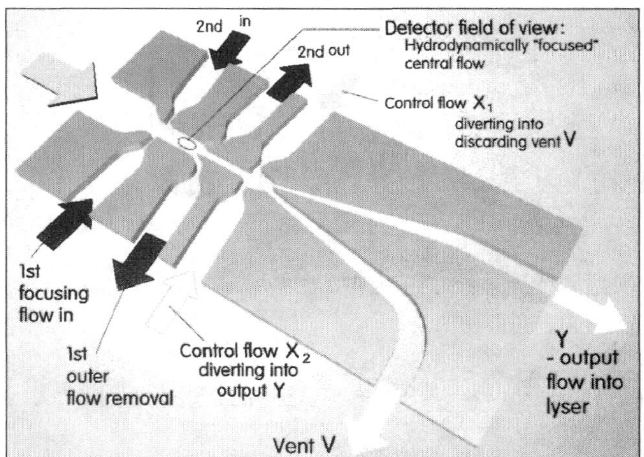

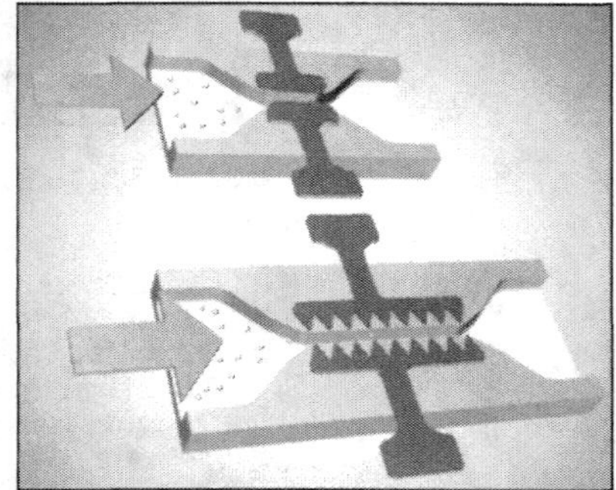

Figure 6.132 (Left) Fluidic sorter device may be thought of as consisting of two stages. In the first stage upstream (at left) the action of the focusing flows from both sides makes the central layer with the sorted cells so thin (Figures 3.119 and 3.120) that the detector can evaluate the detected properties of an individual carried cell. According to the meeting or otherwise of the sorting criterion either control X1 or control X2 is actuated, sending the cell into either Y or V.

Figure 6.133 (Right) Two alternative electroporation devices. They are used for electric lysis of the cell, breaking open its membrane and releasing the cellular material. Top:electrodes supplied with repeated pulses. Bottom: steady applied voltage, a cell experiences pulses as it moves between electrode tips. The flow velocity may be much higher.

An important procedure, which may be actually used in quantitative form for the detection and evaluation step, but mostly serves as a preparatory process, is the *polymerase chain reaction*, (PCR). It is the technique of replicating DNA, so that the number of available copies increases exponentially. It was invented by Kary Mullis in 1983 (sharing Nobel Prize in Chemistry for it in 1993). It usually reproduces only fragments with at most $10 \cdot 10^3$ base pairs (while complete human DNA contains ~ $3 \cdot 10^9$ pairs).

The first step in replication is denaturing, the separation of the two chains of the double helix. A polymerase enzyme then makes a copy, using each strand as a template. The polymerase used is called Taq, named for *Thermus aquaticus*, a bacterium living in a hot spring in Yellowstone National Park, from which the enzyme was isolated. To make a copy of DNA, polymerase requires the presence of other components: a supply of the four nucleotide bases, A, C, G, and T, and a primer, a short sequence of nucleotides to prime the process by making the first few nucleotides of the copy (another primer may be needed to define the end of the copied sequence). The template is mixed with the components in a fluidic mixer, such as one of those described in Section 3.7.

Thermal denaturing requires a temperature of 95°C for a time segment of about 10s. However, the primer cannot bind to the DNA at this temperature, so the mixture is cooled to 57°C at which the primer binds (or anneals) to the ends of the DNA strands. This also takes about 10s. The final step in the reaction is making complete copies of the templates. The temperature required for this part of the process is around 72°C, at which the Taq polymerase works best (this is the temperature of the hot springs where the bacterium was discovered). Taq polymerase has a limited life span, which is partly why it is not advisable to perform more than about 30 amplification cycles. This is not a serious limit, since by then the ideal number of available DNA strands would be 2^{30}, roughly astonishing $1 \cdot 10^9$ (the real number is smaller, in fact the effectiveness of the process decreases after ~ 25 cycles).

The earliest fluidic PCR reactors operated in the batch mode, with the mixture closed in a chamber (Figure 6.134), and an electronic controller operated the heater to perform the necessary

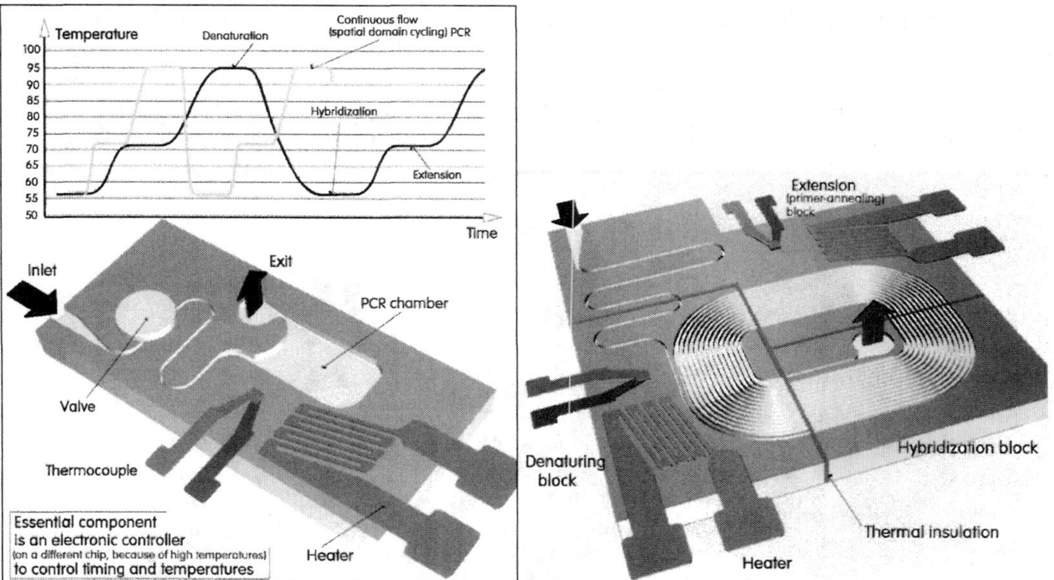

Figure 6.134 (Left) The time-domain polymerase chain-reaction device. It produces a large number of DNA copies by cyclic variation between three temperature levels. With each cycle the number of DNA molecules theoretically doubles.

Figure 6.135 (Right) The spatial-domain polymerase chain-reaction device. The periodic temperature changes of the basic version in Figure 6.134 are replaced by a fluid flowpath passing through three blocks, each kept at a different temperature level (note the heaters and thermocouples).

number of temperature changes, with an ~10s dwell at each temperature level reached. The changes inevitably took some time – which, multiplied by the number of the cycles, resulted in a slow response. Later, time-domain temperature cycling was replaced by a spatial-domain reactor (Figure 6.135), with a continuous flow passing through three regions kept at the three temperature levels. The advantage is a faster overall process, since no time is lost in waiting for the completion of the transition, as shown by the gray curve in the top part of Figure 6.134. Of course, in this layout the whole DNA identification operates continuously [25].

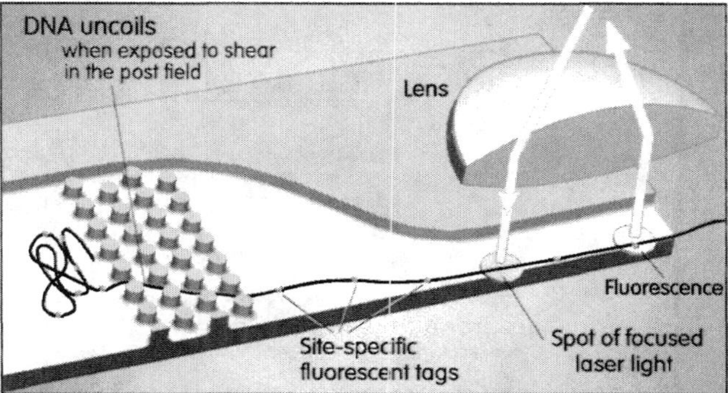

Figure 6.136 Mapping of DNA strands stretched by flow shear, in a device shown here as a one-half section in the axial plane. Laser-induced fluorescence signals are obtained due to the presence of markers that attach to specific locations that are illuminated by laser light. In detection devices such as this, which are capable of mapping a single DNA molecule, chain reaction amplification is unnecessary.

Detection principles include electrocapillary separation, according to Figure 5.90. This used to be almost synonymous with DNA identification, especially with the optical detection of DNA strands marked by fluorescent tags. Recent developments concentrate on the use of electrochemical biosensors, which are less expensive than optical systems with lasers. Also of interest are the detectors (as shown in Figure 6.136) in which fluidic action on the investigated DNA is used, so that it is possible to investigate just a single molecule, making the complexity of PCR superfluous.

6.7 MEDICAL APPLICATIONS

Medicine, at the beginning of the twenty-first century, is under strong economic pressure. The prices of procedures, pharmaceuticals, and instrumentation are increasing while most national health insurance system finances are stretched to the very limit. The solution, at least a partial one, offered by microfluidics is a decentralized prevention and early warning for individuals, making the treatment of illnesses much more effective. Essentially, the idea is to use miniature analytical instruments for convenient, accurate, and inexpensive health monitoring, ideally on a continuous basis, but at least on a regular basis at home.

All the technical aspects of this approach are not yet solved or decided upon. It is not yet known how universal monitoring should be with respect to the innumerable quantity of all possible illnesses. Very probably, the initial effort will be mainly focused on the most common and most fatal illnesses, such as cardiovascular disease and cancer, and on the very widespread ones, such as diabetes. Another question waiting for an answer is the character and frequency of transferring the resultant data to a specialist for evaluation.

There are other aspects yet to be solved. Economic issues dictate that monitoring and drug-delivery devices are mass produced and inexpensive, to the point of being discardable after use (so that there is no need for cleaning or sterilization). The issue of user friendliness is very difficult. Implanting the devices would be technically ideal. This would solve the problem of access to the monitored organs, but it obviously is acceptable only for patients with an already-diagnosed chronic illness.

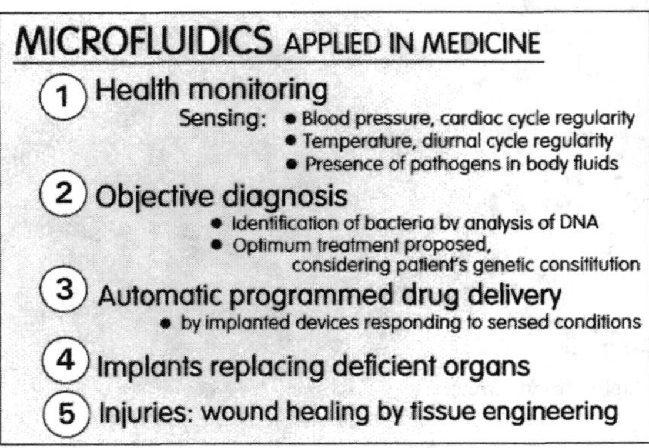

Figure 6.137 A survey of issues discussed—mainly in selected examples—in this chapter.

Apart from monitoring, microfluidics can also contribute substantially to a much better than present-day diagnosis and choice of therapy. Based on the exact analysis of pathogens, the diagnosis is less dependent on subjective opinions, which may be erroneous. Therapy, especially by medication, can be made more efficient with externally controlled, patient-independent, and continuous drug delivery. This again is a suitable domain for fluidics, especially, of course, if the administered pharmaceuticals are in liquid form. A list of these and other opportunities that deserve being mentioned is in Figure 6.137.

6.7.1 Monitoring of the health state

With the number of possibilities so vast, it is the best approach to relate an idea by giving several characteristic examples. Most monitoring devices need an input of a sample; the role of fluidics follows from the fact that most samples—and most that are easily taken—are in liquid form.

6.7.1.1 Flow cytometry

A wealth of data about a person's health may be obtained from measurements on cells suspended in flowing liquids. Cytometry is used to detect cell abnormalities, originally mainly shape deviations of blood cells, but progressively often used also to measure various physical properties. It can also directly detect the presence of bacteria. The procedure typically involves the generation of a very narrow liquid jet, in which the cells are organized in a single file and then come under the detecting or measuring sensor one by one. A typical application is the detection of lung cancer, one of the most prevalent and most lethal cancers, with five-year survival rate less than 15%. As for most cancers, the prognosis depends strongly on the illness stage at the time of diagnosis. Flow cytometry of sputum [50] may provide desirable early stage detection and has the advantage of being noninvasive. It is based on the analysis of epithelial cells present in sputum, by evaluating the number of their aneuploid nuclei—those with an abnormal chromosome number, a condition that increases with tumor malignancy. In a long-term study, sputum cytometry was found to precede a radiological diagnosis by 18–36 months. For the group of patients in which the cancer was detected by cytometry, this was enough to improve the five-year survival rate to 80%.

The presence of a cell in the interrogation area triggers the taking of a picture through a microscope-objective lens on a ccd chip. The first quantity obtained by processing the picture is the cell area - simply the number of all of the gray pixels within the frame [50]. Aneuploidy (changes in chromosome number) was included in the more general concept of DNA index value. This is the integrated magnitude of overall grayscale optical density. Beside taking into account the number of (dark) chromosomes, the index value is also influenced by any extra chromosomal fragments, and the DNA content of individual chromosomes, as well as morphometric features of the nuclei. The value normalized with respect to the magnitude obtained for a normal diploid nuclei is plotted in the diagrammatic examples (Figure 6.138). The cytometer contains mainly the fluidic circuit and the "camera." It is connected to a home computer for picture processing.

6.7.1.2 Glucose sensor

Diabetes mellitus used to be a major cause of death; for example, statistics for Malta in 1927 show that 47.7% of population died there due to diabetes. In 1910, Sir E. A. Sharpey-Schafer found that patients with diabetes were deficient in a hormone, produced by the pancreas, which regulates the uptake of glucose from the blood into cells. He proposed calling it insulin (from the Latin *insula*, island, in reference to the islets of Langerhans in the pancreas). The elevated glucose content is compensated for by excreting excess sugar in urine. This leads to dehydration and poor circulation. Long-term effects include kidney diseases (10–20% of diabetics), blindness (20 000 cases per year in the United States), and nerve damage (60–70%), often leading to amputation

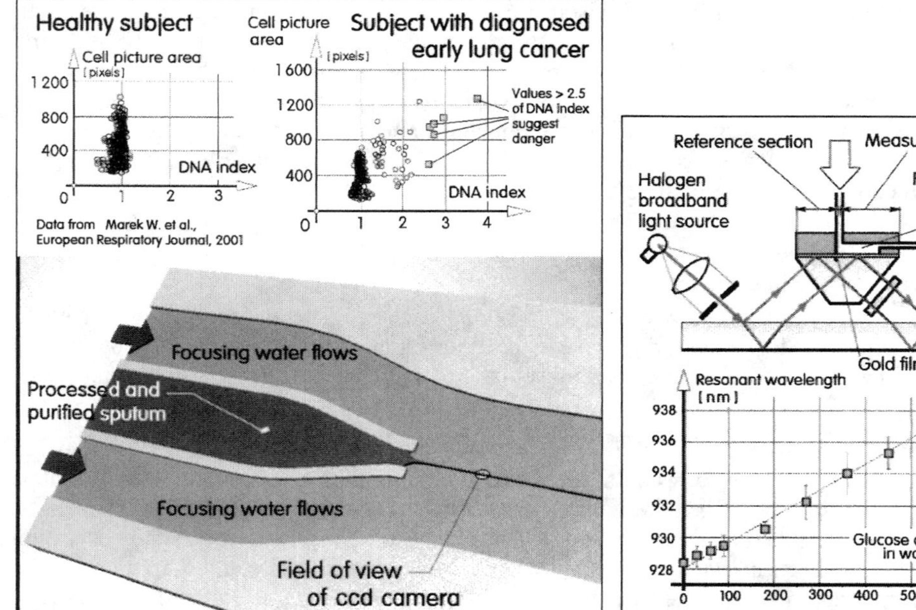

Figure 6.138 (Left) An example of a flow cytometer (bottom) and typical results of measurements of cell optical density (top, after [50]). A 20 x microscope objective projects the cell nucleus picture, triggered by the presence of the cell in the objective lens field of view, on a 1.4 Mpixel, commercially available ccd chip. The typical epithelial cell with a nucleus size 10–12 μm is evaluated with 0.34 μm resolution. The fluidic circuit—sputum processing, purification, and transport—is actually much more complex than is suggested by this picture of the core part.

Figure 6.139 (Right) Optical glucose-concentration sensor based on the surface plasmon resonance phenomenon. The electric fields of the incident photons interact with the electrons in the extremely thin (50 nm) gold film and generate charge-density waves (surface plasmons). In resonance, the intensity of the reflected light is reduced. Even a small change in the optical properties of the liquid shifts the light wavelength at which the resonance occurs. (After [51].)

of extremities. The disease remains incurable, but the administration of insulin obtained from bovine pancreases, a strict diet, and lifestyle modification can help to evade long-term damage. The incidence of diabetes is high. World Health Organization forecasts it to reach 150 million worldwide by 2025. The availability of precise and inexpensive sensors of glucose concentration in blood is urgent. The common principles are the optical sensing (Figure 6.139) and—more promising—the electrochemical method of Figure 6.140. Both employ a nontrivial fluidic circuit operating either directly with blood or with another liquid into which the glucose is transported from blood across a semipermeable membrane. Taking blood samples is unpleasant and may be a potential gate for infections, though microneedles inspired by MEMS microfabrication methods (discussed, for example, in [26]) make it practically painless. The very small needles may not reach the hypodermic depths with sufficient blood flow. The dialysis process can obtain measurable quantities of glucose from the easier-to-access interstitial fluid. There is also glucose in saliva, at a concentration correlated with the one in blood but unfortunately roughly 100 times lower. This noninvasively available source remains a challenge for sensor designers; present devices fail at so low concentrations.

The problem with invasive sampling in general is the necessity of repeated insertion of the probe. The body has a tendency to encapsulate by a separating protein layer any long-term inserted or implanted sensor. Also, some electrochemical method sensor properties tend to drift over a

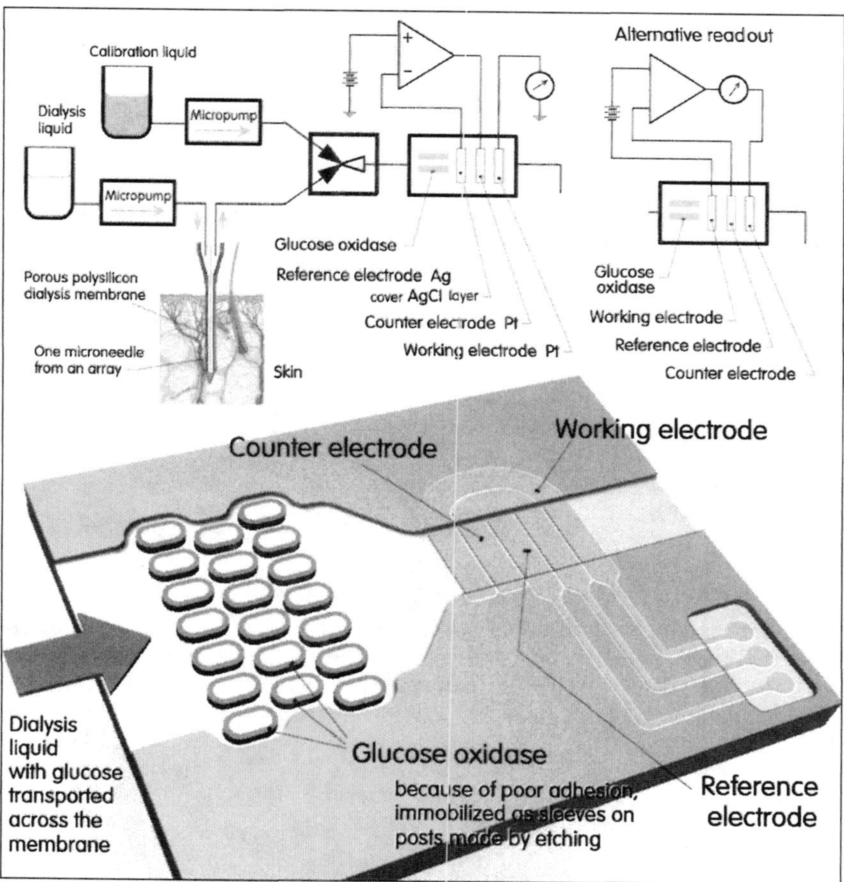

Figure 6.140 A sensor for measuring glucose concentration in blood using the electric potential difference between electrodes in the presence of an enzyme, glucose oxidase, which has to be immobilized inside the flowpath. The actual fluid, instead of the blood, may be a liquid into which the glucose permeates through a dialysis membrane. This may be part of an array of hypodermic microneedles, sometimes made by etching silicon in a variety of shape variants.

longer time period; the solution, complicating the fluidic design, is the use of a calibration liquid (Figure 6.140).

Unfortunately, many variables to be monitored are not externally accessible, and this raises the question of an implantation, as is the case in the following example. Apart from the trauma such a step inevitably brings, from the point of view of fluidics there is the serious problem of the power source for driving the fluid flow and, of course, usually also for powering the electronic circuitry. Batteries, the source mostly used, run out and have to be replaced. This repeats the trauma and is expensive.

6.7.1.3 An implanted blood-pressure sensor

An implanted device for the continuous monitoring of aortic pressure waveforms would be of great benefit for people who have suffered a heart attack. This is part of the grave problem of cardiovascular diseases, with the fatality rate already high and still rising by proportions that deserve to be called an epidemic. Congestive heart failure affects 4.8 million persons in the United States alone. The incurred cost there is over $\$ 38 \cdot 10^9$ annually. Worldwide, it accounts for 5-10%

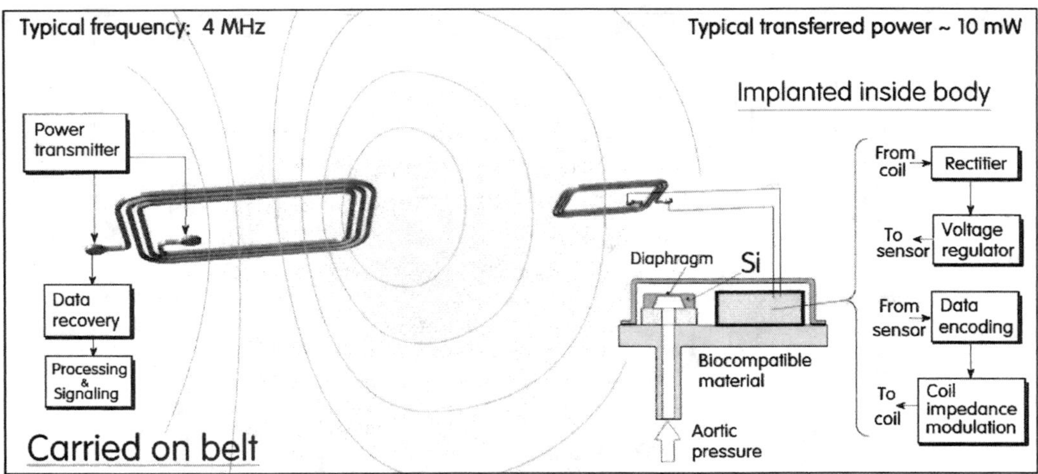

Figure 6.141 An example of a radio-frequency electromagnetic power transfer into an implanted device—here with the monolithic silicone sensor of Figure 5.95—for measuring aortic pressure. The patient has to carry (or have nearby) the power-transmitter/data-recovery device; run-out batteries are exchanged in this external part. Data transfer from the implanted device out is achieved by modulation of the implant's coil impedance, which the outer coil can sense.

of all hospitalizations. A person of age 40 has a 1 in 5 chance of developing this disease. Patients are treated by long-term medication. This is no small problem, since they face a danger of potentially fatal secondary effects (such as pulmonary edema). Decisions about the dosage have to be currently made on the basis of imprecise and subjective, in fact, rudimentary evaluations of symptoms. Of real significance are the variations in aortic pressure wave shapes. Detecting them requires permanent monitoring, which may be, according to Figure 6.141 done by means of an implanted device with a blood pressure sensor. Transfer by an essentially magnetic effect of electromagnetic waves makes unnecessary repeated operations for access to the implanted device. The same principle of power and data transfer may also be used in other applications. Essentially, the power part of the system remains outside the body. Of course, the need to wear it constantly can be unpleasant.

6.7.2 Diagnosis and choice of therapy

Making the correct diagnosis has always been one of the basic problems of medicine. According to published estimates, one-third of patients do not get proper treatment, and one-quarter of them get treatment that is not needed or is even potentially harmful. Diagnoses are mainly based on the evaluation of symptoms and may be very uncertain at the earliest stages before the symptoms fully develop, at which point has already passed the time during which therapy can be the most effective. The ideal of a distributed objective diagnosis process (Figure 6.142), as suggested by the cartoon in Figure 6.143, would not only eliminate subjective errors, but would also provide the correct diagnosis very early. Laboratory techniques are already available, but they are expensive and slow. They are usually implemented at a later stage, while a MEMS device would be instantly available at the slightest suspicion. The technique of DNA identification, discussed in Section 6.6 as means for identifying persons, is in principle applicable to identifying any living organism, including bacteria, viruses, and parasites. The basic principles, as presented in Figure 6.131, are therefore directly applicable to the task of diagnosis, while also taking into the account the classic evaluation of symptoms. Even though the final processing and evaluation of the result is the task of electronics, the importance of the roles of fluidics is obvious: the DNA is extracted, purified, and processed by fluidic devices. There are two possibilities. Ribosomal DNA

DIAGNOSTIC MICROFLUIDICS

(1) **Sensing**: continuous operation or at least frequently repeated

(2) **Assay**: one-time analysis usually biological or biochemical

Typical: detection of antigens or antibodies using the extreme selectivity of immune response to the invasion of foreign molecules into living organisms

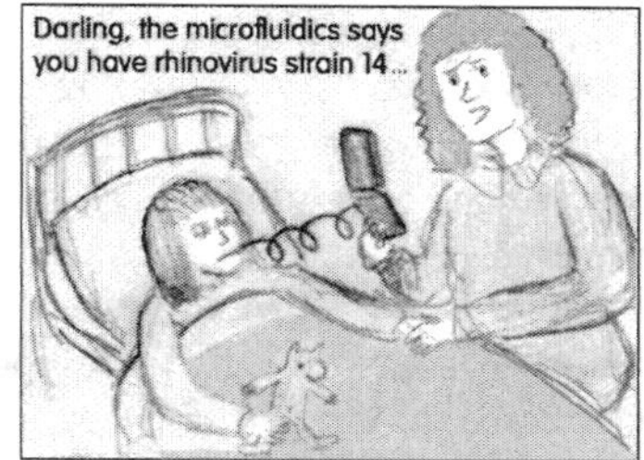

Figure 6.142 (Left) The identification of pathogens is usually done once and not on a continuous-sensing basis. The procedure envisaged as performed by a microfluidic diagnosis device is therefore properly described as an assay.

Figure 6.143 (Right) Although this cartoon is meant as a joke (the users are not expected to be given the results in detailed medical terms) it conveys the basic idea of the distributed domestic healthcare.

tests are sensitive (needing a minimal sample), but not very specific. On the other hand, tests of genomic DNA are highly selective, but their sensitivity is low. A considerable number of instruments have been reported to be in development; so far, as with the identification of persons, none is anywhere near the mature stage needed for practical use. A small disadvantage is that this approach does not distinguish living pathogens from those that are already dead and harmless. Another diagnostic means among the techniques already discussed is "artificial nose" (Figure 6.82). In particular, odor analyzers are proven as capable of diagnosing lung diseases such as sinusitis or pneumonia [27]. The technique considered to be particularly suitable for a small handheld diagnosis instrument is detecting the presence of antibodies against a particular pathogen (Figure 6.144). There is the disadvantage of its inability to distinguish between present and past infections, but this is not a serious problem. More difficult to handle is its dependence on the antigens. These are available commercially and, especially in the amounts needed in microfluidic devices, are not expensive, but they must be kept in a (nondefrosting) refrigerator.

6.7.3 Drug delivery

Fluidics also finds an application in the systems and devices for the treatment of illnesses, in particular the long-term, chronic ones. One of the areas in which it is useful is the controlled

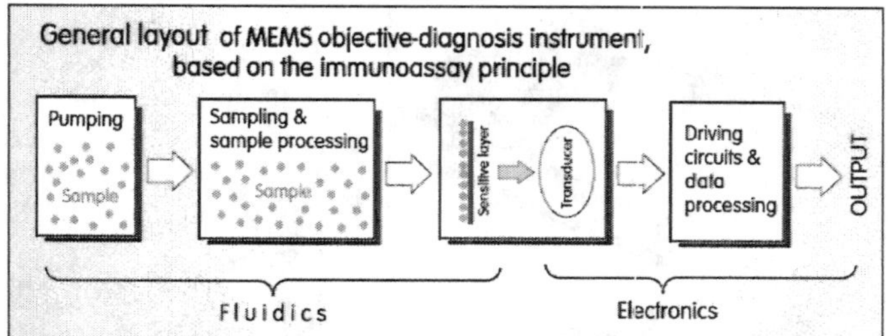

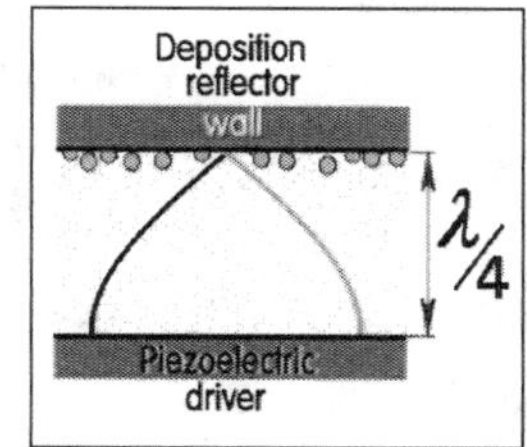

Figure 6.144 (Left) The idea of a typical illness-diagnosis assay. The identification of pathogens is based on detecting the presence of antibodies in body liquids (such as saliva, urine) on a sensitive plate covered with antigens (to which the antibodies may be forced by the method in Figure 6.145), using the selective antibody-antigen reaction. This generates a physically measurable change in the sensitive layer, which is converted by a transducer into an electric output signal.

Figure 6.145 (Right) The deposition of the antibodies on the sensitive layer may be helped—and the immunity reaction thus be made faster—by using the tendency of particles carried with fluid to migrate toward the pressure node of standing waves, as in Figure 5.111. All that is needed is to position a piezoelectric driver opposite of the antigen-covered wall at a distance equal to one-quarter of the wavelength of the generated acoustic waves [34]. Also effective is the deposition by an impinging jet, which (Figure 5.44), may be simultaneously used to excite the transducer reed with the sensitive surface.

delivery of pharmaceuticals into an organism. This remains an important challenge in medicine. Fluidics can secure the continuous release of therapeutic agents over extended time periods in accordance with either a predetermined temporal profile or, better still, in response to a sensor signal monitoring the symptoms.

Again the case of diabetes and its treatment by insulin, already discussed in Section 6.7.1.2, is a good example. Being a protein, insulin cannot be taken orally, as it would be, in effect, digested in the stomach. It is administered by hypodermic injections, an inconvenient and unpleasant, if not painful, process. Unfortunately, it has to be applied often, and in more developed illness several times a day. It is administered to suppress the glucose level in blood reaching and staying for a longer period of time above the value ~140 kg/m^3, which causes irreparable changes in organs (the normal level in healthy subjects is 90–100 kg/m^3). The concentration in blood rises sharply after each meal, which must be eaten in a small portion under strict dietary conditions. If the intake of insulin is not balanced by the glucose input from food, it may cause dangerous states of hypoglycemia. In fact, recent continuous monitoring by glucose sensors (Figures 6.139 and 6.140) revealed quite common occurrences of hypoglycemia for example during the subjects' sleep.

The simpler of the feedback solutions offered by fluidics is based on using an external glucose sensor and an internal insulin-releasing device controlled by a radio-frequency (Figure 6.141) transmitted signal. The protection of the insulin by the walls of the device (Figures 6.146 and also 6.150) during its travel through the stomach (Figure 6.147), makes possible the more convenient method of oral input. The trick in the case of Figure 6.150 is the nearly perfect resistance of the golden membrane to corrosion by body fluids replaced by its fast decomposition when a voltage is applied. Alternatively, a device with a sufficiently large supply of the drug may be also implanted.

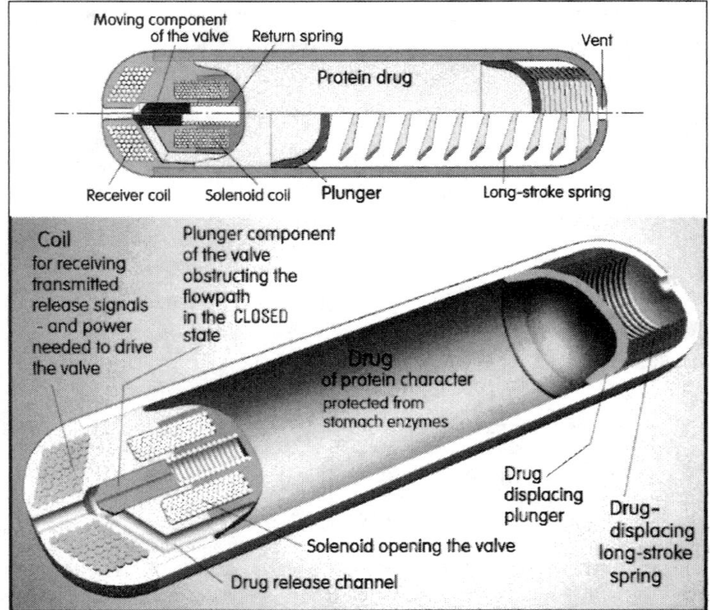

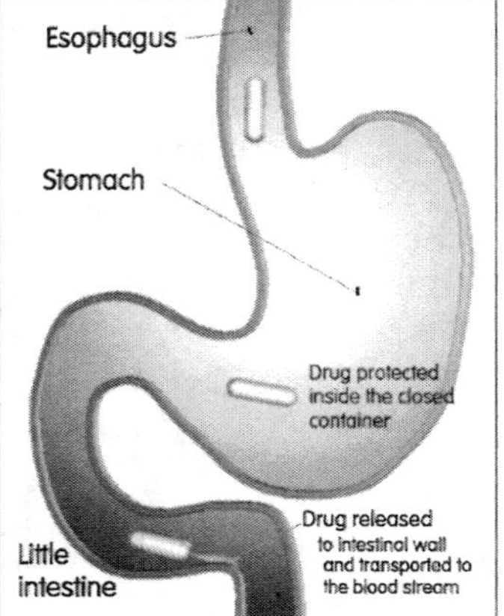

Figure 6.146 (Left) An example of a "smart pill" releasing pharmaceuticals in response to a radio-frequency signal transmitted to the receiver coil according to Figure 6.141. The drug is stored in an elasticity capacitor corresponding to Figure 2.95 (and case F in Figure 2.99).

Figure 6.147 (Right) Use of the pill device (Figure 6.146) to transport a drug of a protein character (such as insulin) through the stomach, where it otherwise would be decomposed.

The improvement in the patient's quality of life is tremendous: no delivery by hypodermic injections and, at least in principle, no diet. What remains is the inconvenience associated with carrying the glucose sensor and the radio-frequency transmitter. This is removed by the totally implanted design, as shown in Figure 6.151. Apart from the need to refill the insulin it contains at periodic intervals, the patient may feel like a nondiabetic, healthy person.

6.7.4 Implanted devices

Implanting devices and thus removing all the inconvenience associated with external therapeutic equipment is often proposed. The idea is, of course, helped by the very small overall size made possible by microfluidics and MEMS in general. However, a number of questions have yet to be answered. Of the technical ones, the most serious are biocompatibility and the provision of driving power. In general, satisfactory biocompatible materials are available for the passive outer walls of the devices. More problematic are the active surfaces, such as dialysis membranes (located inside the needle in Figure 6.140). Transport (mostly selective transfer) across the membrane is hindered by the trend of covering it with a protein layer.

The power needed to drive the devices may be transferred transcutaneously (Figure 6.141) but this eliminates the rationale for the implanting. There are four solutions to the problem of fully autonomous implants:

(1) The least successful of them seems to be the thermal method, using, for example, a thermoelectric effect (Figure 2.24). There is a well-maintained temperature inside the body,

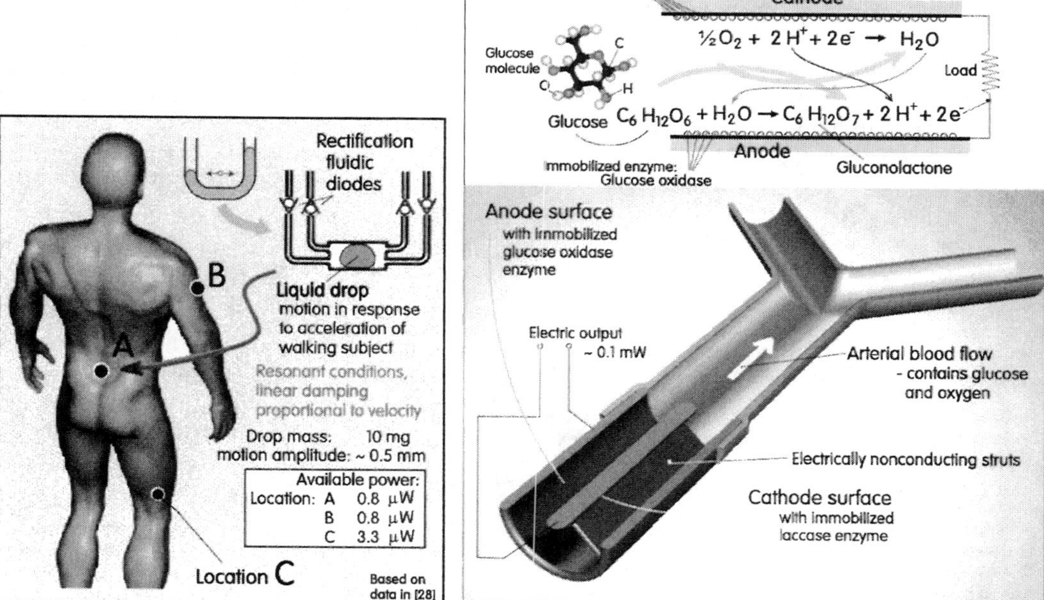

Figure 6.148 (Left) It is, in principle, possible to harvest energy from accelerations accompanying body motions, such as walking, using a fluidic generator. This device was developed from the liquid movement in a translated U-tube (Figure 2.100). However, the obtainable power is low and discontinuous, requiring some form of energy storage.

Figure 6.149 (Right) An enzymatic fuel cell can generate electric power for driving implanted devices using arterial blood flow and its glucose and oxygen contents.

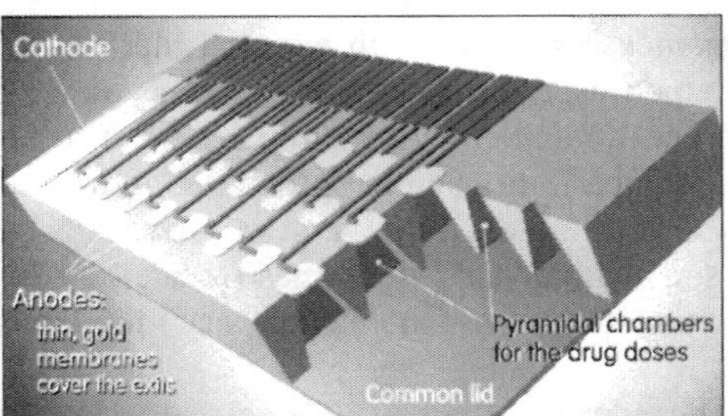

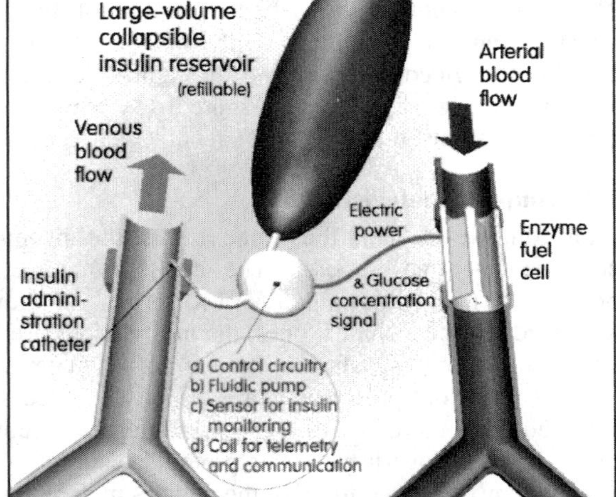

Figure 6.150 (Left) Microfluidics for controlled drug releasing: the pharmaceuticals are allowed to discharge from any of the 30 chambers upon a 1.04 V electric signal which causes the gold membrane to dissolve (converted into soluble gold chloride). The electric power required to dissolve one membrane is ~1 μW applied for 30 s. Note the basic spatio/temporal character of this solution, compared with the purely temporal one in Figure 6.146.

Figure 6.151 (Right) An implanted, electronically controlled, fluidic insulin-control system. As long as there is some insulin in the pouch reservoir, the diabetic can forget about his or her health condition (no need to adhere to any diet).

but the difference relative to the outer environment is small, < 2°C and less in warm climates, and there is no convenient location for the cold ends of the thermocouples (only ear lobes, which are colder than the rest of the body).

(2) Power may be harvested from the inertial effects associated with body motions. This energy source is unfortunately not continuous, needing a complicated energy-storage system. An accelerometric version with a mechanical inertial mass is evaluated in [28]. The fluidic version in Figure 6.148 was developed from use of the liquid motion in the unsteadily translated U-tube resonator (Figure 2.100). The development aimed at miniaturization and elimination of the dependence on orientation with respect to gravity. The result is a configuration with a liquid drop – or its inverse with an air bubble, corresponding to Figure 5.38. In general the available power output is low and unreliable in periods of sleep or immobility.

(3) Stimulated skeletal-muscle-powered generators eliminate discontinuity in energy production by keeping a selected driving muscle, such as the *palmaris longus,* in permanent oscillation by electric stimulation. A piezoelectric material is attached to the tendon of the muscle and bone. The power depends on force and frequency; typically, a ~10 N force at 1 Hz is available. Tests used stacks of lead zirconate titanate plates. Modern piezoelectric polymer materials (Figure 5.61), could provide a much better performance. The generated energy was 500 J per day, of which the muscle stimulator consumed only 0.1 mJ, so the available output was found to suffice for powering prostheses restoring limb function.

In an older study [29], a direct fluidic output was obtained from the electrically stimulated muscle *latissimus dorsi*. The device was basically a cylinder with a piston, connected by a stainless steel bellows, which also acted as the return spring. A flexible sheath formed from a collagen-coated vascular graft was used to prevent tissue growth into the bellows folds. The mass of the device was 91g, and the volume was 26 cm^3. The cylinder was fixed to the rib and the muscle was attached to the piston via its humeral tendon. The typical stroke was 18mm. The power transfer by alternating airflow at a high pressure ~1 kPa and small flowrate amplitudes was chosen to minimize viscous and inertial losses. The output was 1.125 mJ energy per stroke, the device power per muscle mass ~ 6 mW/g, which means that with the 600g *latissimus dorsi* creates more than enough power to drive an artificial heart (typical input < 1.3 W).

(4) A very promising source is the biological fuel cells in which the hydrogen-oxygen reaction in the cell from Figure 6.43 is replaced by the oxidation of glucose from blood and the catalysts are replaced by enzymes. This fuel cell (Figure 6.149), needs no separating membrane and may be placed directly into an artery.

6.7.5 Tissue engineering

The success of transplantations has led to a demand for organs far outpacing (by two to five times, dependent on the organ's character and the recipient's age) the supply from suitable donors. The shortage of available organs is one of the reasons for the widespread effort to grow them artificially in vitro. The growth takes place in bioreactors (Figure 6.152), which, together with their culture medium supply and flow and conditions control, are quite complex fluidic systems. The success of the effort is in inverse proportion to the organ's complexity. Growth of complete organ substitutes, such as kidneys, remains the final goal not yet mastered. What can be done is to separate the production of different tissue types: bones, chondrocytes (cartilage), smooth muscle cells (also the basic structure of blood vessels), vascular endothelial cells, hepatocytes (for the liver), and neurons. Geometrically simple forms with one or two growth degrees of freedom, such as skin or blood vessels, are fabricated rather routinely.

The immediate target is elimination of the rather nonethical final stage of integrating the different tissue types in vivo in laboratory animals. This state increases the demands on the complexity of the bioreactor fluidic control system. It is no longer sufficient just to maintain the physical and chemical conditions. The individual components of the organ require separate growth control, based on individual cell-growth models. The feedback loops mostly use electrochemical

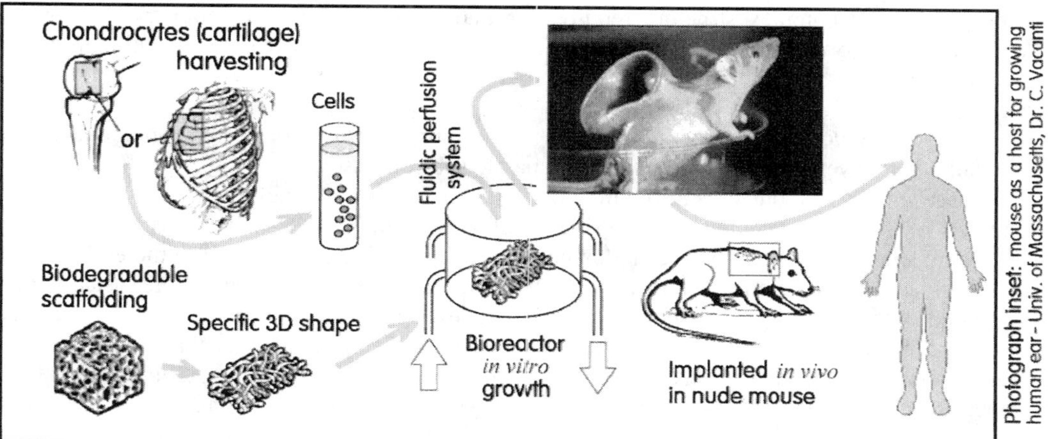

Figure 6.152 The present form of tissue engineering: the organs grow in the bioreactor, supplied by fluidics with nutrients and auxiliary substances, but the final steps require a laboratory animal as a carrier, in this case, to develop a transplantable outer ear.

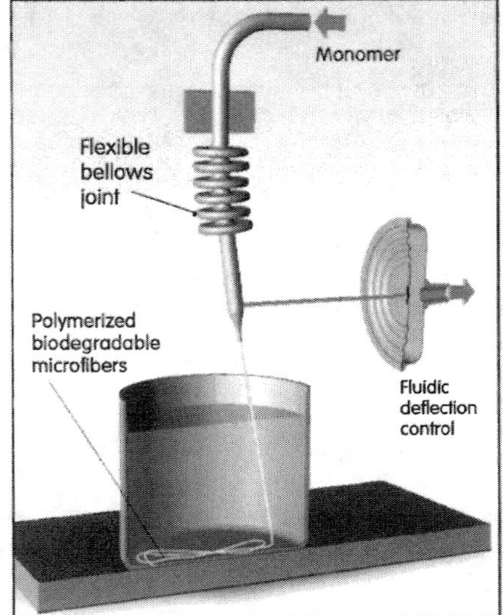

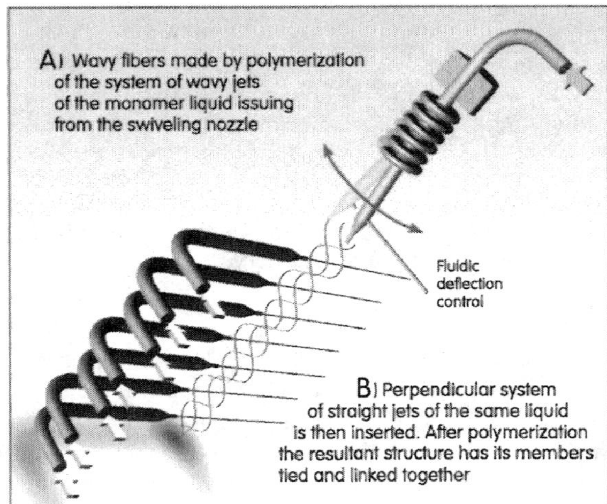

Figure 6.153 (Left) The "scaffolding" that determines the shape of the produced organ (Figure 6.152), is made by several different methods from a polymeric, usually biodegradable material. In this case the shape is built from a jet of monomer liquid polymerizing in contact with the active liquid in the vessel in which the jet issues from a nozzle performing computer-controlled deflection motions.

Figure 6.154 (Right) The essential principle of making the scaffolding as an interlinked, three-dimensionally woven structure from a number of fibers produced by the polymerized liquid jets (in analogy to Figure 6.153).

and optical (noninvasive) sensors as inputs. Also, at the bioreactor exit, the liquids are carefully analyzed for markers of the biosynthetic processes. On the output side of the control system, there are valves governing the culture medium flow distribution among various inlet nozzles. Also the pH and composition of the liquids, consisting mainly of glucose (measured as in Figures 6.139 and 6.140), proteins, antibiotics, and other additives, are appropriately varied.

Organs with a complex structure require, for maintaining the proper geometry, anchorage-dependent cells growing around scaffolding. The progress may be faster by using cell aggregates made in prereactors as the building blocks. The growth processes there are often conditioned by applied mechanical stimuli, which are equally as important as the chemical input for cell differentiation. Endothelial cells lining blood vessels or gastrointestinal tract components are essentially monolayers or sheets also made in separate sub-processes.

The scaffolding, of course, is also built and shaped beforehand. Bones, cartilages, and liver are examples of complex geometries requiring a 3D scaffolding. It usually consists of highly hydrophilic sponges confined by a complex network of fibers. This is essential in the early stages. The cells later create their own matrices. The scaffolding material is therefore usually biodegradable. A typical material is polylactic acid. Bones are grown on ceramic scaffolds derived from coral. Smart scaffolds include locally deposited biochemical markers that encourage cell differentiation; the deposition also involves a computer-controlled fluidics. Building scaffold geometry usually starts with liquid monomers. A typical method of shaping, based on instant polymerization, may also involve rather complex numerically controlled fluidic systems (schematically represented in Figures 6.153 and 6.154. Sometimes the scaffolding (and the final fabrication reactor) also have to be designed to allow an application of the mechanical stimuli to the attached cells.

6.8 AGAINST TERRORISM AND CRIME

Fluidics and microfluidics are also expected to play an important role in making the world and our homes safer. The basic problem of terrorism is its economic aspects, unfortunately favoring the criminals. It is much cheaper to perform a terrorist attack than to prevent terrorists from accomplishing their goal. The crude explosive devices terrorists usually use are inexpensive. Also, a chemical or biological attack can be carried out with minimum cost and effort. On the other hand, maintaining the safety is costly. The primary goal of the antiterrorism war is to eliminate the danger of attacks. This means deploying and maintaining early-warning systems, which cost a large sum of money. The current technologies for the detection of weapons, as well as biological, chemical, and nuclear agents, have to be deployed at a large number of locations, such as airports and their many gates, railway stations, and important public buildings, where they require the constant attention of trained operators. Dogs trained to detect explosives and other dangerous substances are almost always in short supply; they are also expensive, with their long-term training and their need for a dedicated supervisor person. Also the elimination of the consequences of an attack, such as decontamination activities, is expensive. The first responders to a terrorist attack are local law enforcement, fire, and medical personnel. The budget of most localities limits their expenditure on devices that would make all these people less vulnerable.

The emerging technology based on fluidics and microfluidics can change this situation, at least to a certain degree. Sensors capable of detecting the presence of hazardous and prohibited substances may be microfabricated in vast numbers and so inexpensive as to make them deployable almost everywhere. Moreover, these devices are mostly a direct outgrowth of what has been already developed for other purposes. The detection of airborne warfare agents can use the analytical capabilities at the ppb levels of the devices such as a detector of the aromatic volatile organic compounds (Figure 6.42), microchromatographs (Figures 6.27 and 6.28), artificial noses (Figures 6.80 and 6.83), and other devices, such as the bionic odorant sensor (Figure 6.82). Also

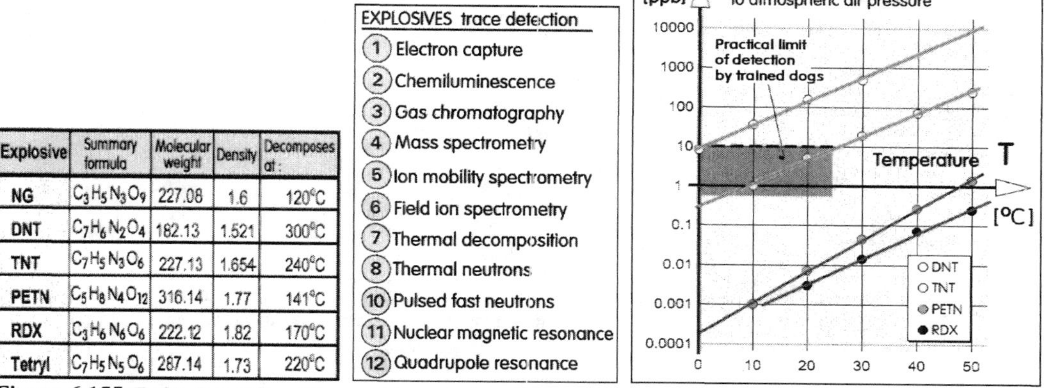

Explosive	Summary formula	Molecular weight	Density	Decomposes at:
NG	$C_3H_5N_3O_9$	227.08	1.6	120°C
DNT	$C_7H_6N_2O_4$	182.13	1.521	300°C
TNT	$C_7H_5N_3O_6$	227.13	1.654	240°C
PETN	$C_5H_8N_4O_{12}$	316.14	1.77	141°C
RDX	$C_3H_6N_6O_6$	222.12	1.82	170°C
Tetryl	$C_7H_5N_5O_6$	287.14	1.73	220°C

Figure 6.155 (Left) Properties of the most important explosives to be detected by the terrorist detection devices (from various sources).

Figure 6.156 (Center) List of detection methods (not exhaustive) discussed in recent scientific journal publications on the possible uses of MEMS in antiterrorist warfare. Most, though not all, methods detect vapors released by the substances.

Figure 6.157 (Right) Ratio of the vapor pressure of various explosives to the atmospheric pressure as a function of temperature. Due to the Dalton law, the plotted pressure ratio corresponds to the vapor concentration. Dinitrotoluene (DNT) is actually not an explosive, but a substance important from the point of view of detection devices, because it is the major impurity in TNT, and its ~25-times higher saturation concentration makes it much easier to discover.

a combination of the already-discussed principles may be useful: the sampling units used in the discovery of new materials (Figures 6.56 and 6.57), may be used for the centralized detection of illicit substances brought in from several locations, the odor sensing of Figures 6.82 and 6.83 may be adapted for the detection of landmines and booby-traps, the interferometer (Figure 5.128), the cantilever sensor (Figure 5.44), and the ion-mobility spectrometer (Figure 6.30) normally used as alternative detectors of mixtures separated by chromatographic means, may be used in parallel or in unusual combinations so as to increase their sensitivity, which is the main problem with the detection of the extremely small trace amounts available, sometimes literally individual molecules remaining on the suit of a terrorist who merely helped in transporting agents. Attacks by contamination of food and water may be instantly discovered by devices based on principles adapted from those presented in Figures 5.121, 6.72, 6.81, and 6.87. The identification of authorized personnel, or, on the other hand, suspects, may use the principles shown above in Figures 6.127 and 6.131.

Terrorism is only one area of the more general problems of crime. Antiterrorism applications pave the way to fighting other crime issues as well by novel microtechnological approaches. These can make crimes more difficult to commit and offenders easier and faster to identify. Smart tagging, for example, by chemical encryption, can suppress document, passport, and banknote fraud. Inexpensive sampling and detection devices may be deployed across the landscape to monitor the airborne propagation of chemical agents or bioaerosols.

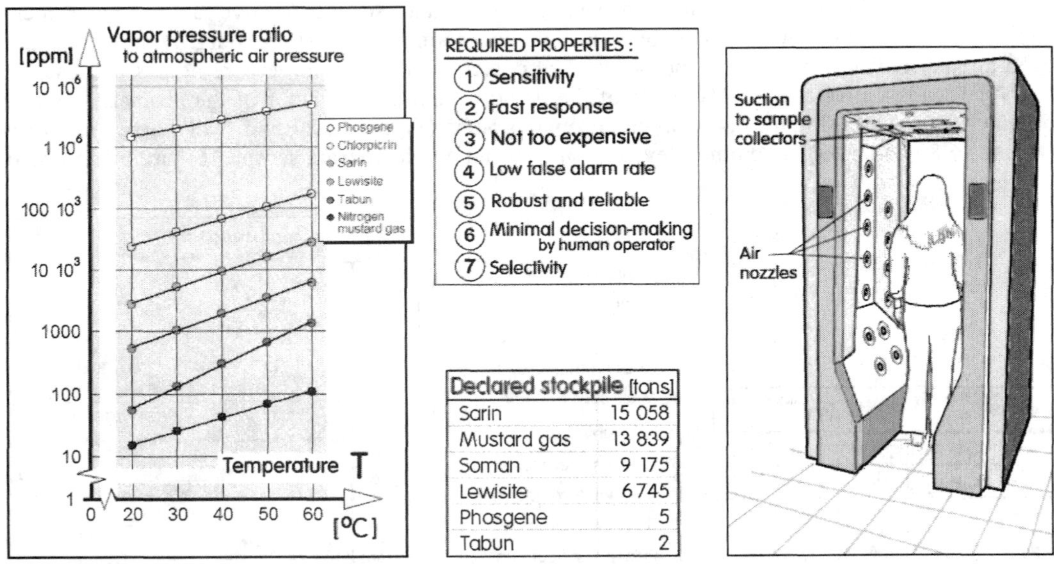

Declared stockpile [tons]	
Sarin	15 058
Mustard gas	13 839
Soman	9 175
Lewisite	6 745
Phosgene	5
Tabun	2

Figure 6.158 (Left) The ratio of the vapor pressure of important chemical warfare agents to the atmospheric pressure, as in Figure 6.156, plotted as a function of temperature. Under the same conditions, the higher vapor pressure of these substances indicate that they are much easier to detect than explosives.

Figure 6.159 (Center-Top) The requirements to be fulfilled by the design of a detector of dangerous materials. The order of importance of individual points depends on the nature of the checkpoint location and the availability of trained operators.

Figure 6.160 (Center-Bottom) Worldwide totals of stored chemical agents reported to the Organisation for the Prohibition of Chemical Weapons in 2000. The probability of the substance getting into the terrorists' hands increases with the stockpile size.

Figure 6.161 (Right) A typical portal used for the detection of illicit-material traces left on persons who recently came into contact with warfare agents or narcotics.

The favored weapon of present-day terrorist is *explosives* (Figure 6.155). Their detection (Figure 6.156), is based on the identification of vapors. The vapor pressure, which determines the vapor concentration in the air, spans 8 orders of magnitude (Figure 6.157). Some of the less volatile explosives may be below the detection capability of dogs (Figure 6.75). Manufacturers are bound by law to tag them, in particular, the plastic explosive RDX and the notorious Semtex (essentially a mixture of RDX and PETN with other additives), by volatile taggants such as the dimethyl-dinitro-butane (DMNB). Figure 6.75 shows its exceptional property to be discovered by dogs.

Because volatility is required for their propagation in atmosphere as well as in action, the *chemical warfare agents* (Figure 6.158), are easier to detect. Even the least volatile mustard gas, made, despite the international ban, in large quantities (Figure 6.160), is comparable in volatility to DNT. Their principles of vapor detection being analogous, it is possible to design dual-purpose devices detecting both explosives and chemical agents. Biological warfare agents require a different approach, such as one of the principles discussed above.

The basic question of detector design is the conflict between the first two requirements listed in Figure 6.159. Sensitivity is essential. It is possible to increase it by preconcentration, according to the principle in Figure 6.29 (which describes accumulation from recirculating water, in the present case it will be extraction from air), but accumulating the diluted substance takes a long time. This makes the device not acceptable for a large-throughput screening, for example, that of persons passing through a checkpoint. Particularly slow are the biological immunoassay methods, which typically require collecting 10^5 antibodies for a reliable detection (typically obtained from 1 500 liters of air); the timescale of the test being tens of minutes.

A typical dual-purpose (chemical agents/explosives) microdetector may approximately correspond to the description in [31]. The components of the sample are separated in a special high-speed, resistively heated chromatographic column. The separation is so fast that the system response time may be only 5 s. The exit from the column is accelerated in a nozzle and directed as an impinging jet on a transducer surface. The typical transduction method uses acoustic waves

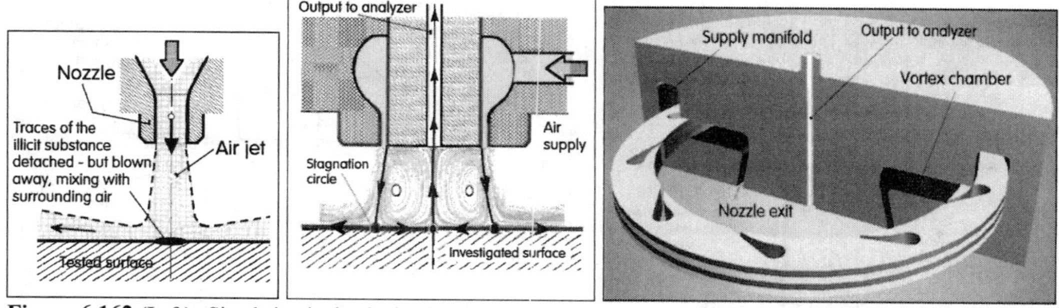

Figure 6.162 (Left) Simple impinging jet issuing from a circular nozzle. The traces of the dangerous substance are released from the surface, but mixed uncontrollably with large amounts of the added air from the jets.

Figure 6.163 (Center) Annular impinging jet surrounds the ring vortex region reaching to the surface. Those traces of the substance that are released from the surface under the vortex are protected from the outer atmosphere and their mixing with the jet air is minimal. Unfortunately, this character of the flow exists only if the nozzle-to-wall distance is short. With longer distances, the vortex (Figure 5.34) does not reach the surface.

Figure 6.164 (Right) An example of the layout of the annular nozzle imparting a rotation to the issuing annular air jet, as in Figure 5.35. The centrifugal force (and the visible outward inclination of the exit slot) counters the tendency of the vortex-ring region to close and makes possible the placement of the nozzle farther away from the impingement wall.

(Figure 5.107). The adsorption of the detected substances to the quartz surface changes the resonant frequency: in [31] the change is 10 Hz per picogram of adsorbed sarin. If the frequency is evaluated every 0.05s, the chromatogram, compared with patterns for illicit substances stored in the device memory, consists of 1 000 data points.

6.8.1 Portal detector and its sample collectors

In the detector design, several lessons were learned from nature, in particular observations about the importance of the exhalation phase in dog's sniffing (Figure 6.76). Similar to the canine nostrils, the fluidic portal detectors, Figure 6.161, also use air jets to release molecules of the detected substances from the surface on which they are carried. This detector type is used not only to discover traces of explosives and chemical agents, but also drugs, sedatives, hypnotics, anxiolytics, opioids, stimulants, hallucinogens, psychedelics, anesthetics, and steroids. In principle, the released substance is available as a vapor component in an air sample, which is processed by a fast, sensitive detector – usually the ion-mobility spectrometer (Figure 6.30), again perhaps with a fast chromatographic separation of the components and computer processing of the chromatogram, applying pattern-recognition analogies with chromatograms of the illicit substances stored in memory. It was found important to impart to the generated jets in the nozzles a considerable momentum so that they impinge on the clothes of the screened persons with sufficient power to ruffle the clothing. This is not as easy as it may sound because of the variability in the screened persons' sizes. The nozzles must be at a position allowing a passage of a large person. If, on the other hand, a person is of a small stature, the nozzles are too far away; to retain the effect, the nozzles have to be large and the jets (Figure 6.162), have to be powerful to counter the loss of their momentum by mixing with the surrounding air. With the large and powerful jets, the tiny amounts of the traces from the clothing are inevitably mixed with a vast

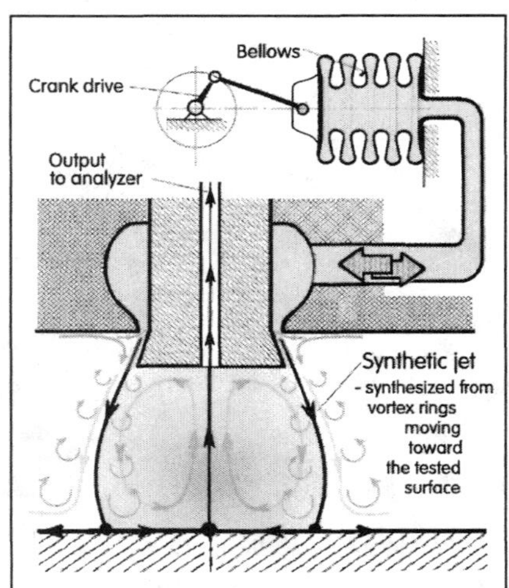

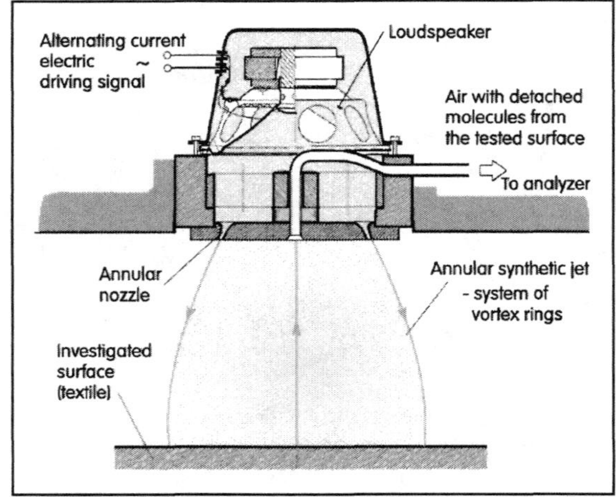

Figure 6.165 (Left) The annular synthetic jet is formed by alternating the outflow from and suction into the nozzle. No jet air is added, so that the trace amounts of substances released from the surface are not diluted. Those released under the nozzle are also separated from the outer atmosphere, as in Figure 6.163. The pulsation produces a quite strong ruffling effect on the clothes. Moreover, experience shows the annular character is retained in longer distances toward the wall.

Figure 6.166 (Right) The symbolic generation of the synthetic jet by the deformed bellows was in performed experimental investigations replaced by the action of a loudspeaker. The formation of a desirable character of the flowfield was found more effective at the achievable, much higher oscillation frequencies.

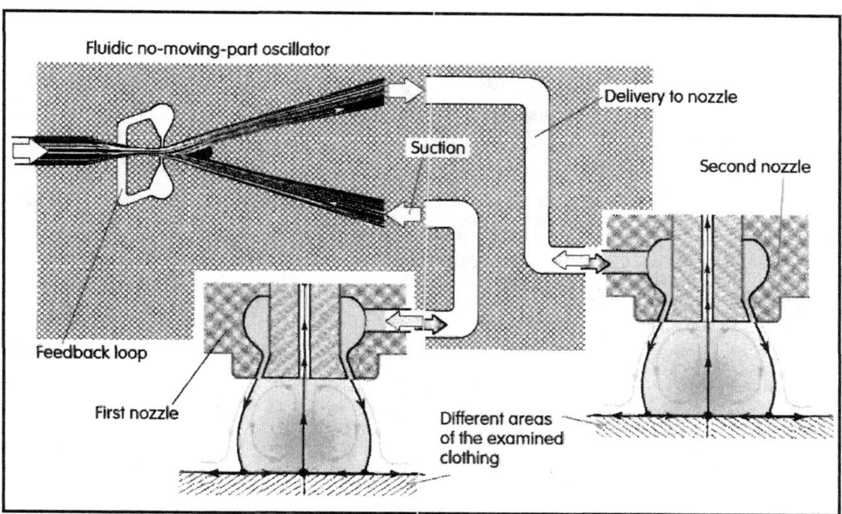

Figure 6.167 The detector with the hybrid-synthetic annular jets generated by the fluidic oscillator, according to Figures 4.15 and, 4.152. The suction flow is caused by the jet-pumping effect of the jet issuing from the main nozzle in the oscillator.

amount of the jet air and become diluted. This decreases the sensitivity of the detection, the key factor in the design. A substantial improvement in sensitivity (which means a faster response, fewer false alarms, and other advantages) is possible by "walling off" the space between the tested surface and the nozzle by the annular jet (Figure 6.163) generated in the sample collectors located so that they replace the passive ruffling jets of Figure 6.161. Unfortunately, the unpredictable distance to the person's clothing surface means that there is a possibility of the nozzle-to-wall distance being very long. In view of the effect shown in Figure 5.34, this may eliminate the vortex ring's reaching up to the wall. (The near-wall part of the flow in Figure 5.34 is effectively the same as in Figure 6.162.) A solution using very large nozzle exit diameters is not welcome, as it leads to high air consumption. A partial solution is provided by the swirling jet (Figures 5.35 and 6.164) [33], but much better is a solution with a synthetic jet (Figures 6.165 and 6.166) or hybrid-synthetic jet (Figure 6.167), the latter preferably with the pure-fluidics, no-moving-part oscillator.

6.8.2 Fluidic decontamination

The primary goal of antiterrorist activities is the prevention of an attack. Nevertheless, it is necessary to consider the possibility of terrorists' success and to have means for the neutralization of nuclear, chemical, and biological agents at the attack site. Since practically all decontaminants are fluids, the task is again a useful opportunity for fluidics. Requirements apart from the biological (sporicide) and chemical action are penetrating capability (important for complex objects, such as mail-sorting machinery, with a high probability of being hit) and chemical nonaggressivity (important when decontaminating for example highly valued objects of art). A fluidic generator of vaporous hydrogen peroxide (decomposing into harmless oxygen and water vapor) may fit the need. Another solution is a fluidic generator of containment foam to cover the objects (and quench possible fires). Available to the first respondents, the device may be analogous to fire extinguishers.

6.9 INTERFACING THE CENTRAL NERVOUS SYSTEM

The development of MEMS has reached the stage at which it is possible to monitor neural signals and even input signals into the neural system. Thanks to the small size of the devices, interfacing is possible at the level of neurons. In this application, fluidics is not the star of the show, but it plays one of the key supporting roles. The most important interaction with neurons takes place in the brain cortex, where neurons, apart from their central cells and long axon fibers, form a dense dentritic structure of synapse fibers with which neurons communicate. There are approximately 30 000 neurons in each cubic millimeter of the human cortex. The largest cells there are the pyramidal neurons with bodies 10–30 μm in diameter, but the dendritic structure is an order of magnitude larger, commensurable with the size of microfabricated devices. Since the details of cortex architecture are not known precisely, the contemporary approach to the interaction is by means of a redundant number of microelectrodes arranged in 3D arrays. The electrodes are on the surface of needlelike micromachined probes (Figures 6.168 and 6.169) with their sharp ends inserted into the cortex. The other ends are fixed on a holder, arranged into a 2D field. In the third dimension the electrodes are distributed along the stem of each probe. The transverse dimension of a probe is usually from 30 μm to 100 μm. The probes are hollow; the internal cavity serves for a microfluidic liquid transport towards the exit hole near the tip. The material of choice for the electrodes (recording and stimulating sites) is iridium oxide, exhibiting a 150 times higher charge delivery capability per surface area than gold, which was used earlier. This is an important factor, because the electrodes have to be small, not only because of the lack of available surface on the needle, but also because of the spatial spread of potential in the tissue, leading to unwelcome averaging. Thin-film conductors, insulated above and below by deposited insulating layers—usually SiO_2, sometimes diamond or silicides of refractory metals—connect the electrodes with

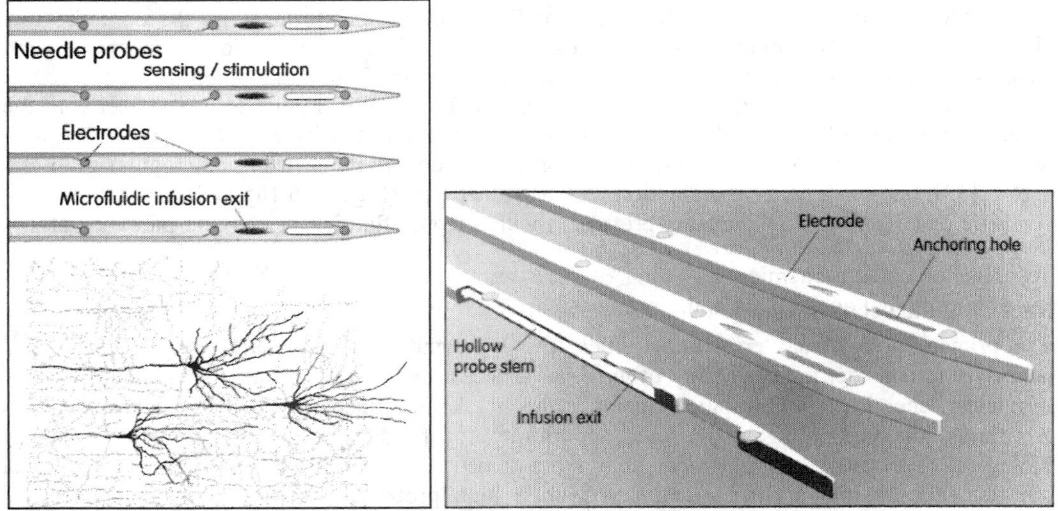

Figure 6.168 (Left) The relative size of neurons in the cortex and part of a typical array of probes with interfacing electrodes.

Figure 6.169 (Right) The probes from Figure 6.168 in detail. The first probe is shown partly cut off to reveal the integral cavity for the transport of fluids (drugs) to the exit near the sharp end of the probe. The anchoring holes allow the ingress of surrounding tissue that anchors and stabilizes the probes in their locations.

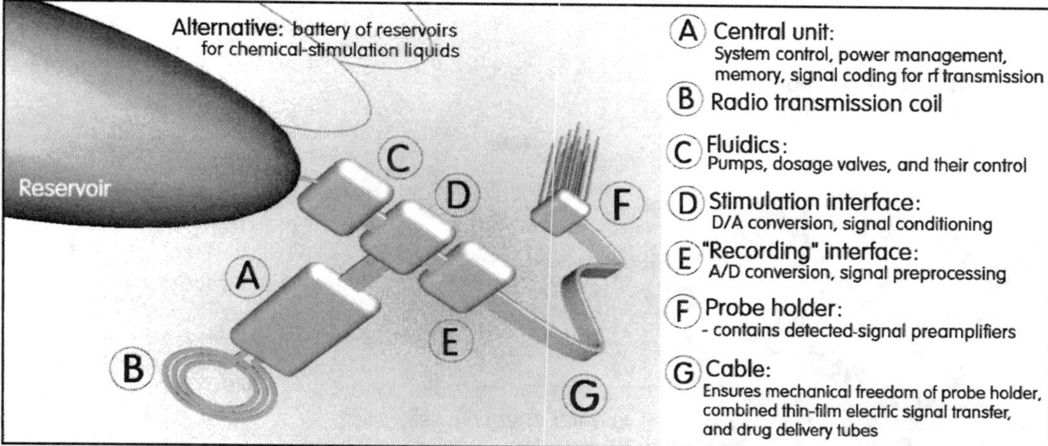

Figure 6.170 An example of a typical implanted device for interfacing the central neural system (recent systems are characterized by more numerous probes forming a more dense array). The array F of the probes is inserted directly into the brain surface. The other parts are connected by the multichannel electro-fluidic cable G, not restraining the free-floating of the brain in the cranial cavity. The drugs infused through the probes are supplied from collapsible (and refillable) reservoirs.

the electronics. The detected signals are first led to preamplifiers, usually microfabricated integrally at the wide base of each needle.

The signal voltages detectable on the electrodes due to the activity of the neurons are usually only from 20 μV up to about 1 mV. (This is in contrast to the micropipette probes capable of reaching the intracellular material, where they detect signals in a range from 10 mV to 100 mV.) When an electrode is used for stimulation, the excitation current depends on how closely the electrode is near the neurons: from 10 μA in the cortex to 300 μA when used to transfer auditory signals in the cochlea. The voltage must remain below the water-decomposition level. The excitation acts in pulses; a typical pulse duration is between 20 and 500 μs. The effect is determined neither by voltage nor current: the decisive factor is the delivered charge.

The insertion of needle probes into the cortex inevitably causes wounds, cellular damage and microhemorrhages. The organism responds first with an immunity reaction and later by the formation of scar tissue. This is unwelcome, walling off the probe from the neurons. The injury response involves a chemical-signaling cascade, and this may be interrupted by administering a suitable chemical agent. This is the primary reason why modern neural implants involve a fluidic subsystem for delivering the agents, as shown in Figure 6.170. This, however, is not the sole reason. A very effective way of stimulating the neurons is chemical excitation. Neuronal ganglia have protein-lined passageways through which specific ions can move along the gradient of concentration or electric field. These ion channels open and close in dependence upon the local conditions of the cell, which are influenced by the injected chemicals. The cells in neuronal systems exhibit firing patterns, or rhythmical changes with timescale on the order of minutes. This is a timescale particularly well controllable by the release of the drugs.

6.9.1 Therapeutic uses

The neural interface devices are effective as neural prostheses. They make it possible—in principle, at least—to treat previously incurable illnesses associated with processes taking place in the brain. Most successful are cochlear prostheses; over 70 000 were implanted worldwide, and

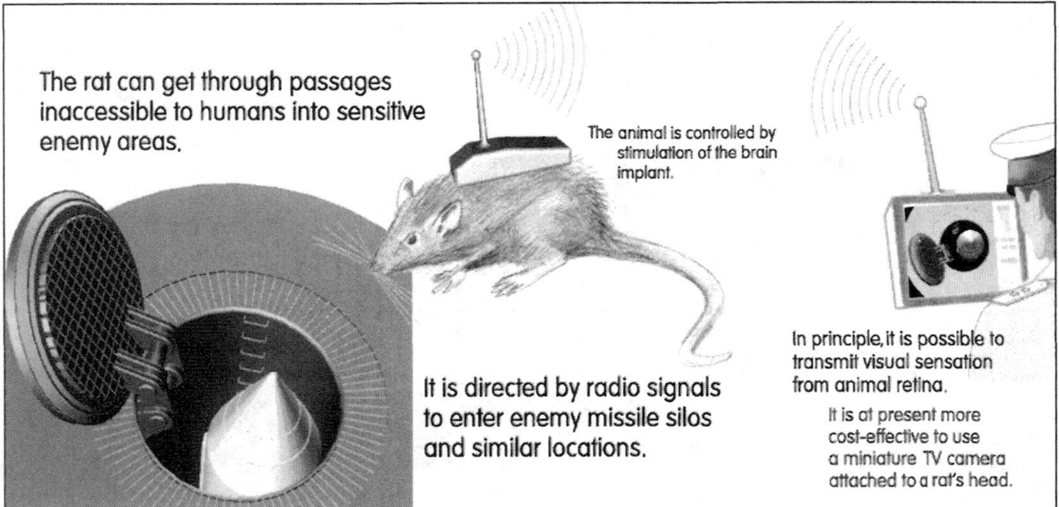

Figure 6.171 The neural interfacing system from Figure 6.170 may be used to remotely control the movement of mammals and evaluate the signals they receive from their senses. Military uses are obvious. The implanting is not of importance; in fact, longer communication distance and shorter timescales may lead to the use of a larger external battery pack as the power source and perhaps even an external attached TV camera as the primary source of the transmitted information.

with them the profoundly deaf can often hear well enough to use the telephone. Retinal implants to remove blindness have received great attention. Also managing severe Parkinson's disease has proved remarkably effective (even though the mechanism by which the deep-brain electrode signals operate is not yet completely understood). Devices for managing severe epilepsy are in development. Experiments for capturing motor control signals from the cortex have been promising to eventually restore at least some moving capability to quadriplegics.

The prosthetic devices combine electric and chemical action. High-density probe arrays with 3D systems of electrodes are used to monitor and record electrical activity of neurons and insert electrical signals to stimulate, and their fluidic channels provide a control of local chemical environment by drug delivery. Typically, the probes are provided with holes for anchoring their location in the surrounding tissue; a certain growth of scar tissue there is desirable. While the fluidic part has to be directly attached, the electronics is divided between the probes, the platform, and an external package – with which the platform maintains a bidirectional radio-frequency telemetry link. The platform is a button-sized implant (size of a few millimeters, thickness < 1 mm). The division places great demands on the flat multichannel interconnecting cable, involving tube(s) for the fluidics, which must allow for free-floating movements of the probe array.

6.9.2 Cyber-animals

Neural interfaces may be implanted to animals. This, after all, is routinely done anyway in the process of testing the system before the application to humans. Inevitably, the idea arose to use remotely controlled animals for defense purposes, mainly as spies (Figure 6. 171). The transmitter has to operate at a much longer distances and this requires a much larger power source. Fortunately, the battery pack as well as aerial may in this application be carried externally. The

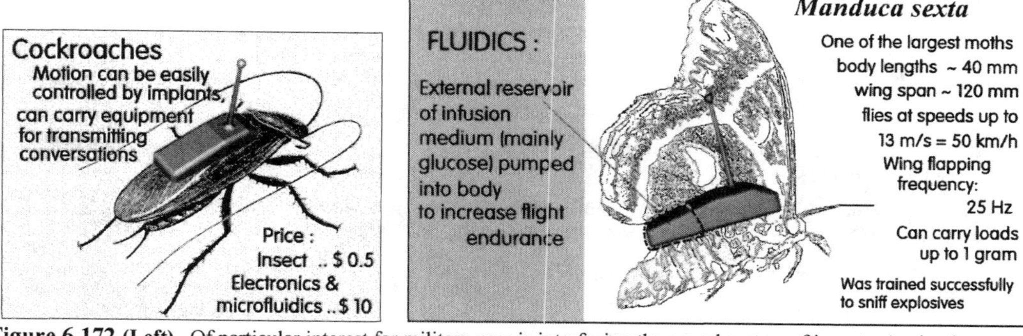

Figure 6.172 (Left) Of particular interest for military uses is interfacing the neural system of insects. Again, the current size of the devices and the longer communication distances favor external attachment rather than implants.

Figure 6.173 (Right) A flying insect is considered particularly suitable for the remotely controlled information gathering. Microfluidic pumping of carbohydrates (simple sugars, mainly glucose) as nutrients into the hemolymph is essential to obtain useful endurance in flight. The infusion is made easier by the open circulatory system of insects.

quality of transmitted visual signals from animal's eyes is generally below the desirable level and a more effective solution is carrying, also externally, a miniature TV camera. More attractive - because of an even lower probability of being discovered (Figure 6.174), is the military use of insects (Figures 6.172, 6.173, 6.174, and 6.175). A disadvantage is the shorter distance the insect can cover. Fluidics offers an improvement of this factor by the possibility of pumping the carbohydrate "fuel" into the hemolymph (the insect "blood") from an external store.

Figure 6.174 (Top) Natural stealth capability, mimicry, of moths is one of their advantages from the point of view of military applications. Manduca sexta sitting in the central part of this picture is hardly recognizable. (Courtesy of Dr. A. Lal, DARPA.)

Figure 6.175 (Bottom) To make the M. sexta in Figure 6.174 recognizable, its contours in this copy of the picture were traced and then filled with contrasting shading.

References

[1] Drost, M., K., et al., "Recent Developments in Microtechnology-Based Chemical Heat Pumps," in Ehrfeld, W., (ed.), *Microreaction Technology: Industrial Prospects*, Springer-Verlag, Berlin, 2000, p. 394.

[2] Ernst, T. C., "Design, Fabrication, and Testing of a Wearable Cooling System," M.Sc. Thesis, Georgia Institute of Technology, February 2005.

[3] Golod, S. V., et al., "Fabrication of Conducting GeSi/Si Micro- and Nanotubes and Helical Microcoils," *Semiconductor Science and Technology*, Vol. 16, 2001, p. 181.

[4] Breuer, K., et al., "Challenges for Lubrication in High Speed MEMS," *NanoTribology*, S. Hsu (ed.), Kluwer Press, 2003.

[5] Vican, J., et al., "Development of a Microreactor as a Thermal Source for Microelectromechanical Systems Power Generation," *Proceedings of the Combustion Institute*, Vol. 29, 2002, p. 909.

[6] Terry, S.C., "A Gas Chromatographic Air Analyser Fabricated on Silicon Wafer Using Integrated Circuit Technology," Ph.D. Thesis, Stanford University, 1975.

[7] Ramamoorthy, R., P. K. Dutta, and S. A. Akbar, "Oxygen Sensors: Materials, Methods, Designs and Applications," *Journal of Materials Science*, Vol. 38, 2003, p. 4271.

[8] Garrigues, S., T. Talou, and D. Nesa, "Comparative Study Between Gas Sensors Array Device, Sensory Evaluiation and GC/MS Analysis for QC in Automotive Industry," *Sensors and Actuators B*, Vol. 103, 2004, p. 55.

[9] Rabl, A. and J. V. Spadaro, "Health Cost of Automobile Pollution," *Revue Francaise d'Allergologie et d'Immunilogie Clinique*, Vol. 40, 2000, p. 55.

[10] "Combinatorial Library Methods and Protocols," English, L. B. (ed.), in *Methods in Molecular Biology*, Vol. 201, Humana Press, Totowa, NJ, 2002.

[11] Zech, T., et al., "Simultaneous Screening of Catalysts in Microchannels: Methodology and Experimental Setup," Ehrfeld, W., (ed.), in *Microreaction Technology: Industrial Prospects*, Springer, Berlin, 2000.

[12] Wilkin, O. M., et al., "High Throughput Testing of Catalysts for the Hydrogenation of Carbon Monoxide to Ethanol," Derouanne, E. G., et al. (eds.), in *Principles and Methods for Accelerated Catalyst Design and Testing*, Kluwer Academic Publishers, the Netherlands, 2002, p. 299.

[13] Pihl, J., M. Karlsson, and D. T. Chiu, "Microfluidic Technologies in Drug Discovery," *Drug Discovery Today*, Vol. 10, No. 20, October 2005, p. 1377.

[14] Tesař, V., et al., "Development of a Microfluidic Unit for Sequencing Fluid Samples for Composition Analysis," *Chemical Engineering Research and Design*, Vol. 82 (A6), June 2004, p. 708.

[15] Tesař, V., "Sampling by Fluidics and Microfluidics," *Acta Polytechnica - Journal of Advanced Engineering*, Vol. 42, No. 2, 2002, p. 41.

[16] Goldberg, D. E., *Genetic Algorithms in Search, Optimization, & Machine Learning,*" Addison-Wesley, Reading, MA, 1989.

[17] Scott, S.M., D. James, and Z. Ali, "Data Analysis for Electronic Nose Systems," *Microchimica Acta*, Springer-Verlag, Berlin, 2006.

[18] Churchill W., "Fifty Years Hence" in *Thoughts and Adventures*, Butterworth, London, 1932.

[19] Benjaminson, M., J. Gilchriest, and M. Lorenz, "In Vitro Edible Muscle Protein Production System. Stage I.: Fish," *Astra Atronautica*, Vol. 51, 2002, p. 879.

[20] Trávníček, Z., T. Vít, and V. Tesař, "Hybrid-Synthetic Jets as the Nonzero-Net-Mass-Flux Synthetic Jets," *Physics of Fluids*, Vol. 18, August 2006, p. 081701-1.

[21] Tesař, V., and Z. Trávníček, "Increasing Heat and/or Mass Transfer Rates in Impinging Jets," *Journal of Visualization*, No. 2, Vol. 8, 2005, p. 91.

[22] Tesař, V., and Z. Trávníček, "Pulsating and Synthetic Impinging Jets," *Journal of Visualization*, No. 3, Vol. 8, 2005, p. 201.

[23] Etude d'Incidence des Cancers à Proximité des Usines d'Incinération d'Ordures Ménagères. Institut de Veille Sanitaire, Département santé environnement, 12, rue du Val d'Osne, 94415 Saint-Maurice, France, November 30, 2006.

[24] Hänel, K., *Biological Treatment of Sewage by the Activated Sludge Process*, Chichester (Ellis Horwood books in water and wastewater technology): Ellis Horwood, 1988.

[25] Zhang, C., et al., "PCR Microfluidic Devices for DNA Amplification," *Biotechnology Advances*, 2005.

[26] Nguyen, N.T., and S. T. Wereley, *Fundamentals and Applications of Microfluidics*, 2nd ed., Artech House, Norwood, MA, 2006.

[27] Preti, G., et al., "Analysis of Lung Air from Patients with Bronchogenic Carcinoma and Controls Using Gas Chromatography, Mass Spectrometry," *Journal of Chromatography B: Biomedical Sciences and Applications*, Vol. 432, 1988, p. 1.

[28] Von Bühren, T. et al., "Optimization of Inertial Micropower Generators for Human Walking Motion," *IEEE Sensors Journal*, Vol. 6, No.1, 2006, p. 28.

[29] Trumble, D. R. and J. A. Magovern, "A Muscle-Powered Energy Delivery System and Means for Chronic in Vivo Testing," *Journ. Applied Physiol.*, Vol. 86, 1999, p. 2106.

[30] Apblett, C., et al., "Bio Micro Fuel Cell Grand Challenge Final Report," Report SAND2005-5734, Sandia National Labs., Albuquerque, N. M., September 2005.

[31] Staples, E. J. and S. Viswanathan, "Ultrahigh-Speed Chromatography and Virtual Chemical Sensors for Detecting Explosives and Chemical Warfare Agents," *IEEE Sensors Journ.*, Vol. 5, No. 4, 2005, p. 622.

[32] Wise, K. D., et al., "Wireless Implantable Microsystems: High-Density Electronic Interfaces to the Nervous System," *Proc. of the IEEE*, Vol. 92, No. 1, 2004, p. 76.

[33] Motchkine, V.S., L. Y. Krasnobaev, and S. N. Bunker, "Cyclone Sampling Nozzle for an Ion Mobility Spectrometer," U.S. Patent No. 6,861,646, March 2005.

[34] Kuznetsova, L. A, and W. T. Coakley, "Applications of Ultrasound Streaming and Radiation Force in Biosensors," *Biosensors and Bioelectronics*, 2007.

[35] Radke, S. M., "A Microfabricated Biosensor for Detecting Foodborne Bioterrorism Agents," *IEEE Sensors Journal*, Vol. 5, No. 4, 2005, p. 744.

[36] Sadik, O. A., A. K. Wanekaya, and S. Andreescu, "Advances in Analytical Technologies for Environmental Protection and Public Safety," *Journal of Environmental Monitoring*, Vol. 6, 2004, p. 513.

[37] Frisk, T., et al., "Fast Narcotics and Explosives Detection Using a Microfluidic Sample Interface," *Proc. of TRANSDUCERS '05*, Seoul, June 2005, p. 2151.

[38] Mlsna, T. E., et al, "Chemicapacitive Microsensors for Chemical Warfare Agent and Toxic Industrial Chemical Detection," *Sensors and Actuators B*, Vol. 116, 2006, p. 192.

[39] Meier, D. C., et al., "Chemical Warfare Agent Detection Using MEMS-Compatible Microsensor Arrays," *IEEE Sensors Journal*, Vol. 5, No. 4, 2005, p. 712.

[40] Lazcka, O., F. J. Del Campo, and F. X. Muñoz, "Pathogen Detection: A Perspective of Traditional Methods and Biosensors," *Biosensors and Bioelectronics*, Vol. 22, 2007, p. 1205.

[41] Prokop, A., "Bioartificial Organs in the Twenty-First Century," *Annals of New York Academy of Sciences*, Vol. 944, 2001, p. 472

[42] Davis, F. and P. J. Higson, "Biofuel Cells—Recent Advances and Applications," *Biosensors and Bioelectronics*, Vol. 22, 2007, p. 1224.

[43] Su, F., K. Chakrabarty, and R. B. Fair, "Microfluidics-Based Biochips: Technology Issues, Implementation Platforms, and Design-Automation Challenges," *IEEE Trans. on Computer-Aided Design of Integrated Circuits and Systems*, Vol. 25, No. 2, 2006, p. 211.

[44] Dario, P., et al, "Micro-Systems in Biomedical Applications," *Journ. Micromech. and Microeng.*, Vol. 10, 2000, p. 235.

[45] Cheung, K. C., and Renaud P., "BioMEMS in Medicine: Diagnostic and Therapeutic Systems," *Proceedings of ESSDERC*, Paper 5.A.1, Grenoble, France, 2005, p. 345.

[46] Balasubramanian, A., B. Bhuva, and F. Haselton, "Si-Based Sensor for Virus Detection," *IEEE Sensors Journal*, Vol. 5, 2005, p. 340.

[47] Gómez-Sjöberg, R., D. T. Morisette, and R. Bashir, "Impedance Microbiology-on-a-Chip: Microfluidics Bioprocessor for Rapid Detection of Bacterial Metabolism," *Journ. of Microelectromechanical Systems*, Vol. 14, 2005, p. 829

[48] Squires, T. M. and S. R. Quake, "Microfluidics: Fluid Physics at the Nanoliter Scale," *Reviews of Modern Physics*, Vol. 77, 2005, p. 97.

[49] Stephan, A., M. Bücking, and H. Steihart, "Novel Analytical Tools for Food Flavours," *Food Research International*, Vol. 33, 2000, p. 199.

[50] Marek, W. et al., "Can Semi-Automated Image Cytometry on Induced Sputum Become a Screening Tool for Lung Cancer? " *European Respiratory Journal*, Vol. 18, 2001, p. 942.

[51] Lam, W.W. et al., "A Surface Plasmon Resonance System for the Measurement of Glucose in Aqueous Solution," *Sensors and Actuators B*, Vol. 105, 2005, p. 138.

Concluding Remarks

Fluidics and in particular microfluidics are new areas of technology, with a rate of development so fast that it is difficult to predict in what direction the progress is going to lead. Whatever the outcome, fluidic systems are here to remain and stay. Chapter 6 presents a number of application examples, some of them already successful and established, others still in their embryonic stage, some bordering on fiction. They show that microfluidics can lead to very profound changes in the world as we know it, the changes ranging from cars to food production and to our very bodies. Like technology in general, the fluidic technology may bring many benefits but may be abused for targets transgressing the moral codes.

This book aims at providing all the basic facts and sufficient details needed for understanding the issues. It does not, however, discuss several themes covered by other books.

Theory and practice of designing fluidic circuits from individual elements is discussed in Chapter 13 "Fluidic Circuits," written by the present author and published in *Microfluidics: History, Theory, and Applications*, ed. W. B. J. Zimmerman (Springer, 2006).

The microfabrication processes used to make the small devices are very well treated in the Chapter 3 "Fabrication Techniques for Microfluidics" in the book by N.-T. Nguyen and S. Wereley in *Fundamentals and Applications of Microfluidics* (Artech House, 2006). Also in this book is a chapter "Fluid Mechanics Theory for Microfluidics," together with the related "Experimental Flow Characterization" for those seeking more information about fluid mechanics aspects of microflows. Readers interested in use of microfluidics for control of external flows (transition into turbulence or flow separation on aircraft bodies and wings) will also find there a discussion of this particular application.

About the Author

Václav Tesař received an Ing. degree in mechanical engineering from the Czech Technical University (ČVUT), Prague, Czech Republic. From 1963 to 1998, he spent teaching fluid mechanics and dynamics of hydromechanic and thermodynamic systems at ČVUT Prague, Professor Tesař progressed gradually from an assistant professor to a docent and later a full professor. He received a C.Sc. degree (the equivalent of a Ph.D.) from ČVUT Prague in 1972. Professor Tesař was also the head of the Department of Fluid Mechanics and Thermodynamics at the Faculty of Mechanical Engineering, ČVUT Prague. His research interests covered the basic principles of fluid mechanics in general, shear flows (boundary layers, jets, wall-jets) with studies and modeling of turbulence, and hydraulic systems, their components, and dynamics of the processes in hydraulic circuits. He was invited in 1985 to stay as a visiting professor at Keio University, Yokohama, Japan. In 1992 he stayed as a visiting professor at Northern Illinois University, DeKalb, Illinois. Professor Tesař published papers on turbulent jets, wall-jets, the Coanda effect of jet attachment to walls, and the applications of these flows to various no-moving-part fluidic flow control devices. He is the inventor of 195 Czech patents, mainly concerning fluidic devices.

In 1998 Professor Tesař joined the University of Sheffield, United Kingdom, as a professor in the Process Fluidics Group at the Department of Chemical and Process Engineering. It was there that he became involved in microfluidics, in particular designing several versions of various pressure-driven microfluidic valves for control of small gas flows.

Recently, Professor Tesař returned to the Czech Republic and is currently employed at the Institute of Thermomechanics of the Academy of Sciences of the Czech Republic. He is the author of four textbooks for mechanical engineering students and of more than 300 research papers in various journals and conference proceedings.

Index

Recent Titles in the Artech House Integrated Microsystems Series

Adaptive Cooling of Integrated Circuits Using Digital Microfluidics, Philip Y. Paik, Krishnendu Chakrabarty, and Vamsee K. Pamula

Fundamentals and Applications of Microfluidics, Second Edition, Nam-Trung Nguyen and Steven T. Wereley

Introduction to Microelectromechanical (MEM) Microwave Systems, Héctor J. De Los Santos

An Introduction to Microelectromechanical Systems Engineering, Nadim Maluf

MEMS Mechanical Sensors, Stephen Beeby et al.

Microfluidics for Biotechnology, Jean Berthier and Pascal Silberzan

Organic and Inorganic Nanostructures, Alexei Nabok

Post-Processing Techniques for Integrated MEMS, Sherif Sedky

Pressure-Driven Microfluidics, Václav Tesař

RF MEMS Circuit Design for Wireless Communications, Héctor J. De Los Santos

Wireless Sensor Network, Nirupama Bulusu and Sanjay Jha

For further information on these and other Artech House titles, including previously considered out-of-print books now available through our In-Print-Forever® (IPF®) program, contact:

Artech House
685 Canton Street
Norwood, MA 02062
Phone: 781-769-9750
Fax: 781-769-6334
e-mail: artech@artechhouse.com

Artech House
46 Gillingham Street
London SW1V 1AH UK
Phone: +44 (0)20 7596-8750
Fax: +44 (0)20 7630-0166
e-mail: artech-uk@artechhouse.com

Find us on the World Wide Web at: www.artechhouse.com